Discrete Mathematics

Mathematical Reasoning and Proof with Puzzles,
Patterns, and Games

Douglas E. Ensley
Shippensburg University

J. Winston Crawley
Shippensburg University

WILEY

John Wiley & Sons, Inc.

To Amy, John, and Jessica, for making life fun. — D. E.

To Margaret, for her unwavering support. — W. C.

VICE PRESIDENT, PUBLISHER	Laurie Rosatone
ASSOCIATE EDITOR	Kelly Boyle
PROGRAM ASSISTANT	Stacy French
SENIOR PRODUCTION EDITOR	Lisa Wasserman
SENIOR DESIGNER	Kevin Murphy
SENIOR ILLUSTRATION EDITOR	Anna Melhorn
COVER PHOTO	©Chuck Carlton/Index Stock

This book was set in LaTex by TechBooks and printed and bound by Sheridan Books, Inc.

This book is printed on acid free paper. ∞

To order books or for customer service please, call 1-800-CALL WILEY (225-5945).

Library of Congress Cataloging in Publication Data:
Ensley, Douglas E.
      Discrete Mathematics: Mathematical Reasoning and Proof with Puzzles, Patterns, and Games/ Douglas E. Ensley. J. Winston Crawley.
p. cm.
Includes bibilographical references and index.x
ISBN-13   978-0-471-47602-3
ISBN-10   0-471-47602-1

1. Logic, Symbolic and mathematical.   2. Mathematical recreations.   3. Computer
      science—Mathematics.
I.   Crawley, J. Winston. II. Title.
QA9.25.E57 2006
511′.1—dc22

2005017417

Printed in the United States of America

11

Preface
for Instructors

This book is written for students who are prepared to make a first departure from the deeply worn path that leads from arithmetic to calculus in the typical mathematics curriculum. Because of its deliberate pace and emphasis on student readability as well as its treatment of engaging topics, this book is suitable for courses taught at several different levels.

- This text originated as notes for an enrichment course in discrete mathematics for talented high school students. This course has been taught for over ten years as part of the core curriculum of the Pennsylvania Governor's School for the Sciences.

- The book has been used for five years for a discrete math course at Shippensburg University. This course is required for students majoring in mathematics or computer science as well as those students majoring in elementary education with a concentration in mathematics.

- In its most recent life, this book has also been used by several instructors as the basis for a "transition" course in a mathematics major program. The gentle introduction to abstraction and the emphasis on written proof have been well-received by students at this level as well.

Discrete mathematics suffers somewhat from having no historically established place in the K-12 mathematics curriculum. According to the National Council of Teachers of Mathematics, discrete math should be emphasized at every level of mathematics. Meanwhile college curriculum committees struggle with what one-semester course ought to be used to address basic discrete math concepts. Since it seems unlikely that there will be a trend in colleges to create a three or four semester sequence in discrete mathematics, perhaps the best one can do is to apply some careful thought and sound pedagogy to what will happen in the crucial one or two semesters.

The approach taken in our book is one of establishing fundamental skills and building connections between basic concepts instead of surveying as many topics as possible in one semester. There is enough variety of topics for the instructor to have some choice for the emphasis of his or her course. However, the instructor might find that a favorite topic is treated only in a specific context instead of more generally. For example, in our book, instead of an entire chapter on binary relations, there are three non-contiguous sections placed at different points in the overall development. By stressing connections between topics, we believe we are enhancing student understanding and avoiding the time-consuming task of thoroughly discussing every aspect of each topic before moving on.

Contents and Organization

A list of mathematical topics that are addressed within the framework of this book is shown below. This list covers most of the content and learning objectives recommended by the ACM/IEEE (for a one-semester course for computer science majors), the MAA CUPM (for math majors), and the CBMS document on the Mathematical Education of Teachers (for future middle and secondary level teachers).

Number sequences	Functions	Markov chains
Truth tables	Binary relations	Eulerian graphs
Propositional logic	Equivalence relations	Planar graphs
Predicate logic	Numerical functions	Hamiltonian graphs
Mathematical proof	Iterated function systems	Traveling salesperson
Proof by induction	Counting techniques	Directed graphs
Pigeonhole principle	RSA cryptography	Adjacency matrices
Modular arithmetic	Recurrence relations	Spanning trees
Representing numbers	Growth of functions	Solving puzzles
Sets	Discrete probability	Analyzing games
Boolean algebra	Expected value	Binary trees
Logic circuits	Matrix arithmetic	

These topics are broadly organized into four parts:

Thinking and writing about mathematics

The first two chapters introduce the fundamental emphasis of the course. Although the *content* is discrete mathematics, the *theme* is learning to think and write mathematically. The material is introduced in the context of puzzles concerning logic and numbers. Although some symbolic logic is covered in the first chapter, it is not directly used in the discussion of proof in Chapter 2. Instead, we build upon the students' innate ability to apply basic logic to their surroundings, and we focus our efforts on attaining a level of comfort with formalization and abstraction. We demonstrate that inductive reasoning and proof are very natural processes. The students practice their newly developed proof-writing skills in connection with a number of mathematical topics, including divisibility properties, rational numbers, sequences, and summations. Chapter 2 ends with applications of these properties of numbers.

Sets, functions, and relations

The next two chapters introduce these important topics. Students are encouraged to apply their problem-solving and proof-writing skills with this new subject matter. An important point in this discussion is the way that abstraction (like Boolean algebra) can actually make certain computations (as with logic circuits) easier instead of harder. Other excursions into iterated functions and growth of functions continue the idea of recursive descriptions and inductive proofs from earlier chapters.

Combinatorics and probability

Chapters 5 and 6 discuss combinatorics, probabilities, and their applications to games of chance. Although mathematical proof does not play as prominent a role in this portion, mathematical reasoning is still paramount. An important example is the emphasis on using the idea of one-to-one correspondence to realize that two seemingly different problems must have the same solution. Excursions into the solution of recurrence relations and the use of matrices to model probability and expected value make connections to other topics in the book.

Graph theory

The fourth part of the text develops graph theory, emphasizing a closer look at puzzles and games. The topic is motivated by the famous Bridges of Königsberg puzzle. Once introduced, graphs provide an opportunity to reinforce several topics from the earlier portions of the book, including binary relations and induction proofs. Alternatively, the classical applications of graph theory can be understood and appreciated independently of any earlier material.

Threads

One innovation in this book is the spiral approach to the development of some important concepts and topics. We achieve this by avoiding as much as possible the fragmentation of the course into disconnected pieces. The book uses several major threads to help weave the core topics into a cohesive whole.

- Throughout the book, we emphasize the application of **mathematical reasoning** to the solution of problems, and we guide the student in thinking about, reading, and writing proofs in a wide variety of contexts.
- An important *content* thread is the focus on (mathematical) **puzzles, games or magic tricks** to engage students with problems that motivate the development of specific topics.

- In addition, we have included a variety of "real world **excursions**" to which these same concepts can be applied, in order to show students the utility of discrete mathematics.

- A primary *conceptual* thread is the use of "**algorithmic thinking**" in solving problems and exploring the connection between related problems.

- Another important feature is the recurrent (so to speak) use of **recursive modeling** and its companion, **mathematical induction**. There are ample opportunities for students to reason recursively and prove inductively throughout the book.

Four Possible Courses

Discrete math courses vary so greatly that it is difficult for one book to serve every purpose. There are four different types of courses for which this book is a good fit. We briefly outline here the core sections that an instructor should consider for each of these courses. Recommendations for emphasis within each section are given in the *Instructor's Guide*.

1. A mixed math, computer science and math education audience requires breadth among applications and conceptual topics. The following sections would provide this experience:

- 1.1–1.5
- 2.1–2.3, 2.5, 2.6
- 3.1–3.5
- 4.1, 4.4, 4.5, 4.7
- 5.1–5.4
- 6.1–6.3
- 7.1, 7.3, 7.5

2. A first-year course for computer science students would want to cover specific computer science applications as well as the logic and proof topics recommended by ACM:

- 1.1–1.5
- 2.1–2.3, 2.5, 2.6
- 3.1–3.5
- 4.1, 4.2, 4.4–4.6, 4.8
- 5.1–5.3, 5.5, 5.6
- 7.1, 7.3, 7.6, 7.7

3. A sophomore or junior level math major transition course might focus on mathematical proof and classical structures, hence the following core sections would be appropriate:

- 1.1–1.6
- 2.1–2.7
- 3.1–3.3
- 4.1–4.5
- 7.1–7.7

4. Advanced high school students might have applications of discrete math as their key interest, so the following sections could provide the core of a course for this audience:

- 1.1–1.4
- 2.1–2.3, 2.6, 2.7
- 3.1–3.5
- 4.1, 4.2, 4.6, 4.7
- 5.1–5.4, 5.6
- 6.1–6.3, 6.6
- 7.1, 7.3, 7.5, 7.7

Aids to Teaching and Learning

Practice problems

Each section contains a number of practice problems distributed throughout. The intent is that students use these to test their understanding of what they have just read. This encourages the students to slow down and think about what they are reading. It also provides positive reinforcement to those who solve the problem correctly and a mid-course correction to those who do not. Detailed solutions for the practice problems are found at the end of each section.

Section reviews

At the conclusion of each chapter, a section review outlines the major terms, concepts, and skills covered by each section within that chapter. In addition to providing a synopsis to the student, these reviews are useful to the instructor in planning which sections to cover.

Exercises

The book has over 1,100 exercises, ranging from routine reinforcement to conceptual challenges. We have tried to include many exercises that are direct applications of the examples in the section, several exercises that make connections to past topics or foreshadow some future developments, and some problems that will challenge and intrigue even the very best students.

Excursions

Each chapter ends with at least one section that leads an excursion into applications of the concepts of the chapter. These independent sections can provide additional applications or individual enrichment, or they can be used as the basis of independent projects.

Appendices

Matrix operations are presented in two different contexts (Markov chains and counting walks in graphs) in this book, and either can be taught without the other. However, each treatment is fairly brief and offers no real chance for directed practice with matrices, so we have included an appendix on basic matrix operations for this purpose. In addition, we have included an appendix explaining the basic rules and terminology for many of the games we use for examples in the book.

Answers and solutions

In addition to the solutions to the practice problems, we give answers or hints in the back of the book for many of the exercises. In the exercise sets themselves, we indicate which problems have answers in the back by highlighting the exercise number in blue. A supplementary *Student Solutions Manual* ISBN 0-471-760978 contains more detailed solutions to those same exercises.

Explore more on the Web.

Electronic resources for students

Every section has examples or exercises supplemented by web-based activities. This supplementary material primarily takes the form of either an online version of the problem or an interactive activity to develop an idea. When an item in the text is connected to an online activity, it is identified with an icon in the margin. The URL for this material and other resources for instructors and students is

http://www.wiley.com/college/ensley

The web-based material related to mathematical proof was funded through NSF grant DUE-0230755.

Resources for instructors

The *Instructor's Guide* located on the instructor companion site at http://www.wiley.com/college/ensley provides more details on how to use this book, including course outlines and an annotated version of the section summaries that appear within the text. In addition, the *Instructor's Solution Manual* ISBN 0-471-488011 provides complete solutions to all the exercises in the book.

Acknowledgements

A textbook combines the clever ideas, generous time, and industrious efforts of many individuals. While the blame for shortcomings and errors always rests on the shoulders of the authors, the credit for the book's successes should be spread generously among all of these people.

For providing feedback from classroom testing, we thank the following professors and their students: Madelaine Bates (Bronx Community College), Debra Borkovitz (Wheelock College), Judith Covington (Louisiana State University, Shreveport), Renee Fister (Murray State University), Catherine Murphy (Purdue University, Calumet), Patricia Oakley (Goshen College), Sharon Robbert (Trinity Christian College), and Madeleine Schep (Columbia College).

In addition, we gratefully acknowledge the following individuals for reviewing our manuscript and generously providing feedback that has greatly improved the book: David Barrington (University of Massachusetts, Amherst), George Davis (Georgia State University), Darin Goldstein (California State University, Long Beach), Jonathan Goldstine (Penn State University), Johannes Hattingh (Georgia State University), Matthew Hudelson (Washington State University), Heather Hulett (University of Wisconsin, La Crosse), Mark Jacobson (University of Northern Iowa), Jay Kappraff (New Jersey Institute of Technology), Theodore Laetsch (University of Arizona), Hong-Jian Lai (West Virginia University), Sheau-Dong Lang (University of Central Florida), Lisa Markus (De Anza College), Jandelyn Plane (University of Maryland, College Park), Jacek Polewczak (California State University, Northridge), and Anne-Louise Radimsky (California State University, Sacramento).

We also thank our patient colleagues at Shippensburg University, James Hamblin, Kate McGivney, Fred Nordai, Cheryl Olsen, and Kim Presser, for their continuous input into the process of developing this material over the past five years. For detailed reviews and technical assistance, we would like to thank John and Amy Ensley, Mark McKibben of Goucher University, Frank Purcell and Elka Block of Twin Prime Editorial, and the entire team at John Wiley and Sons, especially Kelly Boyle, Laurie Rosatone and Lisa Wasserman.

Preface
for Students

"It is not worth the while to go round the world to count the cats in Zanzibar."
– Henry David Thoreau, in *Walden*

Mathematics has often been saddled with the reputation of being a dry academic pursuit characterized by tedious rote. It is easy to lay the blame for this on some traditional mathematics programs, but much of the trouble truly lies with the subject itself. Before the college level, mathematics appears to be a tower of topics, each one requiring mastery before continuation is possible. Certainly no one will understand algebra without first mastering addition and subtraction of numbers. No one will grasp the great utility of trigonometry without understanding something about geometry first. This pyramid of mathematical topics persists to the study of calculus. It is no wonder that calculus is considered by so many outside of mathematical circles to be the very pinnacle of achievement in mathematics.

The most recent national mathematics standards address these concerns, and their recommendations to incorporate many different mathematical topics (to some degree) at all levels might someday change this perception of mathematics, but today, college is likely to be the first place that a student finds out about entirely different kinds of mathematics. There are many courses to take after the calculus sequence which never mention calculus at all. There are even senior level courses which cover new material drawing on no specific mathematical topics the students have ever had. These are not reasons to avoid mathematics in college! On the contrary, you may find that you can enjoy mathematics even if calculus is not your cup of tea.

This discrete mathematics text is designed with two primary goals in mind. The first goal is to expose you to some mathematics that you have never seen before. The topics have not been chosen at random—we will try to see how these new mathematical topics can be applied to real problems and everyday situations. Our second goal is to develop your problem-solving skills. There is no better test of reasoning ability than to understand and give mathematical arguments based upon new material. We hope you will learn to value these skills regardless of your long-term career pursuits.

This book is also designed with your personal intellectual growth valued more highly than your exposure to as many topics as possible. Instead of trying to "go round the world" to learn about discrete mathematics, we focus instead on establishing a solid understanding of each thing we do. In this course, you will be asked to think, explore, write, and discuss in order to make connections between different abstract and concrete mathematical concepts. It is these skills that will be the measure of your success in the course.

The title of this book might give rise to skepticism about the usefulness of this subject. Let us therefore say a few words in defense of puzzles and games. The field of graph theory can be traced to the solution of a popular puzzle concerning crossing bridges in 1735. Today, this is one of the fastest growing areas of mathematics thanks to important ties with computer science and business. Games of chance gave birth to the now ubiquitous fields of probability and statistics. The ability to play chess remains an important test case for intelligence in computers. Mathematicians have shown how certain card tricks are related to the allocation of computer memory. In short, yesterday's games and amusements are among today's deepest mathematical subjects. Looking at it another way, it is also apparent that much of the mathematics that we encounter in our academic and worldly pursuits is not entirely different from "recreational" mathematics. Every math question is a puzzle. Every thought experiment is a game.

The book is organized into four parts. In the first part we will practice thinking and writing about mathematics. The context for most of this activity is in puzzles concerning logic and numbers. This part is built on the premise that logic is a natural function of the human brain, and that the real challenge for developing mathematical thinking rests with attaining comfort with formalization and abstraction and not so much with logic. This part of the course will build some important skills that we will subsequently use.

The second part of the course is an excursion into abstract mathematics via the study of sets and functions. This will primarily give us a setting in which to exercise our problem-solving and proof-writing skills. Another important point in this discussion however is the way that abstraction (like Boolean algebra) can actually make certain computations (as with logic circuits) easier instead of harder. In this part of the course, we also explore the concept of function and its application to problems that are quite different from those in a precalculus setting.

In the third part of the course, we will study probabilities and their applications to games of chance. In order to understand probability, we will have a somewhat lengthy introduction to the mathematical field of combinatorics or counting. The reason for this is simple. A basic premise of probability dictates that the probability of something happening (say rolling a 5 or greater on a toss of a fair die) is simply the ratio of the number of ways for the event to happen to the total number of ways in which any outcome may happen. In the example of rolling a 5 or greater on one toss of a die, there are two ways to have success (rolling a 5 or a 6) and six total equally-likely outcomes from a toss of a die, hence the probability of rolling a 5 or greater is $\frac{2}{6} = \frac{1}{3}$. Of course in real problems, the events are more complicated and the outcomes much more numerous. Therefore, in order to accomplish anything in probability, we will first have to learn some techniques of counting. Along the way, we will be sure to point out that combinatorics is a beautiful and valuable area of mathematics in its own right.

The fourth part of the course will be a closer look at puzzles and games. The route by which we will approach them will be the field of graph theory, which Gottfried Leibniz called the "geometry of position." Once we have seen some of the basic techniques of graph theory we will look in some detail at applications of this important area of mathematics.

There are several threads that run through the course. The two most obvious ones are recursion and applications. Believe it or not, recursive reasoning is not really something new to you. For example, if you have just determined that the sum of the first 100 positive integers is 5,050, and someone asks what the sum of the first 101 positive integers is, you will not start over but will rather quickly add $5050 + 101 = 5151$. The inherent laziness of the human brain that reuses previous information is at the heart of recursive reasoning. The first step in coming to terms with this idea is in realizing that it is a natural form of reasoning, not something invented by your math teacher.

The thread of real world applications is rather overtly presented throughout sections whose main heading is "Excursion." These sections are intended to provide a flavor of application related to some of the ideas developed in the chapter. The intention of this thread is for you to leave the course with the understanding that discrete mathematics is relevant to important problems.

So the big plan is to introduce you to mathematical writing, abstract structures, counting, discrete probability, and graph theory. Along the way you will meet some applications to puzzles and games and subsequently to problems in the real world. This course will probably seem different than most math courses you have had. Be prepared to work and be prepared to play, but most of all, be prepared to think.

"I see you stand like greyhounds in the slips,
Straining upon the start. The games afoot!"
— **William Shakespeare, in** *Henry V*

Contents

1 Puzzles, Patterns, and Mathematical Language   1

1.1 First Examples   1
1.2 Number Puzzles and Sequences   9
1.3 Truth-tellers, Liars, and Propositional Logic   24
1.4 Predicates   40
1.5 Implications   53
1.6 Excursion Validity of Arguments   68

2 A Primer of Mathematical Writing   81

2.1 Mathematical Writing   82
2.2 Proofs About Numbers   98
2.3 Mathematical Induction   110
2.4 More About Induction   122
2.5 Contradiction and the Pigeonhole Principle   132
2.6 Excursion Representations of Numbers   150
2.7 Excursion Modular Arithmetic and Cryptography   166

3 Sets and Boolean Algebra   181

3.1 Set Definitions and Operations   181
3.2 More Operations on Sets   198
3.3 Proving Set Properties   210
3.4 Boolean Algebra   221
3.5 Excursion Logic Circuits   229

4 Functions and Relations   248

4.1 Definitions, Diagrams, and Inverses   249
4.2 The Composition Operation   268
4.3 Properties of Functions and Set Cardinality   283
4.4 Properties of Relations   301
4.5 Equivalence Relations   313
4.6 Numerical Functions in Discrete Math   324
4.7 Excursion Iterated Functions and Chaos   334
4.8 Excursion Growth of Functions   345

5 Combinatorics   368

5.1 Introduction   369
5.2 Basic Rules for Counting   386
5.3 Combinations and the Binomial Theorem   398
5.4 Binary Sequences   408
5.5 Recursive Counting   418
5.6 Excursion Solving Recurrence Relations   423

6 Probability   440

6.1 Introduction   440
6.2 Sum and Product Rules for Probability   448
6.3 Probability in Games of Chance   460
6.4 Expected Value in Games of Chance   466
6.5 Excursion Recursion Revisited   475
6.6 Excursion Matrices and Markov Chains   482

7 Graphs and Trees   505

7.1 Graph Theory   506
7.2 Proofs About Graphs and Trees   519
7.3 Isomorphism and Planarity   533
7.4 Connections to Matrices and Relations   546
7.5 Graphs in Puzzles and Games   567
7.6 Excursion Binary Trees   581
7.7 Excursion Hamiltonian Cycles and the TSP   596

A Rules of the Game   613

Cards   613
Sports   614
Miscellaneous Games   615

B Matrices and Their Operations   618

Matrix Operations   618
Matrix Arithmetic with Technology   620

Selected Answers and Hints   625

References and Further Reading   682

1 | Puzzles, Patterns, and Mathematical Language

Ask five mathematicians what mathematics is, and you might very well get five different answers, but there will no doubt be some underlying themes. Mathematics is about solving problems. Mathematics explains patterns. Mathematics is a set of statements deduced logically from axioms and definitions. Mathematics uses abstraction to model the real world. This introductory chapter lays a foundation for our study of discrete mathematics by addressing all these themes. In the second section, we will examine patterns in number sequences, and in particular, we will discuss the idea of a "recursive model" that will be important throughout the book. The subsequent sections address basic logic particularly as it applies to clarity of language. We are accustomed to people not saying exactly what they mean in conversation, and as a result, we are good at interpreting statements from context, but there is no room for ambiguity in mathematics. We end the chapter with a discussion of validity of the kinds of arguments that one might encounter in life.

1.1 First Examples

In this section we will take a first look at some of the puzzles and games that will occupy us through the weeks ahead. We will not yet worry about the mathematics that lies behind these puzzles and games. The mathematical ideas will confront us soon enough. At this point, you should just immerse yourself in these examples to get a feeling for the type of problems we will be studying in this book. Make some time to perform the magic trick on a roommate, discuss the Josephus problem in class, solve the puzzles, and play the games. This will make the mathematics more meaningful to you as we get to each of the relevant topics behind these examples.

A Magic Trick

At a recent magic show, the following trick was performed. Four playing cards* were stacked face up, with a heart at the bottom, then a club, then a diamond, and then a spade. A spectator held this packet while the blindfolded magician gave these instructions:

1. Turn the spade (the uppermost card) face down.
2. Move any number of cards, one at a time, from the top of the packet to the bottom.
3. Turn over the top two cards as one.
4. Move any number of cards, one at a time, from the top of the packet to the bottom.
5. Turn over the top two cards as one.
6. Move any number of cards, one at a time, from the top of the packet to the bottom.
7. Either turn over the entire stack or do not—your choice.
8. Turn over the topmost card.
9. Turn over the top two cards as one.
10. Turn over the top three cards as one.

At this point, the prestidigitator, while still blindfolded, correctly divined the state of the stack. The club was the only card facing the opposite way from the others!

This trick takes advantage of the fact that the actions that seem like shuffling to the spectator are actually preserving all the properties the magician cares about. We will see how to verify this "invariance of properties" in Section 2.4.

Example 1 *Write down the initial state of the deck (from bottom to top) as* H C D S. *Using uppercase letters to indicate the card is face up and lowercase for face down, we see that after step 1 the deck would be* H C D s. *Starting from here, trace the state of the deck through step 10 if in steps 2, 4, and 6, we move exactly two cards from the top to the bottom, and in step 7 we turn over the packet.*

SOLUTION   Using the notation described, we can show the state of the packet after each step.

Step 1.	H C D s	Step 6.	S d c h
Step 2.	D s H C	Step 7.	H C D s
Step 3.	D s c h	Step 8.	H C D S
Step 4.	c h D s	Step 9.	H C s d
Step 5.	c h S d	Step 10.	H D S c

□

---

* The reader who is unfamiliar with the anatomy of a deck of playing cards should consult Appendix A.

Practice Problem 1   (**Note.** Practice problems allow the reader to stop and try an idea before moving on. The answers to all practice problems are given at the end of the section.) *The following questions refer to the magic trick described in Example 1.*

(a) *Repeat Example 1 but this time in step 7, do not turn over the deck.*

(b) *Repeat Example 1, but this time in steps 2, 4, and 6, move one, two, and three cards (respectively) from the top to the bottom, and in step 7 do not turn over the deck.*

A Matter of Life and Death

In [32], the following legend about the first-century historian Flavius Josephus is recounted:

> *In the Jewish revolt against Rome, Josephus and 39 of his comrades were holding out against the Romans in a cave. With defeat imminent, they resolved that, like the rebels at Masada, they would rather die than be slaves to the Romans. They decided to arrange themselves in a circle. One man was designated as number one, and they proceeded clockwise killing every seventh man. . . . Josephus (according to the story) was among other things an accomplished mathematician; so he instantly figured out where he ought to sit in order to be the last to go. But when the time came, instead of killing himself he joined the Roman side.*

The solution is for Josephus to stand in the 24th position in the circle. It is yet another historical example of how those with a distaste for mathematics quickly become the chaff of evolution. That point aside, the problem rightfully raises the question of how Josephus could have quickly computed this position.

We will refer to this scenario as the "Josephus game" even though it does not sound like fun. We call it a game because it is very similar to methods that children use to decide things on the playground. Perhaps we should imagine the soldiers saying (in Latin, of course):

> *One potato, Two potato, Three potato, Four; Five potato, Six potato, Seven potato, More*

On the "More" of course, someone would be killed, which would be a shame, but it would certainly eliminate that person. This is much like the Josephus game where one skips by eight people between those who are eliminated instead of seven. Another playground example is "eeny-meeny-miney-moe," where the skip number is 16.

Here are some samples of the many questions one can ask about this type of game. We will begin an examination of the first question in the practice problems and exercises of this section. The others are more difficult. In fact, no one knows the answer to the last one.

1. If every second person is killed (instead of every seventh) when there are initially *n* people in the circle, where should Josephus stand?

2. In some versions of the story, Josephus saves a friend by having the friend stand in a position so that the two of them are the last two people left alive. If every second

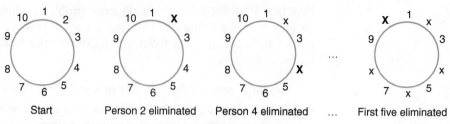

Figure 1-1   The beginning of the Josephus "game."

person is killed (instead of every seventh) when there are initially *n* people in the circle, where should *Josephus's friend* stand?

3. If Josephus is not allowed to move positions but is asked what the "skip number" should be (say Monty Hall is a Roman soldier, for instance), can he always respond in a way so that he will live? What about his friend?

4. Characterize all sets of people in the original circle who can be made to be the last people living by naming an appropriate skip number. In other words, can Josephus save any number of friends in any positions by naming an appropriate skip number?*

The Josephus problem is the basis for introducing recursive modeling in [29] as well as in [18]. In [32] the concept of *Josephus permutation* is introduced so that the entire killing process is captured instead of just the grisly end result. Later in this book, we will explore some aspects of Josephus permutations using recursion and induction.

Explore more on the Web.

Example 2   *If there are 10 people numbered 1 to 10 in a circle, and every other person is eliminated starting with person number 2, which person is the last standing?*

SOLUTION   In Figure 1-1, we use the letter x to represent dead people and a capital **X** to highlight the most recent person eliminated.

In the interest of space, we will write the state of the game in a line rather than draw the circle each time. For example, after the first five people have been eliminated, we can write 1 x 3 x 5 x 7 x 9 **X** instead of drawing the last picture above, and we can still tell from the placement of the **X** that the next person to be eliminated is number 3. If we use this system, the remainder of the game proceeds like this:

● 1 x **X** x 5 x 7 x 9 x
● 1 x x x 5 x **X** x 9 x
● **X** x x x 5 x x x 9 x
● x x x x 5 x x x **X** x

Hence, person number 5 is the last one standing in this game.   ☐

---

* This is given as a "Research problem" in [29].

Practice Problem 2

(a) *If there are nine people numbered 1 to 9 in a circle, and every other person is eliminated starting with person number 2, which person is left?*

(b) *If there are eight people numbered 1 to 8 in a circle, and every other person is eliminated starting with person number 2, which person is left?*

(c) *If there are sixteen people numbered 1 to 16 in a circle, and every other person is eliminated starting with person number 2, which person is left?*

(d) *Explain why the answer to (c) is the same as for (b).*

It's Just a Game

Sporting events present an interesting challenge for mathematical modeling since in sports lies a blend of strategy, skill, and luck that is hard to separate. We will simplify the analysis by using probabilities to simulate all three aspects. For example, if a baseball player bats 0.300 for a season, we will use $\frac{3}{10}$ as the probability of that player getting a hit in any given at-bat. This is certainly an oversimplification—in a given at-bat, the probability of a player getting a hit could be higher or lower depending on who is pitching, how many outs there are, or what he had for breakfast. However, using $\frac{3}{10}$ as if it is the real probability makes our calculations feasible, and it is a reasonable simplification.

We will look at two kinds of sports events in this course, each of which is characterized by the following examples from tennis:

1. In a certain tennis league, the first player to win two sets wins the match. If Player A has a 55% chance of winning a set against Player B, what is the probability that Player A wins the match? If these two players played many matches, what would you expect to be the average number of sets that determine a match?

2. In tennis, each player's score progresses from 0 to 15 to 30 to 40 to *game* with the wrinkle that a final score of *game* to 40 is impossible. A score of 40–40 is called "deuce," and a game can only be won from deuce if one player wins two consecutive points. (This is equivalent to the rule of "winning by two" in other sports.) If Player A has a 60% chance of winning a **point** against Player B, what is the probability that Player A wins the game? If these two players played many games, what would you expect to be the average number of points that determine a game?

The fundamental difference between these two examples is that the first one can be analyzed by exhaustively cataloging all possible matches, but the second one cannot because there are infinitely many possible ways that a game could progress. In Chapter 6, we will see how to handle each type of problem.

Practice Problem 3 *Suppose that Player A and Player B play a tennis match, where Player A has a 55% chance of winning any given set and the first player to win two sets wins the match. (This is called a "best-of-three" match.) One example of how a match could go is, "A wins the first set, B wins the second, and A wins the third." This sequence of events can be represented more briefly by the simple string ABA. Every outcome of a match has a similar representation.*

(a) *Suppose that Player A and Player B play a best-of-three match. List in an organized manner (using the representation described above) all the different ways the match could go.*

(b) *Is every way you listed in Part 1 equally likely to occur? In particular, which do you think is more likely, that A wins in two sets, or that B wins in two sets?*

(c) *Suppose that Player A and Player B play a single game that lasts for eight points. One way this could happen is for the points to be won in this order: ABABABAA. Give two other ways this could happen.*

(d) *Explain why a single game cannot last for exactly seven points.*

(e) *In an organized manner, list all the different ways Player A can win in six or fewer points.*

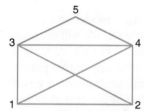

Figure 1-2 A simple envelope.

An Elementary Puzzle

It is not too difficult to draw the picture in Figure 1-2 without lifting your pencil from the paper and without retracing any lines. It is not as easy to decide whether the picture in Figure 1-3 can be drawn in the same way, but it can.

We will see how to make this decision for any picture without resorting to hours of trial and error. Surprisingly, this is related to an eighteenth-century puzzle that was solved by Leonhard Euler. Perhaps even more unexpected is that there are present-day applications of the same idea in many hard problems which have to do with scheduling tasks and constructing networks.

Explore more on the Web.

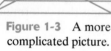 **Example 3** *Starting at the lower left corner of the envelope (position 1 in Figure 1-2), describe a way to draw the envelope of Figure 1-2 without lifting your pencil and without retracing any lines.*

SOLUTION   One possibility is to go up, then across the top, do the "flap," go diagonally to the lower right corner, then up, diagonally to the lower left corner, and finally go across the bottom. Using the numbering in Figure 1-2, we can represent this path by the list of corners we pass through:

$$1, 3, 4, 5, 3, 2, 4, 1, 2$$

□

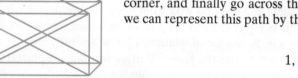

Figure 1-3 A more complicated picture.

Practice Problem 4

(a) *If you have not already done so, find a different solution to Example 3 than the one given.*

(b) *Do you think it is possible to start at the upper left corner (position 3)?*

A Game of Skill?

Play this game with a friend. Start with a 4 × 4 grid as shown in Figure 1-4. Players alternate turns where on each turn a player

Figure 1-4
The grid game.

1. Chooses a row or column that has an empty space, and

2. Places new X's anywhere in that row or column.

X		X	X
X	X	X	
			X
X	X	X	X

X	X	X	
X	X	X	
X			X
X		X	X

Figure 1-5 Grids for Practice Problem 5.

At least one new X must be placed on each turn, and the person to X the last available space wins. Does either player have a foolproof strategy for winning this game?

In Chapter 7, we will see why one player *does* always have a winning strategy in a broad category of similar games. We will also see how this fact does not help in most real-world examples, since the strategy is often too complicated to implement effectively.

Explore more on the Web.

Practice Problem 5

(a) *It's your turn to move, and the board looks like the grid on the left in Figure 1-5. Describe a winning strategy.*

(b) *It's your turn to move, and the board looks like the grid on the right in Figure 1-5. Can you win if your opponent plays intelligently?*

Solutions to Practice Problems

1 We use the same notation as in Example 1, showing the state of the deck after each step.

(a) If in steps 2, 4, and 6, we move exactly two cards from the top to the bottom, and in step 7 we do not turn over the packet, the trick would proceed as follows:

Step 1.	H C D s	Step 6.	S d c h
Step 2.	D s H C	Step 7.	S d c h
Step 3.	D s c h	Step 8.	S d c H
Step 4.	c h D s	Step 9.	S d h C
Step 5.	c h S d	Step 10.	S c H D

(b) If in steps 2, 4, and 6, we move one, two, and three cards (respectively) from the top to the bottom, and in step 7 we do not turn over the deck, the trick would proceed as follows:

Step 1.	H C D s	Step 6.	c h S d
Step 2.	s H C D	Step 7.	c h S d
Step 3.	s H d c	Step 8.	c h S D
Step 4.	d c s H	Step 9.	c h d s
Step 5.	d c h S	Step 10.	c S D H

2 We will use the same notation as in Example 2.

(a) After the first four eliminations we have 1 x 3 x 5 x 7 X 9 and next in line is 1. The next four eliminations will go as follows:

— X x 3 x 5 x 7 x 9
— x x 3 x X x 7 x 9
— x x 3 x x x 7 x X
— x x 3 x x x X x x

(b) After the first four eliminations, we have 1 x 3 x 5 x 7 X, with 3 next in line. Then 3, 7, and 5 go, leaving 1.

(c) After the first eight eliminations, we have 1 x 3 x 5 x 7 x 9 x 11 x 13 x 15 X, with 3 next in line. After the next four eliminations, we have

1 x x x 5 x x x 9 x x x 13 x X x

with 5 next in line. Then 5, 13, and 9 go, leaving 1.

(d) After the first eight eliminations, there are eight people left (1, 3, 5, 7, 9, 11, 13, 15) with the second one next in line for extinction. This is really the same problem as starting with eight people, with the second one the first to go.

3 (a) You can write them out as AA, ABA, ABB, BAA, BAB, BB. Alternatively, the "game tree" in Figure 1-6 provides a handy visualization of the solutions.

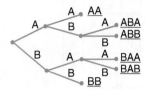

Figure 1-6   A game tree for Practice Problem 3.

First Three Points	Conclusion
AAA	A
	BA
	BBA
AAB	AA
	ABA
	BAA
ABA	AA
	ABA
	BAA
ABB	AAA
BAA	AA
	ABA
	BAA
BAB	AAA
BBA	AAA

Table 1-1   Table for Practice Problem 3

(b) Because A is more likely to win each set, it seems reasonable that A is more likely to win two sets in a row than is B.

	X	X	X	
X	X	X		
X	X		X	
X		X	X	

Figure 1-7   Grid game in Practice Problem 5.

(c) ABBABAAA and BBBAAAAA are two of many possibilities.

(d) Here are the scores that total to seven points with the winner ahead by at least two points: 7–0, 6–1, 5–2. None of these are possible, since in each case the game would have ended when the winner reached four points.

(e) One way to organize is to think about the first three points, which can be either AAA, AAB, ABA, ABB, BAA, BAB, BBA, BBB. For each of these, we list in Table 1-1 all the possible conclusions of the game that result in a win by A.

4 (a) If we use the same notation as in Example 3, a different way to draw the envelope can be represented as

$$1, 2, 4, 5, 3, 1, 4, 3, 2$$

(b) In the graph theory chapter, we will see that this is not possible.

5 (a) By completely filling row 3, you leave two spaces, and your opponent cannot X out both at once.

(b) If you completely fill in any column or any row, your opponent can win. Likewise, if you fill in row 3, column 3 or row 4, column 2, then your opponent has a winning strategy. However, if you fill in row 3, column 2, you have the board in Figure 1-7 and you are guaranteed a win.

Exercises for Section 1.1

Exercises numbered in bold face blue have answers or partial answers in the back of the book.

1. Suppose we use $C(a, b, c, d)$ to mean the result of doing the card trick with a, b, c, and d describing what to do in steps 2, 4, 6, and 7, respectively. For example, $C(2, 2, 2, no)$ would mean to move two cards to the bottom at each of steps 2, 4, and 6, and then to not turn over the deck in step 7. Find the final configuration of the deck for each of the following games:

    (a) $C(1, 1, 1, no)$

    (b) $C(1, 1, 1, yes)$

    (c) $C(3, 3, 3, no)$

    (d) $C(3, 2, 1, yes)$

2. Explain why in the card trick, after each of steps 2–7, the packet of cards contains exactly one card reversed from the others.

3. Suppose we use $J(p, s)$ to mean the Josephus game with p people and a skip amount of s. For example, $J(10, 2)$ means the game with 10 people in which every other person is eliminated, starting with person 2. Play the following Josephus games to determine the winner. Also decide where Josephus's friend should stand—that is,

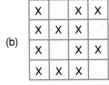 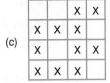

Figure 1-8 Grids for Exercise 12.

who is the last person who would be eliminated before the final survivor?

(a) $J(15, 3)$

(b) $J(32, 2)$

(c) $J(15, 2)$

(d) $J(5, 4)$

4. Suppose a tennis match is a "best of five" rather than a "best of three." Draw the complete game tree. If each of the outcomes is equally likely, what percentage of the time would you expect the match to last five sets?

5. We showed that a tennis game cannot consist of exactly seven points. What other numbers of total points are impossible?

6. Draw a game tree to show the fifteen ways in which Player A can win a tennis game in six or fewer points. (See Practice Problem 3.)

7. Suppose you toss three coins, a nickel, dime, and quarter, and record the results in that order. For example, HTH would mean head on the nickel, tails on the dime, and heads on the quarter.

(a) In a systematic way, list all the different results you could record.

(b) Draw a game tree for the recording of the results.

(c) On the game tree, label each possible result either 0, 1, 2, or 3, indicating how many heads it has. Do you think a person who tosses three coins is more likely to get all three heads, or to get exactly two heads?

8. Suppose you take a true-false test with three questions, and you answer all the questions.

(a) In a systematic way, write all the different answers you could give. Use the representation where, for example, TFF means answering true for Question 1, false for Question 2, and false for Question 3.

(b) Draw a game tree for the different answers.

(c) Suppose the correct answers on the test are TFT. On your game tree, label each possible result either 0, 1, 2, or 3, indicating how many correct answers it has. Do you think a person who guesses is more likely to get exactly one right, or to get exactly two right?

9. Explain the similarity between your answers in Exercises 7 and 8.

10. Explain the similarity between the answers in Exercises 7 and 8 and the list of how the first three points could go in a tennis game between Player A and Player B.

11. In solutions to the envelope puzzle, are there any other places on the envelope where it would be possible to start? After you've investigated this, make a complete list of where you were able to start, and for each of these write down where you finished. What do you notice?

12. Consider the grid game from Practice Problem 5. For each of the grids in Figure 1-8, if it's your turn, can you make a play that guarantees you a win no matter what your opponent does, or can your opponent win no matter what you do? Explain.

1.2 Number Puzzles and Sequences

Guess the Next Number

One reason we focus on puzzles and games in this course is that much of mathematics seems like a puzzle or a game when viewed the right way. Most of you are familiar with the following type of mathematical puzzle.

What is the next number in each of the following lists?

1. 5, 7, 9, 11, 13, _____

2. 1, 9, 17, 25, 33, 41, _____

3. 1, 4, 9, 16, 25, 36, _____
4. 2, 4, 8, 16, 32, 64, _____
5. 1, 2, 6, 24, 120, 720, _____

The key to this type of puzzle is to find a pattern in the sequence of numbers. By "pattern," we mean one of three things:

(i) Each term is related (by arithmetical operations) to previous terms.

(ii) Each term can be described relative to its position in the sequence.

(iii) The sequence merely enumerates a set of integers that the reader must recognize from the scant information given.

Occasionally, a sequence can be described in all three ways, as we see in our first example.

Example 1 *How can the sequence* 1, 3, 5, 7, 9, . . . *be described in each of these ways?*

SOLUTION

(i) Each term is 2 more than the previous term and the first term is 1. When we describe a sequence in this way, we have to describe not only how each term is related to the previous term(s), but also how the sequence begins. For example, if we simply say, "Each term is 2 more than the previous term," that could also describe the sequence 202, 204, 206, 208, ... The stipulation that the first term is 1 guarantees that we have a complete description.

(ii) The nth term is given by the formula $(2n - 1)$. So the first term is $2 \cdot 1 - 1 = 1$, the second term is $2 \cdot 2 - 1 = 3$, and so on.

(iii) The sequence merely enumerates the positive odd integers.

□

In this course, we will be concerned primarily with characterizations of the first two types. One reason we are interested in both types of characterizations is that each has its own particular strengths. The first type of description is often easier to spot and describe, such as in the sequence 1, 1, 2, 3, 5, 8, . . . , where each term is the sum of the two previous terms.

On the other hand, the second type of description means having a formula for calculating each term relative to its position in the sequence, and this makes it easy to calculate a particular term. For example, if the nth term of some sequence is given by the formula $3n - 7$, we can quickly determine that the 100th term in the list is $3(100) - 7 = 293$.

Practice Problem 1 *Consider the sequence* 4, 6, 8, 10, 12, ...

(a) *Describe the sequence in each of the three ways we have mentioned.*

(b) *If you know a certain term of this sequence is 898, what would the next three terms be? Which type of description is most useful for answering this question?*

(c) What is the 1,000th term of the sequence? Which type of description is most useful for answering this question?

Sequences and Sequence Notation

Before we examine some more complicated examples, we will introduce some notation and terminology.

> **Definition**   A **recursive formula** for a sequence is a formula where each term is described in relation to a previous term (or terms) of the sequence. This type of description must include enough information on how the list begins for the recursive relationship to determine every subsequent term in the list. This is sometimes called a **recurrence relation.**

> **Definition**   A **closed formula** for a sequence is a formula where each term is described only in relation to its position in the list.

> **Definition**   [**Sequence notation**] We usually use lowercase letters (a, b, etc.) to name sequences, and we use subscripting to indicate position in a sequence. The notation a_n indicates the nth term of the sequence we are writing as a. We read a_n as "a subscript n," or more usually just "a sub n."

 Example 2 *In the sequence of numbers $1, 3, 5, 7, 9, \ldots$,*

$$a_1 = 1 \text{ means "the first term in the sequence is 1"}$$
$$a_2 = 3 \text{ means "the second term in the sequence is 3"}$$
$$a_3 = 5 \text{ means "the third term in the sequence is 5"}$$
$$a_4 = 7 \text{ means "the fourth term in the sequence is 7"}$$

*and so on. A **closed formula** for this sequence is*

$$a_n = 2n - 1$$

In words, this says, "the nth term of the sequence is given by the formula $2n - 1$."
    *A **recursive formula** for this sequence is*

$$a_1 = 1 \qquad and \qquad a_n = a_{n-1} + 2$$

In words, this says, "the first term in the sequence is 1," and "each term in the sequence can be found by adding 2 to the previous term."

Sequence notation is quite similar to the function notation you are familiar with from algebra courses. When we write $a_n = 2n - 1$, it is similar to writing $f(x) = 2x - 1$. In each case, the formula gives a rule for calculating the result for a particular given number. Just as $f(15)$ would be calculated by substituting 15 for x in the formula $2x - 1$, so a_{15} is found by substituting 15 for n in the formula $2n - 1$. So $a_{15} = 2 \cdot 15 - 1 = 30 - 1 = 29$.

In this notation, we think of the subscripts as *ordinal numbers*. The ordinal numbers are *first, second, third, fourth, . . . , one-hundred-twenty-first*, and so on. They indicate ordering within a list—hence the term ordinal. So when we see a subscript, we can interpret it as a position within the sequence. For example, a_{21} indicates the 21st term in the sequence. Likewise, a_n is the nth term in the sequence, and a_{n-1} would be the $(n-1)^{th}$ term, that is, the term just before the nth term. So when we write $a_n = a_{n-1} + 2$, we can read it as "the nth term in the sequence is equal to the $(n-1)^{th}$ term plus 2." Since the $(n-1)^{th}$ term is the one immediately before the nth term, we sometimes just describe it as "the previous term."

In the preceding discussion, it is implicit that the subscripts start at $n = 1$. We will use this convention throughout our discussion of numerical sequences. For example, when we write the closed formula

$$a_n = 2n - 1$$

it will mean the same as writing

$$a_n = 2n - 1, \quad \text{for integers } n \geq 1$$

If for some reason we decide to begin with a subscript other than 1, we will carefully indicate that fact.

On the other hand, for a recursive formula we explicitly give some initial terms in the sequence, so the formula typically starts with the first term not explicitly given. So when we write the recursive formula

$$a_1 = 1 \quad \text{and} \quad a_n = a_{n-1} + 2$$

we mean that the recursive relationship holds for all values of n beyond the explicitly given term for $n = 1$. That is, this is shorthand for

$$a_1 = 1 \quad \text{and} \quad a_n = a_{n-1} + 2, \quad \text{for integers } n \geq 2$$

The following examples illustrate more about the relationship between closed formulas and recursive formulas for sequences of numbers.

Example 3 *For the sequence whose closed formula is $a_n = 2^n - 1$, do the following:*

1. *List the first five terms of the sequence.*
2. *Calculate the value of the 20th term.*
3. *Give a formula for the $(k+1)^{th}$ term.*
4. *Give a formula for a_{2i-3}.*

SOLUTION

1. $a_1 = 2^1 - 1 = 1$, $a_2 = 2^2 - 1 = 3$, $a_3 = 2^3 - 1 = 7$, $a_4 = 2^4 - 1 = 15$, $a_5 = 2^5 - 1 = 31$.

2. $a_{20} = 2^{20} - 1 = 1{,}048{,}575$.

3. $a_{k+1} = 2^{k+1} - 1$. Just as we find a_{20} by replacing n with 20 in the formula, we find a_{k+1} by replacing n with $k+1$ in the formula.

4. We solve this by replacing n with $2i - 3$ in the formula, so $a_{2i-3} = 2^{2i-3} - 1$.

□

Example 4 *Consider the sequence whose recursive formula is* $a_1 = 11$ *and* $a_n = a_{n-1} + 5$.

1. Write the first five terms of the sequence.
2. Write a recursive formula for the 80th term.
3. Write a recursive formula for the $(k+1)^{\text{th}}$ term.
4. Write a recursive formula for a_{3j-2}.

SOLUTION

1. We can solve the problem by plugging into the recursive formula as follows:

$$a_1 = 11$$
$$a_2 = a_{2-1} + 5 = a_1 + 5 = 11 + 5 = 16$$
$$a_3 = a_{3-1} + 5 = a_2 + 5 = 16 + 5 = 21$$
$$a_4 = a_{4-1} + 5 = a_3 + 5 = 21 + 5 = 26$$
$$a_5 = a_{5-1} + 5 = a_4 + 5 = 26 + 5 = 31$$

Alternatively, once we realize that the recursive formula says, "Each term is equal to the previous term plus 5," we can simply write down the sequence as 11, 16, 21, 26, 31.

2. In the formula $a_n = a_{n-1} + 5$, we replace n with 80, giving $a_{80} = a_{80-1} + 5$, which simplifies to $a_{80} = a_{79} + 5$. In other words, the 80th term is 5 more than the 79th term.

3. We replace n with $k+1$, giving $a_{k+1} = a_{(k+1)-1} + 5$, which simplifies to $a_{k+1} = a_k + 5$. Notice that a_{k+1} is the next term in the sequence after a_k. We can express this formula in words as "the next term is equal to this term plus 5," which is all the original description says.

4. We replace n with $3j - 2$, giving $a_{3j-2} = a_{(3j-2)-1} + 5$, which simplifies to $a_{3j-2} = a_{3j-3} + 5$. $\square$

Discovering Patterns in Sequences

We will now return to the original problems in this section. We will try to do two things with these puzzles. First, we will determine the next number in each sequence. To do this, we will have to discover a pattern for the sequence. We will then try to use the discovered pattern to describe the sequence with either a closed formula or a recursive formula (or perhaps both). Here are the puzzles again:

1. 5, 7, 9, 11, 13, _____
2. 1, 9, 17, 25, 33, 41, _____
3. 1, 4, 9, 16, 25, 36, _____
4. 2, 4, 8, 16, 32, 64, _____
5. 1, 2, 6, 24, 120, 720, _____

Before reading on, try to discover the pattern and determine the next number in each sequence. Then we will discuss the answers.

Explore more on the Web.

 Example 5 *Give both closed and recursive formulas for the sequence that begins*

$$5, 7, 9, 11, 13$$

SOLUTION   It appears that the sequence lists odd numbers in order, so the next number would be 15. We observe that each term is 2 more than the previous term. It is not too hard to use this fact to develop the recursive formula. We just have to remember that the "previous" number in the sequence can be expressed as a_{n-1}. The recursive description is

$$a_1 = 5 \quad \text{and} \quad a_n = a_{n-1} + 2$$

The closed formula requires a bit more thought. One approach that works well is to think of an easier sequence (i.e., one you know a closed formula for) with the same recursive pattern, and compare this sequence to the original. In this case, the sequence $2, 4, 6, 8, \ldots$ has the easy closed formula $b_n = 2n$ and the same recursive pattern, "each term is 2 more than the previous term." In Table 1-2 we line up the sequence whose closed formula we know with the sequence whose closed formula we wish to find. We notice that each term in the bottom row is 3 more than the corresponding term above it. That is, $a_n = b_n + 3$. Therefore, the closed formula for a_n is

n	1	2	3	4	...
b_n	2	4	6	8	...
a_n	5	7	9	11	...

Table 1-2   Table of Values for Example 5

$$a_n = 2n + 3$$

$\square$

 **Example 6**   *Give both closed and recursive formulas for these sequences.*

1. The sequence that begins 1, 9, 17, 25, 33, 41.
2. The sequence that begins 1, 4, 9, 16, 25, 36.
3. The sequence that begins 2, 4, 8, 16, 32, 64.
4. The sequence that begins 1, 2, 6, 24, 120, 720.

SOLUTION

1. The obvious pattern here is that each term is 8 more than the previous term, so the recursive formula is easy to write

$$a_1 = 1 \quad \text{and} \quad a_n = a_{n-1} + 8$$

An easy sequence with this same pattern is $b_n = 8n$, and we can compare this with the original as shown in Table 1-3. Each term in the bottom row is 7 less than the term above it, so the closed formula is $a_n = 8n - 7$.

n	1	2	3	4	...
b_n	8	16	24	32	...
a_n	1	9	17	25	...

Table 1-3   Table of Values for Example 6

2. By now we may be in the habit of looking at differences between successive terms. For the sequence 1, 4, 9, 16, 25, 36, however, those differences change. But there is a pattern in the differences:

$$a_2 = a_1 + \boxed{3}$$
$$a_3 = a_2 + \boxed{5}$$
$$a_4 = a_3 + \boxed{7}$$

$$a_5 = a_4 + \boxed{9}$$
$$a_6 = a_5 + \boxed{11}$$
$$a_7 = a_6 + \boxed{??}$$

This might lead us to conclude that a_7 will be 13 more than a_6. Can we express the difference in terms of n? For $n = 2$ (i.e., when the subscript on the left-hand side of the equation is 2), the number in the box on the right-hand side of the equation is 3. When $n = 3$, the boxed number is 5, and so on. Following this pattern, we conclude that when the subscript on the left-hand side is n, the number in the box has formula $2n - 1$. This leads to our recursive formula for the sequence,

$$a_1 = 1 \quad \text{and} \quad a_n = a_{n-1} + (2n - 1)$$

For the closed formula, we need to find a pattern in the numbers themselves. In this case it helps to recognize that the sequence can be described verbally as "those positive integers that are perfect squares" ($1^2 = 1, 2^2 = 4, 3^2 = 9, \ldots$). This allows us to write the closed formula

$$a_n = n^2$$

3. For the sequence that begins 2, 4, 8, 16, 32, 64, each successive term is twice as large as the previous, so the next term will be 128. The recursive formula is therefore

$$a_1 = 2 \quad \text{and} \quad a_n = 2a_{n-1}$$

For the closed formula, we recognize that the sequence consists of "integers that are a power of 2," beginning at $2 = 2^1, 4 = 2^2, 8 = 2^3$, and we write

$$a_n = 2^n$$

4. When we analyze the sequence that begins 1, 2, 6, 24, 120, 720, we may get stuck for a while. The differences between the terms are 1, 4, 18, 96, 600, and there is no obvious pattern here. Having just worked a problem where each term is the previous term **times** some factor, we look at this to try to detect a pattern:

$$a_2 = a_1 \cdot \boxed{2}$$
$$a_3 = a_2 \cdot \boxed{3}$$
$$a_4 = a_3 \cdot \boxed{4}$$
$$a_5 = a_4 \cdot \boxed{5}$$
$$a_6 = a_5 \cdot \boxed{6}$$
$$a_7 = a_6 \cdot \boxed{??}$$

It appears that a_7 will be a_6 times 7, or 5,040. The recursive formula is

$$a_1 = 1 \quad \text{and} \quad a_n = a_{n-1} \cdot n$$

For the closed formula, notice that $a_1 = 1, a_2 = 1 \cdot 2, a_3 = 1 \cdot 2 \cdot 3, a_4 = 1 \cdot 2 \cdot 3 \cdot 4$, and so on. We write

$$a_n = 1 \cdot 2 \cdot 3 \cdot \ldots \cdot n$$

Note: This "product of the integers from 1 to n" occurs in a variety of mathematical contexts. In fact, mathematicians have created a special notation for it, namely $n!$, which is read as "n factorial." □

Practice Problem 2   *For each sequence, identify the next number in the sequence. Then give either a closed formula, a recursive formula, or a verbal description. Give more than one description when you can.*

(a) 5, 10, 15, 20, 25, _____

(b) 5, 10, 20, 40, 80, _____

(c) 4, 9, 16, 25, 36, _____

Sometimes each term in a sequence is calculated from more than one of the previous terms. A simple example occurs in the famous *Fibonacci numbers*. This sequence was developed by Leonardo Pisano Fibonacci (1170–1250) as a recreational problem about the growth of rabbit populations, but it turns out to have wide-ranging applications. The sequence begins

$$1, 1, 2, 3, 5, 8, 13, 21, 34, \ldots$$

The sequence follows the rule that, starting with the third term, each term is the sum of the two terms that precede it. For example, $3 = 2 + 1, 5 = 3 + 2, 8 = 5 + 3$. If we use F_n to represent the nth term of this sequence, this relationship can be stated succinctly as

$$F_n = F_{n-1} + F_{n-2}, \quad \text{for } n \geq 3$$

where the initial conditions $F_1 = F_2 = 1$ must also be given to exactly describe the sequence. We should note that there *is* a formula that produces the nth Fibonacci number:

$$F_n = \frac{1}{\sqrt{5}}\left(\left(\frac{1+\sqrt{5}}{2}\right)^n - \left(\frac{1-\sqrt{5}}{2}\right)^n\right)$$

but it is not at all obvious from a casual examination of the sequence.

Example 7   *Discover a closed formula for the sequence whose recursive formula is $a_1 = 2$, and $a_n = 3a_{n-1}$.*

SOLUTION   To discover the formula, we first write down the first few terms of the sequence:

$$a_1 = 2, \quad a_2 = 6, \quad a_3 = 18, \quad a_4 = 54$$

In Table 1-4, we compare this to the simpler sequence $b_n = 3^n$, which also has the recursive pattern, "Each term is 3 times the previous term." Observe that each term in the bottom row is 2/3 times the corresponding term in the b_n row. Therefore,

$$a_n = \frac{2}{3} \cdot 3^n = 2 \cdot 3^{n-1}$$

n	1	2	3	4	...
b_n	3	9	27	81	...
a_n	2	6	18	54	...

Table 1-4 Table of Values for Example 7

Perhaps a more obvious connection between the two rows is that each term in the a_n row is twice the *previous* term in the b_n row. This relationship can be literally transcribed, $a_n = 2 \cdot b_{n-1}$, which results in the same closed formula as above. ◻

Discovering closed formulas is much more difficult than finding recursive formulas, so in this course, we will concentrate on how to *verify* a proposed closed formula for a sequence whose recursive description we already know. This will be a focus in the next chapter. Before leaving the topic altogether however, we will look at a classic example of a closed formula that comes from an important type of recursive description.

Example 8 *Discover a closed formula for the sequence whose recursive formula is $s_1 = 1$, and $s_n = s_{n-1} + n$.*

SOLUTION   This recursive description is of the type, "Each term is the previous term plus (something)." If we do not evaluate this sum at each stage, we can see an alternate way to describe s_n:

- $s_1 = 1$
- $s_2 = 1 + 2$
- $s_3 = (1 + 2) + 3$
- $s_4 = (1 + 2 + 3) + 4$

Hence, s_n is the sum of the first n positive integers, and so there is a nice way to look at s_n in order to discover a closed formula for it. We will start by writing s_n as a sum twice, once with terms in increasing order and once with terms in decreasing order:

$$s_n = 1 + 2 + \cdots + (n-1) + n$$
$$s_n = n + (n-1) \cdots + 2 + 1$$

Now by adding the terms aligned in "columns" above, we have

$$2s_n = (n+1) + (n+1) + \cdots + (n+1) + (n+1)$$

Since the right-hand side above consists of $(n+1)$ added to itself n times, we can conclude that

$$s_n = \frac{n \cdot (n+1)}{2}$$

◻

Although discovering a closed formula can be difficult, even when we already know a recursive formula, the converse is rarely true. If we know a closed formula, then finding an alternate recursive description is usually straightforward.

Example 9 *Discover a recursive formula for the sequence whose closed formula is $a_n = 3n + 5$.*

SOLUTION   To discover the formula, we first write down the first few terms of the sequence. Using the formula, we get $a_1 = 3 \cdot 1 + 5 = 8$, $a_2 = 3 \cdot 2 + 5 = 11$, $a_3 = 3 \cdot 3 + 5 = 14$, and so on. We summarize the results by just writing the answers: $8, 11, 14, 17, 20, 23, \ldots$. Then we look for a pattern that relates each term to the previous term, and we see that each one is 3 more than the previous. This leads to the recursive formula

$$a_1 = 8 \quad \text{and} \quad \text{for } n \geq 2, \quad a_n = a_{n-1} + 3$$

We use the subscript $n-1$ to indicate "the previous term," that is, "the term just before the nth term."

☐

Actually finding a recursive formula from a closed formula is often just a matter of doing some algebra. In the previous exercise, since we are *given* the closed formula $a_n = 3n + 5$, we know that $a_{n-1} = 3(n-1) + 5$, and so verifying

$$a_n = a_{n-1} + 3$$

is a simple matter of verifying algebraically that

$$3n + 5 = (3(n-1) + 5) + 3$$

Let's look at an example with a more complicated formula to see this idea in action.

Example 10 *Verify that the sequence given by the closed formula $a_n = 3^n - 2$ satisfies the recursive formula*

$$a_n = 3 \cdot a_{n-1} + 4$$

SOLUTION   Since $a_n = 3^n - 2$, we know that $a_{n-1} = 3^{n-1} - 2$, and so

$$3 \cdot a_{n-1} + 4 = 3 \cdot (3^{n-1} - 2) + 4$$
$$= 3^n - 6 + 4$$
$$= 3^n - 2$$
$$= a_n$$

as desired.

☐

Practice Problem 3   Verify that the sequence given by the closed formula $a_n = n^2 + n$ satisfies the recursive formula

$$a_n = a_{n-1} + 2n$$

Our final examples illustrate some ways we can use what we know about a sequence to learn more about the sequence without "starting from scratch."

Explore more on the Web.

Example 11 *For the sequence whose recursive formula is $a_1 = 11$, and $a_n = a_{n-1} + 5$, suppose someone tells you they have already calculated the 213th term of the sequence, and the answer is 1,071. What is a_{214}?*

SOLUTION   We do not need to start over from a_1 to solve the problem. We simply calculate the 214th term from the 213th. Replacing n with 214 in the recursive formula gives $a_{214} = a_{213} + 5$. Since we know a_{213} is 1,071, we get

$$a_{214} = a_{213} + 5$$
$$= 1{,}071 + 5$$
$$= 1{,}076$$

☐

Explore more on the Web.

Example 12 *For the sequence whose closed formula is $a_n = 2^n - 1$, do the following:*

1. *Calculate the value of the 20th term.*

2. *Someone tells you they have calculated the sum of the first 19 terms, and the answer was $1{,}048{,}555$. What is the sum of the first 20 terms?*

SOLUTION

1. $a_{20} = 2^{20} - 1 = 1{,}048{,}575$.

2. This is similar to the previous example, except that it involves summing up the terms. Again we don't have to start over from the first term; we can reason that the sum of the first 20 terms can be calculated by adding the 20th term to the sum of the first 19 terms. So our answer is $1{,}048{,}555 + 1{,}048{,}575 = 2{,}097{,}130$. We can write this in symbols as

$$a_1 + a_2 + a_3 + \cdots + a_{20} = (a_1 + a_2 + a_3 + \cdots + a_{19}) + a_{20}$$
$$= 1{,}048{,}555 + 1{,}048{,}575$$
$$= 2{,}097{,}130$$

Does it strike you as odd that in this sequence the 20th term is exactly 20 more than the sum of the first 19 terms? Do you think the 21st term will be exactly 21 more than the sum of the first 20 terms? Look at Exercises 19 and 20 below to explore this some more.

Notation for Sums

Example 12 demonstrates an important source of number patterns in mathematics. Summing sequences of numbers is essential in mathematical applications ranging from the analysis of algorithms to integral calculus. We will study sums in more detail in Chapter 2, but for now we will just get acquainted with the traditional notation for sums.

> **Definition**   For a sequence of numbers a_k with $k \geq 1$, we use the notation
>
> $$\sum_{k=1}^{n} a_k$$
>
> to denote the sum of the first n terms of the sequence. This is called *sigma notation* for the sum. In informal situations, we will sometimes write "$a_1 + a_2 + \cdots + a_n$" instead of using sigma notation.

A simple variation on this notation is to write $\sum_{k=m}^{n} a_k$ or $a_m + a_{m+1} + \cdots + a_n$ (where $m \leq n$) whenever we want to sum the numbers in the sequence starting with the mth term and ending with the nth term. This notation is easier to read than it is to write, since writing it requires that we know a closed form for the sequence a_k at the outset, and this is not always the case.

 Example 13 *Write each of the following sums using sigma notation:*

1. *The sum of the first 10 numbers in the sequence $a_k = \frac{1}{k}$ with $k \geq 1$.*
2. $2 + 4 + 8 + 16 + 32 + 64$
3. $2 + 6 + 18 + 54 + 162$
4. $(-4) + (-1) + 2 + 5 + 8 + 11 + 14$

SOLUTION   In each case, we must first determine a closed form for the terms a_k to be summed, and then determine the correct indices of the first term and the last term in the sum.

1. In this case, the form of a_k is given, so we simply have $\sum_{k=1}^{10} \frac{1}{k}$.

2. The terms being summed have closed formula $a_k = 2^k$, the first term in the sum is $a_1 = 2$, and the last term in the sum is $a_6 = 64$. Hence, this sum is represented by $\sum_{k=1}^{6} 2^k$.

3. The terms being summed have closed formula $a_k = 2 \cdot 3^{k-1}$, the first term in the sum is $a_1 = 2$, and the last term in the sum is $a_5 = 162$. Hence, this sum is represented by $\sum_{k=1}^{5} 2 \cdot 3^{k-1}$.

4. The terms being summed have closed formula $a_k = 3k - 7$, the first term in the sum is $a_1 = -4$, and the last term in the sum is $a_7 = 14$. Hence, this sum is represented by $\sum_{k=1}^{7} 3k - 7$.

$\square$

Practice Problem 4   *Evaluate each of the following sums:*

(a) $\sum_{k=1}^{6} (2k-1)$

(b) $\sum_{k=0}^{4} 3^k$

(c) $\sum_{k=3}^{3} k^2$

(d) $\sum_{k=1}^{5} \frac{1}{k(k+1)}$

Sigma notation will be used periodically throughout this book, so it is very important that you become comfortable using it.

Solutions to Practice Problems

1 (a) **Method 1**: The first term is 4, and each term is 2 more than the previous. **Method 2**: The nth term is given by $2(n+1)$. **Method 3**: These are the even integers starting at 4.

(b) From Method 1 above, the next three terms are 900, 902, 904.

(c) From Method 2 above, the $1{,}000^{\text{th}}$ term is $2(1{,}000 + 1) = 2{,}002$.

2 (a) The next number is 30. A closed formula is $a_n = 5n$, a recursive formula is $a_1 = 5$, $a_n = a_{n-1} + 5$, and a description is "multiples of 5."

(b) Next is 160. A closed formula is $a_n = 5 \cdot 2^{n-1}$, and a recursive formula is $a_1 = 5$, $a_n = 2a_{n-1}$.

(c) Next is 49. A closed formula is $a_n = (n+1)^2$, a recursive formula is $a_1 = 4$, $a_n = a_{n-1} + (2n+1)$, and a description is "perfect squares starting at 4."

3 Since $a_n = n^2 + n$, we know that $a_{n-1} = (n-1)^2 + (n-1)$, and so

$$a_{n-1} + 2n = ((n-1)^2 + (n-1)) + 2n$$
$$= (n^2 - 2n + 1) + (n-1) + 2n$$
$$= n^2 + n$$
$$= a_n$$

as desired.

4 (a) $1 + 3 + 5 + 7 + 9 + 11 = 36$

(b) $1 + 3 + 9 + 27 + 81 = 121$

(c) 9

(d) $\frac{1}{2} + \frac{1}{6} + \frac{1}{12} + \frac{1}{20} + \frac{1}{30} = \frac{5}{6}$

Exercises for Section 1.2

1. For each of these sequences, determine the next number in the sequence. Be able to explain how you got it. Then give either a closed formula or a recursive formula. If you can give both, do so.

(a) 2, 4, 6, 8, 10, 12, _____

(b) 4, 9, 16, 25, 36, _____

(c) 2, 5, 10, 17, 26, 37, _____   (HINT. How does this relate to the sequence 1, 4, 9, 16, 25, 36?)

(d) 2, 4, 8, 16, 32, 64, _____

(e) 1, 2, 4, 8, 16, 32, _____

(f) 1, 3, 7, 15, 31, 63,_____
(g) 2, 5, 8, 11, 14, 17,_____
(h) 4, 16, 64, 256, 1,024,_____
(i) 5, 10, 20, 40, 80, 160,_____
(j) 5, 9, 17, 33, 65, 129,_____
(k) 2, 5, 10, 50, 500,_____
(l) 1, 5, 9, 13, 17,_____
(m) 3, 6, 9, 12, 15,_____
(n) 3, 5, 9, 17, 33,_____

2. For each of the following sequences, calculate a_{k-1} and a_{k+1}, and algebraically simplify the expressions:
   (a) $a_n = 5n - 2$
   (b) $a_n = 3n^2 + n$
   (c) $a_n = 2n + 7$
   (d) $a_n = 3^n + 4$
   (e) $a_n = 2^{3n+1} - 1$
   (f) $a_n = \frac{n(2n-1)(n+2)}{6}$

3. If you form the decreasing list $n, n-1, n-2, \ldots$, what will be the kth number in the list? Your answer should involve variables n and k.

4. Here are verbal descriptions of sequences. Use these descriptions to write recursive formulas for the sequences.
   (a) The sequence starts with 2, and each entry is 2 more than the previous entry.
   (b) The sequence starts with 1, and each entry is 6 more than the previous entry.
   (c) The sequence is obtained by starting with 2, and each subsequent entry is 1 more than twice the previous entry.
   (d) The sequence is obtained by starting with 2, and each subsequent entry is the square of the previous entry.

5. For each of these sequences described in English, write out the first five (or more if necessary) terms of the sequence. Use your answer to discover a closed formula for the sequence.
   (a) The sequence starts with 2, and each entry is 2 more than the previous entry.
   (b) The sequence starts with 1, and each entry is 6 more than the previous entry.
   (c) The sequence obtained starting with 2, and each subsequent entry is 1 more than twice the previous entry.
   (d) The sequence obtained starting with 2, and each subsequent entry is the square of the previous entry.

6. For each of the given sequences, use the given closed formula and algebra to check whether the given recursive formula is true:

(a) Given that $a_n = 2^n - 1$ for all $n \geq 1$, is it true that $a_n = 2 \cdot a_{n-1} + 1$?
(b) Given that $a_n = n^2$ for all $n \geq 1$, is it true that $a_n = a_{n-1} + 2n$?
(c) Given that $a_n = 5n + 3$ for all $n \geq 1$, is it true that $a_n = 5 \cdot a_{n-1} - 3$?
(d) Given that $a_n = 1 - \frac{1}{2^n}$ for all $n \geq 1$, is it true that $a_n = a_{n-1} + \frac{1}{2^n}$?

7. For each of these sequences given in closed form, write out the first five (or more if necessary) terms of the sequence. Use your answer to discover a recursive formula for the sequence.
   (a) $a_n = 3n + 1$
   (b) $a_n = 5n - 2$
   (c) $a_n = 2n + 7$
   (d) $a_n = 3^n + 1$
   (e) $a_n = 7n - 6$
   (f) $a_n = 3n^2 + n$
   (g) $a_n = 2^{2n-1} - 1$

8. For each of these sequences given recursively, write out the first five (or more if necessary) terms of the sequence. Use your answer to discover a closed formula for the sequence.
   (a) $a_1 = 5$; $a_n = 5 \cdot a_{n-1}$
   (b) $a_1 = 2$; $a_n = 5 \cdot a_{n-1}$
   (c) $a_1 = 3$; $a_n = 2 \cdot a_{n-1}$
   (d) $a_1 = 4$; $a_n = 4 + a_{n-1}$
   (e) $a_1 = 1$; $a_n = 4 + a_{n-1}$
   (f) $a_1 = 6$; $a_n = 3 + a_{n-1}$

9. For each of these sequences given recursively, write out the first five (or more if necessary) terms of the sequence. Use the idea developed in Example 8 to discover a closed formula for each sequence.
   (a) $a_1 = 5$; $a_n = a_{n-1} + (n+4)$
   (b) $a_1 = 1$; $a_n = a_{n-1} + (2n-1)$
   (c) $a_1 = -3$; $a_n = a_{n-1} + 4n$

10. Give an algebraic expression that describes the nth whole number larger than 1,964.

11. Give an algebraic expression that describes the nth odd three-digit number.

12. Give an algebraic expression that describes the nth even two-digit number.

13. Give an algebraic expression that describes the nth odd perfect square.

14. Form a sequence using the relationship, "a_n is the ones' digit of $(a_{n-1} + a_{n-2})$." Thus, all your terms will be from the set $\{0, 1, 2, \ldots, 9\}$, which means that this sequence will have to repeat itself at some point. Try different pairs of starting numbers from $\{0, 1, 2, \ldots, 9\}$ and see

how long it takes for the sequence to repeat itself. (For example, the sequence 6, 8, 4, 2 follows this rule and would repeat itself if continued.) What is the longest possible such cycle?

15. What does the equation

$$(a+b)^3 = a^3 + 3a^2b + 3ab^2 + b^3$$

say when $a = 2$ and $b = -1$?

16. What does the equation

$$C(n,k) = C(n-1,k) + C(n-1,k-1)$$

say when $n = 2k$?

17. Give an algebraic expression for the nth term in the arithmetic progression

$$a, a+3, a+6, a+9, \ldots$$

18. Give an algebraic expression for the nth term in the arithmetic progression

$$a, a+d, a+2 \cdot d, a+3 \cdot d, \ldots$$

19. For the sequence whose closed formula is $a_n = 2^n$, form the new sequence s_n by the definition

$$s_n = a_1 + a_2 + \cdots + a_n$$

Fill in the blanks in Table 1-5, and make a conjecture about the relationship between the bottom two rows.

20. For the sequence whose closed formula is $a_n = 2^n - 1$, form the new sequence s_n by the definition

$$s_n = a_1 + a_2 + \cdots + a_n$$

Fill in the blanks in Table 1-6, and make a conjecture about the relationship between the bottom two rows.

21. Evaluate each of the following sums:
   (a) $\sum_{k=1}^{7} 3k$
   (b) $\sum_{k=1}^{3} 7k$
   (c) $\sum_{k=1}^{9} 4$
   (d) $\sum_{k=0}^{4} \frac{1}{2^k}$
   (e) $\sum_{k=-1}^{3} 3 - 2k$

n	1	2	3	4	5	6	7	8
a_n	2	4	8					
s_n	2	6	14					

Table 1-5 Table for Exercise 19

n	1	2	3	4	5	6
a_n	1	3				
s_n	1	4				

Table 1-6 Table for Exercise 20

22. Consider the sequence in Practice Problem 3 with recursive description $a_n = a_{n-1} + 2n$ with $a_1 = 2$.
   (a) Show that $a_4 = \sum_{k=1}^{4} 2k$.
   (b) Show that $a_7 = \sum_{k=1}^{7} 2k$.
   (c) Show that $a_1 = \sum_{k=1}^{1} 2k$.

23. Consider the sequence $s_n = \sum_{k=1}^{n} 3$.
   (a) Evaluate s_5 and s_{10}.
   (b) Explain in words why $s_n = s_{n-1} + 3$.
   (c) Give a closed formula for s_n as an algebraic expression using the variable n.

24. Express each of the following sums using sigma notation:
   (a) $4 + 8 + 12 + 16 + 20 + 24 + 28 + 32 + 36$
   (b) $2 + 3 + 4 + \cdots + 16 + 17 + 18$
   (c) $5 + 5 + 5 + 5 + 5 + 5$
   (d) $\frac{1}{2} + \frac{2}{4} + \frac{3}{8} + \frac{4}{16} + \frac{5}{32}$
   (e) $1 + 9 + 17 + 25 + 33 + 41 + 49$

25. Consider the Josephus game from the previous section.
   (a) Suppose we use $J(p,s)$ to mean the Josephus game with p persons and a "skip" amount of s. For example, $J(10,2)$ means the game with 10 people and with the elimination starting out 2, 4, 6. For each value of p from 2 to 12, determine the winner for the $J(p,2)$ game, and fill in Table 1-7.
   (b) If you did this correctly, there should be a discernible pattern in the answers. Describe that

p	Winner of Game $J(p,2)$
2	
3	
4	
5	
6	
7	
8	
9	
10	
11	
12	

Table 1-7 Table for Exercise 25

pattern in words. Then use the pattern to predict the answer for 13, 14, 15, and 16 people.

(c) Predict the answer for 32 people; for 31 people; for 30 people.

26. These are somewhat more challenging number sequences than those given in Exercise 1. For each of these sequences, give a characterization of the sequence. For each, we suggest the type that may be easiest to use, but you may give characterizations of other types.

(a) 1, 9, 25, 49, 81, _____ (Formula for n^{th} term)

(b) 1, 2, 2, 4, 8, 32, 256, _____ (Relate term to previous terms)

(c) 1, 1, 4, 36, 1764, _____ (Relate term to previous terms)

(d) 1, 4, 27, 256, 3125, _____ (Formula for n^{th} term)

(e) 1, 5, 7, 17, 31, 65, _____ (Formula for n^{th} term)

(f) 1, 3, 4, 7, 11, 18, _____ (Relate term to previous terms)

(g) 2, 3, 5, 7, 11, 101, 131, 151, _____ (Recognizable set of integers)

(h) 6, 8, 10, 14, 15, 21, 22, _____ (Recognizable set of integers)

27. Refer to Exercise 25, where we introduced the notation $J(p, s)$ for the Josephus game with p persons and a skip amount of s.

(a) Suppose five people wearing name tags labeled 1, 2, 3, 4, and 5 line up in that order, and play the game with a skip amount of 4. It is easy to verify that Player 1 wins the game. (Try it!) Then answer these questions without actually playing the game.

   i. If five people wearing name tags Anne, Sue, Matt, Tom, and Linda line up in that order and play the game with a skip amount of 4, who will win?

   ii. If five people wearing name tags labeled 5, 6, 1, 2, 3 line up in that order and play the game with a skip amount of 4, who will win?

(b) Suppose six people wearing name tags labeled 1, 2, 3, 4, 5, 6 line up in that order and play the game with a skip amount of 4.

   i. Who is the first person eliminated?

   ii. After the first person is eliminated, who is left (and in what order)?

   iii. Without finishing the game, use the answer to a previous part of this question to tell who will win.

(c) The preceding exercise establishes that, knowing that $J(5, 4)$ is 1, we can easily determine that $J(6, 4)$ is 5. Use this fact, and similar reasoning, to determine $J(7, 4)$.

(d) Suppose you know that $J(15, 4)$ is 13. Use this to calculate $J(16, 4)$.

28. In this exercise, the value of each term depends on a term that appears further back in the sequence, and the rule for a_n depends on whether n is even or odd. Give the first seven terms of each sequence.

(a) $a_1 = 1$ and $a_n = \begin{cases} a_{n/2} + n, & \text{if } n \text{ is even} \\ a_{(n-1)/2} + n, & \text{if } n \text{ is odd} \end{cases}$

(b) $a_1 = 1$ and $a_n = \begin{cases} 3a_{n/2} + 2, & \text{if } n \text{ is even} \\ 3a_{(n-1)/2} + 2, & \text{if } n \text{ is odd} \end{cases}$

(c) $a_1 = 1$ and $a_n = \begin{cases} 3a_{n/2} + n, & \text{if } n \text{ is even} \\ 3a_{(n-1)/2} + n, & \text{if } n \text{ is odd} \end{cases}$

29. In this exercise, we build a sequence whose values are *strings* of A's and B's, similar to those used in Section 1.1 to represent tennis matches. We define the sequence recursively by the following rule:

$$a_1 = B$$
$$a_n = A\, a_{n-1}\, BA$$

That is, a_n consists of the letter A, followed by the string for a_{n-1}, followed by the string BA. For example, a_2 is the string consisting of A followed by a_1 (which is B) followed by BA, making $a_2 = ABBA$.

(a) Calculate terms a_3 through a_6 of this sequence.

(b) Modify the recursive definition to yield the sequence $AB, AABB, AAABBB, AAAABBBB$, and so on. That is, a_n should consist of n occurrences of A followed by n occurrences of B.

(c) Modify the recursive definition to yield the sequence $ABB, AABB, AAABB, AAAABB, AAAAABB$, and so on. That is, a_n should consist of n occurrences of A followed by BB.

30. This exercise combines the ideas in Exercises 28 and 29. Define the sequence as $a_1 = A$ and

$$a_n = \begin{cases} a_{n/2}\, B, & \text{if } n \text{ is even} \\ a_{(n-1)/2}\, A, & \text{if } n \text{ is odd} \end{cases}$$

For example, $a_2 = a_{2/2}\, B = a_1\, B = AB$ and $a_5 = a_{(5-1)/2}\, A = a_2\, A = ABA$.

(a) Write down the first ten strings in this sequence.

(b) Use the results from part (a) to determine a_{17} and a_{21}.

(c) Suppose you already know that $a_{315} = ABBAAABAA$. Give the values of a_{630} and a_{631}.

1.3 Truth-tellers, Liars, and Propositional Logic

Raymond Smullyan (1919–) has written several books of puzzles about conversations held on a fictional island where inhabitants are either truth-tellers or liars. The point of the puzzles is to discern which inhabitant is of which type, using only the statements you hear them make, if possible.

Example 1 *You meet two inhabitants of Smullyan's Island. A says, "Exactly one of us is lying," and B says, "At least one of us is telling the truth." Who (if anyone) is telling the truth?*

To solve this puzzle, we will consider all possibilities for each person's status as a liar or truth-teller. In the table below, we will use p to stand for the phrase, "A is truthful," and q to stand for the phrase, "B is truthful." With this convention, we can organize our consideration of all the possibilities as follows: Either p is true or p is false, and either q is true or q is false. In each case, we can determine the truth or fallacy of what A and B said.

The secret to avoiding dizziness with these puzzles is to **first consider each statement on its own instead of worrying about who said it.** For example, we can tell that the phrase, "At least one of us is telling the truth" is true if either of p or q is true. We can make this conclusion without knowing who made the statement.

In this same way, we can use a table to analyze the truth value of each spoken statement for every possible combination of truth values of the simpler statements p and q.

p	q	Statement 1 Exactly One is Lying	Statement 2 At Least One is Truthful
T	T	F	T
T	F	T	T
F	T	T	T
F	F	F	F

We call this analytic device a **truth table**. It is a valuable tool for analyzing the truth value of any compound statement based on the truth values of its simpler parts, and we will see later in this section that truth tables are useful for far more than solving logic puzzles. Now that we have this representation of the problem, we just need to think about what a solution should look like.

SOLUTION We can think about the meaning of each row in the truth table given above.

- The first row of the table represents the hypothetical situation where both A and B are truth-tellers (because both p and q are true). This cannot be the real situation, however, because Statement 1 is false in this row, so a truth-teller like A could not have said it.

- The second row of the table cannot represent the real situation, since it cannot be that the liar *B* would have made a true statement.
- The third row is ruled out, because in this row the liar *A* has made a true statement.
- In the final row, both *A* and *B* are liars and both statements are false.

We conclude that the fourth row of the truth table describes the only possible real situation, and so the solution to the puzzle is, *Both inhabitants are lying.* □

In this example, we found rows where the truth value describing each inhabitant's type matches the truthfulness of his or her statement. Another way to describe this is to say that the part of the row to the left of the double vertical lines should match the part to the right of those lines. In this way, the use of a truth table helps organize our thinking about the problem, making the search for a solution more routine.

Before we proceed to our next example, let us say a few words about building truth tables. For the two variables *p* and *q* there are a total of four possible combinations. We could list them in some random order, but if we proceed systematically, we can be sure we don't leave any out. Most people do this by thinking analogously to our counting system. When we count from 20 to 39, for example, we write

$$20, 21, 22, 23, \ldots, 29, 30, 31, 32, \ldots, 39$$

The tens' digit stays fixed while we cycle through all the possible ones' digits. Then we go to the next tens digit and repeat the cycle of the ones' digit. Likewise, we wrote the truth table in the order TT, TF, FT, FF—with the right-hand value cycling though its values for T, then again for F.

Those who built "game trees" for exercises in Section 1.1 will be interested to know that this method of listing the rows is actually the same thing, as we see in Figure 1-9. For just two variables this is admittedly not that important, but as we add more variables, it becomes more important to be consistent. For example, for four variables this order would be TTTT, TTTF, TTFT, TTFF, TFTT, TFTF, TFFT, TFFF, FTTT, FTTF, FTFT, FTFF, FFTT, FFTF, FFFT, FFFF.

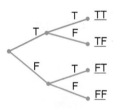

Figure 1-9 The order of rows in a truth table

Example 2 *Suppose you meet three inhabitants of Smullyan's Island and have the following conversation. Can you tell which inhabitants (if any) are lying?*

- A *says, "Exactly one of us is telling the truth."*
- B *says, "We are all lying."*
- C *says, "The other two are lying."*

SOLUTION The analysis of this puzzle is the same once we add a symbol *r* to represent the statement "*C* is truthful." The truth table below begins with TTT and follows the order we discussed earlier. In the table, we mark the row where the truth values of *p*, *q*, and *r* match the truth values of what *A*, *B*, and *C* say, respectively.

p	q	r	Statement 1 Exactly One is Truthful	Statement 2 All are Lying	Statement 3 A and B are Lying
T	T	T	F	F	F
T	T	F	F	F	F
T	F	T	F	F	F
T	F	F	T	F	F
F	T	T	F	F	F
F	T	F	T	F	F
F	F	T	T	F	T
F	F	F	F	T	T

We conclude that the only possible situation is where p is true while q and r are false. Therefore, A is the only inhabitant telling the truth. □

Practice Problem 1 *You meet two inhabitants of Smullyan's Island. A says, "We are both telling the truth." B says, "A is lying." Who (if anyone) is telling the truth?*

In each of the previous examples, exactly one row of the truth table has been consistent, and we can conclude that this one row gives *the solution*. However, it is possible to have situations where more than one row of the table is consistent, in which case we cannot determine the status of all the speakers. On the other hand, if no row is consistent, then we have arrived at a paradox of sorts. Here are the simplest examples of each of these situations.

 Example 3

1. *You meet an inhabitant A, who says, "I am telling the truth." Is she?*
2. *You meet an inhabitant A, who says, "I am lying." Is he?*

SOLUTION   In each part we need only one variable p to represent the statement "A is a truth-teller."

1. The table for this statement looks like this:

p	Statement 1 I am Telling the Truth
T	T
F	F

We cannot conclude anything, since both rows are consistent.

2. The table for this statement looks like this:

p	Statement 1 I am Lying
T	F
F	T

Since neither row is consistent, we have a paradox.

□

The Logic of Propositions

Before we look at more examples, let's agree on some notation to make writing these things a little easier. We will refer to this notation as *propositional logic notation*.

Definition We call a sentence a **proposition** if it is unambiguously true or false. A *propositional variable* is simply a variable name that stands for a proposition.

A *formal proposition* will mean a proposition written using propositional logic notation according to the following rules:

1. Any propositional variable alone is a formal proposition.
2. Given formal propositions p and q, the compound statement $p \wedge q$ is a formal proposition. The proposition $p \wedge q$ stands for, "Both p and q are true," and we read this as "p **and** q."
3. Given formal propositions p and q, the compound statement $p \vee q$ is a formal proposition. The proposition* $p \vee q$ stands for, "Either p or q is true," and we read this as "p **or** q." In mathematics, when we say that either p or q is true, we mean that at least one is true. Saying "or" always allows for the possibility that both are true.
4. Given a formal proposition p, the compound statement $\neg p$ is a formal proposition. The proposition $\neg p$ stands for, "It is not the case that p is true," and we read this as "**not** p." We refer to $\neg p$ as the **negation** of p.

This system allows for the creation of complicated expressions by repeatedly using these basic rules.

Example 4 *Show that the expression $(p \vee q) \wedge \neg(p \wedge q)$ is a formal proposition using the definition above.*

SOLUTION As usual in mathematics, we perform our analysis from inside the parentheses out.

- Since p and q are propositional variables, then they are each formal propositions by (1) in the definition.
- Since p and q are formal propositions, then $p \vee q$ and $p \wedge q$ are formal propositions by (2) and (3) in the definition, respectively.
- Since $p \wedge q$ is a formal proposition, then $\neg(p \wedge q)$ is a formal proposition by (4) in the definition.
- Since $p \vee q$ and $\neg(p \wedge q)$ are formal propositions, then so is $(p \vee q) \wedge \neg(p \wedge q)$ by (2) in the definition. □

---

* One way to keep $\wedge$ and $\vee$ straight in your head is that the word "AND" begins with the letter A, and the $\wedge$ symbol looks sort of like capital A.

It is customary to consider these operations as having precedence rules (much as $+$, $-$, $\cdot$, and/in arithmetic). The "not" operation ($\neg$) has highest precedence, followed by "and" ($\wedge$) and then "or" ($\vee$). For example, the expression $\neg p \wedge \neg q \vee p$ is the same as $((\neg p) \wedge (\neg q)) \vee p$. As with arithmetic operations, use parentheses if you are worried about ambiguity in your expressions.

Practice Problem 2 *You meet two inhabitants of Smullyan's Island, A and B. Using p to represent the proposition "A is truthful" and q to represent the proposition "B is truthful," how would you write each of these statements as formal propositions?*

(a) A is lying.

(b) At least one of us is truthful.

(c) Either B is lying or A is.

(d) Exactly one of us is lying.

We should make a comment here about the meaning of the word "or" in mathematics. This can be a slightly confusing issue since many times in English we use the word "or" to imply a choice must be made. In conversational English, we must rely on context to resolve the issue, and we often have to ask the speaker to clarify the meaning. You may have had conversations like this:

> **Waiter:** Do you want soup or salad?
>
> **You:** Yes, I'll have both.
>
> **Waiter:** No, which one do you want?

In this situation, the waiter meant for the word "or" to convey that a choice was to be made, while perhaps you thought that neither soup nor salad was included in the price of the meal and the waiter is simply asking if you would like an appetizer, in which case it is reasonable to get both things. To avoid this ambiguity of the English language, in mathematics when we say, "*p OR q* is true," we will always mean that either one or both statements are true.

Example 5 *You meet two inhabitants of Smullyan's Island. A says, "Either B is lying or I am," and B says, "A is lying." Who (if anyone) is telling the truth?*

SOLUTION As before we build a truth table to consider all possiblities for the status of the two speakers. Notice that the statement "Either B is lying or A is" is the same as saying, "At least one is lying."

p	q	Statement 1 Either B is Lying or A is	Statement 2 A is Lying
T	T	F	F
★ T	F	T	F
F	T	T	T
F	F	T	T

Only row 2 of the table is consistent, so we can conclude that *A* is truthful and *B* is lying. □

Using the operations ∧, ∨, and ¬, we can write various combinations of propositions in a concise manner. For example, suppose we use *e* to represent the statement "Sue is an English major" and *j* to represent the statement "Sue is a junior." Here are several possible combinations we could write.

Example 6 *How would you write each of these propositions using combinations of e (meaning "Sue is an English major") and j (meaning "Sue is a junior") with the operations ∧, ∨, and ¬?*

1. *Sue is a junior English major.*
2. *Sue is either an English major or she is a junior.*
3. *Sue is a junior, but she is not an English major.*
4. *Sue is neither an English major nor a junior.*
5. *Sue is exactly one of the following: an English major or a junior.*

SOLUTION

1. The sentence implies that *both* conditions are true. That is, she is a junior *and* she is an English major. So we write $j \wedge e$.

2. Although in English there might be some ambiguity about this, in mathematics the word *or* always means either one or both. So we write $j \vee e$.

3. In English when we say "but," it means the same as saying, "and," so the statement is $j \wedge (\neg e)$. Because of the precedence conventions, we do not need the parentheses, and we could simply write $j \wedge \neg e$.

4. There are two ways to think of "neither-nor" in English. The first is that it means the "opposite" (or negation) of either-or. This leads to the statement $\neg(j \vee e)$. Another choice is to reason that "neither this nor that" means "not this and also not that." This leads to the statement $\neg j \wedge \neg e$. These two statements are equivalent.

5. This is sometimes called the **exclusive or** of the two conditions—one or the other is true, but not both. Again, there are two equivalent solutions to the problem. One solution comes from writing the "one or the other is true, but not both" in symbols, as $(j \vee e) \wedge \neg(j \wedge e)$. The other solution comes from reasoning, "She could either be a junior but not an English major, or she could be an English major but not a junior." This gives the symbolic statement $(j \wedge \neg e) \vee (e \wedge \neg j)$. □

Practice Problem 3 *You meet three inhabitants, A, B, and C, of Smullyan's Island. Using p to represent the statement "A is truthful," q to represent the statement "B is truthful," and r to represent the statement "C is truthful," how would you write each of these phrases?*

(a) A and B are lying.

(b) All three are lying.

(c) *One of us is lying.*

(d) *Exactly one of the three is truthful.*

Truth Tables for Formal Propositions

In addition to saving time and space, another reason for creating propositional logic notation is so we can understand the relationship between the truth value of a complex statement and the truth values of its simpler component propositions. We can illustrate this relationship for each of the basic propositional connectives **and**, **or**, and **not** using truth tables.

p	q	$p \wedge q$
T	T	T
T	F	F
F	T	F
F	F	F

p	q	$p \vee q$
T	T	T
T	F	T
F	T	T
F	F	F

p	$\neg p$
T	F
F	T

As in the solutions to the logic puzzles, a truth table shows us the truth value of a compound statement for every possible combination of truth values of its simple components.

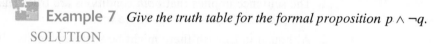 **Example 7** *Give the truth table for the formal proposition $p \wedge \neg q$.*

SOLUTION

p	q	$\neg q$	$p \wedge \neg q$
T	T	F	F
T	F	T	T
F	T	F	F
F	F	T	F

In the solution above, we created a column in the table for $\neg q$ in order to get the column for $p \wedge \neg q$ by simply applying the truth table rule for $\wedge$ to the p column and $\neg q$ column. Specifically, since $A \wedge B$ is true only when both propositions A and B are true, then $p \wedge \neg q$ is true only when both propositions p and $\neg q$ are true, so the p and $\neg q$ columns in the truth table are all we need in order to form the column for $p \wedge \neg q$.

By making the construction of a truth table a formal process, we can easily build tables for complicated expressions without becoming mired in their grammatical or logical muck.

Explore more on the Web.

Practice Problem 4 *Complete the following truth table:*

p	q	$p \vee q$	$\neg (p \vee q)$
T	T		
T	F		
F	T		
F	F		

As the statements get complicated, it is necessary to have a step-by-step procedure for building a truth table. Once again, we will work from inside the parentheses out.

Example 8 *Find the truth table for the formal proposition* $(p \vee q) \wedge \neg(p \wedge q)$

SOLUTION   This is the same proposition we looked closely at in Example 4. We can use the order in which the formation rules were used to get the order in which we should build the truth table. The final column is formed from the two columns that precede it using the truth table rule for $\wedge$.

p	q	$p \wedge q$	$\neg(p \wedge q)$	$p \vee q$	$(p \vee q) \wedge \neg(p \wedge q)$
T	T	T	F	T	F
T	F	F	T	T	T
F	T	F	T	T	T
F	F	F	T	F	F

Negation and Logical Equivalence

As we indicated earlier, we call $\neg p$ the *negation* of p. We now examine this concept in a little more detail for two situations: (1) propositions involving comparisons, and (2) propositions involving the $\wedge$ and $\vee$ connectives.

Example 9 *Let p stand for the proposition "Tammy has more than two children." Express the negated proposition as an English-language sentence.*

SOLUTION   A common mistake is to use "less than" as the negation of "more than." The correct negation is "less than or equal to." We write, "Tammy has less than or equal to 2 children," or "The number of children Tammy has is less than or equal to 2."

If we use the symbol c to indicate how many children Tammy has, then we may write the original proposition p mathematically as "$c > 2$." The negation $\neg p$ can be written as "$c \leq 2$."

Example 10 *In the previous example, if someone tells us that the value of c is actually 3, what does this mean? Which is true, the original proposition p or its negation $\neg p$?*

SOLUTION   This means that Tammy has three children, and we conclude that p is a true statement.

Practice Problem 5 *Let g stand for John's current grade point average. To be admitted to graduate school, a person needs at least a 3.0 grade point average. Use g to express the proposition p, "John's grade point average is high enough to be admitted to graduate school." Then use g to express the negated proposition $\neg p$.*

We now address the issue of forming the negation of propositions involving the $\wedge$ and $\vee$ connectives. In Example 6 we found two different ways to express the proposition "Sue is neither an English major nor a junior," using the symbols e for "Sue is an English major" and j for "Sue is a junior." We stated that the two solutions were "equivalent." To explore exactly what we mean by this, we build a truth table for our first statement "$\neg(j \vee e)$." In the truth table that follows, we calculate the final result, $\neg(j \vee e)$, by first calculating $j \vee e$ and then negating that column.

j	e	$j \vee e$	$\neg(j \vee e)$
T	T	T	F
T	F	T	F
F	T	T	F
F	F	F	T

We see that the statement "$\neg(j \vee e)$" is true precisely when both j and e are false. Next we build a truth table for the second statement "$\neg j \wedge \neg e$." We calculate the final result by first calculating the intermediate values $\neg j$ and $\neg e$.

j	e	$\neg j$	$\neg e$	$\neg j \wedge \neg e$
T	T	F	F	F
T	F	F	T	F
F	T	T	F	F
F	F	T	T	T

We see that the statement "$\neg j \wedge \neg e$" is true precisely when both j and e are false. The two statements ("$\neg(j \vee e)$" and "$\neg j \wedge \neg e$") are true under exactly the same circumstances—they have the same truth table. This is what we mean when we say the statements are equivalent.

> **Definition** Two statements are said to be **logically equivalent** if they have the same truth value for every row of the truth table.

There is another important point to be made—the logical equivalence does not depend on the particular meaning of the symbols j and e in this example. The fact that these two statements are equivalent is just a special instance of a general property. For any statements p and q, it is always true that $\neg(p \vee q)$ is equivalent to $\neg p \wedge \neg q$. This is one of the two facts that have come to be known as DeMorgan's laws.

Proposition 1   *(**DeMorgan's laws**) Let p and q be any propositions. Then*

1. $\neg(p \vee q)$ is logically equivalent to $\neg p \wedge \neg q$.
2. $\neg(p \wedge q)$ is logically equivalent to $\neg p \vee \neg q$.

*In words, to negate a condition containing **and** ($\wedge$) or **or** ($\vee$), negate each part and change the **and** to **or** or vice versa.*

PROOF   The first of these is established in the previous discussion (for propositional variables j and e), and the second is left for Exercise 16. ■

Practice Problem 6   *One way to write, "Sue is not both a junior and an English major" is to write the negation of "Sue **is** both a junior and an English major." This gives the statement "$\neg(j \wedge e)$." Use DeMorgan's laws to write an equivalent statement, and explain what this statement says in words.*

There are several other situations where we can easily see that two statements are logically equivalent. As a simple example, for any statements p and q, it is easy to see that $p \wedge q$ is equivalent to $q \wedge p$. Similarly, we know from our use of ordinary English that any statement p is logically equivalent to its "double negative" $\neg(\neg p)$. To see this formally, we just calculate the truth table:

p	$\neg p$	$\neg(\neg p)$
T	F	T
F	T	F

The fact that the first and third columns are identical establishes the equivalence.

Example 11   *Write the negation of the phrase "Sue is a junior, but she is not an English major" two different ways. What does it mean for this phrase to be false?*

SOLUTION   In an earlier example, we wrote, "Sue is a junior, but she is not an English major," as $j \wedge \neg e$. The simplest way to write the negation is to enclose this in parentheses, preceded by the **not** operation: "$\neg(j \wedge \neg e)$." By DeMorgan's laws and the double negative property, this is equivalent to "$\neg j \vee e$" (change **and** to **or** and negate each part). In words, "Either Sue is *not* a junior, or she *is* an English major." □

Example 12   *Using g to indicate John's score on the recent math test, we can express the statement "John got a B on the test" as "$(g \geq 80) \wedge (g < 90)$." Write the negation as an expression in symbolic logic notation.*

SOLUTION   This example uses both ideas we have considered in the section. To negate the statement, we must change the $\wedge$ to $\vee$ and negate each part. To negate each part, we must take care to properly negate the comparison operations. The solution is "$(g < 80) \vee (g \geq 90)$." In words, "John either got below a B or above a B." □

Practice Problem 7

(a) *Using y to indicate yesterday's high temperature and t for today's high temperature, negate the proposition p written as "$(y < 0) \vee (t < 0)$." Express both p and $\neg p$ as English-language sentences.*

(b) *You meet three inhabitants of Smullyan's Island, A , B and C. Using p to represent the statement "A is truthful," q to represent the statement "B is truthful," and r to represent the statement "C is truthful," how would you write each of these phrases in symbolic logic notation? Use the double negative property and DeMorgan's laws to make your answer as simple as possible.*

   i. *Not all of us are lying. This is the negation of "All of us are lying."*

   ii. *Not one of us is lying. This is the negation of "At least one of us is lying."*

There are many other examples of the logical equivalence of statements that can be established using truth tables. To make these easy to state, we will use the notation $p \equiv q$ to represent the statement "Propositions p and q are logically equivalent."

 Example 13 *Use a truth table to establish the following logical equivalences:*

1. $\neg(\neg p \vee q) \equiv p \wedge \neg q$

2. $p \wedge (p \vee q) \equiv p$

SOLUTION

1. In the table below, because the fifth and eighth columns are identical, we see that $\neg(\neg p \vee q)$ is logically equivalent to $p \wedge \neg q$.

p	q	$\neg p$	$\neg p \vee q$	$\neg(\neg p \vee q)$	p	$\neg q$	$p \wedge \neg q$
T	T	F	T	F	T	F	F
T	F	F	F	T	T	T	T
F	T	T	T	F	F	F	F
F	F	T	T	F	F	T	F

2. The equivalence follows from the fact that, in the table below, the fourth column is the same as the first.

p	q	$p \vee q$	$p \wedge (p \vee q)$
T	T	T	T
T	F	T	T
F	T	T	F
F	F	F	F

□

Explore more on the Web.

Practice Problem 8 *Use a truth table to show that $p \vee (q \wedge r)$ is logically equivalent to $(p \vee q) \wedge (p \vee r)$.*

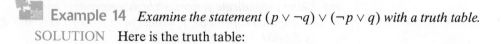 **Example 14** *Examine the statement $(p \vee \neg q) \vee (\neg p \vee q)$ with a truth table.*

SOLUTION   Here is the truth table:

p	q	$\neg p$	$\neg q$	$p \vee \neg q$	$\neg p \vee q$	$(p \vee \neg q) \vee (\neg p \vee q)$
T	T	F	F	T	T	T
T	F	F	T	T	F	T
F	T	T	F	F	T	T
F	F	T	T	T	T	T

$\square$

In this example, the column containing the final result (for the complete statement) consists of all T's. This means that no matter what truth values we assign to the variables p and q, the overall statement is true. It is also possible to have just the opposite situation, where all combinations of truth values yield a false result. This is significant enough to warrant special names for these kinds of propositions.

Definition

1. A **tautology** is a proposition whose value is True for all possible combinations of the truth values of the propositional variables.
2. A **contradiction** is a proposition whose value is False for all possible combinations of the truth values of the propositional variables.

DeMorgan's laws and the double negation property were our first examples of logical equivalence, but there are many more. The following theorem summarizes several that will be important later in this course. The proof of each of these statements consists of simply creating a truth table for the two given propositions and verifying they are the same.

Theorem 2 *Let p, q, and r stand for any propositions. Let t indicate a tautology, and c indicate a contradiction. Then all the logical equivalences shown in Table 1–8 hold.*

PROOF   DeMorgan's laws, the double negative property, one of the absorption properties, and one of the distributive properties are established throughout this section. See the exercises at the end of this section for the rest. ∎

(a)	Commutative	$p \wedge q \equiv q \wedge p$	$p \vee q \equiv q \vee p$
(b)	Associative	$(p \wedge q) \wedge r \equiv p \wedge (q \wedge r)$	$(p \vee q) \vee r \equiv p \vee (q \vee r)$
(c)	Distributive	$p \wedge (q \vee r) \equiv (p \wedge q) \vee (p \wedge r)$	$p \vee (q \wedge r) \equiv (p \vee q) \wedge (p \vee r)$
(d)	Identity	$p \wedge t \equiv p$	$p \vee c \equiv p$
(e)	Negation	$p \vee \neg p \equiv t$	$p \wedge \neg p \equiv c$
(f)	Double negative	$\neg(\neg p) \equiv p$	
(g)	Idempotent	$p \wedge p \equiv p$	$p \vee p \equiv p$
(h)	DeMorgan's laws	$\neg(p \wedge q) \equiv \neg p \vee \neg q$	$\neg(p \vee q) \equiv \neg p \wedge \neg q$
(i)	Universal bound	$p \vee t \equiv t$	$p \wedge c \equiv c$
(j)	Absorption	$p \wedge (p \vee q) \equiv p$	$p \vee (p \wedge q) \equiv p$
(k)	Negations of t and c	$\neg t \equiv c$	$\neg c \equiv t$

Table 1-8   Logical Equivalences

Observe that each property (except (f)) actually consists of two properties. For example, the "commutative" property has two versions, one for the $\wedge$ operation and one for the $\vee$ operation.

Since each of these statements is true under all truth assignments, we can think of these logical equivalences as providing us with a substitution rule of sorts. For example, since $(p \wedge q) \wedge r \equiv p \wedge (q \wedge r)$, we can replace any expression of the form $(p \wedge q) \wedge r$ with the expression $p \wedge (q \wedge r)$ without changing the truth value. This is an instance of the **substitution rule** for propositional logic. We can use this rule along with the facts in Theorem 2 to demonstrate the logical equivalence of two propositions through what appears to be sheer algebraic manipulation.

Example 15 *Verify the logical equivalence $p \vee (\neg p \wedge q) \equiv p \vee q$ using the substitution rule and quoting the appropriate parts of Theorem 2.*

SOLUTION   Each proposition below is equivalent to the previous proposition by the cited part of Theorem 2 and the substitution rule when necessary:

$$
\begin{aligned}
p \vee (\neg p \wedge q) &\equiv (p \vee \neg p) \wedge (p \vee q) \quad &\text{part (c), distributive} \\
&\equiv t \wedge (p \vee q) \quad &\text{part (e), negation} \\
&\equiv (p \vee q) \wedge t \quad &\text{part (a), commutative} \\
&\equiv p \vee q \quad &\text{part (d), identity}
\end{aligned}
$$

In each step, we have replaced part of the expression with an equivalent expression. For example, in the first step we use the second of the two *distributive* properties to replace the entire expression by the equivalent expression. In the second step, we replace $(p \vee \neg p)$ with t by quoting the first of the two *negation* properties. □

Practice Problem 9 *Verify $p \wedge (\neg p \vee q) \equiv p \wedge q$ using the substitution rule and quoting the appropriate parts of Theorem 2.*

Solutions to Practice Problems

1 Here is the truth table:

p	q	Statement 1 We are Both Telling the Truth	Statement 2 A is Lying
T	T	T	F
T	F	F	F
F	T	F	T
F	F	F	T

The third row is the only one that is consistent, so we conclude that A is lying and B is telling the truth.

2 (a) "A is lying" is represented by $\neg p$.
   (b) "At least one of us is truthful" is represented by $p \vee q$.
   (c) "Either B is lying or A is" is represented by $\neg q \vee \neg p$.

   (d) "Exactly one of A and B is lying" is represented by $(p \wedge \neg q) \vee (q \wedge \neg p)$.

3 (a) "A and B are lying" is represented by $\neg p \wedge \neg q$.
   (b) "All are lying" is represented by $\neg p \wedge \neg q \wedge \neg r$.
   (c) "One of us is lying" (take this to mean "at least one of us is lying") is represented by $\neg p \vee \neg q \vee \neg r$.
   (d) "Exactly one is truthful" is represented by $(p \wedge \neg q \wedge \neg r) \vee (q \wedge \neg p \wedge \neg r) \vee (r \wedge \neg p \wedge \neg q)$.

4 The complete truth table is shown below.

p	q	$p \vee q$	$\neg (p \vee q)$
T	T	T	F
T	F	T	F
F	T	T	F
F	F	F	T

5 We write p as "$g \geq 3.0$" and $\neg p$ as "$g < 3.0$."

6   By the second of DeMorgan's laws, "¬(j ∧ e)" is equivalent to "¬j ∨ ¬e." This says that (since she isn't both)
it must be true either that Sue is not a junior or that she
is not an English major.

7   (a) We write ¬p as "($y ≥ 0$) ∧ ($t ≥ 0$)." In English,
p says that the temperature was below 0 at least
one of the two days, and ¬p says it was at or above
0 both days.

(i) "All are lying" is represented by ¬p ∧ ¬q ∧
¬r. The negation is ¬(¬p ∧ ¬q ∧ ¬r). By
DeMorgan's laws, to negate we change **and** to
or and negate each part, giving p ∨ q ∨ r. Observe that the negation of "All are lying" is "At
least one is telling the truth."

(ii) "At least one of us is lying" is represented by
¬p ∨ ¬q ∨ ¬r. The negation is ¬(¬p ∨ ¬q ∨

¬r), which simplifies to p ∧ q ∧ r. The negation
of "At least one of us is lying" is "All are telling
the truth."

8   One way to lay out the truth table is given below.
Because the fifth and eighth columns are identical, the
two expressions at the top of those columns are logically
equivalent.

9   This problem is the "dual" of the previous example—
each ∧ has been changed to ∨ and vice versa. It is not
surprising that the solution mimics the solution to the
example. The duality principle will be addressed further
in Section 3.4.

$$p ∧ (¬p ∨ q) ≡ (p ∧ ¬p) ∨ (p ∧ q)$$ part (c), distributive
$$≡ c ∨ (p ∧ q)$$ part (e), negation
$$≡ (p ∧ q) ∨ c$$ part (a), commutative
$$≡ p ∧ q$$ part (d), identity

p	q	r	$q ∧ r$	$p ∨ (q ∧ r)$	$p ∨ q$	$p ∨ r$	$(p ∨ q) ∧ (p ∨ r)$
T	T	T	T	T	T	T	T
T	T	F	F	T	T	T	T
T	F	T	F	T	T	T	T
T	F	F	F	T	T	T	T
F	T	T	T	T	T	T	T
F	T	F	F	F	T	F	F
F	F	T	F	F	F	T	F
F	F	F	F	F	F	F	F

Truth Table for Practice Problem 8.

Exercises for Section 1.3

1. Solve each of these logic puzzles by using truth tables.
   (a) You come across two inhabitants of Smullyan's Island. A says, "We are both telling the truth," and
   B says, "A is lying." Who if anyone is telling the
   truth?
   (b) You come across three inhabitants of Smullyan's Island. A says, "B or C is lying," B says, "C is lying,"
   and C says, "A and I are both telling the truth." Who
   if anyone is telling the truth?
   (c) You come across three inhabitants of Smullyan's
   Island. A says, "B and C are both lying," B says,
   "Only one of the other two is lying," and C says,
   "At least one of us is lying." Who if anyone is telling
   the truth?

2. Give an example of what two people might say to create a paradox, where each person's individual statement
   does not on its own create a paradox.

3. Suppose you meet two inhabitants A and B of
   Smullyan's Island. Let p represent the statement "A is
   truthful" and q represent the statement "B is truthful."
   Write each of the following in symbolic logic notation:
   (a) A is lying or B is telling the truth.
   (b) Neither A nor B is lying.
   (c) A is truthful but B is not.

4. Upon meeting a third island inhabitant C, you can continue the previous exercise, adding r to represent the
   statement "C is truthful." Use these names along with
   the basic logic operations to write each of the following
   English sentences in symbolic logic notation:
   (a) A is lying and B or C is truthful.
   (b) A and B are lying, or A and C are truthful.
   (c) At least two people are telling the truth.
   (d) Exactly two people are telling the truth.

5. Use the letter f to represent the statement "this person is female," a for "this person is over age 30," and m for "this person is a math major." Use these names along with the basic operations **and** ($\wedge$), **or** ($\vee$), and **not** ($\neg$) to write each of the following English sentences with symbolic logic:

    (a) This person is female but not a math major.

    (b) This person is a female math major over age 30.

    (c) This person is neither female nor a math major.

    (d) Either this person is not female or this person is over age 30.

6. Use the phrases in the previous exercise to write an English statement equivalent to each of the following propositions:

    (a) $\neg f \wedge m$

    (b) $\neg f \wedge \neg a$

    (c) $f \wedge (a \vee m)$

    (d) $(f \wedge a) \vee (\neg f \wedge m)$

7. Use the letter t to represent the statement "Bill is tall," d for "Bill is dark," and h for "Bill is handsome." As in Exercise 5, use these names along with the basic logic operations to write each of the following English sentences in symbolic logic notation:

    (a) Bill is tall, dark, and handsome

    (b) Bill is tall and dark, but not handsome.

    (c) Either Bill is tall or he is handsome, but not both.

    (d) Bill is neither tall nor handsome.

8. Use the phrases in the previous exercise to write an English statement equivalent to each of the following propositions:

    (a) $(t \vee d \vee h) \wedge \neg(t \wedge d \wedge h)$

    (b) $\neg t \wedge \neg d$

    (c) $d \wedge \neg(t \wedge \neg d)$

    (d) $(t \wedge d) \vee (\neg t \wedge d)$

9. Use the letter s to represent the statement "Chris likes to play soccer," r for "Chris likes to read," and p for "Chris likes to eat pizza." As before, use these names along with the basic logic operations to write each of the following English sentences in symbolic logic notation:

    (a) Chris likes pizza but he does not like soccer.

    (b) Chris likes to read and eat pizza, or he likes to play soccer.

    (c) Chris does not like to eat pizza, but he likes to play soccer or read.

    (d) Chris likes to do two of these things but not all three.

10. Suppose x and y indicate particular real numbers. Write conditions that express the following, by using comparisons (such as $x > 0$) and the basic operations of logic.

    (a) Both x and y are positive.

    (b) At least one of x and y is positive.

    (c) Exactly one of x and y is positive.

    (d) Neither x nor y is positive.

11. Complete the following truth tables for the given compound expressions:

(a)

p	q	$\neg p$	$\neg p \vee q$	$p \wedge (\neg p \vee q)$
T	T			
T	F			
F	T			
F	F			

(b)

p	q	$\neg p$	$p \vee q$	$\neg p \wedge (p \vee q)$
T	T			
T	F			
F	T			
F	F			

(c)

p	q	$p \vee q$	$\neg p \vee q$	$(p \vee q) \wedge (\neg p \vee q)$
T	T			
T	F			
F	T			
F	F			

(d)

p	q	r	$q \vee r$	$p \wedge (q \vee r)$
T	T	T		
T	T	F		
T	F	T		
T	F	F		
F	T	T		
F	T	F		
F	F	T		
F	F	F		

(e)

p	q	r	$p \wedge q$	$(p \wedge q) \vee r$
T	T	T		
T	T	F		
T	F	T		
T	F	F		
F	T	T		
F	T	F		
F	F	T		
F	F	F		

12. Rewrite each of the following statements in propositional logic notation, making the meaning of your propositional variables clear, and then create a truth table for the sentence. The first one is done for you as an example.

(ex) *Either the food is terrific, or everyone is tired and hungry at mealtimes.* Let f represent "the food is terrific," t represent "everyone is tired," and h represent "everyone is hungry." Then the given statement can be written as $f \vee (t \wedge h)$, and it has the following truth table:

f	t	h	$t \wedge h$	$f \vee (t \wedge h)$
T	T	T	T	T
T	T	F	F	T
T	F	T	F	T
T	F	F	F	T
F	T	T	T	T
F	T	F	F	F
F	F	T	F	F
F	F	F	F	F

(a) Either everyone is not hungry at mealtime, or everyone is tired and the snackbar makes a profit.

(b) The staff is friendly, or else they are not friendly but very well paid.

(c) The staff is not very well paid, and they are friendly.

(d) You play sports or you play mahjong or nobody knows your name.

(e) You do not play mahjong or you do not play sports, and nobody knows your name.

13. For each of the statements in Exercise 12, obtain a simple symbolic logic expression for the **negation** of that statement, and then rewrite this as a proper English sentence.

14. Use the number variable b to represent Trina's board exam math score and m to represent Trina's math placement test score. To take calculus the first semester, a student must have a board exam math score of at least 600 or a math placement test score of at least 25. Use the variables b and m to express the statement "Trina may take calculus in her first semester." Use the variables b and m to express the negation of this statement as well.

15. Use the number variable a to represent Fred's age and v to represent the number of years Fred has worked at his job. To be eligible for the company's retirement plan, an employee must be at least 62 years of age and have worked for at least 15 years. Use the variables a and v to express the statement "Fred is eligible for the company's retirement plan." Use the variables a and v to express the negation of this statement as well.

16. Use truth tables to verify that $\neg(p \wedge q)$ is logically equivalent to $\neg p \vee \neg q$, establishing the second of DeMorgan's laws in Theorem 2.

17. Use truth tables to verify that $p \vee (p \wedge q)$ is logically equivalent to p, establishing the second absorption property in Theorem 2.

18. Use truth tables to verify that $p \wedge (q \vee r)$ is logically equivalent to $(p \wedge q) \vee (p \wedge r)$, establishing the first distributive property in Theorem 2.

19. Use truth tables to verify the first versions of the commutative and associative properties from Theorem 2.

20. Use truth tables to verify the first versions of the identity, idempotent, and universal bound properties from Theorem 2.

21. Use truth tables to check if each of the given pairs of symbolic logic statements are equivalent.

(a) $p \wedge (\neg p \vee q)$ and $\neg p \wedge (p \vee q)$
(b) $(p \vee q) \wedge (\neg p \vee q)$ and q
(c) $(\neg q \wedge p) \vee (\neg p \wedge q)$ and $\neg p \vee \neg q$
(d) $p \wedge (q \vee r)$ and $(p \wedge q) \vee r$
(e) $(p \vee q) \wedge (q \vee r)$ and $(p \wedge r) \vee q$

22. For each of the following statements, rewrite them in propositional logic notation, making the meaning of your propositional variables clear. Use truth tables to find any pairs of logically equivalent statements.

(a) Jillian likes playing in the sand or volleyball, but she does not like sailing.
(b) Jillian likes playing in the sand, and she likes sailing or volleyball.
(c) Jillian likes playing in the sand and volleyball, or she likes sailing.
(d) Jillian likes playing in the sand and sailing, or she likes volleyball and sailing.

23. Use the double negative property and DeMorgan's laws to rewrite each of the following as an equivalent statement that never has the **not** symbol ($\neg$) outside of a parenthesized expression. The first one is done for you as an example.

(ex) $\neg(p \wedge \neg q)$ is equivalent to $\neg p \vee \neg(\neg q)$ by DeMorgan's law, and this is equivalent to $\underline{\neg p \vee q}$ by the double negative property.

(a) $\neg(\neg p \vee \neg q)$
(b) $\neg(\neg(\neg p))$
(c) $\neg(\neg(\neg p \wedge q))$
(d) $\neg(p \wedge (q \vee \neg p))$

24. By quoting the parts of Theorem 2, verify the following logical equivalences. In each case, start from the left side and use parts of the theorem to change the problem, ending up with the right side. (See Example 15 and the solution to Practice Problem 9.)

(a) $(p \wedge \neg q) \vee p \equiv p$

(b) $(p \vee t) \wedge (p \vee c) \equiv p$

(c) $q \wedge (p \vee r) \equiv (p \wedge q) \vee (q \wedge r)$

25. By using parts of Theorem 2 as algebraic rules, simplify the expression $(p \vee q) \wedge \neg(\neg p \wedge r)$ as much as you can.

1.4 Predicates

In Section 1.3, we worked with propositions such as "Sue is an English major" and "Person A is lying." In each case, the subject is a particular person or other entity. However, in many circumstances the subject is not a particular fixed entity. For example, consider the question "How many students in my discrete math class are English majors?" To answer the question, you would check the truth value of the sentence "_____ is an English major," filling in the blank with the name of each person in the class in turn.

In mathematics, we call a statement of this form a *predicate*. One way to relate to the concept is to think of the meaning of the word *predicate* in an English class. In the English sentence "Mr. Morton wrote Pearl a poem," the predicate is "wrote Pearl a poem," and it tells what Mr. Morton did.* Since predicates do not have subjects, they are not complete sentences. However, with the addition of any subject, you will get some sentence. For example, the predicate "_____ is an English major" can become the propositions "Tom Myers is an English major" or "Mary Johnson is an English major" by adding the appropriate subjects.

In mathematics, it is customary to use a variable rather than an empty blank to represent the missing subject. If we use the variable s (short for *student*), the sentence above would become "s is an English major." Each time we replace s by the name of a student in the class, we get a statement that is either true or false. Sentences of this form, containing one or more variables, are discussed in this section.

Simple Predicates and Their Negations

The preceding discussion leads us to our formal definition.

> *Definition* A *predicate* $P(x)$ is a statement that incorporates a variable x, such that whenever x is replaced by a value, the resulting proposition is unambiguously true or false. If the predicate uses several variables $x_1, x_2, \ldots$, then we will extend the notation to $P(x_1, x_2, \ldots)$ accordingly.

Example 1 *If $P(n)$ is the predicate "n is even," write the proposition that results when the variable n is replaced by each of the values 2, 17, and 240. Which of the resulting statements are true?*

---

* Those of you who are of a certain age will recognize this character from Saturday morning's *Schoolhouse Rock.*

SOLUTION We will write $P(2)$ to denote the proposition resulting from replacing n with 2 in the predicate $P(n)$. In this case, $P(2)$ is the true proposition "2 is even." Similarly, $P(17)$ denotes replacing n by 17, giving us the false statement "17 is even." Finally, $P(240)$ is the true statement "240 is even." ☐

Explore more on the Web.

Practice Problem 1 *In each of the predicates below, replace x with each of the values 2, 23, –5, and 15. Decide if each resulting proposition is true or false.*

(a) $P(x)$ *is the predicate "$x > 15$."*

(b) $Q(x)$ *is the predicate "$x \leq 15$."*

(c) $R(x)$ *is the predicate "$(x > 5) \wedge (x < 20)$."*

Most of what you have learned about propositions has an obvious analogue for predicates. For example, we can combine predicates using the operation symbols $\wedge$ (**and**), $\vee$ (**or**), $\neg$ (**not**) to create new predicates. We can talk about the **negation** of a predicate, and we can make sense of the notion of **equivalent** predicates. For example, here are two equivalent ways to express the negation of the predicate "$x < 0$":

● $\neg(x < 0)$: in English, "x is not less than 0."

● $x \geq 0$: in English, "x is greater than or equal to 0."

These statements are equivalent because, no matter what number is substituted for the variable x, the resulting propositions have the same truth value. In addition, the "double negative" property and DeMorgan's laws apply to predicates just as they do to propositions.

Example 2 *Table 1-9 shows some predicates and their negations. The final example shows a predicate with more than one variable.*

In our next example, we use a simple notation for describing a collection of numbers, written as what we call a *set*. We will study sets in detail in Chapter 3. For now, all you need to know is that our sets are collections of objects, and that the objects in that collection are called *elements* of the set or *members* of the set.

Example 3 *Consider the set $D = \{1, 2, 3, 4, 5, 6, 7, 8, 9, 10\}$. For each predicate below, make a list of the members of the set D that make the statement true. Also,*

Predicate	Negation of Predicate
$x > 5$	$x \leq 5$
$(x > 0) \wedge (x < 10)$	$(x \leq 0) \vee (x \geq 10)$
$\neg(x = 8)$	$x = 8$
$(x \geq 0) \vee (y \geq 0)$	$(x < 0) \wedge (y < 0)$

Table 1-9 Table for Example 2

form the negation of the predicate and list the members of D that make the negation true.

1. $x \geq 8$
2. $(x > 5) \wedge (x \text{ is even})$
3. $x^2 = x$
4. $x > 0$
5. $x > x^2$

SOLUTION   We present the solution in Table 1-10. The negation is true for precisely those numbers in the set D for which the original statement is not true.   □

Truth and Quantifiers

Unlike propositions, we almost never talk about predicates being simply true or false. What is more important is the truth or fallacy of the proposition we get *after* substituting a value for the variable in a predicate. Hence, our discussion of the "truth" of predicates must include the set that these values come from. We will call this set the *domain* of the predicate, and we will never discuss the truth of predicates without making the domain clear.

Given a predicate and domain, the natural question to ask is, "Does the predicate become a true proposition when its variable is replaced by members of the domain?" The three most common answers to this question are **always, sometimes**, or **never**.

1. In Example 3, the answer is "always" for the fourth predicate but not for the others. When the answer is "always," we might use phrases such as "for all members of the set, the statement is true," or "for every element of the set, the statement is true" in describing the situation.

2. In Example 3, the answer is "sometimes" for all but the fifth predicate. When we answer "sometimes" to this question, we are likely to use phrases similar to "there is an element of the set that makes the statement true" or "there exists a member of the set such that the predicate is satisfied."

3. In Example 3, the answer is "never" for the fifth predicate. This is not really necessary to treat separately since "never" is simply the negation of "sometimes." We will address this case in the exercises at the end of this section.

 Example 4 *Here are two statements that use this terminology:*

1. *"For every k that is a member of the set {1, 2, 3, 4, 5}, it is true that k < 20."*

	True for ...	Negation	True for ...
1.	8, 9, 10	$x < 8$	1, 2, 3, 4, 5, 6, 7
2.	6, 8, 10	$(x \leq 5) \vee (x \text{ is odd})$	1, 2, 3, 4, 5, 7, 9
3.	1	$x^2 \neq x$	2, 3, 4, 5, 6, 7, 8, 9, 10
4.	All the members	$x \leq 0$	None of the members
5.	None of the members	$x \leq x^2$	All the members

Table 1-10   Table for Example 3

Predicate	True for These Members of D...	True for At Least One?	True for All?
$x < 0$			
$x > -3$			
$x^2 < x$			
$x^2 \geq x$			

Table 1-11 Table for Practice Problem 2

2. *"There exists a member m of the set $\{-1, 0, 1\}$ with the property that $m^2 = m$."*
   Notice that "there exists" always means "there exists (at least one)."

Explore more on the Web.

Practice Problem 2 *In Table 1-11, list the members of the set $D = \{-1, 0, 1, 2\}$ that make each predicate true. In addition, answer the two questions: (1) Is the predicate true for at least one member of the set? and (2) Is the predicate true for all the members of the set? When the answer to either question is "yes," write the answer as a complete sentence using phrases similar to "there exists" or "for all."*

The phrases "for all," "there exists," "in the set," and so on are used so often that special notation has been developed to make them shorter to write.

Definition

1. The symbol $\in$ indicates membership in a set. For example, "$k \in D$" means that k is a member of the set D.
2. The symbol $\forall$ means "for all" or "for every." Notice that this is an upside-down letter A, and that the letter A is the first letter in the word "all."
3. The symbol $\exists$ means "there is (at least one)" or "there exists (at least one)." Notice that this is a backwards letter E, and that the letter E is the first letter in the word "exists."
4. The symbols $\forall$ and $\exists$ (and the corresponding phrases such as "for all" and "there exists") are called **quantifiers**. When we use quantifiers with a predicate, we refer to the resulting statement as a **quantified predicate**.
5. A **counterexample** is an example illustrating that a "for all" statement is false.

A quantified predicate is unambiguously either true or false. If the domain is small, we can usually determine whether the sentence is true or false by simply substituting each domain element for the variable(s) in the predicate. For larger domains, including infinite sets, it can be more difficult to determine.

Example 5 *Let $D = \{3, 4, 5, 10, 20, 25\}$. Express each sentence using the special symbols $\in$, $\forall$, and $\exists$, and decide whether or not it is true.*

1. *For every n that is a member of D, n < 20.*
2. *For all n in the set D, n < 5 or n is a multiple of 5.*
3. *There is (at least one) k in the set D with the property that k^2 is also in the set D.*
4. *There exists m a member of the set D such that m ≥ 3.*

SOLUTION

1. $\forall n \in D, n < 20$. This statement is false, since 20 and 25 are counterexamples.

2. $\forall n \in D, n < 5$ or n is a multiple of 5. This statement is true.

3. $\exists k \in D$ such that $k^2 \in D$. This statement is true, since $5^2 \in D$. Sometimes we replace the words "such that" with a comma, as in $\exists k \in D, k^2 \in D$. When this is read aloud, words similar to "such that" or "with the property that" are usually added to make it read more like ordinary English.

4. $\exists m \in D$ such that $m \geq 3$, or more simply, just $\exists m \in D, m \geq 3$. This statement is true.

□

Explore more on the Web.

Practice Problem 3 *Translate each symbolic quantified predicate into an English-language sentence about the set $D = \{-2, -1, 0, 1, 2\}$, and then decide whether or not it is true.*

(a) $\forall d \in D, d > -2$

(b) $\exists d \in D, d > -2$

(c) $\forall n \in D, (n > -3) \wedge (n < 3)$

(d) $\forall x \in D, x^2 \leq 4$

(e) $\exists m \in D, m > 10$

When we write a quantified predicate, there is always a set D, called the domain, that we have in mind. If you are being informal, you might not specify the set, or you might specify the set without using the set notation. A useful practice, when faced with a quantified statement written in this less formal style, is to identify the set and rewrite the statement in terms of that set.

The clear specification of the domain is very important in determining whether a quanitified statement is true. The next example illustrates this issue as well as the practical skill of making informal quantified statements more formal.

Example 6 *For each quantified statement, determine the domain D and rewrite the predicate formally in terms of that set D. If the domain is ambiguous, give examples of how different domains can change the truth value of the statement.*

1. *For all x, $x^2 \geq x$.*
2. *∀ even integer m, m ends in the digit 0, 2, 4, 6, or 8.*
3. *There is an integer n whose square root is also an integer.*
4. *Every integer larger than 0 has a square that is larger than 0.*

SOLUTION

1. There are several possibilities. One possibility is that the person writing the predicate was intending to describe a property of the set of all real numbers,

frequently written as $\mathbb{R}$. If we use this notation for the domain, we can write, "$\forall x \in \mathbb{R}, x^2 \geq x$." This statement is false since $x = 0.5$ is a counterexample.

If instead we take the domain to be the set of integers (commonly written $\mathbb{Z}$ after the German word for numbers, *Zahlen*), then we have the *true* statement, "$\forall x \in \mathbb{Z}, x^2 \geq x$."

2. Here D is the set of even integers, and we write

$$\forall m \in D, m \text{ ends in the digit } 0, 2, 4, 6, \text{ or } 8$$

3. Using $\mathbb{Z}$ once again to indicate the set of all integers, we can write, "$\exists k \in \mathbb{Z}, \sqrt{k} \in \mathbb{Z}$."

4. Here the domain D is the set of positive integers, and we write, "$\forall n \in D, n^2 > 0$."

Negating Quantified Statements

In Practice Problem 3, two of the statements are false. As before, when a statement is false, the *negation* of that statement is true. Looking more closely at these two statements in this Practice Problem will help us determine the formal meaning of "negation" for a quantified statement.

Example 7 *For the set $D = \{-2, -1, 0, 1, 2\}$, explain why each of these quantified predicates is false. Use your explanation to write the negation of the statement, first in English and then with more formal notation.*

1. $\forall d \in D, d > -2$
2. $\exists m \in D, m > 10$

SOLUTION

1. If we write $P(d)$ for the predicate "$d > -2$," then $P(-1), P(0), P(1)$, and $P(2)$ are all true. However, $P(-2)$ is false, so we cannot say that *all* the elements of D make $P(d)$ true. That is, *there exists (at least one) element d of D for which P(d) is false*. This negation can be formally written as "$\exists d \in D, \neg P(d)$," or "$\exists d \in D, d \leq -2$."

2. If we write $Q(m)$ for the predicate "$m > 10$," then $Q(-2), Q(-1), Q(0), Q(1)$, and $Q(2)$ are all false, so we cannot find an element of D that makes $Q(m)$ true. That is, *for all elements m of D, Q(m) is false*. This negation can be formally written as "$\forall m \in D, \neg Q(m)$," or "$\forall m \in D, m \leq 10$." $\square$

In words, we can say that the opposite of "true for all" is "false for at least one." Likewise, the opposite of "true for at least one" is "false for all." Using the notation of this section, we can summarize the relationship between a quantified statement and its negation in the following proposition.

Proposition 1 *For any predicates P and Q over a domain D,*

● *The negation of* $\forall x \in D, P(x)$ *is* $\exists x \in D, \neg P(x)$.
● *The negation of* $\exists x \in D, Q(x)$ *is* $\forall x \in D, \neg Q(x)$.

Of course, when we negate the predicates themselves, we use the same ideas we learned about earlier. In particular, there are two important points to remember:

1. Techniques such as the double negative property and DeMorgan's laws apply to predicates.
2. Particular care must be taken when negating comparisons. For example, the negation of a comparison involving $\leq$ will involve $>$.

Explore more on the Web.

Practice Problem 4 *Let $D = \{-1, 0, 1, 2\}$. For each statement, write the negation, and then decide which is true, the original statement or the negation.*

(a) $\forall x \in D, (x \leq 0) \vee (x \geq 2)$

(b) $\exists x \in D, (x < 0) \wedge (x^2 > 0)$

(c) *For all $x \in D, x^2 < x$.*

(d) *There exists $x \in D$ such that $x^2 < x$.*

Multiple Quantifiers and Their Negations

We have said that a predicate can have more than one variable. For example, consider the predicate $P(x, y)$ given as $x + 2y = 3$, where x and y stand for integers. That is, x and y are to be taken from the set $\mathbb{Z} = \{0, \pm 1, \pm 2, \pm 3, \ldots\}$. If we wish to substitute particular values (say, 10 and 4) for the variables, we use the notation $P(10, 4)$. This notation indicates that we replace x by 10 and y by 4 in the predicate, obtaining the (false) proposition $10 + 2 \cdot 4 = 3$. As usual, when we use two variables, we do not imply that the variables *must* indicate different quantities, only that they *can* indicate different quantities. For example, it is perfectly correct to write $P(-5, -5)$ to indicate the statement $(-5) + 2 \cdot (-5) = 3$.

Example 8 *Let $P(x, y)$ be the predicate $x \cdot y = 36$.*

1. *Identify which of the following statements are true: $P(3, 4)$, $P(-9, -4)$, $P(12, -1)$.*
2. *Assuming the domain is the set of integers, find other values for x and y that make the statement $P(x, y)$ true.*

SOLUTION

1. The statements are $3 \cdot 4 = 36$, $(-9) \cdot (-4) = 36$, and $12 \cdot (-1) = 36$. Only the second statement is true, the first and third are false.
2. Here are some additional true statements: $P(36, 1)$, $P(6, 6)$, and $P(-2, -18)$. There are several more that are true. Notice that if the variables x and y are taken from the domain of all real numbers rather than all integers, then there are an infinite number of choices for x and y.

□

When a predicate contains more than one variable, the question of quantifiers becomes more complicated since each variable can be quantified separately. Fortunately, some situations are relatively simple. If all the quantifiers are the same, there is no difficulty in making the proper interpretation.

 **Example 9**

1. *Consider the quantified statement "There exist integers x and y such that $x \cdot y = 36$." Write this using symbolic logic notation, and decide if it is true or not.*

2. *Do the same for the quantified statement "For all integers x and y, it is true that $x \cdot y = 36$."*

SOLUTION

1. We can write this in two equivalent ways:
   - $\exists x \in \mathbb{Z}, \exists y \in \mathbb{Z}, x \cdot y = 36$
   - $\exists x, y \in \mathbb{Z}, x \cdot y = 36$

   The latter is a shortcut that means the same as the former. In either case, we can use $P(x, y)$ to indicate the predicate $x \cdot y = 36$, and write $\exists x \in \mathbb{Z}, \exists y \in \mathbb{Z}, P(x, y)$, or $\exists x, y \in \mathbb{Z}, P(x, y)$. The quantified predicate is true since, for example, $P(-2, -18)$ is true.

2. Similarly, we can write the "for all" quantifiers in two equivalent ways:
   - $\forall x \in \mathbb{Z}, \forall y \in \mathbb{Z}, x \cdot y = 36$
   - $\forall x, y \in \mathbb{Z}, x \cdot y = 36$

   Since $P(10, 2)$ is the false statement "$10 \cdot 2 = 36$," we see that the quantified predicate is false, and we say that $x = 10$, $y = 2$ is a counterexample to the statement $\forall x, y \in \mathbb{Z}, x \cdot y = 36$.

   □

Practice Problem 5 *Write the following as quantified statements using the symbols $\exists$ and $\forall$, and decide whether the statement is true:*

(a) *There are odd integers m and n whose product is 35.*

(b) *There are even integers m and n whose product is 35.*

(c) *For every choice of integers s and t, it is true that $s^2 + t^2 \geq 0$.*

(d) *For every choice of real numbers x, y, and z, $x + y + z > 1$.*

The example and practice problem illustrate two important points. First, the *domain* of the variables is very important. If the domain D is the set of odd integers, the quantified statement $\exists m, n \in D, m \cdot n = 35$ is true. However, the same statement is false if the domain D is the set of even integers. Second, there are some situations where establishing the truth of a quantified predicate is easy, and others where it takes more work.

A more difficult situation occurs when a quantified predicate contains *both* $\exists$ and $\forall$ quantifiers in the same statement. Even with only two variables, there are several different possibilities for the arrangement of the quantifiers and variables. The key issues can be understood by thinking carefully about the difference between these two quantified statements:

- $\forall x, \exists y, P(x, y)$
- $\exists y, \forall x, P(x, y)$

The first thing to note is that, in mathematics, we always interpret these from left to right, while ordinary English is often ambiguous. For example, "For every problem there is a solution," and "There is a solution for every problem," are generally taken to mean the same thing. The listener forms the meaning from the context and, to a certain extent, from the speaker's intonations. In a mathematical setting, we are always more precise in our speaking, writing, and interpretation.

Example 10 *Let $\mathbb{Z}$ indicate the set of all integers. Which of the following quantified predicates statements are true? Explain.*

1. $\forall x \in \mathbb{Z}, \exists y \in \mathbb{Z}, x + 2y = 3$
2. $\forall x \in \mathbb{Z}, \exists y \in \mathbb{Z}, x + y = 15$
3. $\exists y \in \mathbb{Z}, \forall x \in \mathbb{Z}, x + y = 15$

SOLUTION

1. For every possible value of x, you must be able to find a corresponding value for y that makes the predicate "$x + 2y = 3$" true. To see if this is can be done, think of it as a game: Your opponent chooses x, then you try to find a y that makes $P(x, y)$ true.

   (a) The opponent chooses $x = 1$, and you choose $y = 1$, making $P(x, y)$ the true statement $1 + 2 \cdot 1 = 3$.
   (b) The opponent chooses $x = -3$, and you choose $y = 3$, making $P(x, y)$ the true statement $-3 + 2 \cdot 3 = 3$.
   (c) The opponent chooses $x = 1{,}337$, and you choose $y = -667$, making $P(x, y)$ the true statement $1{,}337 + 2 \cdot (-667) = 3$.
   (d) The opponent chooses $x = 0$, and you give up. By algebra you can see that you would have to choose $y = \frac{3}{2}$, but this is not a member of the set $\mathbb{Z}$.

   The quantified statement is false, and $x = 0$ is a counterexample. If x has the value 0, no possible choice for y can make "$x + 2y = 3$" true.

2. Again, think of this as a game. Your opponent chooses a value for x, and you must find a corresponding value for y. If you play the game with an opponent, you may discover a pattern for how you can choose your value for y, once your opponent has chosen his or her value for x. The strategy is "Always choose $y = 15 - x$." This quantified predicate is a true statement.

3. To the casual reader, this may appear the same as the previous example. However, because we read left to right, the rules of the game have changed in a subtle but crucial way. This time, you must go first, so you must try to find a y value that will work no matter how your opponent chooses his or her value for x. Clearly, this is impossible to do—no matter what y you choose, your opponent will have an infinite number of x values to choose from that make the predicate $x + y = 15$ false. For example, if you choose $y = 73$, your opponent only has to make the predicate $x + 73 = 15$ false, and that is easy to do. The quantified predicate is false.

□

Practice Problem 6   *Let $\mathbb{Z}$ indicate the set of all integers and $\mathbb{R}$ the set of real numbers. Which of the following quantified predicates are true? Explain.*

(a) $\forall y \in \mathbb{Z}, \exists x \in \mathbb{Z}, x + 2y = 3$ *(c)* $\exists x \in \mathbb{Z}, \forall y \in \mathbb{Z}, x \cdot y = x$

(b) $\forall x \in \mathbb{R}, \exists y \in \mathbb{R}, x \cdot y = 1$ *(d)* $\exists x \in \mathbb{Z}, \forall y \in \mathbb{Z}, x \cdot y = y$

We now turn to the question of negating predicates that have multiple quantifiers. Actually, we already know how to do this, if we just keep in mind that we read the statements left to right. Recall from Proposition 1 that

The negation of $\forall x \in D, P(x)$ is $\exists x \in D, \neg P(x)$.

The negation of $\exists x \in D, Q(x)$ is $\forall x \in D, \neg Q(x)$.

When there is more than one quantifier, we apply this same process to each quantifier in turn, proceeding from left to right. The following example and practice problem illustrate the process.

Example 11   *Write the negation of each of these statements, simplified so as not to require the $\neg$ symbol to the left of any quantifier.*

1. $\forall x \in \mathbb{Z}, \exists y \in \mathbb{Z}, x + 2y = 3$

2. $\exists x > 0, \forall y > 0, x \cdot y < x$

3. $\exists x \in \mathbb{Z}, \exists y \in \mathbb{Z}, x + y = 13,$ *and* $x \cdot y = 36$

SOLUTION   In each part, we proceed in several steps. First we simply put a negation symbol ($\neg$) in front of the entire statement. Then we apply the negation process to each quantifier in turn and finally to the predicate itself.

1. $\neg(\forall x \in \mathbb{Z}, \exists y \in \mathbb{Z}, x + 2y = 3)$ negation of original statement
    $\exists x \in \mathbb{Z}, \neg(\exists y \in \mathbb{Z}, x + 2y = 3)$ by Proposition 1
    $\exists x \in \mathbb{Z}, \forall y \in \mathbb{Z}, \neg(x + 2y = 3)$ by Proposition 1
    $\exists x \in \mathbb{Z}, \forall y \in \mathbb{Z}, x + 2y \neq 3$ equivalent form of "not equal"

2. $\neg(\exists x > 0, \forall y > 0, x \cdot y < x)$ negation of original statement
    $\forall x > 0, \neg(\forall y > 0, x \cdot y < x)$ by Proposition 1
    $\forall x > 0, \exists y > 0, \neg(x \cdot y < x)$ by Proposition 1
    $\forall x > 0, \exists y > 0, x \cdot y \geq x$ equivalent form of "not less than"

3. $\neg(\exists x \in \mathbb{Z}, \exists y \in \mathbb{Z}, (x + y = 13) \wedge$ negation of original statement
    $(x \cdot y = 36))$
    $\forall x \in \mathbb{Z}, \neg(\exists y \in \mathbb{Z}, (x + y = 13) \wedge$ by Proposition 1
    $(x \cdot y = 36))$
    $\forall x \in \mathbb{Z}, \forall y \in \mathbb{Z}, \neg((x + y = 13) \wedge$ by Proposition 1
    $(x \cdot y = 36))$
    $\forall x \in \mathbb{Z}, \forall y \in \mathbb{Z}, (\neg(x + y = 13) \vee$ by DeMorgan's laws
    $\neg(x \cdot y = 36))$
    $\forall x \in \mathbb{Z}, \forall y \in \mathbb{Z}, (x + y \neq 13) \vee$ equivalent form of "not equal"
    $(x \cdot y \neq 36)$

$\square$

Practice Problem 7 *Negate each quantified predicate. Which is true, the original statement or its negation? The domain for the first statement is $\mathbb{R}$, the set of real numbers.*

(a) $\forall x > 0, \exists y \in \mathbb{R}, (y > x) \wedge (x + y = 2x)$

(b) $\exists x \in \mathbb{Z}, \forall y \in \mathbb{Z}, x \cdot y \le 0$

(c) $\forall x, y, z \in \mathbb{Z}, x^2 + y^2 + z^2 \ge 0$

Solutions to Practice Problems

1 (a) When x is replaced by 2, the statement becomes "$2 > 15$," a false statement. When x is replaced by 23, it becomes "$23 > 15$," which is true. Similarly, "$-5 > 15$" and "$15 > 15$" are both false. Thus, only $P(23)$ is true.

   (b) We see that $Q(2)$, $Q(-5)$, and $Q(15)$ are true. Observe that "$x \le 15$" is true for precisely the values of x for which the predicate "$x > 15$" is false.

   (c) For the listed numbers, only $x = 15$ makes this predicate true. (Only $R(15)$ is true.)

2 The completed table is given as Table 1-12. Here are some sentences that express the true cases shown in this table.

   ● There exists an element x of D such that $x < 0$.

   ● There is an element x in D with the property that $x > -3$.

   ● There is a number in the set D whose square is bigger than or equal to itself.

   ● For every element x of the set D, it is true that $x > -3$.

   ● For all x in D, $x^2 \ge x$.

3 (a) For every d in the set D, $d > -2$. This is false since -2 is a counterexample.

   (b) There is an element d in D such that $d > -2$. This is true.

   (c) For every element n of the set D, $n > -3$ and $n < 3$. This is true.

   (d) For all x in D, $x^2 \le 4$. This is true.

   (e) There exists an m in D with the property that $m > 10$. This is false.

4 Table 1-13 shows each original statement along with its negation.

5 (a) $\exists m, n$ odd integers, such that $m \cdot n = 35$. This is true, since $5 \cdot 7 = 35$.

   (b) $\exists m, n$ even integers, such that $m \cdot n = 35$. This is false. It is possible to prove (we will learn how in Chapter 2) that whenever two even integers are multiplied, the result is even.

   (c) $\forall s, t \in \mathbb{Z}, s^2 + t^2 \ge 0$. This is true. One can prove this by showing that when an integer is squared, the result cannot be negative, and by showing that the sum of two nonnegative numbers is never negative.

   (d) $\forall x, y, z \in \mathbb{R}, x + y + z > 1$. This is false. The choice $x = -2$, $y = 0$, $z = 1$ is a counterexample.

6 (a) This is true. If your opponent chooses a value for y, the pattern $x = 3 - 2y$ can be used to find a corresponding value for x.

   (b) This is false, since $x = 0$ is a counterexample—that is, there is no choice of y that makes $0 \cdot y = 1$ true. However, if we change the domain for x to the set of nonzero real numbers, the statement would be true. If it is clear we are talking about real numbers, this might be written as $\forall x \ne 0, \exists y \in \mathbb{R}, x \cdot y = 1$.

   (c) This is true, since we can choose $x = 0$.

   (d) This is also true, since we can choose $x = 1$.

7 (a) The negation is the true statement $\exists x > 0, \forall y \in \mathbb{R}, (y \le x) \vee (x + y \ne 2x)$.

   (b) The negation is the true statement $\forall x \in \mathbb{Z}, \exists y \in \mathbb{Z}, x \cdot y > 0$.

   (c) The negation is the false statement $\exists x, y, z \in Z, x^2 + y^2 + z^2 < 0$. The original statement is true.

Predicate	True for These Members of D...	True for at Least One?	True for All?
$x < 0$	-1	T	F
$x > -3$	$-1, 0, 1, 2$	T	T
$x^2 < x$	None	F	F
$x^2 \ge x$	$-1, 0, 1, 2$	T	T

Table 1-12 Solution for Practice Problem 2

	Statement	Negation	Which is True?
1.	$\forall x \in D, (x \le 0) \vee (x \ge 2)$	$\exists x \in D, (x > 0) \wedge (x < 2)$	Negation
2.	$\exists x \in D, (x < 0) \wedge (x^2 > 0)$	$\forall x \in D, (x \ge 0) \vee (x^2 \le 0)$	Statement
3.	For all $x \in D, x^2 < x$.	$\exists x \in D$ such that $x^2 \ge x$	Negation
4.	There exists $x \in D$ such that $x^2 < x$.	$\forall x \in D, x^2 \ge x$	Negation

Table 1-13 Table for Practice Problem 4

Exercises for Section 1.4

1. Write each of the following predicates using the simple predicates $x > 0$ and $y > 0$ along with the propositional connectives $\wedge$, $\vee$, and $\neg$:

(a) Both x and y are positive.

(b) At least one of x and y is positive.

(c) Exactly one of x and y is positive.

(d) Neither x nor y is positive.

2. For each of the given values of x and y, determine which predicates from Exercise 1 become true statements.

(a) If $x = 8$ and $y = 3$, which of (a–d) are true?

(b) If $x = -5$ and $y = 0$, which of (a–d) are true?

(c) If $x = 7$ and $y = -7$, which of (a–d) are true?

(d) If $x = 0$ and $y = 0$, which of (a–d) are true?

(e) If $x = -3$ and $y = -10$, which of (a–d) are true?

(f) If $x = -8$ and $y = 2$, which of (a–d) are true?

3. For each predicate given in the first column of Table 1-14, list the members of the set $S = \{1, 2, 3, 4, 5, 6, 7, 8, 9\}$ that make the statement true. (Recall that $\mathbb{Z}$ is the set of all integers.)

4. Which elements of the set $D = \{2, 4, 6, 8, 10, 12\}$ make each of the predicates from Exercise 3 true?

5. Which elements of the set $D = \{2, 4, 6, 8, 10, 12\}$ make the **negation** of each of these predicates true?

(a) $Q(n)$ is the predicate, "$n > 10$."

(b) $R(n)$ is the predicate, "n is even."

Predicate	True for These Members of S ...
x is even	
$n > 5$	
m is even and $m > 5$	
$10 < x^2 + 1 < 25$	
$x/3 \in \mathbb{Z}$	
$10/x \in \mathbb{Z}$	

Table 1-14 Table for Exercise 3

(c) $S(k)$ is the predicate, "$k^2 < 1$."

(d) $T(m)$ is the predicate, "$m - 2$ is an element of D."

6. Based on your answers to Exercise 5, circle the true statements in the list given below.

7. Let $D = \{1, 3, 5, 7, 8, 9\}$. Decide whether each of the following statements is true for all the elements of D. For each that is not, give a counterexample. That is, provide a number in D for which the statement is not true.

(a) x is even and $x > 7$.

(b) x is odd or $x > 7$.

(c) x is not odd and $x \le 7$.

(d) x is odd.

$\forall n \in D, Q(n)$	$\exists n \in D, Q(n)$	$\forall n \in D, \neg Q(n)$	$\exists n \in D, \neg Q(n)$
$\forall n \in D, R(n)$	$\exists n \in D, R(n)$	$\forall n \in D, \neg R(n)$	$\exists n \in D, \neg R(n)$
$\forall k \in D, S(k)$	$\exists k \in D, S(k)$	$\forall k \in D, \neg S(k)$	$\exists k \in D, \neg S(k)$
$\forall m \in D, T(m)$	$\exists m \in D, T(m)$	$\forall m \in D, \neg T(m)$	$\exists m \in D, \neg T(m)$

List for Exercise 6

8. Write each of the following statements using quantifiers and predicates. In each case, you must specify the domain and define the predicates you use.

   (a) Every biology major is required to take geometry.

   (b) There are computer science majors who do not minor in mathematics.

   (c) There is no math major who is required to take a business course.

   (d) There are puzzles that have no solution.

9. Write each of the following statements using quantifiers and predicates. In each case, you must specify the domain and define the predicates you use.

   (a) For every integer n, $2n \neq 9$.

   (b) There exists a triangle T that is equilateral and has perimeter 10.

   (c) Every circle has an integer diameter or an integer area.

   (d) Every two real numbers has an integer in between.

10. Let B be the set of all biology majors and let $G(x)$ be the predicate "x is required to take geometry." Write each statement below using quantifiers over the domain B and the predicate $G(x)$, and then match any statements that are equivalent in meaning.

   (a) There is no biology major who is required to take geometry.

   (b) There is a biology major who is not required to take geometry.

   (c) There is no biology major who is not required to take geometry.

   (d) Every biology major is not required to take geometry.

11. The lesson of Exercise 10 is that a statement of the form "No member of D makes $P(x)$ true" can be formally written as either "$\neg \exists x \in D, P(x)$" or "$\forall x \in D, \neg P(x)$." For each of the following English sentences, specify a domain and a predicate, and write the statement symbolically using both of these forms.

   (a) Friends of Alaina never get tired of playing at the beach.

   (b) No friend of Alaina dislikes doing cartwheels.

   (c) No math course is too hard for Jennica.

   (d) The meals at the camp are never too bad.

12. For each of the following English sentences, specify a domain and a predicate, and write the quantified statement symbolically using either form discussed in Exercise 11.

   (a) Even numbers are never prime.

   (b) Triangles never have four sides.

   (c) There are no integers a and b for which $a^2/b^2 = 2$.

   (d) No square number immediately follows a prime number.

13. Recall that $\mathbb{Z}$ denotes the set of all integers.

   (a) Consider the statement "$\forall x \in \mathbb{Z}, \exists y \in \mathbb{Z}, x + 2y = 3$."

      i. If x is 27, what value can be chosen for y to make the equation true?

      ii. Find two values of x for which it is impossible to find a corresponding y value that makes the equation true.

      iii. Is there a pattern for the x values that serve as counterexamples to the given statement?

   (b) Consider the statement "$\forall y \in \mathbb{Z}, \exists x \in \mathbb{Z}, x + 2y = 3$."

      i. Identify the x value that makes the equation true for each of the following values of y: 3, −10, 0, 17.

      ii. Describe a general strategy for choosing x once your opponent has chosen the value for y.

14. Write the negation of each of the following statements, using Proposition 1 and the rules for negating propositions to simplify each to the point that no $\neg$ symbol occurs to the left of a quantifier. (Recall that $\mathbb{Z}$ denotes the set of all integers, and $\mathbb{R}$ denotes the set of all real numbers.)

   (a) $\forall a \in \mathbb{R}, \forall b \in \mathbb{Z}, a^2 + b \in \mathbb{Z}$

   (b) $\exists y \in \mathbb{R}, \forall x \in \mathbb{R}, x + y = x$

   (c) $\forall x \in \mathbb{Z}, \exists y \in \mathbb{R}, x = 2y$

   (d) $\forall x \in \mathbb{Z}, \exists y \in \mathbb{R}, \frac{x}{y} = 2$

15. For each statement in Exercise 14, is the original statement or its negation true?

16. Write the negation of each of the following statements as an English sentence. You might find it helpful to write a symbolic expression as an intermediate step.

   (a) Every time you roll a "6," you have to take a card.

   (b) There is a day in your life that is better than every other day.

   (c) In every good book, there is a plot twist or a surprise ending.

   (d) Every math course has a topic that everyone finds easy to do.

17. Write the negation of each of the following statements as an English sentence:

   (a) For every integer x, there is an integer y that is bigger than x.

   (b) In every set of integers, there is a smallest number.

   (c) For every positive integer x, there is a positive integer y such that y is smaller than x and y is a factor of x.

18. For each statement in Exercise 17, is the original statement or its negation true?

19. Write the negation of each of the following statements as an English sentence:

 (a) For all odd integers b, there is no real number x such that $x^2 + bx + 15 = 0$.

 (b) For every two real numbers x and y, there is an integer n such that $x < n < y$.

 (c) For every pair of integers that sum to 5, at least one of the numbers must be bigger than 2.

20. For each statement in Exercise 19, is the original statement or its negation true?

1.5 Implications

Many statements in mathematics, as well as in ordinary conversation, refer to the logical connection between two simpler statements rather than to the actual truth of either one. For example,

 (i) If Bob has an 8:00 class today, then it is a Tuesday.
 (ii) If it is raining, then the street is wet.
 (iii) If you are a computer science major, then you must take discrete mathematics.
 (iv) If a real number x satisfies $x^2 > 4$, then $x > 2$.

Definition

1. A statement of the form "if p is true, then q is true" is called an **implication**.
2. We write an implication as $p \rightarrow q$, which is read, "p **implies** q." The $\rightarrow$ operator is taken as having lower precedence than $\wedge$, $\vee$, and $\neg$. For example, the proposition $\neg p \wedge q \rightarrow r$ means $((\neg p) \wedge q) \rightarrow r$.
3. In the statement "if p, then q," we call p the **hypothesis** and q the **conclusion**.

In this definition, p and q can indicate either propositions or predicates. Of our four examples, the first two involve propositions and the last two involve predicates.

When an implication involves predicates, there might be some parts of the sentence that are left unsaid. For example, when the speaker says, "If you are a computer science major, then you must take discrete mathematics," she no doubt has in mind all the students at her particular school. So even though she didn't say so, there is an unspoken domain the speaker has in mind, and there is an unspoken quantifier. Here are more precise versions of examples (iii) and (iv) that make the domain and the quantifier more explicit:

 (iii′) For all students s at this school, if s is a computer science major, then s must take discrete mathematics.
 (iv′) For all real numbers x, if $x^2 > 4$, then $x > 2$.

Example 1 *In each of the following implicational statements, identify the domain D and the predicates $P(x)$ and $Q(x)$ so that the implication is of the form "For all $x \in D$, if $P(x)$, then $Q(x)$." (You may use variables other than x if you wish.)*

1. *If a triangle has three equal sides, then it has three equal angles.*
2. *If an integer ends with a "2," then it is a multiple* of 2.*
3. *If a quadrilateral can be inscribed in a circle, then the opposite angles in the quadrilateral sum to* 180°.

SOLUTION

1. D is the set of all *triangles*, $P(t)$ is "t has three equal sides," and $Q(t)$ is "t has three equal angles."
2. D is the set of *integers*, $P(n)$ is "n ends with 2," and $Q(n)$ is "n is a multiple of 2."
3. D is the set of *quadrilaterals*, $P(d)$ is "d can be inscribed in a circle," and $Q(d)$ is "d has opposite angles summing to 180°."

☐

Practice Problem 1   *As in the preceding example, identify the domain D and the predicates $P(x)$ and $Q(x)$ so that the implication is of the form "For all $x \in D$, if $P(x)$, then $Q(x)$."*

(a) *If a real number x has a real square root, then x is not negative.*
(b) *If a real number x satisfies $x^2 - x = 6$, then $x = 3$.*
(c) *If an integer n is even, then $2^n - 1$ is a multiple of 3.*
(d) *If an integer n ends with a 3, then n is a multiple of 3.*

The Logic of Implications

To help us understand the logic of implicational statements, we begin with a concrete example.

Example 2 †*Trooper Jones walks into the Goldilocks Pub and sees four Boatsville College students (Al, Betty, Cindy, and Dan) enjoying various beverages. She asks the bartender, "Is anyone breaking the drinking law?" The bartender replies, "Everyone in here is obeying the law."*

In front of each person, there is a card which has the person's age on one side and what he or she is drinking on the other side. Trooper Jones sees that the face-up sides of the cards look like Figure 1-10.

The drinking age law states in effect

If you are drinking beer, then you are at least 21 years of age.   (1.1)

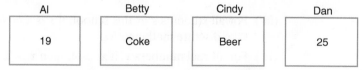

Figure 1-10   Age-drink cards.

---

* The integers resulting from the product of 2 and an integer are the *multiples* of 2.

† This example is based on a psychology experiment carried out by Griggs and Cox [32] in 1982.

1. *Identify a set D and predicates P(x) and Q(x) so that the bartender's statement is of the form "For all $x \in D$, if P(x), then Q(x)."*
2. *Whose cards does Trooper Jones need to turn over to check that everyone is obeying the law?*

SOLUTION

1. *D* is the set of four Boatsville College students currently in the bar, *P(x)* is the predicate "*x* is drinking beer," and *Q(x)* is the predicate "*x* is at least 21 years of age."
2. She should turn over the cards for Al and Cindy. She is looking for a counterexample to the bartender's claim that everyone is obeying the law, and these are the only two students who could possibly be counterexamples. □

This example illustrates the basic logic behind an implication or "if..., then..." statement. It will help to analyze why Trooper Jones does not turn over Betty's card and Dan's.

● *P(Betty)* is the statement "Betty is drinking beer" and *Q(Betty)* is "Betty is at least 21 years of age." Because *P(Betty)* is false, Trooper Jones doesn't care whether *Q(Betty)* is true or not. In ordinary English, Trooper Jones knows that Betty is not drinking beer, so Betty is obeying the law no matter what her age is.

● *P(Dan)* is the statement "Dan is drinking beer" and *Q(Dan)* is "Dan is at least 21 years of age." Because *Q(Dan)* is true, Trooper Jones doesn't care whether *P(Dan)* is true or not. In ordinary English, Trooper Jones knows that Dan is of legal age, so Dan is obeying the law no matter what he is drinking.

If the law is being broken, it must be because someone *is drinking beer* and *is not at least 21 years old*. That is, **the only time that a statement of the form "If *p*, then *q*" is false is when the hypothesis (statement *p*) is true, but the conclusion (statement *q*) is false.** At all other times, we would have to say that the whole statement is true since the law is not being broken. The truth table shown in Table 1-15 sums up our analysis of an implicational statement.

The last two rows of this table are the hardest for most people to swallow. How can a statement of the form "If *p*, then *q*" be true when the statement *p* is false, and *especially* when both statements *p* and *q* are false? For our concrete example, we can see that the last two rows correspond to the situation for Betty, for whom the hypothesis "*x* is drinking beer" is false. Betty would match the third row if she is 23, and the fourth row if she is 18. In either case, though, she is obeying the law, so the implication $p \to q$ is true.

As we will see, this same logic applies to the more abstract setting of mathematical implications.

p	q	$p \to q$
T	T	T
T	F	F
F	T	T
F	F	T

Table 1-15 Truth Table for Implication

 Example 3 *Identify the hypothesis and conclusion for the following statement:*

For every positive integer n, if n is odd, then $n^3 - n$ is divisible by 4.

Do you think the statement is true or false?

n	$n^3 - n$	Divisible by 4?
1	0	Yes, since $0 = (4)(0)$.
3	24	Yes, since $24 = (4)(6)$.
5	120	Yes, since $120 = (4)(30)$.
7	336	Yes, since $336 = (4)(84)$.

Table 1-16 Analysis of the Statement in Example 3

SOLUTION   The domain is the set of positive integers. The hypothesis is the predicate "n is odd," and the conclusion is the predicate "$n^3 - n$ is divisible by 4." To explore the truth of the implication, it is natural to think about examples first, so we do this in Table 1-16. When considering examples, we only listed values for n for which the hypothesis is true. We never considered the values of 2, 4, 6, or 8 for n. This omission is analogous to the Trooper's behavior toward the people whose cards indicated that they were not drinking beer. Whether $n^3 - n$ is divisible by 4 for even values of n has no relevance on the truth of the implication, just as people who are not drinking beer are not breaking the law regardless of their age.

   If we *do* look at these other (even) values of n, we see that sometimes they make the conclusion true and sometimes they make the conclusion false, just as some people who are not drinking beer might be under 21 and some might be 21 or older.

   Table 1-17 shows some examples of this. Since the values of n in Table 1-17 never make the hypothesis true, they cannot be counterexamples to the implication. It turns out that there *are* no counterexamples, and that the statement is therefore true.*

Example 4   *Why do we consider the following statement to be true?*

For all integers n, if 3n = 9, then $n^2 = 9$.

SOLUTION   As in the previous example, each of the component predicates "$3n = 9$" and "$n^2 = 9$" can be true or false depending on the value of n as Table 1-18 illustrates. Once again, when the hypothesis is false, we really don't care about the conclusion. To establish whether the implicational statement is true, we must ask, "Could there possibly be a counterexample—that is, is there a value of n for which the hypothesis is true but the conclusion is false?" Since the only value of n that makes the hypothesis true in this example is $n = 3$, and since this value of n also makes the conclusion true, we conclude that the implicational statement is true.

---

* We will learn how to prove this type of statement in the next chapter.

n	$n^3 - n$	Divisible by 4?
2	6	Not divisible by 4.
4	60	Yes, since $60 = (4)(15)$
6	210	Not divisible by 4.
8	504	Yes, since $504 = (4)(126)$.

Table 1-17 More Analysis of the Statement in Example 3

n	Hypothesis ($3n = 9$)	Conclusion ($n^2 = 9$)
-3	False	True
0	False	False
3	True	True
10	False	False

Table 1-18 Analysis of the Statement in Example 4

 Example 5 *Why do we consider the following statement to be false?*

For all integers n, if $n^2 > 9$, then $n > 3$.

SOLUTION It is possible to find a counterexample, if we remember that "integers" includes both positive and negative numbers. For example, the value $n = -4$ makes the hypothesis ($n^2 > 9$) true while making the conclusion ($n > 3$) false. □

It is only the presence of values of n that make the hypothesis true and the conclusion false, which causes the implicational statement to be false.

> *Summary* For a statement of the form "if **hypothesis**, then **conclusion**" to be *FALSE*, it must be the case that the **hypothesis** is true while the **conclusion** is false. Otherwise, the statement is *TRUE*.
>
> For a quantified statement "$\forall x$, if $P(x)$, then $Q(x)$" to be *FALSE*, it must be the case that at least one value of x is a counterexample—that is, there is at least one value of x that makes the **hypothesis** $P(x)$ true but makes the **conclusion** $Q(x)$ false. Otherwise, the quantified statement is *TRUE*.

Practice Problem 2 *Decide whether each of the following quantified statements is true or false. For each that is false, give a counterexample. Remember that $\mathbb{R}$ is the set of real numbers, and $\mathbb{Z}$ is the set of integers.*

(a) $\forall x \in \mathbb{R}$, *if* $x^2 - 5x + 4 = 0$, *then* $x > 0$.

(b) $\forall n \in \mathbb{Z}$, *if* $n^2 = 1$, *then* $n^3 = 1$.

(c) $\forall$ *positive integers a and b, if a and b are both odd, then* $a + b$ *is also odd.*

(d) $\forall$ *positive integers a and b, if a and b are both odd, then ab is also odd.*

Negating Implications

Since the negation of a statement captures what it means for the statement to be false, we can take our understanding of when an implication is false and turn it into a formal rule for forming the negation of an implicational statement.

Proposition 1   *The negation of the implication* $p \rightarrow q$ *is the statement* $p \wedge (\neg q)$.

PROOF   The proof consists of showing that the propositions $\neg(p \rightarrow q)$ and $p \wedge \neg q$ are logically equivalent using truth tables.

p	q	$p \rightarrow q$	$\neg (p \rightarrow q)$	$\neg q$	$p \wedge \neg q$
T	T	T	F	F	F
T	F	F	T	T	T
F	T	T	F	F	F
F	F	T	F	T	F

Since the columns for $\neg(p \rightarrow q)$ and $p \wedge \neg q$ are identical, we conclude that these statements are logically equivalent. **Notice that the negation of an implication is not an implication!** ∎

If a quantified implicational statement has the form $\forall x \in D, P(x) \rightarrow Q(x)$ for some domain D, then the statement is false if there is a value for x in D that makes the hypothesis $P(x)$ true but the conclusion $Q(x)$ false. Hence, the negation of a quantified implicational statement can itself be expressed as a quantified statement.

Proposition 2   *The negation of the implication* $\forall x \in D, P(x) \rightarrow Q(x)$ *is the statement* $\exists x \in D, P(x) \wedge (\neg Q(x))$.

PROOF   This follows from Proposition 1 of this section and Proposition 1 of Section 1.4. ∎

 Example 6 *Write the negation of each of the following statements:*

1. If Bob has an 8:00 class today, then it is a Tuesday.

2. If Jessica gets chocolate, then she has a happy birthday.

3. For all real numbers x, if $x > 2$, *then* $x^2 > 4$.

4. $\forall$ *real numbers* $x > 0$, *if* $x^2 = 1$, *then* $x^3 = 1$.

SOLUTION   Each negation is given.

1. Bob has an 8:00 class today, and it is not Tuesday.
2. Jessica gets chocolate, but she doesn't have a happy birthday.
3. There exists a real number x such that $x > 2$ but $x^2 \leq 4$.
4. $\exists$ real number $x > 0$, $(x^2 = 1) \wedge (x^3 \neq 1)$. Observe that the domain is still the set of real numbers that are greater than 0.

□

Practice Problem 3   *Write the negation of each of the following statements:*

1. *If you buy the extended warranty, then nothing will go wrong with your television.*
2. *If Christopher gets a flu shot, then he will not get the flu.*
3. *For all triangles t, if t has three equal sides, then t has three equal angles.*
4. *$\forall x \in \{1, 2, 3, 4, 5\}$, x^2 is positive.*

Contrapositives, Converses, and Inverses

We have determined how the truth value of an implicational statement depends on the truth value of its component parts. However, there are many ways to form an implication with a given pair of component statements. Trying different combinations for hypothesis and conclusion is a fairly natural thing to do when considering a new mathematics problem. It is often the case that there are important properties to be studied, and you are first trying to figure out how the properties are related to each other.

Example 7   *Let $P(n)$ stand for the predicate "n ends in a digit 2," and $Q(n)$ for the predicate "n is divisible by 2." We can use these predicates to form many different implications.*

1. *$P(n) \to Q(n)$, that is, "If n ends in a digit 2, then n is divisible by 2."*
2. *$Q(n) \to P(n)$, that is, "If n is divisible by 2, then n ends in a digit 2."*
3. *$\neg P(n) \to \neg Q(n)$, that is, "If n does not end in a digit 2, then n is not divisible by 2."*
4. *$\neg Q(n) \to \neg P(n)$, that is, "If n is not divisible by 2, then n does not end in a digit 2."*

As is often the case, each statement has an implied domain and quantifier—in this case, "For all integers n, ..." Decide if each of the quantified statements is true or false. For each that is false, give a counterexample.

SOLUTION

1. This statement is true.
2. The value $n = 14$ is a counterexample since it makes the hypothesis true (14 is divisible by 2) and the conclusion false (14 does not end in 2).
3. The value $n = 14$ is a counterexample since it makes the hypothesis true (14 does not end in 2) and the conclusion false (14 is divisible by 2).
4. This statement is true.

□

You might have noticed that we used the same counterexample for the second and third statements above. If you think about it, any counterexample to the second statement must be a counterexample to the third statement and vice versa. The same relationship holds between the first and fourth statements as well, so since the first statement has no counterexamples, then neither does the fourth statement. Because of these connections, the relationships illustrated by these four statements have special names in the mathematics literature.

Definition   Consider the implication $\forall x \in D, P(x) \to Q(x)$.

1. The **converse** of the implication is $\forall x \in D, Q(x) \to P(x)$.
2. The **inverse** of the implication is $\forall x \in D, \neg P(x) \to \neg Q(x)$.
3. The **contrapositive** of the implication is $\forall x \in D, \neg Q(x) \to \neg P(x)$.

We have defined the terms in the context of quantified statements, but the same terms can be applied in an obvious way to propositions:

- The converse of $p \to q$ is $q \to p$.
- The inverse of $p \to q$ is $\neg p \to \neg q$.
- The contrapositive of $p \to q$ is $\neg q \to \neg p$.

Since we have a mechanism for proving propositions to be logically equivalent, we can formally establish the relationships mentioned in the discussion following Example 7.

Proposition 3

1. *An implication and its contrapositive are logically equivalent.*
2. *The converse and inverse of an implication are logically equivalent.*
3. *An implication is not logically equivalent to its converse (and thus not to its inverse).*

PROOF   We will prove the first statement by comparing the truth tables of $p \to q$ and $\neg q \to \neg p$.

p	q	$p \to q$	$\neg q$	$\neg p$	$\neg q \to \neg p$
T	T	T	F	F	T
T	F	F	T	F	F
F	T	T	F	T	T
F	F	T	T	T	T

Since the columns for $p \to q$ and $\neg q \to \neg p$ are identical, we conclude that these statements are logically equivalent. The other two parts of this proposition are addressed in Exercise 5 at the end of this section. ∎

Example 8

1. *Give an example of a true implication whose converse is false.*
2. *Give an example of a true implication whose converse is also true.*

SOLUTION

1. The implication "For all integers a and b, if a and b are odd, then $a + b$ is even" is true.* Its converse "For all integers a and b, if $a + b$ is even, then a and b are odd" is false. One counterexample is $a = 2, b = 14$.
2. "If n is even, then n^2 is even" is true and has a true converse. □

Since it is possible for a true implicational statement to have a false converse, but it is also possible for a true statement to have a true converse, we must always treat implications and their converses as two entirely different statements that must each be analyzed on its own merits.

The Language of Implication

One of the frustrations of the English language is that there are many ways to state the same thought. This phenomenon occurs not only in ordinary conversation, but also in mathematical expression. This is one reason mathematicians are so fond of symbolism—a completely symbolic expression of a theorem is less likely to contain the ambiguity inherent in spoken and written language. In this short section, we will give you a few examples of some common sources of confusion in the English language.

One possible source of confusion is our tendency in English to rearrange the parts of an implication without changing its meaning. For example, these sentences convey exactly the same idea:

● I will pass the course if I ace the final.
● If I ace the final, then I will pass the course.

Switching the location of the hypothesis and conclusion within the sentence does not change the meaning. The hypothesis is the part that goes with the word *if*, and in each sentence that is the phrase "I ace the final." Do not confuse this with forming the converse, which involves interchanging the hypothesis and conclusion. Here is a mathematical example:

Statement:	If an integer m ends in the digit 0, then m is a multiple of 5.
Same statement:	An integer m is a multiple of 5 if it ends in the digit 0.
Converse:	If an integer m is a multiple of 5, then m ends in the digit 0.

Another possible source of confusion is the variety of ways in which we can express quantification and implication. We have discussed earlier our tendency to

---

* In the next chapter, we will learn how to prove statements of the type in this example.

leave out domains and quantifiers when we speak. Another issue when formalizing statements is the choice of writing a "for all" statement as an implication or not. Given a statement of the form "$\forall x \in D, Q(x)$" where D is a subset of a natural larger set U, we can equivalently write the statement as "$\forall x \in U, (x \in D) \rightarrow Q(x)$." For now, the choice is a matter of taste, but we will learn to prefer the latter in the next chapter when we discuss formal proofs of this type of statement.

Example 9 *Here is one way to express the idea that every computer science major at this university must take discrete mathematics: "Let D be the set of computer science majors at this university. For all s ∈ D, s must take discrete mathematics."*

1. *Rewrite this as an implication.*
2. *Form the negation of the original statement, and the statement written as an implication.*

SOLUTION

1. Here is one possible solution: The set U of all students at this university is a natural set that includes all of D. So an equivalent implicational statement is "For all $s \in U$, if s is a computer science major, then s must take discrete mathematics."

2. The negation of the original statement is "There exists a computer science major at this university who does not have to take discrete mathematics." For the implicational form, the negation is "There exists a student at this university who is a computer science major but does not have to take discrete mathematics." Both say the same thing, but in a slightly different way. □

Practice Problem 4 *Rewrite each quantified predicate as an implication. Unless otherwise indicated, assume that variables stand for real numbers. Use $\mathbb{R}$ and $\mathbb{Z}$ to indicate the sets of reals and integers, if you wish.*

(a) *$\forall$ even integer m, m ends in the digit 0, 2, 4, 6, or 8.*
(b) *$\forall x > 0, x^2 > x$*
(c) *For every positive odd integer n, $n^3 - n$ is divisible by 4.*

Finally, it is sometimes the case that both an implication and its converse are true. This is the strongest possible relationship between the properties that make up the hypothesis and conclusion of the implication. Because of the importance of this situation, mathematicians have some common ways to convey the idea that an implication and its converse are both true. For example, the statement "If n is even, then n^2 is even" is true, and it has a true converse. A mathematician might say, "If n is even, then n^2 is even, and conversely," or "n is even if and only if n^2 is even."

In general, for any statements p and q (with or without variables), the phrase "p if and only if q" means that both $p \rightarrow q$ and $q \rightarrow p$ are true. For this reason, we use the notation $p \leftrightarrow q$ to express this *biconditional statement*. When considering a statement of this form, it is often easiest to consider each of these two statements separately.

Practice Problem 5   *What are the two implicational statements expressed by the statement "The integer n is a multiple of 10 if and only if n is even"? Is each statement true? If not, give a counterexample.*

Logic Puzzles Revisited

Since now we know how to make truth tables for implicational statements, we end this section with some more logic puzzles, involving the use of implication. Remember to make your truth tables by first considering the statements themselves without worrying about who said them.

Example 10   *You meet two inhabitants. A says, "If B is truthful, then so am I," and B says, "At least one of us is lying." Who (if anyone) is telling the truth?*

SOLUTION   Here is the truth table. Notice that A's statement is $q \to p$, which is false only when q is true but p is false.

	p	q	Statement 1 If B is truthful, then so is A	Statement 2 At least one of us is lying
	T	T	T	F
	T	F	T	T
★	F	T	F	T
	F	F	T	T

We conclude that only B is a truth-teller.

Example 11   *You meet two inhabitants. A says, "If B is truthful, then so am I," and B says, "A is lying." Who (if anyone) is telling the truth?*

SOLUTION   This is similar to the previous example. A's statement (if q, then p) is false only in the situation where q is true but p is false.

	p	q	Statement 1 If B is truthful, then so is A	Statement 2 A is lying
	T	T	T	F
?	T	F	T	F
?	F	T	F	T
	F	F	T	T

In this case there is an ambiguous answer since either of the middle two rows could be solutions. We can conclude that exactly one of A and B is a truth-teller, but we cannot determine which one from what they said.

Example 12 *You meet three inhabitants. A says, "If B is lying, then so is C," B says, "C is truthful," and C says, "At least one of us is lying." Who (if anyone) is telling the truth?*

SOLUTION To build the table for *A*'s statement, we analyze it as being $\neg q \rightarrow \neg r$. The only time it is false is when $\neg q$ is true but $\neg r$ is false. Put another way, the only time Statement 1 is false is when q is false but r is true.

			Statement 1	Statement 2	Statement 3
p	q	r	If *B* is lying, then so is *C*	*C* is truthful	One of us is lying
T	T	T	T	T	F
T	T	F	T	F	T
T	F	T	F	T	T
T	F	F	T	F	T
F	T	T	T	T	T
F	T	F	T	F	T
F	F	T	F	T	T
F	F	F	T	F	T

In this example there are no solutions, so this is a paradox. In other words, given the rules of the problem, no three people could have uttered those particular phrases. $\square$

Solutions to Practice Problems

1 (a) *D* is the set of real numbers, $P(x)$ is "x has a real square root," and $Q(x)$ is "x is not negative."

(b) *D* is the set of real numbers, $P(x)$ is "x satisfies the equation $x^2 - x = 6$," and $Q(x)$ is "$x = 3$."

(c) *D* is the set of integers, $P(n)$ is "n is even," and $Q(n)$ is "$2^n - 1$ is divisible by 3."

(d) *D* is the set of integers, $P(n)$ is "n ends with a 3," and $Q(n)$ is "n is divisible by 3."

2 (a) There are only two values for x that make the hypothesis true ($x = 1$ and $x = 4$), and for both these values the conclusion is also true. Therefore, the "if, then" statement is true.

(b) Although there is a value of n (namely $n = 1$) that makes both the hypothesis and conclusion true, there is also a value ($n = -1$) where the hypothesis is true and the conclusion is false. So the "if, then" statement is false.

(c) This statement is false, and it is easy to find a counterexample—just pick any odd integers for a and b. For example, with $a = 1$ and $b = 7$, we have $a + b = 8$. Since for this choice of a and b the hypothesis is true but the conclusion is false, the original "if, then" statement is false.

(d) Table 1-19 shows some possible values of a and b. The fact that some entries make the conclusion false is irrelevant, since these same values make the hypothesis false as well. It turns out that any time we make the hypothesis true, the conclusion is also true, and so the "if, then" statement is true. In the next chapter we will see how to prove this fact.

3 (a) You buy the extended warranty, and something does go wrong with your television.

(b) Christopher gets the flu shot, but he does get the flu.

(c) There is a triangle t where t has three equal sides but t does not have three equal angles.

(d) $\exists x \in \{1, 2, 3, 4, 5\}$ such that $x^2 \leq 0$.

4 (a) $\forall m \in Z$, if m is even, then m ends in the digit 0, 2, 4, 6, or 8.

(b) $\forall x \in R$, if $x > 0$, then $x^2 > x$.

(c) $\forall n \in Z, (n > 0) \wedge (n \text{ odd}) \rightarrow n^3 - n$ is divisible by 4.

5 The statement "If n is a multiple of 10, then n is even" is a true statement, but the converse "If n is even, then n is a multiple of 10" is false, having $n = 2$ as a counterexample. Hence, the given "if and only if" statement is false.

a	b	Hypothesis a and b are Both Odd	Conclusion ab is Odd
3	5	True	True
3	2	False	False
8	4	False	False
31	11	True	True

Table 1-19 Table for Practice Problem 2(d)

Exercises for Section 1.5

1. Write each statement as a statement of formal propositional logic. That is, assign variable names to the simple phrases, and write the statement using those variables along with the logical connectives $\neg$, $\wedge$, $\vee$, and $\rightarrow$.

   (a) If you don't attend the concert, you will get an F for the course.

   (b) We will go if you go.

   (c) I ate my lunch but I did not eat breakfast.

   (d) If you don't eat your breakfast, you will be hungry.

   (e) It is false that this triangle has both a 30° angle and a 60° angle.

   (f) If a quadrilateral is a square, it has four equal sides and four equal angles.

   (g) If a triangle has either two equal sides or two equal angles, then it is an isosceles triangle.

2. For each statement in Exercise 1, give a truth table for the statement, and use it to explain what conditions would make the statement true and what would make it false.

3. In a certain board game played with a pair of dice, if you roll "doubles" three times in a row, you must place your piece on the board square marked "Jail." After playing the game for two grueling hours, no one ever rolled "doubles" three times in a row. Can we conclude that no piece was ever placed on the "Jail" square during the game? Refer specifically to the hypothesis and conclusion of the rule above in your explanation.

4. Complete the following truth tables for the given compound expressions:

(a)

p	q	$p \wedge q$	$(p \wedge q) \rightarrow q$
T	T		
T	F		
F	T		
F	F		

(b)

p	q	$\neg p$	$p \vee q$	$\neg p \rightarrow (p \vee q)$
T	T			
T	F			
F	T			
F	F			

(c)

p	q	$p \vee q$	$(p \vee q) \rightarrow q$
T	T		
T	F		
F	T		
F	F		

(d)

p	q	$\neg p$	$\neg q$	$\neg q \rightarrow p$	$\neg p \wedge (\neg q \rightarrow p)$
T	T				
T	F				
F	T				
F	F				

(e)

p	q	r	$q \rightarrow r$	$p \wedge (q \rightarrow r)$
T	T	T		
T	T	F		
T	F	T		
T	F	F		
F	T	T		
F	T	F		
F	F	T		
F	F	F		

p	q	r	$p \wedge q$	$p \vee r$	$(p \wedge q) \rightarrow (p \vee r)$
T	T	T			
T	T	F			
T	F	T			
(f) T	F	F			
F	T	T			
F	T	F			
F	F	T			
F	F	F			

5. Use truth tables to check if each of the given pairs of symbolic logic statements are equivalent.

 (a) $p \rightarrow q$ and $q \rightarrow p$ (NOTE: This is a generic implication and its converse.)

 (b) $\neg p \rightarrow \neg q$ and $q \rightarrow p$ (NOTE: This is the inverse and converse of a generic implication $p \rightarrow q$.)

 (c) $p \wedge (p \rightarrow q)$ and $p \wedge q$

 (d) $p \rightarrow (q \rightarrow r)$ and $(p \wedge q) \rightarrow r$

 (e) $(p \vee q) \rightarrow r$ and $(p \rightarrow r) \vee (q \rightarrow r)$

6. It is sometimes useful to realize that the implication $p \rightarrow q$ is logically equivalent to $\neg p \vee q$. This exercise explores this equivalence.

 (a) What combination of truth values for p and q makes the implication $p \rightarrow q$ false?

 (b) Explain how you know that the negation of $\neg p \vee q$ is $p \wedge \neg q$.   (HINT: Refer to Theorem 2 of Section 1.3.)

 (c) What combination of truth values makes $p \wedge \neg q$ true? (This same combination would make $\neg p \vee q$ false.)

 (d) Use a truth table to formally demonstrate that $p \rightarrow q$ is logically equivalent to $\neg p \vee q$.

7. For each of the following statements, rewrite them in propositional logic notation, making the meaning of your propositional variables clear. Use truth tables to find any pairs of logically equivalent statements.

 (a) If Alaina likes basketball, then she likes swimming and gymnastics.

 (b) If Alaina likes gymnastics, then she likes swimming and basketball.

 (c) If Alaina dislikes gymnastics or dislikes swimming, then she dislikes basketball.

 (d) Alaina dislikes basketball or she likes both swimming and gymnastics.

8. Write each of the following predicates using the simple predicates $x > 0$ and $y > 0$ along with the propositional connectives $\wedge$, $\vee$, $\neg$, and $\rightarrow$:

 (a) If x is positive, then y is positive.

 (b) If x is positive, then y is not positive.

 (c) If x is not positive, then y is positive.

 (d) If x is not positive, then y is not positive.

9. For each of the given values of x and y, determine which predicates from Exercise 8 become true statements.

 (a) If $x = 8$ and $y = 3$, which of (a–d) are true?

 (b) If $x = -5$ and $y = 0$, which of (a–d) are true?

 (c) If $x = 7$ and $y = -7$, which of (a–d) are true?

 (d) If $x = 0$ and $y = 0$, which of (a–d) are true?

 (e) If $x = -3$ and $y = -10$, which of (a–d) are true?

 (f) If $x = -8$ and $y = 2$, which of (a–d) are true?

10. Let D be the set $\{1, 3, 5, 7, 8, 10, 11, 12\}$. For each of the following, decide whether it is true for all the elements of D. If it is not, give a counterexample.

 (a) If x is even, then $x > 7$.

 (b) If x is odd, then $x > 7$.

 (c) If x is even, then $x \leq 12$.

 (d) If x is odd, then $x \leq 12$.

 (e) If $x > 12$, then x is even.

11. Let D be the set $\{1, 3, 5, 7, 8, 10, 11, 12\}$. For each of the following, decide whether it is true for all the elements of D. If it is not, give a counterexample.

 (a) If x is even and $x > 7$, then $x < 20$.

 (b) If x is odd, then $x < 10$.

 (c) If $x < 10$ and $x \neq 8$, then x is odd.

 (d) If x is odd or $x < 5$, then $x - 1$ is even.

 (e) If $x > 5$ and $x < 7$, then x is negative.

12. For each of the following statements, describe the predicates P and Q that make the formal statement

$$\text{For every positive integer } n,\ P(n) \rightarrow Q(n)$$

correctly represent the statement.

 (a) If n is even, then $n^2 + n$ is even.

 (b) If n is a multiple of 5, then n has ones' digit of 5.

 (c) If n is prime, then $2^n - 1$ is prime.

13. For each of the following statements, describe the predicates P and Q that make the formal statement

$$\text{There exists an integer } n,\ P(n) \wedge Q(n)$$

correctly represent the statement.

 (a) Some odd numbers n make $2^n - 1$ a multiple of 7.

 (b) At least one multiple of 5 does not have a ones' digit of 0.

 (c) It is possible for a multiple of 3 to have a ones' digit of 7.

14. Express each of the following statements using predicates and the quantifier $\forall$:

 (a) If n ends in 5, then n is a multiple of 5.

 (b) If m ends in 3, then m is a multiple of 3.

(c) If n is a multiple of 5, then $n^2 - 1$ is a multiple of 3.

(d) For every positive real number x, if $x < \sqrt{2}$, then $2/x > \sqrt{2}$.

15. Which of the statements in Exercise 14 are true?

16. For each of the statements in Exercise 14, express the negation using predicates and the quantifier $\exists$.

17. Express each of the following statements using predicates and the quantifiers $\forall$ and $\exists$:

(a) If n is a multiple of 5, then n ends in 5 or n ends in 0.

(b) If n is not a multiple of 3, then $n^2 - 1$ is a multiple of 3.

(c) For all odd integers a and b, there is no real number x such that $x^2 + ax + b = 0$.

(d) For every real number y, if $y \geq 0$, then there exists $x \in \mathbb{R}$ such that $x^2 = y$.

18. Which of the statements in Exercise 17 are true?

19. For each of the statements in Exercise 17, express the negation using predicates and the quantifiers $\exists$ and $\forall$.

20. Give a counterexample to each of the following to show that it is a false implicational statement. If you think the statement is actually true, write one sentence explaining why you think no counterexample exists.

(a) If $n^2 = 4$, then $n^3 = 8$.

(b) If $\sin(x) = 0$, then $\cos(x) = 1$.

(c) If $\cos(x) = 0$, then $\sin(x) = 1$.

(d) If $x^3 = x$, then $x^2 = 1$.

21. Consider the statement "If n is even, then $2^n - 1$ is a multiple of 3." If we let $P(x)$ be the predicate "x is even," and we let $Q(x)$ be the predicate "$2^x - 1$ is a multiple of 3," then we can represent the original statement as

For every natural number n, $P(n) \to Q(n)$.

Create an organized table of your attempts to find a counterexample to this statement. Why do you think that no counterexample exists?

22. A consequence of Proposition 2 is that the negation of the statement $\exists x \in D, P(x) \wedge Q(x)$ is the statement $\forall x \in D, P(x) \to \neg Q(x)$. Use this fact to write the negation of each of the following statements, using careful wording and an "if, then" structure when possible:

(a) There exists a positive integer n such that n is even and $\frac{1}{n} > 1$.

(b) There exist positive integers a and b such that $a - b$ is odd and $a^2 = 2b^2$.

(c) There exist integers a and b such that a and b are positive and $a/b = 1 + b/a$.

(d) There exists a right triangle with perimeter equal to three times the length of one leg.

23. Use the idea in Exercise 22 to write the negation of each of the following statements, using careful wording and an "if, then" structure when possible:

(a) There exist integers m and n such that $m \geq n$ and $m^2 + n^2 = 11$.

(b) There exists a real number z such that $z > 0$ and $z + \frac{1}{z} = 1$.

(c) If x is a positive real number, then there is a positive integer n with $nx > 1$.

(d) If S is a set of three people, then there exist two people in S with the same sex.

24. Which of the original statements in the previous exercise are true?

25. Form the contrapositive of each of these statements:

(a) If you don't attend the concert, you will get an F for the course.

(b) We will go if you go.

(c) If you don't eat your breakfast, you will be hungry.

(d) If a quadrilateral is a square, it has four equal sides and four equal angles.

(e) If a triangle has either two equal sides or two equal angles, then it is an isosceles triangle.

26. Form the converse of each of the statements in Exercise 25.

27. Form the inverse of each of the statements in Exercise 25.

28. Solve each of these logic puzzles by using truth tables.

(a) You come across two inhabitants. A says, "I am lying if B is," and B says, "A is lying if I am." Can you tell who if anyone is telling the truth?

(b) You come across three inhabitants. A says, "If B is lying, then so is C," B says, "If C is lying, then so is A," and C says, "If A is lying, then so is B." Who if anyone is telling the truth?

29. If there was a third person in Example 11, what could she have said that would have determined everyone's truthfulness?

30. Express each of the following using quantified statements over the domain S of all college students and the predicates $C(x)$ meaning "x is a computer science major," and $D(x)$ meaning "x takes discrete mathematics." Which statements are equivalent to one another? Which statements are negations of one another?

(a) Every computer science major takes discrete mathematics.

(b) Some computer science majors take discrete mathematics.

(c) No computer science major takes discrete mathematics.

(d) Not every computer science major takes discrete mathematics.

(e) All computer science majors do not take discrete mathematics.

(f) You must take discrete mathematics if you are a computer science major.

(g) Some computer science majors do not take discrete mathematics.

1.6 Excursion: Validity of Arguments

Charles Sanders Peirce wrote [41]

> ... *Bad reasoning as well as good reasoning is possible; and this fact is the foundation of the practical side of logic.*

Logic has been an important part of intellectual endeavors since the time of the ancient Greeks. Aristotle (384–322 B.C.E.) wrote about the importance of precise reasoning in discourse, and Euclid's famous book *The Elements* has been used in teaching the art of rigorous thought and the skill of persuasive argument for nearly 2,000 years. Since so many decisions in our lives are based on the persuasive use of language and logic, the field of formal logic is an active, important discipline of philosophy. In the next chapter, we will consider the question of mathematical theorems and their proofs. These mathematical proofs are one example of a general concept—using valid arguments to persuade others that something is true. In the current section, we will examine the general notion of arguments in everyday life, and how to detect some common misuses of logic. We begin with an example of a phenomenon that we sometimes see in advertising: *What you think you hear may not be what the advertisement really said.*

Example 1 *A car company advertises, "If you didn't buy from us, you paid too much." Write this as a statement of formal propositional logic, and examine it with a truth table.*

SOLUTION   Let p stand for "You bought a car from us," and q stand for "You paid too much." The statement in the ad is represented by $\neg p \to q$, and its truth table looks like this:

p	q	$\neg p$	$\neg p \to q$
T	T	F	T
T	F	F	T
F	T	T	T
F	F	T	F

Hence, the only circumstances under which the statement is false are if you do not buy your car from this company and you do not pay too much for it. Notice that if you do buy your car from this company, the advertised statement is true regardless of the truth value of statement q. □

The car company that sponsored this ad is probably hoping you hear something different from what they actually say. Two possible misinterpretations are:

- If you paid too much for your car, you didn't buy from us. This is the *converse* of the advertiser's claim.
- If you bought a car from us, you did not pay too much. This is the *inverse* of the advertiser's claim.

As we indicated in the previous section, a statement and its converse (or inverse) do not necessarily have the same truth value. Truth in advertising may force the car company to make true statements in its advertisements—but it is up to the consumer to resist inferring that the inverse or converse is also true.

Why are we so likely to misinterpret the advertiser's claim? One reason is that, in ordinary conversation, we frequently make statements of the form "if p, then q" when our true meaning *does* include both the statement and its converse (or inverse).

Example 2 *When you were a child, your parents said, "If you don't eat your peas, you can't have dessert." You promptly ate your peas and asked for dessert. What statement did you hear? Is it the same statement your parents made? Do you think you misinterpreted their statement?*

SOLUTION   You are acting under the assumption that their statement was "If you *do* eat your peas, you *can* have dessert." This is NOT the same statement. It is the inverse. However, most parents who make a statement such as this really mean both what they actually say *and* its inverse. □

When we are dealing with our parents or our friends, there is perhaps no harm in reading into their statements more than was actually said. However, in a more formal setting such as a debate, a mathematical proof, or a false advertising suit, it is important to establish the exact meaning of each statement, and to analyze whether each successive statement follows logically from the previous statements.

Example 3 *Using p to stand for "You bought a car from us" and q to stand for "You paid too much,"analyze the **inverse** of the advertiser's claim: "If you bought a car from us, you did not pay too much." Under what circumstances is the inverse of this statement false?*

SOLUTION   The statement is written formally as $p \to \neg q$, and it has the following truth table:

p	q	$\neg q$	$p \to \neg q$
T	T	F	F
T	F	T	T
F	T	F	T
F	F	T	T

The first row is the only row where the result is false. Hence, if there is a customer who buys from the dealer and pays too much, then this implicational statement is false. In every other case, the implicational statement is true. □

If the car dealer's ad contained this inverse statement, there would be a good chance that it would be false, and they could be sued for false advertising. The original ad is carefully phrased to avoid false advertising, counting on you the consumer to misinterpret what it says.

Practice Problem 1   *Create the truth tables for each of the following, and relate them to the truth tables for the original claim and the inverse of the claim:*

(a)   The converse of the advertiser's claim: "If you did pay too much, you didn't buy from us."

(b)   The contrapositive of the advertiser's claim: "If you didn't pay too much, you bought from us."

More on the Language of Implication

One obstacle to analyzing the truth of English statements is that English as a language is somewhat complicated and often confusing. One could focus a large part of an entire course on the pitfalls of ambiguity in spoken language. We will not pursue that route since mathematics rarely suffers from this problem. Precision in language is one of the hallmarks of mathematics. Even in mathematics, though, there are a variety of ways to state the basic logical relationships between properties, so it is worth spending some time on the most common ways to express the basic implication relationship.

Proposition 1   *Let p and q stand for statements. The implication "if p is true, then q is true" can be expressed in each of the following ways:*

1. *If p is true, then q is true.*
2. *p is true only if q is true.*
3. *For q to be true, it is sufficient that p is true.*
4. *For p to be true, it is necessary that q is true.*

Example 4   *Each of the following is an equivalent way to express the basic implication "If you live in Pittsburgh, then you live in Pennsylvania:"*

- *You live in Pittsburgh only if you live in Pennsylvania.*
- *It is necessary to live in Pennsylvania in order to live in Pittsburgh.*
- *To live in Pennsylvania, it is sufficient to live in Pittsburgh.*

At least partly because of the possible rearrangements of the clauses within the sentences, implications that are phrased using *only if*, *sufficient*, or *necessary* can be more difficult to understand than those that use the word *if*. Here are some hints that might prove helpful:

- If the order of the clauses is confusing to you, then before doing anything else, rewrite the statement ordered the way with which you are more comfortable.
- A statement that uses "only if" is the converse of the same statement with "if" in its place.

- If a statement uses "sufficient," the condition that is sufficient is the *hypothesis* of the corresponding "if, then" statement.

- If a statement uses "necessary," the condition that is necessary is the *conclusion* of the corresponding "if, then" statement. Hence, a statement that uses "necessary" is the converse of the same statement with "sufficient" in its place.

Practice Problem 2 *Rewrite each of the following sentences in "if, then" form:*

(a) *You will pass the test only if you study for at least four hours.*

(b) *Attending class regularly is a necessary condition for passing the course.*

(c) *In order to be a square, it is sufficient that the quadrilateral have four equal angles.*

(d) *In order to be a square, it is necessary that the quadrilateral have four equal angles.*

(e) *An integer is an odd prime only if it is greater than 2.*

Valid and Invalid Forms of Reasoning

Early in life, children develop reasoning skills. Some of these skills are so fundamental that they seem almost trivial. For example, suppose you tell a child, "Your mother has gone to work, and so has your father." If you later ask, "Where is your mother?" the child will likely state that she is at work. The child has used a form of reasoning that we might write as

Statements p and q are both true.

Therefore, I can conclude that p is true,

where p is the proposition "Mother is at work," and q is the proposition "Father is at work." We can write forms of reasoning like this using two different brief representations:

$$\frac{p \wedge q}{\therefore p} \qquad \text{or} \qquad p \wedge q, \therefore p$$

Similarly, if you tell the child, "Either you must eat your peas or you must eat your carrots," the child will instinctively eat the carrots and announce, "I'm all done." The child has used the form of reasoning represented as follows:

$$\frac{q}{\therefore p \vee q} \qquad \text{or} \qquad q, \therefore p \vee q$$

where p indicates "Peas have been eaten," and q indicates "Carrots have been eaten." In both vertical and horizontal representation formats, we list one or more statements that are presumed to be true, and then we list a conclusion that can be inferred from these premises. Here is another simple example of a valid form of reasoning consisting of two premises followed by the conclusion:

$$\frac{\begin{array}{c} p \\ q \end{array}}{\therefore p \wedge q} \qquad \text{or} \qquad p, q, \therefore p \wedge q$$

This example points out the rather obvious fact that if you know p is true, and you also know q is true, it is valid to conclude that the statement "$p \wedge q$" is also true.

In future examples and exercises, we will explore a variety of other valid forms of reasoning, some trivial and some not so trivial. For now, we will concentrate our attention on some important forms of reasoning related to *implicational statements*.

Example 5 *A parent says, "If you clean your room, we will go play miniature golf." The child cleans her room, and announces, "Let's go golfing." Give a formal description of the reasoning process used by the child.*

SOLUTION We use p for "You clean your room," and q for "We play miniature golf." We can write the given argument structure in vertical or horizontal format as follows:

$$\frac{\begin{array}{c} p \to q \\ p \end{array}}{\therefore q} \qquad \text{or} \qquad p \to q, p, \therefore q$$

In this example, the child has applied the form of reasoning known by the Latin name **modus ponens**, which means "in a manner that asserts." In words, it simply states that if an implication is true, and its hypothesis is also true, then its conclusion must be true.

This direct method of reasoning is related to another valid form of reasoning known as **modus tollens**, which means "in a manner that negates." Simply stated, if an implication statement is true, and its conclusion is false, then its hypothesis must be false. This is represented symbolically as follows:

$$\frac{\begin{array}{c} p \to q \\ \neg q \end{array}}{\therefore \neg p} \qquad \text{or} \qquad p \to q, \neg q, \therefore \neg p$$

Example 6 *Here is an example of a valid argument using modus tollens:*

> *If this quadrilateral is a square, then it has four equal sides.*
> *This quadrilateral does not have four equal sides.*
> *Therefore, this quadrilateral is not a square.*

Notice that with *modus tollens*, one is simply applying the reasoning process of *modus ponens* to the contrapositive statement $\neg q \to \neg p$. That is, the following methods of reasoning are the same because they only differ in the first line, and those lines represent equivalent implicational statements:

$$\frac{\begin{array}{c} p \to q \\ \neg q \end{array}}{\therefore \neg p} \text{ by } modus\ tollens, \qquad \frac{\begin{array}{c} \neg q \to \neg p \\ \neg q \end{array}}{\therefore \neg p} \text{ by } modus\ ponens$$

Of course, not every form of reasoning applied by a person in an argument is a valid form of reasoning. Before we discuss how we tell a valid form from an invalid form of reasoning, we will look at two common erroneous methods of argument. These and other invalid reasoning forms are referred to as **fallacies**. The particular names of the two fallacies below stem from the tendency to confuse an implication with its converse or inverse, as we discussed earlier in this section.

Definition

1. The **converse fallacy** is the invalid reasoning form given by $p \rightarrow q, q, \therefore p$.
2. The **inverse fallacy** is the invalid reasoning form given by $p \rightarrow q, \neg p, \therefore \neg q$.

Example 7 *For each of the following, identify the reasoning as either valid (modus ponens), valid (modus tollens), invalid (converse fallacy), or invalid (inverse fallacy):*

1. *If n is even, then n^2 is even. I know n^2 is even. Therefore, n is even.*
2. *If this triangle has three equal sides, then it has three equal angles. This triangle has three equal sides. Therefore, it has three equal angles.*
3. *If you eat your supper, we will play miniature golf. You didn't eat your supper. Therefore, we will not play miniature golf.*
4. *If I get an A on the final exam, I get an A for the course. I didn't get an A for the course. Therefore, I didn't get an A on the final.*
5. *If I don't get an A on the final exam, I won't get an A for the course. I got an A for the course. Therefore, I got an A on the final.*

SOLUTION

1. Let p stand for "n is even" and q for "n^2 is even." The reasoning follows the form $p \rightarrow q, q, \therefore p$. This is the converse fallacy.

2. Let p be "this triangle has three equal sides" and q "this triangle has three equal angles." We write the reasoning as $p \rightarrow q, p, \therefore q$, which is valid by *modus ponens*.

3. With p as "eat supper" and q as "miniature golf," this is $p \rightarrow q, \neg p, \therefore \neg q$, the inverse fallacy.

4. If p indicates "I get an A on the final exam" and q "I get an A for the course," we have $p \rightarrow q, \neg q, \therefore \neg p$, which is valid by *modus tollens*.

5. We can choose our symbols to make the implication read as $p \rightarrow q$, so p would be "I don't get an A on the final exam" and q would be "I don't get an A for the course." The second sentence, "I got an A for the course," is the negation of q, that is, $\neg q$. Likewise, the third sentence, "I got an A on the final," is $\neg p$. The reasoning is $p \rightarrow q, \neg q, \therefore \neg p$, valid by *modus tollens*. □

Practice Problem 3 *For each of the following, identify the reasoning as either valid (modus ponens), valid (modus tollens), invalid (converse fallacy), or invalid (inverse fallacy):*

(a) *If you don't eat your supper, we won't play miniature golf. You ate your supper. Therefore, we will play miniature golf.*

(b) *If you fail the final, you will fail the course. You passed the final. Therefore, you will pass the course.*

(c) *If you fail the final, you will fail the course. You failed the final. Therefore, you will fail the course.*

(d) *If the sky is red at morning, it will not rain. It rained. Therefore, the sky was not red at morning.*

Analysis of Arguments

In order to use logic effectively in discourse, we need to see how to analyze the validity of persuasive arguments. We have already looked at some specific types of arguments that use *modus ponens* and *modus tollens*, and we have also learned to detect invalid arguments caused by the converse and inverse fallacies. At this point we have all the tools to analyze more complex arguments.

Recall from Section 1.3 that an propositional logic expression is a **tautology** if its truth value is true for all possible combinations of the truth values of the individual components. With this piece of terminology, we can give a formal definition to a **valid argument structure**.

Definition   An *argument structure* is a list of propositions

$$p_1, p_2, \ldots, p_n, \therefore q$$

where statements $p_1, p_2, \ldots, p_n$ are called the *premises* of the argument and the statement q is called the *conclusion* of the argument. Such an argument structure is said to be *valid* if

$$(p_1 \wedge p_2 \wedge \cdots \wedge p_n) \rightarrow q$$

is a tautology, and it is said to be *invalid* otherwise.

To be picky, we should always use the phrase "valid argument structure" rather than "valid argument." Technically, a valid argument is an argument (with actual content) that follows a valid argument structure. For example, $p \rightarrow q, p, \therefore q$ is an argument structure. An argument that follows this structure is "If it is raining, the streets are wet. It is raining. Therefore, the streets are wet." However, many people use the two terms interchangeably.

 Example 8

1. *Show that the modus ponens argument structure is valid.*

2. *Decide whether the following argument is valid. If not, give an assignment of truth values that makes all premises true and the conclusion false.*

   You must pass the final exam or pass all the tests if you pass the course. You failed the final exam and the course. Therefore, you must not have passed all the tests.

SOLUTION

1. Recall that *modus ponens* is formalized as $p \rightarrow q, p, \therefore q$. We build a truth table for the compound expression $((p \rightarrow q) \wedge p) \rightarrow q$:

p	q	$p \rightarrow q$	$(p \rightarrow q) \wedge p$	q	$((p \rightarrow q) \wedge p) \rightarrow q$
T	T	T	T	T	T
T	F	F	F	F	T
F	T	T	F	T	T
F	F	T	F	F	T

Since the final result has all T truth values, the expression $((p \rightarrow q) \wedge p) \rightarrow q$ is a tautology, so the argument structure is valid.

2. Let e stand for passing the final exam, a for passing all the tests, and p for passing the course. Do not be fooled by the placement of the word "if" in the middle of the statement. This means that p is the hypothesis of this statement, so we can write this statement as $p \rightarrow (e \vee a)$.

So we can write the argument as $p \rightarrow (e \vee a), \neg e \wedge \neg p, \therefore \neg a$. The truth table for this statement is shown below. The "final result" column is for the expression $(p \rightarrow (e \vee a)) \wedge (\neg e \wedge \neg p)) \rightarrow \neg a$.

e	a	p	$e \vee a$	$p \rightarrow (e \vee a)$	$\neg e$	$\neg p$	$\neg e \wedge \neg p$	$(p \rightarrow (e \vee a)) \wedge (\neg e \wedge \neg p)$	$\neg a$	Final Result
T	T	T	T	T	F	F	F	F	F	T
T	T	F	T	T	F	T	F	F	F	T
T	F	T	T	T	F	F	F	F	T	T
T	F	F	T	T	F	T	F	F	T	T
F	T	T	T	T	T	F	F	F	F	T
F	T	F	T	T	T	T	T	T	F	F
F	F	T	F	F	T	F	F	F	T	T
F	F	F	F	T	T	T	T	T	T	T

Since there is an F in the final column, the reasoning is faulty. A person who fails the final, passes all the tests, but fails the course demonstrates the fallacy. □

Practice Problem 4

(a) *Show that modus tollens is a valid argument structure.*

(b) *The argument structure $p \vee q, p \rightarrow r, q \rightarrow r, \therefore r$ is called "division into cases." We know one of two things is true, and we prove that no matter which is true, the conclusion r follows. Show that this argument structure is valid.*

Solutions to Practice Problems

1 We use p to stand for "You bought a car from us" and q to stand for "You paid too much."

(a) The converse statement is written formally as $q \rightarrow \neg p$. Here is the truth table. The final result is the same as for the inverse.

p	q	$\neg p$	$q \rightarrow \neg p$
T	T	F	F
T	F	F	T
F	T	T	T
F	F	T	T

(b) The contrapositive statement is written formally as $\neg q \rightarrow p$. Here is the truth table. The final result is the same as for the original statement.

p	q	$\neg q$	$\neg q \rightarrow p$
T	T	F	T
T	F	T	T
F	T	F	T
F	F	T	F

2 (a) If you pass the test, then you studied for at least four hours.

(b) If you pass the course, then you attended class regularly.

(c) If a quadrilateral has four equal angles, then it is a square.

(d) If a quadrilateral is a square, then it has four equal angles.

(e) If an integer is an odd prime, then it is greater than 2.

3 (a) Inverse fallacy

(b) Inverse fallacy

(c) Valid by *modus ponens*

(d) Valid by *modus tollens*

4 (a) *Modus tollens* is formalized as $p \rightarrow q, \neg q, \therefore \neg p$. See the first truth table below, for the compound expression $((p \rightarrow q) \wedge \neg q) \rightarrow \neg p$. The final column entries are all true, so the argument structure is valid.

(b) The second truth table shows $((p \vee q) \wedge (p \rightarrow r) \wedge (q \rightarrow r)) \rightarrow r$:

p	q	$p \rightarrow q$	$\neg q$	$(p \rightarrow q) \wedge \neg q$	$\neg p$	$((p \rightarrow q) \wedge \neg q) \rightarrow \neg p$
T	T	T	F	F	F	T
T	F	F	T	F	F	T
F	T	T	F	F	T	T
F	F	T	T	T	T	T

p	q	r	$p \vee q$	$p \rightarrow r$	$q \rightarrow r$	$(p \vee q) \wedge (p \rightarrow r) \wedge (q \rightarrow r)$	r	$((p \vee q) \wedge (p \rightarrow r) \wedge (q \rightarrow r)) \rightarrow r)$
T	T	T	T	T	T	T	T	T
T	T	F	T	F	F	F	F	T
T	F	T	T	T	T	T	T	T
T	F	F	T	F	T	F	F	T
F	T	T	T	T	T	T	T	T
F	T	F	T	T	F	F	F	T
F	F	T	F	T	T	F	T	T
F	F	F	F	T	T	F	F	T

Exercises for Section 1.6

1. A parent says, "If you don't eat your supper, you can't have dessert." The child eats his supper and the parent says, "No dessert for you." If the child takes his parent to court, does he have a valid case?

2. Each argument is either correct or it has a fallacy. Write the argument in symbols, then determine whether the argument is valid. If it is valid, determine whether it uses *modus ponens* or *modus tollens*. If it isn't valid, identify the fallacy as either the converse or the inverse fallacy.

(a) If both numbers are even, then the sum is even. They are not both even. Therefore, the sum is not even.

(b) If this number is a perfect square, then the equation has a rational solution. The equation has a rational solution. Therefore, this number is a perfect square.

(c) If you didn't buy from us, you paid too much. You did buy from us. Therefore, you didn't pay too much.

(d) If this prime number is even, then it is less than 5. This prime number is even. Therefore, it is less than 5.

(e) If this city is large, then it has large buildings. This city has large buildings. Therefore, it is large.

3. For each of the following, identify the reasoning as either valid (*modus ponens*), valid (*modus tollens*), invalid (converse fallacy), or invalid (inverse fallacy).

(a) If you pass the test, you studied for at least four hours. You studied for at least four hours. Therefore, you will pass the test.

(b) If you pass the course, you attend class regularly. You did not attend class regularly. Therefore, you will not pass the course.

(c) The quadrilateral is a square if it has four equal angles. The quadrilateral has four equal angles. Therefore, it is a square.

4. Recall that you can phrase "if, then" statements with words like "necessary," "sufficient" and "only if." Following the model from the text, convert the following statements to "if, then " statements.

(a) In order to pass this course, it is sufficient to read the book.

(b) In order to pass this course, it is necessary that you read the book.

(c) You will pass this course only if you read the book.

5. Write each of the following as an implication in "if, then" form:

(a) In order for you to get a refund from the store, it is necessary that you have your sales receipt.

(b) To grasp someone's mathematics background, it is sufficient to have them pronounce the name "Euler."

(c) You are legally driving in Pennsylvania only if you are 16 years old.

(d) To understand the history of calculus, it is necessary to study Descartes.

(e) It is sufficient to carry an umbrella to stay dry outside.

(f) Stephen King is fun to read if you like horror stories.

6. Use truth tables to characterize each of the following propositions as a tautology, a contradiction, or neither:

(a) $p \to \neg p$

(b) $(p \to q) \vee (q \to p)$

(c) $(p \wedge q) \vee (q \to \neg p)$

(d) $(p \vee \neg q) \to (q \wedge \neg p)$

7. Use truth tables to characterize each of the following propositions as a tautology, a contradiction, or neither:

(a) $(p \to (q \wedge r)) \vee ((p \wedge q) \to r)$

(b) $((p \to q) \wedge (q \to \neg p)) \to \neg p$

(c) $((p \to q) \wedge (\neg p \to r)) \to (q \vee r)$

8. Some of the valid reasoning forms are so obvious that they come automatically to us. For example, if we know two things are true, then we certainly know that the first is true. We can formalize this as $p \wedge q, \therefore p$. Use a truth table to formally demonstrate the validity of this and the other "obvious" forms of reasoning. For each, give a concrete example of this form of reasoning in daily life.

(a) $p \wedge q, \therefore p$

(b) $p, \therefore p \vee q$

(c) $p, q, \therefore p \wedge q$

(d) $p \vee q, \neg p, \therefore q$

(e) $p \to q, q \to r, \therefore p \to r$

9. Use a truth table to decide whether each of the following argument structures is valid. If it is not, give an assignment of truth values to the propositional variables that makes each premise true and the conclusion false.

(a) $p \to q, q \to p, \therefore p \wedge q$

(b) $p \to (\neg q \wedge r), q, \therefore \neg p$

(c) $p \to q, r \to p, \therefore r \to q$

(d) $p \vee (q \wedge r), p \to r, \therefore \neg q \to r$

(e) $p \wedge (q \vee r), r \to \neg p, \therefore r \to q$

10. Use truth tables to show that each of the following arguments is invalid. Write one sentence explaining a situation (corresponding to an assignment of truth values) that illustrates this.

(a) If Newton is not considered a great mathematician and Leibniz's work is not ignored, then calculus would not be the centerpiece of the modern math curriculum. Newton is considered the greatest mathematician only if Leibniz's work is ignored. Therefore, calculus is the centerpiece of the modern math curriculum and Leibniz's work is not ignored.

(b) If the weather forecast is good, then we have a picnic if and only if we have bread for sandwiches. If we have a picnic, then the weather forecast is good. Therefore, we have bread for sandwiches. (NOTE: The statement "p if and only if q" can be expressed symbolically as $(p \to q) \wedge (q \to p)$.)

(c) If I have a good round of golf, then the wind is calm or the weather is dry. The wind is calm and the weather is dry. Therefore, I have a good round of golf.

(d) Either I read a novel or I both lie on the couch and watch a baseball game on TV. I read a novel only if I lie on the couch. Therefore, if I do not watch a baseball game, then I do not lie on the couch.

11. Use a truth table to show that each of the following arguments is valid:

(a) For our camping trip, we take extra blankets if we take our gas heater, and if we do not take extra blankets, we do not take our air mattress. Therefore, we take our air mattress or gas heater only if we take extra blankets.

(b) Steve votes for a Libertarian candidate if and only if both his wife (Stella) votes for a Democrat and his father (Stan) votes for a Republican. For Stella to vote for a Democrat, it is necessary that Stan not vote for a Republican. Therefore, Steve does not vote Libertarian.

(c) If the prime interest rate goes up, then it is sufficient that unemployment goes down for prices to rise. However, unemployment goes down only if the prime interest rate goes up. Therefore, if prices do not go up, then unemployment does not go down.

(d) Each summer either I visit my family or I take a car trip and take off some time from work. I visit my family only if I do not take off time from work. Therefore, if I take off time from work, then I take a car trip.

Chapter 1 Summary

1.1 First Examples

You should have tried one or more of the games and puzzles in this section:

- Perhaps you have tried the card trick, using different sets of actions by the person following the instructions, to see if the club is always the card facing the opposite way from the others.
- Perhaps you have played the Josephus game with different numbers of people, and different rules for how many people are "skipped" at each step, looking for patterns in the results.
- Perhaps you have looked for organized ways to represent the outcomes of tennis matches, individual tennis games, and other similar situations.
- Perhaps you have tried to find a pattern for what kind of figures can be traced without lifting the pencil from the paper and without retracing any lines.
- Perhaps you have looked for winning strategies in the game played on the 4 × 4 grid.

1.2 Number Puzzles and Sequences

Terms and concepts

- You should recognize the use of the term *sequence* to describe a list of numbers, and feel comfortable with the notation a_n to indicate the nth term of the sequence.
- You should be able to distinguish between a *recursive formula* and a *closed formula* for a sequence.
- You should recognize the *Fibonacci numbers* and their recursive formula.

- You should feel comfortable with *sigma notation*. In particular, you should be able to interpret $\sum_{k=1}^{n} a_k$ as the sum of the first n terms of a sequence.

Working with sequences

- Given a recursive formula for a sequence, you should be able to:
  - Calculate the first several terms of the sequence.
  - Give recursive formulas for specific terms in the sequence (for example, the 20th term, or the $(3k-1)^{\text{st}}$ term).
  - Calculate particular terms of the sequence, given information about earlier terms in the sequence.
- Given a closed formula for a sequence, you should be able to:
  - Calculate the first several terms of the sequence.
  - Give values or formulas for specific terms in the sequence (for example, the 20th term, or the $(3k-1)^{\text{st}}$ term).
- Given the first few terms of a sequence, you should be able to:
  - Discover a pattern in those terms.
  - Apply that pattern to calculate additional terms.
  - Apply that pattern to discover recursive and/or closed formulas for the sequence.
- You should know how to verify that a sequence satisfies a given recursive formula.
- You should be able to convert from sigma notation to the usual summation notation, and back.

1.3 Truth-Tellers, Liars, and Propositional Logic

Terms and concepts

- You should be familiar with *propositions* and the related terms *propositional variable* and *formal proposition*.
- You should understand the concept of the *negation* of a proposition.
- You should understand the meaning of the operations $\wedge$ (*and*), $\vee$ (*or*), and $\neg$ (*not*).
- You should know the terminology "*exclusive or*" and how it differs from the $\vee$ (*or*) operation.
- You should know what it means to say that two statements are *logically equivalent*.
- You should understand the terms *tautology* and *contradiction*.

Operations with propositions

- You should be able to create new propositions from existing ones using the operations $\wedge$ (*and*), $\vee$ (*or*), and $\neg$ (*not*).
- You should be able to interpret and formalize English-language constructions such as "neither-nor" and "exactly one."
- You should be able to form the negation of a proposition using the double negative property and DeMorgan's laws.
- You should be able to apply truth tables to:
  - Analyze the statements made by Smullyan's *Truth-tellers* and *Liars*.
  - Analyze the truth value of compound statements based on the truth value of their components.
  - Determine whether two statements are logically equivalent.
- You should be able to use Theorem 2 to:
  - Simplify compound statements.
  - Verify that two statements are equivalent.

1.4 Predicates

Terms and concepts

- You should be familiar with *predicates* and the notation $P(x)$.
- You should understand the concept of *domain* of a predicate, and the related terms *set*, *element* and *member*, as well as the notation $\in$.

- You should know the purpose of *quantifiers* and understand the notation $\forall$ (*for all*) and $\exists$ (*there exists*).
- You should know the circumstances in which a quantified statement is true, and in particular the meaning of the word *counterexample*.

Operations with predicates

- You should be able to create new predicates from existing ones using the operations $\wedge$ (*and*), $\vee$ (*or*), and $\neg$ (*not*).
- You should be able to form the negation of a predicate using the double negative property and DeMorgan's laws.
- Given a predicate over a specific domain, you should be able to determine the truth value of $P(x)$ for particular elements of the domain.
- Given a quantified statement over a specific domain, you should be able to determine the truth value of the statement.
- You should be able to *negate a quantified statement* both symbolically and in natural language:
  - The negation of $\forall x \in D, P(x)$ is $\exists x \in D, \neg P(x)$.
  - The negation of $\exists x \in D, Q(x)$ is $\forall x \in D, \neg Q(x)$.
- With *multiple quantifiers*, you should know the difference between $\forall x, \exists y, P(x, y)$ and $\exists y, \forall x, P(x, y)$.

1.5 Implications

Terms and concepts

- You should recognize *implications*, whether written as "if p is true, then q is true," as "p implies q," or using the notation $p \rightarrow q$.
- You should be able to identify the *hypothesis* and the *conclusion* of an implication.
- You should understand the meaning of the word *counterexample* as applied to implications.
- You should know the meaning of the terms *converse*, *inverse*, and *contrapositive*.
- You should know the term *biconditional* and its notation $p \leftrightarrow q$.

Interpreting implications

- Given an implication involving predicates, you should be able to identify the domain, the hypothesis, and the conclusion, writing the implication in the form "For all $x \in D$, if $P(x)$, then $Q(x)$."
- You should understand the logic of the implication $p \rightarrow q$.

– You should be able to give a truth table for $p \to q$.

– You should know that the only time $p \to q$ is false is when the hypothesis (statement p) is true but the conclusion (statement q) is false.

- For an implication in the form, "For all $x \in D$, if $P(x)$, then $Q(x)$," you should be able to determine if you think the implication is true, and be able to give a counterexample if it is not true.

- You should be able to interpret a variety of English-language constructions that explicitly or implicitly involve quantified implications:

  – Convert such sentences to formal notation.

  – Recognize which sentences have the same meaning as each other, and which do not.

Operations with implications

- You should be able to negate implications, including quantified implications.

- You should be able to form the converse, inverse, and contrapositive of an implication.

- You should know that the original implication and its contrapositive are logically equivalent, and that the converse and inverse are logically equivalent.

- You should be able to give examples illustrating that a statement and its converse may not have the same truth value.

- You should be able to apply the logic of implications to logic puzzles (Smullyan's truth-tellers and liars).

1.6 Excursion: Validity of Arguments

Terms and concepts

- You should understand the use of the phrases "*if*," "*only if*," "*necessary*," and "*sufficient*," in English-language implications.

- You should know that an *argument structure* has the general form:

$$p_1, p_2, \ldots, p_n, \therefore q$$

and recognize the *premises* of the argument $(p_1, p_2, \ldots, p_n)$ and the *conclusion* of the argument (q).

- You should recognize *modus ponens* and *modus tollens* as two valid argument structures.

- You should recognize the *converse fallacy* and the *inverse fallacy* as two invalid argument structures.

Interpreting arguments

- You should be able to distinguish among a statement, its converse, and its inverse, and apply this to discourse of various types.

- You should know how to analyze what a person has actually said, avoiding the inappropriate substitution of a converse or inverse for the person's statement.

- You should be able to interpret English-language implications stated in a variety of ways (if, only if, necessary, sufficient), converting each to standard "if p, then q" form.

- You should be able to use a truth table to determine whether or not a given argument structure is valid.

- You should be able to analyze English-language arguments using a two-step process:

  – Convert the argument to a formal argument structure.

  – Analyze the validity of that structure. In many situations, you may recognize the structure as:

    * One of the valid argument forms *modus ponens*, *modus tollens*; or
    * One of the invalid argument forms *converse fallacy*, *inverse fallacy*.

2 | A Primer of Mathematical Writing

In the previous chapter, we looked at the nature of deductive reasoning as it relates to everyday experience, and we studied the logical structure of mathematical statements, discussing at length what it means for a statement to be true or false. In this chapter, we will discuss the properties of some common mathematical objects and how to present a mathematical proof about these properties to others.

The subject of our first efforts will be properties of integers, simply because integers are the mathematical objects most familiar to us. We will use these familiar objects to build our skills in writing proofs of mathematical statements. In the third and fourth sections, we will consider the proof technique known as mathematical induction, arguably the most important proof structure in discrete mathematics. In the fifth section, we will study a technique called "proof by contradiction," which applies indirect reasoning in an argument. Finally, in the last two sections we will apply the skills we have developed to explore some problems of interest to the typical student of mathematics, computer science, or education.

It is often the case that the abstract nature of mathematics makes mathematical proofs seem more mysterious than other forms of argument. For this reason, we will make every attempt to separate the logic of our arguments from the abstract nature of the subject matter. We will take time to build an understanding of the mathematical objects and properties before attempting proofs. If we are patient enough, the logical structure of your mathematical writing will seem as natural as an argument with your sister.

2.1 Mathematical Writing

Mathematicians today rely on the notion of symbolic logic to serve as the foundation of the system of reasoning that they bring to bear on the solution of problems. Of course, mathematics was quite successful for many centuries before symbolic logic was introduced, so it is not necessary to study symbolic logic in order to understand formal mathematics. A defining characteristic of mathematics is the thorough understanding of the cause-and-effect relationship between properties of formally defined objects. The demonstration of this understanding is the notion of "proof" around which the mathematics literature is based. In this section we will try to understand how to recognize a proof when we see one, and we will start to develop the skill of writing them ourselves.

Unlike conversational English, most mathematical statements are in the form of implications. That is, since mathematics studies relationships between formal objects, statements that effectively state, "Whenever an object has property P, then it must also have property Q" are commonplace. These are often more succinctly written as "if p, then q" but the nature of the English language allows a wide variety of equivalent forms.

 Example 1 *Write each of the following statements in "if, then" form:*

1. *Whenever n is an even integer, $2n^3 + n$ is divisible by 3.*
2. *For every prime n, $n^2 - n + 41$ is also prime.*
3. *The sum of the interior angles of any triangle is $180°$.*

SOLUTION

1. If an integer n is even, then $2n^3 + n$ is divisible by 3.
2. If a positive integer n is prime, then the number $n^2 - n + 41$ is prime.
3. If P is a triangle, then the sum of the interior angles of P is $180°$.

□

Aside from the variety of different ways to make a mathematical statement, you should glean two important points from the examples above. Not every mathematical statement is necessarily true, and not every mathematical statement is necessarily about numbers.

Practice Problem 1   *Determine which one of the three statements above is false.*

Implications and Their Contrapositives

We begin by highlighting a few of the ideas covered in some detail in Chapter 1, starting with this example first discussed in Section 1.5.

 Example 2 *Trooper Jones walks into the Goldilocks Pub and sees four Boatsville College students enjoying various beverages. In front of each person, there*

Figure 2-1 Age-drink cards.

is a card which has the person's age on one side and what he or she is drinking on the other side. She sees that the face-up sides of the cards look like Figure 2-1.

The drinking age law states in effect:

If you are drinking beer, then you are at least 21 years of age.   (2.1)

Which cards does Trooper Jones need to see turned over to check that everyone is obeying the law?

SOLUTION   If the law is being broken, it must be because someone *is drinking beer* and *is under 21 years of age*. That is, the only time that a statement of the form "if p, then q" is false is when the hypothesis p is true but the conclusion q is false. At all other times, we would have to say that the whole statement is true (i.e., the law is not being broken). Trooper Jones is looking for *counterexamples* to the "if, then" statement given in (2.1). She must turn over the cards that could give counterexamples: the "19" and the "Beer."   □

This example illustrates the basic logic behind an implication or "if, then" statement, which is summarized in Table 2-1. The last two rows of the table are the hardest for most people to swallow. How can a statement of the form "if p, then q" be true when the statement p is false, and *especially* when both statements p and q are false? A visual aid might be useful here.

Example 3 *The diagram in Figure 2-2 shows a way that the Goldilocks Pub can be sure to uphold the Pennsylvania law. It can simply have a separate room for its patrons who are at least 21 years of age, and within that room reserve a table for their patrons who are drinking beer.*

As long as the pub is arranged in this way, the people in the Goldilocks Pub are sure to be obeying the law. The patrons for whom the hypothesis is false (i.e., they are not drinking beer) can be in either room (i.e., the conclusion can be true or false). Notice also that the hypothesis could be false for all the patrons—they could all be enjoying nonalcoholic beverages—and the law would still be obeyed.

Hypothesis (p)	Conclusion (q)	Implication (If p, then q)
You are Drinking Beer	**You are at Least 21**	**You are Obeying the Law**
True	True	True
True	False	False
False	True	True
False	False	True

Table 2-1 Summary Analysis of an Implication

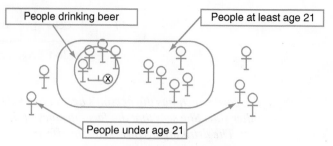

Figure 2-2   A lawful situation.

Most people have no difficulty with this concrete example—certainly Trooper Jones would immediately realize whose cards need to be checked. The same reasoning applies to more abstract settings. It is our contention that it is usually the abstraction in mathematics that causes the confusion, not the deductive logic used. To see this point illustrated, consider the following variation.*

Practice Problem 2   *Suppose someone makes four cards with letters on one side and numbers on the other, and that person then states,* "**If** *one side has a vowel,* **then** *the other side has an odd number." Which of the cards in Figure 2-3 must be turned over to tell whether the person is being truthful?*

In Section 1.5 we also introduced the notion of the *contrapositive* of an implication. Let us review the idea by thinking of equivalent ways of expressing the drinking age law:

$$\text{If you are under 21 years of age, then you are not drinking beer.} \qquad (2.2)$$

To see this is equivalent to our original statement of the law, think of the only situation in which the law is broken—specifically, when someone under the age of 21 drinks beer. This situation makes the hypothesis of (2.2) true and the conclusion false, exactly the situation that makes the implication false as discussed before. Note however that this same situation makes the hypothesis of (2.1) true while making its conclusion false. In other words, the statements (2.2) and (2.1) are false in exactly the same situation and hence true in the same situations—this is what is meant by being "logically equivalent."

In general, the statement "If p is true, then q is true" can always be equivalently expressed by its *contrapositive statement* "If q is not true, then p is not true."

Example 4   *For each implication below, explain the traits a counterexample to the statement must have. Which of the pairs are contrapositives of one another?*

1. *(a)* *If n is even, then $n! + 1$ is prime.*
    (b) *If n is odd, then $n! + 1$ is not prime.*

---

* This experiment, due to P. C. Wason [48], actually precedes the "drinking age" version by almost 20 years. It is now commonly called *Wason's selection task* in the cognitive science literature.

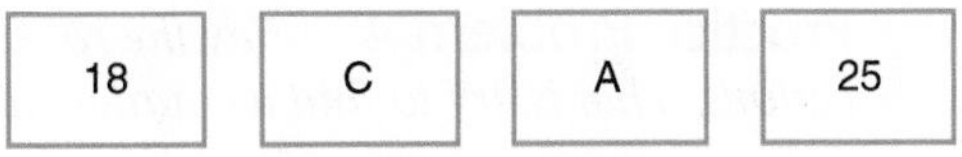

Figure 2-3 Letter-number cards.

2. (a) *If n^2 is even, then n is even.*

   (b) *If n is not even, then n^2 is not even.*

3. (a) *If n is prime, then the number following n is not a perfect square.*

   (b) *If m is perfect square, then the number preceding m is not prime.*

SOLUTION

1. These two statements are *not* contrapositives of each other. We can see this by thinking of the traits a counterexample to each statement must have:

   (a) A counterexample would have to be an even number n for which $n! + 1$ is not prime.

   (b) A counterexample would have to be an odd number n for which $n! + 1$ is prime.

2. These two statements *are* contrapositives of one another. We can also see this by thinking of the traits a counterexample to each statement must have:

   (a) A counterexample would be an integer n where n^2 is even and n is not even.

   (b) A counterexample would be an integer n where n is not even and n^2 is even.

3. These two statements are also contrapositives of one another, but this is clear only after fiddling with the language a bit. We can see this more clearly by thinking of the traits a counterexample to each statement must have:

   (a) A counterexample would have to come from a prime number followed by a perfect square.

   (b) A counterexample would be a perfect square preceded by a prime number.

   □

Practice Problem 3 *Write the contrapositive of each of the following statements. Describe the traits of a counterexample for both the original statement and its contrapositive.*

(a) *If n is divisible by 3, then n^2 is divisible by 9.*

(b) *If n is prime, then 2n is not divisible by 3.*

Mathematical Proofs

Now that we understand the logic of implication statements, we turn to the question "How do we *prove* that an implication is true?" In the small setting encountered by Trooper Jones, the answer is simple. Once Trooper Jones is sure there are no counterexamples, she will be satisfied that the implication is true for all the patrons of the Goldilocks Pub. The same is true in a more abstract settings—if there cannot possibly be a counterexample, then we will know the implication is true for all elements of the domain.

Explore more on the Web.

Practice Problem 4 *Play the role of Trooper Jones for these mathematical implications. That is, try to find a counterexample for each implication.*

(a) *For every integer $n \geq 1$, if n is odd, then $n^2 + 4$ is a prime number.*

(b) *For every positive integer n, if n is odd, then $n^3 - n$ is divisible by 4.*

Perhaps, for the first implication above, "For every integer $n \geq 1$, if n is odd, then $n^2 + 4$ is a prime number," you went through the following thought processes, playing alternately the role of Trooper Jones looking for counterexamples, and the pub's owner trying to show that there are none.

TROOPER JONES: Maybe 1 is a counterexample.

OWNER: No, because $1^2 + 4 = 5$, and 5 is prime.

TROOPER JONES: Let's try 3.

OWNER: Since $3^2 + 4$ is the prime number 13, that doesn't work.

TROOPER JONES: I think 5.

OWNER: Nope, $5^2 + 4$ is the prime number 29.

TROOPER JONES: 7?

OWNER: $7^2 + 4 = 53$, and 53 is prime.

TROOPER JONES: I give up, maybe the implication is true.

OWNER: Thank goodness!

TROOPER JONES: No, wait, let's try 9.

OWNER: Well, $9^2 + 4 = 85$, and 85 can be factored as $5 \cdot 17$.

TROOPER JONES: So 9 is a counterexample! I knew the statement was false!

There are two lessons to be learned from this dialogue:

- Trooper Jones doesn't bother checking even numbers. Even numbers cannot be counterexamples since they don't satisfy the hypothesis of the implication.

- Just because you have not yet found a counterexample doesn't mean there isn't one, unless the domain is finite and you have checked the entire domain.

Table 2-2 shows an abbreviated description of the thought process involved in the second implication of Practice Problem 4, "For every positive integer n, if n

n	$n^3 - n$	Divisible by 4?
1	0	Yes, since $0 = 4 \cdot 0$
3	24	Yes, since $24 = 4 \cdot 6$
5	120	Yes, since $120 = 4 \cdot 30$
7	336	Yes, since $336 = 4 \cdot 84$
9	720	Yes, since $720 = 4 \cdot 180$

Table 2-2 Summary Analysis of Practice Problem 4

is odd, then $n^3 - n$ is divisible by 4." The first column represents the choice of n made by Trooper Jones, and the third column shows the owner checking the truth of the conclusion for this value of n. Once again, the first few chosen numbers satisfying the hypothesis are not counterexamples, but how do we know that in this case, unlike the first practice problem statement, Trooper Jones will *never* find a counterexample? To form a *proof* of the implication, the owner would somehow have to convince Trooper Jones that, no matter what number she chooses, it will not be a counterexample. Since she only chooses numbers that satisfy the hypothesis, this amounts to demonstrating that

> No matter what number is chosen satisfying the hypothesis, it is guaranteed to also satisfy the conclusion.

In the remainder of this section, you will develop your ability to write arguments that will convince even the most skeptical readers that they are wasting their time looking for a counterexample. In Exercise 4e, you will be asked to furnish the proof for the proposition "For every positive integer n, if n is odd, then $n^3 - n$ is divisible by 4."

Proofs as Games

We can think of a mathematical proof as a game between the AUTHOR of the proof and the READER of the proof. Like Trooper Jones, the READER is skeptically trying to show that the statement under consideration is not true, and the AUTHOR's job is to convince the READER that he should quit trying to do this. By being convinced that he is never going to find a counterexample, the READER accepts the AUTHOR's proof.

For any "if, then" statement about an integer n, we can think of the interaction between AUTHOR and READER in terms of the following game:

- READER chooses a value of n that satisfies the hypothesis.
- AUTHOR tries to demonstrate that the conclusion is true for this choice of n.
- If the conclusion is true for this choice of n, then AUTHOR is successful, and it is READER's turn again—the game continues.
- If the conclusion is false for this choice of n, then READER wins.

If the statement under consideration is true, this game will never end. But no one has that kind of time! AUTHOR will write an argument that convinces even the most stubborn READER that the game will never end. This argument is a mathematical proof.

Let's try the game with a simple proposition that comes from the sort of exploration that mathematicians are always doing for fun:

> *Sammy has been playing around with numbers, and has observed that the prime number 3 is followed immediately by the perfect square 4. Intrigued by this, he has tried to find another pair of consecutive integers consisting of a prime number followed by a perfect square. He has come to the conclusion that there is no other such pair, and wants to prove that he is correct.*

Proposition 1   *Other than 3, 4 there is no pair of consecutive integers where the first is a prime number and the second is a perfect square.*

The first important step is to realize that the game cannot be played at all if AUTHOR and READER do not agree on the meaning of the terms in the statement. For this reason, mathematics books make a big deal of giving formal definitions of terms at their outset. There are many different ways to define any given term, so agreeing on a single definition is an important part of the proof process. For the given statement, there are two definitions that need to be given by the textbook so that AUTHOR and READER are using the terms in the same way.

> **Definition** A positive integer n greater than 1 is said to be *prime* if it cannot be factored as $n = a \cdot b$, where both a and b are greater than 1.

> **Definition** A *perfect square* is a positive integer that is equal to z^2 for some positive integer z.

The second observation is that our proposed game requires an "if, then" statement (an implication), and Sammy's proposition is not in that form. However, after some thought Sammy is able to come up with a suitable way to rewrite his proposition.

Theorem 2 *For all integers $n > 4$, if n is a perfect square, then $n - 1$ is not a prime number.*

Here are some sample "plays" of the game for this statement:

R. chooses n $= 4^2 = 16$,	then A. factors $15 = 3 \times 5$,	so 15 is not prime
R. chooses n $= 6^2 = 36$,	then A. factors $35 = 5 \times 7$,	so 35 is not prime
R. chooses n $= 7^2 = 49$,	then A. factors $48 = 16 \times 3$,	so 48 is not prime
R. chooses n $= 10^2 = 100$,	then A. factors $99 = 9 \times 11$,	so 99 is not prime
R. chooses n $= 12^2 = 144$,	then A. factors $143 = 11 \times 13$,	so 143 is not prime
R. chooses n $= 25^2 = 625$,	then A. factors $624 = 13 \times 48$,	so 624 is not prime

It appears that READER cannot win this game, but he will not give up unless we (as AUTHOR) can convince him of the futility of his efforts. We need to look for reasons he should not pick a particular value of n *before* he even bothers to select it. If we can convince READER in advance that no matter what he picks, we will be able to respond, *then* we will have proved Theorem 2.

In this game, AUTHOR's response consists of showing that $n - 1$ is not prime, by factoring it. We should see if we can predict in advance what the factorization will be. The READER is going to choose a number $n = (some\ positive\ number)^2$. Let's give a variable name to the "some positive number," to make it easier to refer to. If we agree to call it m, then each line in the game follows the following pattern:

> READER chooses $n = m^2$,   then AUTHOR tries to factor $n - 1$

Since n is the same as m^2, each line in the game can be seen to follow this slightly revised pattern:

$$\boxed{\text{READER chooses } n = m^2, \quad \text{then AUTHOR tries to factor } m^2 - 1}$$

Will AUTHOR always be able to do this? Perhaps you remember the rule

$$m^2 - 1 = (m-1)(m+1)$$

for factoring the difference of two squares from an algebra course.

With this realization, it appears that AUTHOR will always succeed! Now we just need to write down these observations in a note to READER so that he will stop trying to win the game:

Dear READER,

Every time you choose a perfect square (greater than 4) for n, say, $n = m^2$ (m a positive integer), I can factor $n - 1$. I know this because $n - 1$ is the same as $m^2 - 1$, which factors as $(m-1)(m+1)$. So as long as these two factors, $m-1$ and $m+1$, are both at least 2 (which they are since $n > 4$), this will demonstrate that $n - 1$ is not prime. So give it up and get on with your life.

Your pal,

AUTHOR

Mathematicians do not write proofs of theorems in the friendly style of a letter to READER. For example, here is a formal proof of the theorem as it might appear in a math book:

Theorem 2 (Revisited)   *For all integers $n > 4$, if n is a perfect square, then $n - 1$ is not a prime number.*

PROOF   Let a perfect square $n > 4$ be given. By the definition of a perfect square, $n = m^2$ for some positive integer m. Since $n > 4$, it follows that $m > 2$. Now the number $n - 1 = m^2 - 1$ can be factored as $(m-1)(m+1)$. Since $m > 2$, then both $m - 1$ and $m + 1$ are greater than 1, so $n - 1 = (m-1)(m+1)$ is a factorization of the number $n - 1$ into the product of two positive numbers, each of which is greater than 1. By the definition of prime number, it follows that the number $n - 1$ is not prime.   ∎

These two arguments, the informal and the formal, are really saying the same thing with the same logical structure. The informal proof is certainly less terse, but it includes no fewer details and no less rigorous logic.

Tracing Proofs

Explore more on the Web.

You might imagine READER receiving the letter from AUTHOR. It's a little abstract—it has all these variable names. So READER might attempt to understand the letter by trying it out with some particular numbers. Professional mathematicians frequently do this same thing to help them understand proofs they are encountering for the first time. The process is sometimes called *tracing the proof.*

Perhaps the first number READER tries is 121, which is 11^2. READER realizes the n in the letter is the 121 and the m is the 11. So $n - 1$ is 120, and $(m-1)(m+1)$ is

$n = m^2$	$n - 1$	$(m-1)(m+1)$
$n = (3)^2$	8	$(3-1)(3+1) = (2)(4)$
$n = (4)^2$	15	$(4-1)(4+1) = (3)(5)$
$n = (7)^2$	48	$(7-1)(7+1) = (6)(8)$
$n = (10)^2$	99	$(10-1)(10+1) = (9)(11)$
$n = (12)^2$	143	$(12-1)(12+1) = (11)(13)$
$n = (25)^2$	624	$(25-1)(25+1) = (24)(26)$

Table 2-3 Tracing a Proof

$(11-1)(11+1) = (10)(12)$, and sure enough this factors the 120. Table 2-3 summarizes several more numbers READER tries out to help visualize the letter's contents. This is fairly compelling evidence, since not only do we see that the number $n-1$ *will* factor, we actually see *how* it will factor.

It is important to realize that tracing a proof with particular numbers can help us understand the proof, but it does not really verify that the proof is correct. We must use our reasoning skills to think about the steps of the proof as we read. However, as we will see later in this section, the tracing process *can* sometimes be used to detect flaws in *faulty* proofs.

Simple Proofs About Numbers

Before moving on to other issues, it might help to see more examples of simple proofs. We did not look at these as our first example, because they are *too* easy in the sense that what they are proving does not really seem to *need* proof. It is appropriate to look at them now for further practice in reading and writing formal proofs. These proofs will use familiar terminology like "even," "odd," and "divisible by 4," so we can focus on the actual proof writing. However, we must still agree on the precise formal definitions, even for these easy terms we already know.

Definition

1. An integer n is *even* if it can be written in the form $n = 2 \cdot K$ for some integer K, and an integer m is *odd* if it can be written in the form $m = 2 \cdot L + 1$ for some integer L.

2. An integer n is divisible by 4 if it can be written in the form $n = 4 \cdot M$ for some integer M.

These definitions are all "existence" statements. The phrase "it can be written in the form $n = 2 \cdot K$ for some integer K" is equivalent to the more formal phrase "there exists an integer K such that $n = 2 \cdot K$." As we learned in Section 1.4, to establish the truth of an existence statement, one simply produces a particular value of the domain that makes the predicate true.

These definitions might not be the ones you would have written down if asked to describe the concepts of even and odd integers or of divisibility. For example, you might think of even as meaning that when written in our standard notation, the ones' digit is 0, 2, 4, 6, or 8. This is another perfectly valid definition of the concept of an even integer. As long as the AUTHOR and READER agree that the definition being used is valid, we may use either definition. The definition we have chosen simply makes it easier to write proofs, as we will see shortly.

Example 5 *Use the definitions to show that the following the statements are true:*

1. *72, 0, and −18 are even.*
2. *81 and −15 are odd.*
3. *72 is divisible by 4.*
4. *For any choice of integer n, $4n^2 − 2n$ is even.*

SOLUTION

1. 72 is even since $72 = 2 \cdot 36$, 0 is even since $0 = 2 \cdot 0$, and −18 is even since $-18 = 2 \cdot (-9)$.
2. 81 is odd because $81 = 2 \cdot 40 + 1$, and −15 is odd because $-15 = 2 \cdot (-8) + 1$.
3. Since we can write $72 = 4 \cdot 18$, this shows that 72 is divisible by 4.
4. We must find an integer K such that $4n^2 − 2n$ can be written as $2 \cdot K$. Put another way, we must write $4n^2 − 2n = 2 \cdot$(some integer). Since $4n^2 − 2n = 2 \cdot (2n^2 − n)$, we need only explain how we know that $2n^2 − n$ is an integer. The answer is that n and 2 are integers, and any combination of integers using addition, subtraction, and multiplication always produces another integer. □

The *truly* skeptical reader might be upset by the last line in the solution above since it assumes we agree on a property of numbers without proving it. This property of integer operations (called *closure*) can be established from the formal definition of the integers, but since we are using an informal understanding of the integers, we really cannot prove it here. Instead, we will assume that AUTHOR and READER have agreed that this property of the integers is true.

Closure Property of the Integers: Whenever the operations of addition, subtraction, or multiplication are applied to integers, the result will be an integer.

Practice Problem 5   *Fill in the blanks to make each of the following statements true:*

(a) *240 is even because $240 = 2 \cdot$ _____.*

(b) *−79 is odd because $-79 = 2 \cdot$ _____ $+ 1$.*

(c) *For any choice of integer n, $16n^2 − 44n + 17$ is odd because $16n^2 − 44n + 17 = 2 \cdot$ _____ $+ 1$.*

(d) *For any choice of integer n, $-8n^2 + 12n − 20$ is divisible by 4 because $-8n^2 + 12n − 20 = 4 \cdot$ _____.*

Now here is a proof that uses some of these definitions.

Proposition 3   *The result of summing any odd integer with any even integer is an odd integer.*

PROOF   Let odd integer x and even integer y be given. By the definition of "odd," there is an integer A such that $x = 2 \cdot A + 1$. By the definition of "even," there is an integer B such that $y = 2 \cdot B$. This means that

$$
\begin{aligned}
x + y &= (2 \cdot A + 1) + 2 \cdot B \\
&= 2 \cdot A + 2 \cdot B + 1 \\
&= 2 \cdot (A + B) + 1
\end{aligned}
$$

Since $A + B$ is an integer, the algebraic equations above demonstrate that $x + y$ can be written as 2 times an integer plus 1, so we can conclude from the definition of "odd" that $x + y$ is odd. ■

Although we did not write this proof as a letter to READER, we can still imagine READER following along to try to understand the proof. READER's thoughts might go something like the following example.

Explore more on the Web.

Example 6   *Trace the proof for the particular numbers $x = 17$ and $y = 12$. That is, suppose that READER chooses 17 for x and 12 for y, and show how the proof would read for those numbers.*

SOLUTION   By the definition of "odd," there is an integer A (namely $A = 8$) such that $x = 2 \cdot A + 1$ (i.e., $17 = 2 \cdot 8 + 1$). By the definition of "even," there is an integer B (namely $B = 6$) such that $y = 2 \cdot B$ (i.e., $12 = 2 \cdot 6$). This means that

$$
\begin{aligned}
17 + 12 &= (2 \cdot 8 + 1) + 2 \cdot 6 \\
&= 2 \cdot 8 + 2 \cdot 6 + 1 \\
&= 2 \cdot (8 + 6) + 1
\end{aligned}
$$

This demonstrates that $17 + 12$ can be written as 2 times an integer plus 1, so we can conclude from the definition of "odd" that $17 + 12$ is odd. □

Continuing with more examples, READER might build Table 2-4. So the proof shows exactly how $x + y$ will be written in the form of an odd number given by our formal definition of "odd."

By the way, did you notice that we used two different variable names (A and B)? A common mistake made by students just learning these ideas is to use the same

x	y	A	B	$x + y$	$2 \cdot (A + B) + 1$
17	12	8	6	29	$2 \cdot (14) + 1$
37	8	18	4	45	$2 \cdot (22) + 1$
101	14	50	7	115	$2 \cdot (57) + 1$
−17	84	−9	42	67	$2 \cdot (33) + 1$
51	50	25	25	101	$2 \cdot (50) + 1$

Table 2-4   Tracing the Proof of Proposition 3

variable name for two different things. For example, here is an incorrect version of this same proof.

Proposition 4   *The result of summing any odd integer with any even integer is an odd integer.*

FLAWED ARGUMENT   Let odd integer x, and even integer y be given. By the definition of "odd," there is an integer A such that $x = 2 \cdot A + 1$. By the definition of "even," there is an integer A such that $y = 2 \cdot A$. This means that

$$x + y = (2 \cdot A + 1) + 2 \cdot A$$
$$= 2 \cdot A + 2 \cdot A + 1$$
$$= 2 \cdot (A + A) + 1$$

This demonstrates that $x + y$ can be written as 2 times an integer plus 1, so we can conclude from the definition of "odd" that $x + y$ is odd.   ♮

To see the problem with this proof, imagine again that READER is following along to try to understand the argument. READER's thoughts might go something like this:

I'll choose 17 for x and 12 for y. Then the A would be ... well, I can't tell what it would be. It seems like it would have to be 8, because $17 = 2 \cdot 8 + 1$. But if so, the proof would say that $17 + 12 = 2 \cdot 16 + 1$, which is false. So maybe it needs to be 6, because $12 = 2 \cdot 6$. But then the proof would say that $17 + 12 = 2 \cdot 12 + 1$, which is also false. Hmmm ...

Clearly, READER should be unconvinced by the argument. We summarize this important principle as follows: **In any proof, do not use the same variable to represent two different things.** To be more precise, we use two variables A and B because the two numbers *might* be different. Using two variables does not say that they cannot be the same—for example, when READER chose 51 and 50, both A and B were 25. Using two variables simply says that they *are not known to be the same*—they might be the same, or they might be different.

Practice Problem 6   *Write a proof that the sum of two even integers is even, in the form of a letter from the* AUTHOR *to the* READER.

Here is an example that combines the even/odd concept with the concept of divisibility.

Proposition 5   *If n is even, then n^2 is divisible by 4.*

PROOF   We write this argument as an informal letter from AUTHOR to READER:

Dear READER,

You will never find a counterexample to this one. Choose any even integer and call it n. All I have to do is write your n^2 in the form 4 · (some integer). Here's how I know I'll always be able to do that. I know your n can be written as $2 \cdot k$, with k also an integer—by our agreement on what "even" means. When you square your n, you will get

$$n^2 = (2 \cdot k)^2$$
$$= 4k^2$$
$$= 4 \cdot (k^2)$$

Since k is an integer, so is k^2, and by the definition of "divisible by 4" this shows that your n^2 will always be divisible by 4.

Better luck next time,

AUTHOR ■

Finally, here is an example that illustrates a pitfall and suggests a possible remedy if you get stuck.

Proposition 6   *For all integers n, if n^2 is even, then n is even.*

FLAWED ARGUMENT   *Let n be any integer with n^2 even. We can write $n^2 = 2k$ for some integer k. Our goal is to write $n = 2 \cdot$ (some integer), so we divide both sides by n, getting $n = \frac{2k}{n} = 2 \cdot \left(\frac{k}{n}\right)$. Since $\frac{k}{n}$ is an integer, this proves the result.* ♮

There is a major problem with this attempt at a proof. Although addition, subtraction, and multiplication of integers always produces integers, division does not have that closure property. There is no reason to believe the statement "$\frac{k}{n}$ is an integer." For some integers it might be true, but for others it will not be. When we write a proof, we must be sure that each step is justified.

So once we reach "$n^2 = 2k$ for some integer k," we are stuck. We need to come up with a formula for n, and division does not have the closure property we need. We might try taking a square root, but that is even worse. When you take the square root of an integer, the result is only rarely another integer. What can we do? One possibility is to take an alternative approach based on our earlier discussion.

Proof by Contrapositive

The fact that an implicational statement and its contrapositive are equivalent means that we always have two possible paths to follow when attempting a proof. We can try to prove the theorem as it was originally stated, or we can rewrite the theorem as its contrapositive and prove that instead.

Since sometimes the two statements (the original and the contrapositive) *seem* different, writing the contrapositive often allows us to gain fresh perspective when trying to solve a problem. In some situations, a proof of the contrapositive may be easier than a proof of the original statement. Proposition 6 is a good example of this situation.

Example 7 *Informally prove the statement in Proposition 6, "For all integers n, if n^2 is even, then n is even," to a stubborn* READER.

SOLUTION   According to the discussion above, this statement is equivalent to its contrapositive "If n is not even, then n^2 is not even," or "If n is odd, then

n^2 is odd."* To prove this statement, we imagine a stubborn READER repeatedly choosing odd integers and testing each to see if its square is odd. We might interrupt our READER to offer some advice:

Dear READER,

Any time you choose an odd integer n, we agree that n can be written in the form $2k + 1$ for some integer k. By algebra we can see that

$$n^2 = (2k + 1)^2$$
$$= 4k^2 + 4k + 1$$
$$= 2 \cdot (2k^2 + 2k) + 1$$

This means that n^2 will have the form $2 \cdot$ (an integer) $+ 1$, so n^2 will be odd. So there is no point searching for an odd integer whose square is even.

Hugs and kisses,

AUTHOR

As before, we can turn this into a more formal mathematical proof:

Theorem 7   *For all integers n, if n^2 is even, then n is even.*

PROOF   We prove the contrapositive statement "If n is odd, then n^2 is odd." Let an odd integer n be given. Since n is odd, there is an integer k such that $n = 2k + 1$. It follows that $n^2 = 4k^2 + 4k + 1 = 2 \cdot (2k^2 + 2k) + 1$. Because $2k^2 + 2k$ is an integer, this shows that n^2 is odd. ∎

Solutions to Practice Problems

1   If $n = 41$, then

$$n^2 - n + 41 = 41^2 - 41 + 41 = 41^2$$

which is not prime.

2   A counterexample to the statement "If one side has a vowel, then the other side has an odd integer" must have a vowel on one side and an even integer on the other side. Hence, only the first and third cards (each of which meets one of these two conditions) need to be turned.

3   (a) If n^2 is not divisible by 9, then n is not divisible by 3. A counterexample for the original statement is a value of n where n is divisible by 3 but n^2 is not divisible by 9. A counterexample for the contrapositive is a value of n where n^2 is not divisible by 9 but n is divisible by 3.

   (b) If $2n$ is divisible by 3, then n is not prime. A counterexample for the original statement is a value of

n where n is prime and $2n$ is divisible by 3. A counterexample for the contrapositive is a value of n where $2n$ is divisible by 3 and n is prime.

4   (a) 9 is a counterexample.

   (b) There are no counterexamples.

5   (a) 120

   (b) −40

   (c) $8n^2 - 22n + 8$

   (d) $-2n^2 + 3n - 5$

6   Dear READER:

Suppose you have two even integers that we agree to call x and y. The definition of "even" means that you have agreed there is an integer K such that $x = 2 \cdot K$ and there is an integer L such that $y = 2 \cdot L$. In this case we can use algebra to see that

---

* If at this point you ask, "How do I know that 'not even' is the same as 'odd' for integers?", you have the pedantic inclination to be a first-rate mathematician. It is true that our definitions of even and odd do not guarantee that they are mutually exclusive without some further explanation. That explanation will come in the form of the division theorem in Section 2.2.

$x + y = 2 \cdot K + 2 \cdot L = 2 \cdot (K + L)$. So the sum of x and y is twice an integer (namely, $K + L$), which means that $x + y$ is even, like I told you!

Love always,

AUTHOR

Exercises for Section 2.1*

1. Fill in the blanks in each statement to complete the true statement using the definitions from this section.
   (a) 24 is even because $24 = 2 \cdot$ _____ .
   (b) 123 is odd because $123 = 2 \cdot$ _____ $+ 1$.
   (c) -16 is even because $-16 = 2 \cdot$ _____ .
   (d) -7 is odd because $-7 = 2 \cdot$ _____ $+ 1$.
   (e) $2n^2 + 4n - 6$ is even because $2n^2 + 4n - 6 = 2 \cdot$ _____ .
   (f) $6n + 3$ is odd because $6n + 3 = 2 \cdot$ _____ $+ 1$.
   (g) -72 is divisible by 6 because $-72 = 6 \cdot$ _____ .
   (h) 24 is not prime because $24 =$ _____ .
   (i) 144 is a perfect square because $144 = ($ _____ $)^2$.

2. Decide which of the following you believe to be true and which you believe to be false. If the statement is false, give a specific counterexample. If you believe the statement is true, provide the examples you tried as "evidence."
   (a) For each $n \geq 1$ and $m \geq 1$, if m is odd and n is even, then $m + n$ is divisible by 3.
   (b) For each $n \geq 1$, if n is odd and divisible by 3, then $n^2 - 1$ is divisible by 8.
   (c) For every $n \geq 2$, if n is divisible by 3, then $n(n + 1)(n + 2)$ is divisible by 4.
   (d) For each $n \geq 1$, if n is odd, then $4^n - 1$ is prime.
   (e) For each $n \geq 1$, if n is odd, then $2^n - 1$ is not divisible by 3.
   (f) For each $n \geq 1$ and $m \geq 1$, if m and n are even, then $mn - 1$ is not a perfect square.
   (g) For each $n \geq 1$, if n is not prime, then neither is $2^n - 1$.
   (h) For each $n \geq 1$, if n is divisible by 4, then $3^n + 1$ is divisible by 4.

3. Decide which of the following you believe to be true and which you believe to be false. If the statement is false, give a specific counterexample. If you believe the statement is true, provide the examples you tried as "evidence."
   (a) For each $n \geq 1$ and $m \geq 1$, if mn is a multiple of 4, then either m or n is a multiple of 4.
   (b) For each $n \geq 1$ and $m \geq 1$, if mn is a multiple of 3, then either m or n is a multiple of 3.
   (c) For each $n \geq 1$ and $m \geq 1$, if mn is a perfect square, then either m or n is a perfect square.
   (d) For each $n \geq 1$ and $m \geq 1$, if m is even and n is odd, then either $m + n^2$ or $m^2 + n$ is prime.
   (e) For each $n \geq 1$, if n is even or divisible by 3, then $n^2 + n$ is divisible by 6.
   (f) For each $n \geq 1$, if n is neither even nor divisible by 3, then $n^2 - 1$ is divisible by 6.
   (g) For each $n \geq 1$, if n is divisible by 3, then $n^2 - n + 41$ is prime.
   (h) Given a positive integer $N \geq 10$, form the number N' by removing the ones' digit from N and subtracting this digit from the remaining truncated integer. (For example, if $N = 1{,}309$, then $N' = 130 - 9 = 121$.) The statement to check is: If N' is divisible by 11, then N is divisible by 11.
   (i) Given a positive integer $N \geq 10$, form the number N' by removing the ones' digit from N and adding this digit to the remaining truncated integer. (For example, if $N = 1{,}239$, then $N' = 123 + 9 = 132$.) The statement to check is: If N' is divisible by 7, then N is divisible by 7.

4. Assume that you, the AUTHOR, and the READER can only agree on the definitions given in this section and your knowledge of basic algebra. Write a letter to the READER to convince him of the truth of each of the following statements:
   (a) If n is even, then $3n$ is even.
   (b) If n is even, then $n + 8$ is even.
   (c) If n is even, then $n + 1$ is odd.
   (d) If n is odd, then $n + 1$ is even.
   (e) If n is odd, then $n^3 - n$ is divisible by 4.
   (f) If n is divisible by 4, then $n^3 - n$ is divisible by 4.

---

*Some of the exercises refer to divisibility by 3, by 6, and so on. This is defined analogously to the definition of divisibility by 4.

(g) If n is divisible by 3, then n^2 is divisible by 9.

5. Fill in the missing steps in each of the following formal proofs:

(a) Proposition The sum of two odd integers is even.
Proof. Let m and n be odd integers. This means there is an integer K such that $m = 2K + 1$ and there is an integer L such that $n = 2L + 1$, and so

$$m + n = (\underline{}) + (\underline{})$$
$$= 2 \cdot (\underline{})$$

Since $\underline{}$ is an integer, this means that $m + n$ is even. ∎

(b) Proposition If n is even, then n^2 is even.
Proof. Let n be an even integer. This means that we can write $n = \underline{}$ for some integer m. This, in turn, means that $n^2 = 2 \cdot (\underline{})$, so n^2 is even, because $\underline{}$ is an integer. ∎

(c) Proposition Every odd perfect square can be written in the form $4k + 1$, where k is an integer.
Proof. Let s be an odd perfect square. So $s = n^2$ for some integer n, and n^2 is odd. By the contrapositive of the previous exercise, n is odd. Since n is odd, there is an integer L such that $n = \underline{}$. This means

$$s = n^2 = (\underline{})^2$$
$$= \underline{}$$

So $s = 4(\underline{}) + 1$, where $\underline{}$ is an integer, as desired. ∎

6. Here is a proof that the sum of two even integers is even:
Proof. Let even integers x and y be given. Then there is an integer K such that $x = 2 \cdot K$, and there is an integer L such that $y = 2 \cdot L$. So $x + y = 2 \cdot K + 2 \cdot L = 2 \cdot (K + L)$. Since $K + L$ is an integer, this shows that $x + y$ is even. ∎
As we did in Example 6, trace this proof for each of the following choices of x and y:
(a) $x = 18$, $y = 20$
(b) $x = 34$, $y = 8$
(c) $x = 18$, $y = 18$
(d) $x = -24$, $y = -40$

7. Refer to Exercise 6 above. Explain why the proof must use two different variables K and L. Does using two different letters mean that the two integers must be different, or just that they might be different?

8. Write a proof in the form of a letter from the AUTHOR to the READER of the statement "If n is odd, then $3n^2 + 1$ is divisible by 4."

9. Write a proof for each of the following statements in the form of a letter from the AUTHOR to the READER:
(a) The product of two odd integers is always odd.

(b) The product of an odd integer and an even integer is always even.
(c) If one integer is even and another is divisible by 3, then their product is divisible by 6.

10. Write a proof for each of the following statements:
(a) If n is even, then the product of n and its successor is even. (NOTE: The successor of an integer n is the integer $n + 1$.)
(b) If n is odd, then the product of n and its successor is even.
(c) The product of any two consecutive integers is even. (HINT: Refer the READER to the first two parts of this problem, and explain why this proves the proposition.)

11. Change your informal proof to a formal proof for each part of Exercise 9.

12. Which of the following pairs of statements are contrapositives of one another? In each case, describe the traits that a counterexample to each statement must have.
(a) i. If I root for the Braves, my brother roots for the other team.
ii. If I do not root for the Braves, my brother roots for the other team.
(b) i. If n is a positive even integer, then $3^n + 1$ is divisible by 5.
ii. If $3^n + 1$ is an odd integer, then n is not divisible by 5.
(c) i. If you do not do some math problems every night, you will not be good at mathematics.
ii. If you are good at mathematics, you do math problems every night.
(d) i. If you like computers, you would love computer science.
ii. If you do not like computers, you would not like computer science.
(e) i. If n is a prime greater than 3, then $n + 1$ is not a perfect square.
ii. If m is a perfect square greater than 4, then $m - 1$ is not prime.

13. For each of the following, write the contrapositive statement, and then prove the original statement by proving its contrapositive:
(a) If $m^2 + n^2 \neq 0$, then $m \neq 0$ or $n \neq 0$.
(b) If $3n$ is odd, then n is odd.
(c) If $m + n$ is odd, then m or n must be even. (HINT: Have you already proved this?)
(d) If mn is even, then m or n must be even. (HINT: Have you already proved this?)

14. Write a proof for each of the following statements in the form of a letter from the AUTHOR to the READER:

   (a) If n is an even integer greater than 2, then $2^n - 1$ is not prime.

   (b) Every odd perfect square can be written in the form $8k + 1$.   (HINT: Exercise 10c can help.)

   (c) 7 is the only prime number that precedes a perfect cube.*   (HINT: How does the polynomial $x^3 - 1$ factor?)

   (d) Every even perfect square is a multiple of 4. (HINT: Use Theorem 7.)

2.2 Proofs About Numbers

The primary goal of mathematical writing is clarity. In their profession, mathematicians have given up flowery adjectives or interesting metaphors in order to bring about a clarity in exposition as absolute as the vacuum of space. There are two side effects of this: (1) knowing standard definitions is essential for understanding, and (2) mathematics is often boring to read. No math textbook will compare a function to a summer day or a number to a bird in flight, so to understand what you are reading, you must know what the abstract objects are, and be prepared to be told of their interesting properties in perhaps not-so-interesting words.

   This section is intended to expose you to some of these standard mathematical definitions still within the familiar environment of integers and fractions. Along the way you will continue practicing your logical analysis of statements and your written explanations of why they are true or false. The trick will be to hold on to the basic processes we have discussed as the subject matter moves deeper into the world of abstract mathematics.

Divisibility

In the previous section, we relied entirely on our familiarity with division of integers to make sense of statements about numbers. We will now develop these ideas with slightly more abstract definitions in order to build our skills at writing coherent mathematical arguments about formally defined concepts. We begin with a general definition of divisibility.

> **Definition**   An integer n is *divisible by* a nonzero integer k if there is an integer q (called the quotient) such that $n = k \cdot q$. Equivalent ways to say this are "k *divides* n," "k is a *factor* of n," or even "n is a *multiple* of k."

Proposition 1   *If the integers m and n are both divisible by 3, then the number $m + n$ is also divisible by 3.*

   PROOF   Let m and n be integers, each divisible by 3. This means that there are integers K and L such that $m = 3K$ and $n = 3L$. Hence, $m + n = 3K + 3L = 3(K + L)$. Since $K + L$ is an integer, this shows that $m + n$ is divisible by 3.   ∎

> *__Definition__   A number x is a perfect cube if $x = y^3$ for some integer y.

Proposition 2   *If the integers m and n are both divisible by 3, then the number $m \cdot n$ is divisible by 9.*

PROOF   Let m and n be integers, each divisible by 3. This means that there are integers K and L such that $m = 3K$ and $n = 3L$. Hence, $m \cdot n = 3K \cdot 3L = 9(K \cdot L)$. Since $K \cdot L$ is an integer, this shows that $m \cdot n$ is divisible by 9. ∎

Not every proof has to use the definition of divisiblity. It is typical in mathematics to use results that have been previously established when constructing new arguments.

Proposition 3   *If the integer n is divisible by 3, then the number $n^2 + 3n$ is divisible by 9.*

PROOF   Let an integer n that is divisible by 3 be given. By Proposition 1, $n + 3$ is divisible by 3, and so by Proposition 2, $n \cdot (n + 3)$ is divisible by 9. Since $n^2 + 3n = n \cdot (n + 3)$, this establishes the fact that $n^2 + 3n$ is divisible by 9. ∎

Explore more on the Web.

Practice Problem 1   *Emulate Propositions 1 and 2 to write proofs of the following two statements:*

(a) *If m and n are divisible by 4, then m + n is divisible by 4.*

(b) *If m is divisible by 4 and n is any even integer, then $m \cdot n$ is divisible by 8.*

Another habit of mathematicians is to generalize results so that they apply to many situations. This practice makes the mathematical results more universally useful, but at the same time the results become more abstract and hence sometimes harder to understand. For example, Proposition 1 can be generalized as follows:

Proposition 4   *For any nonzero integer d, if the integers m and n are both divisible by d, then the number m + n is also divisible by d.*

Compare the following proof of this proposition to the proof of Proposition 1. The only difference is that instead of using the particular integer 3, the proof uses the variable d that can stand for *any* nonzero integer.

PROOF   Let the integer $d \neq 0$ be given, and let m and n be integers, each divisible by d. This means that there are integers K and L such that $m = d \cdot K$ and $n = d \cdot L$. Hence, $m + n = d \cdot K + d \cdot L = d \cdot (K + L)$. Since $K + L$ is an integer, this shows that $m + n$ is also divisible by d. ∎

Rational Numbers

In the proofs of the previous section, we mentioned the *closure property* of the set of integers with respect to the operations of addition, subtraction, and multiplication. This means that if we start with integers and use only those operations, we will always get integer results. As soon as we allow division, however, we have to acknowledge that the set is no longer closed. For example, the result of dividing the integer 5 by the integer 8 certainly exists, but it is not an integer. We will refer to such fractional numbers as rational numbers.

Definition

1. A real number r is *rational* if there exist integers a and b ($b \neq 0$) with $r = \frac{a}{b}$. Rational numbers (also called *fractions*) can be expressed in many equivalent ways (e.g., $\frac{1}{2} = \frac{2}{4} = \frac{3}{6} = \cdots$). It is always possible to choose the integers a and b with no common divisors* greater than 1.

2. A real number is *irrational* if it is not rational.

Proofs about rational numbers are often similar to the proofs we have been writing in the sense that the result you are trying to establish can be viewed as an "existence proof." For example, to show that a particular number is even, you must show that there exists an integer k such that your number is equal to $2 \cdot k$. Put another way, you must show that your number can be written as $2 \cdot$ (some integer). Similarly, to show that a particular number is a rational number, you must show that it can be written as $\frac{\text{some integer}}{\text{some nonzero integer}}$. Let's look at an example of a proof of this type using the formal definition of "rational."

Proposition 5   *For any rational number r, the number $r + 1$ is also rational.*

PROOF   Let r be a rational number. Then we may write $r = \frac{a}{b}$ for some integers a and b with $b \neq 0$. It follows that

$$r + 1 = \frac{a}{b} + 1$$
$$= \frac{a}{b} + \frac{b}{b}$$
$$= \frac{a + b}{b}$$

Since $a + b$ and b are integers and $b \neq 0$, by the definition of rational numbers this shows that $r + 1$ is a rational number.   ■

The style of this proof is similar to those we have seen before. We use the hypothesis and the definition of rational numbers to write r in "fraction form" as $\frac{a}{b}$. Our goal is to use this to show that there exist integers c and d with the property that $r + 1 = \frac{c}{d}$. The proof simply shows that the integers $c = a + b$ and $d = b$ satisfy the conditions.

The set of rational numbers has all the familiar closure properties (for addition, subtraction, and multiplication) enjoyed by the set of integers as well as some additional closure properties. For example, the average of two rational numbers is rational, as is the quotient provided the denominator is not zero. These properties will be developed in the exercises, using proofs that are similar to the one given in Proposition 5.

As in Section 2.1, when faced with a statement to prove, we are free to prove the contrapositive statement instead. It might be easier to do this if the conclusion of the original implication is a negative condition, like the property that a number is irrational. In general, it is easier to prove a number *has* a particular form than to prove

---

* Two integers with no common divisor greater than 1 are said to be *relatively prime*.

a number does *not* have that form. With this in mind, think about the contrapositive statement in completing the following problem.

Practice Problem 2   *Prove that for any real number x, if 2x is irrational, then x is also irrational.*

Proving by Cases

We continue the discussion of proofs about properties of integers by using an argument structure called "proof by cases." This technique is logically unremarkable—the READER agrees that one of several cases must hold and then the AUTHOR illustrates that in each one of these cases the same conclusion follows. Here are two examples that use the fact that every integer must be either even or odd.

Proposition 6   *For any integer n, $n^2 + n$ is even.*

   PROOF   (Informal version)

Dear READER,

   Any time you choose an integer n, we both know it will either be even or odd. The proof is slightly different for those two situations, so I'll address them one at a time in separate paragraphs.
   If your number is even, then I can write $n = 2 \cdot L$ for some integer L. From this, I can substitute and do a little algebra to get

$$n^2 + n = (2L)^2 + (2L) = 4L^2 + 2L = 2(2L^2 + L)$$

Since you will agree $2L^2 + L$ is an integer (adding and multiplying integers always gives an integer), this shows that $n^2 + n$ is even, but only for this case where your choice of n is even.
   Perhaps the number you have chosen is odd. If so, then I can write $n = 2 \cdot M + 1$ for some integer M. In this case, I can also substitute and do a little algebra to get

$$\begin{aligned} n^2 + n &= (2M + 1)^2 + (2M + 1) \\ &= 4M^2 + 4M + 1 + 2M + 1 \\ &= 4M^2 + 6M + 2 \\ &= 2(2M^2 + 3M + 1) \end{aligned}$$

Since $2M^2 + 3M + 1$ is an integer, this demonstrates that $n^2 + n$ is also even in the event that your choice of n is odd.
   So you can see that no matter how you choose your integer n, whether it is even or odd, I'll be able to demonstrate that $n^2 + n$ is even.

Your good friend,

AUTHOR

   PROOF   (Formal version) Let n be given. Since every integer is either even or odd, we have two cases to consider for n:

● **Case 1:** Suppose n is even. Then $n = 2 \cdot L$ for some integer L, and

$$n^2 + n = (2L)^2 + (2L)$$
$$= 4L^2 + 2L = 2(2L^2 + L)$$

Since $2L^2 + L$ is an integer, we have that $n^2 + n$ is even in this case.

- **Case 2:** Suppose n is odd. Then $n = 2 \cdot M + 1$ for some integer M, and

$$n^2 + n = (2M + 1)^2 + (2M + 1)$$
$$= 4M^2 + 4M + 1 + 2M + 1$$
$$= 4M^2 + 6M + 2 = 2(2M^2 + 3M + 1)$$

Since $2M^2 + 3M + 1$ is an integer, $n^2 + n$ is even in this case.

Thus, in either case we have shown that $n^2 + n$ is even. ■

In the preceding proof, we used two different variables L and M, but we could have used the same variable. Because the two cases are really separate little "proofs within a proof," they are completely independent of each other. The next example illustrates the alternative approach. Notice that for any particular integer n chosen by the READER, only one of the cases will apply, so as one reads through the proof, there will only be one value given to k even though the letter appears in two different parts of the proof. Exercise 5 looks at this issue in more detail.

Proposition 7   *Every perfect square is either a multiple of 4 or of the form* $4q + 1$ *for some integer q.*

PROOF   Let a perfect square n be given. We will give the name m to the integer that has $m^2 = n$. We know that every integer is either even or odd, so we have two cases to consider:

- **Case 1:** Suppose m is even. Then $m = 2k$ for some integer k, and hence

$$n = m^2 = (2k)^2 = 4k^2$$

Since k^2 is an integer, this establishes that n is a multiple of 4 in this case.

- **Case 2:** Suppose m is odd. Then $m = 2k + 1$ for some integer k, and hence

$$n = m^2 = (2k + 1)^2 = 4(k^2 + k) + 1$$

establishing that n is of the form $4 \cdot q + 1$, where in this case, $q = k^2 + k$. ■

The Division Theorem

In the above argument we stated matter-of-factly that "we know every integer is either even or odd." While this remark most likely did not elicit any gasps of amazement, it is worth stopping for a moment to think about what that statement means. It has to do with the way we prefer to define even and odd as well as the way we learn to do division in elementary school. Specifically, *whenever you divide an integer n by 2, you get some quotient q and a remainder r that is either 0 or 1.*

We can state this more formally by writing that every integer n can be written in the form

$$n = 2 \cdot q + r$$

for some integer q and where r must be 0 or 1. In other words, either $n = 2 \cdot q$ or $n = 2 \cdot q + 1$ for some integer q. This is precisely what we mean when we say, "We know that every integer is either even or odd."

This way of thinking about even and odd numbers explains why we have resisted the temptation to simply define "odd" as meaning "not even," or vice versa. We have seen the benefits of this point of view in writing proofs—if we know a number is odd, it is more useful to be able to say something positive about it (e.g., "it can be written as $2k + 1$") rather than something negative about it (e.g., "it cannot be written as $2k$"). But the *most* important thing about this point of view about even and odd numbers is that it can be generalized for any divisor. Specifically,

Whenever you divide an integer n by a positive integer d, you get a unique integer quotient q and a unique *remainder r from the set* $\{0, 1, \ldots, d - 1\}$.

Example 1 *Investigate the above statement for various integers n divided by $d = 5$.*

SOLUTION In Table 2-5, we give the quotient and remainder, as well as the multiplicative way to write the relationship, for various values of n. These examples support the claim that when an integer is divided by 5, the only possible remainders are 0, 1, 2, 3, and 4. □

This fact, using the multiplicative style from the example to express it, is called the *division theorem*. It is also called *division algorithm* in some contexts. You will be asked to prove this through exercises at appropriate times in this book. The important thing now is to remember what it means in terms of elementary school division.

Theorem 8 (Division Theorem) *For all integers a and b (with b > 0), there is an integer q (called "the quotient when a is divided by b") and an integer r (called "the remainder when a is divided by b") such that*

1. $a = b \cdot q + r$, *and*
2. $0 \le r < b$.

Furthermore, q and r are the only two integers satisfying both these conditions.

---

n	1	23	49	0	-13
Quotient	0	4	9	0	-3
Remainder	1	3	4	0	2
Equation $n =$	$0 \cdot 5 + 1$	$4 \cdot 5 + 3$	$9 \cdot 5 + 4$	$0 \cdot 0 + 0$	$-3 \cdot 5 + 2$

Table 2-5 Five Remainders when Dividing by 5

PROOF   The existence of values for q and r is the focus of Exercise 14 in Section 2.4, and the uniqueness of these values is addressed in Exercises 24 and 25 at the end of the current section and in Exercises 7 and 8 in Section 2.5. ■

Since this is not written the way we are accustomed to seeing division with a remainder, it might be helpful to consider one more example before moving on to proofs that use this rather abstract-looking result.

Example 2 *Discuss $73 \div 5$ in terms of quotients and remainders, and connect the way you understand the answer to the formal statement of the division theorem.*

SOLUTION   In elementary school, we would have said that $73 \div 5$ yields a quotient of 14 with a remainder of 3. Perhaps later in our education we might have given the answer as $14\frac{3}{5}$, and today we might just say the answer is 14.6. The division theorem takes us back to our original understanding of division.

More to the point, the formal division theorem is most like the way we would have checked our elementary school answer. To *check* that answer, we would have multiplied the *quotient* 14 by the *divisor* 5 and added the *remainder* 3. If we got the *dividend* 73, we knew we had the right answer. Thus, our "check" consists of checking that

$$73 = 5 \cdot 14 + 3$$

This is exactly the way the division theorem writes the result of dividing 73 by 5. □

In this book, when we use the division theorem in a proof, we almost always have a particular divisor in mind. Propositions 6 and 7 are good examples of this. In each of these proofs, we use cases based on whether a number is even or odd. This is exactly the same thing as using cases for the remainder of the number when divided by 2. When we do have a particular divisor in mind, we can think about the division theorem in a slightly less abstract form.

Example 3 *Suppose we have "division by 5" in mind, just as we did in Example 1. Then the division theorem says that for any integer a, we can find a quotient q and a remainder r, where*

$$a = 5 \cdot q + r \qquad and \qquad r \text{ is either } 0, 1, 2, 3, \text{ or } 4$$

A less formal way to phrase this is to say that, for any integer a, we can find a quotient q so that a can be written as one of the following:

$$a = 5 \cdot q, \quad a = 5 \cdot q + 1, \quad a = 5 \cdot q + 2, \quad a = 5 \cdot q + 3, \quad or \quad a = 5 \cdot q + 4$$

Here is an example of how one might use the division theorem in a proof in the same way we used our understanding of evens and odds in the proofs of Propositions 6 and 7.

Proposition 9 *If n is any integer not divisible by 5, then n has a square that is of either the form $5k + 1$ or $5k + 4$. (For example, $13^2 = 5 \cdot 33 + 4$ and $9^2 = 5 \cdot 16 + 1$.)*

PROOF   Let an integer n not divisible by 5 be given. By the division theorem, when n is divided by 5, it leaves a remainder of 0, 1, 2, 3, or 4. But since n is not divisible by 5, we know the remainder must be 1, 2, 3, or 4. That is, one of the following cases must be true:

- **Case 1:** It might be that $n = 5q + 1$ for some integer q. In this case, $n^2 = 25q^2 + 10q + 1 = 5 \cdot (5q^2 + 2q) + 1 = 5 \cdot$ (an integer) $+ 1$.
- **Case 2:** It might be that $n = 5q + 2$ for some integer q. In this case, $n^2 = 25q^2 + 20q + 4 = 5 \cdot (5q^2 + 4q) + 4 = 5 \cdot$ (an integer) $+ 4$.
- **Case 3:** It might be that $n = 5q + 3$ for some integer q. In this case, $n^2 = 25q^2 + 30q + 9 = 5 \cdot (5q^2 + 6q + 1) + 4 = 5 \cdot$ (an integer) $+ 4$.
- **Case 4:** It might be that $n = 5q + 4$ for some integer q. In this case, $n^2 = 25q^2 + 40q + 16 = 5 \cdot (5q^2 + 8q + 3) + 1 = 5 \cdot$ (an integer) $+ 1$.

Therefore, n^2 is of the form $5 \cdot k + 1$ or $5 \cdot k + 4$ for some integer k in every possible case. ∎

We conclude this section with a final word on remainders. The operation that returns the remainder of a division problem is not only useful in mathematics proofs but it is also very common in computer programming, especially when a program needs to "branch" to handle cases separately.

> **Definition**   In computer languages as well as in mathematics, the term **mod** is used to describe the remainder when one integer is divided by another. Thus, we write $a \bmod b = r$ to mean that r is the remainder when a is divided by b (i.e., $a = b \cdot q + r$ and $0 \le r < b$). Some computer languages (like C, $C++$, and *Java*, e.g.) write this operation as $a \% b$ rather than as $a \bmod b$.

 Example 4 *Compute each of the following:*

1. 73 mod 5
2. −22 mod 6
3. 8 mod 11
4. *Let n be any integer. What is* $(4n^2 - 12n + 3) \bmod 4$?
5. *Let n be any integer. What is* $(4n^2 - 12n + 9) \bmod 4$?

SOLUTION

1. $73 \bmod 5 = 3$, since $73 = 5 \cdot 14 + 3$.
2. $-22 \bmod 6 = 2$, since $-22 = 6 \cdot (-4) + 2$.
3. $8 \bmod 11 = 8$, since $8 = 11 \cdot (0) + 8$.
4. Let n be any integer. Then $(4n^2 - 12n + 3) \bmod 4 = 3$, since $4n^2 - 12n + 3$ can be written as $4(n^2 - 3n) + 3$.

5. Let n be any integer. Then $(4n^2 - 12n + 9) \bmod 4 = 1$, since

$$4n^2 - 12n + 9 = 4n^2 - 12n + 8 + 1$$
$$= 4(n^2 - 3n + 2) + 1$$

$\square$

Practice Problem 3

(a) *For each of the following pairs of integers a and b, write $a = b \cdot q + r$ where $0 \le r < b$:*

   i. $a = 17, b = 3$

   ii. $a = 100, b = 12$

   iii. $a = -19, b = 4$

   iv. $a = 6, b = 13$

(b) *Calculate the following:*

   i. 17 mod 3

   ii. 100 mod 12

   iii. -19 mod 4

   iv. 6 mod 13

   v. $(25q^2 + 10q + 1) \bmod 5$

   vi. $(25q^2 + 30q + 9) \bmod 5$

(c) *Complete the following sentence: "The division theorem with $b = 7$ tell us that, for any integer a, we can find a quotient q so that a can be written as one of the following:* _____."

The mod notation is convenient for mathematicians, too. For example, if we use this terminology, Proposition 9 can be more concisely written as "If $n \bmod 5 \ne 0$, then $n^2 \bmod 5$ is either 1 or 4."

To test your understanding of the division theorem, proof by cases, and the mod notation, try this problem before moving on to the section exercises.

Explore more on the Web.

Practice Problem 4

Prove that every integer not divisible by 3 has a square that is of the form $3K + 1$. Also write the statement of the theorem using mod *notation.*

Solutions to Practice Problems

1 (a) *Proof.* Let integers m and n, both divisible by 4, be given. This means that there is an integer K such that $m = 4K$, and there is an integer L such that $n = 4L$. It follows that

$$m + n = 4K + 4L = 4(K + L)$$

Since $K + L$ is an integer, the sum $m + n$ is divisible by 4. $\blacksquare$

(b) *Proof.* Let integers m and n, with m divisible by 4 and n even, be given. This means that there is an integer X such that $m = 4X$, and an integer Y such

that $n = 2Y$. It follows that

$$m \cdot n = (4X) \cdot (2Y) = 8 \cdot (X \cdot Y)$$

and so, since $X \cdot Y$ is an integer, the product $m \cdot n$ is divisible by 8. $\blacksquare$

2 *Proof.* We prove the contrapositive "For any real number x, if x is rational, then $2x$ is also rational." Let rational number x be given and write $x = \frac{a}{b}$, where a and b are integers and $b \ne 0$. Then

$$2x = 2 \cdot \frac{a}{b} = \frac{2a}{b}$$

Since $2a$ and b are integers and $b \neq 0$, this shows that $2x$ is rational. ∎

3 (a) $17 = 3 \cdot 5 + 2$, $100 = 12 \cdot 8 + 4$, $-19 = 4 \cdot (-5) + 1$, and $6 = 13 \cdot (0) + 6$

(b) $2, 4, 1, 6, 1$, and 4

(c) $a = 7 \cdot q$, $a = 7 \cdot q + 1$, $a = 7 \cdot q + 2$, $a = 7 \cdot q + 3$, $a = 7 \cdot q + 4$, $a = 7 \cdot q + 5$, or $a = 7 \cdot q + 6$

4 **Proposition** If $n \bmod 3 \neq 0$, then $n^2 \bmod 3 = 1$.
Proof. Let n be an integer not divisible by 3. By the division theorem, every integer n must be of the form $3q$, $3q + 1$ or $3q + 2$, but since this n is not divisible by 3, we have just two cases to consider:

● **Case 1:** If $n = 3q + 1$, then
$$n^2 = (3q + 1)^2 = 9q^2 + 6q + 1$$
$$= 3(3q^2 + 2q) + 1$$
so n^2 is of the form $3 \cdot (\text{an integer}) + 1$.

● **Case 2:** If $n = 3q + 2$, then
$$n^2 = (3q + 2)^2 = 9q^2 + 12q + 4$$
$$= 3(3q^2 + 4q + 1) + 1$$
so n^2 is of the form $3 \cdot (\text{an integer}) + 1$. ∎

Exercises for Section 2.2

1. Fill in the blanks to create multiplicative equations in the style of the division theorem. (NOTE: To be completely consistent with the division theorem, there is only one way to fill in each blank!)

(a) $73 = \underline{\hspace{2cm}} \cdot 6 + \underline{\hspace{2cm}}$

(b) $187 = \underline{\hspace{2cm}} \cdot 11 + \underline{\hspace{2cm}}$

(c) $-1{,}234 = \underline{\hspace{2cm}} \cdot 15 + \underline{\hspace{2cm}}$

(d) $-24 = \underline{\hspace{2cm}} \cdot 4 + \underline{\hspace{2cm}}$

(e) $1{,}000 = \underline{\hspace{2cm}} \cdot 7 + \underline{\hspace{2cm}}$

(f) $(9k^2 + 5) = \underline{\hspace{2cm}} \cdot 3 + \underline{\hspace{2cm}}$

2. Compute each of the following:

(a) $73 \bmod 6$

(b) $187 \bmod 11$

(c) $-1{,}234 \bmod 15$

(d) $-24 \bmod 4$

(e) $1{,}000 \bmod 7$

(f) $(9k^2 + 5) \bmod 3$

3. Compute each of the following:

(a) $55 \bmod 6$

(b) $55 \bmod 5$

(c) $6 \bmod 55$

(d) $-55 \bmod 6$

(e) $(6(k^3 - 3k^2 - 2k + 8) + 5) \bmod 6$

(f) $(9k^4 - 3k^2 - 30k + 11) \bmod 3$

(g) $10(9k + 1) \bmod 9$

(h) $(7(5k + 2^{n-1}) - 2^n) \bmod 5$. (HINT: Write 2^n as $2 \cdot 2^{n-1}$.)

4. Provide a counterexample to each of the following statements about integers that is false. You do not need to prove the statements you believe to be true.

all False

(a) If $(a \% c) = (b \% c)$, then $a = b$.

(b) If $(a \% b) = c$, then $((a + 1) \% b) = c + 1$

(c) If $a < b$, then $(a \% c) + (b \% c) = ((a + b) \% c)$.

(d) If $b < c$, then $(a \% b) + (a \% c) = ((a \% (b + c))$.

5. For the values of n given in Table 2-6, find the values of m and k described in the proof of Proposition 7.

6. Fill in the missing details in the following proofs:

(a) **Proposition** If 3 divides b and b divides c, then 3 divides c.
Proof. Let integers b and c be given, and assume that 3 divides b and b divides c. This means that $b = 3 \cdot k$ for some integer k and $c = m \cdot b$ for some integer m.
$\vdots$
Since $\underline{\hspace{2cm}}$ is an integer, this establishes that 3 divides c. ∎

(b) **Proposition** If n^3 is even, then so is n.
Proof of the contrapositive. Let an odd integer n be given. Then we may choose an integer m so that $n = 2m + 1$.
$\vdots$
Hence, n^3 is odd, completing the proof. ∎

(c) **Proposition** If 3 divides $4^{n-1} - 1$, then 3 divides $4^n - 1$.
Proof. Let the integer n be given, and assume that 3 divides $4^{n-1} - 1$. This means that $4^{n-1} - 1 = 3k$ for some integer. So $4^{n-1} = 3k + 1$, and
$$4^n = 4(4^{n-1})$$
$$= 4(\underline{\hspace{2cm}})$$
$$= 3(\underline{\hspace{2cm}}) + 1$$

n	4	25	49	144	225	1,024
m	2	5	7	12	15	32
k	1	2	3	6	7	16

Table 2-6 Table for Exercise 5

$4k^2 = _n$, for even

$4(k^2 + k) + 1 = _n$, for odd

Therefore, $4^n - 1 = 3(\underline{\hspace{2cm}})$. Since $\underline{\hspace{3cm}}$ is an integer, this means that 3 divides $4^n - 1$. ∎

7. Prove each of the following propositions:

⚡ (a) If a divides b and a divides c, then a divides $b + c$.

(b) If a divides b and a divides c, then a^2 divides $b \cdot c$.

(c) If a divides b and c divides d, then ac divides bd.

(d) If c divides a and x is any nonzero integer, then cx divides ax.

(e) If 9 divides $10^{n-1} - 1$, then 9 divides $10^n - 1$.

(f) If 6 divides $n^3 - n$, then 6 divides $(n+1)^3 - (n+1)$. (HINT: Multiply out $(n+1)^3 - (n+1)$ and simplify.)

8. Fill in the blanks to complete the proof of the following statement.

Proposition The sum of two rational numbers is a rational number. (That is, if a and b are rational numbers, then $a + b$ is a rational number.)

Proof. Let rational numbers a and b be given. Since $\underline{\hspace{2cm}}$, we know that $a = \frac{x}{y}$ for some integers x and y with $y \neq 0$. Likewise, since $\underline{\hspace{2cm}}$, we know that $b = \frac{z}{w}$ for some integers z and w with $w \neq 0$. From the rules for adding fractions, we know that

$$a + b = \frac{?}{yw}$$

We know that $\underline{\hspace{2cm}}$ and yw are both integers, and $yw \neq 0$ because $\underline{\hspace{2cm}}$. Hence, we know that $a + b$ is rational, by the definition of "rational." ∎

9. Using Exercise 8 as a guide, prove that the difference between any two rational numbers is a rational number.

10. Using Exercise 8 as a guide, prove that the product of any two rational numbers is a rational number.

11. The average of two numbers a and b is the number $\frac{a+b}{2}$. Prove that the average of two rational numbers is a rational number.

12. Prove that every integer is a rational number.

13. The following propositions have to do with number sequences defined recursively as we saw in Section 1.2:

(a) Suppose the sequence a_n is defined by the recurrence relation $a_n = a_{n-1} + 2n$.

   i. Prove that if a_{n-1} is even, then so is a_n.

   ii. Write out the first five terms of this sequence with the additional assumption that $a_1 = 10$. Is the statement you proved in (i) true for this sequence?

   iii. Write out the first five terms of this sequence with the additional assumption that $a_1 = 7$. Is the statement you proved in (i) true for this sequence?

(b) Suppose the sequence b_n is defined by the recurrence relation $b_n = 3b_{n-1} + 2b_{n-2}$.

   i. Prove that if b_{n-2} is a multiple of 3, then so is b_n.

   ii. Write out the first eight terms of this sequence with the additional assumption that $b_1 = 0$ and $b_2 = 1$. Is the statement you proved in (i) true for this sequence?

   iii. Write out the first eight terms of this sequence with the additional assumption that $b_1 = 1$ and $b_2 = 2$. Is the statement you proved in (i) true for this sequence?

(c) Suppose the sequence c_n is defined by the recurrence relation $c_n = c_{n-1} + 2^{n-1}$.

   i. Prove that if $c_{n-1} < 2^{n-1}$, then $c_n < 2^n$.

   ii. Write out the first five terms of this sequence with the additional assumption that $c_1 = 0$. Is the statement you proved in (i) true for this sequence?

   iii. Write out the first five terms of this sequence with the additional assumption that $c_1 = 3$. Is the statement you proved in (i) true for this sequence?

14. Fill in the missing details in the following proofs:

⚡ (a) **Proposition** If n^2 is divisible by 3, then so is n. (HINT: Check each of the three possible cases for remainders when n is divided by 3.)

Proof. Considering the contrapositive statement, let an integer n be given that is not divisible by 3. By the division theorem, when any integer is divided by 3, it leaves a remainder of 0, 1, or 2. That is, one of the following cases must be true:

● **Case 1:** It might be that $n = 3q$ for some integer q. However, for this particular integer n, we know this case does not happen, because $\underline{\hspace{2cm}}$.

● **Case 2:** It might be that $n = 3q + 1$ for some integer q. In this case,

$$n^2 = \underline{\hspace{2cm}}$$
$$= 3 \cdot (\underline{\hspace{2cm}}) + \underline{\hspace{2cm}}$$

● **Case 3:** It might be that $n = 3q + 2$ for some integer q. In this case,

$$n^2 = \underline{\hspace{2cm}}$$
$$= 3 \cdot (\underline{\hspace{2cm}}) + \underline{\hspace{2cm}}$$

Thus, in every case that satisfies the hypothesis, we see that n^2 is not divisible by 3, completing the proof. ∎

(b) **Proposition** Every prime number greater than 3 has a square that is of the form $12k + 1$. (For example, $7^2 = 12 \cdot 4 + 1$ and $19^2 = 12 \cdot 30 + 1$.)

Proof. Let a prime number n greater than 3 be given. By the division theorem, when n is divided by 6, it leaves a remainder of 0, 1, 2, 3, 4, or 5. That is, one of the following cases must be true:

- **Case 1:** It might be that $n = 6q$ for some integer q. But we know this case does not happen for this n, since _____.
- **Case 2:** It might be that $n = 6q + 1$ for some integer q. In this case,
$$n^2 = \underline{\qquad}$$
$$= 12 \cdot (\underline{\qquad}) + 1$$
- **Case 3:** It might be that $n = 6q + 2$ for some integer q. This case does not happen for this n, since _____.
- **Case 4:** It might be that $n = 6q + 3$ for some integer q. This case does not happen for this n, since _____.
- **Case 5:** It might be that $n = 6q + 4$ for some integer q. This case does not happen for this n, since _____.
- **Case 6:** It might be that $n = 6q + 5$ for some integer q. In this case,
$$n^2 = \underline{\qquad}$$
$$= 12 \cdot (\underline{\qquad}) + 1$$

Therefore, n^2 is of the form $12 \cdot$ (an integer) $+ 1$ in the only possible cases. ∎

15. Prove that if n is divisible by 3 and n is divisible by 4, then n is divisible by 12. (HINT: Write $n = 12q + r$ and consider the cases.)

16. Prove that if n is not divisible by 3, then $n^2 + 2$ is divisible by 3.

17. Prove that if n is even, then $n^3 + 2n$ is divisible by 4.

18. Use the previous three exercises to prove that if n is even, then $n^3 + 2n$ is divisible by 12.

19. Prove that if n^2 is divisible by 5, then n is divisible by 5. (HINT: Consider the contrapositive.)

20. Prove that the sum of any three consecutive perfect cubes is divisible by 9. (HINT: Agree with your READER to denote the three perfect cubes $(n-1)^3$, n^3, and $(n+1)^3$, and be prepared to use Exercise 16.)

21. Prove that no perfect square ends with the digit 2. (HINT: The ones' digit of a number is mathematically the same as the remainder when that number is divided by 10. So write $n = 10q + r$ and calculate n^2.)

22. Prove that for every integer n, $n^5 - n$ is divisible by 5. (HINT: Write $n = 5q + r$.)

23. Prove that if $2^{3m-3} \bmod 7 = 1$, then $2^{3m} \bmod 7 = 1$. (HINT: $2^{3m} = 2^3 \cdot 2^{3m-3}$.)

24. Fill in the blanks in the proof below:
Proposition If one person says they can write a number n in the form $5A + B$, where A and B are integers with $0 \le B \le 4$, and another person says they can write the same number in the form $5C + D$, where C and D are integers with $0 \le D \le 4$, show that $B = D$ and $A = C$.
Proof. Let A, B, C, and D be given such that $5A + B = 5C + D$ with $0 \le B \le 4$ and $0 \le D \le 4$. Since _____, $D - B$ is divisible by 5. But
$$\underline{\qquad} \le D - B \le \underline{\qquad}$$
and _____ is the only number in this interval that is divisible by 5, so we can conclude that $B = D$. From this it follows that $5(A - C) = 0$ because _____, from which it follows that $A = C$ as well. ∎

25. Emulate Exercise 24 to prove the following proposition:
Proposition If one person says they can write a number n in the form $7A + B$, where A and B are integers with $0 \le B \le 6$, and another person says they can write the same number in the form $7C + D$, where C and D are integers with $0 \le D \le 6$, show that $B = D$ and $A = C$.

26. We discussed in this section that the rather obvious fact that every integer is either even or odd follows from the division theorem. Let's see how we can use this starting point to prove something less obvious.
(a) What is the contrapositive of the statement "If n^2 is even, then so is n"? Prove it. (You may use the fact that if an integer is not even, then it must be odd.)
(b) Prove that every perfect square is of the form $4k$ or $4k + 1$ for some integer k.
(c) Prove that in Pythagorean triples $a^2 + b^2 = c^2$, if c is even, then so are both a and b.

27. If a and b are both odd integers, show that the polynomial $x^2 + ax + b$ cannot be factored.

28. Prove that no perfect square is of the form $3k + 2$.

29. Prove that if x is an integer greater than 5 such that both $x - 1$ and $x + 1$ are prime, then x is divisible by 6. (NOTE: When two consecutive odd numbers are both prime, they are called twin primes.)

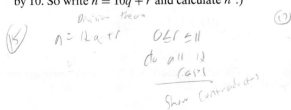

30. Notice that each of

$$1 \cdot 2 + 2 = 4$$
$$1 \cdot 7 + 2 = 9$$
$$2 \cdot 7 + 2 = 16$$

is a perfect square. That is, the product of each pair of integers from the set $\{1, 2, 7\}$ is 2 less than a perfect square. Show that the set cannot be extended to look like $\{1, 2, 7, c\}$ and still have this property. Specifically you should explain why not all of $1 \cdot c + 2, 2 \cdot c + 2$, and $7 \cdot c + 2$ can be perfect squares for any value of c.*

(HINT: Consider the four possible forms $c = 4 \cdot q + r$ from the division theorem, and make use of Proposition 7.)

31. The Fermat numbers are defined by the rule $F_n = 2^{2^n} + 1$. Prove that for all $n \geq 0$, $F_{n+1} = (F_n - 1)^2 + 1$. Explain why it follows that F_n always divides $F_{n+1} - 2$.

32. It has been proven (by Lagrange in 1770) that every positive integer can be written as a sum of four perfect squares. (For example, $23 = 3^2 + 3^2 + 2^2 + 1^2$.) Show that no integer of the form $8n + 7$ can be written as the sum of three perfect squares.

2.3 Mathematical Induction

The word "induction" usually refers to a type of reasoning that is fundamental to the experimental or observational sciences. It is the inference of an event from past events. For example, concluding that the sun will rise tomorrow, based on our experience that it has risen every day before, is an instance of inductive reasoning. It cannot be proven, but it sits well with our experience. Since mathematics is an exact science, this type of reasoning is not mathematical reasoning. So what does the phrase "mathematical induction" mean?

Mathematical induction does have something in common with inductive reasoning in the experimental sciences. When using mathematical induction, we will, for example, use the fact that we know a property is true of the integers 1 through 999 to prove that the same property is true for the integer 1,000. In a way, this is similar to the inductive reasoning of the experimental scientist. The big difference, however, is that we will *prove* the truth for the next integer, not merely *infer* it from a pattern.

Imagine, for example, that one could conclusively prove that "if the sun rises today, then it will rise tomorrow." Then one would have to accept the inference that the sun will rise tomorrow without a doubt. In fact, one would then *prove* that the sun will rise every day forever. This is analogous to what we would like to do in mathematics.

Inductive reasoning in mathematics is not only valid, it is quite natural. For example, suppose that a teacher's idea of a good arithmetic assignment is to have students sum all of the integers from 1 to 100 inclusive. Imagine that you are a student of this teacher, but you have heard about this assignment from your friends in an earlier class. Your friends faithfully tell you that the earlier class all arrived at the answer of 5,050 to the sum $1 + 2 + 3 + \cdots + 100$. When you arrive in your class, your stomach sinks when you see a *different problem* on the board:

$$1 + 2 + 3 + \cdots + 101 = ?$$

What can you do? Of course, you could just start adding up the numbers, but hopefully you will realize that this new problem could have been written as

$$1 + 2 + 3 + \cdots + 100 + 101 = ?$$

---

* This result appeared as the three-page paper "The set of numbers $\{1, 2, 7\}$" in the *Proceedings of the Calcutta Mathematics Society* (v. **72**, pp. 195–97, 1980).

That is, the sum of all the integers from 1 to 101 is the same as the sum from 1 to 100 but with one additional integer added on at the end. Since you are sure your friends' answer was correct, you figure that

$$1 + 2 + 3 + \cdots + 100 + 101 = (1 + 2 + 3 + \cdots + 100) + 101$$
$$= (5{,}050) + 101$$
$$= 5{,}151$$

and get the correct answer, not from a clever method or lightning calculation, but rather from simply using relevant information that has already been carefully checked. In fact, presented in this way, it would seem silly *not* to use the previous information to answer the question at hand.

Induction as a Game

In Section 1.2 we played with number sequences in a game of "What's next?" At the time, we saw that some sequences had descriptions based on both *recurrence relations* (i.e., recursive formulas) and closed formulas. For example, consider the sequence given by the recurrence relation

$$a_k = a_{k-1} + (2k - 1), \quad \text{for } k \geq 2 \text{ and } a_1 = 1$$

If we calculate the first few terms of this sequence, we notice a pattern:

$$a_1 = 1, \quad a_2 = 4, \quad a_3 = 9, \quad a_4 = 16, \quad a_5 = 25, \ldots$$

It *looks like* $a_n = n^2$ for all $n \geq 1$, but how can we be sure that this pattern continues forever?

As with our proofs in Section 2.1, it is AUTHOR's job to convince a skeptical READER that the pattern is true for every positive integer n. So AUTHOR builds Table 2-7 and shows it to READER.

Of course, READER is skeptical as usual and wants to play the game from Section 2.1. As before, the game consists of READER choosing an integer and AUTHOR demonstrating the truth for that number. Now READER is no dummy, and so he does not choose a number already in the table. Suppose READER says, "I choose $n = 50$." Because the sequence of numbers is only described recursively, AUTHOR has to work for a while to compute a_{50}. The benefit is that she has to generate $a_1, a_2, \ldots, a_{49}$ along the way, and so she can check every value of the sequence up to a_{50} as well. In doing this, her table now looks like Table 2-8.

n	Formula for a_n	Value Calculated for a_n	n^2	Is $a_n = n^2$?
1	$a_1 = 1$	1	1	Yes
2	$a_2 = a_1 + (2 \cdot 2 - 1)$	$1 + 3 = 4$	4	Yes
3	$a_3 = a_2 + (2 \cdot 3 - 1)$	$4 + 5 = 9$	9	Yes
4	$a_4 = a_3 + (2 \cdot 4 - 1)$	$9 + 7 = 16$	16	Yes
5	$a_5 = a_4 + (2 \cdot 5 - 1)$	$16 + 9 = 25$	25	Yes

Table 2-7   Checking the First Few Values

n	Formula for a_n	Value Calculated for a_n	n^2	Is $a_n = n^2$?
1	$a_1 = 1$	1	1	yes
2	$a_2 = a_1 + (2 \cdot 2 - 1)$	$1 + 3 = 4$	4	yes
$\vdots$	$\vdots$	$\vdots$	$\vdots$	$\vdots$
49	$a_{49} = a_{48} + (2 \cdot 49 - 1)$	$2{,}304 + 97 = 2{,}401$	2,401	yes
50	$a_{50} = a_{49} + (2 \cdot 50 - 1)$	$2{,}401 + 99 = 2{,}500$	2,500	yes

Table 2-8   After the First 50 Values

But READER is still skeptical, and he will not give up the game. AUTHOR realizes she must find a pattern in what she is doing if she hopes to win.

We can see the pattern if we examine the last row of our table for $n = 50$. It basically states that

$$a_{50} = a_{49} + (2 \cdot 50 - 1) = 2{,}401 + 99 = 2{,}500$$

The "$a_{50} = a_{49} + (2 \cdot 50 - 1)$" part comes from the recursive formula for the sequence, $a_k = a_{k-1} + (2 \cdot k - 1)$. Then the "2,401" part is simply the value calculated for a_{49} in the previous row of the table, which has already been shown to be equal to 49^2. So the formula for a_{50} can really be written as

$$a_{50} = a_{49} + (2 \cdot 50 - 1) = 49^2 + (2 \cdot 50 - 1)$$

This can be generalized as

$$a_m = a_{m-1} + (2 \cdot m - 1) = (m - 1)^2 + (2 \cdot m - 1)$$

We can see this statement is valid for the same reason the statement about a_{50} was valid. The first step, $a_m = a_{m-1} + (2 \cdot m - 1)$, is simply the recursive definition of the sequence. For the second step, we simply replace a_{m-1} with $(m - 1)^2$ because we know that by the time we get to the step that does the calculation for a_m, we will already have done the calculation and verified the proposed closed formula for everything in the sequence from a_1 to a_{m-1}. AUTHOR is ready to write her letter.

Dear READER,

You have already seen that my guess for a closed formula, $a_n = n^2$, is correct for $n = 1, 2, 3, 4$, and 5. You should feel free to check it for as many more values of n as you wish. Let m be the number of the first row that has not yet been checked. Table 2-9 shows the mth row and the one just before it. Keep in mind that the $(m - 1)^{\text{th}}$ row is the last one that you *did* check. To compute the value of a_m, we must use the recursive description, $a_m = a_{m-1} + (2 \cdot m - 1)$. Since we are now filling in row m, we already know that $a_{m-1} = (m - 1)^2$. Substituting this into the formula for a_m, we get

$$a_m = (m - 1)^2 + (2 \cdot m - 1)$$

and a little algebra will show that $(m - 1)^2 + (2 \cdot m - 1)$ simplifies to $(m^2 - 2m + 1) + (2m - 1) = m^2$. This means that $a_m = m^2$ must be true, so we have completed row m of our table.

n	Formula for a_n	a_n simplified	n^2	Is $a_n = n^2$
1	$a_1 = 1$	1	1	Yes
2	$a_2 = a_1 + (2 \cdot 2 - 1)$	$1 + 3 = 4$	4	Yes
3	$a_3 = a_2 + (2 \cdot 3 - 1)$	$4 + 5 = 9$	9	Yes
$\vdots$	$\vdots$	$\vdots$	$\vdots$	$\vdots$
$m-1$	$a_{m-1} = a_{m-2} + (2 \cdot (m-1) - 1)$	$(m-1)^2$	$(m-1)^2$	Yes
m	$a_m = a_{m-1} + (2 \cdot m - 1)$			???

Table 2-9   Illustrating the General Case

I know this is a little more complicated than the letters I've written for our earlier games, but this argument should convince you that no matter how many statements you check, the very next statement must also be true. A little thought will tell you that since the first five statements were explicitly checked, this means that all the statements must be true.

Your exhausted friend,

AUTHOR

Explore more on the Web.

Practice Problem 1   *Try out* AUTHOR'S *reasoning in this proof by answering the following questions:*

(a) *Here is the row in the table for $n = 29$:*

n	Formula for a_n	Value Calculated for a_n	n^2	Is $a_n = n^2$?
29	$a_{29} = a_{28} + (2 \cdot 29 - 1)$	29^2	29^2	Yes

Perform the calculations indicated in AUTHOR'S *letter for $m = 30$. Does the result verify that $a_{30} = 30^2$?*

(b) *Here is the row in the table for $n = 73$:*

n	Formula for a_n	Value Calculated for a_n	n^2	Is $a_n = n^2$?
73	$a_{73} = a_{72} + (2 \cdot 73 - 1)$	73^2	73^2	Yes

Perform the calculations indicated in AUTHOR'S *letter for $m = 74$. Does the result verify that $a_{74} = 74^2$?*

(c) *Imagine a different sequence with the recursive definition*

$$a_n = a_{n-1} + 2 \cdot n, \quad \text{with } a_1 = 2$$

AUTHOR *sets up Table 2-10 for this sequence to investigate her suspicion that the sequence really has the closed formula $a_n = n(n + 1)$.*

   Rewrite the paragraph of AUTHOR'S *letter that begins "To compute the value of $a_m, \ldots$" so it applies to this different sequence.*

n	Formula for a_n	Value Calculated for a_n	$n(n+1)$	Is $a_n = n(n+1)$?
1	$a_1 = 2$	2	$1 \cdot 2$	Yes
2	$a_2 = a_1 + 2 \cdot 2$	$2 + 4 = 6$	$2 \cdot 3$	Yes
$\vdots$	$\vdots$	$\vdots$	$\vdots$	$\vdots$
$m-1$	$a_{m-1} = a_{m-2} + 2 \cdot (m-1)$	$(m-1)m$	$(m-1)m$	Yes
m	$a_m = a_{m-1} + 2 \cdot m$	?	$m(m+1)$	???

Table 2-10   Table for Practice Problem 1

Closed Formulas for Sums

Although READER is stubborn, in the face of flawless logic, he is a good sport. He admits that there is no counterexample, and he is ready for a new challenge. The first example we gave to illustrate inductive thinking involved the sum of the numbers:

$$\underbrace{1 + 2 + 3 + \cdots + n}_{n \text{ terms}}$$

This can be equivalently expressed using the *sigma notation* introduced in Section 1.2 as

$$\sum_{i=1}^{n} i = 1 + 2 + 3 + \cdots + n$$

Now suppose that AUTHOR has a friend who has taken this class before who tells her a shortcut formula for computing this sum:

$$\sum_{i=1}^{n} i = \frac{n \cdot (n+1)}{2}$$

This seems too good to be true, so fresh from her victory in the previous game, AUTHOR sets up a similar table and begins to explore. She immediately realizes that each new sum can be calculated from the sum in the previous row, and uses parentheses to indicate this reasoning. Here is this table through row 34:

n	$\sum_{i=1}^{n} i$ or $1 + 2 + \cdots + n$	Simplified *sum*	$\frac{n(n+1)}{2}$	Is *sum* $= \frac{n(n+1)}{2}$?
1	$\sum_{i=1}^{1} i = 1$	1	$\frac{1 \cdot 2}{2} = 1$	Yes
2	$\sum_{i=1}^{2} i = 1 + 2$	$1 + 2 = 3$	$\frac{2 \cdot 3}{2} = 3$	Yes
3	$\sum_{i=1}^{3} i = (1+2) + 3$	$3 + 3 = 6$	$\frac{3 \cdot 4}{2} = 6$	Yes
4	$\sum_{i=1}^{4} i = (1+2+3) + 4$	$6 + 4 = 10$	$\frac{4 \cdot 5}{2} = 10$	Yes
5	$\sum_{i=1}^{5} i = (1+2+3+4) + 5$	$10 + 5 = 15$	$\frac{5 \cdot 6}{2} = 15$	Yes
$\vdots$	$\vdots$	$\vdots$	$\vdots$	$\vdots$
34	$\sum_{i=1}^{34} i = (1+2+\cdots+33) + 34$	$561 + 34 = 595$	$\frac{34 \cdot 35}{2} = 595$	Yes

*Explore more on
the Web.*

Practice Problem 2   *Answer the following questions about subsequent rows of this table:*

(a) *What will row 35 of the table above look like?*

(b) *Assuming that the pattern continues, row 66 will be as shown below. What will row 67 look like?*

n	$\sum_{i=1}^{n} i$ or $1 + 2 + \cdots + n$	Simplified *sum*	$\frac{n(n+1)}{2}$	Is *sum* $= \frac{n(n+1)}{2}$?
66	$\sum_{i=1}^{66} i = (1 + 2 + \cdots + 65) + 66$	$2{,}145 + 66 = 2{,}211$	$\frac{66 \cdot 67}{2} = 2{,}211$	Yes

As before, AUTHOR looks for a pattern in what she is doing. She visualizes two rows, the last one that she has already checked, and the next one to be checked. Once again, she labels the one she is about to check as row m:

n	$\sum_{i=1}^{n} i$ or $1 + 2 + \cdots + n$	Simplified *sum*	$\frac{n(n+1)}{2}$	Is *sum* $= \frac{n(n+1)}{2}$?
$\vdots$	$\vdots$	$\vdots$	$\vdots$	$\vdots$
$m - 1$	$\sum_{i=1}^{m-1} i = 1 + 2 + \cdots + (m-1)$	$\frac{(m-1) \cdot m}{2}$	$\frac{(m-1) \cdot m}{2}$	Yes
m	$\sum_{i=1}^{m} i = (1 + 2 + \cdots + (m-1)) + m$	???	$\frac{m(m+1)}{2}$	???

Now she uses the fact that the sum up to m is the same as the sum to $m - 1$ plus one more number (m), and *the fact that row $m - 1$ has already been checked*, to write

$$\sum_{i=1}^{m} i = 1 + 2 + \cdots + m$$

$$= (1 + 2 + \cdots + (m-1)) + m$$

$$= \frac{(m-1) \cdot m}{2} + m \quad \text{(from row } m - 1)$$

$$= \frac{(m-1) \cdot m}{2} + \frac{2m}{2} \quad \text{(common denominator)}$$

$$= \frac{m(m-1) + 2m}{2}$$

$$= \frac{m(m - 1 + 2)}{2} \quad \text{(factor out } m)$$

$$= \frac{m(m+1)}{2}$$

She is now ready to write her letter, which will have to wait until a bit later in the section. However, before we move on, we should check some examples to see that her logic is sound for any possible positive integer.

 Example 1   *Apply the AUTHOR's logic to these specific situations:*

1. *Having checked rows 1 to 97, and in particular knowing that $1 + 2 + \cdots + 97 = \frac{97 \cdot 98}{2}$, verify that the conclusion is true for $m = 98$. That is, verify that $1 + 2 + \cdots + 98 = \frac{98 \cdot 99}{2}$.*

2. *Do the same for $m = 146$. That is, use the fact that $1 + 2 + \cdots + 145 = \frac{145 \cdot 146}{2}$ to verify that $1 + 2 + \cdots + 146 = \frac{146 \cdot 147}{2}$.*

3. *Do the same for $m = 1{,}341$.*

SOLUTION

1. $1 + 2 + \cdots + 98 = (1 + 2 + \cdots + 97) + 98 = \frac{97 \cdot 98}{2} + 98 = \frac{97 \cdot 98}{2} + \frac{2 \cdot 98}{2} = \frac{98 \cdot (97 + 2)}{2} = \frac{98 \cdot 99}{2}$

2. $1 + 2 + \cdots + 146 = (1 + 2 + \cdots + 145) + 146 = \frac{145 \cdot 146}{2} + 146 = \frac{145 \cdot 146}{2} + \frac{2 \cdot 146}{2} = \frac{146 \cdot (145 + 2)}{2} = \frac{146 \cdot 147}{2}$

3. $1 + 2 + \cdots + 1{,}341 = (1 + 2 + \cdots + 1{,}340) + 1{,}341 = \frac{1{,}340 \cdot 1{,}341}{2} + 1{,}341 = \frac{1{,}340 \cdot 1{,}341}{2} + \frac{2 \cdot 1{,}341}{2} = \frac{1{,}341 \cdot (1{,}340 + 2)}{2} = \frac{1{,}341 \cdot 1{,}342}{2}$ □

Formal Proofs by Mathematical Induction

Now that we have informally discussed this form of reasoning, let us move toward formalizing the underlying principle as we will use it in our proofs. First of all, there are specific types of statements to which this form of reasoning can be applied. The underlying issue is that there must be some way to order an infinite family of statements so that there is a first statement, a second statement, and so forth and so that for every statement proved, there is a next statement to consider. We can formally guarantee this by insisting that our statements be given in a way that makes it easy to "index" them with positive integers.

> **Definition**   A *statement about the positive integers* is a predicate $P(n)$ with the set of positive integers as its domain. That is, when any positive integer is substituted for n in statement $P(n)$, the result is a proposition that is unambiguously either true or false.

For example, "n is even," "$2^n - 1$ is prime," and "every single-elimination tournament with n players must have $n - 1$ games" are all statements about the positive integers. On the other hand, "$n^2 + 1$," for example, is not a statement—when one substitutes 7 for n one gets "50," which is, of course, a number and neither true nor false.

Example 2 *Which of the following are statements about the positive integers, according to the definition given above?*

1. *$n^2 + n$ is even.*

2. *$100 - n$*

3. *$100 - n > 83$*

4. *John has fewer than n apples in his refrigerator.*

SOLUTION   All but #2 are valid statements. When we substitute 12, for example, into "$100 - n$," we get "$100 - 12$," which is neither true nor false.   □

Example 3 *For each of the predicates below, write the corresponding sentences when $n = 2$ and $n = 30$, and decide which are true propositions.*

1. $E(n)$ *is the statement* "$n^2 + n$ *is even.*"
2. $G(n)$ *is the statement* "$100 - n > 83$.*"
3. $S(n)$ *is the statement* "$1 + 2 + \cdots + n = \frac{n(n+1)}{2}$.*"

SOLUTION

1. $E(2)$ is the statement "$2^2 + 2$ is even." $E(30)$ is the statement "$30^2 + 30$ is even." Both are true statements.
2. $G(2)$ is the statement "$100 - 2 > 83$." This is true. $G(30)$ is the statement "$100 - 30 > 83$." This is false.
3. $S(2)$ is the statement "$1 + 2 = \frac{2(2+1)}{2}$." $S(30)$ is the statement "$1 + 2 + \cdots + 30 = \frac{30(30+1)}{2}$." Using our $\sum$ notation, we could also write "$\sum_{i=1}^{30} i = \frac{30(30+1)}{2}$." Both $S(2)$ and $S(30)$ are true.

□

Practice Problem 3

(a) $P(n)$ is "If there are n students in the class, the room will be too small." What is $P(35)$? What is $P(m-1)$?

(b) $S(n)$ is "$1 + 4 + 9 + \cdots + n^2 = \frac{n(n+1)(2n+1)}{6}$." Rewrite $S(n)$ using $\sum$ notation. Then write $S(1)$, $S(2)$, $S(3)$, and $S(4)$, and decide if they are true. Also write $S(m-1)$ and simplify it.

(c) Consider the sequence defined recursively as $a_1 = 11$, and $a_k = a_{k-1} + 4$, and let $R(n)$ be the statement "$a_n = 4n + 7$." Write $R(1)$, $R(2)$, $R(3)$, and $R(4)$, and decide if they are true. Also write $R(m-1)$ and simplify it.

The simple idea of having the statements we wish to prove given in some order allows us to use inductive reasoning. This form of reasoning is central to many concepts in discrete math, so it is of the utmost importance in this course that everyone have the correct mental picture of how inductive reasoning works. Before we give some examples, we will state the underlying principle. This principle is stated in the terminology of a generic predicate $P(n)$ over the domain of positive integers.

The Principle of Mathematical Induction　　Let $P(n)$ be a statement about the positive integers. If one can prove that

 (i) $P(1)$ is true, and that
 (ii) for every integer $m \geq 2$, whenever $P(1), P(2), \ldots, P(m-1)$ have all been checked to be true, it follows that $P(m)$ is true,

then we can conclude that $P(n)$ is true for every positive integer n.

This precise statement might seem to be a long way from the intuition in the discussion leading up to it, but it is a very natural principle. Imagine a long line of dominoes numbered with the positive integers in sequence, and statement $P(n)$ is "Domino n falls." In this analogy, condition (ii) essentially says that all the dominoes are spaced so that if all previous dominoes fall, they knock down the next one. Clearly, if that condition holds of all the dominoes and someone tips the first domino (which is condition (i)), then we can reasonably conclude that all the dominoes will fall.

Note that technically there is no reason for the first statement to correspond to $n = 1$—all that matters is that there is a first statement that is checked explicitly to be true. In reality that statement might be called $P(0)$, $P(5)$, or something else. The conclusion of mathematical induction is that every statement is true from that point on.

To illustrate the principle stated in this way, we revisit an earlier example and give a formal proof.

Proposition 1   *For every positive integer n,*

$$\sum_{i=1}^{n} i = \frac{n(n+1)}{2}$$

PROOF   (By induction on n) Let $S(n)$ be the predicate "$\sum_{i=1}^{n} i = \frac{n(n+1)}{2}$." It is easy to check that statement $S(1)$, which states "$1 = \frac{1(1+1)}{2}$," is true. Now let the integer $m \geq 2$ be given such that the statements $S(1), \ldots, S(m-1)$ have already been checked to be true. We now consider statement $S(m)$:

$$\sum_{i=1}^{m} i = 1 + 2 + 3 + \cdots + m$$
$$= (1 + 2 + 3 + \cdots (m-1)) + m$$
$$= \frac{(m-1)m}{2} + m, \quad \text{since } S(m-1) \text{ is true}$$
$$= \frac{(m-1)m}{2} + \frac{2m}{2}$$
$$= \frac{m((m-1)+2)}{2}$$
$$= \frac{m(m+1)}{2}$$

This shows that $S(m)$ is true, completing the induction.   ■

Explore more on the Web.

Practice Problem 4   *Let $P(n)$ be the statement "$\sum_{i=1}^{n} 2^{i-1} = 2^n - 1$." By imitating Proposition 1, give a logically convincing reason that $P(n)$ is true for every positive integer n.*

Our final example of this section revisits the problem on recursively defined sequences in the first letter from AUTHOR to READER in this section. We give a semiformal proof, midway between the "Letter to READER" and the terse proof of the preceding proposition.

Proposition 2   *Let the sequence of numbers a_n be given by the recurrence relation $a_k = a_{k-1} + (2k - 1)$ for $k \geq 2$ and $a_1 = 1$. Show that $a_n = n^2$ for all $n \geq 1$.*

PROOF   (By induction on n) Consider the statement "$a_n = n^2$." We are given that $a_1 = 1$. Clearly, the first statement $a_1 = 1^2$ is true. We check a few of the initial statements, then imagine we are about to check statement m. That is, we imagine that everything in Table 2-11 above the line has been checked. Now we simply calculate a_m according to the recursive formula, *using the fact that row $m-1$ has already been checked*,

$$
\begin{aligned}
a_m &= a_{m-1} + (2 \cdot m - 1) \\
&= (m-1)^2 + (2m-1), \quad \text{using "} a_{m-1} = (m-1)^2 \text{"} \\
&= (m^2 - 2m + 1) + (2m - 1) \\
&= m^2 - 2m + 1 + 2m - 1 \\
&= m^2
\end{aligned}
$$

We therefore have shown that $a_m = m^2$, which is just what we need to complete row m of our table. We have thus shown how knowing the truth of statements up to "$a_{m-1} = (m-1)^2$" (and a little algebra) leads to knowing the truth of the next statement, "$a_m = m^2$." This establishes the truth of all statements "$a_n = n^2$" for all $n \geq 1$ by mathematical induction.   ■

Mathematical induction gives us a rigorous way to show that a number sequence, for which we know a recursive description, has a particular closed description, as well. See Exercise 3 for some practice with this idea.

Explore more on the Web.

Practice Problem 5   *Let the sequence of numbers a_n be given by the recurrence relation $a_k = a_{k-1} + 2k$ for $k \geq 2$ and $a_1 = 2$. Show that $a_n = n(n+1)$ for all $n \geq 1$.*

Strong Versus Weak Induction

According to many mathematics textbooks, the form of induction we have presented in this section is called *strong induction* because we are explicitly imagining having all previous statements $P(1), P(2), \ldots, P(m-1)$ available for use when we are ready to prove statement $P(m)$. In our actual sample proofs, you might have noticed that even though we had all the previous statements available, we only *used* the immediately prior statement $P(m-1)$. In the proof, if we had assumed only that statement

n	a_n	"$a_n = n^2$"	True?
1	$a_1 = 1$	"$a_1 = 1^2$"	Yes
2	$a_2 = a_1 + (2 \cdot 2 - 1) = 1 + 3 = 4$	"$a_2 = 2^2$"	Yes
3	$a_3 = a_2 + (2 \cdot 3 - 1) = 4 + 5 = 9$	"$a_3 = 3^2$"	Yes
4	$a_4 = a_3 + (2 \cdot 4 - 1) = 9 + 7 = 16$	"$a_4 = 4^2$"	Yes
5	$a_5 = a_4 + (2 \cdot 5 - 1) = 16 + 9 = 25$	"$a_5 = 5^2$"	Yes
$\vdots$	and so on, up to	$\vdots$	$\vdots$
$m-1$	$a_{m-1} = a_{m-2} + 2 \cdot ((m-1) - 1)$	"$a_{m-1} = (m-1)^2$"	Yes
m	$a_m = a_{m-1} + (2 \cdot m - 1)$	"$a_m = m^2$"	???

Table 2-11   Proof of Proposition 2

$P(m-1)$ had been checked, then the proof would still be valid, but some math textbooks would consider this a use of *weak induction*. These names are unfortunate since the two argument structures are equally valid. In this book, we will use only the so-called strong form of induction. This allows us to maintain our mental image of the READER checking all statements starting from the first one. In addition, we will encounter proofs (Exercise 12 in this section, e.g.) that cannot be done with the weak form, and it is simpler to consistently use the form that is more widely applicable.

Solutions to Practice Problems

1 (a) $a_{30} = a_{29} + (2 \cdot 30 - 1) = 29^2 + (2 \cdot 30 - 1) = 841 + 59 = 900$, and this is the same as 30^2.

   (b) $a_{74} = a_{73} + (2 \cdot 74 - 1) = 73^2 + (2 \cdot 74 - 1) = 5{,}329 + 147 = 5{,}476$, and this is the same as 74^2.

   (c) The formula for a_m is given by $a_m = a_{m-1} + 2 \cdot m$. By the time I'm filling in row m, all the rows from 1 to $m-1$ will already be done, so I'll already know that a_{m-1} is equal to $(m-1)m$. Substituting this for a_{m-1}, I'll get $a_m = (m-1)m + 2m$, and a little algebra will show that this simplifies to $(m-1)m + 2m = m^2 - m + 2m = m^2 + m = m(m+1)$.

2 We show both answers in Table 2-12.

3 (a) $P(35)$ is "If there are 35 students in the class, the room will be too small." $P(m-1)$ is "If there are $m-1$ students in the class, the room will be too small."

   (b) If we use $\sum$ notation, $S(n)$ is "$\sum_{i=1}^{n} i^2 = \frac{n(n+1)(2n+1)}{6}$." $S(1)$ is "$1 = \frac{1(1+1)(2\cdot1+1)}{6}$." $S(2)$ is "$1 + 4 = \frac{2(2+1)(2\cdot2+1)}{6}$." $S(3)$ is "$1 + 4 + 9 = \frac{3(3+1)(2\cdot3+1)}{6}$." $S(4)$ is "$1 + 4 + 9 + 16 = \frac{4(4+1)(2\cdot4+1)}{6}$." All four

are true. $S(m-1)$ is "$1 + 4 + 9 + \cdots + (m-1)^2 = \frac{(m-1)(m)(2m-1)}{6}$," or "$\sum_{i=1}^{m-1} i^2 = \frac{(m-1)(m)(2m-1)}{6}$."

   (c) $R(1)$ is "$a_1 = 4 \cdot 1 + 7$." $R(2)$ is "$a_2 = 4 \cdot 2 + 7$." $R(3)$ is "$a_3 = 4 \cdot 3 + 7$." $R(4)$ is "$a_4 = 4 \cdot 4 + 7$." Since (according to the recursive definition) the sequence begins with $11, 15, 19, 23$, all four are true. $R(m-1)$ is "$a_{m-1} = 4 \cdot (m-1) + 7$."

4 So $P(n)$ is "$2^0 + 2^1 + 2^2 + \cdots + 2^{n-1} = 2^n - 1$." Table 2-13 shows the inductive reasoning for the proof. Here is the calculation that uses $P(m-1)$ to verify $P(m)$:

$$2^0 + 2^1 + 2^2 + \cdots + 2^{m-1}$$
$$= \left(2^0 + 2^1 + 2^2 + \cdots + 2^{m-2}\right) + 2^{m-1}$$
$$= (2^{m-1} - 1) + 2^{m-1}, \quad \text{(since P(m-1)}$$
$$\text{was already checked)}$$
$$= 2^{m-1} + 2^{m-1} - 1$$
$$= 2 \cdot 2^{m-1} - 1$$
$$= 2^m - 1$$

5 Consider the statement "$a_n = n(n+1)$." Table 2-14 shows the inductive reasoning for the proof. Here is the

n	sum $(= 1 + 2 + \cdots + n)$	sum	$\frac{n(n+1)}{2}$	Is sum $= \frac{n(n+1)}{2}$?
35	$(1 + 2 + \cdots + 34) + 35$	$595 + 35 = 630$	$\frac{35 \cdot 36}{2} = 630$	Yes
67	$(1 + 2 + \cdots + 66) + 67$	$2{,}211 + 67 = 2{,}278$	$\frac{67 \cdot 68}{2} = 2{,}278$	Yes

Table 2-12   Solution to Practice Problem 2

n	$P(n)$	$P(n)$ Simplified	True?
1	"$2^0 = 2^1 - 1$"	"$1 = 2 - 1$"	Yes
2	"$2^0 + 2^1 = 2^2 - 1$"	"$1 + 2 = 4 - 1$"	Yes
3	"$2^0 + 2^1 + 2^2 = 2^3 - 1$"	"$1 + 2 + 4 = 8 - 1$"	Yes
4	"$2^0 + 2^1 + 2^2 + 2^3 = 2^4 - 1$"	"$1 + 2 + 4 + 8 = 16 - 1$"	Yes
5	"$2^0 + 2^1 + \cdots + 2^4 = 2^5 - 1$"	"$1 + 2 + \cdots + 16 = 32 - 1$"	Yes
$\vdots$	and so on, up to	$\vdots$	$\vdots$
$m-1$	"$2^0 + 2^1 + \cdots + 2^{m-2} = 2^{m-1} - 1$"	"$1 + 2 + \cdots + 2^{m-2} = 2^{m-1} - 1$"	Yes
m	"$2^0 + 2^1 + \cdots + 2^{m-1} = 2^m - 1$"	"$1 + 2 + \cdots + 2^{m-1} = 2^m - 1$"	???

Table 2-13   Solution to Practice Problem 4

n	a_n	"$a_n = n(n+1)$"	True?
1	$a_1 = 2$	"$a_1 = 1 \cdot (1+1)$"	Yes
2	$a_2 = a_1 + 2 \cdot 2 = 2 + 4 = 6$	"$a_2 = 2 \cdot (2+1)$"	Yes
3	$a_3 = a_2 + 2 \cdot 3 = 6 + 6 = 12$	"$a_3 = 3 \cdot (3+1)$"	Yes
4	$a_4 = a_3 + 2 \cdot 4 = 12 + 8 = 20$	"$a_4 = 4 \cdot (4+1)$"	Yes
5	$a_5 = a_4 + 2 \cdot 5 = 20 + 10 = 30$	"$a_5 = 5 \cdot (5+1)$"	Yes
$\vdots$	and so on, up to	$\vdots$	$\vdots$
$m-1$	$a_{m-1} = a_{m-2} + 2 \cdot (m-1)$	"$a_{m-1} = (m-1)m$"	Yes
m	$a_m = a_{m-1} + 2 \cdot m$	"$a_m = m(m+1)$"	???

Table 2-14 Solution to Practice Problem 5

calculation that verifies the truth of "$a_m = m(m+1)$."

$$a_m = a_{m-1} + 2 \cdot m$$
$$= (m-1)m + 2m, \quad \text{using "}a_{m-1} = (m-1)(m)\text{"}$$
$$= m^2 - m + 2m$$
$$= m^2 + m$$
$$= m(m+1)$$

Exercises for Section 2.3

1. **(a)** If $P(n)$ is "$n^2 + 1$ is prime," write $P(1)$, $P(2)$, and $P(12)$. Which if any are true? Also, write $P(m-1)$.

   (b) If $L(n)$ is "$n^2 < 2^n$," write $L(1)$, $L(2)$, $L(3)$, $L(4)$, $L(5)$, and $L(6)$. Which if any are true? Also, write $L(m-1)$.

   (c) If $S(n)$ is "$\sum_{i=1}^{n} i^2 = \frac{n(n+1)(2n+1)}{6}$," write $S(1)$, $S(2)$, $S(3)$, $S(4)$, $S(5)$, and $S(6)$. Which if any are true? Also, write $S(m-1)$ and simplify it.

   (d) If $P(n)$ is "n can be written as the product of exactly two different prime numbers," write $P(4)$, $P(5)$, and $P(6)$. Which if any are true? Also, write $P(m-1)$.

   (e) If $J(n)$ is "If the Josephus game of Section 1.1 starts with 2^n people, and every second person is killed, then the person in position 1 is the last survivor," write $J(1)$, $J(2)$, $J(3)$, and $J(4)$. Which if any are true? Also, write $J(m-1)$.

2. For the sequence given recursively by $a_k = 2a_{k-1} - 3$ for $k \geq 2$, where $a_1 = 4$, someone has guessed a closed form and used the predicate $R(n)$ for the statement "$a_n = 2^{n-1} + 3$."

   (a) Are $R(1)$, $R(2)$, $R(3)$, and $R(4)$ true?

   (b) Write $R(8)$ and $R(9)$.

   (c) Suppose that Amanda has already verified $R(8)$. Show how to verify $R(9)$ without starting all over.

3. In Proposition 2, we showed how to argue that a recursively defined number sequence is equally well described by a given closed formula. For each of the sequences defined recursively below, prove that the given closed formula is correct. As part of each proof, verify the given statement for at least $n = 1$, $n = 2$, $n = 3$, and $n = 4$.

   (a) Show that the sequence defined by $a_k = a_{k-1} + 4$ for $k \geq 2$, where $a_1 = 1$, is equivalently described by the closed formula $a_n = 4n - 3$.

   (b) Show that the sequence defined by $a_k = a_{k-1} + (k+4)$ for $k \geq 2$, where $a_1 = 5$, is equivalently described by the closed formula $a_n = \frac{n(n+9)}{2}$.

   (c) Show that the sequence defined by $a_k = a_{k-1} + k^2$ for $k \geq 2$, where $a_1 = 1$, is equivalently described by the closed formula $a_n = \frac{n(n+1)(2n+1)}{6}$.

   (d) Show that the sequence defined by $a_k = 2a_{k-1} + 1$ for $k \geq 2$, where $a_1 = 1$, is equivalently described by the closed formula $a_n = 2^n - 1$.

   (e) Show that the sequence defined by $a_k = a_{k-1} + \frac{1}{(k)(k+1)}$ for $k \geq 2$, where $a_1 = \frac{1}{2}$, is equivalently described by the closed formula $a_n = \frac{n}{n+1}$.

   (f) Show that the sequence defined by $a_k = 2a_{k-1} - 3$ for $k \geq 2$, where $a_1 = 4$, is equivalently described by the closed formula $a_n = 2^{n-1} + 3$.

4. For each of the sequences defined recursively below, use mathematical induction to prove that the given closed formula is correct.

(a) Show that the sequence defined by $b_k = b_{k-1} + 2^k$ for $k \geq 2$, where $b_1 = 4$, is equivalently described by the closed formula $b_n = 2^{n+1}$.

(b) Show that the sequence defined by $b_k = 3b_{k-1}$ for $k \geq 2$, where $b_1 = 6$, is equivalently described by the closed formula $b_n = 2 \cdot 3^n$.

(c) Show that the sequence defined by $b_k = 3b_{k-1} + 2$ for $k \geq 2$, where $b_1 = 2$, is equivalently described by the closed formula $b_n = 3^n - 1$.

(d) Show that the sequence defined by $b_k = 4b_{k-1} + 3$ for $k \geq 2$, where $b_1 = 3$, is equivalently described by the closed formula $b_n = 2^{2n} - 1$.

(e) Show that the sequence defined by $b_k = 2b_{k-1} + k$ for $k \geq 2$, where $b_1 = 3$, is equivalently described by the closed formula $b_n = 3 \cdot 2^n - n - 2$.

5. Guess a closed formula, in terms of n, for the sequence given recursively by $a_1 = 1$, and for $k \geq 2$, $a_k = a_{k-1} + (2k - 3)$. Use induction to prove your guess is correct. HINT: Write the first eight terms of the sequence. How do the terms relate to the sequence $1, 4, 9, 16, 25, 36, \ldots$?

6. Suppose $S(n)$ is the statement $\sum_{i=1}^{n} \frac{i(i+1)}{2} = \frac{n(n+1)(n+2)}{6}$. Rewrite $S(n)$ without using sigma notation. Write $S(1)$, $S(2)$, $S(3)$, and $S(4)$ and simplify each to see if it is true. Write $S(m-1)$ without using sigma notation.

7. If $G(n)$ is the statement "$\sum_{i=2}^{2n} \frac{1}{i} \geq \frac{n}{2}$," write $G(n)$ without using sigma notation. Then write $G(1)$, $G(2)$, $G(3)$, and $G(4)$ without using sigma notation.

8. Use induction to prove each of the following. As part of your proof, write and verify each statement for at least $n = 1$, $n = 2$, $n = 3$, and $n = 4$.

(a) $\sum_{i=1}^{n} (2 \cdot i - 1) = n^2$ for each $n \geq 1$.

(b) $\sum_{i=1}^{n} (2 \cdot i + 4) = n^2 + 5n$ for each $n \geq 1$.

(c) $\sum_{i=1}^{n} (2^i - 1) = 2^{n+1} - n - 2$ for each $n \geq 1$.

(d) $2(\sum_{i=1}^{n} 3^{i-1}) = 3^n - 1$ for each $n \geq 1$.

(e) $\sum_{i=1}^{n} \frac{1}{2^i} = 1 - \frac{1}{2^n}$ for each $n \geq 1$.

(f) $\sum_{i=1}^{n} \frac{1}{(i)(i+1)} = \frac{n}{n+1}$ for all $n \geq 1$

(g) $\sum_{i=1}^{n} \frac{1}{(2i-1)(2i+1)} = \frac{n}{2n+1}$ for all $n \geq 1$.

9. Use mathematical induction to prove each of the following:

(a) Prove by induction that for all positive integers n,

$$1 + 3 + 6 + 10 + \cdots + \frac{n(n+1)}{2}$$
$$= \frac{n(n+1)(n+2)}{6}$$

(b) Prove by induction that for all natural numbers $n \geq 1$,

$$1(3) + 2(4) + 3(5) + \cdots + n(n+2)$$
$$= \frac{n(n+1)(2n+7)}{6}$$

10. Write each of the equations in Exercise 9 using sigma notation.

11. Guess a formula in terms of n, for

$$(1)(2^1) + (2)(2^2) + (3)(2^3) + \cdots + (n)(2^n)$$

and use induction to prove that your guess is correct.

12. Show that the sequence defined by $a_k = a_{k-1} + 2a_{k-2}$ for $k \geq 3$, where $a_1 = 1$ and $a_2 = 2$, is equivalently described by the closed formula $a_n = 2^{n-1}$. (Notice that because the recursive formula starts at $k = 3$ you will have to at least check the statements for $n = 1$ and $n = 2$ before considering the induction step. Also notice that when you verify $a^m = 2^{m-1}$, you will need to use the closed formulas for a_{m-1} and a_{m-2} that have already been verified.)

13. Let a real number $x \neq 1$ be given. Prove by induction that for all positive integers n,

$$1 + x + x^2 + \cdots + x^n = \frac{x^{n+1} - 1}{x - 1}$$

14. Explain how Exercise 13 implies that when x is a number between 0 and 1, the more terms of

$$1 + x + x^2 + x^3 + \cdots$$

you add, the closer the sums get to $\frac{1}{1-x}$.

2.4 More About Induction

In this section we explore induction in a little more depth, illustrating the wide variety of problems for which induction is an effective tool. Even though our proofs in this section will be written more formally, keep in mind that the formal exposition is simply a compressed version of the reasoning process we outlined in the preceding section.

You may have noticed that both types of problems in the preceding section, sums and recursive sequences, had a certain common "feel." In both situations, the calculations to verify statement $P(m)$ used the result from $P(m-1)$ and some simple algebra. Our first objective in this section is to explore this specific connection.

Sums as Recursive Sequences

Consider the proof from Proposition 2 of Section 2.3, that the recursive sequence $a_1 = 1, a_k = a_{k-1} + (2k - 1)$ has closed formula $a_n = n^2$. In proving "$a_m = m^2$," we wrote

$$
\begin{aligned}
a_m &= a_{m-1} + (2m - 1) \\
&= (m - 1)^2 + (2m - 1), \quad \text{using "} a_{m-1} = (m - 1)^2 \text{"} \\
&= m^2 - 2m + 1 + 2m - 1 \\
&= m^2
\end{aligned}
$$

The first step used the relationship between a_m and a_{m-1}, while the second step used the fact that "$a_{m-1} = (m - 1)^2$." Now consider the proof from Proposition 1 of Section 2.3, that the sum $1 + 2 + 3 + \cdots + n$ has the closed formula $\frac{n(n+1)}{2}$. In proving the statement $S(m)$, we wrote

$$
\begin{aligned}
1 + 2 + 3 + \cdots + m &= (1 + 2 + 3 + \cdots + (m - 1)) + m \\
&= \frac{(m - 1)(m)}{2} + m, \quad \text{using } S(m - 1) \\
&= \frac{m^2 - m}{2} + \frac{2m}{2} \\
&= \frac{m^2 + m}{2} \\
&= \frac{m(m + 1)}{2}
\end{aligned}
$$

Once again, the first step established the relationship between statements $S(m - 1)$ and $S(m)$, and the second step used the information. It is no coincidence that this is exactly what happened in the other proof. Sums are recursive by nature—the sum of 101 numbers is simply the sum of the first 100 numbers plus the last number. Let's see how to make this connection explicit. In the examples that follow, we use s_n rather than a_n for the sequences to suggest that s_n is defined by a sum.

Example 1 *Consider the sum $\sum_{i=1}^{n}(2i - 1)$, which is the same as $1 + 3 + 5 + \cdots + (2n - 1)$. Let us use the notation s_n to denote this sum. For example, s_5 means $1 + 3 + 5 + 7 + 9$. Find a recursive description of s_n.*

SOLUTION   The key to the solution is realizing that s_{n-1} means the sum of the first $n - 1$ terms, and that to sum the first n terms, we sum the first $n - 1$ terms and then add the final term (i.e., the term when $i = n$). In symbols,

$$
\begin{aligned}
s_n &= \sum_{i=1}^{n}(2i - 1) \\
&= 1 + 3 + 5 + \cdots + (2n - 1) \\
&= (1 + 3 + 5 + \cdots + (2n - 3)) + (2n - 1) \\
&= s_{n-1} + (2n - 1)
\end{aligned}
$$

So $s_n = s_{n-1} + (2n - 1)$ is the recurrence relation. However, this by itself does not determine the sequence of numbers—we must also provide a starting value. This comes from the somewhat silly case, $s_1 = 1$, since the sum $\sum_{i=1}^{1}(2i - 1)$ is 1. □

Practice Problem 1 *Let $s_n = \sum_{i=1}^{n}(4i - 1)$. Find a recursive description of s_n.*

If we write a sum recursively as in Example 1, it makes it somewhat easier to write an induction proof about the sum. In Example 1, we showed that the sum $s_n = \sum_{i=1}^{n}(2i - 1)$ is described by the recurrence relation $s_n = s_{n-1} + (2n - 1)$ with $s_1 = 1$, but in Proposition 2 of Section 2.3, we proved that the sequence with this recursive description satisfies the closed formula $s_n = n^2$. So in a roundabout way, we have proved

$$\sum_{i=1}^{n}(2i - 1) = n^2$$

Perhaps it would be clearer to see another example of this process all in one place.

Example 2 *Show that $\sum_{i=1}^{n}\frac{2}{3^i} = 1 - \frac{1}{3^n}$ for every positive integer n.*

PROOF   (By induction on n) Let s_n represent the sum $\sum_{i=1}^{n}\frac{2}{3^i}$. Then s_n has the recursive description $s_n = s_{n-1} + \frac{2}{3^n}$ with $s_1 = \frac{2}{3}$. So instead of the original statement, we will prove that

$$s_n = 1 - \frac{1}{3^n}$$

It is easy to see that the first statement, "$s_1 = 1 - \frac{1}{3}$," is true. Now let a positive integer m be given such that the statements up to "$s_{m-1} = 1 - \frac{1}{3^{m-1}}$" have been checked to be true. Now considering the next term s_m, we see that

$$s_m = s_{m-1} + \frac{2}{3^m}, \quad \text{from the recurrence relation}$$

$$= \left(1 - \frac{1}{3^{m-1}}\right) + \frac{2}{3^m}, \quad \text{since "}s_{m-1} = 1 - \frac{1}{3^{m-1}}\text{"}$$

$$= 1 - \frac{3}{3^m} + \frac{2}{3^m}$$

$$= 1 - \frac{1}{3^m}$$

So we have verified the truth of the statement

$$s_m = 1 - \frac{1}{3^m}$$

which is the correct closed form for the next term.   ■

Practice Problem 2 *Emulate Example 2 to show that $\sum_{i=1}^{n}(4i - 1) = n \cdot (2n + 1)$ for every positive integer n.*

There is no logical necessity in rewriting a sum as a recurrence relation, but it is an intriguing idea. It turns out that finding nice formulas for sums is really just a special case of finding closed formulas for recurrence relations. There has been a great deal of work done in mathematics to develop tools to do this. You will get an introduction to some of these ideas in Section 5.6. For now, understanding how to *prove* a closed formula is correct is an essential first step.

Other Uses of Induction

There are times when it is either very difficult or very messy to find a closed formula for a recursively defined sequence of numbers. In many of these cases, we might not even need a closed formula, but we do need to establish that the sequence has some particular property. Mathematical induction gives us the tool to prove properties of a recursively defined sequence of numbers directly without finding a closed formula first.

Example 3 *Let the number sequence $\{a_n\}$ be defined by the recurrence relation $a_n = 2a_{n-1} + a_{n-2}$, where $a_1 = 5$ and $a_2 = 10$. Use induction to show that for all $n \geq 3$, $a_n < 3^n$.*

PROOF   (By induction on n) Consider the statement "$a_n < 3^n$." The first several terms of the sequence described are

$$5, 10, 25, 60, \ldots$$

so it is easy to verify that "$a_3 < 3^3$" and "$a_4 < 3^4$."

Now let $m \geq 5$ be given such that all these inequalities up to "$a_{m-1} < 3^{m-1}$" have already been checked, and now consider the next statement:

$$
\begin{aligned}
a_m &= 2a_{m-1} + a_{m-2}, && \text{by definition of the sequence} \\
&< 2 \cdot 3^{m-1} + a_{m-2}, && \text{since "}a_{m-1} < 3^{m-1}\text{" is true} \\
&< 2 \cdot 3^{m-1} + 3^{m-2}, && \text{since "}a_{m-2} < 3^{m-2}\text{" is true} \\
&< 2 \cdot 3^{m-1} + 3^{m-1}, && \text{since } 3^{m-2} < 3^{m-1} \\
&= 3 \cdot 3^{m-1} \\
&= 3^m
\end{aligned}
$$

This shows that the statement "$a_m < 3^m$" is true, completing the induction.   ∎

You may have noticed that in this example we first checked the statement for $n = 3$ instead of for $n = 1$. This is simply because the statement being proved, "$a_n < 3^n$," is only asserted true for $n \geq 3$—in other words, "from $n = 3$ on." In terms of our domino analogy, this is like having the "dominoes" properly spaced but then knocking down the third one first. If this happens, then our conclusion is that they all fall down from the third one on. This is still a useful thing to know as the previous example attests.

The other item to notice in the previous example is that we checked the statement for **both** $n = 3$ **and** $n = 4$ before taking on the "inductive step." While it is certainly harmless to show more initial steps than necessary in any of our inductive proofs, in this case it is actually necessary in the proof. To see why, notice that the proof requires we know something about both a_{m-1} and a_{m-2}. If we tried to start the inductive step at $m = 4$, we would be using the statements "$a_3 < 3^3$" and "$a_2 < 3^2$." However "$a_2 < 3^2$" was **not** checked, and, in fact, it is not even true! We need to check "$a_3 < 3^3$" and "$a_4 < 3^4$" by hand, and *then* prove the induction step for $m \geq 5$. This raises the general issue that it is not always sufficient to show only one initial statement for an inductive proof to be logically sound.

One of the number sequences we encountered in Section 1.2 was the sequence of *Fibonacci numbers*. These numbers are the source of a great many problems in science and recreational mathematics. Most of us know the Fibonacci numbers as the sequence obtained by starting with 1 and 1, and then using the sum of two successive numbers in the sequence to get the next number in the sequence. So the sequence looks like this:

$$1, 1, 2, 3, 5, 8, 13, 21, 34, 55, 89, \ldots$$

Formally, this means that the Fibonacci numbers are defined with the recurrence relation

$$F_n = F_{n-1} + F_{n-2}, \quad \text{for all } n \geq 3 \tag{2.3}$$

where $F_1 = 1$ and $F_2 = 1$.

Example 4 *Show that the Fibonacci number F_{3n} is always even. (This means that F_k is even whenever k is divisible by 3.)*

PROOF (By induction on n) Consider the statement "F_{3n} is even." Since we wrote out some of the Fibonacci numbers above, we can quickly see that $F_3 = 2$, $F_6 = 8$, and $F_9 = 34$, so the statements "F_3 is even," "F_6 is even," and "F_9 is even" are each true. Now let $m \geq 2$ be given, and assume that we have already checked that the Fibonacci numbers $F_3, F_6, \ldots, F_{3(m-1)}$ are all even. The next statement under consideration is $P(m)$, which states that "F_{3m} is even." With careful use of the recurrence relation (2.3), we can see that

$$\begin{aligned} F_{3m} &= F_{3m-1} + F_{3m-2} \\ &= (F_{3m-2} + F_{3m-3}) + F_{3m-2} \\ &= F_{3m-3} + 2F_{3m-2} \end{aligned}$$

So $F_{3m} = F_{3m-3} + 2F_{3m-2}$. It has already been checked that $F_{3m-3} = F_{3(m-1)}$ is even, and $2F_{3m-2}$ is certainly even since it is twice an integer. Since the sum of two even numbers is even, we can conclude that the next statement "F_{3m} is even" is true. This completes the induction. ∎

Example 5 *Show that the Fibonacci numbers satisfy the following for all $n \geq 1$:*

$$\sum_{i=1}^{n}(F_i)^2 = \underbrace{(F_1)^2 + (F_2)^2 + \cdots + (F_n)^2}_{n \text{ terms}} = (F_n)(F_{n+1})$$

PROOF (By induction on n) Let $P(n)$ be the statement "$\sum_{i=1}^{n}(F_i)^2 = (F_n)(F_{n+1})$." Since $P(1)$ states that "$(F_1)^2 = (F_1)(F_2)$," and by definition $F_1 = F_2 = 1$, then $P(1)$ is true.

Now let the positive integer m be given such that statements $P(1), \ldots, P(m-1)$ have already been checked to be true—in particular, we already know for sure that $\sum_{i=1}^{m-1}(F_i)^2 = (F_{m-1})(F_m)$. Now note that

$$\begin{aligned} \sum_{i=1}^{m}(F_i)^2 &= (F_1)^2 + \cdots + (F_m)^2 \\ &= (F_1)^2 + \cdots + (F_{m-1})^2 + (F_m)^2 \\ &= \left(\sum_{i=1}^{m-1}(F_i)^2\right) + (F_m)^2 \end{aligned}$$

$$= (F_{m-1})(F_m) + (F_m)^2, \quad \text{by statement } P(m-1)$$
$$= (F_m)(F_m + F_{m-1})$$
$$= (F_m)(F_{m+1}), \quad \text{by the Fibonacci recurrence relation (2.3)}$$

So we have shown that $\sum_{i=1}^{m}(F_i)^2 = (F_m)(F_{m+1})$. That is, $P(m)$ is true, as desired. ∎

Practice Problem 3 *Emulate Example 3 to show that $F_n < 2^n$ for all positive integers n.*

We have seen how induction can be used to prove that every number in a sequence has a particular property. The common theme has been the use of the recursive description of the sequence to show that the property is preserved from one term to the next. This general idea can be extended to situations where we start with a closed formula for the sequence. In these cases, there are often perfectly good noninduction proofs of the same properties, but an induction proof might be more elegant or offer a new insight into the property at hand.

*Explore more on
the Web.*

Example 6 *Show that $n^3 + 2n$ is divisible by 3 for all positive integers n.*

PROOF (By induction on n) Let $D(n)$ be the statement "$n^3 + 2n$ is divisible by 3." Statement $D(1)$ states that "$1^3 + 2(1)$ is divisible by 3," which is true. Now let the positive integer m be given such that statements $D(1), \ldots, D(m-1)$ have already been checked to be true—in particular, we already know that $(m-1)^3 + 2(m-1)$ is divisible by 3. This means that there is an integer K so that $(m-1)^3 + 2(m-1) = 3K$. This can be simplified algebraically to say $m^3 - 3m^2 + 3m - 1 + 2m - 2 = 3K$, or $m^3 - 3m^2 + 5m - 3 = 3K$. Note that

$$m^3 + 2m = (m^3 - 3m^2 + 5m - 3) + (3m^2 - 3m + 3)$$
$$= (3K) + (3m^2 - 3m + 3), \quad \text{using the simplified statement } D(m-1)$$
$$= 3K + 3m^2 - 3m + 3$$
$$= 3(K + m^2 - m + 1)$$

That is, $m^3 + 2m = 3(K + m^2 - m + 1)$, so "$m^3 + 2m$ is divisible by 3" is a true statement. This is statement $D(m)$, as desired. ∎

It might not be clear from the examples we have seen up to this point, but mathematical induction can be applied to *any* situation where there is a sequence of statements about the positive integers, even when there is no "formula" involved.

Example 7 *Show that in a single-elimination basketball tournament with n teams there must be $n-1$ games played to determine a champion.*

PROOF (By induction on n) Let $T(n)$ be the statement "In a single-elimination basketball tournament with n teams there must be $n-1$ games played to determine a champion." Certainly in any tournament with 1 team, no games need to be played. Hence, the statement $T(1)$ is true. Also in any tournament with two teams, only one game needs to be played. Hence, the statement $T(2)$ is true. Let m be given such that statements $T(1), \ldots, T(m-1)$ have already been shown to be true—in particular, we already know that in any tournament with $m - 1$ teams, $m - 2$ games are required. Now let any tournament with m teams be given. After one game has been played in this tournament, one team is eliminated, leaving $m - 1$ teams still "alive." Previously checked statement $T(m - 1)$ told us that to determine a champion from among these remaining $m - 1$ teams, exactly $m - 2$ games are required. So to determine a champion in the tournament involving m teams, we need the first game **plus** the other $m - 2$ games for a total of $m - 1$ games. This means that $T(m)$ is true, as desired.   ■

Example 8   *In the Josephus problem from Section 1.1, we placed a number of people in a circle and eliminated every other person until only one person was left. The puzzle is to find the correct place to stand so that you are the last person surviving. If the game starts with 2^n people (numbered $1, \ldots, 2^n$) and every second person is eliminated starting with Person 2, where should you stand?*

SOLUTION   **Claim:** You should stand in position 1.

Proof by induction on n. Let $J(n)$ be the statement "If this game starts with 2^n people, then the person in position 1 is the last survivor." If the game is played out with two people, then since the Person 2 is eliminated first, Person 1 is the last survivor. That is, $J(1)$ is true. It is also easy to check $J(2)$: If the game is played with $2^2 = 4$ people, then the elimination order is 2, 4, 3, 1.

Now let m be given such that statements $J(1), \ldots, J(m - 1)$ have already been checked to be true—in particular, we know that when the game starts with 2^{m-1} people, then Person 1 is the last survivor. Now consider the game starting with 2^m people. After one time around the circle, players in the even positions are eliminated, leaving only the people in the odd positions $1, 3, 5, 7, \ldots, 2^m - 1$ alive. This leaves half of the original 2^m people, which amounts to 2^{m-1} people, remaining. We know from statement $J(m - 1)$ that in this game starting with 2^{m-1} people, the person in the first position (which is Person 1) will be the last survivor. Hence, Person 1 survives the game with 2^m players, completing the induction.   □

Example 9   *In the magic trick that opens Section 1.1, there are three ways in which the packet of four cards can be mixed. The packet can be cut, the top two cards can be turned over as one, or the entire packet can be turned over together. Recall that the packet starts off with one card (the spade) facing the "wrong way" from the rest. Prove that for all $n \geq 0$, after n of these shuffles, there will still be one card facing the wrong way from the rest.*

PROOF   (By induction on n) Let $P(n)$ be the statement "After n of these shuffles, there will still be one card facing the wrong way from the rest." The first statement is $P(0)$, which refers to the packet before any shuffle has been performed. Certainly at this point, there is one card (the spade) facing the wrong way, so statement $P(0)$ is true. Now let $m \geq 1$ be given such that statements $P(0), P(1), \ldots, P(m - 1)$ have all been checked to be true.

Once the packet has undergone m shuffles, we know that after the first $m - 1$ of these shuffles, there is still one card turned the wrong way from the others, by statement $P(m - 1)$. The final (m^{th}) shuffle must be one of the three given types, so we consider each possibility as a separate case.

Case 1: If the mth shuffle cuts the packet, then no cards are turned over, so the one card that was turned the wrong way before this last shuffle will still be facing the wrong way after the shuffle.

Case 2: If the mth shuffle reverses the entire packet, then *all* the cards are turned over, so the one card that was facing the wrong way before this last shuffle will still be turned the wrong way after the shuffle.

Case 3: If the mth shuffle flips the top two cards as one, then things become a bit more complicated. If the card that is facing the wrong way before this final shuffle is among the top two in the packet, then the mth shuffle creates a new, single wrong-way card in the same position in the packet. On the other hand, if the card that is facing the wrong way before this final shuffle is among the *bottom* two in the packet, then the mth shuffle creates a new, single wrong-way card in a different position among the bottom two cards in the packet.

In each case, after the mth shuffle, the packet still has a single wrong-way card. That is, statement $P(m)$ is true, completing the induction. ∎

We end this section with a result in mathematics that is so important it is called the *fundamental theorem of arithmetic*. This important idea was known to the ancient Greeks, so it is easy for us to take it for granted. Since its proof is a simple application of mathematical induction, this course is a good place to see why it is true.

Theorem 1 (Fundamental Theorem of Arithmetic) *Every integer greater than 1 can be expressed as the product of a list of prime numbers.*[*]

PROOF   (By induction on n) Let $P(n)$ be the statement "n can be written as the product of a list of prime numbers." It is easy to check that $P(2)$ and $P(3)$ are true, since 2 and 3 *are* prime numbers themselves. It might seem silly to think of 2 as being the product of the prime numbers in the list "2," but technically it is. Let's look at a few more statements before moving on to the induction step:

- $P(4)$ is true since 4 is the product of the prime numbers in the list "2,2."
- $P(5)$ is true since 5 is the product of the prime numbers in the list "5."
- $P(6)$ is true since 6 is the product of the prime numbers in the list "2,3."

Now let $m \geq 7$ be given such that $P(2), P(3), \ldots, P(m-1)$ have all been checked to be true, and we are now considering statement $P(m)$. Either the number m is prime or it is not.

- **Case 1:** If m is prime, then technically m is the product of the prime numbers in the list "m," so $P(m)$ is true in this case.
- **Case 2:** If m is not prime, then by the definition of "prime," we know that $m = a \times b$ for some positive integers a and b, each strictly between 1 and m.

---

[*] The usual theorem also states that this product of prime numbers is unique—that is, a number cannot be written as the product, say, 5×7, and also as some other product using primes other than 5 and 7. We will not prove uniqueness here.

Since we already have checked statement $P(a)$, we know that a can be written as the product of a list of primes, and since we already have checked statement $P(b)$, we know that b can be written as the product of a list of primes. If we take the list of primes whose product is a along with the list of primes whose product is b, we will have a list of primes whose product is $a \times b = m$. Hence, $P(m)$ is true in this case.

Since $P(m)$ is true in either case, this completes the induction step. ■

Solutions to Practice Problems

1 $s_n = \sum_{i=1}^{n}(4i - 1)$ is the same as $s_n = 3 + 7 + 11 + \cdots + (4n - 1)$. Since $s_{n-1} = 3 + 7 + \cdots + (4(n-1) - 1)$ and $s_n = 3 + 7 + \cdots + (4(n-1) - 1) + (4n - 1)$, we conclude that $s_n = s_{n-1} + (4n - 1)$. This along with the fact that $s_1 = 3$ is enough to determine all the s_n.

2 *Proof by induction on n.* Consider the statement "$s_n = n \cdot (2n + 1)$." The first statement says, "$s_1 = 1 \cdot 3$," which is true. Now suppose that statements "$s_1 = 1 \cdot 3$" "$s_2 = 2 \cdot 5$," ... , "$s_{m-1} = (m - 1) \cdot (2(m - 1) + 1)$" have all been checked for a given $m \geq 2$. Then we know

$$
\begin{aligned}
s_m &= s_{m-1} + (4m - 1), \quad \text{from Practice Problem 1} \\
&= (m - 1) \cdot (2(m - 1) + 1) + (4m - 1), \\
&\quad \text{since } s_{m-1} = (m - 1) \cdot (2(m - 1) + 1) \\
&= (m - 1) \cdot (2m - 1) + (4m - 1) \\
&= 2m^2 - 3m + 1 + 4m - 1 \\
&= 2m^2 + m \\
&= m(2m + 1)
\end{aligned}
$$

This shows that the next statement "$s_m = m \cdot (2m + 1)$" is true, completing the induction step. ■

3 *Proof by induction on n.* Consider the statement "$F_n < 2^n$." We can easily check that $F_1 < 2^1$ and $F_2 < 2^2$, so the first two statements are true. Now suppose we have checked all the statements from "$F_1 < 2^1$" up to "$F_{m-1} < 2^{m-1}$," for a given $m \geq 3$. Then

$$
\begin{aligned}
F_m &= F_{m-1} + F_{m-2} \\
&< 2^{m-1} + 2^{m-2}, \quad \text{since } F_{m-1} < 2^{m-1} \text{ and } F_{m-2} < 2^{m-2} \\
&< 2^{m-1} + 2^{m-1}, \quad \text{since } 2^{m-2} < 2^{m-1} \\
&= 2 \cdot 2^{m-1} \\
&= 2^m
\end{aligned}
$$

This establishes that the next statement "$F_m < 2^m$" is true, completing the induction. ■

Exercises for Section 2.4

1. Write each of the following sequences as a recurrence relation (with sufficient initial values specified):

   (a) $a_n = 2(3^0 + 3^1 + 3^2 + \cdots + 3^{n-1})$ for each $n \geq 1$.

   (b) $b_n = \frac{1}{2^1} + \frac{1}{2^2} + \cdots + \frac{1}{2^n}$ for each $n \geq 1$.

   (c) $c_n = \frac{1}{(1)(2)} + \frac{1}{(2)(3)} + \frac{1}{(3)(4)} + \cdots + \frac{1}{(n)(n+1)}$ for all $n \geq 1$.

   (d) $d_n = \frac{1}{(1)(3)} + \frac{1}{(3)(5)} + \frac{1}{(5)(7)} + \cdots + \frac{1}{(2n-1)(2n+1)}$ for all $n \geq 1$.

2. The following refer to the sequences in the previous exercise. Use induction for each proof.

   (a) Show that $a_n = 3^n - 1$ for all positive integers n.

   (b) Show that $b_n = 1 - \frac{1}{2^n}$ for all positive integers n.

   (c) Show that $c_n = \frac{n}{n+1}$ for all positive integers n.

   (d) Show that $d_n = \frac{n}{2n+1}$ for all positive integers n.

3. Rewrite the proof in Example 6 using the following outline:

   (a) Define the number sequence $\{g_n\}$ with the closed formula $g_n = n^3 + 2n$. Use algebra to show that g_m is the same thing as $g_{m-1} + 3(m^2 - m + 1)$.

   (b) Use the relationship from the previous part to explain that if g_{m-1} is divisible by 3, then g_m is divisible by 3.

   (c) Using this relationship, prove by induction that g_n is divisible by 3 for all integers $n \geq 1$.

3. Use mathematical induction to prove each of the following:

   (a) Use induction to prove that for each integer $n \geq 1$, $n^2 - n$ is even.

   (b) Use induction to prove that for each integer $n \geq 1$, $n^3 - n$ is divisible by 3.

   (c) Use induction to prove that for each integer $n \geq 1$, $n^5 - n$ is divisible by 5.

5. Use mathematical induction to prove that for all integers $n \geq 0$, $10^n - 1$ is divisible by 9.

6. Use mathematical induction to prove that for all integers $n \geq 2$, $2^{3n} - 1$ is not prime.

7. We define the Pell sequence by the initial values $p_1 = 1$ and $p_2 = 2$ along with the recurrence relation

$$p_n = 2p_{n-1} + p_{n-2}$$

Prove by induction on n that

$$2p_n^2 + (-1)^n = (p_{n+1} - p_n)^2$$

(This establishes that the left-hand side of the above equation is always a perfect square, thus solving Problem 602 from the *College Mathematics Journal*, May 1997.)

8. The following problems refer to the Fibonacci numbers defined before Example 4:
   (a) Show that for all $n \geq 2$, $F_n < 2^{n-1}$.
   (b) Show that for all $n \geq 1$,
   $$F_2 + F_4 + \cdots + F_{2n} = F_{2n+1} - 1$$
   (c) Show that for all $n \geq 1$,
   $$F_1 + F_3 + \cdots + F_{2n-1} = F_{2n}$$
   (d) Show that for all $n \geq 1$,
   $$F_1 + F_2 + \cdots + F_n = F_{n+2} - 1$$

9. The following problems refer to the Fibonacci numbers defined before Example 4:
   (a) Prove by induction that for all $n \geq 1$, F_{4n} is divisible by 3.
   (b) Prove by induction that for all $n \geq 1$, F_{5n} is divisible by 5.
   (c) Show that $F_{n+2} \geq \phi^n$ for all $n \geq 1$, where $\phi = \frac{1+\sqrt{5}}{2}$. (HINT: First check that ϕ satisfies the equation $\phi^2 = \phi + 1$. This is the property of ϕ that is important for this problem.)

10. This problem refers to the Josephus problem from Section 1.1 and Example 8.
   (a) In the Josephus problem, if the game starts with 2^n people and every second person is eliminated, where should Josephus's friend stand? Prove your answer by induction on n.
   (b) Use the answer in Example 8 to explain where Josephus should stand in the Josephus game starting with n people where every second person is eliminated.
   (c) Use the answer to part (a) above to explain where Josephus's friend should stand if the game starts with n people and every second one is eliminated.

11. For the magic trick that opens Section 1.1 (and revisited in Example 9), prove that after any n shuffles, there is one card between the "club" and the wrong-way card.

12. Prove by induction on $n \geq 1$ that the product of n odd integers is an odd integer.

13. (a) Prove by induction on $n \geq 1$ that the sum of n rational numbers is a rational number.
   (b) Prove by induction on $n \geq 1$ that the average of n rational numbers is a rational number.

14. The following statements prove the existence of the quotient and remainder in Theorem 8 of Section 2.2, the division theorem.
   (a) Prove by induction on $n \geq 0$ that there exist integers q and r such that $n = 3 \cdot q + r$ and $0 \leq r \leq 2$. (HINT: Use statement $P(m-3)$ in trying to prove statement $P(m)$.)
   (b) Prove by induction on $n \geq 0$ that there exist integers q and r such that $n = 5 \cdot q + r$ and $0 \leq r \leq 4$.
   (c) Let the positive integer k be given. Prove by induction on $n \geq 0$ that there exist integers q and r such that $n = k \cdot q + r$ and $0 \leq q \leq k - 1$.

15. Suppose you have an unlimited supply of 3-cent and 8-cent stamps at your disposal. Show by induction that you can just use combinations of these stamps to make n cents in postage for any $n \geq 14$.

16. Suppose you have an unlimited supply of 5-cent and 8-cent stamps at your disposal. Find the smallest value of N so that the statement "You can use just combinations of these stamps to make n cents in postage for any $n \geq N$" is true, and prove the statement by induction.

17. Prove that for every $n \geq 1$,
$$\sum_{i=2}^{2^n} \frac{1}{i} \geq \frac{n}{2}$$

18. Nicole Oresme (1323–1382) used the result in the previous problem to argue that the sum
$$\frac{1}{1} + \frac{1}{2} + \frac{1}{3} + \frac{1}{4} + \frac{1}{5} + \cdots + \frac{1}{k}$$
can be made larger than any fixed number if k is taken to be large enough. Informally explain Oresme's argument. In the language of modern calculus, this means that the *infinite* sum $\sum_{k=1}^{\infty} \frac{1}{k}$, called the *harmonic series*, diverges.

19. Find the error in each of the following "proofs":
   (a) **Proposition** All sets are infinite.
   Flawed Argument. Let $P(n)$ be the statement "Every set has at least n elements." We will prove by induction on n that $P(n)$ is always true, from which the proposition will follow.
   First note that $P(0)$ says, "Every set has at least 0 elements," which is true. Now let $m \geq 1$ be given, and assume we have verified that $P(0), P(1), \ldots, P(m-1)$ are all true. Let a set S be given, and

choose an element $a \in S$. Applying the induction hypothesis $P(m-1)$ to the set $S - \{a\}$ tells us that $S - \{a\}$ has at least $m-1$ elements. Since S has one more element than $S - \{a\}$, it follows that S has at least m elements. Since S is any given set, this shows then that every set has at least m elements. That is, $P(m)$ is true.   ♯

(b) Proposition Define the number sequence $\{a_n\}$ by the recurrence relation $a_n = a_{n-2} + 2$, where $a_1 = 3$ and $a_2 = 2$. For all $n \geq 1$, a_n is odd.

Flawed Argument. Let $P(n)$ be the statement "a_n is odd." First note that $P(1)$ says, "a_1 is odd," which is true since we are given the value $a_1 = 3$. Now assume that we have verified $P(1), \ldots, P(m-1)$ (where $m \geq 2$), and we are now considering $P(m)$. The recurrence relation tells us that $a_m = a_{m-2} + 2$ and $P(m-2)$ tells us that a_{m-2} is odd; hence, a_m is odd. That is, $P(m)$ is true, completing the induction step.   ♯

(c) Proposition All horses are the same color.

Flawed Argument. Let $P(n)$ be the statement "Every corral containing n horses has all horses the same color." We will prove by induction on n that $P(n)$ is always true, from which the proposition will follow.

First note that $P(1)$ says, "Every corral containing one horse has all horses the same color," which is true. Now assume that we have verified $P(1), P(2), \ldots, P(m-1)$ are all true (where $m \geq 2$), and we are now considering $P(m)$. To do this, we must allow the reader to pick any corral containing m horses she wishes. If we designate one of the horses "Ed," we can have Ed temporarily removed, leaving us with a corral containing $m-1$ horses. Since we have already checked that $P(m-1)$ is true, we know that every horse in this corral is the same color. Now if we return Ed and remove a different horse, whom we will designate as "Silver," we will be led to the same conclusion. Hence, the corral of horses without Ed has horses all of the same color, and the corral of horses without Silver has horses all of the same color. Since these two groups overlap, we conclude that all horses in the original corral of m horses are of this same color. Since this reasoning can be applied to any corral of m horses supplied by the reader, this means that $P(m)$ is true.   ♯

2.5 Contradiction and the Pigeonhole Principle

The main emphasis in this chapter has been proving statements of the form "For all x in some domain of interest, if this hypothesis is satisfied, then this conclusion is true." An example of this type of statement is "For all integers n, if n is even, then n^2 is even." In the notation of Chapter 1, these statements are quantified predicates of the form $\forall x \in D, P(x) \rightarrow Q(x)$.

At this point, we have three major tools in our arsenal of proof techniques for proving statements of this type:

1. *Direct proof.* We imagine that the READER of the proof has selected an element of the domain for which the hypothesis is true, and the AUTHOR must demonstrate to the READER that the theorem's conclusion must be true for that element. The term *direct proof* comes from the fact that we start by assuming the hypothesis, and logically move "forward" until we reach the conclusion.

2. *Proof by contrapositive.* To use this technique, we form the contrapositive of the theorem. We then do a direct proof of the contrapositive statement. Since a theorem and its contrapositive are equivalent, this also proves the original theorem. This is sometimes called an *indirect proof* because we have not directly proved the original theorem, but have instead obtained the proof in an indirect manner.

3. *Mathematical induction.* This technique applies to theorems about the positive integers, or more generally to theorems about all integers greater than or equal to some fixed value. The heart of the proof is typically a direct proof that assumes $P(1), P(2), \ldots, P(m-1)$ as the hypothesis, and proves the conclusion $P(m)$.

In this section we examine a truly indirect proof technique known as *proof by contradiction*. We also study an important example of a theorem that is usually proved using this technique.

Proof by Contradiction

One way to think about direct proof and proof by contrapositive is that they both demonstrate, in different ways, that **there cannot possibly be a counterexample to the theorem**. Thus, we know the theorem must be true for all elements of the domain.

These proof techniques are based on the two properties that a potential counterexample to a given implication must possess:

(i) It must make the hypothesis of the implication true.

(ii) It must make the conclusion of the implication false.

In a direct proof, we show the READER that if she chooses an element satisfying property (i), that same element cannot satisfy property (ii). In this way, we show that she cannot possibly find a counterexample.

When we prove the contrapositive, we show the READER that if she chooses an element satisfying property (ii), that same element cannot satisfy property (i). Once again, the reader will be unable to find a counterexample.

In a proof by contradiction, we allow the READER to imagine she *has* found a counterexample, and we show that this cannot be so because it would lead to a "contradiction"—that is, to a statement we know to be false. In so doing, we are essentially showing that properties (i) and (ii) are logically incompatible.

Explore more on the Web.

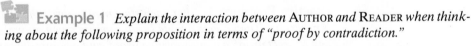

Example 1 *Explain the interaction between AUTHOR and READER when thinking about the following proposition in terms of "proof by contradiction."*

Proposition *If n is an odd integer, then $n^2 + n$ is even.*

SOLUTION Remember that the READER sets out looking for counterexamples to the given statement. The method of "proof by contradiction" asks the READER what will happen (i.e., follow deductively) if he **does** happen to find a counterexample. So the AUTHOR might write the following:

Dear READER,

I appreciate your efforts to find a counterexample to the given statement, but did you ever stop to think what will happen if you find one? Suppose you **do** find an example (let's agree to call it n) that makes the hypothesis true and the conclusion false. This means that your n is an odd integer n that makes $n^2 + n$ an odd integer. Because of our definition of "odd," this means that $n = 2K + 1$ for some integer K and $n^2 + n = 2L + 1$ for some integer L. Since your example would create these two equations, then substituting the first equation into the second

$$2L + 1 = n^2 + n$$
$$= (2K + 1)^2 + (2K + 1)$$
$$= 4K^2 + 6K + 2$$

will give us the equation $2L + 1 = 4K^2 + 6K + 2$. This might not seem so bad as written, but a little algebra will turn this into the equation

$$L - 2K^2 - 3K = \frac{1}{2}$$

Since L and K are integers, we know that $L - 2K^2 - 3K$ is an integer, but we know that $\frac{1}{2}$ is definitely *not* an integer. This is clearly nonsense! However, it is a logically valid deduction if the example you are looking for really exists. I think we will both sleep better at night knowing that there is no counterexample to the given statement.

Only concerned with your well-being,
AUTHOR

   It is useful to examine how each type of proof would begin, and how it would end. Table 2-15 shows the outline of three different forms of proof for the proposition "If n is an odd integer, then $n^2 + n$ is even." For a direct proof, we start by assuming the hypothesis is true, and eventually show that the conclusion must be true. For a contrapositive proof, we use the same approach for the contrapositive statement "If $n^2 + n$ is odd integer, then n is even." For a proof by contradiction, we assume that there is a counterexample, and we try to show how a false statement follows from this assumption. In both the direct and contrapositive proofs, the AUTHOR has well-defined starting and ending points. With a proof by contradiction, by contrast, the AUTHOR has more information to start with, but no particular goal toward which to work. This is a typical trade-off when developing a proof by contradiction. You may assume two things (the hypothesis is true **and** the conclusion is false), but in exchange you lose the advantage of having a well-established goal.

Practice Problem 1   *For each statement, rewrite the statement in "if, then" form, if necessary. Then create a table similar to that given above, showing how to start and end a direct proof, a contrapositive proof, and a contradiction proof for the statement.*

(a) *If n^2 is odd, then n is odd.*

(b) *Even perfect squares are always divisible by 4.*

Direct	Contrapositive
Let n be an odd integer.	Let n be given such that $n^2 + n$ is odd.
.	.
.	.
.	.
We conclude that $n^2 + n$ must be even.	We conclude that n is even.

Contradiction
Suppose there is an integer n such that n is odd **and** $n^2 + n$ is odd.
.
.
.
We infer a false statement.
This is a contradiction, so we know there cannot be a counterexample.

Table 2-15   The Structure of Proofs

For the statement proved in Example 1, it turns out that a direct proof is very short and simple. (Refer to Exercise (10b) in Section 2.1.) You might wonder why anyone would prefer the proof by contradiction. It is not really a matter of preference, but more a matter of the thought processes by which one discovers the proof.

It is usually a good idea to try to find a direct proof, or a proof of the contrapositive statement, before resorting to proof by contradiction. However, there are some theorems for which proof by contradiction may be considered as the best alternative. One situation occurs when both the statement and its contrapositive have "negative" conclusions. In general, it can be difficult to prove that something does not happen. Here is an example of a proposition that we have *not* proved before.

Example 2 *Why might you try a proof by contradiction for the proposition "If n has the form* $4K + 3$ *for some integer K, then* $n^2 - n$ *is not divisible by* 4"?

SOLUTION   Table 2-16 shows the starting and ending points for a direct proof and a proof of the contrapositive. In both direct and contrapositive proofs, the goal contains the word *not*. That does not mean we cannot do such a proof, but it does suggest considering a proof by contradiction first. □

Now let's see the proof discovery process as well as the formal version of a "proof by contradiction" for this proposition.

Proposition 1   *If n has the form* $4K + 3$ *for some integer K, then* $n^2 - n$ *is not divisible by* 4.

PROOF   (Informal version)

Dear READER,

In your search for a counterexample to the above statement, suppose you actually find one. Let's agree to call it *n* and talk about what properties *n* must have. Since *n* is a counterexample to the given statement, we agree that *n* makes the hypothesis true and the conclusion false. In other words, $n = 4K + 3$ for some integer *K*, **and** $n^2 - n = 4L$ for some integer *L*. Substituting the first equation into the second, we see that

$$4L = n^2 - n$$
$$= (4K + 3)^2 - (4K + 3)$$
$$= 16K^2 + 20K + 6$$

Direct	Contrapositive
Let *n* be an integer with the form $4K + 3$.	Let *n* be given such that $n^2 - n$ is divisible by 4.
⋮	⋮
We conclude that $n^2 - n$ is not divisible by 4.	We conclude that *n* does not have the form $4K + 3$.

Table 2-16   Solution to Example 2

So your value of n creates the equation $16K^2 + 20K + 6 = 4L$, which can be rewritten as

$$L - 4K^2 - 5K = \frac{3}{2}$$

The trouble with this is that $L - 4K^2 - 5K$ is an integer but $\frac{3}{2}$ is not. So you see that if you *do* find a counterexample, it will lead (through simple algebra) to an equation showing that $\frac{3}{2}$ is an integer. Since this is nonsense, it must be the case that no counterexample to the original statement exists.

Always failing to never be your friend,

AUTHOR

PROOF   (Formal version)

Suppose that a counterexample n to the above statement exists. This means that n has the form $4K + 3$, and $n^2 - n$ **is divisible** by 4. In other words, $n = 4K + 3$, and $n^2 - n = 4L$ for some integers K and L. Combining these two facts leads us to say

$$\begin{aligned} 4L &= n^2 - n \\ &= (4K+3)^2 - (4K+3) \\ &= 16K^2 + 20K + 6 \end{aligned}$$

The equation $4L = 16K^2 + 20K + 6$ can be rearranged to say

$$L - 4K^2 - 5K = \frac{3}{2}$$

from which it follows that $\frac{3}{2}$ is an integer, which is a contradiction. Therefore, no counterexample to the original statement exists.

Another situation that suggests trying proof by contradiction occurs when the original "if, then" statement and its contrapositive seem equally difficult to prove. For example, the statement "If $x > 0$, then $\frac{1}{x} > 0$" has contrapositive "If $\frac{1}{x} \le 0$, then $x \le 0$," which is essentially the same.

Proposition 2   *For every real number x, if $x > 0$, then $\frac{1}{x} > 0$.*

PROOF   Suppose there is a counterexample to this statement. That is, a real number x exists such that $x > 0$ and $\frac{1}{x} \le 0$. Since $x > 0$, multiplying by x on both sides[*] of the inequality

$$\frac{1}{x} \le 0$$

gives us the inequality

$$\frac{1}{x} \cdot x \le 0 \cdot x \quad \text{or} \quad 1 \le 0$$

Certainly, "$1 \le 0$" is nonsense, so we have arrived at a contradiction. Therefore, no counterexample exists for the original statement.

---

[*] We are, of course, assuming that the READER acknowledges this is a valid rule for inequalities.

If you ask yourself, "How did the AUTHOR of the proof know what to do to get a contradiction?" you have uncovered the difficulty with a proof by contradiction—one never knows in advance what the contradiction will look like. This is one reason we suggest first trying to find a direct proof or a contrapositive proof—at least then you have a definite goal to work toward.

Existence and Nonexistence Proofs

Most of our theorems to this point have been *implications*, characterized as "For all x in some domain D, if $P(x)$, then $Q(x)$." We have emphasized this type of theorem because it is such a frequently encountered form. However, some mathematical statements assert the existence or nonexistence of some mathematical object with a particular property. An *existence proof* is a proof of a theorem characterized as "There exists x in some domain D such that $P(x)$." Likewise, the proof of a theorem characterized as "There does *not* exist an x in some domain D such that $P(x)$" is called a *nonexistence proof.*

Example 3 *Determine which of the following statements are true. Give a convincing argument for your answer:*

1. *There exists a positive real number r satisfying the equation $r^5 - r^4 + r^3 - r^2 + r - 1 = 0$.*
2. *There exists a positive integer s satisfying the equation $6s^2 - 5s + 1 = 0$.*
3. *There exists a positive rational number t satisfying the equation $t^2 - 2 = 0$.*
4. *There exists a smallest positive integer.*

SOLUTION   Only statements #1 and #4 are true.

1. This is a true statement. If we test the value $r = 1$ in the equation, we find that $1^5 - 1^4 + 1^3 - 1^2 + 1 - 1 = 0$, so this shows that the positive real number $r = 1$ satisfies the equation.

2. This is false because the quadratic formula tells us that the only real solutions to the equation $6s^2 - 5s + 1 = 0$ are $\frac{1}{2}$ and $\frac{1}{3}$, neither of which is an integer value. Hence, there are no integers with the desired property.

3. This is a false statement. We will address the formal justification for this in Theorem 4.

4. This is a true statement since we know that the number 1 **is** the smallest positive integer.

□

To prove the true statements above, in each case, we simply supplied an example of a number with the desired property. This type of proof is called a *constructive proof* of an "existence statement," and it is very hard to take issue with, even for the most critical READER.

Existence proofs are actually not new to you. We have seen one major theorem whose proof is an existence proof, the division theorem (Theorem 8 of Section 2.2). The proof of the existence of a quotient and remainder with the desired properties was addressed in Exercise 14 of Section 2.4. In addition, you have done small existence proofs within almost every proof in the first two sections of the chapter. For example, consider this proof from Section 2.2.

Proposition 3   *If the integers m and n are both divisible by 3, then the number m + n is also divisible by 3.*

   PROOF   *Let m and n be integers, each divisible by 3. This means that there are integers K and L such that $m = 3K$ and $n = 3L$. Hence, $m + n = 3K + 3L = 3(K + L)$. Since $K + L$ is an integer, this shows that $m + n$ is divisible by 3.* ■

To show that $m + n$ is divisible by 3, we had to produce an integer Q such that $m + n = 3Q$. The last step of the proof shows that the integer $Q = K + L$ works. The definitions of even, odd, divisibility, and rational all involve the existence of integers with certain properties. As a result, proofs where the conclusion involves any of these concepts usually contain a constructive existence proof.

Nonexistence proofs, on the other hand, can be more challenging, and may best be handled using proof by contradiction. Consider the following classic result:

Theorem 4   *The real number $\sqrt{2}$ is irrational. That is, there does not exist a rational number r such that $r^2 = 2$.*

The proof, given below, is a quintessential example of a *nonexistence* proof. This category of proof is another that is frequently attacked using proof by contradiction. In a direct proof, the "goal" would be to show that something does *not* exist, and it is not clear how one could reach such a goal. We would need to show that "for every rational number r, $r^2 \neq 2$." We can find rational numbers whose square is very close to 2—for example, $(\frac{35355339}{25000061})^2 \approx 1.999990233$. How can we show directly that r^2 will never be exactly 2? Rather than try, we will show that a counterexample to the theorem would lead to a contradiction. The contradiction will center on the following idea that we first saw in the definition of a rational number.

> **Definition**   Two integers are said to be *relatively prime* if they have no common divisor greater than 1.

   PROOF   (Theorem 4) Suppose to the contrary that this proposition is not true. That is, there *does* exist a rational number r such that $r^2 = 2$. Since r is a rational number, we know that there exist integers a and b such that $r = \frac{a}{b}$. Moreover, we know that these two integers can be chosen to be relatively prime.

   Now since $r^2 = 2$, we have $(\frac{a}{b})^2 = 2$. Using algebra, we rewrite this as $a^2 = 2b^2$. From this ensues the following chain of reasoning:

● Since $a^2 = 2b^2$, we know that a^2 is even, which means that a is even.[*]

● Since a is even, we know that $a = 2K$ for some integer K. This means $a^2 = (2K)^2 = 4K^2$.

---

[*] In Theorem 7 of Section 2.1, we proved the proposition "If n^2 is even, then n is even."

- Since $a^2 = 2b^2$ and $a^2 = 4K^2$, it follows that $b^2 = 2K^2$. From this, we know that b^2, and hence b, is even.
- Since a and b are both even, then they have 2 as a common divisor.
- Remember that a and b are relatively prime.

The final two statements are contradictory, so the assumption that the proposition is false must be wrong. We conclude that the proposition is true. ■

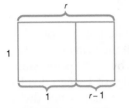

Figure 2-4   A golden rectangle.

The same argument structure can be used to prove that other important numbers are not rational. For example, in art and architecture, the *golden ratio* is considered to represent the ideal proportion of height to width for a rectangle. Formally, a rectangle is in this proportion[*] if the removal of an appropriate square leaves a smaller rectangle that is similar to (i.e., in the same proportions as) the original rectangle. As Figure 2-4 illustrates, the implication is that the ratio of 1 to r is the same as the ratio of $r - 1$ to 1. That is, $\frac{1}{r} = \frac{r-1}{1}$. It can be shown that the positive real number satisfying this equation is approximately 1.62, but we can show that the very relationship defining this value of r can be used to prove that r is irrational.

Proposition 5   *There does not exist a rational number r satisfying $\frac{r-1}{1} = \frac{1}{r}$.*

PROOF   Suppose to the contrary that there *is* such a rational number r. That means we can write $r = \frac{a}{b}$, where a and b are relatively prime integers. Now $\frac{r-1}{1} = \frac{1}{r}$ can be rewritten as $\frac{a}{b} - 1 = \frac{b}{a}$, which implies that $a^2 - ab = b^2$, or $a^2 - b^2 = ab$. This implies that $(a - b)(a + b) = ab$.

Since a and b are relatively prime, they cannot both be even. Thus, one of the following must be true: (1) a is even and b is odd; or (2) b is even and a is odd; or (3) both are odd. We can proceed by cases.[†]

- **Case 1:** If a is even and b is odd, then $(a - b)$ and $(a + b)$ are odd, from which it follows that $(a - b)(a + b)$ is odd. However, since a is even, we know ab is even. We conclude that $(a - b)(a + b) \neq ab$, contradicting our earlier calculation that $(a - b)(a + b) = ab$.
- **Case 2:** A similar contradiction arises if a is odd and b is even.
- **Case 3:** If a and b are both odd, then ab is odd, but $(a - b)$ and $(a + b)$ are both even, which means that $(a - b)(a + b)$ is even. We conclude that $(a - b)(a + b) \neq ab$, contradicting our earlier calculation that $(a - b)(a + b) = ab$.

In every case, a contradiction arises. Hence, our original assumption that $\frac{r-1}{1} = \frac{1}{r}$ is satisfied by a rational number r must be incorrect. ■

---

[*] Such a rectangle is naturally called a *golden rectangle*.

[†] In these cases, we have omitted some details, such as the fact that the product of two odd numbers is always odd. These details are of the type you learned about in Section 2.1 so we will not clutter up this proof with them.

More Classic Proofs by Contradiction

In Book IX of Euclid's *Elements*, a series of proofs about prime numbers is given.[*] In Greek mathematics all numbers represent geometric measurements, so it is fairly interesting that the notion of "prime number" was considered important at all. The following classic example of using "proof by contradiction" is Proposition 20 in Book IX of the *Elements*.

Theorem 6   *There are an infinite number of prime numbers.*

PROOF   Suppose to the contrary that there are only a finite number of prime numbers. Then we can form the number x by multiplying all the primes together. Even though x is a large number, it will be evenly divisible by every prime number. This means that the number $x + 1$ will leave a remainder of 1 when divided by any prime number. Hence, we have produced a number that is not divisible by any prime number. This is a contradiction to the fundamental theorem of arithmetic (Theorem 1 in the previous section) that follows deductively from the assumption that there are only finitely many prime numbers. Hence, that assumption must be false, which means that there are, in fact, infinitely many prime numbers. ∎

There is a specific kind of proof by contradiction that is related to mathematical induction. Since we have already devoted two sections to mathematical induction, we will not discuss it much more here. We will just note that another way to express the principle of mathematical induction is to state that any nonempty set of positive integers will always have a smallest number in the set. This fact is called the *well-ordering principle* of the set of positive integers, and it is used in the following proof of a property that is fundamental to the study of number theory.

Theorem 7   *For integers a and b, define the set $S_{a,b}$ to be the set of all integers of the form $au + bv$, where $u, v \in \mathbb{Z}$. (For example, $S_{6,8}$ includes numbers like 20 and 4, since $20 = 2 \cdot 6 + 1 \cdot 8$ and $4 = -2 \cdot 6 + 2 \cdot 8$.) If c is the smallest positive integer in $S_{a,b}$, then every number in $S_{a,b}$ is a multiple of c.*

PROOF   (By contradiction) Let integers a and b be given, and let $c = au + bv$ (for $u, v \in \mathbb{Z}$) be the smallest positive integer in the set $S_{a,b}$. Assume that the theorem is not true, and let d be the smallest positive counterexample to the statement. That is, $d = ax + by$ for some $x, y \in \mathbb{Z}$, but d is not divisible by c. Since d is not divisible by c, we know from the division theorem that $d = q \cdot c + r$ for $1 \le r < c$. But in this case,

$$r = d - q \cdot c$$
$$= (ax + by) - q(au + bv)$$
$$= a(x - qu) + b(y - qv)$$

---

[*] Ronald Calinger's *Classics of Mathematics* is a very worthwhile book full of original sources from the history of math, including translations of these particular propositions from Book IX of Euclid's *Elements*.

This means that $r \in S_{a,b}$, but r is a positive number less than c, so this is a contradiction to the fact that c is the smallest positive number in $S_{a,b}$. ∎

How Not to Use Proof by Contradiction

Some instructors complain, "Once I show my students proof by contradiction, they want to use it for every proof they do!" Some of these instructors have even taken the extreme point of view that one should *never* use contradiction, since there is surely a direct or contrapositive proof if you only look hard enough.

Why is this a complaint? Isn't proof by contradiction a valid form of proof? The answer is "Yes, certainly!" However, there are some situations where a relatively small rewrite yields a more elegant and straightforward proof. This process sometimes disguises the hard work that was done in discovering a proof, but it is still worthwhile if brevity and logical clarity are among your proof-writing goals.

How can a student just learning about proof-writing possibly avoid this complaint? First, you can always try first to find a direct or contrapositive proof. Second, when you do use contradiction, you can recognize two dead giveaways that a rewrite is possible and not difficult. Both situations apply to proving an "if, then" proposition. To help you understand the discussion, Table 2-17 summarizes how each of the three types of proof begins and ends. Now suppose you write a perfectly correct proof by contradiction, but in your proof, the contradiction you reach is "The conclusion is true, and that contradicts our assumption that the conclusion is false." In this case,

- The first step of your proof includes the assumption that the "hypothesis is true"; and
- The last step of your proof is the inference that the "conclusion is true."

If you look closely, buried in your proof you will find a direct proof of the proposition, since this is exactly how a direct proof begins and ends.

On the other hand, suppose you write a perfectly correct proof by contradiction, but in your proof, the contradiction you reach is "The hypothesis is false, and that contradicts our assumption that the hypothesis is true." In this case,

- The first step of your proof includes the assumption that the "conclusion is false"; and
- The last step of your proof is the inference that the "hypothesis is false."

	Direct	**Contrapositive**	**Contradiction**
Start	Hypothesis is true.	Conclusion is false.	Hypothesis is true **and** Conclusion is false.
Goal	Conclusion is true.	Hypothesis is false.	Some contradiction.

Table 2-17   Summary of Proof Techniques

If you look closely, buried in your proof you will find a contrapositive proof of the proposition, since this is how a contrapositive proof begins and ends.

It will help to see examples of each of these types of overuse of the contradiction technique. Notice how much shorter and direct the rewrite is in each case.

Example 4 *The following is written as "proof by contradiction." Rewrite it as a direct proof or contrapositive proof instead.*

Proposition *If n is an odd integer, then 3n is odd.*

PROOF *Suppose you have found a counterexample to this statement. Let's agree to call it m. Since m is a counterexample, we must have m odd while 3m is even. That is, $m = 2K + 1$ for some integer K, which implies that $3m = 3(2K + 1) = 2(3K + 1) + 1$. Since $3K + 1$ is an integer, this means that 3m is odd. Hence, a counterexample m must make 3m even and 3m odd at the same time, which clearly cannot happen. Therefore, no counterexample to the original statement exists.* ∎

SOLUTION　Since the contradiction is that $3m$ is odd, which is the same as the conclusion of the original proposition, we suspect that we can write a direct proof.

Proof. Let an odd integer n be given. This means $n = 2k + 1$ for some integer K, which implies that $3n = 3(2K + 1) = 2(3K + 1) + 1$. Since $3K + 1$ is an integer, this means that $3n$ is odd. □

Practice Problem 2　*The following is written as "proof by contradiction." Rewrite it as a direct or contrapositive proof instead.*

Proposition　*If n^2 is not divisible by 4, then n is odd.*

PROOF　*Suppose you have found a counterexample to this statement. Let's agree to call it m. Since it is a counterexample, it must make the hypothesis true and the conclusion false. That is, m^2 is not divisible by 4 **and** $m = 2K$ for some integer K. The second equation implies*

$$m^2 = (2K)^2$$
$$= 4K^2$$

*Since K^2 is an integer, this implies that m^2 is divisible by 4. But m was originally chosen so that m^2 is **not** divisible by 4, so we have a contradiction. From this, we conclude that the original assumption there is a counterexample must be wrong. Hence, the original statement is true.* ∎

The Pigeonhole Principle

So far we have seen some classic results traditionally proved using the "proof by contradiction" technique. The next example is an instance of a more general result that can be proved and used in many applications. It is called the *pigeonhole principle*[*], and it will be the focus of the remainder of this section.

---

[*] The original version involves placing letters into the pigeonholes of a desk. Although the analogy is out of date, the name persists.

Example 5   *Prove that if 29 tennis balls are given out to four players, then (at least) one player gets eight or more of the balls.*

SOLUTION   (Proof by contradiction) Suppose to the contrary that this statement is false. This means that we can give 29 balls to four players, and each of the four players gets seven or fewer balls. Use the variables a, b, c, and d to represent the number of balls given to the four players. Since each player gets seven or fewer balls, we have $a \leq 7, b \leq 7, c \leq 7, d \leq 7$. Since there are 29 balls, we also have

$$a + b + c + d = 29$$

But $a \leq 7, b \leq 7, c \leq 7$, and $d \leq 7$ together imply that

$$a + b + c + d \leq 7 + 7 + 7 + 7 = 28$$

Since it is impossible that $a + b + c + d = 29$ **and** $a + b + c + d \leq 28$, we have reached a contradiction. This establishes the claim.   □

The generalization of this simple fact is one version of the pigeonhole principle.

Theorem 8   *(Pigeonhole Principle, General Version) If $m \cdot n + 1$ objects are distributed among n different boxes, then there must be some box containing at least $m + 1$ objects.*

PROOF   (By contradiction) Suppose it is possible to distribute $m \cdot n + 1$ objects into n different boxes, such that each box contains m or fewer objects. Label the n boxes $1, 2, 3, \ldots, n$. After the objects have been distributed among the boxes, define $x_1, x_2, \ldots, x_n$ by the following rule:

$$x_i = \text{ the number of objects in box } i$$

Since each object can go into only one box, we know that

$$x_1 + x_2 + x_3 + \cdots + x_n = m \cdot n + 1$$

On the other hand, we have $x_1 \leq m, x_2 \leq m, x_3 \leq m, \ldots, x_n \leq m$. However, this implies that

$$x_1 + x_2 + x_3 + \cdots + x_n \leq \underbrace{m + m + m + \cdots + m}_{n \text{ times}}$$

and thus

$$x_1 + x_2 + x_3 + \cdots + x_n \leq m \cdot n$$

The statements $x_1 + x_2 + x_3 + \cdots + x_n = m \cdot n + 1$ and $x_1 + x_2 + x_3 + \cdots + x_n \leq m \cdot n$ form a contradiction, so we conclude that no counterexample can exist, completing the proof.   ■

In light of our recent discussion, you might wonder whether "proof by contradiction" is necessary for this statement. Exercise 38 asks you to show that it is not.

Box number	0	1	2	3	4	5
Ones' digit	0	1 or 9	2 or 8	3 or 7	4 or 6	5

Table 2-18   Boxes for Example 6

Practice Problem 3   *(Basic pigeonhole principle) Use the previous proof as a guide to prove the following statement: "If n + 1 objects are distributed among n boxes, then some box must contain more than one object."*

This "basic" version of the pigeonhole principle is the form most commonly encountered in a discrete math course. Applying it is simply a matter of defining a rule for placing objects into boxes so that the conclusion "some box contains more than one object" matches the desired conclusion. Here is a typical example.

Example 6   *Show that among any seven positive integers, there are two whose sum or difference is divisible by 10.*

SOLUTION   We define six "boxes" so that each integer is placed in one box based on that number's ones' digit, as shown in Table 2-18. Now given any seven integers, when placed among these six boxes, some box will contain at least two integers by the basic pigeonhole principle. If these two integers have the same last digit, then their difference will be divisible by 10. Otherwise, by the definition of the boxes, the two numbers will have a sum that is divisible by 10.   □

Explore more on the Web.

Example 7   *Here are two different choices for the seven integers referenced in Example 6. Trace through the proof for the example by showing the placement of these integers into boxes and discussing the result.*

1. *The numbers chosen are {13, 15, 28, 30, 1, 46, 58}.*
2. *The numbers chosen are {27, 86, 50, 35, 11, 44, 108}.*

SOLUTION

1. The numbers are placed into boxes as shown in Figure 2-5. In this case, there are two numbers in the third box. These two numbers have the same ones' digit; hence, their difference, $58 - 28$, is divisible by 10.
2. They are placed into boxes as shown in Figure 2-6. In this case, there are two numbers in the fifth box. These two numbers do not have the same ones'

Figure 2-5   Illustration of boxes for Example 6.

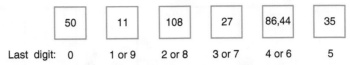

Figure 2-6 Illustration of boxes for Example 6.

digit; hence (because of the way the boxes are defined), their sum, $86 + 44$, is divisible by 10. □

The fascination that mathematicians have with the pigeonhole principle can be attributed to its relative simplicity and its utility in proving things in a variety of mathematical areas.

Proposition 9   *(Number theory) Given any four positive integers, some pair of them will have a difference divisible by* 3.

PROOF   Let four positive integers be given. Place these numbers into boxes labeled 0, 1, 2 according to the rule: A number x goes into the box labeled i if i is the remainder when x is divided by 3. By the basic pigeonhole principle, some box (let's call its label d) contains at least two numbers. Let's call these numbers a and b. Since a and b are in the box labeled d, then (by the division theorem) this means that $a = 3K + d$ for some integer K and $b = 3L + d$ for some integer L. In this case,

$$a - b = (3K + d) - (3L + d)$$
$$= 3 \cdot (K - L)$$

Since $K - L$ is an integer, this means that the difference between a and b is divisible by 3. ■

Explore more on the Web.

Practice Problem 4   *Prove that for any eleven positive integers, some pair of them will have a difference divisible by* 10.

Proposition 10   *(Geometry) Given any five points placed in a unit square, there must be two that are within $\frac{\sqrt{2}}{2}$ of each other.*

PROOF   Divide the square into four equal quadrants as shown by the dashed lines in Figure 2-7.

By the basic pigeonhole principle (thinking of the small squares as "boxes"), we know that no matter how the points are distributed, some small square will contain at least two points. Since the diagonal of each small square has length $\frac{\sqrt{2}}{2}$, these two points must be within this distance of each other. ■

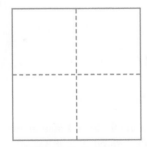

Figure 2-7 Four quadrants of a square.

Example 8   *Figure 2-8 illustrates two different choices of points within the unit square for Proposition 10. Discuss what the proof of the proposition tells you in each case.*

SOLUTION   In the first choice, points C and D are those produced by the proof. In the second, the points are A and E. Note that the proof does not necessarily produce the two points closest together. □

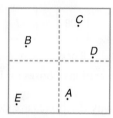
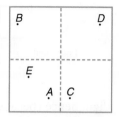

Figure 2-8   Two possible choices for five points.

Solutions to Practice Problems

1 (a) In Table 2-19, we give the starting and ending point for each type of proof of the statement "If n^2 is odd, then n is odd."

   (b) We rewrite the statement as "If n is even and n is a perfect square, then n is divisible by 4." Table 2-20 on page 147 shows the starting and ending points for a proof of this statement.

2 Since the contradiction is that m^2 is divisible by 4, the negation of the proposition's hypothesis, we try to find a proof by contrapositive. The contrapositive of the original statement is "If n is even, then n^2 is divisible by 4." To prove this, let an even integer n be given. This means that $n = 2k$ for some integer k, which implies that $n^2 = 4k^2$. Since k^2 is an integer, we conclude that n^2 is divisible by 4.

3 *Proof by contradiction.* Suppose there is a counterexample. Then $n + 1$ objects can be distributed into n boxes, with each box getting no more than one object. Label the n boxes $1, 2, 3, \ldots, n$. After the objects have been distributed among the boxes, define $x_1, x_2, \ldots, x_n$ by the following rule:

$$x_i = \text{ the number of objects in box } i$$

Since each object can go into only one box, we know that

$$x_1 + x_2 + x_3 + \cdots + x_n = n + 1$$

but since each x_i is no more than 1, we have

$$x_1 + x_2 + x_3 + \cdots + x_n \leq \underbrace{1 + 1 + 1 + \cdots + 1}_{n \text{ times}}$$

$$x_1 + x_2 + x_3 + \cdots + x_n \leq n$$

This is a contradiction, so there is no counterexample, and the theorem is true. ∎

4 *Proof.* Let 11 positive integers be given. Place these numbers into boxes labeled $0, 1, 2, 3, \ldots, 8, 9$ according to the following rule: A number x goes into the box labeled i if i is the remainder when x is divided by 10 (i.e., i is the ones' digit of x). By the basic pigeonhole principle, some box (let's call its label d) contains at least two numbers. Let's call these numbers a and b. Since a and b are in the box labeled d, then (by the division theorem) this means that $a = 10K + d$ for some integer K and $b = 10L + d$ for some integer L. In this case,

$$a - b = (10K + d) - (10L + d)$$
$$= 10 \cdot (K - L)$$

Since $K - L$ is an integer, this means that the difference between a and b is divisible by 10. ∎

	Direct	**Contrapositive**	**Contradiction**
Start	n^2 is odd.	n is even.	n^2 is odd, and n is even.
Goal	n is odd.	n^2 is even.	Some contradiction.

Table 2-19   The Structure of Proof 1 in Practice Problem 1

Exercises for Section 2.5

1. Fill in the details to complete the following proof:
⚡ **Proposition**   If an integer n has the form $3K + 1$, then it does not have the form $9L + 5$.

Proof. Suppose a counterexample to this statement does exist. Let's agree to call it m. Since m is a counterexample, it must make the hypothesis of the original statement true while making the conclusion false.

	Direct	Contrapositive	Contradiction
Start	n is even and n is a perfect square.	n is not divisible by 4.	n is even and n is a perfect square, and n is not divisible by 4.
Goal	n is divisible by 4.	n is odd or n is not a perfect square.	Some contradiction.

Table 2-20   The Structure of Proof 2 in Practice Problem 1

That is, $m =$ _____ for some integer K and $m =$ _____ for some integer L. Combining these equations gives us

$$\underline{\hspace{3cm}} = \underline{\hspace{3cm}}$$

from which it follows that $K - 3L = \frac{4}{3}$. Since _____ is an integer, this is nonsense. Therefore, _____. ∎

2. Fill in the details to complete the following proof:

⚡ **Proposition**   For all real numbers x and y, either $x \geq \frac{x+y}{2}$ or $y \geq \frac{x+y}{2}$.

Proof.   Suppose there is a counterexample to this statement. That is, real numbers x and y exist such that _____ and _____. It follows from this that

$$x + y < \underline{\hspace{4cm}}$$
$$= x + y$$

Therefore, these x and y would have to satisfy $x + y < x + y$. Clearly, no numbers can do this, so we have a contradiction. Hence, there is no counterexample to the original statement. ∎

3. Earlier in this chapter we proved the following using the contrapositive statements. Prove them now using "proof by contradiction."
   (a) If n^2 is even, then n is even.
   (b) If n^2 is odd, then n is odd.

4. Use "proof by contradiction" to prove that an even perfect square cannot have the form $4k + 2$.   (HINT: Use the previous exercise.)

5. Use the previous two exercises to prove that the sum of two odd perfect squares is not a perfect square. (This shows that no right triangle with integer sides can have both legs of odd length.)

6. Use "proof by contradiction" to explain why no integer can be both odd and even.

7. Use "proof by contradiction" to explain why it is impossible for a number n to be of the form $5K + 3$ and of the form $5L + 1$ for integers K and L.

8. Using the previous two proofs as models, develop an alternative approach to proving the uniqueness of the values q and r in the division theorem, Theorem 8 of Section 2.2.

9. Using the proof of Proposition 5 as a model, show that there are no relatively prime integers a and b satisfying $a^2 = 2b^2$. (This gives an alternate proof that $\sqrt{2}$ is irrational.)

10. Suppose we have already established the "sign rules" for multiplication of nonzero real numbers: (i) the product of a positive number and a positive number is positive; (ii) the product of a negative number and a positive number is negative; and (iii) the product of a negative number and a negative number is positive. Use the appropriate one(s) and "proof by contradiction" to prove the corresponding rules for division. (Interpret $x \div y = z$ to mean the same thing as $y \times z = x$.)
   (a) A positive number divided by a positive number is a positive number.
   (b) A negative number divided by a positive number is a negative number.
   (c) A negative number divided by a negative number is a positive number.

11. Use "proof by contradiction" to prove that if a is a rational number, then $a + \sqrt{2}$ is irrational.   (HINT: You will need the result in Exercise 9 from Section 2.2.)

12. Use "proof by contradiction" to prove that if a is a rational number and b is an irrational number, then $a + b$ is irrational.

13. Carefully write the contrapositive of the statement in the previous exercise, and then prove it without using "proof by contradiction."

14. Prove each of the following existence statements:
   (a) There exists a positive rational number r such that $6r^2 + 11r = 35$.
   (b) There exists a positive integer n such that $\frac{1}{\sqrt{n}} < 0.001$.
   (c) There exists a positive integer n such that $n \bmod 7 = 1$ and $n \bmod 3 = 2$.

15. Prove that $\sqrt{5}$ is irrational.   (HINT: Use Exercise 19 of Section 2.2.)

16. Does the fact that $\sqrt{2}$ is irrational have to be proved by contradiction? Suppose that your READER has agreed with the truth of Propositions 1 and 2 below. How would you prove Proposition 3 to her without using "proof by contradiction?"
    - **Proposition 1**: r is rational if and only if there exist relatively prime integers a and b such that $r = a/b$.
    - **Proposition 2**: If $(a/b)^2 = 2$, then a and b have 2 as a common divisor.
    - **Proposition 3**: If $r^2 = 2$, then r is not rational.

17. Prove that any positive real number r satisfying $r - \frac{1}{r} = 5$ must be irrational.

18. Not every nonexistence proof is best handled using proof by contradiction. Consider the proposition "There does not exist a largest positive integer."

    (a) The proposition is the negation of "There exists an integer n such that n is the largest integer." We know that the negation of a "there exists" statement can be written as a "for all" statement. Write the proposition in this form.

    (b) Prove the proposition you wrote in part (a).

19. The following proof is written using "proof by contradiction." Rewrite the proof so that "proof by contradiction" is not used.
    Proposition   For all integers n, if $5n + 4$ is odd, then n is odd.
    Proof.   Suppose there is a counterexample to this statement. Let's agree to call it n. This number must make the hypothesis true (i.e, $5n + 4$ is odd) and the conclusion false (i.e., n is even). Since n is even, then $n = 2K$ for some integer K, and since $5n + 4$ is odd, then $5n + 4 = 2L + 1$ for some integer L. Combining these equations gives us

$$2L + 1 = 5n + 4$$
$$= 5(2K) + 4$$
$$= 10K + 4$$

    The final equation $2L + 1 = 10K + 4$ can be rewritten as $L - 5K = 3/2$. Since $L - 5K$ must be an integer, this is a contradiction. Therefore, no counterexample can exist.   ∎

20. The following proof is written using "proof by contradiction." Rewrite the proof so that "proof by contradiction" is not used.
    Proposition For all real numbers x and y, if $x^2 + y^2 \neq 0$, then $x \neq 0$ or $y \neq 0$.
    Proof.   Suppose there is a counterexample to this statement. That is, real numbers x and y exist such that $x^2 + y^2 \neq 0$ and it is not true that "$x \neq 0$ or $y \neq 0$." This is the same as saying that $x^2 + y^2 \neq 0$ and $x = 0$ and $y = 0$. Of course, $x = 0$ and $y = 0$ implies that $x^2 + y^2 = 0^2 + 0^2 = 0$. Therefore, these x and y would have to sat-

isfy $x^2 + y^2 \neq 0$ on the one hand and $x^2 + y^2 = 0$ on the other hand. Clearly, no numbers can do both, so we have a contradiction. Therefore, no counterexample exists for the original statement.   ∎

21. Refer to Theorem 7. Prove that for all nonzero integers a and b, the smallest positive integer c in the set $S_{a,b}$ is the greatest common divisor of a and b.   (HINT: First show c divides both a and b, and then show that every common divisor of a and b must also divide c.)

22. Prove the following statement, using either the contrapositive statement or a proof by contradiction: "If the average age of four children is 10 years old, then (at least) one child is at least 10 years old."

23. Prove the following statement, using either the contrapositive statement or a proof by contradiction: "If the average net weight of seven boxes of cereal is 17 ounces, then at least one of the boxes has a net weight of at least 17 ounces."

24. Prove the following statement, using either the contrapositive statement or a proof by contradiction: "If 10 real numbers have an average of 89.63, and one of the numbers is less than 89.63, then one of the numbers must be greater than 89.63."

25. The previous three exercises are related to a slightly different form of the pigeonhole principle, often called the *average version of the pigeonhole principle*: "If z is the average of the collection of numbers

$$x_1, x_2, x_3, \ldots, x_n$$

then at least one number in the list is at least z." Prove this statement, using either the contrapositive statement or a proof by contradiction.

26. Use the average version of the pigeonhole principle from Exercise 25 to directly prove (a) the basic version of the pigeonhole principle and (b) the general version of the pigeonhole principle.

27. Prove the following statement, using either the contrapositive statement or a proof by contradiction: "If integers x, y and z satisfy $x + y + z \geq 11$, then either $x \geq 4$, $y \geq 4$ or $z \geq 5$."

28. Prove the following, using either the contrapositive statement or a proof by contradiction: "In a collection of numbers $x_1, x_2, x_3, \ldots, x_n$, if $x_1 + x_2 + x_3 + \cdots + x_n > \frac{n(n+1)}{2}$, then there is a value of i for which $x_i > i$."

    (a) By emulating the proof of Theorem 8.

    (b) By proving the contrapositive statement.

29. Prove the statement in Example 5 by considering the contrapositive statement and not using "proof by contradiction."

30. Prove the basic pigeonhole principle (Practice Problem 3) by considering the contrapositive statement (and not using "proof by contradiction").

31. Garrison Keillor signs off his news from Lake Wobegon by saying, "...where every child is above average." Your little brother says, "I don't get it." Explain to him why this is funny.

32. Use the pigeonhole principle to prove each of the following statements about numbers:

    (a) Given any seven integers, there will be two that have a difference divisible by 6.

    (b) Given any five integers, there will be two that have a sum or difference divisible by 7.

33. Fill in the details to complete the proof below:

    Proposition Given any five integers, there will be three for which the sum of the squares of those integers is divisible by 3.

    Proof. Let five integers be given. Think of two boxes, one labeled "divisible by 3" and one labeled "not divisible by 3." By the distribution version of the pigeonhole principle (with $n =$ _____ and $m =$ _____), we conclude that there are at least three of the numbers in one box. Let's refer to these three numbers as a, b, and c, and consider two cases based on which box they are in.

    ● **Case 1:** If a, b, c are in the box labeled "divisible by 3," then $a^2 + b^2 + c^2$ is divisible by 3 because ...

    ● **Case 2:** If a, b, c are in the box labeled "not divisible by 3," then by Practice Problem 4 from Section 2.2, a^2 can be written in the form _____, b^2 can be written in the form _____, and c^2 can be written in the form _____. Hence, $a^2 + b^2 + c^2$ is divisible by 3 because ...

    In either case, $a^2 + b^2 + c^2$ is divisible by 3, completing the proof. ∎

34. Use the pigeonhole principle to prove each of the following statements about numbers:

    (a) Given any seven integers, there will be four for which the sum of the squares of those integers is divisible by 4.

    (b) Show that among any 52 integers, there are two whose sum or difference is divisible by 100.

35. Use the pigeonhole principle to prove each of the following geometric statements:

    (a) Whenever five points are placed inside an equilateral triangle with sides of length 1, there must be at least two points within $\frac{1}{2}$ of each other.

    (b) Whenever nine points are placed inside a cube with sides of length 1, there must be at least two points within $\frac{\sqrt{3}}{2}$ of each other.

    (c) Let O denote the origin of the plane. Whenever five points (other than O) are placed on the plane, there will exist two points P and Q such that $\angle POQ$ is an acute angle.

36. For every integer $n \geq 1$, show that some number in $\{\pi, 2\pi, 3\pi, \ldots, n\pi\}$ is within $\frac{1}{n}$ of an integer using the following steps:

    ● Define n boxes into which we can place real numbers x that are between 0 and 1 according to the rule shown in Table 2-21.

    ● Each multiple of π is placed into these boxes according to its fractional part. (For example, the fractional part of $2\pi \approx 6.283$ is 0.283.)

    (a) What does it mean if one of these multiples is in Box 1?

    (b) If no number from $\{\pi, 2\pi, 3\pi, \ldots, n\pi\}$ is in Box 1, what does the pigeonhole principle say will happen?

    (c) If two multiples of π are in the same box, what does this mean about the difference between these multiples of π?

    (d) How does this prove the original statement?

37. Generalize the previous exercise to show that for any real number r and for every integer $n \geq 1$, some number in the set $\{r, 2r, 3r, \ldots, nr\}$ is within $\frac{1}{n}$ of an integer.

38. Write a proof of Theorem 8 that does not use "proof by contradiction."

39. Prove that in any gathering of six people, there are either three people who are mutual friends (i.e., each pair are friends) or there are three people who are mutual strangers (i.e., each pair do not know each other). This is the easiest result in a field of mathematics called Ramsey theory.

Box 1	Box 2	Box 3	$\cdots$	Box $n-1$	Box n
$0 \leq x < \frac{1}{n}$	$\frac{1}{n} \leq x < \frac{2}{n}$	$\frac{2}{n} \leq x < \frac{3}{n}$	$\cdots$	$\frac{n-2}{n} \leq x < \frac{n-1}{n}$	$\frac{n-1}{n} \leq x < 1$

Table 2-21 Definition of Boxes in Exercise 36

2.6 Excursion: Representations of Numbers

An important mathematical issue that is relevant for both computer scientists and mathematics teachers is the nature of numbers and numerals. To see that numbers and numerals are not the same thing, we need only consider the abundant examples of numeral systems other than our own Hindu-Arabic system. Almost every culture has developed or adapted some system of numerals to express a number of objects. The earliest systems usually resemble something like the tally marks that children might use to keep score in a game. The Roman numeral system is the best known system that is clearly related to tally marks. The Roman numeral V and the Hindu-Arabic numeral 5 represent the number of fingers on a person's hand. Hence, a *numeral* is a representation system for a number, and a *number* is actually an abstract notion of quantity.

Although we are all familiar with Roman numerals, few of us would ever consider trying to do arithmetic with them. If asked how many years there were between Super Bowls XIV and XXXI, we would all convert these two Roman numerals into regular Hindu-Arabic numerals and subtract with our familiar procedure, complete with regrouping in this case. It is not strictly necessary to have a numeral system to do arithmetic, but over the last eight hundred[*] years, the versatility of the Hindu-Arabic system in performing arithmetic procedures has made it the most common system in the world. In this section, we will refer to this common system as the *base ten*, or *decimal*, numeral system.

Decimal and Binary Numerals

There are many properties of arithmetic that make computations simple. Some of them really are properties of the operations, but others are actually properties of the decimal numeral system itself.

Example 1 *Which of the following properties of multiplication are true no matter what numeral system is used?*

1. *(The identity property) For all $a \in \mathbf{Z}$, $a \cdot 1 = a$.*
2. *(The commutative property) For all $a, b \in \mathbf{Z}$, $a \cdot b = b \cdot a$.*
3. *(The distributive property) For all $a, b, c \in \mathbf{Z}$, $a \cdot (b + c) = a \cdot b + a \cdot c$.*
4. *(Shifting) For all $a \in \mathbf{Z}$, $10 \cdot a$ adds a "0" on the right end of a.*

SOLUTION   Only the fourth property depends on the numeral system. Just try to make sense of that rule with Roman numerals! □

So if the shifting rule above is not really a property of multiplication, then how can we prove that it is true? To address this, we need to first give a formal definition of the decimal representation of a positive integer.

---

[*] The first "Western" book in which the Hindu-Arabic numbers were used was the book on arithmetic *Liber Abaci* published in A.D. 1202 by none other than our old friend Fibonacci. This same book includes the famous Fibonacci number sequence.

> **Definition**   Given a positive integer X, the *decimal representation* for X is a string consisting of digits from $\{0, 1, 2, 3, 4, 5, 6, 7, 8, 9\}$ that looks like $d_n d_{n-1}, \cdots, d_2 d_1 d_0$, where
>
> $$X = \sum_{i=0}^{n} d_i \cdot 10^i$$
> $$= d_n \cdot 10^n + d_{n-1} \cdot 10^{n-1} + \cdots + d_2 \cdot 10^2 + d_1 \cdot 10^1 + d_0 \cdot 10^0$$

Example 2 *Write the positive integer 24,317 in expanded form as shown in the definition. Multiply the expanded form by 10 and simplify.*

SOLUTION $24{,}317 = 2 \cdot 10^4 + 4 \cdot 10^3 + 3 \cdot 10^2 + 1 \cdot 10^1 + 7 \cdot 10^0$. *When we multiply by* 10, *we get*

$$10 \cdot (2 \cdot 10^4 + 4 \cdot 10^3 + 3 \cdot 10^2 + 1 \cdot 10^1 + 7 \cdot 10^0)$$
$$= 10 \cdot 2 \cdot 10^4 + 10 \cdot 4 \cdot 10^3 + 10 \cdot 3 \cdot 10^2 + 10 \cdot 1 \cdot 10^1 + 10 \cdot 7 \cdot 10^0$$
$$= 2 \cdot 10^5 + 4 \cdot 10^4 + 3 \cdot 10^3 + 1 \cdot 10^2 + 7 \cdot 10^1 + 0 \cdot 10^0$$
$$= 243{,}170$$

Proposition 1   *Multiplying a decimal numeral by 10 shifts the digits one place to the left and places a "0" digit on the end.*

PROOF   The proof simply repeats the steps of Example 2, but in general rather than for a particular integer. The base ten representation of a number x is given by $d_n d_{n-1} \ldots d_2 d_1 d_0$, which means

$$x = \sum_{i=0}^{n} d_i \cdot 10^i$$
$$= d_n \cdot 10^n + d_{n-1} \cdot 10^{n-1} + \cdots + d_2 \cdot 10^2 + d_1 \cdot 10^1 + d_0 \cdot 10^0$$

When this is multiplied by 10, we get

$$10x = \sum_{i=0}^{n} d_i \cdot 10^{i+1}$$
$$= d_n \cdot 10^{n+1} + d_{n-1} \cdot 10^n + \cdots + d_2 \cdot 10^3 + d_1 \cdot 10^2 + d_0 \cdot 10^1$$
$$= d_n \cdot 10^{n+1} + d_{n-1} \cdot 10^n + \cdots + d_2 \cdot 10^3 + d_1 \cdot 10^2 + d_0 \cdot 10^1 + 0 \cdot 10^0$$

Hence, $10 \cdot x$ has the base ten representation $d_n d_{n-1} \ldots d_2 d_1 d_0 0$—that is, the base ten representation of x with the digits shifted left and a 0 placed at the end.   ■

Properties of decimal numerals are hard to think about critically since it is difficult for us to think about the concept of a number like 112 without picturing the decimal numeral 112. To overcome this difficulty, we should work in other numeral systems in which we cannot take the processes or properties for granted. We will continue using a place value system—that is, a system in which there are digits whose positions reflect their values—but we will change the base to numbers other than 10. Not only does this help us understand properties of our decimal system better, it is also of practical value for computer science applications.

> **Definition**   The *base two* (a.k.a. *binary*) *representation* of a positive integer X is a string consisting of digits from $\{0, 1\}$ that looks like $b_n b_{n-1} \cdots b_2 b_1 b_0$, where
>
> $$X = \sum_{i=0}^{n} b_i \cdot 2^i$$
> $$= b_n \cdot 2^n + b_{n-1} \cdot 2^{n-1} + \cdots + b_2 \cdot 2^2 + b_1 \cdot 2^1 + b_0 \cdot 2^0$$
>
> We will refer to binary digits as *bits*, a term coined by Princeton University statistician and computer pioneer John W. Tukey (1915–2000).

When our discussion leaves some doubt as to the base being used, we will simply write the base as a subscript. For example, $(1001011)_{ten}$ represents "one million, one thousand and eleven," while the binary numeral $(1001011)_{two}$ represents a much smaller number, as we see in the next example.

 Example 3

1. *What decimal numeral represents the same number as the binary numeral $(1001011)_{two}$?*
2. *If we label the number in the first part as x, what is the binary representation for $2x$? For $2x + 1$?*

SOLUTION

1. For decimal numbers, the place values are ones, tens, hundreds, and so on— that is, $10^0, 10^1, 10^2$, and so on. For binary numbers, the place values follow the same pattern, but using powers of 2 rather than powers of 10. From Table 2-22, we conclude that $(1001011)_{two}$ represents

$$\mathbf{1} \cdot 2^6 + \mathbf{0} \cdot 2^5 + \mathbf{0} \cdot 2^4 + \mathbf{1} \cdot 2^3 + \mathbf{0} \cdot 2^2 + \mathbf{1} \cdot 2^1 + \mathbf{1} \cdot 2^0$$
$$= \mathbf{1} \cdot 64 + \mathbf{0} \cdot 32 + \mathbf{0} \cdot 16 + \mathbf{1} \cdot 8 + \mathbf{0} \cdot 4 + \mathbf{1} \cdot 2 + \mathbf{1} \cdot 1$$
$$= (75)_{ten}$$

2. As we did for decimal numerals in Example 2, we have

$$2x = 2 \cdot (\mathbf{1} \cdot 2^6 + \mathbf{0} \cdot 2^5 + \mathbf{0} \cdot 2^4 + \mathbf{1} \cdot 2^3 + \mathbf{0} \cdot 2^2 + \mathbf{1} \cdot 2^1 + \mathbf{1} \cdot 2^0)$$
$$= \mathbf{1} \cdot 2^7 + \mathbf{0} \cdot 2^6 + \mathbf{0} \cdot 2^5 + \mathbf{1} \cdot 2^4 + \mathbf{0} \cdot 2^3 + \mathbf{1} \cdot 2^2 + \mathbf{1} \cdot 2^1 + \mathbf{0} \cdot 2^0$$
$$= (10010110)_{two}$$

Bit	1	0	0	1	0	1	1
Place value	2^6	2^5	2^4	2^3	2^2	2^1	2^0
In words	Sixty-fours	Thirty-twos	Sixteens	Eights	Fours	Twos	Ones

Table 2-22   Place Values for a Binary Number

and $2x + 1$ will be

$$2x = 2 \cdot (\mathbf{1} \cdot 2^6 + \mathbf{0} \cdot 2^5 + \mathbf{0} \cdot 2^4 + \mathbf{1} \cdot 2^3 + \mathbf{0} \cdot 2^2 + \mathbf{1} \cdot 2^1 + \mathbf{1} \cdot 2^0) + 1$$
$$= \mathbf{1} \cdot 2^7 + \mathbf{0} \cdot 2^6 + \mathbf{0} \cdot 2^5 + \mathbf{1} \cdot 2^4 + \mathbf{0} \cdot 2^3 + \mathbf{1} \cdot 2^2 + \mathbf{1} \cdot 2^1 + \mathbf{1} \cdot 2^0$$
$$= (10010111)_{two}$$

The binary numeral for $2x$ is formed by shifting x's bits to the left and placing a "0" on the right; for $2x + 1$ we place a "1" on the right after doing the shift. □

Example 4 *Suppose we have found the binary representations for all natural numbers up to and including* 18 *in Table 2-23. Use this information to find the binary representation for* 19.

SOLUTION   By the division theorem (applied to the division problem $19 \div 2$), we can write $19 = 2 \cdot 9 + 1$. If we look in the chart, we find that the base two representation of 9 is 1001. Hence, the representation for 19 can be found by placing the remainder (1) onto the right side of the representation for 9, which gives us 10011. □

Practice Problem 1   *Using the chart in Table 2-23, find the binary representations for the numbers* 25, 39, *and* 50.

It might not be clear that our definition allows every number to have a binary representation, so we will take the opportunity to use induction to prove that this is so. We conclude (since every natural number has a binary representation) that the binary numeral system is just as good as the decimal system for representing numbers. Example 4 and Practice Problem 1 illustrate the crucial step of the proof.

Proposition 2   *Every natural number has a binary representation.*

PROOF   (By induction) Let $P(n)$ be the statement "n has a binary representation." Since $(0)_{two}$ represents 0, and $(1)_{two}$ represents 1, we have checked statements $P(0)$ and $P(1)$ without even trying. Now let $m \geq 2$ be given such that $P(0), P(1), \ldots, P(m-1)$ have been checked. That is, every number up to and including $m - 1$ has a binary representation. We next must consider the number m. By the division theorem, there are integers q (the quotient) and r (the remainder) such that (1) $m = 2 \cdot q + r$ and (2) r is from the set $\{0, 1\}$. Since $q < m$, then statement $P(q)$ has already been checked, so we know that q has a binary representation: let's refer to it as $b_k b_{k-1} \cdots b_2 b_1 b_0$. By this, we mean that $q = \sum_{i=0}^{k} b_i \cdot 2^i$. Now

n	Base Two	n	Base Two	n	Base Two	n	Base Two	n	Base Two
0	0	4	100	8	1000	12	1100	16	10000
1	1	5	101	9	1001	13	1101	17	10001
2	10	6	110	10	1010	14	1110	18	10010
3	11	7	111	11	1011	15	1111	19	??

Table 2-23   Binary Numerals for Numbers 1 through 18

$$m = 2 \cdot q + r$$
$$= 2 \cdot \left(\sum_{i=0}^{k} b_i \cdot 2^i \right) + r$$
$$= \left(\sum_{i=0}^{k} b_i \cdot 2^{i+1} \right) + r$$

In the above expressions, we see that multiplying q by 2 shifts the bits for q to the left and adds a "0" on the right—this is the same effect that multiplying a decimal number by 10 has in Proposition 1. Hence, when r is added, this essentially places r in the ones place. This means that the binary representation for m is $b_k b_{k-1} \cdots b_2 b_1 b_0 r$. ∎

Not only does this proposition prove that a base two representation always exists, it actually provides instructions for finding it. We can summarize the steps as follows, given the natural number n:

1. Write $n = 2q + r$ (i.e., divide n by 2, finding the quotient q and the remainder r).
2. Write down the base two representation for q.
3. Append r to the right of the string you wrote in step 2.

We used this algorithm in Example 4 and Practice Problem 1, where we already had a table of binary values. The question arises, how do we proceed in general? If we don't already know the binary representation for q, what do we do? The answer is that we apply this same process to q. To make this easier to do, we can rewrite the process as a repeating process as shown below. It is important to notice that the instructions have the unusual feature of writing the digits of the base two number **from right to left**.

Algorithm for Writing a Number in Base Two

- Input a natural number n.
- While $n > 0$, do the following:
   Divide n by 2 and get a quotient q and remainder r.
   Write r as the next (right-to-left) digit.
   Replace the value of n with q, and repeat.

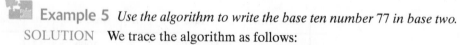

Example 5 *Use the algorithm to write the base ten number 77 in base two.*

SOLUTION   We trace the algorithm as follows:

- We begin the algorithm with $n = 77$
- $77 \div 2 = 38$ with remainder 1, so we write **1**, and let $n = 38$.
- $38 \div 2 = 19$ with remainder 0, so we write **0**, and let $n = 19$.
- $19 \div 2 = 9$ with remainder 1, so we write **1**, and let $n = 9$.

- $9 \div 2 = 4$ with remainder 1, so we write **1**, and let $n = 4$.
- $4 \div 2 = 2$ with remainder 0, so we write **0**, and let $n = 2$.
- $2 \div 2 = 1$ with remainder 0, so we write **0**, and let $n = 1$.
- $1 \div 2 = 0$ with remainder 1, so we write **1**, and let $n = 0$.
- Since $n = 0$, we quit.

Thus, the numeral formed by the algorithm is 1001101, so this is the base two numeral. To check, we can compute

$$1 \cdot 2^6 + 0 \cdot 2^5 + 0 \cdot 2^4 + 1 \cdot 2^3 + 1 \cdot 2^2 + 0 \cdot 2^1 + 1 \cdot 2^0 = 64 + 8 + 4 + 1 = 77$$

$\square$

Practice Problem 2   *Write the base ten numbers 37 and 125 in base two.*

Why should we care about properties of numbers represented in different numeral systems? Just as with the shifting property in Proposition 1, there are properties of arithmetic that can only really be understood in terms of our decimal numeral system. An interesting one is the well-known test for divisibility by 9:

A number is divisible by 9 if and only if the sum of its digits is divisible by 9.

As you will see in the exercises, this statement is not true when the number is written as a numeral using a base other than ten. We will show how a number is related to the sum of its decimal digits, and we will leave the actual explanation of the rule above as a practice problem.

Proposition 3   *If the digits of the decimal representation of a positive integer x sum to s, then $x - s$ is divisible by 9.*   (HINT: *Exercise 5 from Section 2.4 proves that $10^n - 1$ is divisible by 9 for all $n \geq 0$.)*

PROOF   The base ten representation of a number x is given by the string of digits $d_n d_{n-1} \cdots d_2 d_1 d_0$, which means

$$x = \sum_{i=0}^{n} d_i \cdot 10^i$$
$$= d_n \cdot 10^n + d_{n-1} \cdot 10^{n-1} + \cdots + d_2 \cdot 10^2 + d_1 \cdot 10^1 + d_0 \cdot 10^0$$

On the other hand, if we use the letter s to stand for the sum of these digits, then we have

$$s = \sum_{i=0}^{n} d_i = d_n + d_{n-1} + \cdots + d_2 + d_1 + d_0$$

We can put these two facts together to get

$$x - s = \sum_{i=0}^{n} d_i \cdot 10^i - \sum_{i=0}^{n} d_i$$
$$= \sum_{i=0}^{n} d_i \cdot \left(10^i - 1\right)$$

Since $10^i - 1$ is divisible by 9 for each of the values of i in the above sum, it follows that $x - s$ is divisible by 9.   ■

Example 6 *Trace the above proof with the value $x = 1,934$.*

SOLUTION   Using the expanded base ten representation, we have $x = 1,934 = 1 \cdot 10^3 + 9 \cdot 10^2 + 3 \cdot 10^1 + 4 \cdot 10^0$ and the sum of the digits is $s = 1 + 9 + 3 + 4$, so

$$
\begin{aligned}
x - s &= \left(1 \cdot 10^3 + 9 \cdot 10^2 + 3 \cdot 10^1 + 4 \cdot 10^0\right) - (1 + 9 + 3 + 4) \\
&= 1 \cdot \left(10^3 - 1\right) + 9 \cdot \left(10^2 - 1\right) + 3 \cdot \left(10^1 - 1\right) + 4 \cdot \left(10^0 - 1\right) \\
&= 1 \cdot 999 + 9 \cdot 99 + 3 \cdot 9 + 4 \cdot 0
\end{aligned}
$$

Since each term in the sum is divisible by 9, then $x - s$ is divisible by 9.   □

Proposition 4   *A natural number is divisible by 9 if and only if the sum of its (decimal) digits is divisible by 9.*

PROOF   Since this statement has an "if and only if" form, we must actually prove two things:

- **Claim 1:** If a natural number X is divisible by 9, then the sum of its digits is divisible by 9.

  Proof. Let a natural number X that is divisible by 9 be given. This means that $X = 9K$ for some integer K. We will use S to denote the sum of the decimal digits of X. Proposition 3 tells us that $X - S$ is divisible by 9, which means that $X - S = 9L$ for some integer L. Combining these facts tells us that

  $$
  \begin{aligned}
  S &= X - (X - S) \\
  &= 9K - 9L \\
  &= 9(K - L)
  \end{aligned}
  $$

  Since $K - L$ is an integer, we can conclude that S is divisible by 9.
- **Claim 2:** If the sum of the digits of a natural number X is divisible by 9, then X is divisible by 9.

  Proof. See the practice problem below.

  ■

Practice Problem 3   *Prove that if the sum of the digits of a natural number X is divisible by 9, then X is divisible by 9.*

Numbers in Other Bases

We have discussed our usual decimal representation (using the digits 0 through 9), and the binary representation (using the digits 0 and 1). These ideas can be generalized to other bases. For example, in base five we would use the digits $0, 1, 2, 3, 4$, and place values would be powers of 5. In general, a natural number x has the base b representation

$$
x = d_n d_{n-1} \cdots d_2 d_1 d_0
$$

where each of the digits is from the set $\{0, 1, 2, \ldots, b - 1\}$. The exact meaning of this place value numeral is

$$x = \sum_{i=0}^{n} d_i \cdot b^i$$

 Example 7 *Write the base ten number for each of the following:*

1. $(412)_{five}$

2. $(2046)_{seven}$

3. $(4011)_{three}$

SOLUTION

1. $4 \cdot 5^2 + 1 \cdot 5^1 + 2 \cdot 5^0 = 4 \cdot 25 + 1 \cdot 5 + 2 \cdot 1 = 107$
2. $2 \cdot 7^3 + 0 \cdot 7^2 + 4 \cdot 7^1 + 6 \cdot 7^0 = 2 \cdot 343 + 4 \cdot 7 + 6 \cdot 1 = 720$
3. 4 *is not a legal digit for a base three numeral.*

☐

Proposition 5   *Let any integer* $b \geq 2$ *be given. Every natural number has a representation in base b.*

PROOF   The proof is left as Exercise 23. It is an induction proof almost identical to that for Proposition 2. ■

Since the proof for a general base b is almost identical to the proof for base two, you should not be surprised to find out that the proof also yields an algorithm for conversion to base b. Moreover, that algorithm is essentially the same, but we divide by b rather than by 2 in each step.

 Example 8 *Write the base ten number* 1,964 *in base eight.*

SOLUTION   We trace the algorithm (dividing by 8 rather than 2) as follows:

- We begin the algorithm with $n = 1,964$ and $i = 0$.
- $1,964 \div 8 = 245$ with remainder 4, so we write **4**, and let $n = 245$.
- $245 \div 8 = 30$ with remainder 5, so we write **5**, and let $n = 30$.
- $30 \div 8 = 3$ with remainder 6, so we write **6**, and let $n = 3$.
- $3 \div 8 = 0$ with remainder 3, so we write **3**, and let $n = 0$.
- Since $n = 0$, we quit.

Thus, the numeral $d_3 d_2 d_1 d_0$ formed by the algorithm is 3,654, so this is the base eight numeral. To check, we can compute

$$3 \cdot 8^3 + 6 \cdot 8^2 + 5 \cdot 8^1 + 4 \cdot 8^0 = 1,964$$

☐

Practice Problem 4   *Write the base ten number* 1,992 *in base five.*

Hexadecimal Numerals in Computer Science

Binary representation are the most important in computer science, but if you look at machine-level addresses and values, you will often see mysterious strings of digits and letters instead of zeroes and ones. When you see these strings, you are looking at a higher representation of the binary numerals that represent the lowest level of machine representation. This higher representation is called the *hexadecimal*, or base sixteen, system, and we will see shortly how these numbers stand in for binary numerals.

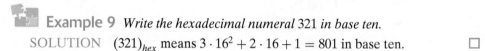 **Example 9** *Write the hexadecimal numeral 321 in base ten.*

SOLUTION $(321)_{hex}$ means $3 \cdot 16^2 + 2 \cdot 16 + 1 = 801$ in base ten. □

Hexadecimal (base sixteen) is strange in that its base is larger than our decimal base of ten. This means that the digits from Proposition 5 must be from the set of numbers from zero to 15. But we cannot use 15 as a single digit, for example, since $(15)_{hex}$ represents 1 in the sixteens' place and 5 in the ones' place, which is the decimal numeral 21. To overcome this difficulty, we introduce some symbols whose use as digits will not cause this confusion. We will use the letters A, B, C, D, E, and F to represent the numbers 10, 11, 12, 13, 14, and 15, respectively.

Example 10 *Write the decimal numeral 8,940 as a hexadecimal numeral.*

SOLUTION We can simply follow the algorithm.

- We begin the algorithm with $n = 8,940$.
- $8,940 \div 16 = 558$ with remainder 12, so we write **C** and set $n = 558$.
- $558 \div 16 = 34$ with remainder 14, so we write **E** (to the *left* of the previous digit) and set $n = 34$.
- $34 \div 16 = 2$ with remainder 2, so we write **2** (to the *left* of the previous digit) and set $n = 3$.
- $2 \div 16 = 0$ with remainder 2, so we write **2** (to the *left* of the previous digit) and set $n = 0$.
- Since $n = 0$, we quit.

We conclude that the hexadecimal numeral $22EC$ represents the decimal numeral 8,940. □

Practice Problem 5 *Which of the following (decimal) numbers make an English word when you write their hexadecimal representations?*

(a) *4,013*

(b) *65,261*

(c) *700,638*

As we mentioned before, the main benefit of hexadecimal numerals is that they actually stand in for binary numerals when the latter would be long and cumbersome. Before we continue that discussion, we recap what we know so far about binary numerals and hexadecimal digits in Table 2-24. Note that the set of hexadecimal digits corresponds exactly to the set of four-bit binary numerals. With current technology, a common computer configuration has internal words consisting of 32-bit binary numerals. If you look at a binary numeral of this length, it can be rather daunting:

Decimal	0	1	2	3	4	5	6	7	8	9	10	11	12	13	14	15
Binary	0000	0001	0010	0011	0100	0101	0110	0111	1000	1001	1010	1011	1100	1101	1110	1111
Hexadecimal	0	1	2	3	4	5	6	7	8	9	A	B	C	D	E	F

Table 2-24 Binary-Hexadecimal Conversion

$$10110100010010100011001110010111$$

By breaking it into eight four-bit pieces, it is a little easier to read:

$$1011 \quad 0100 \quad 0100 \quad 1010 \quad 0011 \quad 0011 \quad 1001 \quad 0111$$

and when we realize that each of these four-bit pieces corresponds to a single hexadecimal digit, the representation improves even more:

$$\underbrace{1011}_{B}\underbrace{0100}_{4}\underbrace{0100}_{4}\underbrace{1010}_{A}\underbrace{0011}_{3}\underbrace{0011}_{3}\underbrace{1001}_{9}\underbrace{0111}_{7}$$

Hence, the hexadecimal representation of

$$(10110100010010100011001110010111)_{two}$$

is $(B44A3397)_{hex}$. Since this gives us a quick and easy procedure for changing between hexadecimal and binary numerals, it is clear why, even when the real representation of an address or value is stored as a binary numeral, a human being would prefer to look at the hexadecimal representation.

 **Example 11** *Convert the binary numeral* $(111011110110010101111)_{two}$ *to its hexadecimal representation.*

SOLUTION Since every block of four bits in the 21-bit string 111011110110010101111 represents a single hexadecimal digit, we will need to pad the left side of the binary representation with three leading 0's in order for the total number of bits to be a multiple of 4. This will not change the value of the numeral.

$$\underbrace{0001}_{1}\underbrace{1101}_{D}\underbrace{1110}_{E}\underbrace{1100}_{C}\underbrace{1010}_{A}\underbrace{1111}_{F}$$

By blocking the bits in fours and converting each block of four bits to the corresponding hexadecimal digit, we see that the hexadecimal representation will be $1DECAF$.   □

Practice Problem 6   *Convert the hexadecimal numeral* $(2FACED)_{hex}$ *to binary and then to decimal.*

Some Mathematical Applications of Binary Numbers

Characterization of the solution to the Josephus problem

Recall the Josephus problem (as described in [32]) from Section 1.1 of this book:

> *In the Jewish revolt against Rome, Josephus and 39 of his comrades were holding out against the Romans in a cave. With defeat imminent, they resolved that, like the rebels at Masada, they would rather die than be slaves to the Romans. They*

decided to arrange themselves in a circle. One man was designated as number one, and they proceeded clockwise killing every seventh man. . . . Josephus was among other things an accomplished mathematician; so he instantly figured out where he ought to sit in order to be the last to go. But when the time came, instead of killing himself he joined the Roman side.

For a binary numeral $b_k b_{k-1} \ldots b_2 b_1 b_0$, we define the "cyclic left shift" to be the result of moving the leftmost bit b_k to the rightmost place. For example, the cyclic left shift of the binary numeral 1010011 is the binary numeral 0100111. If we form Table 2-25 showing the result of performing this operation on the binary numerals that represent the first 10 positive integers, a familiar pattern emerges. If possible, we would like to find some algebraic "closed formula" that will directly relate the numbers in the top row to the numbers in the bottom row of this table. If we think about what the cyclic left shift does arithmetically, this is a fairly simple process that we can study in a specific example before we state it more generally. Let's use the example $n = 11$ since it would be the next value to be entered into the table above.

- $(11)_{ten} = (1011)_{two}$
- Removing the leftmost 1 from the binary numeral 1011 would give us 011. This is the same thing as subtracting from 11 the largest power of 2 that is less than or equal to 11. In this case, $11 - 2^3 = 3$, and 011 is the binary numeral for 3.
- Placing this 1 on the right of the remaining string is actually done in two steps:
  1. First shifting the string to the left. In our example, 011 shifted one place to the left would give us 0110. This is equivalent to multiplying the number for 011 by 2. In this case, $2 \cdot 3 = 6$, and 0110 really is the binary numeral representing 6.
  2. After shifting these bits one place to the left, we can simply add 1 to put a 1 digit on the rightmost end of the string, to give us 0111. Adding 1 to 6 gives us 7, which is indeed the number represented by the binary numeral 0111.
- Hence, the result of a cyclic left shift on the binary numeral representing 11 is the binary numeral representing 7.

This example can be generalized to give us a formula for the result of a cyclic left shift.

Proposition 6   *Given a positive integer number n, find the largest value of k for which $2^k \le n$, and the number in the bottom row will be given by the formula $2 \cdot (n - 2^k) + 1$.*

PROOF   Given any positive integer n and the number k as defined above,

Value of n	1	2	3	4	5	6	7	8	9	10
Binary representation of n	1	10	11	100	101	110	111	1000	1001	1010
Cyclic left shift	1	01	11	001	011	101	111	0001	0011	0101
Decimal after shift	1	1	3	1	3	5	7	1	3	5

Table 2-25   Cyclic Left Shift of Binary Numerals

- Removing the leftmost 1 from the binary numeral for n is the same thing as subtracting 2^k from n, where 2^k is the largest power of 2 that is less than n.
- Placing this 1 on the right of the remaining string is done by first shifting the string to the left (which we have seen is done by multiplying the number by 2) and adding 1.

■

Example 12 *Test the formula for $n = 10$, $n = 20$, and $n = 30$, and verify (using the binary numbers) that it works for these values.*

SOLUTION   For $n = 10$, the formula predicts that the cyclic left shift will yield the number $2 \cdot (10 - 2^3) + 1 = 5$, which is correct according to the table above. For $n = 20$, the formula predicts that the cyclic left shift will yield the number $2 \cdot (20 - 2^4) + 1 = 9$, which is correct because $(20)_{ten} = (10100)_{two}$, and the cyclic left shift on $(10100)_{two}$ yields $(01001)_{two} = (9)_{ten}$. For $n = 30$, the formula predicts that the cyclic left shift will yield the number $2 \cdot (30 - 2^4) + 1 = 29$, which is correct because $(30)_{ten} = (11110)_{two}$, and the cyclic left shift on $(11110)_{two}$ yields $(11101)_{two} = (29)_{ten}$. $\square$

Theorem 7   *In the Josephus game with n players in which every second person is eliminated, the number of the last remaining person is $2 \cdot (n - 2^k) + 1$, where k is the largest integer for which $2^k \le n$.*

PROOF   In Example 8 of Section 2.4, we proved by induction that for every integer $m \ge 0$, "In the Josephus game with 2^m players in which every second person is eliminated, the first person passed over (person 1 in the given labeling of the circle) will be the last one left at the end." Given this result, it is easy to find a formula for the last person left at the end given any starting number n. Let a positive integer n be given, and let k be the largest number such that $2^k \le n$. Eliminate people $2, 4, 6, \ldots$ until 2^k people remain. You will have eliminated exactly $n - 2^k$ people. This number must be less than half the total you started with, so all the eliminated people have even position numbers. Hence, the list of people eliminated looks like

$$2, 4, 6, \ldots, 2 \cdot \left(n - 2^k\right)$$

At this point, it looks like the game starting with 2^k people in a circle, and the first one you are going to let live is the very next one, person $2 \cdot \left(n - 2^k\right) + 1$. By the example cited above, this same person will be the last one left at the end of this game.   ■

Corollary 8   *In the Josephus game with n players in which every second person is eliminated, the number of the last remaining person can be found by performing a cyclic left shift on the binary representation of n.*

Perfect shuffles

The so-called *perfect shuffle* really refers to two separate permutations (rearrangements) that can be performed on a packet of cards. In order to make this shuffle fairly easy to do, we will use only one suit from our deck, so assume we have a packet of 13 cards. The basic premise of the perfect shuffle is that the packet is divided in half and then the cards are perfectly interwoven, with the original top card either staying on top (called an *out-shuffle*) or moving to the second position from the top (called an *in-shuffle*). See Figure 2-9 for clarification. The only further issue is the meaning of "divide in half" for a packet containing an odd number of cards. We will follow the convention of magicians by always dividing our packet so that the resulting shuffle has the larger half straddling the smaller half.

The perfect shuffle is a component of many card tricks performed by sleight-of-hand artists, and it is certainly the magic trick most thoroughly studied by mathematicians. Brent Morris's book [39] offers a rich amalgamation of math results and magic history. In particular, using the notation **I** for in-shuffle and **O** for out-shuffle on a packet of 13 cards, Morris provides the following basic facts, which can be verified by the reader on at least the examples shown in Figure 2-9.

Proposition 9   *Assume the card positions are labeled 0, 1, 2, 3, 4, 5, 6, 7, 8, 9, 10, 11, 12, with 0 denoting the top card.*

*1. The card initially in position p is in position **I**(p) after the in-shuffle, where*

$$\mathbf{I}(p) = (2p + 1) \bmod 13 \tag{2.4}$$

*2. The card initially in position p is in position **O**(p) after the out-shuffle, where*

$$\mathbf{O}(p) = (2p) \bmod 13 \tag{2.5}$$

Because each of these shuffles corresponds to simple arithmetic operations on binary numerals, we can use the binary representations of position numbers to predict the movements of individual cards. As an example, we will show how to obtain a sequence of shuffles to move the top card of the packet of 13 to any desired position in the packet. Use $6 = (110)_{two}$ as an example. As we go through the shuffles, we will keep track of the original top card's position using its binary representation.

1. The position $(0)_{two}$ card (the top) is moved to position $(1)_{two}$ via an **I** shuffle.
2. The position $(1)_{two}$ card is moved to position $(11)_{two}$ via an **I** shuffle.
3. The position $(11)_{two}$ card is moved to position $(110)_{two}$ via an **O** shuffle.

Hence, after three appropriate shuffles, the card that was originally on top can be moved to position six (which is seventh from the top) in the packet.

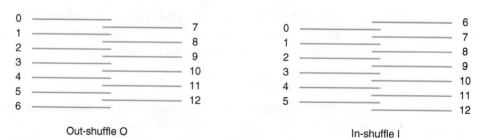

Out-shuffle O   In-shuffle I

Figure 2-9   The out- and in-shuffles with an odd number of cards.

Theorem 10 *In a packet of 13 cards, to move a card from position 0 (the top) to position k, one must do in-shuffles (**I**) and out-shuffles (**O**) according to the binary number for k, using an **O** shuffle for each 0 and an **I** shuffle for each 1, reading the binary number from left to right.*

PROOF   Simply emulate the example above for a given position number n. ∎

Practice Problem 7   *(Do a magic trick!) Get a packet of 13 cards of the same suit with the ace on top. Ask a friend to name a number between 1 and 10. Do the appropriate sequence of in-shuffles and out-shuffles* to move the top card to that position in the pack. Have your friend take the pack and move cards from top to bottom, counting each card until his or her chosen number has been reached. The top card on the packet is now the ace!*

Solutions to Practice Problems

1   $25 = 2 \cdot 12 + 1$, so since $(12)_{ten} = (1100)_{two}$, then adding a "1" digit to the right side of this gives us $(25)_{ten} = (11001)_{two}$. $39 = 2 \cdot 19 + 1$, so since $(19)_{ten} = (10011)_{two}$, then adding a "1" digit to the right side of this gives us $(39)_{ten} = (100111)_{two}$. $50 = 2 \cdot 25 + 0$, so since $(25)_{ten} = (11001)_{two}$, then adding a "0" digit to the right side of this gives us $(50)_{ten} = (110010)_{two}$.

2   100101 and 1111101.

3   *Proof.*   Let a natural number X be given such that the sum S of the decimal digits of X is divisible by 9. This means that $S = 9k$ for some integer k. Proposition 3 tells us that $X - S$ is divisible by 9, which means that $X - S = 9m$ for some integer m. Combining these facts tells us that

$$X = (X - S) + S$$
$$= 9m + 9k$$
$$= 9(m + k)$$

Since $m + k$ is an integer, we can conclude that S is divisible by 9. ∎

4   We trace the algorithm with $b = 5$ as follows:

- We begin the algorithm with $n = 1,992$.
- $1,992 \div 5 = 398$ with remainder 2, so we write 2 and set $n = 398$.
- $398 \div 5 = 79$ with remainder 3, so we write 3 (to the left of the previous digit) and set $n = 79$.
- $79 \div 5 = 15$ with remainder 4, so we write 4 (to the left of the previous digit) and set $n = 15$.
- $15 \div 5 = 3$ with remainder 0, so we let $d_3 = 0$ (to the left of the previous digit) and set $n = 3$.
- $3 \div 5 = 0$ with remainder 3, so we let $d_4 = 3$, $n = 0$, and $i = 5$.
- Since $n = 0$, we quit.

We conclude that $(1992)_{ten} = (30432)_{five}$.

5   They all do (sort of)!

(a) $(4013)_{ten} = (FAD)_{hex}$

(b) $(65261)_{ten} = (FEED)_{hex}$

(c) $(700638)_{ten} = (AB0DE)_{hex}$

6   Using the table

$$\underbrace{0010}_{2}\ \underbrace{1111}_{F}\ \underbrace{1010}_{A}\ \underbrace{1100}_{C}\ \underbrace{1110}_{E}\ \underbrace{1101}_{D}$$

and dropping the unnecessary leading zeroes, we see that

$$(2FACED)_{hex} = (101111101011001110 1101)_{two}$$

Using the meaning of the hexadecimal digits, we can compute

$$(2FACED)_{hex} = 2 \cdot (16)^5 + 15 \cdot (16)^4 + 10 \cdot (16)^3 +$$
$$12 \cdot (16)^2 + 14 \cdot (16)^1 + 13 \cdot (16)^0$$
$$= (3124461)_{ten}$$

---

* If you cannot master the perfect shuffle with a small packet, here is a way to fake it. Divide the packet as evenly as possible and give one pack to your friend. Holding your packets face *down*, take turns dealing your cards off the *bottom* into a single pile on the table. You will have to give some thought to the division of the packet so that the person with the bigger pack always starts the deal and the deal results in the correct shuffle, but a little practice should make this possible.

Exercises for Section 2.6

1. Write the following decimal numerals using their binary representations:
   (a) 35
   (b) 125
   (c) 123
   (d) 1,024

2. Write the following decimal numerals using their base five representations:
   (a) 35
   (b) 125
   (c) 123
   (d) 1,024

3. Write the following decimal numerals using their base eight* representations:
   (a) 35
   (b) 125
   (c) 123
   (d) 1,024

4. Write the following numerals, each given in the indicated bases, in their base ten representations:
   (a) $(35)_{eight}$
   (b) $(125)_{eight}$
   (c) $(1010)_{two}$
   (d) $(1024)_{five}$

5. Use the shortcut method of Example 11 to convert each of the following numerals from hexadecimal to binary or vice versa, whichever is appropriate:
   (a) $(DAD)_{sixteen}$
   (b) $(1F0B)_{sixteen}$
   (c) $(EFEF0707)_{sixteen}$
   (d) $(11001)_{two}$
   (e) $(1011000011011110)_{two}$
   (f) $(11001010111111101111000000001101)_{two}$

6. The method of Example 11 can be adapted to convert between base eight (octal) numerals and binary representations using Table 2-26 for individual octal digits. Use this adapted method to convert each of the following numerals from octal to binary or vice versa, whichever is appropriate:
   (a) $(17)_{eight}$
   (b) $(2005)_{eight}$
   (c) $(24601)_{eight}$
   (d) $(11001)_{two}$
   (e) $(1011000011011110)_{two}$
   (f) $(11001010111111011110101)_{two}$

7. Use the ideas from Exercises 5 and 6 to convert each of the following numerals from octal to hexadecimal or vice versa, whichever is appropriate. (HINT: As an intermediate step, you should produce the equivalent binary numeral in each case.)
   (a) $(17)_{eight}$
   (b) $(1245)_{eight}$
   (c) $(13570246)_{eight}$
   (d) $(DAD)_{sixteen}$
   (e) $(111111)_{sixteen}$
   (f) $(1A2B3C4D)_{sixteen}$

8. Prove that a decimal number is divisible by 3 if and only if the sum of its digits is divisible by 3.

9. What can the ones digit be of a perfect square written in base eight?

10. What can the ones digit be of a perfect square written in base five?

11. Which of the following decimal numbers is divisible by 9?
   (a) 765
   (b) 1,234
   (c) 123,678
   (d) 1,909,876

12. Prove that if a number n is represented as $d_4d_3d_2d_1d_0$ in base b, then the product $n \cdot b$ is represented as $d_4d_3d_2d_1d_00$ in base b.

13. Demonstrate that the following rule is not true of binary numerals: "If a number is divisible by 3, then the sum of that number's binary digits is divisible by 3."

Decimal	0	1	2	3	4	5	6	7
Binary	000	001	010	011	100	101	110	111
Octal	0	1	2	3	4	5	6	7

Table 2-26 Binary-Octal Conversion

---

*Base eight is often called *octal*.

14. Demonstrate that the converse of the rule in Exercise 13 is also not true for binary numerals.

15. Demonstrate that the following rule is not true of base eight (octal) numerals: "If a number is divisible by 3, then the sum of that number's octal digits is divisible by 3."

16. Demonstrate that the converse of the rule in Exercise 15 is also not true for base eight numerals.

17. Prove the following rule for base eight (octal) numerals: "A number is divisible by 7 if and only if the sum of its octal digits is divisible by 7."

18. Refine the proof of Proposition 4 to show that $x \bmod 9 = s \bmod 9$ when x is any natural number and s is the sum of the decimal digits of x.

19. Hexadecimal comes up frequently in computer science applications. For example, colors on a Web page are typically expressed using RGB values. If you open "Notepad" in Windows (or whatever text editor your computer might have) and type the following single line in a file that is then saved as "colortest.html," you will see colors when you open this file with your browser:

```
< BODY BGCOLOR = #777700 >
```

(a) Try using the RGB value FF0000, 00FF00, and 0000FF to see all red, all green, and all blue.

(b) Can you make the background of your screen yellow?

(c) How many different colors is it possible to express in this way?

20. Write the following decimal numbers in base twelve. (You will need to invent symbols for the numbers 10 and 11 for your system.)

(a) 7,369

(b) 1,427

(c) 553

(d) 14

21. Here is a magic trick that uses binary numbers. Choose any number you see on any of the cards in Figure 2-10. Notice that most numbers are on several cards. For each of the five cards, if your number is on the card, circle the number in the upper left corner of that card. Now sum the circled numbers, and the sum will reveal your chosen number! Why does this trick work?

22. Design a magic trick like the one above taking advantage of base three representations of numbers.

23. Prove Proposition 5.

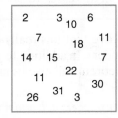

Figure 2-10 Binary magic cards.

2.7 Excursion: Modular Arithmetic and Cryptography

In Section 2.2, we examined properties of integers related to divisibility and remainders. The point there was to provide some familiar mathematical territory as the starting point for our journey into formal definitions and mathematical writing. It turns out that the ideas of divisibility and remainders actually can be used to formulate a whole new system of arithmetic with properties of interest in their own right. It might be surprising to some that there can be "whole new systems of arithmetic" out there, but it is nothing to be worried about. The new systems (yes, there are more than one) have many connections to the arithmetic we are familiar with.[*]

We first establish a new notion of "equivalence" for integers where we use the notation

$$a \equiv_n b$$

to denote that a and b have the same remainder on division by the positive integer n. In the notation of Section 2.2, this means that $a \bmod n = b \bmod n$.

This connects our discussion with the division theorem, but to be honest, that might not be the most useful way of thinking about this relationship. We give a less intuitive but more useful definition instead.

Definition For all integers a, b and n with $n \geq 1$, $a \equiv_n b$ means that $a - b$ is divisible by n. In this case, we say that a and b are *congruent* mod n.

We will first prove that this really does mean the same thing as the statement above about remainders. In addition to giving us two ways to think about this sort of equivalence, this theorem will give us more practice in mathematical writing.

Theorem 1 *For integers a, b and n with n $\geq$ 1, a mod n = b mod n if and only if n divides a − b.*

  PROOF   To establish this, we must prove each of the following propositions:

1. **Proposition.** If a and b have the same remainder on division by n, then n divides $a - b$.
   Proof. Let integers a, b and $n \geq 1$ be given, and suppose that a and b have the same remainder r on division by n. This means that for some integers q_1 and q_2, we have $a = n \cdot q_1 + r$ and $b = n \cdot q_2 + r$. But this means that

$$a - b = (n \cdot q_1 + r) - (n \cdot q_2 + r)$$
$$= n \cdot (q_1 - q_2)$$

   which means that n divides $a - b$.

2. **Proposition.** If n divides $a - b$, then a and b have the same remainder on division by n.

---

[*] The general study of these systems is the starting point for the mathematical field of *abstract algebra*.

Proof. Let a, b and $n \geq 1$ be given, and suppose that n divides $a - b$. By the division theorem (Theorem 8 of Section 2.2), there are integers q_1 and r_1, q_2 and r_2 such that

$$a = n \cdot q_1 + r_1, \quad \text{with } 0 \leq r_1 < n$$
$$b = n \cdot q_2 + r_2, \quad \text{with } 0 \leq r_2 < n$$

So we have

$$a - b = (n \cdot q_1 + r_1) - (n \cdot q_2 + r_2)$$
$$= n \cdot (q_1 - q_2) + (r_1 - r_2)$$

Now the fact that $a - b$ is divisible by n implies from this last equation that $r_1 - r_2$ is divisible by n. Because each of r_1 and r_2 is from the set $\{0, 1, 2, \ldots, n - 1\}$, this can only happen if $r_1 - r_2 = 0$. This means that a and b have the same remainder on division by n. ∎

The right way to think of this is to imagine the n being fixed in advance, and having the $\equiv_n$ equivalence applied to pairs of integers. In this way, the integers sort themselves into coherent cohorts as the following example illustrates.

Example 1 *Let $n = 6$. All numbers in the same row of the table below are congruent mod 6. For example, 20 and 2 are in the same row because both have remainder 2 on division by 6, **and** because $20 - 2 = 18$ is evenly divisible by 6. Notice that every integer will appear in one of the six rows because the division theorem guarantees that every integer when divided by 6 will have a remainder from $\{0, 1, 2, 3, 4, 5\}$, and there is one row of numbers for each of these values.*

$$\ldots, -18, -12, -6, \mathbf{0}, 6, 12, 18, \ldots$$
$$\ldots, -17, -11, -5, \mathbf{1}, 7, 13, 19, \ldots$$
$$\ldots, -16, -10, -4, \mathbf{2}, 8, 14, 20, \ldots$$
$$\ldots, -15, -9, -3, \mathbf{3}, 9, 15, 21, \ldots$$
$$\ldots, -14, -8, -2, \mathbf{4}, 10, 16, 22, \ldots$$
$$\ldots, -13, -7, -1, \mathbf{5}, 11, 17, 23, \ldots$$

In this example, we consider all the integers in the same row as being *equivalent* in this new sense. This is certainly a different meaning of equivalence from the usual one, but we still have not seen how this constitutes a new system of arithmetic. Let's experiment a bit with the table above to see what "mod 6 arithmetic" is all about.

Example 2 *We will refer to each row in the previous table by the remainder associated with the numbers in that row. Investigate the result of adding or multiplying numbers, one from row 2 and the other from row 3.*

SOLUTION Since there are many numbers in row 2 and many numbers in row 3, we should try addition and multiplication with different choices of representatives to see what difference it makes. A few such trials are shown in Table 2-27. It seems that no matter which two representatives from row 2 and row 3 we choose, their sum will be in row 5 and their product will be in row 0. Thus, we can sensibly say that "row 2 plus row 3 equals row 5" and "row 2

Choice from Row 2	Choice from Row 3	Sum?	Row?	Product?	Row?
8	9	17	5	72	0
8	−15	−7	5	−120	0
2	−15	−13	5	−30	0
14	3	17	5	42	0

Table 2-27 Solution for Example 2

times row 3 equals row 0," or more simply,

$$2 + 3 \equiv_6 5$$
$$2 \cdot 3 \equiv_6 0$$

This gives us our new system of arithmetic on the numbers $\{0, 1, 2, 3, 4, 5\}$. □

We can extend this example to build entire addition and multiplication tables for our new mod 6 arithmetic.

Example 3 *Figure 2-11 shows the addition and multiplication tables for* mod 6 *arithmetic.*

Practice Problem 1 *Give the complete multiplication table for* mod 5 *arithmetic. In what way are the rows in your table different from the rows in the* mod 6 *multiplication table in Example 3?*

The key to these tables being well-defined is the observation in Example 2 that the row number of the "output" of a sum or product does not depend on which representatives are used for the "inputs." Since our main purpose here is to practice mathematical writing, we will prove this behavior below and in Exercise 2 at the end of this section.

Theorem 2 *Consider an integer $n \geq 2$. For all integers a, b, c, and d, if $a \equiv_n b$ and $c \equiv_n d$, then*

1. $(a \cdot c) \equiv_n (b \cdot d)$, and
2. $(a + c) \equiv_n (b + d)$.

PROOF   Let a, b, c, d and $n \geq 2$ be given, and assume that $a \equiv_n b$ and $c \equiv_n d$. (We will prove the first conclusion here and leave the proof of the second conclusion to Exercise 2.) By Theorem 1, both $a - b$ and $c - d$ are divisible by n. This

+	0	1	2	3	4	5
0	0	1	2	3	4	5
1	1	2	3	4	5	0
2	2	3	4	5	0	1
3	3	4	5	0	1	2
4	4	5	0	1	2	3
5	5	0	1	2	3	4

·	0	1	2	3	4	5
0	0	0	0	0	0	0
1	0	1	2	3	4	5
2	0	2	4	0	2	4
3	0	3	0	3	0	3
4	0	4	2	0	4	2
5	0	5	4	3	2	1

Figure 2-11   Arithmetic mod 6.

means that $a - b = K \cdot n$ and $c - d = L \cdot n$ for some integers K and L. Rewriting these equations as $a = b + K \cdot n$ and $c = d + L \cdot n$, respectively, allows us to substitute and expand:

$$a \cdot c = (b + K \cdot n)(d + L \cdot n)$$
$$= b \cdot d + (b \cdot L + d \cdot K + n \cdot K \cdot L) \cdot n$$

This means that $a \cdot c - b \cdot d = $ (an ugly integer) $\cdot n$, so $a \cdot c - b \cdot d$ is divisible by n. We conclude (using Theorem 1 again) that $(a \cdot c) \equiv_n (b \cdot d)$. ∎

This fact alone can be used for some amusing math puzzles. A little later on we will see a more practical side to these ideas.

n	3^n	Remainder mod 10
1	3	3
2	9	9
3	27	7
4	81	1

Table 2-28 Small Powers of 3 in mod 10 Arithmetic

Example 4 *What is the ones digit of the rather large number $3^{2,005}$?*

SOLUTION   This is equivalent to asking, "What is the remainder when $3^{2,005}$ is divided by 10?" or equivalently, "What number from the set $\{0, 1, 2, \cdots, 9\}$ is congruent to $3^{2,005}$ mod 10?" To answer the latter question, we first observe a pattern in Table 2-28.

Proposition 3 uses mathematical induction to show that for every integer $n \geq 1$, $3^{4n} \equiv_{10} 1$. From this it follows that $3^{2,004} \equiv_{10} 1$, and so

$$3^{2,005} = 3 \cdot 3^{2,004}$$
$$\equiv_{10} 3 \cdot 1, \quad \text{by Theorem 2}$$
$$= 3$$

Hence, $3^{2,005}$ has a 3 in the ones digit. □

Here is the promised proof of the assertion from the previous example.

Proposition 3   *For all positive integers n, $3^{4n} \equiv_{10} 1$.*

PROOF   Consider the statement "$3^{4n} \equiv_{10} 1$." The table in Example 4 illustrates that the first statement, "$3^{4 \cdot 1} \equiv_{10} 1$," is true. Now let an integer $m \geq 1$ be given such that all the statements up to "$3^{4 \cdot (m-1)} \equiv_{10} 1$" have been checked to be true. Now consider the next statement:

$$3^{4m} = 3^4 \cdot 3^{4m-4}$$
$$\equiv_{10} 1 \cdot 1, \quad \text{by Theorem 2 and the facts that } 3^4 \equiv_{10} 3^{4(m-1)} \equiv_{10} 1$$
$$= 1$$

This proves the statement "$3^{4 \cdot m} \equiv_{10} 1$," completing the induction. ∎

Practice Problem 2   *Use mathematical induction to show that for all positive integers n, $2^{4n} \equiv_{10} 6$.*

We will now get back to the general idea of a "new system of arithmetic." In arithmetic, once we have a way to add and multiply, it is only natural to ask whether we can subtract and divide. Since our modular relationship is originally defined on all integers (including the negative ones), the notion of subtraction works out as one would expect—namely, subtracting a from b is the same as adding $-a$ to b. Division is more complicated, however. We will take the idea that division should amount to multiplying by a "reciprocal" and see how far we can get.

Example 5 *In Example 4, we saw that $3^3 \equiv_{10} 7$ and $3^4 \equiv_{10} 1$. A particular side effect of this is the fact that*

$$3 \cdot 7 \equiv_{10} 1$$

Just as 2 and $\frac{1}{2}$ are reciprocals in normal "rational number arithmetic" because their product is 1, it makes sense that we can call 3 and 7 reciprocals in mod 10 arithmetic. Hence, dividing by 3 is the same as multiplying by 7 when doing arithmetic mod 10.

This fact has the practical advantage of allowing us to do some algebra in our new system of arithmetic.

Example 6 *Solve the equation $3 \cdot x \equiv_{10} 4$ for x.*

SOLUTION   By Theorem 2, multiplying on both sides of the equation by 7 gives us

$$7 \cdot 3 \cdot x \equiv_{10} 7 \cdot 4 \qquad \text{or} \qquad 1 \cdot x \equiv_{10} 8$$

So any integer $x \equiv_{10} 8$ solves the given equation.   □

Practice Problem 3   *For each of the following, find an integer for x that satisfies the equation, if possible:*

(a) $2 \cdot x \equiv_{25} 3$

(b) $6 \cdot x \equiv_{15} 4$

(c) $4 \cdot x \equiv_{71} 3$

The second equation in the previous practice problem shows that one cannot always do division mod n. In this example, this happens because 6 has no reciprocal mod 10. This is equivalent to saying, "No multiple of 6 can have a remainder of 1 when divided by 10," which can be proved simply in terms of even and odd numbers. This problem arises from the fact that 6 and 10 have a common divisor other than 1.

To avoid this issue, we can just stick to arithmetic mod p for prime numbers p, and then all nonzero numbers have reciprocals, which allows us to do division. Therefore, in the exercises we will look mostly at mod p arithmetic for prime numbers p.

The fact that reciprocals always exist in arithmetic mod p can be proved in many ways. We prefer to see it as a corollary to the following useful theorem. This theorem is not proved here, but it is a consequence of the fact that every number can be factored into a product of prime numbers in a unique way.

Theorem 4   *Given a prime number p and integers a and b, if $a \cdot b$ is a multiple of p, then a is a multiple of p or b is a multiple of p.*

As our final proof of this section, we will show that the property of prime numbers given in the previous theorem does, in fact, ensure that all numbers have multiplicative reciprocals in mod p arithmetic.

Proposition 5   *Let p be prime and a be a number from $\{1, 2, \ldots, p-1\}$. It follows from Theorem 4 that there is a number b in $\{1, 2, \ldots, p-1\}$ such that $b \cdot a \equiv_p 1$.*

PROOF   For the given prime p and number a from $\{1, 2, \ldots, p-1\}$, we can form the following list of numbers:

$$1 \cdot a, 2 \cdot a, 3 \cdot a, \ldots, (p-1) \cdot a$$

Theorem 4 guarantees that none of the numbers in this list is divisible by p. That is, none of the numbers in this list is congruent to 0 mod p. Exercise 12 at the end of this section will show that this implies that no two numbers in this list are congruent to each other mod p.

Now the conclusion follows if we think about the situation carefully. The list above contains values congruent (mod p) to **different** numbers in $\{1, 2, \ldots, p-1\}$, and so each of the values $1, 2, \ldots, p-1$ is represented once in the list. In particular, one of the numbers in the list is congruent to 1 mod p.

Since all numbers in the list are of the form $b \cdot a$, this means that some number of the form $b \cdot a$ is congruent to 1 mod p, completing the proof. ■

This proof might be unsatisfying in that it logically concludes something exists (the mod p *reciprocal* of a in this case) without giving any clue at all about how you might find it! There are several effective ways to find a reciprocal, one of which derives from the following classic result of number theory.

Theorem 6 (Fermat)   *If p is a prime number, then for any positive integer $a < p$,*

$$a^{p-1} \equiv_p 1$$

PROOF   Let prime p and positive integer $a < p$ be given. In the proof of Proposition 5, we said that every value in the list

$$1 \cdot a, 2 \cdot a, 3 \cdot a, \ldots, (p-1) \cdot a$$

is congruent mod p to a different value from the set $\{1, 2, \ldots, p-1\}$, so the product of the numbers in the list is congruent mod p to the product of the numbers in $\{1, 2, \ldots, p-1\}$. That is,

$$(1 \cdot a)(2 \cdot a)(3 \cdot a) \cdots ((p-1) \cdot a) \equiv_p (p-1)!$$

which is the same thing as

$$(p-1)! \cdot a^{p-1} \equiv_p (p-1)!$$

Since $(p-1)! \not\equiv_p 0$, we can multiply on each side of this equivalence by the reciprocal of $(p-1)!$ to get the desired conclusion, $a^{p-1} \equiv_p 1$. ■

Corollary 7 *Given a prime number p and a number a from* $\{1, 2, \ldots, p - 1\}$*, the number* a^{p-2} mod *p is the* mod *p reciprocal of a.*

Example 7 *Letting* $p = 17$ *and* $a = 4$*, we can compute*

$$4^{p-2} = 4^{15}$$
$$= 1{,}073{,}741{,}824$$
$$\equiv_{17} 13$$

And it is easy to check that $4 \cdot 13 = 52 \equiv_{17} 1$*, so 13 is the reciprocal of 4 in* mod 17 *arithmetic.*

RSA Cryptosystem

"Encryption" means to make a message readable only to an intended receiver. For years, the science of cryptography was associated only with military secrets and national security. However, cryptography is now a part of everyday life. There is some level of encryption in most computer systems, from credit card transactions to simple everyday e-mail. Consequently, many cryptologists continue to develop new encryption schemes that are effective but easily implemented.

In the remainder of this section we will discuss a popular encryption scheme and see the mathematics behind it. The RSA system of cryptography is named for Ronald Rivest, Adi Shamir, and Leonard Adleman, who published a paper about it in 1977. It uses the strange idea that the information necessary to encrypt a message can be made public while the information needed to decrypt the message remains private. Somewhere at the heart of such a system, there must be a mathematical operation that is easy to compute but difficult to reverse. Surprisingly, the operation in the RSA system is simply the multiplication of large numbers.

We will first give a general description of the encryption/decryption process, and then we will look at the mathematics behind the system.

1. We first must translate our message into a number. We will use the following simple system:
   - Break the message into strings of four letters each. (The last string might be smaller.) For example, the message "PLEASE COME HERE" would be broken down into the strings PLEA, SECO, MEHE, and RE.
   - Each small word is changed into a number using the simple concatenation of numbers from Table 2-29.

A	B	C	D	E	F	G	H	I	J	K	L	M
01	02	03	04	05	06	07	08	09	10	11	12	13
N	**O**	**P**	**Q**	**R**	**S**	**T**	**U**	**V**	**W**	**X**	**Y**	**Z**
14	15	16	17	18	19	20	21	22	23	24	25	26

Table 2-29 Changing Letters to Numbers

- So the string PLEA becomes the number 16,120,501 (i.e., the concatenation of the number 16 for P, 12 for L, 05 for E, and 01 for A), and similarly,

    The string SECO becomes the number 19,050,315,

    The string MEHE becomes the number 13,050,805, and

    The string RE becomes the number 1,805.

2. Each of these four numbers will be encrypted separately and be sent in order, so the receiver will need to decrypt them, change them back into strings, and then put the strings together to form the original message.

3. Prime numbers p and q are chosen so that their product $n = p \cdot q$ is more than eight digits long, and hence each number to be encrypted is between 0 and n. For this example we will use the primes $p = 5{,}021$ and $q = 25{,}013$ giving us a product of $n = 125{,}590{,}273$.

4. The value of n is made public, but the values of p and q are eventually erased, shredded, and burned, because the security of the encyphering scheme is based on the fact that it is hard (i.e., it takes a long time) to recover p and q from knowing n. Imagine how long it would take you to factor the value $n = 125{,}590{,}273$ into the product of primes p and q by hand. If our p and q values had each been over 100 digits long, as they are in secure systems, then even the fastest computers would struggle.

5. Before p and q are destroyed, we need them for a couple of other tasks to make our encryption scheme ready to use:

    - First we compute the number $k = (p-1) * (q-1)$. This also will have to be destroyed a bit later. In our example, $k = 5{,}020 \cdot 25{,}012 = 125{,}560{,}240$.

    - Next we need another prime number, this time between 1 and k, that is *not* a factor of k. We will call this number e, and it will be the publicly available encryption key. For our example, let's pick an easy number like $e = 7$.

    - We will use the k and e to find positive integers d and v so that the equation

$$d * e - v * k = 1$$

    is true. This is the only computationally difficult step, and it has to do with what is called the Euclidean algorithm. We will not discuss this in this course, but rather we will give you these values when you need them. In our example, with $k = 125{,}560{,}240$ and $e = 7$, the equation is

$$107{,}623{,}063 \cdot 7 - 6 \cdot 125{,}560{,}240 = 1$$

    That is, $d = 107{,}623{,}063$ and $v = 6$.

6. Only the numbers n, e, and d are kept—the others are disposed of. If this is to be your public key code, you should publish your value of n and e for the general public. **The value of d is your decryption key, and you should not share it with anyone.** Notice that if anyone were able to factor n, they would be able to compute k and then find d just as you did above. The security of the whole system depends on n being hard to factor.

7. The point is that now if I have a number M to send you as a message, I can encrypt it by computing $C = M^e \bmod n$, since your values of n and e are publicly available.

When you receive the encrypted message C, *you* compute C^d mod n, and you will see M. No one else knows d, and to find it, they would have to be able to factor the rather large number n.

Example 8 *Let's see how I would send you the message "PLEASE COME HERE" once I have looked you up in my public key book to find out that your public keys are $n = 125{,}590{,}273$ and $e = 7$.*

SOLUTION

1. First the message is broken down into the four numbers to be sent: 16,120,501; 19,050,315; 13,050,805; and 1,805.

2. For each of these numbers M, I compute M^7 mod 125590273 with a computer, as shown in Table 2-30.

3. I send you the encrypted numbers: 106,118,249; 108,681,457; 30,461,159; 107,623,063.

M	M^7 mod 125590273
16,120,501	106,118,249
19,050,315	108,681,457
13,050,805	30,461,159
1,805	59,000,375

Table 2-30 Plaintext to Cyphertext

To decrypt my message, you must compute M^d mod 125590273 for each of these four numbers, using your secret decryption key $d = 107{,}623{,}063$ and a friendly computer, as shown in Table 2-31.

Notice that the numbers we got back in the end are precisely the original ones before encryption. The method works! □

The Mathematics Behind RSA

We will now see why this public-key cryptography works. The basis for the RSA system is the following consequence of Theorem 6, called *Fermat's little theorem*:

Theorem 8 *For any prime p and any positive integer $a < p$,*

$$a^p \bmod p = a$$

In order to show that the RSA scheme really works, we need to show that with all the values selected as in the example, we will always get back the original message M after it has been encrypted and then decrypted.

Proposition 9 *For n, e, and d selected as in the previous discussion of the RSA system, we will have for any message M (between 1 and n),*

$$(M^e)^d \bmod p = M$$

M	$M^{107623063}$ mod 125590273
106,118,249	16,120,501
108,681,457	19,050,315
30,461,159	13,050,805
59,000,375	1,805

Table 2-31 Cyphertext to Plaintext

PROOF   Fermat's theorem tells us that $M^{p-1} \bmod p = 1$, from which it is straightforward to infer that $M^k \bmod p = 1$ since

$$M^k \bmod p = M^{(p-1)(q-1)} \bmod p$$
$$= \left(M^{p-1}\right)^{q-1} \bmod p$$
$$= 1^{q-1} = 1$$

Similarly, we know that $M^k \bmod q = 1$, and so (with a bit more thought) since $n = p \cdot q$, we know that

$$M^k \bmod n = 1 \qquad (2.6)$$

By applying this to the encryption/decryption scheme, we find that

$$(M^e)^d = M^{e \cdot d}$$
$$= M^{1+k \cdot v}$$
$$= M \cdot (M^k)^v$$

But we know from (2.6) that $M^k \bmod n = 1$, so

$$M \cdot (M^k)^v \bmod n = M \cdot 1^v = M$$

So successively encrypting and decrypting will return a message unchanged. ■

Solutions to Practice Problems

1 Figure 2-12 shows the mod 5 multiplication. One way in which this differs from the mod 6 multiplication table is that the one here has no 0 entries except for those that come from multiplying by zero.

2 Consider the statement "$2^{4n} \equiv_{10} 6$." It is easy to check that the first statement "$2^4 \equiv_{10} 6$" is true. Suppose we have checked all the statements up to the statement "$2^{4(m-1)} \equiv_{10} 6$," for some given integer $m \geq 2$. Then it follows that

$$2^{4m} = 2^{4m-4} \cdot 2^4$$
$$= 2^{4(m-1)} \cdot 2^4$$
$$\equiv_{10} 6 \cdot 2^4, \qquad \text{since } 2^{4(m-1)} \equiv_{10} 6$$
$$\equiv_{10} 6$$

This shows that $2^{4m} \equiv_{10} 6$, completing the induction.

3 The second equation has no solution.

(a) $2 \cdot x \equiv_{25} 3$ for any $x \equiv_{25} 14$; in particular, $x = 14$ will do it.

(b) $6 \cdot x \equiv_{15} 4$ has no solution for x because $6x - 4$ cannot be divisible by 15 since it cannot be divisible by 3. (Look at the remainder when $6x - 4$ is divided by 3.)

(c) $4 \cdot x \equiv_{71} 3$ for any $x \equiv_{71} 54$; in particular, $x = 54$ will do it.

×	0	1	2	3	4
0	0	0	0	0	0
1	0	1	2	3	4
2	0	2	4	1	3
3	0	3	1	4	2
4	0	4	3	2	1

Figure 2-12 Multiplication in mod 5 arithmetic.

Exercises for Section 2.7

1. Each of the following statements is false. Provide a counterexample to each.

(a) If $a \cdot b \equiv_n 0$, then $a \equiv_n 0$ or $b \equiv_n 0$.

(b) For every $b \not\equiv_n 0$, there is an integer a such that $a \cdot b \equiv_n 1$.

(c) If $a^3 \equiv_n 0$, then $a^2 \equiv_n 0$.

(d) If $a \equiv_n 1$ and $b \equiv_m 1$, then $a \cdot b \equiv_{m \cdot n} 1$.

(e) We will say that 5 is a perfect square mod n if there is some integer a such that $a^2 \equiv_n 5$. If 5 is a perfect square mod n, then 5 must divide $n(n-1)$.

(f) If $a^2 \equiv_n a$, then either $a \equiv_n 0$ or $a \equiv_n 1$.

(g) If $a \equiv_n b$, then $2^a \equiv_n 2^b$.

2. Fill in the missing details in the following proofs:

(a) **Proposition** For any integer $n \neq 0$ and for all integers a, b, c, and d, if $a \equiv_n b$ and $c \equiv_n d$, then $a + c \equiv_n b + d$.

Proof. Let a, b, c, d, and n be given, and assume that $a \equiv_n b$ and $c \equiv_n d$. By Theorem 1, both $a - b$ and $c - d$ are divisible by n.

$$\vdots$$

Since $(a + c) - (b + d)$ is divisible by n, we conclude (by Theorem 1 again) that $a + c \equiv_n b + d$.

(b) **Proposition** For every integer $n \geq 0$, if n is divisible by 4, then $2^n \equiv_5 1$.

Proof by induction. Let $P(k)$ be the statement "$2^{4k} \equiv_5 1$." We first check the first few statements:

● $P(0)$ states, _____, which is true.

● $P(1)$ states, _____, which is true.

● $P(2)$ states, _____, which is true.

Now assume that for some given $m \geq 1$, we have established all the statements $P(0), P(1), \ldots, P(m-1)$, and we are now considering $P(m)$.

$$2^{4m} = \underline{\hspace{2cm}}$$
$$\equiv_5 16 \cdot \underline{\hspace{2cm}}$$
$$\equiv_5 1 \cdot 1 \text{ by Theorem } \underline{\hspace{2cm}}$$
$$\text{and } P(m-1)$$
$$\equiv_5 1$$

Since $2^{4m} \equiv_5 1$, this establishes $P(m)$, completing the induction.

(c) **Proposition** For every integer n, $n^3 \equiv_3 n$.

Proof. Let the integer n be given. By the division theorem, either $n \equiv_3 0$, $n \equiv_3 1$, or $n \equiv_3 2$, so we can simply address each possible case:

● If $n \equiv_3 0$, then $n^3 \equiv_3$ _____ by Theorem _____.

● If $n \equiv_3 1$, then $n^3 \equiv_3$ _____ by Theorem _____.

● If $n \equiv_3 2$, then $n^3 \equiv_3$ _____ by Theorem _____.

Hence, in every possible case, $n^3 \equiv_3 n$. ■

3. Make complete multiplication and addition tables for mod 7 arithmetic.

4. Make complete multiplication and addition tables for mod 8 arithmetic.

5. Find reciprocals of each of the values $1, 2, 3, \ldots, 10$ in mod 11 arithmetic.

6. Find reciprocals of each of the values $1, 2, 3, \ldots, 12$ in mod 13 arithmetic.

7. What is the remainder when $2^{1,000,000}$ is divided by 19? (HINT: First find a so that $2^a \equiv_{19} 1$.)

8. Let x be any positive integer and s_x be the sum of the (base ten) digits of x. Prove that $x \equiv_9 s_x$. (HINT: Use Proposition 3 from Section 2.6.)

9. Suppose that the number $2^{1,000,000}$ is written out and its digits are summed, then this resulting number is written out and its digits are summed, and this process is repeated until there is only a one-digit number result. What is this single digit? (HINT: Use the previous exercise.)

10. Find all solutions for x in each of the following equations:

(a) $2x + 3 \equiv_{15} 7$

(b) $3x - 1 \equiv_8 7$

(c) $x^2 - x \equiv_{11} 1$

(d) $2x - 1 \equiv_{20} 6$

(e) $x^2 \equiv_{17} -1$

(f) $x^2 - 5 \equiv_{23} 0$

11. Find all primes p less than 100 for which the equation $x^2 + 1 \equiv_p 0$ has an integer solution. Make a conjecture about what these values of p all have in common.

12. Show that if none of the numbers in the list $1 \cdot a, 2 \cdot a, \ldots, (p-1) \cdot a$ are congruent to 0 mod p, then no two numbers in the list are congruent to each other mod p. (HINT: If two of the numbers are congruent to each other, what do you know about their difference?)

13. Prove that for each natural number n, n^5 has the same ones digit as n does.

14. If we want to send a message one letter at a time, we can use the RSA system with small primes like $p = 11$ and $q = 7$, and just send the letters A through Z as numbers 1 through 26, respectively.

(a) If I use the encryption key $e = 43$, what is the smallest positive number you can use for the decryption key?

(b) Using public keys $n = 11 \cdot 7 = 77$ and $e = 43$, I send you the following message:

   41 26 26 69 01 69 41 64 26 61

Decipher the message using the value of d you found in the previous exercise.

(c) Give two reasons why you would not want to send a sensitive message using this scheme.

15. Suppose I want to send a message one letter at a time, using the RSA system with the primes $p = 5$ and $q = 7$, and just send the letters A through Z as numbers 1 through 26, respectively.

(a) If I use the encryption key $e = 11$, what is the smallest positive number you can use for the decryption key?

(b) If I use the encryption key $e = 13$, what is the smallest positive number you can use for the decryption key?

(c) If I use the encryption key $e = 17$, what is the smallest positive number you can use for the decryption key?

(d) Using public keys $n = 5 \cdot 7 = 35$ and $e = 17$, I send you the following message:

   24 10 33 23 10 20 24 12 21 04 23 23
   10 17

Decipher the message using the value of d you found in part (c).

16. Here is a message to you in which I am using an RSA system with public key $n = 2,773$ and encryption key $e = 157$.

   0245 2040 1698 1439 1364 1758 0946 0881
   1979 1130

I have broken my original message into pairs of characters and converted these pairs to numbers as we did in the text. For example, the word "MATH" would be broken into "MA" and "TH" that would be converted to the numbers 1,301 and 2,008, respectively, and encrypted. Figure out the decryption key d, decrypt the message, and answer the question that it asks.

17. Define a *k-pseudoprime* as a number $n > 1$ for which $k^{n-1} \equiv_n 1$, and define a *Carmichael number* as a number $n > 1$ that is a k-pseudoprime for all positive k which have no factors greater than 1 in common with n.

(a) Find all three-digit 2-pseudoprimes that are not prime.

(b) Find all three-digit 3-pseudoprimes that are not prime.

(c) Find all three-digit 5-pseudoprimes that are not prime.

(d) Find the smallest Carmichael number.

The next few exercises involve systems of more than one equation of modular arithmetic.

18. A single number satisfies many different congruences using modular arithmetic. For example, the number 16 satisfies all the following:

$$16 \equiv_2 0 \quad 16 \equiv_4 0 \quad 16 \equiv_6$$
$$16 \equiv_3 1 \quad 16 \equiv_5 1 \quad 16 \equiv_7$$

Some of these are redundant. For example, if a number n satisfies $n \equiv_6 4$, we know that $n = 6k + 4$ for some integer k. From this form (specifically since $n = 2(3k + 2)$), we can see that if n is divided by 2, there will be a 0 remainder, and also (since $n = 3(2k + 1) + 1$) that if n is divided by 3, there will be a remainder of 1. Hence, we have proved the following:

Proposition If $n \equiv_6 4$, then $n \equiv_2 0$ and $n \equiv_3 1$.

Now you prove each of the following:

(a) If $n \equiv_6$ 5, then $n \equiv_2 1$ and $n \equiv_3 2$.

(b) If $n \equiv_{20}$ 7, then $n \equiv_4 3$ and $n \equiv_5 2$.

(c) If $n \equiv_{30} 11$, then $n \equiv_2 1$, $\quad n \equiv_3 2$, and $n \equiv_5 1$

19. Another interesting fact is that each of the above statements also has a converse that is true. In other words, complete information about mod 2 and mod 3 determines mod 6 information for any given number. Let's see how to prove this:

Proposition If $n \equiv_2 0$ and $n \equiv_3 2$, then $n \equiv_6 2$.

Proof. Let an integer n be given, and assume that $n \equiv_2 0$ and $n \equiv_3 2$. The division theorem leaves us only six possibilities for the remainder when n is divided by 6.

● **Case 0:** Suppose $n \equiv_6 0$. Then reasoning as in (a) above, we can conclude that $n \equiv_2 0$ and $n \equiv_3 0$. The latter of these conditions contradicts the assumption that $n \equiv_3 2$.

● **Case 1:** Suppose $n \equiv_6 1$. Then reasoning as in (a) above, we can conclude that $n \equiv_2 1$ and $n \equiv_3 1$. Both these conditions contradict the assumptions that $n \equiv_2 0$ and $n \equiv_3 2$.

● **Case 2:** Suppose $n \equiv_6 2$. Then reasoning as in (a) above, we can conclude that $n \equiv_2 0$ and $n \equiv_3 2$. This does not create a contradiction with the given information.

● **Case 3:** Suppose $n \equiv_6 3$. Then reasoning as in (a) above, we can conclude that $n \equiv_2 1$ and $n \equiv_3 0$. Both these conditions contradict the assumptions that $n \equiv_2 0$ and $n \equiv_3 2$.

● **Case 4:** Suppose $n \equiv_6 4$. Then reasoning as in (a) above, we can conclude that $n \equiv_2 0$ and $n \equiv_3 1$. The latter of these conditions contradicts the assumption that $n \equiv_3 2$.

- **Case 5:** Suppose $n \equiv_6 5$. Then reasoning as in (a) above, we can conclude that $n \equiv_2 1$ and $n \equiv_3 2$. The first of these conditions contradicts the assumption that $n \equiv_2 0$.

The only case that does not create a contradiction is Case 2. Hence, it must be the case that $n \equiv_6 2$. ∎

For each of the following, find the smallest positive integer x that satisfies all the given congruences. If you believe that no such an integer exists, explain why.

(a) $x \equiv_{10} 3$ and $x \equiv_7 1$
(b) $x \equiv_{10} 7$ and $x \equiv_4 3$
(c) $x \equiv_{10} 3$ and $x \equiv_4 2$
(d) $x \equiv_4 3$, $x \equiv_5 4$, and $x \equiv_6 5$
(e) $x \equiv_4 2$, $x \equiv_5 3$, and $x \equiv_6 1$
(f) $x \equiv_{10} 4$, $x \equiv_{16} 2$, and $x \equiv_{24} 18$

20. The Chinese remainder theorem states that when n and m have no positive common divisors other than 1, the two congruences $x \equiv_n a$ and $x \equiv_m b$ can be satisfied by the same number x. Use facts from this section to prove this is true in the special case when m and n are prime numbers.

Chapter 2 Summary

2.1 Mathematical Writing

Terms and concepts

- You should be familiar with *implications* and the circumstances in which an implication is false.
- You should be familiar with the term *counterexample*, and you should be able to find counterexamples to false implications about simple properties of numbers.
- You should be able to identify the *hypothesis* and *conclusion* of an implication, even when the statement is written informally.
- You should be able to write the *contrapositive* of an implicational statement, and you should understand that it is equivalent to the original statement.
- You should know the formal definitions of *even* and *odd* as they will be used in proofs in this text.

Proofs

- You should be able to write a proof for a simple implication involving basic properties of numbers like "even" and "odd."
- You should be able to prove an implication by forming its contrapositive, then proving that contrapositive implication.
- You should be able to trace a proof for a simple implication involving basic properties of numbers.

2.2 Proofs About Numbers

Terms and concepts

- You should know the formal definition of *divisible by* as it will be used in proofs in this text.

- You should understand the *division theorem* and be able to identify quotients and remainders for any division problem.
- You should know the definition of *rational number* as it will be used in proofs in this text.
- You should be able to use the *mod* operation on integers.

Proofs

- You should be able to write a proof for a statement whose hypothesis or conclusion states that a particular integer is divisible by another.
- You should be able to correctly use the division theorem to define *cases* for a proof.
- You should be able to write a proof for a statement whose conclusion states that a particular integer has a particular form.
- You should be able to write a proof for a statement whose hypothesis or conclusion states that a particular real number is rational.

2.3 Mathematical Induction

Terms and concepts

- You should be familiar with a *predicate over the natural numbers*, and you should be able to write statements like $P(3)$ or $P(m-1)$ given a specific statement of the form $P(n)$. There is more discussion of predicates in general in Section 1.4.
- You should be comfortable with *sigma notation*. There is more discussion of sigma notation in Section 1.2.
- You should understand the *principle of mathematical induction*.

Proofs

- You should be able to give an informal "table-oriented" proof by mathematical induction, for these situations:
  - Given a recursive description of a sequence of integers, you should be able to prove that the terms of the sequence also satisfy a given closed formula.
  - Given a summation (whether in sigma notation or not), you should be able to prove that the summation satisfies a given closed formula.
- You should also be able to write these proofs by mathematical induction using the formal textbook style.

2.4 More About Induction

Skills and proofs

- You should gain exposure to using mathematical induction in a variety of contexts.
- In particular, you should be familiar with the use of mathematical induction in some of these settings:
  - Given a summation, write a recursive description for s_n, the sum of the first n terms of the sum, and prove a given formula for s_n using induction. This ties together the two types of problems in Section 2.3.
  - Write proofs concerning the *Fibonacci numbers* and other *second order* recurrences.
  - Write induction proofs involving inequalities satisfied by recursively defined sequences.
  - Use mathematical induction to establish divisibility properties.
  - Prove statements that revisit some problems from Section 1.1.

2.5 Contradiction and the Pigeonhole Principle

Terms and concepts

- You should understand the idea of *proof by contradiction* and how it differs from *direct proof* and from *proof of the contrapositive statement*.
- You should be familiar with some classic proofs by contradiction:
  - The real number $\sqrt{2}$ is irrational.
  - There are an infinite number of primes.
- You should be familiar with the various forms of the *pigeonhole principle*: the *general* version, the *basic* version, and perhaps the *average* version which is introduced in the exercises.

Skills

- For a given implication, you should be able to identify how a proof by contradiction would begin.
- You should be able to correctly negate an existence statement. This revisits a topic first appearing in Section 1.4.
- You should be able to recognize instances where a contradiction proof can be converted to a direct proof or to a proof of the contrapositive statement.
- You should be able to apply the *pigeonhole principle* to appropriate problems.

Proofs

- You should be able to write a proof by contradiction for a statement about basic properties of numbers.
- You should be able to prove that a particular real number is irrational using either contradiction or contraposition.
- You should be able to prove the various versions of the *pigeonhole principle* using either contradiction or contraposition.

2.6 Excursion: Representation of Numbers

Terms, concepts, and skills

- You should be familiar with the terms *binary*, *octal*, *decimal* and *hexadecimal* as relating to various *bases* for *numerals*.
- You should be able to convert numbers between decimal and any other base.
- You should be able to convert directly between *binary*, *octal* and *hexadecimal* without converting to decimal as an intermediate step.
- You should be able to prove simple statements about representations of numbers such as rules for divisibility.
- At the end of this section we learn about some applications of binary numerals to specific problems like the Josephus problem and perfect shuffles of cards. You should be able to follow these examples.

2.7 Excursion: Modular Arithmetic and Cryptography

Terms, concepts, and skills

- You should be able to do basic computations using the mod n equivalence, denoted $\equiv_n$ in the section.

- You should be able to construct the addition and multiplication tables for mod n arithmetic for any positive integer n.
- You should understand the statement and proof of *Fermat's little theorem*.

- You should know that multiplicative inverses (of nonzero elements) do not always exist in mod n arithmetic unless n is prime.
- You should understand how to encrypt and decrypt simple messages using the *RSA system*.

3 | Sets and Boolean Algebra

In the previous chapter, we introduced the fundamentals of mathematical proof-writing, and we practiced both writing about and reasoning with numbers. In this chapter, we study sets and their generalization to a more abstract structure, Boolean algebra. In addition to providing another setting in which to practice mathematical reasoning and writing, these topics will illustrate an important attribute of higher mathematics. Moving from the concrete to the abstract can often make computational problems easier for those willing to adapt to the more abstract setting.

The chapter begins by introducing the general operations on, and properties of, sets. After we become comfortable with the formal terminology and notation, we will practice our proof techniques using sets as a new context for mathematical writing. The concept of "set" naturally leads to the abstract concept of a *Boolean algebra*, which resolves some connections among set operations, logical connectives from the first chapter, and the study of circuits in computer science. The chapter concludes with a practical technique for simplifying circuits that is based on the abstract properties of Boolean algebra.

3.1 Set Definitions and Operations

Sets are among the fundamental building blocks of mathematics. They also have a simple logical structure, so they provide good practice to strengthen our reasoning and writing abilities. We will work with the informal definition of a *set* as a simple collection of objects, called the *members* or *elements* of the set. It has been well established that this loose definition can lead to paradoxes.* To avoid this, we will take the pragmatic point of view of most mathematicians. In practice, we will work only within a small number of well-understood sets, so we will not worry about the paradoxical set constructions that are better discussed in a course on the foundations of mathematics.

Definition   Here are some of the common sets of numbers we will use:

- $\mathbb{N}$ is the set of natural numbers—these are numbers that can answer counting problems ($\mathbb{N} = \{0, 1, 2, 3, \ldots\}$).
- $\mathbb{Z}$ is the set of integers ($\mathbb{Z} = \{\ldots, -3, -2, -1, 0, 1, 2, 3, \ldots\}$).
- $\mathbb{Q}$ is the set of rational numbers—these are characterized as ratios of integers such as $\frac{1}{2}$ or $\frac{-17}{4}$ or $\frac{3}{1}$.
- $\mathbb{R}$ is the set of real numbers—these can be thought of as decimal numbers with possibly unending strings of digits after the decimal point.

At times variations of these basic sets are used, so it's good to adopt consistent notation for the most common circumstances:

$\mathbb{R}^+$ is the set of positive real numbers.
$\mathbb{R}^{\geq 0}$ is the set of nonnegative real numbers.
$\mathbb{Q}^+$ is the set of positive rationals.
$\mathbb{Q}^{\geq 0}$ is the set of nonnegative rationals.
$\mathbb{Z}^+$ is the set of positive integers.
$\mathbb{Z}^{\geq 0}$ is the same as $\mathbb{N}$.

 Example 1   *List three numbers that are*

1. *Integers but not natural numbers.*
2. *Rational numbers but not integers.*
3. *Real numbers but not rational numbers.*

SOLUTION

1. -100, -5, and -1 are all integers, but none are natural numbers.
2. $\frac{1}{5}$, $\frac{-117}{23}$, and 2.12 (since $2.12 = \frac{53}{25}$) are all rational numbers, but none are integers.
3. In Section 2.5 we proved (between examples and exercises) that the real numbers $\sqrt{2} \approx 1.414213562\ldots$, $\sqrt{3} \approx 1.732050808\ldots$, and the golden ratio $\frac{1+\sqrt{5}}{2} \approx 1.618033988\ldots$ are not rational numbers. □

---

* The most famous of these is *Russell's paradox*. See Exercise 33 for details.

Subsets

The sets in the above definition are clearly related to one another. For example, every rational number is also a real number. To make relationships like this an integrated part of our mathematical language, we need some new terminology and notation:

Definition

1. The notation $x \in A$ means "x is an *element* of A," which means that x is one of the members of the set A.

2. A is a *subset* of B (written $A \subseteq B$) if every element in A is also an element in B. Formally, this means that for every x, if $x \in A$, then $x \in B$.

3. A is equal to B (simply written $A = B$) means that A and B have exactly the same members. This is expressed formally by saying, "$A \subseteq B$ and $B \subseteq A$."

4. A set that contains no elements is called an *empty set*, and is denoted by $\{\ \}$ or $\emptyset$.

5. For any given discussion, all the sets will be subsets of a larger set called the *universal set* or *universe*, for short. We commonly use the letter U to denote this set.

For example, since 2 is a natural number, we can write $2 \in \mathbb{N}$. Since 2 is also a positive real number, we can also write $2 \in \mathbb{R}^+$. With this notation, we can express relationships between some of the common sets of numbers described above. For example,

- $\mathbb{N} \subseteq \mathbb{Z}$ conveys the fact that every natural number is also an integer.
- $\mathbb{Z} \subseteq \mathbb{Q}$ conveys the fact that every integer is also a rational number since, for example, $3 = \frac{3}{1}$.
- $\mathbb{Q} \subseteq \mathbb{R}$ conveys the fact that every rational number is also a real number.

Since many problems involve sets of numbers of one type or another, frequent choices for the universal set are $\mathbb{R}$, the set of all real numbers, or $\mathbb{Z}$, the set of all integers. We will be explicit about the universal set whenever it is not clear from the context of a problem.

To describe small sets, we can simply list all the elements within braces. This is called the *roster method* for set description.

 Example 2

1. $\{1, 2, 3, 4, 5\}$ *is a set containing exactly five positive integers. Notice that* $\{1, 2, 3, 4, 5\} \subseteq \mathbb{N}$.

2. *These three sets are all the same: the set* $\{1, 2, 3, 4, 5\}$, *the set* $\{2, 4, 1, 5, 3\}$, *and the set* $\{1, 1, 2, 2, 3, 3, 4, 4, 5, 5, 5, 5\}$. *Each consists of the numbers 1, 2, 3, 4, and 5. The order in which we list the members is irrelevant. Listing the elements more than once does not change the set.*

3. $\{2, 4\} \subseteq \{1, 2, 3, 4, 5\}$ *is true, but* $\{1, 2, 3, 4, 5\} \subseteq \{2, 4\}$ *is false.* 1 *is a counterexample to the statement "If* $x \in \{1, 2, 3, 4, 5\}$, *then* $x \in \{2, 4\}.$"

4. $\emptyset \subseteq \{1, 2, 3\}$ *is true, since there is no counterexample to the statement "If* $x \in \emptyset$, *then* $x \in \{1, 2, 3\}.$" *The empty set is a subset of every set.*

5. *{Joe, Tom, Sue, Mary} is a set containing four names. (Here the universal set might be the set of all names of people.)*

6. $\{(1, 3), (2, 5), (3, 7)\}$ *is a set of ordered pairs such as one might plot on graph paper in an algebra class.*

7. $\{\{3, 4\}, \{5, 6, 7\}\}$ *is a strange-looking set, but it is completely legitimate. This set contains two members, each of which is itself a set. This is really no more shocking than having a box containing two smaller boxes or an envelope containing two smaller envelopes. We will see more examples like this in the next section.*

Practice Problem 1   *Which of these statements are true?*

(a) *{abba, aabba, ababa} is a legal set.*

(b) $\{4, 1, 2, 3\} \subseteq \{1, 2, 3, 4, 3, 2, 1\}$.

(c) $\{6, 8, 10\} \subseteq \{1, 2, 3, 4, 5, 6, 7, 8, 9, 10\}$.

(d) $\mathbb{Q} \subseteq \mathbb{N}$.

(e) $\emptyset \subseteq \{abba, aabba, ababa\}$.

It is often impossible, or at best impractical, to list every element in a set. To describe larger sets, we usually use *set builder notation*, which actually has two somewhat different forms. Here are some examples to illustrate the first of these ideas.

 Example 3

1. *To describe the even natural numbers, we can write* $\{x : x \in \mathbb{N} \text{ and } x \text{ is even}\}$. *This is read, "The set of x such that x is an element of* $\mathbb{N}$ *and x is even." Here are two other ways to describe the same set:*
    - $\{x \in \mathbb{N} : x \text{ is even}\}$.
    - $\{x \in \mathbb{N} : x = 2k \text{ for some } k \in \mathbb{N}\}$.

2. $\{x \in \mathbb{R} : -2 < x \le 2\}$ *can also be written as the interval* * $(-2, 2]$.

3. $\{n \in \mathbb{N} : n \text{ has exactly two positive divisors}\}$ *is better known as the set of prime numbers.*

4. $\{x \in \mathbb{R} : x^2 + 1 = 0\}$ *contains no elements at all, so* $\{x \in \mathbb{R} : x^2 + 1 = 0\} = \{\}$.

In each of these examples, there is a universe U given and the description of the set has the form

$$\{x \in U : x \text{ is } \underline{\hspace{2cm}}\}$$

---

* See Exercises 20 and 21 at the end of this section for more on interval notation.

which is read, "The set of all x in U such that x is _____." Since the large blank contains some property of x that determines whether a particular element of U is or is not to be included in the set, we will refer to these as *property descriptions* for set-builder notation.

Practice Problem 2   *Write each of the following in set-builder notation using a property description:*

(a) *The set of even integers*

(b) *The set of real numbers bigger than 10*

(c) *The set of rational numbers whose square is less than 2*

In the first example above, we see that the defining "property" for x is really a matter of the form of x. For example, when describing the set of even natural numbers, the property "is even" is the same as "has the form $2k$ for some $k \in \mathbb{N}$." Hence, we can describe this set based on the form of its elements as follows:

$$\{2k : k \in \mathbb{N}\}$$

We read this, "All numbers of the form $2k$ where k is from $\mathbb{N}$." We will refer to this as a *form description* within our set-builder notation.

Example 4 *Write each of the following sets in set-builder notation using a form description:*

1. *The set of integers that are multiples of 3*

2. *The set of perfect square integers*

3. *The set of natural numbers that end with a 1*

4. $\mathbb{Q}$

SOLUTION

1. $\{3k : k \in \mathbb{Z}\}$

2. $\{m^2 : m \in \mathbb{N}\}$ or $\{m^2 : m \in \mathbb{Z}\}$ (They describe the same set.)

3. $\{10k + 1 : k \in \mathbb{N}\}$

4. $\{\frac{a}{b} : a \in \mathbb{Z} \text{ and } b \in \mathbb{Z}^+\}$

□

Explore more on the Web.

Practice Problem 3   *Each of the following sets is written using a property description. For each one, write five elements of the set and then give a "form description" of the whole set.*

(a) $\{x \in \mathbb{Z} : x = 5k \text{ for some } k \in \mathbb{Z}\}$

(b) $\{y \in \mathbb{N} : y \text{ is one more than a perfect square}\}$

(c) $\{r \in \mathbb{Q} : r = \frac{a}{2k+1} \text{ for some } a \in \mathbb{Z} \text{ and } k \in \mathbb{N}\}$

New Sets from Old

Everyone is familiar with the many ways a scientific calculator can operate on a number or numbers to produce some other related number. For example, the $+$ button takes two numbers and returns the sum of these numbers, and the e^x button takes a single number and returns the value of the constant e raised to the power x. Similarly, there are many ways to build new sets from given sets. Just as we learn simple arithmetic before more complicated numerical functions, in this section we will begin with the simplest set operations, and proceed to more complex operations in the next section.

Definition Given two sets A and B of elements from a universal set U,

1. The *intersection* of A and B (written $A \cap B$—think of "$\cap$ for intersection") is the set that contains those elements common to both A and B. In set-builder notation, we write

$$A \cap B = \{x \in U : x \in A \text{ and } x \in B\}$$

2. The *union* of A and B (written $A \cup B$—think of "$\cup$ for Union) is the set that contains those elements in either set A or B. In set-builder notation, we write

$$A \cup B = \{x \in U : x \in A \text{ or } x \in B\}$$

3. The *difference* of A and B (written $A - B$) is the set that contains those elements in A which are not in B. In set-builder notation, we write

$$A - B = \{x \in U : x \in A \text{ and } x \notin B\}$$

Example 5 *Let $U = \mathbb{N}$. For the sets $A = \{1, 2, 3, 4, 5\}$, $B = \{2, 4, 6, 8, 10\}$, and $C = \{8, 10, 12\}$, find (1) $A \cap B$, (2) $A \cup B$, (3) $A - B$, (4) $B - A$, and (5) $A \cap C$.*

SOLUTION

1. $A \cap B = \{2, 4\}$
2. $A \cup B = \{1, 2, 3, 4, 5, 6, 8, 10\}$
3. $A - B = \{1, 3, 5\}$
4. $B - A = \{6, 8, 10\}$
5. $A \cap C$ does not contain any elements—that is, $A \cap C = \{\}$. □

The last example illustrates an important relationship between sets. Since A and C have no common elements (i.e., $A \cap C = \{\}$), we can refer to them as being *nonoverlapping*, *mutually exclusive*, or *disjoint*. We will use the latter of these terms later when we discuss counting and probability, so it is worth making it a formal definition now.

> **Definition**   Sets A and B are *disjoint* if $A \cap B = \emptyset$.

Explore more on the Web.

Practice Problem 4   *Let $U = \mathbb{N}$. For the sets $A = \{1, 3\}$, $B = \emptyset$, and $C = \{1, 2, 3, 4, 5\}$:*

(a)   What is $A \cap B$?

(b)   What is $A \cup B$?

(c)   What is $A \cup C$?

(d)   Is A a subset of B?

(e)   Is B a subset of A?

(f)   Give an example of two disjoint sets D and E whose union is C.

As we mentioned earlier, to avoid some well-known paradoxes we will work only within well-understood sets. Although we do not always mention it, there is always a "*universal set*" so that every set we are discussing is a subset of that universal set. (We usually use the notation U for this universal set.) Under this assumption, for a given set A it makes sense to talk about the set containing those elements that are *not* in A. We call this set the *complement* of A.

> **Definition**   Given a set A with elements from the universe U, the *complement* of A (written A') is the set that contains those elements of the universal set U which are not in A. That is, $A' = U - A$.

Example 6   *For the universal set $U = \{1, 2, 3, 4, 5, 6, 7, 8, 9, 10\}$ and the sets $A = \{1, 2, 3, 4, 5\}$ and $B = \{2, 4, 6, 8, 10\}$, find each of the following:*

1. *A'*

2. *B'*

3. *$A \cap B'$*

4. *U'*

SOLUTION

1. $A' = \{6, 7, 8, 9, 10\}$

2. $B' = \{1, 3, 5, 7, 9\}$

3. $A \cap B' = \{1, 2, 3, 4, 5\} \cap \{1, 3, 5, 7, 9\} = \{1, 3, 5\}$. This is exactly the same thing as $A - B$.

4. $U' = \emptyset$. Since U is the entire universal set, "the set that contains those elements of the universal set U which are not in U" is a set with no elements in it.

Practice Problem 5   *For the universal set* $U = \{1, 2, 3, 4, 5, 6, 7, 8, 9, 10\}$ *and the sets* $A = \{1, 2, 3, 4, 5\}$ *and* $B = \{2, 4, 6, 8, 10\}$, *calculate the following:*

(a) $(A \cap B)'$

(b) $A' \cup B'$. *Is this the same as* $(A \cap B)'$?

(c) $\emptyset'$

Each of the notational conventions for describing sets has a particular form when it comes to expressing the result of these set operations. As an example, we look specifically at set-builder notation below.

Example 7   *Given the sets* $A = \{3k : k \in \mathbb{N}\}$ *and* $B = \{y \in \mathbb{N} : y \text{ is prime}\}$, *express each of the following using set-builder notation:*

1. $A \cap B$

2. $A \cup B$

3. $A - B$

4. $B - A$

SOLUTION

1. $A \cap B = \{x \in \mathbb{N} : x = 3\}$ (= $\{3\}$ via the roster method)
2. $A \cup B = \{z \in \mathbb{N} : z \text{ is prime or a multiple of } 3\}$
3. $A - B = \{x \in \mathbb{N} : x \neq 3 \text{ and } x \text{ is a multiple of } 3\}$ or $\{3k : k \in \mathbb{N} \text{ and } k \geq 2\}$)
4. $B - A = \{x \in \mathbb{N} : x \neq 3 \text{ and } x \text{ is prime}\}$

$\square$

Properties and Venn Diagrams

When we study operations on numbers, we take advantage of many properties of and relationships between these operations to allow us to write expressions in different ways. For example, the distributive property allows us to factor $5x + 15$ as $5 \cdot (x + 3)$, and we know that this is useful in the algebraic context of solving equations. Properties like the commutative property, the associative property, and the distributive property are clearly important for performing algebraic manipulations and generally understanding numerical functions.

The set operations that we have just introduced have many of these same properties, as well as some properties that the numerical operations definitely do not have.

Proposition 1   *For sets A, B, and C, the following properties hold:*

Commutative property for $\cap$	$A \cap B = B \cap A$
Commutative property for $\cup$	$A \cup B = B \cup A$
Associative property for $\cap$	$(A \cap B) \cap C = A \cap (B \cap C)$
Associative property for $\cup$	$(A \cup B) \cup C = A \cup (B \cup C)$
Distributive property of $\cap$ *over* $\cup$	$A \cap (B \cup C) = (A \cap B) \cup (A \cap C)$
Distributive property of $\cup$ *over* $\cap$	$A \cup (B \cap C) = (A \cup B) \cap (A \cup C)$

We will compose a more extensive list of these properties in Theorem 6 of Section 3.3, and we will see some striking similarities between the set properties given in that theorem and the properties for elementary logic given in Theorem 2 of Section 1.3. In the section on Boolean algebra, we will discover that this is not a coincidence.

In the remainder of the current section, we will discuss how we might assess a new statement about sets whose truth is in question. To do this, we will use a visualization technique called a *Venn diagram*.

A Venn diagram is simple. We use a large square to represent the universe U, and then add circles within the square to represent sets. We imagine that elements of the universe in a set S fall inside the circle for S and elements not in the set S fall outside of that circle. When the sets are small, we can actually draw the whole picture without taxing our imaginations too much.

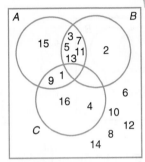

Figure 3-1 Venn diagram for Example 8.

Example 8 *Let* $U = \{1, 2, 3, \ldots, 14, 15, 16\}$, $A = \{1, 3, 5, 7, 9, 11, 13, 15\}$, $B = \{2, 3, 5, 7, 11, 13\}$, *and* $C = \{1, 4, 9, 16\}$. *Draw the Venn diagram for these sets showing all elements of the universe* U.

SOLUTION The rectangle in Figure 3-1 denotes the universe U, so all 16 elements from the universe are shown within the rectangle. Numbers within the labeled circles indicate the elements of the respective sets. Notice that the picture has empty regions where there are no set elements. So, for example, the fact that $B \cap C = \{\}$ is evident from the picture. □

While this visualization tool is helpful for many aspects of the study of sets, we will use it mainly to analyze statements about sets like those in Proposition 1. To investigate a *statement* about generic sets using Venn diagrams, we try to draw the most general situation for two or three sets so that we can easily see how a proposition might fail. Since the sets in such a statement are not specifically given, we use shading to indicate regions in the picture that correspond to the results of the set operations within the proposed statement. This process is best illustrated with an example.

Example 9 *Use Venn diagrams to illustrate the truth of the distributive property,* $A \cap (B \cup C) = (A \cap B) \cup (A \cap C)$.

SOLUTION We do this by considering the two sides of the equation separately, in each case thinking about how to "build up" the set described. On the left-hand side, the set $A \cap (B \cup C)$ is built by first computing $B \cup C$ and then $A \cap (B \cup C)$. Figure 3-2 shows how to draw these two steps.

On the right-hand side, the set $(A \cap B) \cup (A \cap C)$ is built by first computing $A \cap B$ and $A \cap C$, and *then* forming $(A \cap B) \cup (A \cap C)$. Figure 3-3 shows how to draw these three steps. Since the diagram for $A \cap (B \cup C)$ is the same as the diagram for $(A \cap B) \cup (A \cap C)$, we have our visual evidence that the statement is true. □

Explore more on the Web.

Practice Problem 6 *Draw a Venn diagram to illustrate each side of the statement* $A - (B \cup C) = (A - B) \cap (A - C)$. *Use your diagram to decide if the statement is true.*

Venn diagrams do not exactly constitute formal proofs since they are based less on deductive logic than on visual intuition. They are also limited since they cannot be easily applied to situations involving more than three sets. These diagrams can,

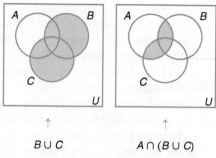

$$B \cup C \qquad\qquad A \cap (B \cup C)$$

Figure 3-2 Venn diagram for the
left-hand side of Example 9.

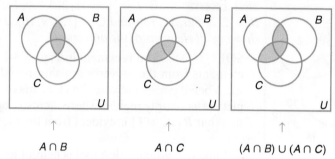

$$A \cap B \qquad\qquad A \cap C \qquad\qquad (A \cap B) \cup (A \cap C)$$

Figure 3-3 Venn diagram for the right-hand side of
Example 9.

however, be quite useful in determining whether we believe a proposition, and in finding counterexamples for false statements.

Example 10 *Find a counterexample to the following statement:*

$$A \cup (B - C) = (A \cup B) - C$$

SOLUTION   On the left-hand side, the set $A \cup (B - C)$ is built up from computing $B - C$ and then $A \cup (B - C)$, as shown in Figure 3-4.

On the right-hand side, the set $(A \cup B) - C$ is built up from computing $A \cup B$ and then $(A \cup B) - C$, as shown in Figure 3-5. We notice that in the two final diagrams above, the $A \cap C$ region is shaded in the diagram for $(A \cup B) - C$ but not shaded in the diagram for $A \cup (B - C)$. Hence, choosing sets A, B, and C so that $A \cap C$ is nonempty should provide a counterexample. Letting $A = \{1, 2\}$, $B = \{3, 4\}$, and $C = \{2, 5\}$, we see that $(A \cup B) - C = \{1, 3, 4\}$ and $A \cup (B - C) = \{1, 2, 3, 4\}$, which are not equal sets.   □

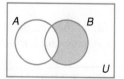

Figure 3-6   A Venn
diagram for two sets.

It should be noted that if only sets A and B are involved in a statement about sets, we can form a "two-set Venn diagram" by simply leaving out the set C in the diagram. For example, the Venn diagram for $A' \cap B$ is shown in Figure 3-6.

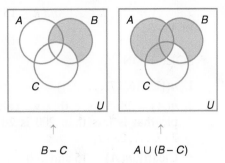

$$\uparrow \qquad\qquad \uparrow$$
$$B - C \qquad\qquad A \cup (B - C)$$

Figure 3-4 Venn diagram for the
left-hand side of Example 10.

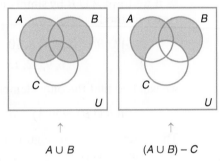

$$\uparrow \qquad\qquad \uparrow$$
$$A \cup B \qquad\qquad (A \cup B) - C$$

Figure 3-5 Venn diagram for the
right-hand side of Example 10.

*Explore more on
the Web.*

Practice Problem 7 *Draw a two-set Venn diagram for each side of the following
equation. Use the result to give a specific example to illustrate that the statement is false.*

$$(A \cup B) - B = A$$

The Inclusion-Exclusion Principle

Because sets provide a common structure for grouping together mathematical objects, a natural question for a given set A is "How many elements does A have?" We
will see in later chapters that this simple question is central to some fairly advanced
mathematics. In the present chapter, we will merely address some basic questions so
that we can build on the answers in the pages to come.

> **Definition** If A is a finite set, we use the notation $n(A)$ to indicate the
> number of elements in the set A.

 Example 11 *Find each of the following:*

1. $n(\{15, 16, 17, \ldots, 22\})$
2. *Let $A = \{m \in \mathbb{Z}^+ : m < 200 \text{ and } m \text{ is a multiple of } 7\}$. Find $n(A)$.*

3. Let $A = \{2k : k \in \mathbb{Z}^+ \text{ and } k \leq 15\}$ and $B = \{3k : k \in \mathbb{Z}^+ \text{ and } k \leq 10\}$. Find $n(A \cup B)$.

SOLUTION

1. $n(\{15, 16, 17, \ldots, 22\}) = 8$
2. $n(A) = 28$. To see this, we calculate $200/7 = 28.57$, so the largest multiple that is less than 200 is $28 \cdot 7 = 196$. So $A = \{7, 14, \ldots, 196\} = \{1 \cdot 7, 2 \cdot 7, \ldots, 28 \cdot 7\}$.
3. Clearly, $n(A) = 15$ and $n(B) = 10$. A little thought tells us that

$$A \cap B = \{6k : k \in \mathbb{Z}^+ \text{ and } k \leq 5\} = \{6, 12, 18, 24, 30\}$$

so if we form $A \cup B$ by simply combining the elements of A and B (via the roster method)

$$A \cup B = \{2, 4, 6, 8, 10, 12, 14, 16, 18, 20, 22, 24, 26, 28, 30,$$
$$3, 6, 9, 12, 15, 18, 21, 24, 27, 30\}$$

we will find that the only elements listed twice are those in $A \cap B$. Hence,

$$n(A \cup B) = n(A) + n(B) - n(A \cap B) = 15 + 10 - 5 = 20$$

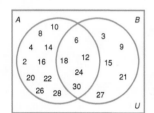

Figure 3-7 Venn diagram for Example 11.

The Venn diagram in Figure 3-7 illustrates this relationship among A, B, $A \cap B$, and $A \cup B$. ◻

The last example illustrates an important principle about the size of the union of sets, known as the *inclusion-exclusion principle*. We state the two-set and three-set versions as a theorem, and we leave the more general statement to be explored by the reader in the exercises.

Theorem 2 (The Inclusion-Exclusion Principle) *Let sets A, B, and C be given.*

1. $n(A \cup B) = n(A) + n(B) - n(A \cap B)$
2. $n(A \cup B \cup C) = n(A) + n(B) + n(C) - n(A \cap B) - n(A \cap C) - n(B \cap C) + n(A \cap B \cap C)$

The following informal proof for the two-set case will be sufficient for our purposes: *When you add $n(A)$ and $n(B)$, anything that is in both sets has been counted twice, so you need to subtract $n(A \cap B)$ to adjust for this overcount.* At the end of this section, Exercise 24 asks you to give a similar explanation of the three-set version. Exercise 25 then asks you to discover the correct four-set version of the principle.

Practice Problem 8 *For $A = \{n \in \mathbb{N} : 200 \leq n \leq 500 \text{ and } n \text{ is a multiple of } 7\}$ and $B = \{n \in \mathbb{N} : 200 \leq n \leq 500 \text{ and } n \text{ is a multiple of } 5\}$, calculate the following:*

(a) $n(A)$
(b) $n(B)$
(c) $n(A \cap B)$
(d) $n(A \cup B)$

Solutions to Practice Problems

1 (a) True. The universal set for this example might be the set of all strings consisting of the letters a and b.

   (b) True. Both sets contain exactly the numbers $1, 2, 3$, and 4, so these sets are equal. Any set is a subset of itself.

   (c) True. Every number in the set $\{6, 8, 10\}$ is also in the set $\{1, 2, 3, 4, 5, 6, 7, 8, 9, 10\}$.

   (d) False. The number $x = 1/2$ provides a counterexample, since it is in the set $\mathbb{Q}$ but not in the set $\mathbb{N}$.

   (e) True. It is impossible to find a counterexample, since a counterexample would have to be an element of the set on the left side, and $\{\}$ has no elements. Hence, the empty set is a subset of any set.

2 There are many correct ways to represent each set. We give only one for each.

   (a) The set of even integers is $\{x \in \mathbb{Z} : x = 2y \text{ for some } y \in \mathbb{Z}\}$.

   (b) The set of real numbers bigger than 10 is $\{x \in \mathbb{R} : x > 10\}$.

   (c) The set of rational numbers whose square is less than 2 is $\{x \in \mathbb{Q} : x^2 < 2\}$.

3 (a) $\{0, 5, 10, -5, -10, \ldots\} = \{5k : k \in \mathbb{Z}\}$

   (b) $\{1, 5, 10, 101, 1025, \ldots\} = \{k^2 + 1 : k \in \mathbb{N}\}$

   (c) $\{\ldots, -\frac{13}{15}, 0, -5, \frac{2}{3}, \ldots\} = \{\frac{a}{2k+1} : a \in \mathbb{Z}, k \in \mathbb{N}\}$

4 (a) Since B contains no elements, there cannot be any elements that A and B have in common, so $A \cap B = \emptyset$; that is, $A \cap \emptyset = \emptyset$.

   (b) $\{1, 3\}$. Observe that $A \cup \emptyset = A$.

   (c) $\{1, 2, 3, 4, 5\}$. $A \cup C = C$ because $A \subseteq C$.

   (d) No, because the element 1 is in A but not in B.

   (e) Yes. There are no possible counterexamples. The empty set is a subset of every set.

   (f) There are several ways to do this. One is to choose $D = \{1\}$ and $E = \{2, 3, 4, 5\}$. Another is to choose $D = \{1, 2, 3, 4, 5\}$ and $E = \emptyset$.

5 (a) $(A \cap B)' = \{2, 4\}' = \{1, 3, 5, 6, 7, 8, 9, 10\}$

   (b) $A' \cup B' = \{6, 7, 8, 9, 10\} \cup \{1, 3, 5, 7, 9\} = \{1, 3, 5, 6, 7, 8, 9, 10\}$. Yes, this is the same as $(A \cap B)'$.

   (c) $\emptyset' = U$ (since $\emptyset$ has nothing in it, "the set that contains those elements of the universal set U which are not in $\emptyset$" contains all the elements of U).

6 On the left-hand side, the set $A - (B \cup C)$ is built up from first computing $B \cup C$, and then $A - (B \cup C)$. Figure 3-8 shows how to draw these two steps.

   On the right-hand side, the set $(A - B) \cap (A - C)$ is built up from first computing $A - B$ and $A - C$, and then forming $(A - B) \cap (A - C)$. Figure 3-9 on page 194 shows how to draw these two steps.

   Since the diagram for $A - (B \cup C)$ is the same as the diagram for $(A - B) \cap (A - C)$, we believe that the statement is true.

7 If we let $A = \{1, 2\}$ and $B = \{1, 3\}$, then $A \cup B = \{1, 2, 3\}$ and so

$$(A \cup B) - B = \{2\} \neq A$$

8 (a) $n(\{203, 210, \ldots, 497\}) = n(\{29 \cdot 7, 30 \cdot 7, \ldots 71 \cdot 7\}) = 43$

   (b) $n(\{200, 205, \ldots, 500\}) = n(\{40 \cdot 5, 41 \cdot 5, \ldots 100 \cdot 5\}) = 61$

   (c) $n(\{n \in \mathbb{N} : 200 \leq n \leq 500 \text{ and } n \text{ is a multiple of } 35\}) = n(\{210, 245, \ldots, 490\}) = n(\{6 \cdot 35, 7 \cdot 35, \ldots, 14 \cdot 35\}) = 9$

   (d) $43 + 61 - 9 = 95$

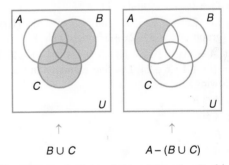

Figure 3-8   Left-hand side of Practice Problem 6.

Exercises for Section 3.1

1. Given the universal set $U = \{1, 2, 3, 4, 5, 6, 7, 8, 9, 10\}$ and sets $A = \{2, 4\}$, $B = \{1, 2, 8\}$, and $C = \{1, 2, 5, 6, 10\}$, find each of the following:

   (a) $A \cap C$

   (b) $A \cup B$

   (c) C'

   (d) $B - C$

   (e) $C \cap B'$

2. Let $A = \{x \in \mathbb{N} : x = 4k \text{ for some } k \in \mathbb{N}\}$, $B = \{y \in \mathbb{Z}^+ : 2y \text{ is a perfect square}\}$, and $C = \{z \in \mathbb{Z} : z^2 < 1,000\}$. List five elements in each of the following sets:

   (a) $A \cup (B \cap C)$

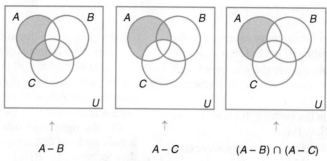

Figure 3-9 Right-hand side of Practice Problem 6.

(b) $(A \cup B) \cap C$

(c) $A \cap (B \cup C)$

(d) $A \cap (A \cup C)$

3. Fill in the blanks to make each sentence true.

   (a) The set of even integers is $\{x \in \mathbb{Z} : x =$ _____$\}$.

   (b) The set of integers that are powers of 2 is $\{x \in \mathbb{N} :$ $x =$ _____$\}$.

   (c) The set of odd natural numbers is $\{x \in$ _____ $: x =$ _____$\}$.

   (d) The set of odd perfect squares is $\{x \in \mathbb{N} : x =$ _____$\}$.

4. List five elements of each of the following sets:

   (a) $\{3n : n \in \mathbb{N}\}$

   (b) $\{m^2 : m \in \mathbb{Z} \text{ and } -5 < m < 5\}$

   (c) $\{3 - 2r : r \in \mathbb{Q} \text{ and } 0 \leq r \leq 5\}$

   (d) $\{3 - 2x + x^2 : x \in \mathbb{R} \text{ and } 0 \leq x \leq 5\}$ (HINT: A graph of $y = 3 - 2x + x^2$ might help.)

5. Let $A = \{3x + 1 : x \in \mathbb{N}\}$, $B = \{2x + 1 : x \in \mathbb{Z}^+\}$, and $C = \{6x + 1 : x \in \mathbb{Z}\}$. List five elements in each of the following sets:

   (a) $A \cap B$

   (b) $A \cup C$

   (c) $B - A$

   (d) $A - C$

6. Let $U = \{0, 1, 2, 3, 4, 5, 6\}$. Match each of the sets on the left with the appropriate set on the right. Not every set on the right will be used.

(a) $\{2k + 2 : k \in U\}$	(A)	$\{-2, 2, 6, 10, 14, 18, 22\}$
(b) $\{2m : m \in U\}$	(B)	$\{0, 4, 8, 12, 16, 20, 24\}$
(c) $\{n : n - 2 = 4L$ for	(C)	$\{4, 8, 12, 16, 20, 24, 28\}$
some $L \in U\}$		
(d) $\{4p : p \in U\}$	(D)	$\{0, 2, 4, 6, 8, 10, 12\}$
	(E)	$\{2, 6, 10, 14, 18, 22, 26\}$
	(F)	$\{2, 4, 6, 8, 10, 12, 14\}$

7. Fill in the blanks to make each sentence true.

   (a) The set of even integers is $\{2x : x \in$ _____$\}$.

   (b) The set of perfect cubes is $\{$_____$: x \in \mathbb{Z}\}$.

   (c) The set of natural numbers that end with a 7 is $\{$_____$: x \in$ _____$\}$.

   (d) The set of rational numbers between -1 and 1 is $\{\frac{a}{b} \in \mathbb{Q} :$ _____ $< a <$ _____$\}$.

8. Fill in the blanks with either the symbol $\subseteq$ or the symbol $\not\subseteq$ (which denotes "is not a subset of") to make the statement true. If you believe the first set is *not* a subset of the second set, give an example of an element that supports your answer.

   (a) $\mathbb{N}$ _____ $\mathbb{Z}^+$

   (b) $\mathbb{Z}$ _____ $\mathbb{Q}^+$

   (c) $\mathbb{Q}$ _____ $\mathbb{R}^+$

   (d) $\mathbb{R}$ _____ $\mathbb{R}^{\geq 0}$

   (e) $\mathbb{Z}$ _____ $\mathbb{Q}$

   (f) $\mathbb{R}$ _____ $\mathbb{Q}$

9. In each of the following problems, three sets are described. One of the sets is not the same as the other two. In each case, find the set that is not like the others.

   (a) $A = \{a + b : a \in \mathbb{N}, b \in \mathbb{N}\}$
   $B = \{a - b : a \in \mathbb{N}, b \in \mathbb{N}\}$
   $C = \mathbb{N}$

   (b) $A = \{x^2 : x \in \mathbb{N}\}$
   $B = \{x^2 : x \in \mathbb{Z}\}$
   $C = \{x^2 : x \in \mathbb{R}\}$

   (c) $A = \{4z + 2 : z \in \mathbb{Z}\}$
   $B = \{4z - 2 : z \in \mathbb{Z}\}$
   $C = \{2z + 4 : z \in \mathbb{Z}\}$

10. Write each of the following sets using a "form description" instead of a "property description." (For example, the property description $\{x \in \mathbb{Z} : x \text{ is even}\}$ can be written using the form description $\{2k : k \in \mathbb{Z}\}$.)

    (a) $\{x \in \mathbb{Z} : x \text{ is odd}\}$

    (b) $\{x \in \mathbb{Q} : x = 2^m \text{ for some } m \in \mathbb{Z}\}$

(c) $\{x \in \mathbb{N} : x \text{ is twice a perfect square}\}$

(d) $\{x \in \mathbb{Z} : x \text{ is the product of two consecutive integers}\}$

11. List five elements of each of the following sets. In each case, also state which of the sets $\mathbb{N}, \mathbb{Z}, \mathbb{Q}$, or $\mathbb{R}$ would be an appropriate universe for the set.

   (a) $\{\frac{2a}{2b+1} : a \in \mathbb{Z}, b \in \mathbb{Z}\}$

   (b) $\{a^2 + 1 : a \in \mathbb{N}\}$

   (c) $\{a^b : a \in \mathbb{Z}^+, b \in \mathbb{Z}\}$

   (d) $\{a^2 + b^2 : a \in \mathbb{N}, b \in \mathbb{N}\}$

   (e) $\{\sqrt{a} : a \in \mathbb{N}\}$

12. For each of the following, determine what the given elements have in common and then write the set-builder notation description of a set that includes them. (Note that there are many correct answers, but the best answers are the simplest ones to describe.)

   (a) $\{1, 3, 5, 7, 9, \ldots\}$

   (b) $\{11, 33, 55, 77, \ldots\}$

   (c) $\{1, 9, 17, 25, \ldots\}$

   (d) $\{1, 2, 4, 8, 16, \ldots\}$

   (e) $\{\frac{1}{2}, \frac{-11}{4}, \frac{7}{8}, \frac{113}{64}, \frac{-1}{1,024}, \ldots\}$

13. Decide if each of the following statements about finite sets is true or false. For any false ones, provide a specific counterexample to back up your answer. (You do not need to prove the true statements.)

   (a) If $A \subseteq B$, then $n(A) \le n(B)$.

   (b) If $n(A) \le n(B)$, then $A \subseteq B$.

   (c) If $A \subseteq B$ and $A \ne B$, then $n(A) < n(B)$.

   (d) If $n(A) < n(B)$, then $A \subseteq B$.

14. Sometimes you will see the notation $A \subset B$ or $A \subsetneq B$ to denote the fact that A is a subset of B but $A \ne B$. (We read $A \subset B$ as "A is a proper subset of B.") Write a few sentences explaining how to remember the difference between the set notation $A \subset B$ and $A \subseteq B$, using an analogy to the numerical notation $x < y$ and $x \le y$.

15. For the universal set $U = \{1, 2, 3, 4, 5, 6, 7, 8, 9, 10\}$ and sets $A = \{1, 3, 5\}$, $B = \{1, 2, 3, 4\}$, and $C = \{1, 2, 5, 6, 10\}$, verify that the following set properties hold by calculating each side of the equation separately.

   (a) The distributive property of $\cap$ over $\cup$:

$$A \cap (B \cup C) = (A \cap B) \cup (A \cap C)$$

   (b) The distributive property of $\cup$ over $\cap$:

$$A \cup (B \cap C) = (A \cup B) \cap (A \cup C)$$

   (c) One of DeMorgan's laws:

$$(A \cup B)' = A' \cap B'$$

(d) The other one of DeMorgan's laws:

$$(A \cap B)' = A' \cup B'$$

(e) One of the absorption properties:

$$A \cap (A \cup B) = A$$

(f) The other absorption property:

$$A \cup (A \cap B) = A$$

16. Use Venn diagrams to verify each of the following properties of our set operations:

   (a) The distributive property of $\cup$ over $\cap$:

$$A \cup (B \cap C) = (A \cup B) \cap (A \cup C)$$

   (b) One of DeMorgan's laws:

$$(A \cup B)' = A' \cap B'$$

   (c) The other one of DeMorgan's laws:

$$(A \cap B)' = A' \cup B'$$

   (d) One of the absorption properties:

$$A \cap (A \cup B) = A$$

   (e) The other absorption property:

$$A \cup (A \cap B) = A$$

17. Use Venn diagrams to check whether or not you believe these properties are true. For those that are not true, give a specific counterexample.

   (a) $A \cap (B \cup C) = (A \cap B) \cup C$

   (b) $(B \cup C) - A = (B - A) \cup (C - A)$

   (c) $A - (B \cap C) = (A - B) \cup (A - C)$

   (d) $(A - B) \cup (B - C) \subseteq (A - C)$

   (e) $(B \cap C) - A = B \cap (C - A)$

   (f) If $A \subseteq B$, then $A - B = B - A$. (HINT: Since $A \subseteq B$, draw the circle for A inside the circle for B.)

   (g) If $A \subseteq B$, then $A = A \cup B$.

18. Given $A = \{2x : x \in \mathbb{Z}\}$ and $B = \{3y : y \in \mathbb{Z}\}$, describe each of the following using the simplest set-builder notation possible:

   (a) $A \cap B$

   (b) $A - B$

   (c) $B - A$

   (d) $A \cup B$

   (e) $\mathbb{Z} - A$

19. A partial sum of the harmonic series is a sum of the form $\frac{1}{1} + \frac{1}{2} + \frac{1}{3} + \cdots + \frac{1}{n}$ for some positive integer n. $A = \{x \in \mathbb{Q} : x \text{ (in lowest terms) has an even denominator}\}$, $B = \{x \in \mathbb{Q} : x \text{ is a partial sum of the harmonic series}\}$, and $C = \{x \in \mathbb{Q} : 1 < x < 2\}$ are given.

(a) What does it mean in English if $B \subseteq A$?

(b) What does it mean in English if $B \subseteq C$?

(c) What does it mean in English if $A \cap C = \emptyset$?

(d) Write in notation involving A, B, or C and set operations the following claim: "Every rational number between 1 and 2 has an even denominator (when in lowest terms)."

(e) Write in notation involving A, B, or C and set operations the following claim: "If a rational number between 1 and 2 has an even denominator (when in lowest terms), then that number is a partial sum of the harmonic series."

20. The notation $[a, b]$ (called *interval notation*) is used to describe a *closed interval* of real numbers. That is, $[a, b] = \{x \in \mathbb{R} : a \le x \le b\}$. Rewrite these sets using either interval notation, set-builder notation, or a simple list of elements.

(a) $[-5, 3.2] \cap \mathbb{Z}$

(b) $[-6.1, 4] \cap [3.1, 7.2]$

(c) $[0, \pi] \cap [4, 10]$

21. There are several other variations on the *interval notation* from Exercise 20.

- $(a, b) = \{x \in \mathbb{R} : a < x < b\}$
- $[a, b) = \{x \in \mathbb{R} : a \le x < b\}$
- $(a, b] = \{x \in \mathbb{R} : a < x \le b\}$
- $[a, \infty) = \{x \in \mathbb{R} : a \le x\}$
- $(-\infty, b] = \{x \in \mathbb{R} : x \le b\}$

Rewrite the following sets using either interval notation, set-builder notation, or a familiar name of the set:

(a) $[0, 10] - [3, 12]$

(b) $(0, 10] - [5, 10)$

(c) $(-5, \infty) \cup (-\infty, 2]$

(d) $(-\infty, 2) \cap (-\infty, 0]$

(e) $\mathbb{Q} \cap [0, \infty)$

(f) $\mathbb{Z} \cap [0, \infty)$

22. Express each of the following sets as an interval or the union of intervals:

(a) $\{3x : x \in [0, 4]\}$

(b) $\{3 - 2x : x \in [0, 4]\}$

(c) $\{3 - 2x + x^2 : x \in [0, 6]\}$ (HINT: A graph of $y = 3 - 2x + x^2$ might help.)

(d) $\{x \in \mathbb{R} : 3 - 2x + x^2 \in [3, 6]\}$ (HINT: A graph of $y = 3 - 2x + x^2$ might help.)

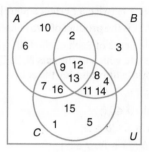

Figure 3-10   Sets for Exercises 23 and 24.

23. See Figure 3-10.

(a) Calculate $n(A)$, $n(B)$, and $n(C)$, where

$$A = \{2, 6, 7, 9, 10, 12, 13, 16\},$$
$$B = \{2, 3, 4, 8, 9, 11, 12, 13, 14\}$$
$$C = \{1, 4, 5, 7, 8, 9, 11, 12, 13, 14, 15, 16\}$$

(b) For the sets in part (a), find the following, using (where appropriate) the two-set inclusion-exclusion counting principle.

i. $n(A \cap B)$ and $n(A \cup B)$

ii. $n(A \cap C)$ and $n(A \cup C)$

iii. $n(B \cap C)$ and $n(B \cup C)$

iv. $n(A \cap B \cap C)$

v. $n(A \cup B \cup C)$

24. See Figure 3-10 for the Venn diagram of the sets A, B, and C in Exercise 23. Based on the two-set inclusion-exclusion principle, you might expect that $n(A \cup B \cup C)$ would be $n(A) + n(B) + n(C) - n(A \cap B) - n(A \cap C) - n(B \cap C)$—that is, add up the three sets and then subtract the overlaps. However, this does not match your answer in Exercise 23(b). Why is this formula not correct? In particular, how many times does this formula count each element in $A \cap B \cap C$?

25. State the four-set version of the inclusion-exclusion principle, and invent an example to illustrate that you are correct.

26. Let $T = \{x \in \mathbb{N} : x \le 1{,}000\}$.

(a) Use the inclusion-exclusion property to find the size of the set

$$\{n \in T : n \text{ is a multiple of 2 or 3}\}$$

(b) Use the inclusion-exclusion property to find the size of the set

$$\{n \in T : n \text{ is a multiple of 2 or 3 or 5}\}$$

(c) Use your answer to the previous part to find the size of the set

$$\{n \in T : n \text{ is a multiple of neither 2 nor 3 nor 5}\}$$

27. A certain club is forming a recruitment committee consisting of five of its members. They have calculated that there are 8,568 different ways to form such a committee. The club has two members named Jack and Jill. They have calculated that 2,380 of the potential committees have Jack on them, 2,380 have Jill, 1,820 have Jack but not Jill, 1,820 have Jill but not Jack, and 560 have both Jack and Jill.

(a) How many committees have either Jack or Jill?

(b) How many committees have neither Jack nor Jill?

(c) Jack and Jill are car-pooling, so they insist that if either one is on the committee, the other person must also be on the committee. How many committees meet this condition?

(d) Jack and Jill have had a fight. Jack says, "If Jill is on the committee, I won't be." Jill says, "Likewise." How many of the committees meet this condition?

28. In Mrs. Smith's science class, there are a total of 25 students. Of these, 15 are female, and 15 are honor students. Draw a Venn diagram for a specific situation consistent with this information, and then answer the following questions:

(a) What is the smallest possible number of female students who are on the honor roll?

(b) What is the largest possible number of female students who are on the honor roll?

29. In Mr. Jones' math class, all 40 students must complete activities at three stations during the current marking period. So far, 6 students have completed stations A and B, 10 students have completed stations A and C, and 15 students have completed stations B and C. Everyone in class has completed at least one station, and for each of the three stations, five students have completed only that one station. Draw a Venn diagram for this situation, and use it to determine how many students have completed all three stations.

30. Students at Apollo Elementary School receive ribbons at the end of each year at a school-wide awards ceremony. This year 120 students receive gold stars for perfect attendance, 180 receive certificates for participating in the science fair, and 80 students receive blue ribbons for outstanding grades. Of these, 40 students who receive the attendance star receive no other award, 50 students who receive the science fair certificate re-

ceive no other award, and 10 students who receive the blue ribbon receive no other awards. In addition, 10 students receive all three awards and 65 students receive no awards. Draw a Venn diagram for this situation, and determine how many students attend the school this year.

31. Students at Apollo Elementary School receive ribbons at the end of each year at a school-wide awards ceremony. Last year 100 students received gold stars for perfect attendance, 200 received certificates for participating in the science fair, and 50 students received blue ribbons for outstanding grades. In addition, 10 students received all three awards and 80 students received no awards. Draw a Venn diagram for a specific situation consistent with this information, and then answer the following questions:

(a) What was the smallest possible number of children in the school last year?

(b) What was the largest possible number of children in the school last year?

(c) If in addition we know that a total of 100 children received exactly one of the three awards, what was the exact number of students in the school last year?

32. For each of the following, determine if the statement is true or false. If false, give a specific counterexample. If true, explain informally (using Venn diagrams if you wish) why you think so.

(a) If A and B are disjoint, then $A - B = A$.

(b) If A and B are disjoint, then $n(A \cup B) = n(A) + n(B)$.

(c) If A and B are disjoint, then $n(B - A) = n(B) - n(A)$.

(d) If $A \subseteq B$, then $n(B - A) = n(B) - n(A)$.

(e) If $A \subseteq B$, then $A - B = B - A$.

33. Discuss the following paradox attributed to Bertrand Russell. Let S be defined as the set of all sets X for which $X \notin X$.

(a) Is $S \in S$?

(b) Is $S \notin S$?

(c) If $\mathcal{C}$ denotes the collection of all possible sets, is $\mathcal{C}$ itself a set? (HINT: The previous parts are related to this question.)

3.2 More Operations on Sets

So far we have seen how to build new sets by combining or restricting the elements in existing sets. For example, if A and B are sets of *numbers*, then using any of the operations discussed so far, the resulting set will also be a set of *numbers*. We will now see that there are other ways to build new sets from old that involve creating completely new types of objects.

Cartesian Products

We first consider sets of pairs of objects. Anyone who has drawn a graph for an algebra class is already familiar with ordered pairs of numbers. In algebra, we sometimes plot individual points and sometimes graph "smooth" lines or curves. In either case, we are creating a visual representation of a set of ordered pairs, as the following example illustrates.

Example 1 *In Figure 3-11, the graph on the left illustrates the set of points* $\{(1, 2), (2, 5)\}$, *while the graph on the right illustrates the set of all ordered pairs* (x, y) *of real numbers, where the y-coordinate is always one less than three times the x-coordinate.*

These examples show that we need notation and some language for working with sets of ordered pairs.

> **Definition**   Given sets A and B, we define $A \times B = \{(a, b) : a \in A$ and $b \in B\}$. (We read $A \times B$ as "A cross B" and call (a, b) "the ordered pair a, b.") We often refer to $A \times B$ as the *Cartesian* * *product* of A and B. In the common special case that both coordinates are taken from the same set, we often write A^2 instead of $A \times A$.

With this notation, the set of points on the right in Example 1 could be described by

$$\{(x, y) \in \mathbb{R}^2 : y = 3x - 1\}$$

which is closer to the way we usually would describe the line in an algebra class.

When it is possible to draw a coordinate axis as in Example 1 with the horizontal axis labeled with elements of A and the vertical axis labeled with elements of B, then we can produce a picture like the one in that example. In particular, this is always possible to do when the elements of A and B are numbers and when A and B are small sets. We will refer to the picture we get as the *Cartesian graph* of the set of ordered pairs. We will use the adjective "Cartesian" faithfully because of an unfortunate confluence of math terminology that can lead to confusion in Chapter 7—it is nothing to worry about at the moment.

---

* The adjective "Cartesian" pays homage to René Descartes (1596–1650), a French mathematician and philosopher who greatly contributed to the notion that geometry could be studied using coordinates.

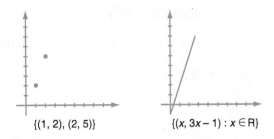

$\{(1, 2), (2, 5)\}$ $\{(x, 3x - 1) : x \in \mathsf{R}\}$

Figure 3-11 The Cartesian graph of a set of ordered pairs.

Practice Problem 1 *Draw the Cartesian graph for each of the following sets of ordered pairs:*

(a) $\{(1, 2), (1, 3), (2, 2), (2, 4), (3, 1)\}$

(b) $\{(x, y) \in \mathbb{R}^2 : y = 2x - 1\}$ *(It is only possible to draw a portion of the graph for this set.)*

(c) $\{(x, y) \in \mathbb{N}^2 : y = x^2\}$ *(It is only possible to draw a portion of the graph for this set.)*

Of course, sets of ordered pairs are relevant beyond the confines of a high school algebra course. In particular, we have already seen them (in spirit at least) in some of our earlier work in this book, and we will continue to bump into them in the future.

 Example 2 *Express each of the sets described below using efficient set notation.*

1. *In Chapter 1, when encountering two inhabitants A and B of the Island of Liars and Truthtellers, we found ourselves considering all possibilities for what type of person each was. That is, we formed (by way of rows in a truth table) the set of all pairs of the form*

   (type of A, type of B)

2. *In Chapters 1 and 2, we described a sequence like*

   $$2, 4, 8, 16, 32, \ldots$$

using the notation $a_1 = 2$, $a_2 = 4$, $a_3 = 8$, and so forth. To avoid the subscripted letter a, we could have used ordered pairs where the first coordinate tells the position of a term in the sequence and the second coordinate describes the value of that term:

   $$(1, 2), (2, 4), (3, 8), (4, 16), (5, 32), \ldots$$

3. *In Chapter 5, we will develop procedures for counting the size of increasingly complex sets. An early example asks how many ways one can order a sandwich and a drink from a restaurant that numbers sandwich choices 1, 2, 3, and labels drink choices A, B, C, D. The set of all possible choices includes ordered pairs like these:*

   $$(1, A), (1, C), (2, A), (3, D), \ldots$$

SOLUTION

1. If we use $V = \{$Truthteller, Liar$\}$, then the given set can be simply expressed as $V \times V = V^2$.

2. $\{(n, 2^n) : n \in \mathbb{Z}^+\}$

3. $\{1, 2, 3\} \times \{A, B, C, D\}$

Before moving on to look at other set structures, we will investigate the size of the Cartesian product of two sets. This is obviously relevant to the last part of the preceding example.

Example 3 *Let $A = \{2, 4, 6, 8\}$ and $B = \{1, 2, 3, 4, 5\}$. List all elements in $A \times B$ in an orderly way. Will $B \times A$ have more elements than $A \times B$? Are they the same set? Draw the Cartesian graph of both sets.*

SOLUTION   In each case, we list the elements of the new set in a tabular format that allows us to see a pattern. Table 3-1 shows $A \times B$ followed by $B \times A$. Because a 4×5 table and a 5×4 table have the same size, we can see that $A \times B$ and $B \times A$ are sets of the same size. The Cartesian graph, shown in Figure 3-12, makes the relationship between the size of $A \times B$ and the size of $B \times A$ even more apparent. To see that $A \times B$ and $B \times A$ are not the same set, we simply note that there are elements of $A \times B$, like $(2, 1)$ and $(8, 5)$, for example, that are not elements of $B \times A$.

Elements of A $\downarrow$	Elements of $B \rightarrow$				
	1	2	3	4	5
2	$(2, 1)$	$(2, 2)$	$(2, 3)$	$(2, 4)$	$(2, 5)$
4	$(4, 1)$	$(4, 2)$	$(4, 3)$	$(4, 4)$	$(4, 5)$
6	$(6, 1)$	$(6, 2)$	$(6, 3)$	$(6, 4)$	$(6, 5)$
8	$(8, 1)$	$(8, 2)$	$(8, 3)$	$(8, 4)$	$(8, 5)$

Elements of B $\downarrow$	Elements of $A \rightarrow$			
	2	4	6	8
1	$(1, 2)$	$(1, 4)$	$(1, 6)$	$(1, 8)$
2	$(2, 2)$	$(2, 4)$	$(2, 6)$	$(2, 8)$
3	$(3, 2)$	$(3, 4)$	$(3, 6)$	$(3, 8)$
4	$(4, 2)$	$(4, 4)$	$(4, 6)$	$(4, 8)$
5	$(5, 2)$	$(5, 4)$	$(5, 6)$	$(5, 8)$

Table 3-1 $A \times B$ and $B \times A$ in Example 3

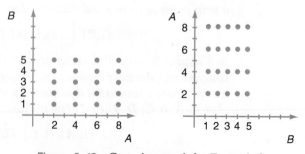

Figure 3-12   Cartesian graph for Example 3.

Practice Problem 2   *For $A = \{1, 2\}$ and $B = \{2, 4, 6, 8\}$, use the table approach above to list all the elements of the set $A \times B$. What is $n(A \times B)$?*

The table and Cartesian graphs lead to an obvious observation about the size of the Cartesian product of any two sets A and B. We will step through one possible proof of this fact in the exercises.

Theorem 1   *For all finite sets A and B,*

$$n(A \times B) = n(A) \cdot n(B)$$

PROOF   See Exercise 20 at the end of this section.   ■

We have seen that the Cartesian product of sets A and B is a set of ordered *pairs*. There is no reason to restrict ourselves to *pairs* of elements when it might be natural to have more than two objects tied together in this way. We can easily modify the language of *ordered pair* to include *ordered triples* or *ordered quadruples*, and the analogy is fairly clear. At some point, we will no longer know common English words for describing the number of elements we would like to convey, so mathematicians adopt a convention that is as descriptive as it is funny to say.

Definition   For any integer $n \geq 3$, the structure $(x_1, x_2, \ldots, x_n)$ is called an *n* - tuple.*

This generalization of ordered pair leads to the obvious generalization of the Cartesian product.

Definition   Given sets $S_1, S_2, \ldots, S_n$, we define the Cartesian product

$$S_1 \times S_2 \times \cdots \times S_n$$

as the set of all *n* - tuples $(x_1, x_2, \ldots, x_n)$ such that $x_1 \in S_1$, $x_2 \in S_2$, and so on. As before, in the event that the sets S_1, S_2, and so on are actually all the same set, we will often use the notational shorthand

$$S^n = \underbrace{S \times S \times \cdots \times S}_{n \text{ terms}}$$

Example 4   *Express each of the sets described below using efficient set notation.*

1. In Chapter 1, when analyzing a proposition with four propositional variables p, q, r, and s, we examined (via rows in a truth table) all possible quadruples of the form

(*truth value of p, truth value of q, truth value of r, truth value of s*)

---
* You can pronounce this as "tupple" or "toople." In either case, mathematicians will know what you mean, while nonmathematicians will just crack up.

2. *The familiar* 12×12 *multiplication table we learn in elementary school conveys a relationship among three pieces of information, two factors and their product. Hence, instead of using the table format, we could describe the information in the table as a list* $(2, 3, 6)$, $(5, 1, 5)$, *and so on, where the third number is product of the first two numbers.*

SOLUTION

1. This set can be expressed as $\{T, F\} \times \{T, F\} \times \{T, F\} \times \{T, F\}$ or equivalently as $\{T, F\}^4$.

2. If we use S to denote the set $\{1, 2, 3, \ldots, 11, 12\}$, then we can express the 12×12 multiplication table using the "property description"

$$\{(a, b, c) \in S \times S \times \mathbb{N} : a \cdot b = c\}$$

or the "form description"

$$\{(a, b, a \cdot b) : (a, b) \in S \times S\}$$

$\square$

Practice Problem 3   *List five elements of each of the following sets:*

(a) $\{T, F\}^3$

(b) *The set*

$$S = \{s \in \{0, 1\}^5 : s \text{ consists of three 0's and two 1's in some order}\}$$

(c) *At a restaurant that serves sandwiches from the set* $\{1, 2, 3, 4, 5\}$, *side orders from the set* $\{x, y, z\}$, *and drinks from the set* $\{A, B, C, D\}$, *the set of all possible ways to order one of each type of item can be described as*

$$\{1, 2, 3, 4, 5\} \times \{x, y, z\} \times \{A, B, C, D\}$$

To finish this discussion and to follow up on our earlier work on the size of the Cartesian product of sets, we present the following generalization of Theorem 1.

Theorem 2   *For any finite sets* $S_1, S_2, \ldots, S_k$,

$$n(S_1 \times S_2 \times \cdots \times S_k) = n(S_1) \cdot n(S_2), \cdots, n(S_k)$$

PROOF   See Exercise 23 at the end of this section.   ■

Sets of Sets

We will now look at another common construction with sets. Namely, we will consider the possibility that the elements of a set can themselves be sets. For example, the notation $\{1, \{3, 4\}, \{1, 2\}\}$ simply describes a set with three elements, one of which is a number (namely, 1) and the other two of which are sets (namely, $\{3, 4\}$ and $\{1, 2\}$).

If this seems strange, imagine a school supply box containing three objects: a pencil and two pencil boxes, with each containing pencils. The notion that a set can contain other sets as elements is really no different than the idea that a box can contain other boxes. However, many people confuse the basic set terminology like "element" and "subset" when these types of sets are used. The following example is intended to help clarify some of these potential difficulties.

Example 5 *Answer each of the following questions about sets:*

1. *Is $\{1, 2\} \subseteq \{\{1, 2\}, \{1, 3, 4\}\}$?*
2. *Is $\{1, 2\} \in \{\{1, 2\}, \{1, 3, 4\}\}$?*
3. *What is $n(\{\{1, 2\}, \{2, 4\}, \{1, 3, 4\}\})$?*

SOLUTION

1. No. The number 1 is an element of the set $\{1, 2\}$, but the two elements of $\{\{1, 2\}, \{1, 3, 4\}\}$ are $\{1, 2\}$ and $\{1, 3, 4\}$—while these sets themselves contain 1 as an element, there is no doubt that neither of these sets is actually equal to 1.
2. Yes. As mentioned above, the set $\{1, 2\}$ is an element of the set $\{\{1, 2\}, \{1, 3, 4\}\}$.
3. $n(\{\{1, 2\}, \{2, 4\}, \{1, 3, 4\}\}) = 3$ since the set $\{\{1, 2\}, \{2, 4\}, \{1, 3, 4\}\}$ contains three elements, each of which happens to be a set. □

Practice Problem 4 *The set $S = \{1, 2, 3, \{1, 2\}, \{1, 3, 4\}\}$ is strange since some of its elements are numbers and some of its elements are sets. List the five elements of S. What element of S is also a subset of S?*

Once we accept the legitimacy of a set of sets, it makes sense to define an operation that produces the set of all subsets of a given set.

Definition $\mathcal{P}(A) = \{S : S \subseteq A\}$ (We read $\mathcal{P}(A)$ as "the power set of A.")

Example 6 *Let $A = \{1, 2, 3, 4\}$, $B = \{2, 4, 6, 8\}$, and $C = \{1, 2, 3, 4, 5\}$.*

1. *List all elements in $\mathcal{P}(A)$ in an orderly way.*
2. *How much larger do you expect $\mathcal{P}(B)$ to be?*
3. *How much larger do you expect $\mathcal{P}(C)$ to be?*

SOLUTION

1. Table 3-2 illustrates a systematic way to build up the answer one row at a time. Each row contains everything in the previous row, plus all the new sets (i.e., those with the new element). To get the new sets, we simply insert the new element into each of the old sets.

$\mathcal{P}(\{1\})$	∅	{1}						
$\mathcal{P}(\{1,2\})$	∅	{1}						
	{2}	{1, 2}						
$\mathcal{P}(\{1,2,3\})$	∅	{1}	{2}	{1, 2}				
	{3}	{1, 3}	{2, 3}	{1, 2, 3}				
$\mathcal{P}(\{1,2,3,4\})$	∅	{1}	{2}	{1, 2}	{3}	{1, 3}	{2, 3}	{1, 2, 3}
	{4}	{1, 4}	{2, 4}	{1, 2, 4}	{3, 4}	{1, 3, 4}	{2, 3, 4}	{1, 2, 3, 4}

Table 3-2 Building the Power Set Recursively

2. We expect that $\mathcal{P}(\{2, 4, 6, 8\})$ will have the same size as $\mathcal{P}(\{1, 2, 3, 4\})$ since the number of subsets of a four-element set should be the same regardless of what the four elements are.

3. Using this same pattern, we expect $\mathcal{P}(\{1, 2, 3, 4, 5\})$ to contain 32 subsets—the 16 we already have from $\mathcal{P}(\{1, 2, 3, 4\})$, and each of those same 16 subsets but with the element 5 inserted.

□

The solution to Example 6 suggests a method for a proof of the following theorem.

Theorem 3   *For any finite set A, if $k = n(A)$, then $n(\mathcal{P}(A)) = 2^k$.*

PROOF   See Exercise 24 at the end of this section.   ∎

Before moving on, let's look at one more practice problem that combines the two main ideas from this section.

Practice Problem 5   *Let $A = \{1, 2, 3\}$ and $B = \{1, 3, 5\}$.*

1. *Which is the larger set, $\mathcal{P}(A \cup B)$ or $\mathcal{P}(A) \cup \mathcal{P}(B)$?*
2. *Let $C = \{x, y, z\}$. Which is the larger set, $C \times (A \cap B)$ or $(C \times A) \cap (C \times B)$?*

Partitions of a Set

There are many instances in mathematics where a particular type of set whose elements are sets is relevant. The basic idea is even present in problems being solved every day by children on playgrounds everywhere.

Example 7   *Alison, Billie, Chris, David, Ellen, and Fred want to play three-on-three basketball. How can we address the question "How many ways can they divide themselves into teams?" using the idea of sets of sets?*

SOLUTION   A basketball game can be represented as a set containing two basketball teams, each of which is a set of three players. Here are three examples of basketball games represented in this way:

● $\{\{A, B, C\}, \{D, E, F\}\}$

- $\{\{A, B, F\}, \{D, E, C\}\}$
- $\{\{F, C, B\}, \{D, E, A\}\}$ □

Each of the examples above is indeed a set of sets, but they are special in a couple of ways. First, all six children want to play, so all the letters A, B, C, D, E, F must be used. Second, no child can be on both teams. These two properties are the defining characteristics of what we will call a *partition* of the set of children.

Definition   For a set A, a *partition of A* is a set $S = \{S_1, S_2, S_3, \ldots\}$ of subsets of A (each set S_i is called *a part of S*) such that

1. For all i, $S_i \neq \emptyset$. That is, each part is nonempty.
2. For all i and j, if $S_i \neq S_j$, then $S_i \cap S_j = \emptyset$. That is, different parts have nothing in common.
3. $S_1 \cup S_2 \cup S_3 \cup \cdots = A$. That is, every element in A is in some part.

Practice Problem 6   *The six children in Example 7 want to form three teams of two children each for a doubles table tennis tournament. Use partitions to list all the ways they can do this.*

Example 8 *For each of the following requirements, find a partition of the set $\{1, 2, 3, 4, 5, 6\}$ that satisfies it:*

1. *Every part has the same size.*
2. *No two parts have the same size.*
3. *There are as many parts as possible.*
4. *There are as few parts as possible.*

SOLUTION   For the first two parts, there is more than one answer.

1. $\{\{1, 3\}, \{2, 4\}, \{5, 6\}\}$
2. $\{\{2\}, \{3, 6\}, \{4, 1, 5\}\}$
3. $\{\{1\}, \{2\}, \{3\}, \{4\}, \{5\}, \{6\}\}$
4. $\{\{1, 2, 3, 4, 5, 6\}\}$ □

We can also have partitions of our (infinite) universal sets $\mathbb{Z}$, $\mathbb{Q}$, and so on. Some of these partitions are connected to concepts we have already studied, and some are important for higher mathematics.

Example 9 *Verify that each of the following are partitions of the given set:*

1. *Let $A = \{2k : k \in \mathbb{Z}\}$ and $B = \{2k + 1 : k \in \mathbb{Z}\}$. Show that $\{A, B\}$ is a partition of $\mathbb{Z}$.*

2. Let $A = \{3k : k \in \mathbb{Z}\}$, $B = \{3k+1 : k \in \mathbb{Z}\}$, and $C = \{3k+2 : k \in \mathbb{Z}\}$. *Show that* $\{A, B, C\}$ *is a partition of* $\mathbb{Z}$.

3. For a rational number r, let $P_r = \{(a, b) \in \mathbb{Z} \times \mathbb{Z}^+ : a/b = r\}$. *Show that* $\{P_r : r \in \mathbb{Q}\}$ *is a partition of* $\mathbb{Z} \times \mathbb{Z}^+$.

SOLUTION   In each case, we verify the three properties of a partition given by the definition:

1. (1) Both A and B are nonempty since $0 \in A$ and $1 \in B$. (2) $A \cap B = \emptyset$ since any number x in both A and B would have to satisfy $x = 2K$ and $x = 2L+1$ for some $K \in \mathbb{Z}$ and $L \in \mathbb{Z}$, which in turn would imply that $K - L = \frac{1}{2}$, which is impossible for integers K and L. (3) The division theorem (for division by 2) from Section 2.2 tells us that any integer must be in A or B.

2. (1) All three sets are nonempty since $0 \in A$, $1 \in B$, and $2 \in C$. (2) $A \cap B = \emptyset$ since any number x in both A and B would have to satisfy $x = 3K$ and $x = 3L+1$ for some $K \in \mathbb{Z}$ and $L \in \mathbb{Z}$, which in turn would imply that $K - L = \frac{1}{3}$, which is impossible for integers K and L. A similar argument shows that $A \cap C = \emptyset$ and $B \cap C = \emptyset$. (3) The division theorem (for division by 3) from Section 2.2 tells us that any integer must be in A, B, or C.

3. (1) Let $r \in \mathbb{Q}$ be given. Since $r = \frac{a}{b}$ for some $a \in \mathbb{Z}$, $b \in \mathbb{Z}^+$ by definition of $\mathbb{Q}$, so $(a, b) \in \mathbb{Z} \times \mathbb{Z}^+$. Hence, P_r is nonempty. (2) Let $r \neq s \in \mathbb{Q}$ be given. $P_r \cap P_s = \emptyset$, since any pair (a, b) in both sets would have to satisfy $r = \frac{a}{b}$ and $s = \frac{a}{b}$, which would in turn mean that $r = s$, a contradiction. (3) Since for any $(a, b) \in \mathbb{Z} \times \mathbb{Z}^+$, we know that $\frac{a}{b} \in \mathbb{Q}$, so it follows that $(a, b) \in P_{a/b}$. Hence, every element of $\mathbb{Z} \times \mathbb{Z}^+$ is in one of the parts.   □

Another common way to define a partition of a set A is to describe which elements of A should belong to the same part of the partition. This is difficult to write down with our set-builder notation right now, but we will come back to this issue in Chapter 4 when we have a richer vocabulary for writing the description. The danger in the type of description below is that it is easy to accidentally describe something that is not actually a partition.

Example 10 *For each description, write down the partition of the set* $A = \{1, 2, \ldots, 8, 9\}$ *described, if possible.*

1. Elements $a, b \in A$ are in the same part if and only if $a - b$ is even.
2. Elements $a, b \in A$ are in the same part if and only if a^2 and b^2 have the same digit in the ones place.
3. Elements $a, b \in A$ are in the same part if and only if $a + b$ is a prime number.

SOLUTION

1. The partition is $\{\{1, 3, 5, 7, 9\}, \{2, 4, 6, 8\}\}$.
2. The partition is $\{\{1, 9\}, \{2, 8\}, \{3, 7\}, \{4, 6\}, \{5\}\}$.
3. There is no such partition. To see why, consider the numbers 2, 3, and 5. According to the description, 2 and 3 should be in the same part (since $2 + 3 = 5$ is prime), 2 and 5 should be in the same part (since $2 + 5 = 7$ is

prime), but 3 and 5 should *not* be in the same part (since $3 + 5 = 8$ is not prime). This is impossible. ☐

More on the Size of Sets

We close this section with an important connection between two of the structures we have discussed. This connection will be developed much more in Chapter 5.

If you have been paying attention, you may have noticed that the results about the size of Cartesian products and the size of the power set are identical in some cases.

- For $A = \{1, 2, \ldots, k\}$, Theorem 3 tells us that $n(\mathcal{P}(A)) = 2^k$.
- For $S = \{0, 1\}$, Theorem 2 tells us that $n(S^k) = 2^k$.

Could this be a coincidence? Of course not. To see the connection, we will list the elements in the two relevant sets when k has a very manageable value of 3.

Example 11 *Find a direct connection between the sets $\mathcal{P}(\{1, 2, 3\})$ and $\{0, 1\} \times \{0, 1\} \times \{0, 1\}$.*

SOLUTION In Table 3-3, each set on the left describes which coordinates are 1's in the corresponding ordered triple on the right. ☐

Practice Problem 7 *Table 3-4 maintains the same correspondence that we saw in Example 11. Fill in the missing entries.*

In some books, this correspondence is taken one step further, leading the authors to use the notation 2^A instead of $\mathcal{P}(A)$ for the power set of A. Given the above correspondence, this choice of notation is very appropriate.

$\mathcal{P}(\{1, 2, 3\})$	$\{0, 1\}^3$
$\{\}$	$(0, 0, 0)$
$\{3\}$	$(0, 0, 1)$
$\{2\}$	$(0, 1, 0)$
$\{2, 3\}$	$(0, 1, 1)$
$\{1\}$	$(1, 0, 0)$
$\{1, 3\}$	$(1, 0, 1)$
$\{1, 2\}$	$(1, 1, 0)$
$\{1, 2, 3\}$	$(1, 1, 1)$

Table 3-3
Correspondence for
Example 11

$\mathcal{P}(\{1, 2, 3, 4\})$	$\{2, 4\}$		$\{1, 3\}$		
$\{0, 1\}^4$		$(0, 1, 0, 0)$		$(1, 1, 0, 0)$	$(1, 1, 0, 1)$

Table 3-4   Complete for Practice Problem 7

Solutions to Practice Problems

1 Figure 3-13 on page 208 shows the three Cartesian graphs in order from left to right.

2 $A \times B = \{(1, 2), (1, 4), (1, 6), (1, 8), (2, 2), (2, 4), (2, 6), (2, 8)\}$, so $n(A \times B) = 8$.

3 (a) $\{T, F\}^3 = \{(T, T, T), (T, T, F), (T, F, T), (T, F, F), (F, T, T), \ldots\}$

   (b) $S = \{(0, 0, 0, 1, 1), (0, 0, 1, 0, 1), (1, 1, 0, 0, 0), (0, 1, 0, 1, 0), (1, 0, 0, 0, 1), \ldots\}$

(c) $(1, x, A), (1, y, D), (2, x, A), (3, y, B), (2, x, C), \ldots$ are a few of the elements.

4 $\{1, 2\}$ is an element of the set $\{1, 2, 3, \{1, 2\}, \{1, 3, 4\}\}$. Since 1 and 2 are also *elements* of $\{1, 2, 3, \{1, 2\}, \{1, 3, 4\}\}$, then $\{1, 2\}$ is a subset of $\{1, 2, 3, \{1, 2\}, \{1, 3, 4\}\}$ as well.

5 Let $A = \{1, 2, 3\}$ and $B = \{1, 3, 5\}$.

   (a) $\mathcal{P}(A) \cup \mathcal{P}(B)$ consists of all sets that are subsets of A or subsets of B, while $\mathcal{P}(A \cup B)$ consists of all subsets of $A \cup B$. The latter includes more items since,

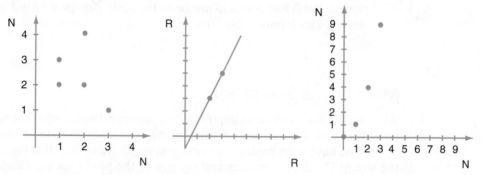

Figure 3-13   Cartesian graphs for Practice Problem 1.

for example, $\{1, 2, 5\}$ is a subset of $A \cup B$ while not being a subset of either A or B itself.

(b) Let $C = \{x, y, z\}$. $C \times (A \cap B)$ and $(C \times A) \cap (C \times B)$ both describe all ordered pairs with the first ele-

ment from C and the second from $\{1, 3\}$, so they are identical sets.

6   The complete list of all 15 partitions is given below:

7   Table 3-5 has been completed.

$$\{\{A, B\}, \{C, D\}, \{E, F\}\} \quad \{\{A, B\}, \{C, E\}, \{D, F\}\} \quad \{\{A, B\}, \{C, F\}, \{D, E\}\}$$
$$\{\{A, C\}, \{B, D\}, \{E, F\}\} \quad \{\{A, C\}, \{B, E\}, \{D, F\}\} \quad \{\{A, C\}, \{B, F\}, \{D, E\}\}$$
$$\{\{A, D\}, \{B, C\}, \{E, F\}\} \quad \{\{A, D\}, \{B, E\}, \{C, F\}\} \quad \{\{A, D\}, \{B, F\}, \{C, E\}\}$$
$$\{\{A, E\}, \{B, C\}, \{D, F\}\} \quad \{\{A, E\}, \{B, D\}, \{C, F\}\} \quad \{\{A, E\}, \{B, F\}, \{C, D\}\}$$
$$\{\{A, F\}, \{B, C\}, \{D, E\}\} \quad \{\{A, F\}, \{B, D\}, \{C, E\}\} \quad \{\{A, F\}, \{B, E\}, \{C, D\}\}$$

List for Practice Problem 6.

$\mathcal{P}(\{1,2,3,4\})$	$\{2,4\}$	$\{2\}$	$\{1,3\}$	$\{1,2\}$	$\{1,2,4\}$
$\{0,1\}^4$	$(0,1,0,1)$	$(0,1,0,0)$	$(1,0,1,0)$	$(1,1,0,0)$	$(1,1,0,1)$

Table 3-5   Solution for Practice Problem 7

Exercises for Section 3.2

1. Given the universal set $U = \{1, 2, 3, 4, 5, 6, 7, 8, 9, 10\}$ and sets $A = \{2, 4\}$, $B = \{1, 2, 8\}$, and $C = \{1, 2, 5, 6, 10\}$, find each of the following:
   (a) $A \times B$
   (b) $(A \times B) - (A \times A)$
   (c) $\mathcal{P}(B)$
   (d) $\mathcal{P}(B \cap C)$
   (e) $\mathcal{P}(B) - \mathcal{P}(B \cap C)$

2. In each of the following problems, three sets are described. In each case, decide which set is not like the others.

   (a) $A = \{(x, 3x) : x \in \mathbb{N}\}$
   $B = \{(y/3, y) : y \in \mathbb{N}\}$
   $C = \{(z - 1, 3z - 3) : z \in \mathbb{Z}^+\}$
   (b) $A = \{(x, y, x + y) : x, y \in \mathbb{Z}\}$
   $B = \{(x, y, x - y) : x, y \in \mathbb{Z}\}$
   $C = \{(x, y, x + y) : x, y \in \mathbb{N}\}$

3. Determine the size of each of the following sets:
   (a) $\{S \in \mathcal{P}(\{1, 2, 3\}) : n(S) \geq 2\}$
   (b) $\{S \in \mathcal{P}(\{1, 2, 3\}) : S \cap \{1, 2\} \neq \emptyset\}$
   (c) $\{S \in \mathcal{P}(\{1, 2, 3, 4\}) : S \cap \{1, 4\} = \emptyset\}$

4. List five elements of each of the following sets:
   (a) $\{(X, Y) \in \mathcal{P}(\{1, 2, 3\}) \times \mathcal{P}(\{1, 2, 3\}) : X \subseteq Y\}$

(b) $\{(X, Y) \in \mathcal{P}(\{1, 2, 3\}) \times \mathcal{P}(\{1, 2, 3\}) : X \cap Y = \emptyset\}$

(c) $\{(X, Y) \in \mathcal{P}(\{1, 2, 3\}) \times \mathcal{P}(\{1, 2, 3\}) : n(X) = n(Y)\}$

5. Let $A = \{4k : k \in \mathbb{Z}\}$, $B = \{4k + 1 : k \in \mathbb{Z}\}$, and $C = \{4k + 2 : k \in \mathbb{Z}\}$. Show that $\{A, B, C\}$ is **not** a partition of $\mathbb{Z}$.

6. Let $A = \{2k : k \in \mathbb{Z}\}$, $B = \{3k + 1 : k \in \mathbb{Z}\}$, and $C = \{6k + 5 : k \in \mathbb{Z}\}$. Show that $\{A, B, C\}$ is **not** a partition of $\mathbb{Z}$.

7. Let $A = \{2k : k \in \mathbb{Z}\}$, $B = \{3k + 1 : k \in \mathbb{Z}\}$, $C = \{4k + 2 : k \in \mathbb{Z}\}$, and $D = \{5k + 3 : k \in \mathbb{Z}\}$. Show that $\{A, B, C, D\}$ is **not** a partition of $\mathbb{Z}$

8. Let $A = \{3k : k \in \mathbb{Z}\}$, $B = \{4k + 1 : k \in \mathbb{Z}\}$, $C = \{4k + 3 : k \in \mathbb{Z}\}$, and $D = \{6k + 1 : k \in \mathbb{Z}\}$. Show that $\{A, B, C, D\}$ is **not** a partition of $\mathbb{Z}$.

9. Let $A = \{2k : k \in \mathbb{Z}\}$, $B = \{4k + 1 : k \in \mathbb{Z}\}$, and $C = \{4k + 3 : k \in \mathbb{Z}\}$. Explain why $\{A, B, C\}$ is a partition of $\mathbb{Z}$.

10. Each of these statements is false. Find a counterexample for each.

    (a) For all sets A, B, and C, $A \times (B \cup C) = (A \times B) \cap (A \times C)$.

    (b) For all sets A and B, $A \times (B - A) = A \times B$.

    (c) For all sets A and B, $\mathcal{P}(A) \cup \mathcal{P}(B) = \mathcal{P}(A \cup B)$.

    (d) For all sets A and B, if $A - B = \emptyset$, then $A = B$.

11. Identify which of the following statements is true. For each false statement, provide a counterexample. You do not have to prove the true statements.

    (a) For all sets A, B, and C, if $A \subseteq B$, then $(A \times C) \subseteq (B \times C)$.

    (b) For all sets A and B, $(A \cup B) \times (A - B) = A^2 - B^2$.

    (c) For all sets A, B, and C, $A \times (B \times C) = (A \times B) \times C$.

12. Which is typically larger, $\mathcal{P}(A \times B)$ or $\mathcal{P}(A) \times \mathcal{P}(B)$? Give an example of sets A and B for which these are the same size.

13. Calculate the size of these sets, given $A = \{1, 3, 5\}$ and $B = \{2, 4, 6, 8\}$. This can be done using only the facts that $n(A) = 3$ and $n(B) = 4$. You do not need to list all elements of the new sets.

    (a) $\mathcal{P}(A)$

    (b) $\mathcal{P}(B)$

    (c) $A \times B$

    (d) $\mathcal{P}(A \times B)$

    (e) $\mathcal{P}(A) \times \mathcal{P}(B)$

14. Generalize the previous problem. That is, using the letters $k = n(A)$ and $m = n(B)$, find an expression for each of the following in terms of k and m:

    (a) $n(\mathcal{P}(A))$

    (b) $n(A \times B)$

    (c) $n(\mathcal{P}(A \times B))$

    (d) $n(\mathcal{P}(A) \times \mathcal{P}(B))$

15. How many different basketball games can be played in Example 7?

16. Which of the following are partitions of the set $A = \{1, 2, 3, 4, 5, 6, 7, 8\}$? For those that are not, explain why not.

    (a) $\mathcal{S} = \{1, 2, \{3, 4, 5\}, \{6, 7, 8\}\}$

    (b) $\mathbf{P} = \{\{1, 5\}, \{6, 7, 2\}, \{4, 3, 5\}, \{8\}\}$

    (c) $\Pi = \{\{1, 8\}, \{4, 3, 5\}, \{7, 2\}\}$

    (d) $\Gamma = \{\{4, 2, 3\}, \{5, 1, 8\}, \{6, 7\}\}$

17. Give a partition of $A = \{1, 2, 3, 4, 5, 6, 7, 8\}$ satisfying each of the following criteria:

    (a) Every part has the same size.

    (b) There are exactly three parts, all having different sizes.

    (c) Even numbers are all in the same part.

    (d) Prime numbers are all in the same part.

18. Using the definition of $\times$, explain why $A \times \emptyset = \emptyset$ for any given set A.

19. Let $A = \{1, 2, 3, 4, 5\}$. Use mathematical induction to prove that for all integers $n \geq 1$, if $B = \{1, 2, \ldots, n\}$, then the number of elements in $A \times B$ is $5 \cdot n$.

20. Let $A = \{1, 2, \ldots, k\}$. Use mathematical induction to prove that for all integers $n \geq 1$, if $B = \{1, 2, \ldots, n\}$, then the number of elements in $A \times B$ is $k \cdot n$.

21. The sets $A \times B \times C$ and $(A \times B) \times C$ are very similar but not equal. Give an example of these sets for a specific choice of A, B, and C, and explain the differences between the two sets.

22. Explain informally why (for $k \geq 3$) the sets $S_1 \times S_2 \times \cdots \times S_{k-1} \times S_k$ and $(S_1 \times S_2 \times \cdots \times S_{k-1}) \times S_k$ have the same size.

23. Use the previous exercise along with Theorem 1 to prove by induction on $k \geq 2$ that for all sets $S_1, S_2, \ldots, S_k$,

$$n(S_1 \times S_2 \times \cdots \times S_k) = n(S_1) \cdot n(S_2) \cdot \ldots \cdot n(S_k)$$

24. Use mathematical induction to *prove* that for every $n \geq 1$, if a set has n elements, then its power set has 2^n elements.

3.3 Proving Set Properties

The set operations *union*, *intersection*, and *complement* have many important properties, some of which were illustrated in the examples and exercises of the previous section. Some of those properties are "obvious." For example, if you think about what intersection of sets ($\cap$) means, then it is not hard to see that $A \cap B \subseteq A$ and that $A \cap B = B \cap A$. Our main purpose in this section is to learn how to create careful proofs of theorems about sets.

In particular, in this section we will learn how to prove that one set is a subset of another set, and how to prove that two sets are equal. The techniques we use apply equally well to "obvious" properties as they do to the less intuitive properties.

Element-Wise Proofs

We begin by learning how to prove that one set is a subset of another set, using a proof technique called an *element-wise proof*. Here is a simple concrete example of an element-wise proof.

Example 1 *For the sets $B = \{2, 4, 6, 8, 10\}$ and $C = \{2, 4\}$, explain why $C \subseteq B$ is true.*

SOLUTION   *The only elements of C are 2 and 4, and each of those elements is also an element of B.*   □

To prove that C is a subset of B, you must convince the READER that each element of C has to be an element of B. The terminology *element-wise* comes from the fact that the proof deals with the individual elements of the sets.

> **To show that one set is a subset of another, we establish that every element of the first set must also be an element of the second set.**

Here is a slightly more abstract example. The sets are too large for us to list all the elements, so we are forced to think instead about the properties the elements have.

Example 2 *Let A be the set $\{0, 10, 20, 30, 40, \ldots\}$ and let B be the set $\{\ldots, -6, -4 - 2, 0, 2, 4, 6, \ldots\}$. That is, $A = \{10k : k \in \mathbb{N}\}$, and $B = \{k \in \mathbb{Z} : k$ is even\}. Explain why $A \subseteq B$ is true.*

SOLUTION   Because the sets are infinite, we cannot list all the elements of A and verify that each is in B. Instead, we imagine a READER who is trying to find a counterexample to the if, then statement "If $x \in A$, then $x \in B$," and we write this letter:

Dear READER,

As you search for a counterexample, you are choosing numbers that make the hypothesis true and hoping to discover that the conclusion is false. To make the hypothesis true, you must choose an element of A—let's agree to call it x. Since $x \in A$, I know that x has the form $10k$ for some $k \in \mathbb{N}$.

To convince you that this value cannot make the conclusion false, I need to show that x is in B. To show $x \in B$, I need to show x is even. Since you agree that $x = 10 \cdot k$, I know I can write $x = 2 \cdot (5 \cdot k)$. Since $5 \cdot k$ is also an integer, this shows that your x is even, which means that your x is in the set B.

This is why there is no counterexample to the if, then statement, and hence you must agree that the statement $A \subseteq B$ is true.

Hoping we're still friends,

AUTHOR

$\square$

Our next example is more abstract. We will not just prove a statement about particular sets, but we will prove that a property holds for any choice of sets. We give two versions of the proof, the first a "letter to the READER," and the second a more concise version of the same proof.

Proposition 1 $A \cap B \subseteq A$.

PROOF   (Informal version)

Dear READER,

No matter how you choose your sets A and B, I'll be able to show that there is no counterexample to the statement "If x is an element of $A \cap B$, then x is an element of A."

If your choice of A and B makes $A \cap B = \emptyset$, then there can be no counterexample because you cannot make the hypothesis true. So we will focus on the case where $A \cap B \neq \emptyset$.

Choose an element of $A \cap B$, and let's agree to call that element x. Since we agree that this x is in $A \cap B$, by the definition of the intersection operation $\cap$, I know that two things are true: x is in A and x is in B. Notice that in particular, we have agreed that this x is in the set A.

Therefore, no matter which element of $A \cap B$ you choose, I will know it is also an element of A. But this is exactly what we mean when we say that $A \cap B \subseteq A$.

Looking forward to the next challenge,

AUTHOR

■

In general, when we prove the subset relationship $S \subseteq T$, we will not separately address the case where $S = \emptyset$. The above reasoning establishes the general fact that *the empty set is a subset of any set*; hence, there is no need to address this possibility in every proof. We see this approach right away with the formal proof of Proposition 1.

PROOF   (Formal version)

Let sets A and B be given. Let $x \in A \cap B$ be given. By the definition of $\cap$, we know that $x \in A$ and $x \in B$. In particular, we know that $x \in A$. Thus, every element of $A \cap B$ is also an element of A, so $A \cap B \subseteq A$.   ■

In a way, this proof simply explains in detail how our careful set definitions logically support the obvious fact that $A \cap B \subseteq A$. Intuitively, everything that is in the overlap of sets must be in the first set, so this result is not surprising. The proof simply explains the result using careful mathematical language and the deductive argument structure that we learned about in the previous chapter.

Practice Problem 1   *Another obvious property of any sets A and B is that $A \subseteq A \cup B$.*

(a) *Write a letter to the* READER *to convince her that this property is true.*
(b) *Write a more formal proof of the property.*
(c) *Explain what the property says in nonmathematical terms.*

The following proposition gives us more practice with the technique of element-wise set proofs, but it differs from the previous proposition in a crucial way. The READER is not choosing *just any* given sets A and B as before, but instead she is choosing sets that satisfy the particular property $A \cup B \subseteq B$. When proving this type of statement, we will be using even more of the ideas about proofs that we learned about in the previous chapter.

Proposition 2   *If $(A \cup B) \subseteq B$, then $A \subseteq (A \cap B)$.*

PROOF   (Detailed version) To help make the proof clear, we will number the paragraphs.

1. Suppose that the READER has chosen sets A and B satisfying the hypothesis of the theorem (i.e., $(A \cup B) \subseteq B$). We must convince him that $A \subseteq (A \cap B)$. This means we must demonstrate that every element of A is also an element of $A \cap B$.

2. To do this, we invite the READER to choose any element of A, and we agree to call that element x.

3. Since we know that $x \in A$, the definition of union tells us that $x \in A \cup B$. (Remember that to be in the union, an element just has to be in either one of the sets.)

4. From paragraph 3 we know that $x \in A \cup B$. From paragraph 1 we know that $(A \cup B) \subseteq B$. By the definition of subset, everything in $A \cup B$ is also in B. Therefore, $x \in B$.

5. From paragraph 2 we know that $x \in A$. From paragraph 4 we know that $x \in B$. So $x \in A \cap B$, by the definition of intersection.

6. We have demonstrated that every element of A is also an element of $A \cap B$, so the proof is complete. ∎

PROOF   (Less detailed version) Let sets A and B be given satisfying $A \cup B \subseteq B$, and let $x \in A$ be given. (We must show that $x \in A \cap B$.) Since $x \in A$, clearly $x \in A \cup B$. But since $A \cup B \subseteq B$, this implies that $x \in B$. Thus, $x \in A$ and also $x \in B$, so $x \in A \cap B$. Therefore, $A \subseteq A \cap B$. ∎

Practice Problem 2 *Show that if $A \subseteq B$ and $B \subseteq C$, then $A \subseteq C$. Be as formal or informal in your writing as you would like, but be sure to include every logical detail.*

The next proposition and practice problem together establish an important property of intersection and union operations, the "distributive property": $A \cap (B \cup C) = (A \cap B) \cup (A \cap C)$. (Notice the similarity to the familiar distributive property for the multiplication and addition operations on numbers: $a \cdot (b + c) = (a \cdot b) + (a \cdot c)$.)

In order to prove this property, we first need a formal definition of the meaning of "=" for sets. The intuitive meaning of two sets being *equal* is that they contain the same elements. In order to build on the terms and proof techniques that we already know, we will use the following informal strategy to prove set equality:

> **To show that two sets are equal, we show that each is a subset of the other.**
> *This often requires two separate proofs!*

Why is this? Well, sets being equal means that they contain the same elements. So every element of the first must be an element of the second, and every element of the second must be an element of the first. We take this explanation as our formal definition of set equality.

> **Definition** Two sets A and B are *equal*, written $A = B$, if $A \subseteq B$ and $B \subseteq A$.

From this point of view, the distributive property for sets, written

$$A \cap (B \cup C) = (A \cap B) \cup (A \cap C)$$

is actually equivalent to the **two** separate properties

$$A \cap (B \cup C) \subseteq (A \cap B) \cup (A \cap C) \text{ and } (A \cap B) \cup (A \cap C) \subseteq A \cap (B \cup C)$$

These properties will be proven separately, one as the proposition below and the other as the practice problem that immediately follows. Notice that the proof of the proposition uses the "division into cases" technique that we first encountered in Section 2.2.

Proposition 3 $A \cap (B \cup C) \subseteq (A \cap B) \cup (A \cap C)$.

PROOF Let sets A, B, and C be given, and let $x \in A \cap (B \cup C)$ be given. By the definition of intersection ($\cap$), we know that $x \in A$ and $x \in B \cup C$. By the definition of union ($\cup$), either $x \in B$ or $x \in C$, so we consider each possibility as a separate case. (Remember that we know that $x \in A$ for sure in either case.)

● If it happens that the given x is in B, we now know that $x \in A$ and $x \in B$, from which it follows by the definition of intersection ($\cap$) that $x \in A \cap B$. This is enough (by Practice Problem 1) to guarantee that $x \in (A \cap B) \cup (A \cap C)$.

● If, on the other hand, it happens that the given x is in C, then we know that $x \in A$ and $x \in C$, from which it follows by the definition of intersection $(\cap)$ that $x \in A \cap C$. This is enough (by Practice Problem 1) to guarantee that $x \in (A \cap B) \cup (A \cap C)$.

Since in either case we have the desired result, $x \in (A \cap B) \cup (A \cap C)$, we can conclude that $A \cap (B \cup C) \subseteq (A \cap B) \cup (A \cap C)$. ■

Explore more on the Web.

Practice Problem 3   *Prove that $(A \cap B) \cup (A \cap C) \subseteq A \cap (B \cup C)$.*

We are finally ready to conclude that the distributive property is true of the set intersection and union operations. All the hard work has already been done.

Theorem 4 (Distributive Property)   *For all sets A, B, and C, $A \cap (B \cup C) = (A \cap B) \cup (A \cap C)$.*

PROOF   To show that $A \cap (B \cup C) = (A \cap B) \cup (A \cap C)$, we must show that each is a subset of the other. In Proposition 3 we proved that $A \cap (B \cup C) \subseteq (A \cap B) \cup (A \cap C)$, and in Practice Problem 3 we proved that $(A \cap B) \cup (A \cap C) \subseteq A \cap (B \cup C)$. ■

Our final example of an element-wise set proof involves an "if and only if" statement in order to show yet another level of complexity in proof structure. In Section 1.6, we saw that a statement of the form "p if and only if q" is called a *biconditional statement* because it is equivalent to the statement "if p, then q, **and** if q, then p." Therefore, we will prove a statement of the form "p if and only if q" by writing **two** proofs, one for the statement "if p, then q," and another for the statement "if q, then p."

Proposition 5   *$A \subseteq B$ if and only if $A \cap B = A$.*

PROOF   We must actually prove two different propositions:

1. **Claim:** If $A \subseteq B$, then $A \cap B = A$. Let sets A and B be given satisfying the hypothesis $A \subseteq B$. We wish to show that $A \cap B = A$, so we will need to show that both (a) $A \cap B \subseteq A$ and (b) $A \subseteq A \cap B$.

   (a) We proved in Proposition 1 that the statement $A \cap B \subseteq A$ is true of all sets.

   (b) To show that $A \subseteq A \cap B$, we start by letting any $x \in A$ be given. Since the sets A and B satisfy the hypothesis $A \subseteq B$, it follows that $x \in B$, and consequently (since x is in A and in B) that $x \in A \cap B$. Hence, $A \subseteq A \cap B$.

   Since we have shown that $A \subseteq A \cap B$ and $A \cap B \subseteq A$, we conclude (under the hypothesis that $A \subseteq B$) that $A \cap B = A$.

2. **Claim:** If $A \cap B = A$, then $A \subseteq B$. Let sets A and B be given such that $A \cap B = A$. Let any $x \in A$ be given. Since $A \cap B = A$, it follows that $x \in A \cap B$, and consequently that $x \in B$. Since we have shown that any x in A must also be in B, we conclude that $A \subseteq B$. ■

The result in Proposition 5 establishes an important connection between the $\subseteq$ relationship and the $\cap$ operation. In particular, it gives us a way to express the subset relationship as an equation involving set intersection.

Proving New Properties from Old

We will next see how proofs about sets can be based on algebraic properties of set equations. This means that once we establish some basic properties, we can then prove new properties using these earlier results. This will allow us to avoid using an element-wise proof structure that clearly gets more cumbersome with every additional layer of complexity in the statements being proven.

So once we have a core of basic rules, we can use this type of proof technique to derive even more results without having to use the "element-wise proof" structure at all. Theorem 6 below is a particularly useful collection because it consists of set properties with a flavor of algebraic manipulation rules. Theorem 4 above, for example, shows up as one of the versions of the distributive property in Theorem 6. You will be asked to prove some of the other properties in the exercises, while we focus the remainder of this section on using these properties to create proofs about sets using algebraic manipulation.

Theorem 6 *For sets A, B, and C, the universal set U, and the empty set $\emptyset$, the properties shown in Table 3-6 hold.*

The two-column organization of Theorem 6 is worth taking a moment to discuss. Note that most of the properties listed have two very similar versions, one given in each column. In fact, the two versions are exactly the same with some standard changes.

Example 3 *For each of properties (a), (b), (c), (g), and (j), verify that interchanging the symbols $\cup$ and $\cap$ changes the first version of the property into the second version. What additional changes need to be made in the other properties?*

SOLUTION Interchanging $\cap$ and $\cup$ **and** interchanging U and $\emptyset$ will turn *every* first-column version into the second-column version. □

(a)	Commutative	$A \cap B = B \cap A$	$A \cup B = B \cup A$
(b)	Associative	$(A \cap B) \cap C = A \cap (B \cap C)$	$(A \cup B) \cup C = A \cup (B \cup C)$
(c)	Distributive	$A \cap (B \cup C) = (A \cap B) \cup (A \cap C)$	$A \cup (B \cap C) = (A \cup B) \cap (A \cup C)$
(d)	Identity	$A \cap U = A$	$A \cup \emptyset = A$
(e)	Negation	$A \cup A' = U$	$A \cap A' = \emptyset$
(f)	Double negative	$(A')' = A$	
(g)	Idempotent	$A \cap A = A$	$A \cup A = A$
(h)	DeMorgan's laws	$(A \cap B)' = A' \cup B'$	$(A \cup B)' = A' \cap B'$
(i)	Universal bound	$A \cup U = U$	$A \cap \emptyset = \emptyset$
(j)	Absorption	$A \cap (A \cup B) = A$	$A \cup (A \cap B) = A$
(k)	Complements of U and $\emptyset$	$U' = \emptyset$	$\emptyset' = U$
(l)	Complement and negation	$A - B = A \cap B'$	

Table 3-6 Properties of Set Operations

The above observation leads us to formalize a principle that will not only account for the organization of Theorem 6, but more generally will give us a "free" result for every one we prove.

> **Definition**   For any expression involving the set operations $\cap$, $\cup$, $\varnothing$, U, and $'$, the *dual expression* is obtained by interchanging every $\cap$ and $\cup$ and by interchanging every U and $\varnothing$.

> **Theorem 7 (Duality Principle)**   *For any true equation of two set expressions, the equation obtained by replacing each expression by its dual expression is also true.*

We will address more on duality in the exercises for this section. We complete the section by illustrating how Theorem 6 can be used to prove other properties about sets.

In our first example below, we are to show that two sets are equal, where one of the sets is given by a fairly complicated expression and the other is much simpler. For this type of problem, a common technique is to start with the more complicated expression and simplify it as much as possible, justifying each step with properties from Theorem 6.

 Example 4   *Assume the properties of Theorem 6 are all true. Use them to prove that for all sets A and B, $A \cup (A' \cap B) = A \cup B$. What is the dual equation that must also be true by the duality principle?*

SOLUTION   We start with $A \cup (A' \cap B)$, simplifying until we obtain $A \cup B$. Here are the steps:

$$
\begin{aligned}
A \cup (A' \cap B) &= (A \cup A') \cap (A \cup B) &&\text{(c, distributive)} \\
&= U \cap (A \cup B) &&\text{(e, negation)} \\
&= (A \cup B) \cap U &&\text{(a, commutative)} \\
&= A \cup B &&\text{(d, identity)}
\end{aligned}
$$

The dual of this equation is $A \cap (A' \cup B) = A \cap B$. By the duality principle, we can conclude that this is also true.   □

Explore more on the Web.

Practice Problem 4   *Which properties of Theorem 6 justify each step of the following proof that $A \cap (B' \cap A)' = A \cap B$? What is the dual equation that must also be true?*

$$
\begin{aligned}
A \cap (B' \cap A)' &= A \cap ((B')' \cup A') &&\underline{\hspace{2cm}} \\
&= A \cap (B \cup A') &&\underline{\hspace{2cm}} \\
&= (A \cap B) \cup (A \cap A') &&\underline{\hspace{2cm}} \\
&= (A \cap B) \cup \varnothing &&\underline{\hspace{2cm}} \\
&= A \cap B &&\underline{\hspace{2cm}}
\end{aligned}
$$

In the next example of this proof technique, both sides of the proposed equation are fairly complicated, so we will simplify each side as much as possible, hoping that they simplify to the same thing. (It is worth noting in this example that we are implicitly using the associative property of "union" when we write expressions such as $A' \cup B \cup C'$ rather than $(A' \cup B) \cup C'$ or $A' \cup (B \cup C')$. We do this simply to avoid an unnecessary level of parentheses in our expresssions.)

Example 5 *Assume the properties of Theorem 6 are all true. Use them to prove that for all sets A, B, and C,*

$$(A \cap B) \cup (A' \cup C)' = (A \cup B) \cap (A \cup B') \cap (A' \cup B \cup C')$$

SOLUTION First, we take the expression on the left-hand side, $(A \cap B) \cup (A' \cup C)'$, and we simplify it as much as possible:

$$
\begin{aligned}
(A \cap B) \cup (A' \cup C)' &= (A \cap B) \cup ((A')' \cap C') && \text{(h, DeMorgan's)} \\
&= (A \cap B) \cup (A \cap C') && \text{(f, double negative)} \\
&= A \cap (B \cup C') && \text{(c, distributive)}
\end{aligned}
$$

Next, we take the expression on the right-hand side, $(A \cup B) \cap (A \cup B') \cap (A' \cup B \cup C')$, and again we simplify it as much as possible.

$$
\begin{aligned}
&(A \cup B) \cap (A \cup B') \cap (A' \cup B \cup C') \\
&= (A \cup (B \cap B')) \cap (A' \cup B \cup C') && \text{(c, distributive)} \\
&= (A \cup \emptyset) \cap (A' \cup B \cup C') && \text{(e, negation)} \\
&= A \cap (A' \cup B \cup C') && \text{(d, identity)} \\
&= (A \cap A') \cup (A \cap (B \cup C')) && \text{(c, distributive)} \\
&= \emptyset \cup (A \cap (B \cup C')) && \text{(e, negation)} \\
&= A \cap (B \cup C') && \text{(d, identity)}
\end{aligned}
$$

Because both sides of the original equation simplify to the same expression, namely $A \cap (B \cup C')$, we conclude that they are equal, meaning the original equation is true. □

Practice Problem 5 *Using properties of Theorem 6, simplify each side as much as possible to show that the following expression is true:*

$$(A \cap B) \cup (A' \cup B)' = ((A \cap A)' \cap (A' \cup B))'$$

What is the dual equation?

Finally, we will examine the proof of an if, then statement relating two particular relationships between arbitrary sets A and B. In this case, we follow the same reasoning as with all direct proofs of if, then statements. We invite the READER to form an example satisfying the hypothesis, and we demonstrate (using properties from Theorem 6) that such an example must also satisfy the conclusion.

Proposition 8   *If $A \subseteq B$, then $A \cup B = B$.*

PROOF   Let sets A and B be given such that $A \subseteq B$. By Proposition 5, this means that $A \cap B = A$. Since these are the same set, taking the union of each with B will result in the same set. That is,

$$A \cup B = (A \cap B) \cup B$$

But we can use properties from Theorem 6 to simplify this as follows:

$$
\begin{aligned}
(A \cap B) \cup B &= B \cup (A \cap B) && \text{by commutativity} \\
&= B \cup (B \cap A) && \text{by commutativity} \\
&= B && \text{by absorption}
\end{aligned}
$$

Since $A \cup B = (A \cap B) \cup B$ and $(A \cap B) \cup B = B$, we conclude that $A \cup B = B$. ∎

Solutions to Practice Problems

1 (a) The text of the letter might go something like this. You choose the two sets A and B. To show you that $A \subseteq A \cup B$, I have to convince you that every element of A is also in $A \cup B$. But if you choose an element (call it x) in set A, I'll automatically know it's in $A \cup B$. Why? To be in the union, all you have to do is be in either one of the two sets, and your x certainly satisfies that condition.

(b) Let sets A and B be given, and let $x \in A$ be given. By the definition of union, to establish that $x \in A \cup B$, we need only show either that $x \in A$ or that $x \in B$. Since we do know that $x \in A$, this shows that $x \in A \cup B$.

(c) In plain terms, the property $A \subseteq (A \cup B)$ means that joining together A with any set B produces a set that must contain all the elements of A.

2 Suppose that the READER has chosen sets A, B, and C satisfying the hypothesis of the theorem (i.e., $A \subseteq B$ and also $B \subseteq C$). We must convince him that $A \subseteq C$. This means we must demonstrate that every element of A is also an element of C.

So we suppose the READER has chosen a particular element x of A, and we must convince him that x is also an element of C. We reason as follows:

Since x is an element of A, and since every element of A is also an element of B (because $A \subseteq B$), x must be an element of B.

Now we know that x is an element of B and that every element of B is also an element of C (because $B \subseteq C$), and therefore x must be an element of C.

We have shown that no matter which element x of A is chosen, x must also be an element of C. This is precisely what we mean by the statement $A \subseteq C$, and so we have proved that $A \subseteq C$.

3 Let sets A, B, and C be given, and let $x \in (A \cap B) \cup (A \cap C)$ be given. By the definition of $\cup$, either $x \in A \cap B$ or $x \in A \cap C$.

- If it happens that $x \in A \cap B$, we know that $x \in A$ and $x \in B$ (from the definition of $\cap$). Since $x \in B$, the definition of $\cup$ tells us that $x \in B \cup C$. Now because $x \in A$ and $x \in B \cup C$, we conclude from the definition of $\cap$ that $A \cap (B \cup C)$.

- If, on the other hand, it happens that $x \in A \cap C$, we know that $x \in A$ and $x \in C$ (from the definition of $\cap$). Since $x \in C$, the definition of $\cup$ tells us that $x \in B \cup C$. Now because $x \in A$ and $x \in B \cup C$, we conclude from the definition of $\cap$ that $A \cap (B \cup C)$.

Thus, in either case we have the desired result $x \in A \cap (B \cup C)$, and we conclude that $(A \cap B) \cup (A \cap C) \subseteq A \cap (B \cup C)$.

4 (h) DeMorgan's, (f) double negative, (c) distributive, (e) negation, and (d) identity. The dual of the proven equation is $A \cup (B' \cup A)' = A \cup B$, which we can conclude is also true by the duality principle.

5 The left side simplifies to

$$
\begin{aligned}
(A \cap B) \cup (A' \cup B)' &= (A \cap B) \cup ((A')' \cap B') \\
&= (A \cap B) \cup (A \cap B') \\
&= A \cap (B \cup B') \\
&= A \cap U \\
&= A
\end{aligned}
$$

The right side simplifies to

$$
\begin{aligned}
((A \cap A)' \cap (A' \cup B))' &= ((A)' \cap (A' \cup B))' \\
&= (A' \cap (A' \cup B))' \\
&= (A')' \\
&= A
\end{aligned}
$$

The expressions are the same set. The dual equation, which is also true by the duality principle, is

$$(A \cup B) \cap (A' \cap B)' = ((A \cup A)' \cup (A' \cap B))'$$

Exercises for Section 3.3

1. Decide which of the following statements are true. For each true statement, explain why in a letter from the AUTHOR to the READER. For each false statement, give a specific example to support your answer.

   (a) $\{1, 3, 5, 7, 9\} \subseteq \{k \in \mathbb{N} : k \text{ is odd}\}$

   (b) $\{k \in \mathbb{N} : k \text{ ends in } 0\} \subseteq \{k \in \mathbb{N} : k \text{ is divisible by } 5\}$

   (c) $\{k \in \mathbb{N} : k \text{ is prime}\} \subseteq \{k \in \mathbb{N} : k \text{ is odd}\}$

   (d) $\{x \in \mathbb{R} : x < 1\} \subseteq \{x \in \mathbb{R} : x^2 < 1\}$

2. Prove each of the following statements about specific sets:

   (a) $\{4m : m \in \mathbb{Z}\} \subseteq \{2n : n \in \mathbb{Z}\}$

   (b) $\{4m + 1 : m \in \mathbb{Z}\} \subseteq \{2n - 1 : n \in \mathbb{Z}\}$

   (c) $\mathbb{Z} \subseteq \mathbb{Q}$ (Use the definition of $\mathbb{Q}$ given in Section 3.1 in your proof.)

   (d) $(\{2n + 1 : n \in \mathbb{Z}\} \cap \{3m + 1 : m \in \mathbb{Z}\}) \subseteq \{6k + 1 : k \in \mathbb{Z}\}$

   (e) $(\{2n + 1 : n \in \mathbb{Z}\} \cap \{5m + 4 : m \in \mathbb{Z}\}) \subseteq \{10k + 9 : k \in \mathbb{Z}\}$

3. Prove each of the following statements about specific sets:

   (a) If a is divisible by b, then $\{a \cdot m : m \in \mathbb{Z}\} \subseteq \{b \cdot n : n \in \mathbb{Z}\}$.

   (b) $(\{n^2 - 1 : n \in \mathbb{Z}\} \cap \{2k : k \in \mathbb{Z}\}) \subseteq \{4m : m \in \mathbb{Z}\}$

   (c) $(\{p : p \text{ is a prime number}\} \cap \{k^2 - 1 : k \in \mathbb{N}\}) = \{3\}$

   (d) $(\{a^2 + b^2 : a, b \in \mathbb{N}\} \cap \{3k : k \in \mathbb{N}\}) = \{9m : m \in \mathbb{N}\}$

4. Fill in the missing steps in each of the following element-wise proofs:

   (a) **Proposition**   For all sets A and B, $A \cap B \subseteq B$.
   Proof.   Let A and B be given. Let $x \in A \cap B$ be given.
   . . .
   So $x \in B$. Therefore, $A \cap B \subseteq A$.   ■

   (b) **Proposition**   For all sets A and B, $B \subseteq A \cup B$.
   Proof.   Let A and B be given. Let $x \in B$ be given.
   . . .
   So $x \in A \cup B$. Therefore, $B \subseteq A \cup B$.   ■

   (c) **Proposition***   If $A \subseteq B$, then $A \cup B \subseteq B$.

Proof.   Let A and B be given, and assume that $A \subseteq B$. Let $x \in A \cup B$ be given. We consider the two possible cases, either $x \in A$ or $x \in B$.
. . .
So in either case $x \in B$. Therefore, $A \cup B \subseteq B$.   ■

5. Fill in the missing steps in the following element-wise proof.
   Proposition   If $A \subseteq B$ and $A \subseteq C$, then $A \subseteq (B \cap C)$.
   Proof.   Let A, B, and C be given, and assume that $A \subseteq B$ and $A \subseteq C$. Let $x \in A$ be given.
   . . .
   So $x \in B \cap C$. Therefore, $A \subseteq (B \cap C)$.   ■

6. Exercise 5 gives us the following strategy for proving statements of the form "$A \subseteq (B \cap C)$." *Do two separate proofs, one to show $A \subseteq B$ and another to show $A \subseteq C$, and then cite Exercise 5 to justify the final conclusion.* Use this strategy to prove the following statements:

   (a) $\{10n - 1 : n \in \mathbb{Z}\} \subseteq (\{2k + 1 : k \in \mathbb{Z}\} \cap \{5m + 4 : m \in \mathbb{Z}\})$

   (b) $\{6k + 1 : k \in \mathbb{Z}\} \subseteq (\{2n + 1 : n \in \mathbb{Z}\} \cap \{3m + 1 : m \in \mathbb{Z}\})$

   (c) $\{x \in \mathbb{R} : x^2 < 4\} \subseteq (\{x \in \mathbb{R} : x < 2\} \cap \{x \in \mathbb{R} : x > -2\})$

7. Fill in the missing steps in the following element-wise proof:
   Proposition   If $A \subseteq C$ and $B \subseteq C$, then $A \cup B \subseteq C$.
   Proof.   Let A, B, and C be given, and assume that $A \subseteq C$ and $B \subseteq C$. Let $x \in A \cup B$ be given. We consider the two possible cases, either $x \in A$ or $x \in B$.
   Case 1: Suppose $x \in A$.
   . . .
   Therefore, $x \in C$.
   Case 2: Suppose $x \in B$.
   . . .
   Therefore, $x \in C$.
   So in either case, $x \in C$. Therefore, $A \cup B \subseteq C$.   ■

8. Exercise 7 gives us the following strategy for proving statements of the form "$(A \cup B) \subseteq C$." *Do two separate proofs, one to show $A \subseteq C$ and another to show $B \subseteq C$, and then cite Exercise 7 to justify the final conclusion.* Use this strategy to prove the following statements:

---

*Exercises 4(b) and 4(c) together establish that if $A \subseteq B$, then $A \cup B = B$.

(a) $(\{4k+1 : k \in \mathbb{Z}\} \cup \{4m+3 : m \in \mathbb{Z}\}) \subseteq \{2n+1 : n \in \mathbb{Z}\}$

(b) $(\{6k+1 : k \in \mathbb{Z}\} \cup \{6m-1 : m \in \mathbb{Z}\}) \subseteq \{2n+1 : n \in \mathbb{Z}\}$

(c) $(\{x \in \mathbb{R} : x < -3\} \cup \{x \in \mathbb{R} : x > 3\}) \subseteq \{x \in \mathbb{R} : x^2 > 9\}$

(d) $(\{x \in \mathbb{R} : x < -1\} \cup \{x \in \mathbb{R} : x > 3\}) \subseteq \{x \in \mathbb{R} : x^2 - 2x - 3 > 0\}$ (HINT: It might help to know that algebraically $x^2 - 2x - 3 = (x-1)^2 - 4$.)

9. Fill in the missing steps in each of the following element-wise proofs:

(a) **Proposition** $A \cup (B \cap C) \subseteq (A \cup B) \cap (A \cup C)$.
Proof. Let A, B, and C be given, and let $x \in A \cup (B \cap C)$ be given. We consider the two possible cases, either $x \in A$ or $x \in B \cap C$.
$\cdots$
So in either case, $x \in (A \cup B) \cap (A \cup C)$. Therefore, $A \cup (B \cap C) \subseteq (A \cup B) \cap (A \cup C)$. ∎

(b) **Proposition*** $(A \cup B) \cap (A \cup C) \subseteq A \cup (B \cap C)$.
Proof. Let A, B, and C be given, and let $x \in (A \cup B) \cap (A \cup C)$ be given. Then $x \in A \cup B$ and also $x \in A \cup C$. We consider two possibilities, either $x \in A$ or $x \notin A$.
$\cdots$
So in either case, $x \in A \cup (B \cap C)$. Therefore, $(A \cup B) \cap (A \cup C) \subseteq (A \cup B) \cap (A \cup C)$. ∎

10. Give element-wise proofs of the following properties from Theorem 6:

(a) (Commutative property) $A \cup B = B \cup A$

(b) (Distributive property) $A \cap (B \cup C) = (A \cap B) \cup (A \cap C)$

(c) (Absorption property) $A \cap (A \cup B) = A$

(d) (DeMorgan's law) $(A \cup B)' = A' \cap B'$

11. Give element-wise proofs for the following:

(a) If $A \cup B = B$, then $A \cap B = A$.

(b) If $A \cap B = A$, then $A \cup B = B$.

(c) If $A \cap B = A$ and $B \cap C = B$, then $A \cap C = A$.

(d) If $A \cup B = B$ and $B \cup C = C$, then $A \cup C = C$.

12. When proving a statement of the form "$S = \emptyset$," one effective strategy is to use proof by contradiction. Specifically, begin with the assumption that there *is* some element $x \in S$, and argue that some absurdity results. Try this type of reasoning on the following statements to be proven:

(a) $\{2k+1 : k \in \mathbb{N}\} \cap \{4k : k \in \mathbb{N}\} = \emptyset$

(b) $(\{3k+1 : k \in \mathbb{Z}\} \cap \{6k+5 : k \in \mathbb{Z}\}) = \emptyset$

(c) $(\{(x, y) \in \mathbb{R} \times \mathbb{R} : x^2 - 2x - 3 = y\} \cap \{(x, y) \in \mathbb{R} \times \mathbb{R} : x - 6 = y\}) = \emptyset$

(d) If $A \subseteq B$, then $A \cap B' = \emptyset$.

13. Give element-wise proofs for the following:

(a) If $A \cap B = A$, then $A' \cup B = U$.

(b) If $A \subseteq B$, then $B' \subseteq A'$.

(c) If $A \subseteq B$, then $A \cup (B - A) = B$.

14. Prove the following by quoting the appropriate parts of Theorem 6:

(a) $(A \cup U) \cap (A \cup \emptyset) = A$

(b) $A \cap (A' \cup B) = A \cap B$

(c) $A \cup (A' \cap B) = A \cup B$

(d) $(A \cup B) \cap (B \cup C) = (A \cap C) \cup B$

(e) $(A \cup B) \cap (A' \cap C)' = A \cup (B \cap C')$

15. Form the dual of each equation in the previous exercise.

16. It turns out that if we have given an elementwise proof of parts (a), (c), (d), and (e) of Theorem 6, all the other parts can be proved by reference to those four parts. Justify each step of the following proofs by quoting parts (a), (c), (d), and (e) of the theorem. Once you have proved a property, you can use it or its dual in later proofs.

(a) Prove part (g), idempotent: $A \cap A = A$

$$A = A \cap U \qquad \underline{\hspace{2cm}}$$
$$= A \cap (A \cup A') \qquad \underline{\hspace{2cm}}$$
$$= (A \cap A) \cup (A \cap A') \qquad \underline{\hspace{2cm}}$$
$$= (A \cap A) \cup \emptyset \qquad \underline{\hspace{2cm}}$$
$$= A \cap A \qquad \underline{\hspace{2cm}}$$

(b) You can use parts (a), (c), (d), (e), and (g). Prove part (i), universal bound: $A \cup U = U$

$$A \cup U = (A \cup U) \cap U \qquad \underline{\hspace{2cm}}$$
$$= (A \cup U) \cap (A \cup A') \qquad \underline{\hspace{2cm}}$$
$$= A \cup (U \cap A') \qquad \underline{\hspace{2cm}}$$
$$= A \cup (A' \cap U) \qquad \underline{\hspace{2cm}}$$
$$= A \cup A' \qquad \underline{\hspace{2cm}}$$
$$= U \qquad \underline{\hspace{2cm}}$$

(c) Now you can use parts (a), (c), (d), (e), (g), and (i) to prove part (j), absorption: $A \cap (A \cup B) = A$

$$A \cap (A \cup B) = (A \cup \emptyset) \cap (A \cup B) \qquad \underline{\hspace{2cm}}$$
$$= A \cup (\emptyset \cap B) \qquad \underline{\hspace{2cm}}$$
$$= A \cup (B \cap \emptyset) \qquad \underline{\hspace{2cm}}$$
$$= A \cup \emptyset \qquad \underline{\hspace{2cm}}$$
$$= A \qquad \underline{\hspace{2cm}}$$

---

*Exercises 9(a) and 9(b) together establish that $A \cup (B \cap C) = (A \cup B) \cap (A \cup C)$.

17. We did not prove the duality principle, but we can get an idea about why it is true by considering the proofs in the previous exercise.

   (a) Prove that $A \cup A = A$ by taking the proof of part (a) above and replacing *every line* with its dual. How do the reasons for each step change?

   (b) Prove that $A \cap \emptyset = \emptyset$ by taking the proof of (b) above and replacing *every line* with its dual. How do the reasons for each step change?

   (c) Prove that $A \cup (A \cap B) = A$ by taking the proof of (c) above and replacing *every line* with its dual. How do the reasons for each step change?

18. Prove the following by quoting the parts of Theorem 6:

   (a) If $A \cup B = B$, then $A \cap B = A$.

   (b) If $A \cap B = A$, then $A \cup B = B$.

   (c) If $A \cap B = A$, then $A' \cup B = U$.

19. Prove the following by quoting the parts of Theorem 6:

   (a) If $A' \cup B = U$, then $A \cap B' = \emptyset$.

   (b) If $A \cup B = B$, then $A \cup (B \cap A') = B$.

(c) If $A \cap B = A$, then $B \cap (B \cap A')' = A$.

20. Prove each of the following statements:

   (a) $\mathcal{P}(A \cap B) \subseteq \mathcal{P}(A) \cap \mathcal{P}(B)$ (Use the strategy from Exercise 6.)

   (b) $\mathcal{P}(A) \cup \mathcal{P}(B) \subseteq \mathcal{P}(A \cup B)$ (Use the strategy from Exercise 8.)

   (c) $A \times \emptyset = \emptyset$ (Use the strategy from Exercise 12.)

21. Prove each of the following statements:

   (a) If $A \subseteq B$, then $(A \times C) \subseteq (B \times C)$.

   (b) If $A \subseteq B$, then $\mathcal{P}(A) \subseteq \mathcal{P}(B)$.

22. Prove each of the following statements:

   (a) $(A \cap B) \times C = (A \times C) \cap (B \times C)$

   (b) $(A \cup B) \times C = (A \times C) \cup (B \times C)$

23. Prove each of the following statements:

   (a) $\mathcal{P}(A) \cap \mathcal{P}(B) = \mathcal{P}(A \cap B)$

   (b) $\mathcal{P}(A) \cup \mathcal{P}(B) \subseteq \mathcal{P}(A \cup B)$

   (c) Give an example to show that the statement $\mathcal{P}(A) \cup \mathcal{P}(B) = \mathcal{P}(A \cup B)$ is not necessarily true.

3.4 Boolean Algebra

In the remainder of this chapter, we will consider the important notion of Boolean algebra as a thread uniting concepts about "logic systems" (from Section 1.6) with concepts about "set systems" (from Section 3.3), and we will see how in this case the use of *abstraction* actually makes it easier to discover and understand new properties of these more concrete systems. We conclude the chapter with a traditional computer science application of Boolean algebra to logic circuit design issues.

One of the most remarkable properties of the study of mathematics is that we frequently discover essentially identical patterns in widely diverse subject matter. In fact, we have already encountered an important example of this phenomenon. Take a look at this theorem from Section 1.6, which describes properties of the logical connectives *and*, *or*, and *not*.

Theorem 2 (Section 1.6) again   *Let p, q, and r be propositions, and let t indicate a tautology and c a contradiction. The logical equivalences shown in Table 3-7 on page 222 hold.*

Compare this with Theorem 6 of the previous section which describes properties of the set operations intersection, union, and complement.

Theorem 6 (Section 3.3) again   *For sets A, B, and C, the universal set U, and the empty set Ø, the properties shown in Table 3-8 on page 222 hold.*

(a)	Commutative	$p \wedge q \equiv q \wedge p$	$p \vee q \equiv q \vee p$
(b)	Associative	$(p \wedge q) \wedge r \equiv p \wedge (q \wedge r)$	$(p \vee q) \vee r \equiv p \vee (q \vee r)$
(c)	Distributive	$p \wedge (q \vee r) \equiv (p \wedge q) \vee (p \wedge r)$	$p \vee (q \wedge r) \equiv (p \vee q) \wedge (p \vee r)$
(d)	Identity	$p \wedge t \equiv p$	$p \vee c \equiv p$
(e)	Negation	$p \vee \neg p \equiv t$	$p \wedge \neg p \equiv c$
(f)	Double negative	$\neg(\neg p) \equiv p$	
(g)	Idempotent	$p \wedge p \equiv p$	$p \vee p \equiv p$
(h)	DeMorgan's laws	$\neg(p \wedge q) \equiv \neg p \vee \neg q$	$\neg(p \vee q) \equiv \neg p \wedge \neg q$
(i)	Universal bound	$p \vee t \equiv t$	$p \wedge c \equiv c$
(j)	Absorption	$p \wedge (p \vee q) \equiv p$	$p \vee (p \wedge q) \equiv p$
(k)	Negations of t and c	$\neg t \equiv c$	$\neg c \equiv t$

Table 3-7 Logical Equivalences

(a)	Commutative	$A \cap B = B \cap A$	$A \cup B = B \cup A$
(b)	Associative	$(A \cap B) \cap C = A \cap (B \cap C)$	$(A \cup B) \cup C = A \cup (B \cup C)$
(c)	Distributive	$A \cap (B \cup C) = (A \cap B) \cup (A \cap C)$	$A \cup (B \cap C) = (A \cup B) \cap (A \cup C)$
(d)	Identity	$A \cap U = A$	$A \cup \emptyset = A$
(e)	Negation	$A \cup A' = U$	$A \cap A' = \emptyset$
(f)	Double negative	$(A')' = A$	
(g)	Idempotent	$A \cap A = A$	$A \cup A = A$
(h)	DeMorgan's laws	$(A \cap B)' = A' \cup B'$	$(A \cup B)' = A' \cap B'$
(i)	Universal bound	$A \cup U = U$	$A \cap \emptyset = \emptyset$
(j)	Absorption	$A \cap (A \cup B) = A$	$A \cup (A \cap B) = A$
(k)	Complements of U and $\emptyset$	$U' = \emptyset$	$\emptyset' = U$

Table 3-8 Set Equivalences

These theorems are remarkably similar. In fact, simply making the following mechanical changes:

- From p, q, r to A, B, C
- From $\wedge$ to $\cap$
- From $\vee$ to $\cup$
- From $\neg$ to $'$
- From $\equiv$ to $=$
- From t to U
- From c to $\emptyset$

translates the first theorem into the second. Similarly, any logical expression can be converted to a related expression from set theory (and vice versa), as the next examples illustrate.

Example 1 *Convert these logical expressions/equivalences to set theory notation, using sets A, B, and C from the universal set U:*

1. $\neg(p \wedge q) \vee (\neg q \wedge r)$

2. $(p \vee \neg q) \wedge t \equiv \neg(\neg p \wedge q)$

SOLUTION

1. $(A \cap B)' \cup (B' \cap C)$
2. $(A \cup B') \cap U = (A' \cap B)'$

Example 2 *Convert these set theory expressions/equalities to logical notation, using logical variables p, q, and r:*

1. $(A \cup B) \cap (C' \cup U)$
2. $(A \cap B') \cup \emptyset = (A' \cup B)'$

SOLUTION

1. $(p \vee q) \wedge (\neg r \vee t)$
2. $(p \wedge \neg q) \vee c \equiv \neg(\neg p \vee q)$

□

Practice Problem 1

(a) *Convert* $(\neg(p \vee c) \wedge (\neg q \wedge t)) \vee p$ *to set notation, using sets A and B.*
(b) *Convert* $(A' \cup \emptyset) \cap (B \cup C)' = A' \cap B' \cap C'$ *to logical notation, using logical variables p, q, and r.*

Here are the solutions to two earlier examples and exercises that further illustrate the similarities between properties of logical expressions and properties of sets.

Example 3 *Verify the logical equivalence* $p \vee (\neg p \wedge q) \equiv p \vee q$ *by quoting Theorem 2 of Section 1.6.*

SOLUTION

$$\begin{aligned}
p \vee (\neg p \wedge q) &\equiv (p \vee \neg p) \wedge (p \vee q) &&\text{(c, distributive)} \\
&\equiv t \wedge (p \vee q) &&\text{(e, negation)} \\
&\equiv (p \vee q) \wedge t &&\text{(a, commutative)} \\
&\equiv p \vee q &&\text{(d, identity)}
\end{aligned}$$

□

Example 4 *Verify the set equality* $A \cup (A' \cap B) = A \cup B$ *by quoting Theorem 6 of Section 3.3.*

SOLUTION

$$\begin{aligned}
A \cup (A' \cap B) &= (A \cup A') \cap (A \cup B) &&\text{(c, distributive)} \\
&= U \cap (A \cup B) &&\text{(e, negation)} \\
&= (A \cup B) \cap U &&\text{(a, commutative)} \\
&= A \cup B &&\text{(d, identity)}
\end{aligned}$$

□

In the mid-1800s the English mathematician George Boole investigated systems with properties such as these, giving rise to the structure that today is called Boolean algebra. It is customary in describing a Boolean algebra to use lowercase letters, to

	Logical	Sets	Boolean Algebra
Variables	p, q, r	A, B, C	a, b, c
Operations	$\wedge$ $\vee$ $\neg$	$\cap$ $\cup$ $'$	$\cdot$ $+$ $'$
Special elements	c t	$\emptyset$ U	0 1

Table 3-9   Connections between Logic, Sets, and Boolean Algebra

use $\cdot$ and $+$ for the operations,* and to use 0 and 1 for the special elements. It is also customary to consider $\cdot$ to have higher precedence than $+$, reducing the need for parentheses. This results in the list of properties satisfied by any Boolean algebra shown in Table 3-10. You should compare this list to the two preceding theorems, observing the relationships summarized in Table 3-9.

Any logical expression or expression of set theory can be written using Boolean algebra notation.

Example 5 *Write* $(p \vee \neg q) \wedge t \equiv \neg(\neg p \wedge q)$ *and* $(A \cap B') \cup \emptyset = (A' \cup B)'$ *using Boolean algebra notation, with variables a and b.*

SOLUTION   *The logical equivalence translates as* $(a + b') \cdot 1 = (a'b)'$ *and the set equality as* $ab' + 0 = (a' + b)'$*. Observe that we have omitted some of the $\cdot$ operations and removed unnecessary parentheses in these solutions.* □

Practice Problem 2 *Convert* $(\neg(p \vee c) \wedge (\neg q \wedge t)) \vee p$ *and* $(A' \cup \emptyset) \cap (B \cup C)' = A' \cap B' \cap C'$ *to Boolean algebra notation, using variables a, b, and c.*

Because the three systems have exactly the same properties, a proof that two expressions are equal in any one of the systems automatically shows that the corresponding equalities hold in the other two systems. This is true because a proof in any one of the three would use exactly the same properties, in exactly the same order, as in the other two settings. We illustrate this by giving a Boolean algebra version of the proofs in Examples 3 and 4.

Example 6 *Verify the Boolean algebra equality* $a + a'b = a + b$ *by quoting the properties of a Boolean algebra.*

SOLUTION

$$
\begin{aligned}
a + (a'b) &= (a + a')(a + b) && \text{(c, distributive)} \\
&= 1 \cdot (a + b) && \text{(e, negation)} \\
&= (a + b) \cdot 1 && \text{(a, commutative)} \\
&= a + b && \text{(d, identity)}
\end{aligned}
$$

□

---

* We frequently omit the $\cdot$ operation, writing ab for $a \cdot b$ when no confusion can result.

(a)	Commutative	$a \cdot b = b \cdot a$	$a + b = b + a$
(b)	Associative	$(a \cdot b) \cdot c = a \cdot (b \cdot c)$	$(a + b) + c = a + (b + c)$
(c)	Distributive	$a \cdot (b + c) = (a \cdot b) + (a \cdot c)$	$a + (b \cdot c) = (a + b) \cdot (a + c)$
(d)	Identity	$a \cdot 1 = a$	$a + 0 = a$
(e)	Negation	$a + a' = 1$	$a \cdot a' = 0$
(f)	Double negative	$(a')' = a$	
(g)	Idempotent	$a \cdot a = a$	$a + a = a$
(h)	DeMorgan's laws	$(a \cdot b)' = a' + b'$	$(a + b)' = a' \cdot b'$
(i)	Universal bound	$a + 1 = 1$	$a \cdot 0 = 0$
(j)	Absorption	$a \cdot (a + b) = a$	$a + (a \cdot b) = a$
(k)	Complements of 1 and 0	$1' = 0$	$0' = 1$

Table 3-10 Properties of a Boolean Algebra

Notes:

1. It turns out that to verify that any given system is a Boolean algebra, one needs only check that properties (a), (c), (d), and (e) are true. The other seven properties can be proven if we know these four hold.

2. One of the advantages of the Boolean algebra notation is that some of the properties are analogous to familiar properties of ordinary algebra. For example, the first of the two distributive properties is the same as what we are familiar with. This makes some symbolic manipulations easier for us than they would be using either the logical symbols or the set symbols.

Duality

Every property of a Boolean algebra, except for property (f), occurs in two forms. For example, we have these two negation properties:

$$a + a' = 1$$
$$a \cdot a' = 0$$

We say these properties are "*duals* of each other." If we start with the first negation property, $a + a' = 1$, and change $+$ to $\cdot$ and 1 to 0, we get the second negation property.

> **Definition** The *dual* of a Boolean algebra expression is the expression obtained by interchanging the roles of $+$ and $\cdot$, and also interchanging the roles of 1 and 0.

 Example 7 *Form the dual of these expressions/equalities:*

1. $(a + b)(c' + 1)$

2. $ab' + 0 = (a' + b)'$

SOLUTION

1. $(a \cdot b) + (c' \cdot 0)$. *Observe that we can drop some parentheses and $\cdot$ symbols, obtaining $ab + c' \cdot 0$.*
2. $(a + b') \cdot 1 = (a'b)'$. *Be careful to insert the required parentheses when converting $\cdot$ to $+$ as you form the dual.*

□

Practice Problem 3 *Form the dual of the expression $(a + 0)'(b' \cdot 1) + a$ and of the equality $(a' + 0)(b + c)' = a'b'c'$.*

Because all the properties occur in dual pairs (and (f) is self-dual because it contains no 1, 0, +, or ·), we obtain the following principle:

Theorem 1 **(Duality)** *For every true equality in a Boolean algebra, the "dual" of that property (given by swapping +/· and swapping 0 / 1) is also true.*

Why? Because the same proof that proved the original equality could be used to prove the dual property. Consider the following example.

Example 8 *Verify the Boolean algebra equality $a(a' + b) = a \cdot b$.*
SOLUTION (Version 1)

$$
\begin{aligned}
a \cdot (a' + b) &= a \cdot a' + a \cdot b &&\text{(c, distributive)} \\
&= 0 + a \cdot b &&\text{(e, negation)} \\
&= a \cdot b + 0 &&\text{(a, commutative)} \\
&= a \cdot b &&\text{(d, identity)}
\end{aligned}
$$

(Version 2) *This is the dual of the equality established in Example 6.* □

Observe that the Version 1 proof quotes exactly the same properties, in exactly the same order, and for exactly the same reasons, as the proof in Example 6. Informally, this is the reason the duality principle applies. It is worth noting that sometimes it is easier to prove the dual of a theorem than it is to prove the theorem as stated, simply because some properties of Boolean algebra more closely mirror properties of "ordinary" algebra. For example, the distributive property $a(b + c) = ab + ac$ comes more naturally to us than the dual distributive property $a + bc = (a + b)(a + c)$.

Explore more on the Web.

Practice Problem 4 *Here is a proof of an equality in Boolean algebra. Write out the proof of the dual equality, and for each step tell which Boolean algebra property is being applied.*

$$(a+b)(ab')' = (a+b)(a'+b'')$$
$$= (a+b)(a'+b)$$
$$= (b+a)(b+a')$$
$$= b+aa'$$
$$= b+0$$
$$= b$$

Solutions to Practice Problems

1 (a) $((A \cup \emptyset)' \cap (B' \cap U)) \cup A$

    (b) $(\neg p \vee c) \wedge \neg (q \vee r) = \neg p \wedge \neg q \wedge \neg r$

2 $(a+0)'(b'1) + a$; $(a'+0)(b+c)' = a'b'c'$

3 $((a1)' + (b'+0))a$; $(a'1) + (bc)' = a' + b' + c'$

4 $ab + (a+b')' = ab + a'b''$ DeMorgan's

                $= ab + a'b$ double negative

                $= ba + ba'$ commutative

                $= b(a+a')$ distributive

                $= b \cdot 1$ negation

                $= b$ identity

Exercises for Section 3.4

1. Rewrite these logical equivalences in Boolean algebra notation:

    (a) $(p \wedge \neg q) \vee p \equiv p$

    (b) $\neg(p \wedge \neg q) \equiv \neg p \vee (p \wedge q)$

    (c) $\neg(\neg p \wedge q) \wedge (p \vee q) \equiv p$

    (d) $\neg(p \vee (\neg p \wedge q)) \equiv \neg p \wedge \neg q$

    (e) $\neg(p \wedge q) \vee (p \vee q) \equiv t$

    (f) $(p \vee r) \wedge (q \vee r) \equiv (p \wedge q) \vee r$

2. Rewrite each of these set equalities in Boolean algebra notation. (HINT: Remember that $A - B$ is the same as $A \cap B'$.)

    (a) $A \cap (B' \cup A) = A$

    (b) $(A \cup B) \cap C' = (A \cap C') \cup (B \cap C')$

    (c) $(A \cup B) \cap (A' \cap C)' = A \cup (B \cap C')$

    (d) $(A \cap B') \cup A = A$

    (e) $(A - B)' = A' \cup (A \cap B)$

    (f) $(A \cap B) - (B \cap C) = (A \cap B) - C$

3. Justify each step in the following proofs by quoting one of the properties of a Boolean algebra:

    (a) **Claim:** For all a, $(a+1)(a+0) = a$.
        Proof.

$$(a+1)(a+0) = 1 \cdot (a+0) \quad \underline{\hspace{2cm}}$$
$$= (a+0) \cdot 1 \quad \underline{\hspace{2cm}}$$
$$= a+0 \quad \underline{\hspace{2cm}}$$
$$= a \quad \underline{\hspace{2cm}} \quad \blacksquare$$

    (b) **Claim:** For all a and b, $a(a'+b) = ab$.
        Proof.

$$a(a'+b) = a \cdot a' + a \cdot b \quad \underline{\hspace{2cm}}$$
$$= 0 + a \cdot b \quad \underline{\hspace{2cm}}$$
$$= a \cdot b + 0 \quad \underline{\hspace{2cm}}$$
$$= a \cdot b \quad \underline{\hspace{2cm}} \quad \blacksquare$$

    (c) **Claim:** For all a and b, $a + a'b = a + b$.
        Proof.

$$a + b = (a+b) \cdot 1 \quad \underline{\hspace{2cm}}$$
$$= 1 \cdot (a+b) \quad \underline{\hspace{2cm}}$$
$$= (a+a') \cdot (a+b) \quad \underline{\hspace{2cm}}$$
$$= a \cdot (a+b) + a' \cdot (a+b) \quad \underline{\hspace{2cm}}$$
$$= a + a' \cdot (a+b) \quad \underline{\hspace{2cm}}$$
$$= a + a' \cdot a + a' \cdot b \quad \underline{\hspace{2cm}}$$
$$= a + a \cdot a' + a' \cdot b \quad \underline{\hspace{2cm}}$$
$$= a + 0 + a' \cdot b \quad \underline{\hspace{2cm}}$$
$$= a + a' \cdot b \quad \underline{\hspace{2cm}} \quad \blacksquare$$

4. Prove the following identities in a Boolean algebra, justifying each step by quoting one of the properties of a Boolean algebra:

    (a) $(a+b)(b+c) = ac + b$

    (b) $ab + bc = (a+c)b$

    (c) $(a+b)(a'c)' = a + bc'$

    (d) $ab + (a'+c)' = a(b+c')$

5. Fill in the blanks in each proof below with the properties of Boolean algebra that make each equality true:

(a) **Claim:** If $a + b = b$ then $a + (b \cdot a') = b$.
Proof. Let a and b be given such that $a + b = b$. Then

$$
\begin{aligned}
b &= a + b & \text{since } b = a + b \\
&= (a + b) \cdot 1 & \underline{\hspace{2cm}} \\
&= (a + b) \cdot (a + a') & \underline{\hspace{2cm}} \\
&= a + (b \cdot a') & \underline{\hspace{2cm}} \\
&= a + b \cdot a' & \underline{\hspace{2cm}} \quad \blacksquare
\end{aligned}
$$

(b) If $ab = a$ then $b(ba')' = a$.
Proof. Let a and b be given such that $ab = b$. Then

$$
\begin{aligned}
b(ba')' &= b(b' + (a')') & \underline{\hspace{2cm}} \\
&= b(b' + a) & \underline{\hspace{2cm}} \\
&= b \cdot b' + b \cdot a & \underline{\hspace{2cm}} \\
&= 0 + b \cdot a & \underline{\hspace{2cm}} \\
&= b \cdot a + 0 & \underline{\hspace{2cm}} \\
&= b \cdot a & \underline{\hspace{2cm}} \\
&= a \cdot b & \underline{\hspace{2cm}} \\
&= a & \text{since } a \cdot b = b \quad \blacksquare
\end{aligned}
$$

6. For a Boolean algebra, prove each of the following:
   (a) If $a + b = b$, then $ab = a$.
   (b) If $ab = a$, then $a' + b = 1$.
   (c) If $a' + b = 1$, then $ab' = 0$.
   (d) If $ab' = 0$, then $a + b = b$.

   NOTE: Proving these four statements establishes that the statements (i) $a + b = b$, (ii) $ab = a$, (iii) $a' + b = 1$, and (iv) $ab' = 0$ are equivalent. That is, if one of them is true, then all four of them are true.

7. Rewrite the two propositions you proved in Exercise 5 as theorems of set theory. (Use the notation $F - G$ in place of $F \cap G'$ when you write the theorems.) Which do you suppose is easier, the Boolean algebra proof of these theorems or an element-wise proof in the style of the previous section?

8. In a Boolean algebra, we can define a relationship $\leq$ by saying $a \leq b$ means that $a \cdot b = a$. By quoting properties of a Boolean algebra, verify the following properties this relationship has:
   (a) $a \leq a$
   (b) If $a \leq b$ and $b \leq a$, then $a = b$.
   (c) If $a \leq b$ and $b \leq c$, then $a \leq c$.

9. Consider the set of all divisors of 30, $S = \{1, 2, 3, 5, 6, 10, 15, 30\}$. Define operations $\cdot$ and $+$ by $a \cdot b = \gcd(a, b)$ and $a + b = \mathrm{lcm}(a, b)$. (By *gcd* we mean the greatest common divisor, and by *lcm* we mean the least common multiple.)
   (a) Fill in the entries in Table 3-11 to show the result of doing the operations.
   (b) There is an element of S that acts as an identity for multiplication. (If we call this element u, it will be true that for any element a, $a \cdot u = a$.) Which element has this property? 30
   (c) There is an element of S that acts as an identity for addition. (If we call this element z, it will be true that for any element a, $a + z = a$.) Which element has this property? 1
   (d) For each element a of the set, define a' as the element that satisfies both $a \cdot a' = z$ and $a + a' = u$, where z and u are the elements determined in the previous steps. Calculate a' for each element of S.

10. Repeat Exercise 9 for the divisors of 70.

d) $1' = 30$
$2' = 15$
$3' = 10$
$5' = 6$
$30' = 1$

$\cdot$	1	2	3	5	6	10	15	30
1	1	1	1	1	1	1	1	1
2	1	2	1	1	2	2	1	2
3	1	1	3	1	3	1	3	3
5	1	1	1	5	1	5	5	5
6	1	2	3	1	6	2	3	6
10	1	2	1	5	2	10	5	10
15	1	1	3	5	3	5	15	15
30	1	2	3	5	6	10	15	30

$+$	1	2	3	5	6	10	15	30
1	1	2	3	5	6	10	15	30
2	2	2	6	10	6	10	30	30
3	3	6	3	15	6	30	15	30
5	5	10	15	5	30	10	30	30
6	6	6	6	30	6	30	30	30
10	10	10	30	10	30	10	30	30
15	15	30	15	15	30	30	15	30
30	30	30	30	30	30	30	30	30

Table 3-11 Addition and Multiplication Tables

b∕a	1	2	3	5	6	10	15	30
1								
2								
3								
5								
6								
10								
15								
30								

Table 3-12 Table for Exercise 12

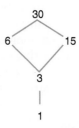

Figure 3-14 Lattice of divisors of 30.

(d) Repeat part 12(c) for the divisors of 70.

13. Starting with only properties (a), (c), (d), and (e) of a Boolean algebra, prove the following properties. Once you have proved a property, it can be used for the later proofs.

    (a) Property (g) (HINT: Start with $a = a \cdot 1$.)

    (b) Property (i) (HINT: Start with $a + 1 = (a + 1) \cdot 1$ and use the fact that $1 = a + a'$.)

    (c) Property (j) (HINT: Start with $a(a + b) = (a + 0)(a + b)$.)

14. It turns out that for any elements x and y of a Boolean algebra, if $x + y = 1$ and $xy = 0$, then y must be equal to x'. Thus, to prove DeMorgan's law

$$(a \cdot b)' = a' + b'$$

all we have to do is verify that

$$(a \cdot b) + (a' + b') = 1$$
$$\text{and that}$$
$$(a \cdot b) \cdot (a' + b') = 0$$

Do this. You may quote any property of Boolean algebra that you wish except DeMorgan's law, of course, since that is the one you are proving.

11. Refer to Exercise 9. Do you think this set with these operations ($\cdot$, $+$, and $'$) forms a Boolean algebra? To decide, you need to decide if properties (a), (c), (d), and (e) are true. What elements play the role of 1 and 0?

12. Refer to Exercise 9. Define the predicate L by saying $L(a, b)$ means that $a \cdot b = a$.

    (a) In Table 3-12, place a checkmark to indicate those pairs where $L(a, b)$ is true.

    (b) Describe in words what relationship a and b must have for $L(a, b)$ to be true.

    (c) Complete the visualization in Figure 3-14 showing the truth value of the L predicate among divisors of 30. To do this, we place a at a level below b and draw a line joining them whenever $L(a, b)$ is true.

3.5 Excursion: Logic Circuits

One of the fascinating things about mathematics is that applications of an idea are frequently discovered many years, even centuries, after the mathematical idea is developed. Boolean algebra provides an excellent example of this phenomenon. George Boole developed his ideas in the mid-1800s, and in the mid-1900s these same ideas were used in the construction of the electrical circuits that make up a computer.

    We will give an overview of how Boolean algebra relates to computer circuits. Data are stored and manipulated in a computer as binary (base two) numbers, that is, as strings of 1's and 0's. For example, on many computers the letter "c" is stored as the string 01000011, and the number 75 as 0000000001001011. Each of the numbers in the string is referred to as a bit (binary digit). Individual bits are represented by voltages, with two different voltage levels representing the 0 value and the 1 value.

These bits are combined using complicated circuits, to allow for operations such as integer arithmetic. For example, a device called the arithmetic-logic unit (ALU) can, among other things, combine two strings that represent numbers to obtain a third string that represents the sum of those numbers. Given the string 0000000001001011 (representing 75) and the string 0000000000000011 (representing 3), it creates the string 0000000001001110 (representing 78).

How is this accomplished? The answer is that simple circuits called gates are used to combine one or more single bits into a single bit answer. These answers are then combined (using more gates) to obtain further answers. We will not examine in detail the various ways in which these combinations can be used, for example, to do addition. However, we will learn about the individual gates, and we will see how to use combinations of these gates to obtain any desired output from a given set of inputs. In essence, we will see how to use these gates to build truth tables.

Logic Gates

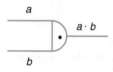

Figure 3-15 The **and** gate.

One of the gates that is used to combine bits is called the **and** gate. Figure 3-15 shows the symbolic representation of a two-input **and** gate.

The wires labeled a and b contain an "input" voltage that is either the voltage that represents the bit value 1, or the voltage that represents the bit value 0. The resulting "output" voltage, labeled $a \cdot b$, is given by this "truth table":

a	b	$a \cdot b$
0	0	0
0	1	0
1	0	0
1	1	1

Observe that if we think of 1 as meaning true and 0 as meaning false, this is exactly the truth table for the logical *and* operation. This is the motivation for calling this an **and** gate. It is also possible to build **and** gates with more than two inputs, where the rule is "The output is 1 if and only if all the inputs are 1." We also note that we frequently omit the $\cdot$ operation, writing ab for $a \cdot b$.

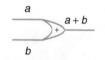

Figure 3-16 The **or** gate.

Similarly, one can build **or** gates with two or more inputs, where the rule is "The output is 1 if and only if at least one of the inputs is 1." For two inputs, Figure 3-16 shows the symbolic representation. Here is the truth table:

a	b	$a + b$
0	0	0
0	1	1
1	0	1
1	1	1

Again with 0 for false and 1 for true, this is the truth table for the logical *or* operation—hence, the name **or** gate.

The final basic building block is the **inverter**, which takes a single input and forms its negation or complement. Figure 3-17 shows the symbolic representation. Here is the truth table:

Figure 3-17 The **not** gate.

a	a'
0	1
1	0

With 0 meaning false and 1 meaning true, this is the same as the logical *not* operation, and an inverter is also called a **not** gate.

These simple building blocks can be combined in an unlimited number of ways.

 Example 1 *Consider the circuit in Figure 3-18. What output will be obtained for the different combinations of input? That is, what is the truth table for this circuit?*

SOLUTION   Perhaps the easiest way to answer this question is to label each of the outputs from the gates, working our way from left to right, as shown in Figure 3-19 on page 232.

We can then build the truth table as follows:

a	b	a'	$a' + b$	$a \cdot (a' + b)$
0	0	1	1	0
0	1	1	1	0
1	0	0	0	0
1	1	0	1	1

Observe that the output from this complicated circuit (a two-input **or** gate, a two-input **and** gate, and a **not** gate) is the same as what we would get with a circuit consisting of just a two-input **and** gate. We say that the two circuits are equivalent because their outputs are the same.   □

Explore more on the Web.

Practice Problem 1   *Build the truth table for the circuit shown in Figure 3-20 on page 232.*

Simplifying Circuits

An important question to be considered in this section is how we can build a circuit with a desired output, and how we can accomplish this with as few gates as possible. Part of the answer to the latter question comes from realizing that the expressions which represent the output from circuits form a Boolean algebra, where "=" means equivalent circuits. To verify this, we only need to verify these four properties:

(a)	Commutative	$a \cdot b = b \cdot a$	$a + b = b + a$
(c)	Distributive	$a \cdot (b + c) = (a \cdot b) + (a \cdot c)$	$a + (b \cdot c) = (a + b) \cdot (a + c)$
(d)	Identity	$a \cdot 1 = a$	$a + 0 = a$
(e)	Negation	$a + a' = 1$	$a \cdot a' = 0$

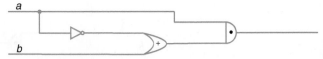

Figure 3-18   The circuit for Example 1.

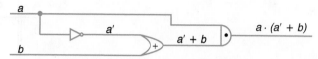

Figure 3-19 The solution to Example 1.

Two circuits are equivalent if they give the same output for all possible combinations of input; hence, we can establish these equivalences with truth tables.

Example 2 *Show* $a + (b \cdot c) = (a + b) \cdot (a + c)$.

SOLUTION   Here are the truth tables for the expressions $a + (b \cdot c)$ and $(a + b) \cdot (a + c)$:

a	b	c	$b \cdot c$	$a + b \cdot c$	a	b	c	$a + b$	$a + c$	$(a + b) \cdot (a + c)$
0	0	0	0	0	0	0	0	0	0	0
0	0	1	0	0	0	0	1	0	1	0
0	1	0	0	0	0	1	0	1	0	0
0	1	1	1	1	0	1	1	1	1	1
1	0	0	0	1	1	0	0	1	1	1
1	0	1	0	1	1	0	1	1	1	1
1	1	0	0	1	1	1	0	1	1	1
1	1	1	1	1	1	1	1	1	1	1

The fact that the two circuits give the same output for all possible combinations of input is exactly what we mean when we write $a + (b \cdot c) = (a + b) \cdot (a + c)$. □

It is easy to verify that the other three properties also hold. Thus, these combinations of simple gates constitute a Boolean algebra. We may therefore use all the properties of a Boolean algebra to simplify the expression for a given circuit. We could have anticipated that the $a(a' + b)$ circuit would have the same output as the $a \cdot b$ circuit, because we had previously done this simplification using the properties of a Boolean algebra:

$$
\begin{aligned}
a(a' + b) &= aa' + ab &\quad \text{part (c)} \\
&= 0 + ab &\quad \text{part (e)} \\
&= ab + 0 &\quad \text{part (a)} \\
&= ab &\quad \text{part (d)}
\end{aligned}
$$

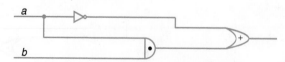

Figure 3-20 The circuit for Practice Problem 1.

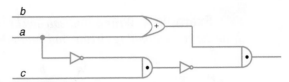

Figure 3-21 The circuit for Example 3.

Example 3 *Write the Boolean algebra expression for the circuit shown in Figure 3-21, and simplify as much as you can.*

SOLUTION The expression is $(a + b)(a'c)'$. We simplify this as follows:

$$
\begin{aligned}
(a + b)(a'c)' &= (a + b)((a')' + c') & \text{part (h)} \\
&= (a + b)(a + c') & \text{part (f)} \\
&= a + bc' & \text{part (c)}
\end{aligned}
$$

$\square$

Observe that the original circuit contained one two-input **or** gate, two two-input **and** gates, and two **not** gates. The circuit for the simplified expression has one two-input **or** gate, one two-input **and** gate, and one **not** gate.

Practice Problem 2 *Write the Boolean algebra expression for the circuit shown in Figure 3-22, and simplify as much as you can.*

The next example illustrates a truth table for three variables, and shows a general way to convert from a truth table to a Boolean algebra expression. The Boolean algebra expression obtained in this manner is not the simplest, but there are standard methods for simplifying the expression. We will consider this topic in more detail in the next section.

It is worth noting that, as we did for logical expressions, we list the truth table in a systematic manner, starting with a row of 0's and ending with a row of 1's. Just as in counting, we cause the right most position to vary most frequently. It would certainly be possible to start with all 1's. This would correspond to starting with all true for the logical expressions. We have chosen to begin with all 0's because that makes the rows progress just as numbers in the binary number system: 000, 001, 010, 011, 100, 101, 110, 111.

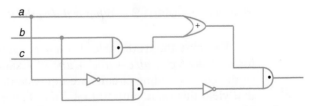

Figure 3-22 The circuit for Practice Problem 2.

Example 4 *Write a Boolean algebra expression that gives the following result, draw the corresponding circuit, and count the gates that are used:*

a	b	c	Result
0	0	0	0
0	0	1	0
0	1	0	1
0	1	1	0
1	0	0	1
1	0	1	1
1	1	0	1
1	1	1	1

SOLUTION   We present a technique that yields a so-called *sum-of-products* expression. The answer will be the sum of several terms of the general form xyz, where the first factor is either a or a', the second factor is either b or b', and the third factor is either c or c'. If we use terms of this form, it is easy to build an expression that has a 1 exactly where we want it. For example, the factor $a'bc'$ will have a 1 in the row where a is 0, b is 1, and c is 0. That is, $a'bc'$ has a 1 in row 3 and 0 in the other rows. We apply this same idea for each of the other four rows where we want a 1:

$a'bc'$ has a 1 in row 3, and 0 in the other rows

$ab'c'$ has a 1 in row 5, and 0 in the other rows

$ab'c$ has a 1 in row 6, and 0 in the other rows

abc' has a 1 in row 7, and 0 in the other rows

abc has a 1 in row 8, and 0 in the other rows

If we combine these five results with the logical *or* operation, we will get a 1 in rows 3, 5, 6, 7, 8 (each row where at least one of these results was 1), and 0 everywhere else. This is the desired result. So the answer is

$$a'bc' + ab'c' + ab'c + abc' + abc$$

The corresponding circuit is shown in Figure 3-23. It uses five three-input **and** gates, one five-input **or** gate, and six **not** gates.

NOTE: In Figure 3-23, we adopt some frequently used shorthands: (1) we indicate inverters (**not** gates) by a small circle placed immediately adjacent to the next **and** gate or **or** gate; and (2) we do not include the part of the circuitry that splits each incoming signal.   □

Practice Problem 3   *What is the truth table for $a + bc'$?*

Examine the truth tables for $a + bc'$ you just did in the practice problem and for $a'bc' + ab'c' + ab'c + abc' + abc$ of Example 4. The truth tables are identical, so the corresponding circuits are equivalent. We have now seen three expressions that yield this same output: $(a + b)(a'c)'$, $a + bc'$, and $a'bc' + ab'c' + ab'c + abc' + $

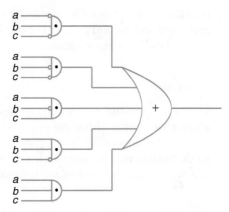

Figure 3-23 A more complicated
circuit.

abc. Simplifying $(a + b)(a'c)'$ to $a + bc'$ was not difficult. Simplifying $a'bc' + ab'c' + ab'c + abc' + abc$ is more tedious, but it can be done:

$$
\begin{aligned}
a'bc' + ab'c' + ab'c + abc' + abc &= a'bc' + ab'(c' + c) + ab(c' + c) \\
&= a'bc' + ab'(c + c') + ab(c + c') \\
&= a'bc' + ab'(1) + ab(1) \\
&= a'bc' + ab' + ab \\
&= a'bc' + a(b' + b) \\
&= a'bc' + a(b + b') \\
&= a'bc' + a(1) \\
&= a'bc' + a \\
&= a + a'bc' \\
&= (a + a')(a + bc') \\
&= (1)(a + bc') \\
&= (a + bc')(1) \\
&= a + bc'
\end{aligned}
$$

We finish this section with a visual method for simplifying sum-of-products expressions.

Karnaugh Maps

In the remainder of this section, we introduce a graphical technique for simplifying sum-of-products expressions containing up to four variables. We will name the variables x, y, z, and w. The technique is due to Maurice Karnaugh, and can actually be extended to expressions containing up to six variables. We use a visualization referred to as a *Karnaugh map*. The use of Karnaugh maps gives the simplest possible sum-of-products expression. It is sometimes possible to reduce the number of gates further if we do not use a sum-of-products expression. For example, $xy' + xz'$ requires two two-input **and** gates, one two-input **or** gate, and two **not** gates. We can rewrite this as $xy' + xz' = x(y' + z') = x(yz)'$, which can be built with two two-input **and** gates and one **not** gate. In the section, we restrict our attention to finding minimal sum-of-products expressions.

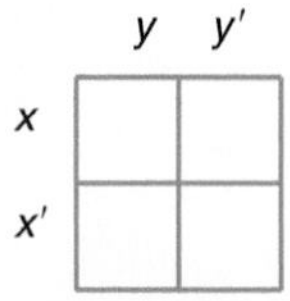

Two variables

The technique and its underlying logic are relatively easy to visualize for two variables, so we begin with some two-variable examples. The basic idea is to use a grid to lay out all the possible terms (products of the two variables or their complements) so that terms we can simplify are in adjacent squares. When only two variables are involved in the expression, we lay out the grid as in Figure 3-24.

Figure 3-24
The Karnaugh map grid for two variables.

Example 5 *Perhaps the simplest example is $xy + xy'$, which can be simplified to $x(y + y') = x \cdot 1 = x$. In the Karnaugh map we place checkmarks in the boxes that correspond to the two terms xy and xy', as shown in the grid on the left in Figure 3-25. The existence of a rectangular shape in the checkmarks is our signal that simplification is possible. We put an oval around the rectangle, as shown in the grid on the right in Figure 3-25, and we recognize that the presence of a rectangle of "area 2" indicates that one variable will not appear in the simplification. Which variable will not appear? The one that appears in both complemented and noncomplemented form (y in this example) since when we simplify, the y and y' will sum to 1 and thus no longer appear.*

$$xy + xy' = x(y + y')$$
$$= x \cdot 1$$
$$= x$$

Practice Problem 4 *Use a Karnaugh map to simplify $xy' + x'y'$.*

Figure 3-26
The Karnaugh map for Example 6.

Example 6 *Simplify this more complicated expression: $xy + xy' + x'y$.*

SOLUTION   In the Karnaugh map shown in Figure 3-26, we can identify both a horizontal rectangle and a vertical rectangle, so we circle them both. A single checkmark is allowed to appear in more than one rectangle. The expression for the horizontal rectangle is x, for the vertical rectangle is y, and the final answer is the sum of the expressions for the two rectangles: $x + y$. □

To see that this gives the same answer we could get by doing the algebra, we first of all realize that property (g) (the idempotent property) allows us to write xy as $xy + xy$. If we omit the steps that just apply the commutative property, the algebraic simplification goes as follows:

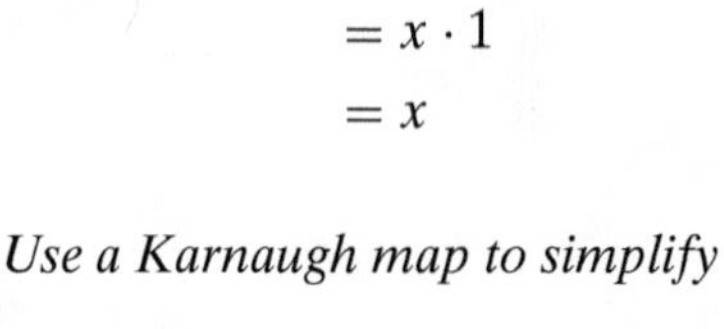

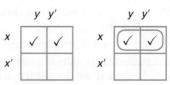

Figure 3-25   The Karnaugh map for Example 5.

$$xy + xy' + x'y = xy + xy + xy' + x'y$$
$$= xy + xy' + xy + x'y$$
$$= x(y + y') + (x + x')y$$
$$= x \cdot 1 + 1 \cdot y$$
$$= x + y$$

Figure 3-27 The Karnaugh map with no simplification.

The horizontal rectangle provides a visualization that $xy + xy'$ will simplify to x, and the vertical rectangle is a visualization that $xy + x'y$ will simplify to y.

Practice Problem 5 *Use a Karnaugh map to simplify $xy' + x'y + x'y'$.*

Before moving on to more complex expressions, note that sometimes no simplification can be done. For example, the Karnaugh map for the expression $xy + x'y'$ is shown in Figure 3-27. Since the only rectangles we can circle are 1×1 rectangles, we conclude that the original expression is already simplified as much as possible.

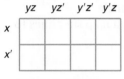

Figure 3-28 The Karnaugh map grid for three variables.

Three variables

If a sum-of-products expression involves three variables, we use the Karnaugh map form shown in Figure 3-28.

Along the top of this picture, labels that are side by side (like yz and yz') differ in exactly one of the two variables. This is the key in all Karnaugh maps to using rectangles to simplify an expression.

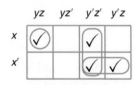

Figure 3-29 The Karnaugh map for Example 7.

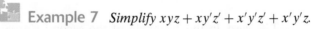

Example 7 *Simplify $xyz + xy'z' + x'y'z' + x'y'z$.*

SOLUTION   The Karnaugh map with the terms checked and the rectangles circled is shown in Figure 3-29. Since we can completely cover the checked squares with three rectangles, the simplified expression will contain three terms:

● The 1×1 rectangle does not remove any variables. Its term is xyz.
● The vertical circle of "area 2" removes one variable (x). We reason that x appears in the rectangle both as x and x', so it will not appear in the answer. The term is $y'z'$.
● The horizontal circle of "area 2" removes one variable (z). We reason that z appears in the rectangle both as z and z', so it will not appear in the answer. The term is $x'y'$.

The final answer is $xyz + y'z' + x'y'$.   □

Practice Problem 6 *Simplify $xyz' + x'yz + x'yz' + xy'z'$.*

It is no surprise that the three-variable problem is more complicated than the two-variable problem. The larger Karnaugh map used for the three-variable problem creates some issues that we have not had to deal with before. For example, it is now relevant that only a rectangle whose area is a power of 2 corresponds to a collection

	yz	yz'	$y'z'$	$y'z$
x	✓	✓	✓	✓
x'				

	yz	yz'	$y'z'$	$y'z$
x		✓	✓	
x'		✓	✓	

Figure 3-30 The Karnaugh maps for Example 8.

of terms that simplify. In particular, a simplifying rectangle has sides of length 1, 2 or 4. *No simplifying rectangle has a side of length 3.*

Example 8 *Find the simplest sum-of-products expression for the Karnaugh maps shown in Figure 3-30.*

SOLUTION In each, we circle a rectangle of area 4. Such a rectangle will cause two variables not to appear in the answer, and once again the general rule is that any variable appearing in the rectangle both complemented and uncomplemented will not be part of the term for that rectangle.

Using this rule of thumb, we see that the first Karnaugh map simplifies to x. Here is the algebra that the Karnaugh map captures:

$$\begin{aligned} xyz + xyz' + xy'z' + xy'z &= xy(z+z') + xy'(z+z') \\ &= xy \cdot 1 + xy' \cdot 1 \\ &= xy + xy' \\ &= x(y+y') \\ &= x \cdot 1 \\ &= x \end{aligned}$$

Likewise, in the rectangle of the second Karnaugh map, both x and x' boxes are checked, and y and y' boxes are checked, so x and y are not in the term for that rectangle. Thus, the solution for that map is z'. □

	yz	yz'	$y'z'$	$y'z$
x	✓	✓		
x'	✓	✓	✓	✓

Figure 3-31 The Karnaugh map for Practice Problem 7.

Practice Problem 7 *Find the simplest sum-of-products expression for the Karnaugh map shown in Figure 3-31.* (HINT: *Use a* 1×4 *rectangle and a* 2×2 *rectangle.)*

	yz	yz'	$y'z'$	$y'z$
x				
x'	✓			✓

Figure 3-32 A complication with Karnaugh maps.

A second complication with the three-variable problem is that checkmarks which do not appear to be side by side nevertheless can be simplified. For example, consider the Karnaugh map shown in Figure 3-32. Algebraically, the expression does simplify as

$$\begin{aligned} x'yz + x'y'z &= x'z(y+y') \\ &= x'z \cdot 1 \\ &= x'z \end{aligned}$$

	yz	yz'	$y'z'$	$y'z$
x				
x'	✓			✓

Figure 3-33 A rectangle can "wrap-around".

For the Karnaugh map visualization to give us this simplification, we merely have to visualize the map as if it were the label on a tin can, so that the left and right columns would be adjacent to each other. Then we can circle the area 2 rectangle as in Figure 3-33, and our usual rule of thumb gives the proper answer.

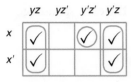

Figure 3-34 A solution to Example 9.

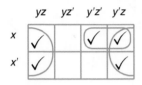

Figure 3-35 A better solution to Example 9.

Figure 3-36 Karnaugh map for Practice Problem 8.

Figure 3-37 Karnaugh map for $xy + xy'z' + x'z$.

Example 9 *Simplify* $xyz + xy'z' + xy'z + x'yz + x'y'z$.

SOLUTION   The Karnaugh map is shown in Figure 3-34.

This choice of rectangles gives $yz + xy'z' + y'z$. However, this is **not** the simplest sum-of-products expression. We should follow this guideline when we choose our rectangles:

> **Choose rectangles so that the number of rectangles is as small as possible and each individual rectangle is as large as possible (but remembering that sides of length 3 are not allowed).**

Following this guideline, we obtain the better choice of rectangles for Example 9 shown in Figure 3-35. The area 4 rectangle in the first and last columns has both x and x' checked, and also both y and y' checked, so the term is z. The area 2 rectangle has both z and z' checked, and the term is xy'. The final answer is $z + xy'$.   ☐

Practice Problem 8   *Choose rectangles for the Karnaugh map shown in Figure 3-36.*

Before we proceed to four variables, we make one last note. Suppose the expression to be simplified is $xy + xy'z' + x'z$. The question is, what box(es) do we check to correspond to the xy in the problem? What about the $x'z$? To answer this, we reason using the reverse of the process of going from Karnaugh map to term:

> **For a term in which one or more variables do not appear, think of those variables as appearing both complemented and uncomplemented, and check all the corresponding boxes.**

Applying this principle to the expression $xy + xy'z' + x'z$, for term xy we check both xyz and xyz', and similarly for $x'z$. Figure 3-37 shows the Karnaugh map for this expression.

Four variables

When the expression involves four variables, we use a Karnaugh map form as shown in Figure 3-38. Observe that again each label differs from the one on either side only in one of the two variables. As for the three-variable map, the last column differs from the first column in only one variable, and we again think of the first and last columns as being adjacent. In addition, observe that the top and bottom row labels likewise differ in only one variable. We may therefore consider the top and bottom rows as being adjacent.

	zw	zw'	z'w'	z'w
xy				
xy'				
x'y'				
x'y				

Figure 3-38 Karnaugh map grid for four variables.

Example 10 *Simplify* $xyz'w' + xy'z'w' + xy'z'w + x'y'z'w' + x'yz'w'$.

SOLUTION   Figure 3-39 shows the Karnaugh map on page 240.

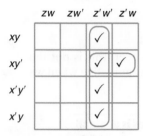

Figure 3-39 *Karnaugh map for Example 10.*

We circle an area 4 rectangle and an area 2 rectangle. The corresponding terms are found as follows:

In the area 4 rectangle:	In the area 2 rectangle:
x appears as x and x'	x appears only as x
y appears as y and y'	y appears only as y'
z appears only as z'	z appears only as z'
w appears only as w'	w appears as w and w'
The term is $z'w'$	The term is $xy'z'$

The final answer is $z'w' + xy'z'$.

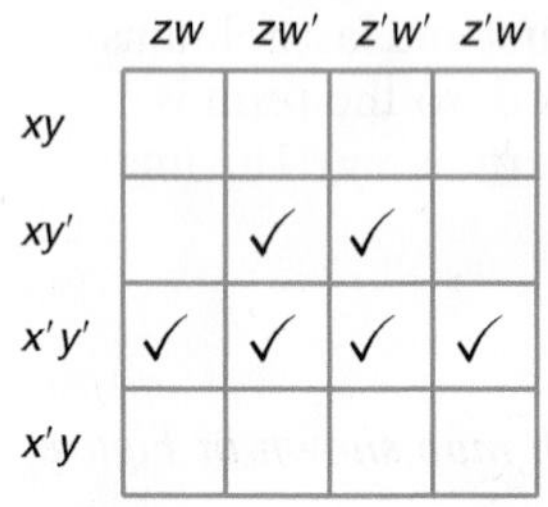

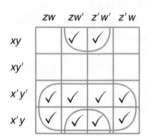

Figure 3-40 Karnaugh map for Practice Problem 10.

Practice Problem 9 *Write the expression that corresponds to the Karnaugh map in Figure 3-40.*

Example 11 *Simplify* $xyzw' + xyz'w' + x'y'zw + x'y'zw' + x'y'z'w' + x'y'z'w + x'yzw + x'yzw' + x'yz'w' + x'yz'w.$

SOLUTION Figure 3-41 shows the Karnaugh map.

Observe that we do not just use an area 2 rectangle for the top two checkmarks. Because the top and bottom rows are adjacent, we can combine them with the two checkmarks on the bottom row to obtain a rectangle of area 4. Again, we go through the reasoning in detail:

Figure 3-41 Karnaugh map for Example 11.

In the area 8 rectangle:	In the area 4 rectangle:
x appears only as x'	x appears as x and x'
y appears as y and y'	y appears only as y
z appears as z and z'	z appears as z and z'
w appears as w and w'	w appears only as w'
The term is x'	The term is yw'

The final answer is $x' + yw'$.

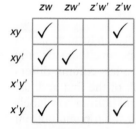

Figure 3-42 Karnaugh map for Example 12.

Example 12 *Circle the rectangles for the Karnaugh map shown in Figure 3-42.*

SOLUTION The grid on the left in Figure 3-43 shows the most obvious solution, but it is not the best solution. We have forgotten here that the top and bottom rows are adjacent, and also the left and right columns are adjacent. Thus, the four corners form a 2×2 rectangle of area 4, and we have the better solution on the right in Figure 3-43.

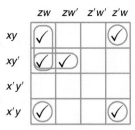

Figure 3-43 Solution for Example 12.

Practice Problem 10 *Give the expression that corresponds to the better selection of rectangles in Example 12.*

Example 13 *Give the Karnaugh map for the expression $y'w' + yz + x'yw'$.*

SOLUTION As we did in the three-variable problem, we handle the $y'w'$ term by realizing that since x is missing, we will have both x and x' appearing in the Karnaugh map, and that since z is missing, we will have both z and z' appearing. Just this term will cause the checkmarks shown in the leftmost grid in Figure 3-44. For the yz term we reason that since x is missing, we will have both x and x' appearing, and since w is missing, we will have both w and w' appearing. This gives the checkmarks shown in the center grid of Figure 3-44. Finally, for the $x'yw'$ term we reason that we will have both z and z' appearing, as shown in the rightmost grid of Figure 3-44. The complete final answer is given in Figure 3-45. □

Figure 3-45 Complete Karnaugh map for Example 13.

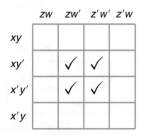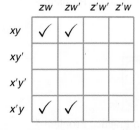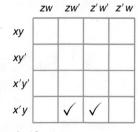

Figure 3-44 The three terms in Example 13.

Practice Problem 11 *Give the corresponding simplified expression for the Karnaugh map in Example 13.*

Solutions to Practice Problems

1 The expression is $a' + ab$. Here is the truth table:

a	b	a'	ab	$a' + ab$
0	0	1	0	1
0	1	1	0	1
1	0	0	0	0
1	1	0	1	1

2 $(a + bc)(a'b)' = (a + bc)(a + b') = a + (bc)b' = a + cbb' = a + c0 = a + 0 = a$

3

a	b	c	$b \cdot c'$	$a + b \cdot c'$
0	0	0	0	0
0	0	1	0	0
0	1	0	1	1
0	1	1	0	0
1	0	0	0	1
1	0	1	0	1
1	1	0	1	1
1	1	1	0	1

4 The Karnaugh map, with a vertical rectangle of area 2 circled, is shown in Figure 3-46. This time only the y' will appear in the final answer (the rectangle is in the y' column), and the x and x' will add up to 1. The answer is y'.

5 Figure 3-47 shows the Karnaugh map. The vertical rectangle represents y', and the horizontal rectangle x'. The answer is $x' + y'$.

Figure 3-46 Solution to Practice Problem 4.

Figure 3-47 Solution to Practice Problem 5.

Exercises for Section 3.5

1. To show that the output from circuits made up of **and**, **or**, and **not** gates forms a Boolean algebra, where "$=$" means equivalent circuits, we need to verify the commutative, distributive, identity, and negation properties (properties (a), (c), (d), and (e)). We used a truth table to verify the equivalence $a + (b \cdot c) = (a + b) \cdot (a + c)$, which is one of the two distributive laws. Do the same for the following:

(a) $a \cdot (b + c) = (a \cdot b) + (a \cdot c)$ (the other distributive law)

(b) $a \cdot 1 = a$ (one of the identity laws)

(c) $a \cdot a' = 0$ (one of the negation laws)

Figure 3-48 Solution to Practice Problem 6.

6 Figure 3-48 shows the Karnaugh map with the terms checked and the rectangles circled. The checked squares can be entirely covered by two rectangles. (It would be possible also to put in a vertical rectangle around the xyz' and the $x'yz'$, but we get a simpler answer if we use as few rectangles as possible.) The top rectangle has both y and y', so the y terms drop out and the term is xz'. For the bottom rectangle z appears both as z and as z', so there is no z in the term—the term is $x'y$. The final answer is $xz' + x'y$.

7 $x' + y$ (The x' is the 1×4 rectangle along the bottom. The y is the 2×2 rectangle in the first two columns.)

8 There are two equally good solutions shown in Figure 3-49.

9 $x'y' + y'w'$

10 $yw + xy'z$

11 The expression $y'w' + yz + x'w'$ comes from the rectangles shown in Figure 3-50.

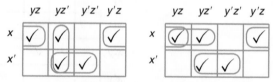

Figure 3-49 Solution to Practice Problem 8.

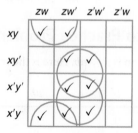

Figure 3-50 Solution to Practice Problem 11.

2. By quoting the properties of a Boolean algebra, justify each step in the simplification

$$a'bc' + ab'c' + ab'c + abc' + abc = a + bc'$$

that we did in this section.

3. Give the truth table and draw the circuit corresponding to the following Boolean algebra expressions. For the circuits, you may use the shorthand notation developed in Example 4.

(a) $a + b(a + b)$

(b) $(a + b')(b + a'b)$

(c) $(ab + c')(b' + c)$

(d) $(a'b')'(a + c)'$

(e) $abc + abc' + ab'$

(f) $abc + abc' + ab'c' + ab'c$

4. For each expression of Exercise 3, simplify the expression as much as you can by using the laws of Boolean algebra.

5. Write a sum-of-products Boolean algebra expression for each of the following truth tables:

a	b	c	Result
0	0	0	0
0	0	1	0
0	1	0	0
(a) 0	1	1	0
1	0	0	1
1	0	1	0
1	1	0	0
1	1	1	1

a	b	c	Result
0	0	0	1
0	0	1	1
0	1	0	1
(b) 0	1	1	1
1	0	0	0
1	0	1	0
1	1	0	0
1	1	1	0

a	b	c	Result
0	0	0	0
0	0	1	1
0	1	0	0
(c) 0	1	1	1
1	0	0	0
1	0	1	1
1	1	0	0
1	1	1	1

6. For each expression of Exercise 5, simplify the expression as much as possible using Boolean algebra.

7. Write a sum-of-products Boolean algebra expression for each of the following truth tables:

a	b	c	Result
0	0	0	1
0	0	1	0
0	1	0	0
(a) 0	1	1	1
1	0	0	1
1	0	1	0
1	1	0	0
1	1	1	1

a	b	c	Result
0	0	0	1
0	0	1	1
0	1	0	1
(b) 0	1	1	1
1	0	0	1
1	0	1	1
1	1	0	1
1	1	1	1

a	b	c	Result
0	0	0	0
0	0	1	0
0	1	0	1
(c) 0	1	1	0
1	0	0	0
1	0	1	1
1	1	0	0
1	1	1	1

8. For each of the following pairs of circuits, determine whether or not they are equivalent. If they are not equivalent, give an input that demonstrates they are not.

   (a) $ab + ab'$ and a

   (b) $a'b + ab'$ and ab

   (c) $a'bc + a'b'c$ and $a'c$

   (d) $abc + ab'c' + a'b'c' + a'b'c$ and $abc + b'c' + a'b'$

   (e) $ab + ac + a'c'$ and $ac + bc + a'c'$

9. Find the expression for each of these Karnaugh maps. (HINT: In each case the answer consists of a single product.)

(a)

	yz	yz'	y'z'	y'z
x				
x'		✓	✓	

(b)

	yz	yz'	y'z'	y'z
x	✓			✓
x'				

(c)

	yz	yz'	y'z'	y'z
x	✓			✓
x'	✓			✓

(d)

	zw	zw'	z'w'	z'w
xy				
xy'			✓	✓
x'y'				
x'y				

(e)

	zw	zw'	z'w'	z'w
xy	✓	✓		
xy'				
x'y'				
x'y	✓	✓		

(f)

	zw	zw'	z'w'	z'w
xy	✓			✓
xy'	✓			✓
x'y'	✓			✓
x'y	✓			✓

10. Find a minimal expression for each of these Karnaugh maps.

(a)

	zw	zw'	z'w'	z'w
xy		✓		
xy'		✓	✓	✓
x'y'	✓	✓	✓	
x'y			✓	

(b)

	zw	zw'	z'w'	z'w
xy		✓		
xy'		✓	✓	
x'y'		✓		
x'y	✓	✓		✓

(c)

	zw	zw'	z'w'	z'w
xy	✓			✓
xy'		✓	✓	
x'y'	✓		✓	
x'y	✓	✓	✓	✓

11. Find a minimal expression for these expressions:
- (a) $xy + xyz + x'y'z' + x'yzt'$
- (b) $xyz + xy'z' + xy'z + x'yz + x'yz'$
- (c) $xyz + xy'z' + xy'z + x'yz + x'yz' + x'y'z$
- (d) $xyz + xy'z + x'yz + x'yz' + x'y'z' + x'y'z$

12. For each column (a), (b), and (c) in the truth table below, give a Karnaugh map for the corresponding sum-of-products expression, and simplify the expression. To

save space, the 16 possible values of the four variables x, y, z, and w are presented in two tables, each with eight rows.

13. Find a minimal expression for each of the following:
- (a) $xy + x'y + x'y'$
- (b) $x + x'yz + xy'z'$
- (c) $y'z + y'z'w' + z'w$
- (d) $y'zw + xzw' + xy'z'$

x	y	z	w	(a)	(b)	(c)	x	y	z	w	(a)	(b)	(c)
0	0	0	0	0	0	0	1	0	0	0	0	1	1
0	0	0	1	0	0	0	1	0	0	1	1	1	0
0	0	1	0	1	1	1	1	0	1	0	1	1	1
0	0	1	1	0	0	0	1	0	1	1	1	0	0
0	1	0	0	1	1	1	1	1	0	0	0	1	0
0	1	0	1	1	0	1	1	1	0	1	1	0	1
0	1	1	0	1	1	1	1	1	1	0	1	1	0
0	1	1	1	1	1	1	1	1	1	1	0	1	1

Truth Table for Exercise 12.

Chapter 3 Summary

3.1 Set Definitions and Operations

Terms and concepts

- You should recognize the notation for frequently used sets of numbers, including:
  - $\mathbb{N}$ (natural numbers), $\mathbb{Z}$ (integers), $\mathbb{Q}$ (rational numbers), and $\mathbb{R}$ (real numbers).
  - $\mathbb{R}^+, \mathbb{R}^{\geq 0}, \mathbb{Q}^+, \mathbb{Q}^{\geq 0}, \mathbb{Z}^+, \mathbb{Z}^{\geq 0}$
- You should know what it means to be an *element* of a set (written $x \in A$), and what it means to be a *subset* of a set (written $A \subseteq B$).
- You should recognize the *empty set*, whether denoted by $\{\}$ or $\emptyset$, and you should recognize the use of U to denote the *universal set*.
- You should understand the notation used for various set operations: $A \cap B$ for *intersection*, $A \cup B$ for *union*, $A - B$ for *difference*, A' for *complement*. You should know that *disjoint sets* are sets whose intersection is the empty set.
- You should be familiar with important properties of set operations: *commutative*, *associative*, *distributive*.

- You should be able to use and interpret the notation $n(A)$ used to indicate the number of elements in the finite set A.

Describing and working with sets

- You should be able to identify elements of the standard sets $\mathbb{N}$, $\mathbb{Z}$, and so on.
- You should understand that two sets are equal when each is a subset of the other.
- You should be aware that sets may contain elements that are not numbers. For example, we may work with sets of strings, sets of students, or sets of ordered pairs.
- You should be able to work with sets described using set-builder notation, as in these examples:
  - $\{x \in \mathbb{N} : x = 2k$ for some $k \in \mathbb{N}\}$ – using *property descriptions*
  - $\{2k : k \in \mathbb{N}\}$ – using *form descriptions*
- You should be able to determine whether or not one set is a subset of another set. The given sets may be described in a variety of ways, including set-builder notation.

Set operations and set size

- You should be able to build new sets from existing sets using various combinations of the set operations intersection, union, difference, and complement.
- You should be able to use Venn diagrams to illustrate and investigate properties of the set operations.
- You should be able to determine the size of a set described in various ways, including those described using set-builder notation.
- You should be able to apply the *Inclusion-Exclusion Principle*, for two and three sets, to determine the size of the union of sets.

3.2 More Operations on Sets

Terms and concepts

- You should recall the notion of *ordered pair* from high school algebra. For example, (3, 5) is read as "the ordered pair 3, 5."
- You should recognize the notation $A \times B$ ("A cross B") for the *Cartesian product* of A and B, and the special notation A^2 used to indicate $A \times A$.
- You should be able to work with the generalization of these concepts:
  - the n - tuple $(x_1, x_2, \ldots, x_n)$
  - the Cartesian product of n sets, written as $S_1 \times S_2 \times \cdots \times S_n$
  - the use of the shorthand S^n
- You should recognize the use of $\mathcal{P}(A)$ to denote the "*power set of the set A.*"
- You should know what is meant by a *partition* of a set, and understand the use of the phrase "*a part of S*" in this context.

Describing and working with Cartesian products

- Given two or more sets, you should be able to calculate, and answer questions about, their Cartesian product.
- You should be able to give the *Cartesian graph* of small sets of ordered pairs, or of sets of ordered pairs described using algebraic equations.
- You should be able to relate Cartesian products to earlier concepts (for example, the truth tables of Chapter 1).

Describing and working with sets of sets

- You should be able to answer questions about sets whose elements are also sets.

- You should be able to calculate the complete power set for small sets, and to answer questions about the power set for any given set.
- For a given set, you should be able to find partitions of that set satisfying given conditions.
- Given a set of subsets, you should be able to determine if it represents a partition; and if not, identify which property or properties of a partition are not satisfied.

Set operations and set size

- You should be able to combine the operations of this section (Cartesian product, power set) with those of the previous section (intersection, etc.).

Set size

- You should be able to calculate the size of the Cartesian product of sets, and of the power set of a set.
- You should be able to combine these two calculations with set size calculations from the previous section of the text.
- You should understand, for small sets, the relationship between the size of the power set and the size of the Cartesian product $\{0, 1\}^k$.

3.3 Proving Set Properties

Terms and concepts

- You should recognize that an *element-wise proof* of a set property involves choosing an arbitrary element of one set and showing that it must be an element of some other set.
- For an expression involving the set operations $\cap$, $\cup$, $\emptyset$, U, and $'$, you should know what is meant by the *dual expression*, and you should be able to write that dual expression.
- You should know and be able to apply the *duality principle* (if a property is true for sets, so is its dual property).
- You should recognize set properties that are similar to properties of numbers: *commutative, associative, distributive, identity, double negative.*
- You should also realize that sets have other properties numbers do not satisfy: *negation, idempotent, DeMorgan's laws, universal bound, absorption, complements of U and $\emptyset$, complement and negation.*

Element-wise proofs

- You should be able to give element-wise proofs to show that one set is a subset of another:

- For particular sets – for example, $\{10k : k \in \mathbb{N}\} \subseteq \{k \in \mathbb{Z} : k$ is even$\}$.
  - For set properties – for example, $A \cap B \subseteq A$.
  - For set properties that are true with additional hypotheses – for example, "If $(A \cup B) \subseteq B$, then $A \subseteq (A \cap B)$."
- You should know (and be able to apply) that we prove two sets are equal by proving that each is a subset of the other.
- You should know how to prove that a set is empty.

Proofs using previously established properties

- You should be able to use the properties in Theorem 6 to prove additional properties of sets.
- For each property you prove, you should be able to state the dual property that is also true.

3.4 Boolean Algebra

Terms and concepts

- You should recognize the connections among *sets*, *logical connectives*, and *Boolean algebra*, and be able to translate expressions and properties written in one of these forms to either of the other forms.
- You should recognize properties of a Boolean algebra, and their relationship to similar properties for sets and for logical connectives: *commutative, associative, distributive, identity, negation, double negative, idempotent, DeMorgan's laws, universal bound, absorption, complements of 1 and 0, complement and negation.*
- For an expression involving the Boolean algebra operations $\cdot$, $+$, 0, 1, and $'$, you should know what is meant by the *dual expression*, and you should be able to write that dual expression.
- You should know and be able to apply the *duality principle* (if a given property is true, so is its dual property).

Proofs

- You should be able to use the properties of a Boolean algebra to prove additional properties.
- For each property you prove, you should be able to state the dual property that is also true.

3.5 Excursion: Logic Circuits

Terms and concepts

- You should understand, and be able to draw, the standard *logic gates* that make up circuits: the *and gate*, the *or gate*, and the *not gate* (or *inverter*).
- You should understand the terminology "*sum-of-products*" as it applies to Boolean expressions for circuits.
- You should know that *Karnaugh maps* are graphical tools to assist in simplifying sum-of-products expressions.

Circuits, Boolean algebra expression, and truth tables

- Given a circuit, you should be able to:
  - Give the corresponding Boolean algebra expression that describes the output for that circuit.
  - Give a truth table for the circuit.
- Given a truth table that represents the desired output for a circuit, you should be able to write a "sum-of-products" expression that yields that output.
- Given a Boolean algebra expression, you should be able to draw the corresponding circuit.
- For a given circuit, you should be able to use Boolean algebra properties to simplify the expression that corresponds to the circuit.

Karnaugh maps

- For 2, 3, or 4 variables, you should be able to create an empty Karnaugh map diagram.
- For any sum-of-products expression, you should be able to check the appropriate boxes in the Karnaugh map.
- For any Karnaugh map with boxes checked, you should be able to locate a minimal number of maximum-size rectangles that cover the checked boxes, and use this information to write a minimal sum-of-products expression for the map.
- You should be able to combine these steps, using Karnaugh maps to simplify arbitrary sum-of-products expressions.

4 | **Functions and Relations**

Everyone has experience with mathematical functions from their high school days, but that experience is often limited to studying functions that model physical motion or time-dependent data (like parabolas, lines, or trigonometric functions). Discrete math makes use of some of these functions as well, but we also have a second goal regarding functions. We would like to learn something about functions as abstract mathematical objects. In this pursuit, we will encounter a broader class of mathematical objects, called binary relations. Binary relations will not only provide us with a fresh perspective on the nature of functions, they will also be important for applications in this chapter as well as later in the book.

A distinguishing feature of discrete mathematics is that it deals primarily with integers. In the previous chapter, we saw this feature in that most of our examples and applications focused on either finite sets (i.e., sets whose size is a natural number) or the set of natural numbers itself. In this same vein, our functions and relations will primarily involve finite sets or the set of natural numbers. In the ultimate twist of irony, we will see that functions and relations can actually be thought of as sets, so much of the material from the previous chapter is relevant here once again.

The general study of functions and relations typically includes the notions of "inverse" and "composition," so we will stress these operations initially. In order to make connections with the previous and subsequent chapters, we will then study the relationship between certain properties of a function and the sizes of the sets on which that function is defined, and then investigate the notion of equivalence relations. A key concept throughout these early sections is the "arrow diagram" of a function. Aside from its visual appeal, this structure is important because it will show up later as the notion of a directed graph in Chapter 7. The next section will introduce (or reinforce) specific numerical functions that arise frequently in discrete mathematics, including the floor, the ceiling, and the logarithm. The chapter ends with two excursions that can be tackled independently of one another. Both excursions deal with recursive thinking, some proofs by induction, and the numerical functions that have been used in the earlier sections.

4.1 Definitions, Diagrams, and Inverses

We begin our discussion with functions, where some familiar examples will help us flesh out the notation and terminology we will be using.

Notation and Terminology of Functions

> **Definition** The notation $f : A \to B$ is used for a function, simply called f, with a set of *inputs* A (called the *domain*), and a set B (called the *codomain*) that includes all the *outputs*. The function f associates with each input in A one and only one output in B.

We use the notation $f : A \to B$ and say, "f is a function from A to B." If the rule associates to the element a of the domain the element b of the codomain, we write $f(a) = b$, and say, "f maps a to b," "the value of f at a is b," or "f of a equals b."

Example 1 *Suppose $f : \mathbb{N} \to \mathbb{N}$ is defined by the rule $f(x) = 2x + 1$. We can think of this as meaning, "Given an input $x \in \mathbb{N}$, f maps x to the output value $2x + 1 \in \mathbb{N}$." Is every element of the codomain an output of one and only one input into the function?*

SOLUTION No, there are codomain elements like $0 \in \mathbb{N}$ that are not the value of f at any input value $a \in \mathbb{N}$—that is, there is no $a \in \mathbb{N}$ for which $2a + 1 = 0$. This does not affect the fact that f is a function. □

Example 2 *Suppose $f : \mathbb{Z} \to \mathbb{Z}$ is defined by the rule $f(x) = x^2$. We think of this as meaning, "Given an input $x \in \mathbb{Z}$, the value of f at x is $x^2 \in \mathbb{Z}$." Is every element of the codomain an output of one and only one input into the function?*

SOLUTION No, there are codomain elements like $1 \in \mathbb{Z}$ for which there are two input values $a \in \mathbb{Z}$ with $a^2 = 1$. In particular, both $a = 1$ and $a = -1$ satisfy this requirement. This does not affect the fact that f is a function. □

People are most familiar with functions that serve to describe some computation to be performed on a number to get a new number. However, we use other types of functions all the time without really being aware of it.

 Example 3

1. *When you look up a phone number in a directory, you are using a function whose inputs are names and whose outputs are phone numbers.*

2. *When you type a document in word processing software, your computer uses a function whose inputs are keystrokes and whose outputs are symbols displayed on your computer screen.*

3. *When we add numbers, we are really using a function whose inputs are pairs of numbers and whose outputs are the numbers resulting from the sum of the pair.*

We have also encountered functions in this book starting from the very first pages.

Example 4 *The actions on the packet of four cards in the magic trick in Example 1 of Section 1.1 are each a function that takes a packet of cards as input and returns a rearranged packet of cards as output. In that example, we refer to the cards using the letters C, S, H, D. For this example, we will not allow the cards to be flipped. Hence, HCDS refers to the following packet of cards, from bottom to top: heart, club, diamond, spade. Explore the cutting action as a function.*

SOLUTION   Let κ denote the action of cutting a single card from top to bottom. Then we can list the output for each of the 24 different packets of cards that we take as inputs:

$$\kappa(HCDS) = SHCD, \quad \kappa(SHCD) = DSHC, \quad \kappa(DSHC) = CDSH, \quad \kappa(CDSH) = HCDS$$
$$\kappa(HCSD) = DHCS, \quad \kappa(DHCS) = SDHC, \quad \kappa(SDHC) = CSDH, \quad \kappa(CSDH) = HCSD$$
$$\kappa(HSCD) = DHSC, \quad \kappa(DHSC) = CDHS, \quad \kappa(CDHS) = SCDH, \quad \kappa(SCDH) = HSCD$$
$$\kappa(HSDC) = CHSD, \quad \kappa(CHSD) = DCHS, \quad \kappa(DCHS) = SDCH, \quad \kappa(SDCH) = HSDC$$
$$\kappa(HDCS) = SHDC, \quad \kappa(SHDC) = CSHD, \quad \kappa(CSHD) = DCSH, \quad \kappa(DCSH) = HDCS$$
$$\kappa(HDSC) = CHDS, \quad \kappa(CHDS) = SCHD, \quad \kappa(SCHD) = DSCH, \quad \kappa(DSCH) = HDSC$$

$\square$

Example 5 *The truth tables from Section 1.3 are functions that take as input the truth values of the variables and return the truth value of a compound expression as the single output. What are the domain and codomain of the function given by the truth table shown in Table 4-1?*

p	q	$p \wedge q$
T	T	T
T	F	F
F	T	F
F	F	F

Table 4-1   Truth Table for Example 5

SOLUTION   The inputs are actually *pairs* of truth values. The set of inputs can be represented as the Cartesian product

$$\{T, F\} \times \{T, F\} = \{(T, T), (T, F), (F, T), (F, F)\}$$

and the outputs are simply truth values from the set $\{T, F\}$. The given table explains the rule for this function perfectly well, but if we would like to state it in words, we could say, "To the input value (x, y), associate as output the truth value of the statement 'p and q,' where p has truth value x and q has truth value y."

In summary, to completely describe a function, we must do four things:

1. Give the function a name. f, g, and h are popular names for functions, but it's always okay to be creative and descriptive.
2. Describe the domain.
3. Describe the codomain.
4. Describe the rule.

There are many ways to describe the rule for a given function. In some of the previous examples, we gave formulas describing an arithmetic operation to be performed to find the output to associate with each input. In others, we gave a table that showed every possible input with its associated output. To illustrate the connection between these different descriptions, we will look closely at a single function whose rule is described in several different ways.

Example 6   *Give several alternative ways to describe the rule for the following function:*

- **Name:** f
- **Domain**: $\{1, 2, 3, 4, 5\}$
- **Codomain**: $\mathbb{N}$
- **Rule**: *To each number in the domain, associate the square of the number.*

SOLUTION   The original description of the rule is in the form of a simple English sentence. The following are equivalent descriptions of this same rule:

1. We can use the sentence above to describe the rule.
2. We can use an algebraic formula to describe the rule: $f(x) = x^2$.
3. We can describe the rule by listing all the combinations of inputs and outputs, as ordered pairs: *f consists of the ordered pairs of numbers* $(1, 1)$, $(2, 4)$, $(3, 9)$, $(4, 16)$, $(5, 25)$.
4. Table 4-2 lists the inputs and outputs in table form.

The table representation is especially nice when the domain of the function is small. We can make this representation even more visual by using a picture to show

Input:	1	2	3	4	5
Output:	1	4	9	16	25

Table 4-2   Table Listing of the Function in Example 6

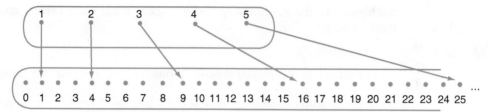

Figure 4-1   Arrow diagram for Example 6.

the mapping of inputs to outputs. In Figure 4-1, we use an arrow *from* each input value *to* its associated output value. We will refer to this kind of picture as an *arrow diagram*.

Notice that in Example 6, the codomain is not a finite set, so we can only show part of that set in our diagram. When both the domain and codomain are small enough, we will show the entire diagram.

 Example 7   *Show the arrow diagrams for the following functions:*

1. *Let f be the function with domain $\{a, b, c\}$ and codomain $\{1, 2, 3\}$ defined by the set of ordered pairs $\{(a, 2), (b, 3), (c, 1)\}$.*

2. *Let $S = \{a, b, c\}$, and consider the function $n : \mathcal{P}(S) \to \{0, 1, 2, 3\}$, where $n(A)$ is the number of elements in the set A.*

SOLUTION   The diagrams are shown in Figure 4-2. In the arrow diagram, each function has *exactly one* arrow beginning at each point in the domain.   □

Practice Problem 1   *For the set $X = \{a, b, c\}$, define the function $C : \mathcal{P}(X) \to \mathcal{P}(X)$ with the rule $C(A) = X - A$. (Recall that $X - A$ is the set of elements from X that are not also in A.) Complete the arrow diagram for this function shown in Figure 4-3.*

Binary Relations

So far we have defined functions as rules acting on the elements of specified sets. We have seen that there are several ways for describing the rule for a function, but one of them warrants a closer look. When we describe a rule by listing input-output pairs, we are actually adopting a mathematically sound point of view from which *a*

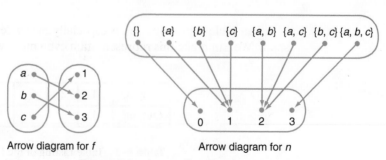

Arrow diagram for *f*   Arrow diagram for *n*

Figure 4-2   Arrow diagrams for Example 7.

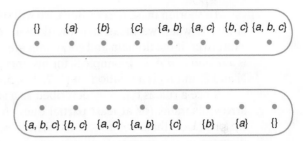

Figure 4-3 Diagram for Practice Problem 1.

function is just another kind of set. We will see that even when a rule fails to describe a function, it can still have important mathematical properties and applications worthy of our attention.

In general, a set of ordered pairs is a very simple structure for storing information in a way that captures a basic relationship between data. This type of structure is called a *binary relation*, and it will provide a generalization of our notion of function. Formally, a binary relation is nothing more than a subset of the Cartesian product $A \times B$ for specified sets A and B.

Definition   A *binary relation R* consists of three components: a domain A, a codomain B, and a subset of $A \times B$ called the "rule" for the relation. We will often omit the adjective "binary" since we will not address other kinds of relations in this book.

 Example 8   *The following are examples of binary relations:*

1. *Relation R_1*
   - **Domain:** *The set $\mathbb{S}$ of all students at your college this semester.*
   - **Codomain:** *The set $\mathbb{C}$ of all classes offered at your college this semester.*
   - **Rule:** (x, y) *is in R_1 if student x is enrolled in class y this semester.*
2. *Relation R_2*
   - **Domain:** *The set $A = \{1, 2, 3, 4, 5, 6\}$.*
   - **Codomain:** *The same set A.*
   - **Rule:** (x, y) *is in R_2 if $x - y$ is an even integer.*
3. *Relation R_3*
   - **Domain:** *The set $\mathbb{N}$ of natural numbers.*
   - **Codomain:** *The set $\mathbb{Z}$ of integers.*
   - **Rule:** $R_3 = \{(x, y) \in \mathbb{N} \times \mathbb{Z} : x = y^2\}$, *or equivalently,* $R_3 = \{(y^2, y) : y \in \mathbb{Z}\}$.
4. *Relation R_4*
   - **Domain:** *The set $\mathbb{Z}$ of integers.*
   - **Codomain:** *The set $\mathbb{N}$ of natural numbers.*
   - **Rule:** $R_4 = \{(x, y) \in \mathbb{N} \times \mathbb{Z} : y = x^2\}$, *or equivalently,* $R_4 = \{(x, x^2) : x \in \mathbb{N}\}$.

To shorten these descriptions, we will sometimes refer to R as *a relation between A and B* when we mean that the domain of R is A and the codomain of R is B. Moreover, if the domain and codomain are the same set A, we will simply say that R is *a relation on A*. In Example 8, for instance, we can say that R_3 is a relation between $\mathbb{N}$ and $\mathbb{Z}$ and R_2 is a relation on $\{1, 2, 3, 4, 5, 6\}$.

Since a rule is typically described as a set of ordered pairs, it is easy to produce arrow diagrams (or at least partial ones) for any binary relation in the same way we produced them for functions. When we want to further emphasize the arrow diagram or the grammatical structure of a binary relation, there is an alternative notation called *infix notation* with which we write "$a\,R\,b$" (and read "a is R-related to b" or just "a is related to b" when the relation is made clear from context) instead of "$(a, b) \in R$." The next example illustrates when this notation makes sense.

 Example 9 *Draw an arrow diagram for the relation L defined by*

- **Domain**: $A = \{1, 2, 3, 4, 5\}$
- **Codomain**: $B = \{2, 3, 5, 7\}$
- **Rule**: $L = \{(1, 2), (1, 3), (1, 5), (1, 7), (2, 3), (2, 5), (2, 7), (3, 5),$
  $(3, 7), (4, 5), (4, 7), (5, 7)\}$. *We can use set-builder notation to write*

$$L = \{(x, y) \in A \times B : x < y\}$$

or we can use infix notation to write

$$1\,L\,2, 1\,L\,3, 1\,L\,5, 1\,L\,7, 2\,L\,3, 2\,L\,5, 2\,L\,7, 3\,L\,5, 3\,L\,7, 4\,L\,5, 4\,L\,7, 5\,L\,7$$

The latter list of infix pairs clearly reflects the list of statements

$$1 < 2, 1 < 3, 1 < 5, 1 < 7, 2 < 3, 2 < 5, 2 < 7, 3 < 5, 3 < 7, 4 < 5, 4 < 7, 5 < 7$$

that defines the relation L.

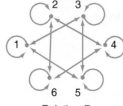

Relation L

Figure 4-4 Diagram for Example 9.

SOLUTION   The arrow diagram, shown in Figure 4-4, contains an arrow from a to b exactly when $(a, b) \in L$, or equivalently, when $a\,L\,b$ is true.   □

When the domain and codomain of a relation are the same set, we can simplify the arrow diagram by using just one set of nodes to represent both domain and codomain. We will call this the *one-set arrow diagram* and use the phrase *two-set arrow diagram* to refer to diagrams like those in the previous examples.

Relation R_2

Figure 4-5 Diagram for Example 10.

Example 10 *Draw the one-set arrow diagram for relation R_2 from Example 8.*

SOLUTION   To keep the arrow diagram relatively uncluttered, we will often use "double arrows" to indicate edges in both directions between a pair of nodes. The result is shown in Figure 4-5.   □

Practice Problem 2   *Draw the one-set arrow diagram for the relation R described below:*

- **Domain**: $A = \{1, 2, 3, 4, 5, 6, 7, 8, 9, 10\}$.
- **Codomain**: *The same set A.*

Explore more on the Web.

- **Rule**: $R = \{(1, 2), (2, 4), (3, 6), (4, 8), (5, 10)\}$. *Alternate ways to describe the rule for R include*

$$x\,R\,y \quad \text{if and only if} \quad y = 2x$$

or

$$R = \{(x, 2x) : x \in \{1, 2, 3, 4, 5\}\}$$

Draw part of the arrow diagrams for relations R_3 and R_4 from Example 8 as well.

When is a Relation a Function?

We saw at the beginning of this section that one way of describing the rule for a function is to specify a set of ordered pairs. Although not just *any* set of ordered pairs meets the defining criteria for a function, we can rewrite our definition to reflect that a function is simply a special kind of binary relation.

> **Definition**   A *function F from A to B* is a binary relation with domain A and codomain B with the property that for every $x \in A$, there is exactly one element $y \in B$ for which $(x, y) \in F$.

Example 11 *Which of the relations in Example 8 are actually functions?*

SOLUTION   We go through the list of relations to explain why each one is or is not a function.

1. Relation R_1 has just one problem—most students take more than one class each semester.
2. Relation R_2 includes pairs $(1, 3)$ and $(1, 5)$, violating the definition of a function.
3. Relation R_3 includes pairs $(1, 1)$ and $(1, -1)$, violating the definition of a function. There are also domain elements (like 2) that are not the first coordinate of any ordered pair in R_3.
4. Relation R_4 is a function.

$\square$

   We see that in practice, to be a function, a relation must satisfy two conditions: (1) Every element of the domain must have *something* associated to it, and (2) no element of the domain can have *two* things associated to it.

Practice Problem 3   *Which of the following relations are actually functions? For each relation that is not a function, give a specific way in which it violates the definition of a function.*

(a) *Let A be the set of letters in the alphabet and B be the set of people in your math class, and let R be the relation from A to B with the rule "$\alpha\,R\,P$ if and only if the first name of person P begins with letter α."*

(b) *Let R be the relation on $\mathbb{R}$ given by the rule "$x\,R\,y$ if and only if $x \cdot y = 1$."*

(c) Let $A = \{1, 4, 9, 16, 25\}$, and define the relation R from A to $\mathbb{N}$ with the rule $(x, n) \in R$ if $n^2 = x$.

(d) Let R be the relation whose domain is $\{1, 2, 3, 4, 5\}$, whose codomain is $\mathbb{N}$, and whose rule is given by $R = \{(1, 12), (2, 4), (3, 4), (2, 9), (5, 25)\}$.

The arrow diagram of a relation gives us a more visual way to determine whether the relation is a function.

 Example 12 *Show the arrow diagram for each of the following relations. Which are functions? How can you tell from the arrow diagram whether a relation is a function?*

1. *Let f be the relation with domain $\{a, b, c\}$ and codomain $\{1, 2, 3\}$ defined by the set of ordered pairs $\{(a, 2), (b, 3), (c, 1)\}$.*

2. *Let $\mathbf{V}$ be the relation from $\{1, 2, 3, 4, 5\}$ to $\{a, e, i, o, u\}$ defined so that $(x, \alpha) \in \mathbf{V}$ if α is a vowel in the English word for the number x.*

SOLUTION   In its arrow diagram, a function will have *exactly one* arrow beginning at each point in the domain. In the diagram for f shown in Figure 4-6, each input has exactly one arrow beginning at that input, so f is a function. In the diagram for $\mathbf{V}$, three different inputs (1, 4, and 5) have more than one arrow, which means they are mapped to more than one output value, so $\mathbf{V}$ is not a function. □

Explore more on the Web.

Practice Problem 4   *Using sets $A = \{1, 2, 3, 4\}$ and $B = \{1, 2, 3\}$, draw arrow diagrams for each of the following:*

(a) *A function from A to B*

(b) *A relation between A and B that is not a function because some domain element is not associated with any codomain element*

(c) *A relation between A and B that is not a function because some domain elements are associated with more than one codomain element*

Inverse Relations

In practice, a rule for a function is typically interpreted as an "action" on domain elements that results in codomain elements, while a rule for a relation is viewed

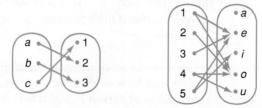

Figure 4-6   Diagram for Example 12.

simply as a relationship between domain elements and codomain elements that does not "favor" one type over the other. For example, the relation $R = \{(x, 2x) : x \in \mathbb{Z}\}$ on $\mathbb{Z}$ can be thought of as the action of doubling each domain element to find the associated codomain element. On the other hand, the following relations R and S describe very similar relationships between men and women:

● Relation R

   Domain: The set $\mathbb{M}$ of all males on the planet.

   Codomain: The set $\mathbb{F}$ of females on the planet.

   Rule: (x, y) is in R if y is the sister of x.

● Relation S

   Domain: The set $\mathbb{F}$ of females on the planet.

   Codomain: The set $\mathbb{M}$ of all males on the planet.

   Rule: (x, y) is in S if x is the sister of y.

The only difference between R and S is the interchanged roles of domain and codomain sets and the order of the elements in the ordered pairs. Specifically, given two people $a \in \mathbb{M}$ and $b \in \mathbb{F}$,

$$(a, b) \in R \quad \text{if and only if} \quad (b, a) \in S$$

If we call the pair (b, a) the *reverse* of the pair (a, b), then we can describe S as the set of all reversed pairs from relation R. In this case, we will say that the relation R is the *inverse* of relation S.

Definition   Given a relation R with domain A and codomain B, the relation R^{-1} (read "R inverse") with domain B and codomain A is called the *inverse of R* and is defined so that

$$(x, y) \in R \quad \text{if and only if} \quad (y, x) \in R^{-1}$$

Note that this definition also tells us that the inverse of R^{-1} is R, so we could simply say that these two relations are inverses of each other.

Example 13   *Recall the following from Example 8:*

● *Relation R_3*

   Domain: *The set $\mathbb{N}$ of natural numbers*

   Codomain: *The set $\mathbb{Z}$ of integers*

   Rule: $R_3 = \{(x, y) \in \mathbb{N} \times \mathbb{Z} : x = y^2\}$, *or equivalently,* $R_3 = \{(y^2, y) : y \in \mathbb{Z}\}$

● *Relation R_4*

   Domain: *The set $\mathbb{Z}$ of integers*

   Codomain: *The set $\mathbb{N}$ of natural numbers*

   Rule: $R_4 = \{(x, y) \in \mathbb{N} \times \mathbb{Z} : y = x^2\}$, *or equivalently,* $R_4 = \{(x, x^2) : x \in \mathbb{N}\}$

Show that R_3 and R_4 are inverses of one another.

SOLUTION   The domain of R_3 is the codomain of R_4 and vice versa, so we only have to check the relationship between the rules for the two relations. Consider a pair $(x, y) \in R_3$. This means that $x = y^2$, which in turns means that the *reversed* pair (y, x) is in R_4. Similarly, every pair in R_4 has its reverse in R_3. Therefore, R_3 and R_4 are inverses of one another.   □

The arrow diagrams of relations that are inverses of one another have a very simple relationship with each other. Since the arrow diagram of a relation R consists of an arrow from a to b whenever the pair $(a, b) \in R$, it follows that to find the arrow diagram for the inverse of a relation, one must simply reverse the arrows in the diagram for the original relation.

Example 14   *Draw the arrow diagram for the inverse of the relation E shown in Figure 4-7.*

SOLUTION   The relation E can be best described as follows:

- **Domain:** The set $A = \{1, 2, 3\}$.
- **Codomain:** $\mathcal{P}(A)$, the power set of A.
- **Rule:** $x \, E \, y$ if and only if $x \in y$.

This means that the inverse E^{-1} can be described as follows:

- **Domain:** $\mathcal{P}(A)$, the power set of A.
- **Codomain:** The set $A = \{1, 2, 3\}$.
- **Rule:** The reverse of the rule for E. Namely, "$(x, y) \in E^{-1}$ means that $y \in x$."

The arrow diagram is shown in Figure 4-8.   □

Practice Problem 5   *Draw the arrow diagrams for the inverses of the relations R from Practice Problem 2 and L from Example 9.*

In the special case where the relation has the same domain as codomain, the relationship between the arrow diagrams of the relation and its inverse appears even stronger.

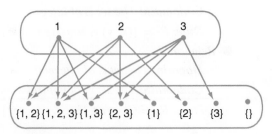

Figure 4-7   Arrow diagram for relation E.

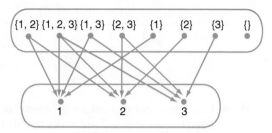

Figure 4-8 Arrow diagram for relation E^{-1}.

Example 15 *For the relation R described by the arrow diagram in Figure 4-9 on the left, draw the arrow diagram for R^{-1}.*

- **Domain:** *The set $S = \{A, B, C, D, E, F\}$ of men's basketball teams in a certain college athletic conference.*
- **Codomain:** *The same set S.*
- **Rule:** *$(x, y) \in R$ (or equivalently, "$x\,R\,y$") means "x beat y this year," where R is given by the arrow diagram below on the left.*

SOLUTION   The arrow diagram for R^{-1} is shown in Figure 4-9 on the right. □

Inverse Functions

We have already noted that when considering a relation that is actually a function, we can think of it as performing an operation or action on domain elements to create an element from the codomain. Under this interpretation, we can think of an inverse function as "undoing the action" of the original function. It is very natural when you think about it: Once you have a function to perform some operation or action, it is inevitable that you will need to create another function to "undo" the operation or action of the first function. This process has been a theme throughout your mathematical life.

- Once you learn to add, it is inevitable that you will have to subtract. For example, "adding 3" and "subtracting 3" are inverses of each other.
- Once you learn to multiply, it is inevitable that you will have to divide. For example, "doubling" and "halving" are inverses of each other.

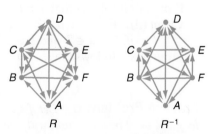

Figure 4-9 Arrow diagrams for Example 15.

a	-5	-1	0	7	$\cdots$	b	-2	2	3	10	$\cdots$
$f(a)$	-2	2	3	10	$\cdots$	$g(b)$	-5	-1	0	7	$\cdots$

Table 4-3 Table for Example 16

● Once you learn to set up problems involving x^2 in algebra, you soon thereafter need to compute square roots to solve these problems. "Finding the square" and "taking the square root" are inverse functions.

So we have seen inverse functions at many different times in our lives, perhaps without ever noticing the recurring theme. Let's see what inverses mean within the context of our function notation so that we can work toward a formal definition of inverse functions.

Example 16 *Let $f : \mathbb{Z} \to \mathbb{Z}$ be a function with the rule $f(x) = x + 3$, and let $g : \mathbb{Z} \to \mathbb{Z}$ be a function with the rule $g(y) = y - 3$. Compare the actions of f and g for several input values.*

SOLUTION We present these results in Table 4-3. We can see that $g(b) = a$ exactly when $f(a) = b$ for every a and b mentioned in the tables, but this is far from a formal proof. □

This is hardly surprising, perhaps to the point that it seems like a waste of time. Indeed, it is much more efficient to show symbolically that any choice of values from $\mathbb{Z}$ must be inserted into the table in this way.

Proposition 1 *Let $f : \mathbb{Z} \to \mathbb{Z}$ be a function with the rule $f(x) = x + 3$, and let $g : \mathbb{Z} \to \mathbb{Z}$ be a function with the rule $g(y) = y - 3$. Then for all $a \in \mathbb{Z}$ and for all $b \in \mathbb{Z}$, $f(a) = b$ if and only if $g(b) = a$.*

NOTE: The definition of inverse functions involves the biconditional "if and only if," so we need to actually write two proofs: one to prove (i) "if $f(a) = b$, then $g(b) = a$," and a second to prove (ii) "if $g(b) = a$, then $f(a) = b$." Exercise 19 explores what can go wrong if we don't include both proofs.

PROOF
Claim (i) For all $a, b \in \mathbb{Z}$, if $f(a) = b$, then $g(b) = a$.
  Proof. Let $a \in \mathbb{Z}$ and $b \in \mathbb{Z}$ be given such that $f(a) = b$. That is, $a + 3 = b$. From this it follows that $a = b - 3$, and hence $a = g(b)$.
Claim (ii) For all $a, b \in \mathbb{Z}$, if $g(b) = a$, then $f(a) = b$.
  Proof. Let $a \in \mathbb{Z}$ and $b \in \mathbb{Z}$ be given such that $g(b) = a$. That is, $b - 3 = a$. From this it follows that $b = a + 3$, and hence $b = f(a)$. ■

Practice Problem 6 *Let $f : \mathbb{R} \to \mathbb{R}$ be a function with the rule $f(x) = 2x$, and let $g : \mathbb{R} \to \mathbb{R}$ be a function with the rule $g(y) = y/2$. Complete Table 4-4 to show the actions of f and g for several input values, then write a formal proof that $f(a) = b$ if and only if $g(b) = a$ for all $a, b \in \mathbb{R}$.*

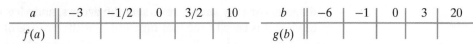

a	-3	$-1/2$	0	$3/2$	10	b	-6	-1	0	3	20
$f(a)$						$g(b)$					

Table 4-4 Table for Practice Problem 6

From these examples and observations, we can make the following formal definition:

> **Definition** Function $f : A \to B$ and function $g : B \to A$ are *inverses* of each other if $f(a) = b$ if and only if $g(b) = a$ for all $a \in A$ and $b \in B$. In this case, we refer to f as *the inverse of g* and to g as *the inverse of f*. In this case, we will often use the notation f^{-1} to mean the function that is the inverse of function f.

The formal definition gives us a strategy for finding the inverse for a given function if we are given the function rule in terms of an algebraic relationship.

Example 17 *Find a function g that is the inverse of the function $f : \mathbb{Q} \to \mathbb{Q}$ with rule $f(x) = \frac{2}{5}x - 2$.*

SOLUTION According to the definition, we must have $g : \mathbb{Q} \to \mathbb{Q}$ such that

$$f(a) = b \quad \text{if and only if} \quad g(b) = a$$

so suppose we have $a, b \in \mathbb{Q}$ satisfying $f(a) = b$. This means that $\frac{2}{5}a - 2 = b$, from which it follows (solving for a) that $a = \frac{5}{2}b + 5$. Hence, we can let g have the rule $g(y) = \frac{5}{2}y + 5$, and it can be shown that f and g are inverses of each other. We could say this more concisely by simply writing, "$f^{-1}(y) = \frac{5}{2}y + 5$." $\qquad\square$

It is even easier to see the relationship between inverse functions in the context of arrow diagrams.

Example 18 *Which of the functions f_1, f_2, or f_3 is the inverse of the function g shown in Figure 4-10?*

SOLUTION In terms of arrow diagrams, in order for f and g to be inverses of each other, we must have an arrow in f pointing from x to y whenever there is an arrow in g pointing from y to x, and vice versa. In other words, **if f and g are inverses, then the arrow diagram for f can be obtained from the arrow**

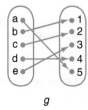
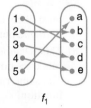
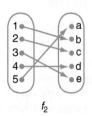
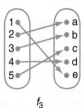

Figure 4-10 Arrow diagrams for Example 18.

diagram for g by simply reversing the arrows, and vice versa. From this point of view, it is easy to see that function f_2 is the inverse of g. That is, $f_2 = g^{-1}$. □

Practice Problem 7   *Draw arrow diagrams for the inverse of functions f_1 and f_3 in Example 18.*

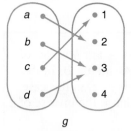

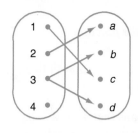

g

Figure 4-11   Arrow diagram for Example 19.

We can see from our examples that when a function's rule is described as an algebraic formula, then the inverse function's rule might be derived using algebra, and when a function is presented by an arrow diagram, then the inverse function can be found by reversing the arrows. Unfortunately, this is an oversimplification. Remember that *a function is a special kind of relation*, and there is no reason why reversing the rule for a function will necessarily produce another function, as the following example illustrates.

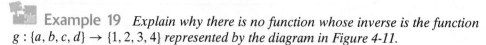 **Example 19**   *Explain why there is no function whose inverse is the function $g : \{a, b, c, d\} \to \{1, 2, 3, 4\}$ represented by the diagram in Figure 4-11.*

SOLUTION   It is easy enough to reverse the arrows as in Figure 4-12, but will the result be the arrow diagram of a *function*?

The answer is "no" for two reasons. First, since no arrow in g points to 4, when the arrows are reversed, there is no arrow starting at 4. Second, since two arrows in g point to 3, when we reverse the arrows, there are two arrows starting at 3. Remember that the one requirement for a function is that *for every element of the domain there is exactly one element of the codomain associated with it.* Therefore, there is no *function* with domain $\{1, 2, 3, 4\}$ and codomain $\{a, b, c, d\}$ that has the (reversed) arrow diagram above. □

Reversing the arrows

Figure 4-12   Solution to Example 19.

Summary

A *binary relation* is simply any set of ordered pairs with its first element from the *domain A* and second element from the *codomain B*. In the language of sets, this means that a binary relation is just a subset of $A \times B$. In this context, a *function* is a binary relation with an extra restriction: For each domain element x, a function has exactly one pair with x as its first coordinate. The inverse of a binary relation can always be obtained by reversing the pairs in the relation, but a side effect of this is that the *inverse* of a relation that is a function can result in a relation that is not a function. We will study this specific issue later in this chapter.

Solutions to Practice Problems

1   The arrow diagram is given in Figure 4-13.

2   The three diagrams are shown in Figure 4-14.

3   Only the third rule is a function for sure.

(a) It is possible, but very unlikely, that this is a function. Many letters (like probably x or q) will not be the first initial of anyone in your class, and there are most likely letters that may be associated with more than one person with that first initial.

(b) This is "almost" a function. The problem is that this rule does not associate anything with the number 0. The same rule for a relation on the set $\mathbb{R} - \{0\}$ is a function.

{}
{a}
{b}
{c}
{a, b}
{a, c}
{b, c}
{a, b, c}
{a, b, c}
{b, c}
{a, c}
{a, b}
{c}
{b}
{a}
{}

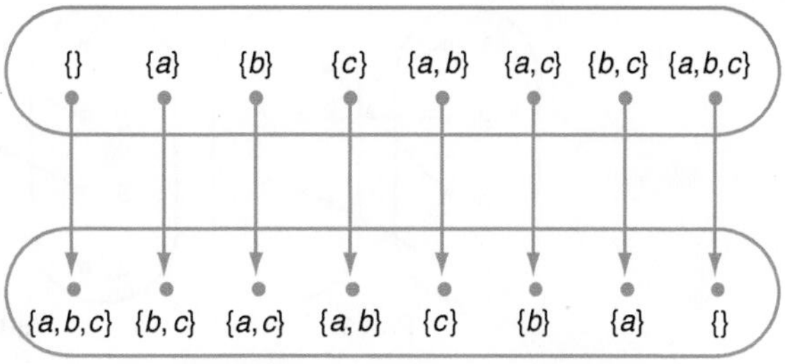

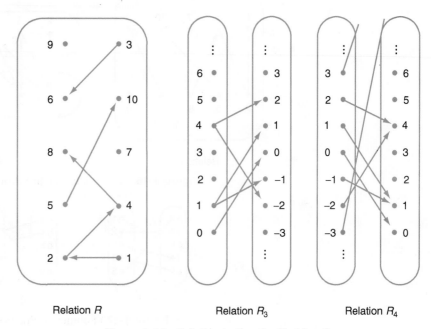

Figure 4-13 Solution to Practice Problem 1.

Relation R
Relation R_3
Relation R_4

Figure 4-14 Solution to Practice Problem 2.

(c) The rule is equivalently described by

$$R = \{(1, 1), (4, 2), (9, 3), (16, 4), (30, 5)\}$$

which satisfies the criteria to be a function from A to $\mathbb{N}$.

(d) There are two problems—the domain element 2 is associated with two codomain values (4 and 9), and the domain element 4 has no associated codomain element.

4 There are many possible solutions. In Figure 4-15, we give one possible solution for each part.

5 The two arrow diagrams are shown in Figure 4-16.

6 Table 4-5 shows the completed table, so we only need to write the proof.

Claim For all $a, b \in \mathbb{R}$, $f(a) = b$ if and only if $g(b) = a$.

Proof. Let $a, b \in \mathbb{R}$ be given such that $f(a) = b$. This means $2a = b$, from which it follows that $a = b/2$, and hence $a = g(b)$, as desired.

Now let $a, b \in \mathbb{R}$ be given such that $g(b) = a$. This means $b/2 = a$, from which it follows that $b = 2a$ and so $b = f(a)$, as desired. ■

7 The diagrams in Figure 4-17 show the inverse of f_1 and the inverse of f_3.

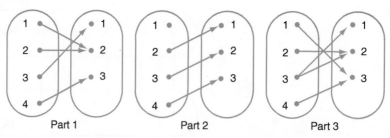

Part 1   Part 2   Part 3

Figure 4-15   Solution to Practice Problem 4.

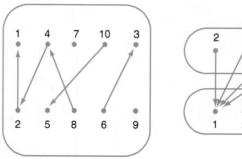

Relation R^{-1} Relation L^{-1}

Figure 4-16   Solution to Practice Problem 5.

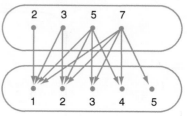

$f_1{}^{-1}$ $f_3{}^{-1}$

Figure 4-17   Solution to Practice
Problem 7.

Exercises for Section 4.1

1. For each of the following functions, draw a portion of the arrow diagram showing at least five elements of the domain and codomain:

   (a) $f : \mathbb{N} \to \mathbb{N}$ with $f(n) = 2n$

   (b) $g : \mathbb{N} \to \mathbb{Z}$ with $f(m) = 13 - 3m$

   (c) $f : \mathbb{N} \to \mathbb{Q}$ with $f(x) = 2 - \frac{1}{x+1}$

   (d) $g : \mathbb{N} \to \mathbb{Q}$ with $g(z) = z + \frac{1}{z^2+1}$

2. Sometimes in the description of a function, a rule is given but the domain is not clearly specified. That leaves it up to the reader to guess what was intended. For each of the following rules, provide a sensible domain and codomain:

   (a) $g(z) = 2z + \frac{1}{z+1}$

   (b) $g(t) = 3t + 1 + \sqrt{t}$

   (c) $f(x) = \sqrt{2x + 1}$

   (d) $f(t) = \frac{1}{\sqrt{t+1}}$

   (e) $g(x) = \frac{1}{x^2+1}$

3. Let $f : \mathcal{P}(\{1, 2, 3\}) \to \mathcal{P}(\{1, 2, 3\})$ be the function with the rule $f(A) = A - \{3\}$. Complete the two-set arrow diagram shown in Figure 4-18.

4. In the previous exercise, draw a single-set arrow diagram for the function f.

5. Let $f : \mathcal{P}(\{1, 2, 3, 4\}) \to \mathcal{P}(\{1, 2, 3, 4\})$ be the function with the rule $g(A) = A - \{4\}$. Complete the single-set arrow diagram in Figure 4-19.

a	-3	$-1/2$	0	$3/2$	10
$f(a)$	-6	-1	0	3	20

b	-6	-1	0	3	20
$g(b)$	-3	$-1/2$	0	$3/2$	10

Table 4-5   Solution for Practice Problem 6

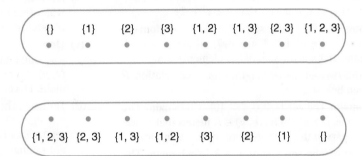

Arrow diagram for f

Figure 4-18   Diagram for Exercise 3.

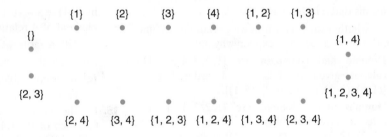

Arrow diagram for g

Figure 4-19   Diagram for Exercise 5.

6. Draw the complete arrow diagram for the following function n:
   - **Domain**: The set $\mathcal{P}(\{a, b, c, d\})$.
   - **Codomain**: The set $\{0, 1, 2, 3, 4\}$.
   - **Rule**: To each input A (a subset of $\{a, b, c, d\}$), associate the size of the set, and denote this number by $n(A)$. For example, $n(\{b, d\}) = 2$.

7. Draw a partial arrow diagram for the following function σ. (This is the Greek letter "sigma.")
   - **Domain**: The set $\mathbb{Z}^+$ of positive integers.
   - **Codomain**: The set $\mathbb{Z}$ of all integers.
   - **Rule**: To each input n in $\mathbb{Z}^+$, associate the sum of the positive numbers that evenly divide into n and denote this number by $\sigma(n)$. For example, here are some of the infinitely many ordered pairs that describe the function: $(6, 12)$, $(8, 15)$, $(20, 42)$.

8. Draw the two-set arrow diagram for each relation R described below:

   (a) Domain: The set $A = \{0, 1, 2, 3, 4, 5, 6\}$. **Codomain:** The same set A. **Rule:** $(x, y) \in R$ means that the number $x + y$ leaves a remainder of 0 or 4 when divided by 7.

   (b) Domain: The set $A = \{1, 2, 3, \ldots, 9, 10\}$. **Codomain:** The set $A = \{11, 12, 13, \ldots, 19, 20\}$. **Rule:** $(x, y) \in R$ means that $y < x^2 < 2y$.

   (c) Domain: The power set, $\mathcal{P}(\{1, 2, 3\})$. **Codomain:** The set $B = \{0, 1, 2, 3, 4, 5, 6, 7\}$. **Rule:** $(S, n) \in R$ means that n is the sum of the elements in S.

9. Draw the two-set arrow diagram for each relation R described below:

   (a) Domain: The set $\{1, 2, 3, \ldots, 10\}$. **Codomain:** The set $\{a, e, i, o, u\}$. **Rule:** $(n, v) \in R$ means that vowel v appears in the Spanish word for the number n.

   (b) Domain: The set $A = \{0, 1, 2, 3\}$. **Codomain:** The set B of all possible strings one can make using three characters, each of which is a 0 or a 1. ($B = \{101, 001, 100, 111, \ldots\}$ is called the set of *binary strings* of length 3.) **Rule:** $(n, s) \in R$ if the n is the number of 1's in s.

   (c) Domain: The power set, $\mathcal{P}(\{1, 2, 4, 8\})$. **Codomain:** The set $B = \{0, 1, 2, \ldots, 14, 15\}$. **Rule:** $(S, n) \in R$ means that n is the sum of the elements in S.

10. Decide whether each of the following relations is a function. If it is not a function, explain why not.

   (a) Domain and **codomain** are $\{1, 2, 3, 4, 5\}$. The **rule** is given by this set of ordered pairs: $\{(1, 5), (2, 3), (3, 3), (4, 2), (5, 1)\}$.

   (b) Domain and **codomain** are $\{1, 2, 3, 4, 5\}$. The **rule** is given by this set of ordered pairs: $\{(1, 5), (2, 3), (3, 3), (1, 2), (4, 1)\}$.

   (c) Domain: $\mathbb{Q}$. **Codomain:** $\mathbb{Z}$. **Rule:** $(r, z) \in R$ means that the rational number r can be written as a fraction with numerator z.

11. Decide whether each of the following relations is a function. If it is not a function, explain why not.

   (a) Domain: The set of all finite strings consisting of a's and b's, with at least one of each. **Codomain:** $\mathbb{Z}$. **Rule:** $(s, z) \in R$ means that z is the number of a's minus the number of b's in the string s. For example, $(bbaabab, -1) \in R$.

   (b) Domain: $\mathbb{N}$. **Codomain:** The set of all binary strings. **Rule:** $(n, s) \in R$ means that string s begins with n consecutive 0's.

   (c) Domain: $\mathbb{Z}$. **Codomain:** The set M of all people who were alive at midnight, December 31, 1999. **Rule:** $(z, p) \in R$ means that person p is z years old at midnight, December 31, 1999.

   (d) Domain: The set M of all people who were alive at midnight, December 31, 1999. **Codomain:** The same set M. **Rule:** $(x, y) \in R$ means that persons x and y are siblings.

12. Which of the arrow diagrams in Figure 4-20 represents a function whose domain and codomain are the set $\{1, 2, 3, 4, 5, 6\}$?

13. Which of the arrow diagrams in Figure 4-21, *when the arrows are reversed*, represents a function whose domain and codomain are the set $\{1, 2, 3, 4, 5, 6\}$?

14. Let $S = \{1, 2, 3\}$, and define a function $c : \mathcal{P}(S) \to \mathcal{P}(S)$ by $c(A) = S - A$. (That is, c maps a set A to the complement of A relative to the universe S.)

   (a) Fill in Table 4-6 indicating the function values.

   (b) Draw the arrow diagram for c.

   (c) If c is invertible, give a formula for the inverse function.

15. For the function $f : \mathbb{Q} \to \mathbb{Q}$ with the rule $f(x) = 2x + 6$, which of the following is the rule for the function $g : \mathbb{Q} \to \mathbb{Q}$ that is the inverse of f?

   (a) $g(y) = 2y - 6$

   (b) $g(y) = \frac{1}{2}y + 3$

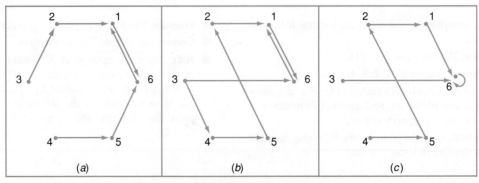

Figure 4-20 Diagrams for Exercise 12.

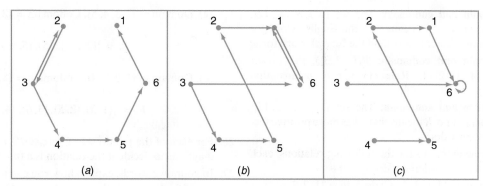

Figure 4-21 Diagrams for Exercise 13.

(c) $g(y) = \frac{1}{2}y - 3$

(d) $g(y) = \frac{1}{2}y + 6$

16. Fill in the missing steps to complete the following proof that the function $f : \mathbb{Q} \to \mathbb{Q}$ with rule $f(x) = 3x + 9$ and the function $g : \mathbb{Q} \to \mathbb{Q}$ with rule $g(y) = \frac{y}{3} - 3$ are inverses of each other.

   Claim (i) For all $a, b \in \mathbb{Q}$, if $f(a) = b$, then $g(b) = a$.
   Proof. Let $a \in$ _____ and $b \in$ _____ be given such that $f(a) = b$. That is, _____ $= b$. From this it follows that ____, and hence $a = g(b)$.
   Claim (ii) For all $a, b \in \mathbb{Q}$, if $g(b) = a$, then $f(a) = b$.
   Proof. Let $a \in$ _____ and $b \in$ _____ be given such that _____. That is, _____ $= a$. From this it follows that _____, and hence $b = f(a)$.

17. For each of the following functions with domain and codomain $\mathbb{Q}$, find the rule for the inverse function f^{-1}. Prove your answer is correct in each case.

   (a) $f(x) = 3x - 6$

   (b) $f(x) = 2x + 8$

   (c) $f(x) = \frac{1}{3}x - 1$

   (d) $f(x) = x + 5$

18. For each of the following functions with domain and codomain $\mathbb{R}$, find the rule for the inverse function g. Prove your answer is correct in each case.

   (a) $f : \mathbb{R} \to \mathbb{R}$ with $f(x) = 2x - 4$ for all $x \in \mathbb{R}$.

   (b) $f : \mathbb{R} \to \mathbb{R}$ with $f(x) = 3x + 5$ for all $x \in \mathbb{R}$.

   (c) $f : \mathbb{R} \to \mathbb{R}$ with $f(x) = \frac{2}{7}x - \frac{1}{3}$ for all $x \in \mathbb{R}$.

   (d) $f : \mathbb{R} \to \mathbb{R}$ with $f(x) = \sqrt[3]{x} - 1$ for all $x \in \mathbb{R}$.

19. We have stated that when proving a pair of functions are inverses of one another, we must prove two parts of

a biconditional statement. This exercise explores why both proofs are required. Let $f : \mathbb{Q}^{\geq 0} \to \mathbb{R}$ be given by the rule $f(x) = \sqrt{x}$, and let $g : \mathbb{R} \to \mathbb{Q}^{\geq 0}$ be given by the rule $g(x) = x^2$.

   (a) Prove that for all $a \in \mathbb{Q}^{\geq 0}$ and $b \in \mathbb{R}$, if $f(a) = b$, then $g(b) = a$.

   (b) Give a counterexample to the statement "For all $a \in \mathbb{Q}^{\geq 0}$ and $b \in \mathbb{R}$, if $g(b) = a$, then $f(a) = b$."

   (c) Explain in plain English why f and g are not inverses of each other.

20. Prove that the function $f : (1, \infty) \to (1, \infty)$ defined by the rule $f(x) = \frac{x+1}{x-1}$ is its own inverse.

21. Draw the arrow diagram for the *inverse* of each relation given in Exercise 8.

22. Draw the arrow diagram for the *inverse* of each relation given in Exercise 9.

23. For each relation R described below, draw the one-set arrow diagrams for R and R^{-1}.

   (a) Domain and codomain: The set $A = \{0, 1, 2, 3, 4, 5, 6\}$. **Rule:** $(x, y) \in R$ means that the number $x + y$ leaves a remainder of 0 or 4 on division by 7.

   (b) Domain and codomain: The set $A = \{0, 1, 2, \ldots, 10\}$. **Rule:** $(x, y) \in R$ means that the number $x^2 - y$ is divisible by 11.

   (c) Domain and codomain: The set $A = \{0, 1, 2, \ldots, 10\}$. **Rule:** $(x, y) \in R$ means that $x^2 < y$.

24. For each relation R described below, draw the one-set arrow diagrams for R and R^{-1}.

A	$\emptyset$	$\{1\}$	$\{2\}$	$\{3\}$	$\{1, 2\}$	$\{1, 3\}$	$\{2, 3\}$	$\{1, 2, 3\}$
$c(A)$								

Table 4-6 Table for Exercise 14

(a) **Domain and codomain:** The set $\{1, 2, 3, \ldots, 12\}$. **Rule:** $(x, y) \in R$ means that the English word for x has one fewer letters than the English word for y.

(b) **Domain and codomain:** $\mathcal{P}(\{1, 2, 3\})$, the power set of $\{1, 2, 3\}$. **Rule:** $(x, y) \in R$ means that $x \subseteq y$.

(c) **Domain and codomain:** The set $\{1, 2, 3, \ldots, 12\}$. **Rule:** $(x, y) \in R$ means that x has more positive factors than y does.

25. Give arrow diagrams for the following relations, each given as a set of ordered pairs:

(a) Domain $= \{4, 5, 8, 9\}$. Codomain $= \{1, 2, 3, 5\}$.

$$R_a = \{(4, 5), (5, 3), (8, 1), (9, 2)\}$$

(b) Domain $= \{1, 2, 3\}$. Codomain $= \{6, 7, 8, 9\}$.

$$R_b = \{(1, 6), (2, 7), (3, 8)\}$$

(c) Domain $= \{2, 3, 4, 5\}$. Codomain $= \{2, 5, 6\}$.

$$R_c = \{(2, 2), (3, 5), (5, 6)\}$$

(d) Domain $= \{1, 2, 3, 4\}$. Codomain $= \{5, 6, 7\}$.

$$R_d = \{(1, 5), (2, 5), (3, 6), (4, 7)\}$$

26. For each of the relations in Exercise 25, use the arrow diagrams to decide if the relation is a function.

27. In your own words, explain how you can tell from a list of ordered pairs whether the relation with that list as its rule is a function.

28. Explain how you can tell from a list of ordered pairs whether the *inverse* of the relation with that list as its rule is a function.

4.2 The Composition Operation

The idea of composition of functions and relations is a very basic one. Mathematics is largely about how complex concepts, structures, and properties can be built logically out of simpler ones. Since functions and relations are fundamental structures in mathematics, it stands to reason that combinations of two or more of these structures could be important.

Composition of Functions

Before being swept away by a formal definition, let us consider an example of how composition naturally arises in the English language. One rule that relates pairs of people is the "husband of" relation. Another rule of this type is the "mother of" relation. These relations can be combined to give two distinct meanings:

● The relation "mother of the husband of" associates a woman with her mother-in-law.

● The relation "husband of the mother of" associates any person with his or her father or stepfather.

The preposition "of" naturally ties together English clauses in the same way that composition ties together mathematical functions, as we see in our formal definition.

Definition   If $f : A \rightarrow B$ and $g : B \rightarrow C$, then we can build a new function called $(g \circ f)$ that has domain A and codomain C, and that follows the rule $(g \circ f)(x) = g(f(x))$. We call $(g \circ f)$, read "g of f," the *composition* of g with f.

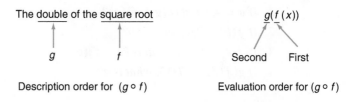

Figure 4-22 How to read $(g \circ f)$.

Example 1 *Given the function $f : \mathbb{R}^{\geq 0} \to \mathbb{R}$ defined by the rule $f(x) = \sqrt{x}$, and the function $g : \mathbb{R} \to \mathbb{R}$ defined by the rule $g(y) = 2 \cdot y$, describe the domain, codomain, and rule for the function $(g \circ f)$.*

SOLUTION Since the codomain of f matches the domain of g, the composition $(g \circ f) : \mathbb{R}^{\geq 0} \to \mathbb{R}$ makes sense. The rule for $(g \circ f)$ can be stated in a couple of different ways:

1. Since the rule for f can be described as "computing the square root of" and the rule for g can be described as "doubling," the rule for $(g \circ f)$ can be described as "computing the double of the square root of."

2. Algebraically,

$$(g \circ f)(x) = g(f(x)) = g(\sqrt{x}) = 2 \cdot \sqrt{x}$$

so we see that $(g \circ f)$ maps an input x to the output $2 \cdot \sqrt{x}$.

□

This example demonstrates a quirk about composing functions that we illustrate in Figure 4-22. When describing the rule for the function $(g \circ f)$ in English, it is natural to describe the action of g on the result of the action of f. In other words, we would typically mention the functions in the same order (left to right) they are written in the notation $(g \circ f)$. However, to actually *evaluate* the particular value, say, $(g \circ f)(9)$, we must first compute $f(9) = 3$, and then double that result to get 6. This means that to evaluate $(g \circ f)(x)$, we must use the functions in the reverse order (right to left) from how they are written. This evaluation order is consistent with the usual rule in algebra that one works from the innermost parentheses to the outermost parentheses.

There is an easy parallel of this situation in our English-language example. When we describe "the mother of the husband of" relationship, we are not thinking about any particular person, but as soon as we try to apply this relationship to a real person, we must think from right to left. For example, to find the mother of the husband of Melissa, we must first discover that Melissa's husband is Matt, and then find out that Matt's mother is Michelle.

Here is an exercise and practice problem that will seem a bit like a puzzle. Playing "mathematics detective" will give you some practice with the order of evaluation in the composition of functions.

Example 2 *Given the function $f : \mathbb{N} \to \mathbb{Q}^{+}$ with the rule $f(n) = \frac{n}{n+3}$ and the function $g : \mathbb{Q}^{+} \to \mathbb{Q}$ with the rule $g(r) = \frac{1}{r+1}$, answer the following questions about the composite function $(g \circ f)$:*

1. If $a = 4$, what is $(g \circ f)(a)$?
2. If $f(b) = 7/8$, what is b?
3. If $f(c) = 9/10$, what is $(g \circ f)(c)$?
4. If $g(f(d)) = 7/13$, what is d?

SOLUTION

1. $(g \circ f)(4) = g(f(4)) = g(4/7)$ and

$$g(4/7) = \frac{1}{\frac{4}{7}+1} = \frac{7}{11}$$

2. If $f(b) = 7/8$, solving the equation $\frac{b}{b+3} = 7/8$ gives us $b = 21$.
3. If $f(c) = 9/10$, then $(g \circ f)(c) = g(f(c)) = g(9/10)$ and

$$g(9/10) = \frac{1}{\frac{9}{10}+1} = \frac{10}{19}$$

Notice that we never need to know the value of c to find this.

4. If $g(f(d)) = 7/13$, then this means that $\frac{1}{f(d)+1} = \frac{7}{13}$, which can be solved to tell us that $f(d) = 6/7$. This, in turn, leads to the equation $\frac{d}{d+3} = 6/7$, which has as its solution $d = 18$.

□

Practice Problem 1 *Given the function $g : \mathbb{N} \to \mathbb{N}$ with the rule $g(z) = 2z + 1$ and the function $f : \mathbb{Z} \to \mathbb{Z}$ with the rule $f(m) = 5m - 7$, fill in the missing values in Table 4-7.*

So we see that when our functions have a context, either from the real world or from specified mathematical operations, the composition of functions is a perfectly natural thing. When the functions are merely abstract objects, we rely on tools like our arrow diagrams to help us form the proper mental picture of composition.

Example 3 *Figure 4-23 shows functions $g : \{a, b, c, d\} \to \{1, 2\}$ and $h : \{1, 2\} \to \{X, Y, Z\}$. Draw the arrow diagram of the composition $(h \circ g)$: $\{a, b, c, d\} \to \{X, Y, Z\}$.*

SOLUTION The diagram for $(h \circ g) : \{a, b, c, d\} \to \{X, Y, Z\}$ is found by thinking about the "evaluation order" of the function. That is, we take each input in the oval for $\{a, b, c, d\}$, trace its g arrow to the "intermediate" set $\{1, 2\}$, and then continue along the h arrow from that element of $\{1, 2\}$ into the final output set shown by the oval containing $\{X, Y, Z\}$. (It often helps to see the two diagrams merged together as in Figure 4-24 to show them sharing the

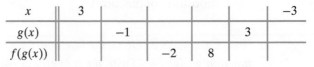

x	3					−3
$g(x)$		−1			3	
$f(g(x))$			−2	8		

Table 4-7   Table for Practice Problem 1

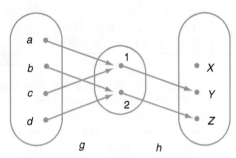

Figure 4-23 Diagrams for Example 3.

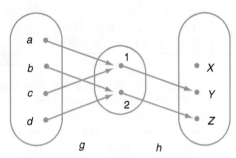

Figure 4-24 Another look at the diagrams for Example 3.

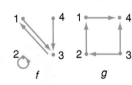

Figure 4-25 Solution for Example 3.

intermediate set $\{1, 2\}$.) For example, we can trace one such input like this:

$$\underbrace{a \underset{g}{\mapsto} 1 \underset{h}{\mapsto} Y}_{h \circ g}$$

Figure 4-25 shows the complete arrow diagram for $(h \circ g)$. We do not show the intermediate set $\{1, 2\}$ in this diagram since that set is neither the domain nor the codomain of the new function $(h \circ g)$. □

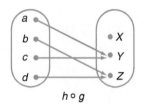

Figure 4-26
Diagrams for Example 4.

Example 4 *Consider the functions f and g with the set $\{1, 2, 3, 4\}$ as domain and codomain, and whose rules are given by the arrow diagrams in Figure 4-26. (The "loop" at 2 represents an arrow that starts at 2 and also ends at 2—this means that $f(2) = 2$.) Make a table to illustrate the rule for the function $g \circ f$.*

SOLUTION To calculate $(g \circ f)(1)$, we start at 1 in the f diagram, and follow the arrow to $f(1) = 3$. Then in the g diagram we start at 3 and follow the arrow to $g(3) = 2$. Thus, $(g \circ f)(1) = g(f(1)) = g(3) = 2$. We can find all the outputs of $g \circ f$ in this way.

- $(g \circ f)(1) = g(f(1)) = g(3) = 2$
- $(g \circ f)(2) = g(f(2)) = g(2) = 1$
- $(g \circ f)(3) = g(f(3)) = g(1) = 4$
- $(g \circ f)(4) = g(f(4)) = g(3) = 2$

□

Explore more on the Web.

Practice Problem 2   *For the functions f and g in Example 4, give arrow diagrams for $g \circ f$, $f \circ g$, and $f \circ f$.*

Inverse Functions Revisited

The composition operation on functions gives us another way to express the inverse relationship between functions. To understand this new perspective on inverse functions, we need to first understand the concept of an identity function.

> **Definition**   For a given set A, the *identity function** on A is the function $\iota_A : A \to A$ with the rule $\iota_A(x) = x$ for all $x \in A$. If the set A is clear from the context of a problem, we will often omit A and just use ι. Note that $\iota_A = \{(x, x) : x \in A\}$ for when we wish to consider ι_A as a binary relation.

In plain terms, ι is a function that performs no action at all, or equivalently, it is a relation that only associates each element with itself. Our first notion of inverse functions was built on the idea of one function undoing the action of the other. Another way to say this is that the successive actions of the two functions, in the end, do nothing at all. Let's revisit the pair of functions from Example 17 from Section 4.1 to see this in action.

Example 5 *Let $f : \mathbb{Q} \to \mathbb{Q}$ be the function with the rule $f(x) = \frac{2}{5}x - 2$, and let $g : \mathbb{Q} \to \mathbb{Q}$ be the function with the rule $g(y) = \frac{5}{2}y + 5$.*

SOLUTION   Let $x \in \mathbb{Q}$ be given. Then

$$
(g \circ f)(x) = g(f(x)) = g\left(\frac{2}{5}x - 2\right)
$$
$$
= \frac{5}{2}\left(\frac{2}{5}x - 2\right) + 5
$$
$$
= (x - 5) + 5
$$
$$
= x
$$

Now let $y \in \mathbb{Q}$ be given. Then

$$
(f \circ g)(y) = f(g(y)) = f\left(\frac{5}{2}y + 5\right)
$$
$$
= \frac{2}{5}\left(\frac{5}{2}y + 5\right) - 2
$$
$$
= (y + 2) - 2
$$
$$
= y
$$
□

It might seem silly to check both $f \circ g$ and $g \circ f$, but it is possible for one of them to be equal to the identity function but not the other.

---

* The traditional symbol for the identity function is ι, the lowercase Greek letter "iota."

Example 6 *Let $A = \{1, 2, 3, 4\}$ and consider the functions $f : A \to (A \times A)$ with the rule $f(a) = (a, a)$ and $g : (A \times A) \to A$ with the rule $g(x, y) = x$. Show that $g \circ f = \iota_A$. Explain why $f \circ g \neq \iota_{A \times A}$.*

SOLUTION

Claim $g \circ f = \iota_A$.

PROOF   Note that $(g \circ f) : A \to A$, so $(g \circ f)$ and ι_A have the same domain and codomain. We must only check then that their rules agree. Let $a \in A$ be given. We can evaluate

$$(g \circ f)(a) = g(f(a)) = g(a, a) = a$$

to see that $g \circ f$ agrees with the rule for ι_A for every element of their domains. Hence, these two functions are the same.

To see why $f \circ g \neq \iota_{A \times A}$, we can simply consider the example $(1, 2) \in A \times A$, and evaluate

$$(f \circ g)(1, 2) = f(g(1, 2)) = f(1) = (1, 1)$$

Hence, $(f \circ g)(1, 2) \neq (1, 2)$, while $\iota_{A \times A}(1, 2) = (1, 2)$, so the rules for these functions do not agree. ∎

We are ready for the formal statement of this equivalent way to express the inverse relationship between functions. The proof of the theorem is given in the form of "fill-in-the-blank" exercises at the end of this section.

Theorem 1   *Functions $f : A \to B$ and $g : B \to A$ are inverses of each other if and only if $f \circ g = \iota_B$ and $g \circ f = \iota_A$.*

PROOF   See Exercises 11 and 12. ∎

Under this interpretation, we finally see the sense in using the notation f^{-1} to denote the inverse of function f. At least we can understand it by analogy to the more familiar use of the -1 exponent in arithmetic. Think about the parallels between the following statements:

- 2^{-1} is the inverse of 2 under the multiplication operation $\times$. This means that

$$2 \times 2^{-1} = 2^{-1} \times 2 = 1$$

- f^{-1} is the inverse of f under the composition operation $\circ$. This means that

$$f \circ f^{-1} = f^{-1} \circ f = \iota$$

Practice Problem 3   *Use Theorem 1 to show that $f : \mathbb{Q} \to \mathbb{Q}$ with the rule $f(x) = \frac{7x - 10}{5}$ and $g : \mathbb{Q} \to \mathbb{Q}$ with the rule $g(y) = \frac{5y + 10}{7}$ are inverses of each other.*

Composition of Binary Relations

Binary relations can be combined to create new relations in much the same way that we combined functions earlier in this section. Sometimes the actual meaning of the new relation is somewhat elusive, so our examples and exercises will emphasize the interpretation of composition of relations.

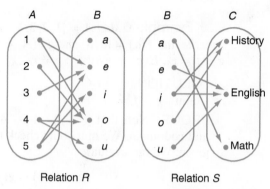

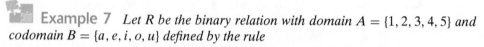

Relation R   Relation S

Figure 4-27 Diagrams for Example 7.

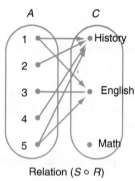

Relation (S ∘ R)

Figure 4-28 Solution to Example 7.

Example 7 *Let R be the binary relation with domain $A = \{1, 2, 3, 4, 5\}$ and codomain $B = \{a, e, i, o, u\}$ defined by the rule*

$$R = \{(x, y) : y \text{ is a vowel in the English word for } x\}$$

Let S be the binary relation with domain $B = \{a, e, i, o, u\}$ and codomain $C = \{math, English, history\}$ defined by the rule

$$S = \{(x, y) : x \text{ is a vowel in the word } y\}$$

Use an analogy to composition of functions to draw an arrow diagram for the composition $S \circ R$.

SOLUTION Figure 4-27 shows the arrow diagrams for the relations R and S. By following the arrows from the set A to the set B and then on to the set C, we can form the diagram for the new relation $(S \circ R)$ with domain A and codomain C as shown in Figure 4-28. □

Interpreting the composition of relations can sometimes be tricky and sometimes confusing. In the previous example, the rule can be described as "$(x, y) \in (S \circ R)$ if the English word for the number x has a common vowel with the word y." This leads to our formal definition of the composition of relations.

Definition Given the relation R_1 with domain A and codomain B and the relation R_2 with domain B and codomain C, we define the new relation $(R_2 \circ R_1)$ with domain A, codomain C, and the rule "$(a, c) \in (R_2 \circ R_1)$ if there is an element $b \in B$ such that $(a, b) \in R_1$ and $(b, c) \in R_2$."

This new relation is called *the composition of R_2 with R_1*, and we read the notation $(R_2 \circ R_1)$ as "R_2 of R_1."

When the relations express "real relationships," their composition has a natural interpretation.

Example 8 *Let S be the relation "is a sibling of," and C be the relation "is a child of," where both relations have the set of all people as domain and codomain. That is, we will have $(x, y) \in S$ if x is a sibling of y, and we will have $(a, b) \in C$ if a is a child of b. What is the natural interpretation of the relations $(S \circ C)$ and $(C \circ S)$?*

SOLUTION   The best approach is to use the formal definition of these compositions:

- According to the formal definition, $(a, c) \in (S \circ C)$ if there is a person b such that $(a, b) \in C$ and $(b, c) \in S$. In words, $(a, c) \in (S \circ C)$ if a is a child of a person b who is a sibling of c—that is, a is a child of a sibling of c.
- According to the formal definition, $(x, z) \in (C \circ S)$ if there is a person y such that $(x, y) \in S$ and $(y, z) \in C$. In words, $(x, z) \in (C \circ S)$ if x is a sibling of a person y who is a child of z—that is, x is a sibling of a child of z.

The two relations $(S \circ C)$ and $(C \circ S)$ are definitely different. The relation $(S \circ C)$ expresses (part of) the niece/nephew relationship, while $(C \circ S)$ is not very different from the relation C itself.   □

Practice Problem 4   *For the relations described below, draw an arrow diagram for (DistributedBy ∘ Stocks), and give a natural interpretation of this relation:*

- **Domain of Stocks**: *The set A of grocery stores in town, A = {Big Food, Food King, Leviathan, Big Market, Super Food, Grocelot}.*
- **Codomain of Stocks**: *The set B of brands of cereal*

  $$B = \{Krunchies, Prime\text{-}O's, Great\ Mornin', Quotidians\}$$

- **Rule for Stocks**: *$(a, b) \in$ Stocks if store a stocks cereal brand b, as given by the arrow diagram on the left in Figure 4-29.*
- **Domain of DistributedBy**: *The set B of brands of cereal*

  $$B = \{Krunchies, Prime\text{-}O's, Great\ Mornin', Quotidians\}$$

- **Codomain of DistributedBy**: *The set C of distribution companies, C = {Tom's Trucks, National Distributing, Interstate Amalgamated, Allied Foods}.*
- **Rule for DistributedBy**: *$(b, c) \in S$ if cereal brand b is distributed by company c, as given by the arrow diagram on the right in Figure 4-29.*

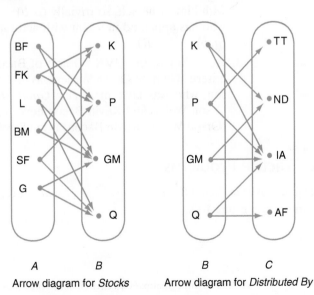

A   B   B   C

Arrow diagram for *Stocks*   Arrow diagram for *Distributed By*

Figure 4-29   Diagrams for Practice Problem 4.

x	3	−1	0	1	1	−3
$g(x)$	7	−1	1	3	3	−5
$f(g(x))$	28	−12	−2	8	8	−32

Table 4-8 Solution for Practice Problem 1

The previous practice problem is an example of an "inventory" problem that is common in computer science courses on databases. In that context, the composition of relations is similar to the *join* operation for relational databases. In fact, the University of Virginia Computer Science Department uses a movie database* for an application of "composition of relations" made famous by a parlor game based on the 1993 movie *Six Degrees of Separation*.

Example 9 *Let A denote the set of all actors who have ever appeared in a commercially released motion picture, and define the relation R on A by the rule $(x, y) \in R$ if actors x and y appeared in the same film. Answer the following questions about this relation (KB stands for Kevin Bacon):*

1. *What does it mean in plain English for $(x, y) \in R \circ R$?*
2. *Is $R \subseteq (R \circ R)$? Prove your answer is correct.*
3. *Using your own knowledge of movies or UVA's "Oracle of Bacon," find an actor X such that $(X, KB) \in R \circ R$ but $(X, KB) \notin R$.*

The "Kevin Bacon Game" is examined in more detail in Exercise 22.

SOLUTION

1. $(x, y) \in R \circ R$ means that there is an actor z who appeared in a movie with actor x and appeared in a movie with actor y.
2. **Claim:** $R \subseteq (R \circ R)$.
   Proof. Let $(a, b) \in R$ be given. This means that actors a and b appeared together in a film. It is of course also true that actor b appeared in some film with him or herself, so trivially $(b, b) \in R$. So there is an actor (namely b) who has appeared in a film with actor a and in a film with actor b; hence, $(a, b) \in (R \circ R)$.
3. According to the UVA "Oracle of Bacon," Orson Welles meets the given criteria. Namely, Orson Welles was in *A Safe Place* (1971) with Jack Nicholson, who was later in *A Few Good Men* (1992) with Kevin Bacon, but Orson Welles and Kevin Bacon never appeared in a film together. Hence, (Orson Welles, Kevin Bacon) is in $(R \circ R)$ but not in R itself.

□

Solutions to Practice Problems

1 The complete table is given in Table 4-8.
2 The diagrams are given in Figure 4-30.

3 *Proof.* Let $x \in \mathbb{Q}$ be given. Then
$$(g \circ f)(x) = g(f(x)) = g\left(\frac{7x - 10}{5}\right)$$

---

* Accessible through http://www.cs.virginia.edu/oracle/ as of June 2005.

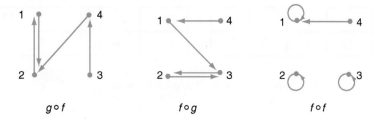

$g \circ f$ $\qquad$ $f \circ g$ $\qquad$ $f \circ f$

Figure 4-30 Solutions to Practice Problem 2.

$$= \frac{5 \cdot \frac{7x-10}{5} + 10}{7}$$
$$= \frac{(7x - 10) + 10}{7}$$
$$= x$$

Now let $y \in \mathbb{Q}$ be given. Then

$$(f \circ g)(y) = f(g(y)) = f\left(\frac{5y+10}{7}\right)$$
$$= \frac{7 \cdot \frac{5y+10}{7} - 10}{5}$$

$$= \frac{(5y+10) - 10}{5}$$
$$= y$$

Therefore, f and g are inverses of each other. ■

4 Figure 4-31 shows the arrow diagram for the relation (*DistributedBy* ∘ *Stocks*). $(a, c) \in$ (*DistributedBy* ∘ *Stocks*) means that there is a brand of cereal $b \in B$ such that store a stocks brand b, which is distributed by company c. So the natural interpretation is "Stocks something distributed by." So this relation tells us which distributors each store does business with for its cereal sales.

Exercises for Section 4.2

1. All functions in this problem have the set of real numbers $\mathbb{R}$ as their domain and codomain.
   (a) If $f(x) = 2x + 1$ and $g(y) = y^2 - 1$, what is $(f \circ g)(z)$?
   (b) If $f(x) = 3x - 2$ and $(f \circ g)(y) = 12y + 7$, what is $g(y)$?
   (c) If $g(y) = 2y - 1$ and $(f \circ g)(z) = 6z - 1$, what is $f(x)$?

2. All functions in this problem have the set of real numbers $\mathbb{R}$ as their domain and codomain.
   (a) If $f(x) = 2x + 1$ and $g(y) = y^2 - 1$, what is $(g \circ f)(z)$?
   (b) If $f(x) = 3x - 2$ and $(g \circ f)(z) = 9z^2 - 9z$, what is $g(y)$?
   (c) If $g(y) = 2y - 1$ and $(g \circ f)(z) = 4z^2 - 1$, what is $f(x)$?

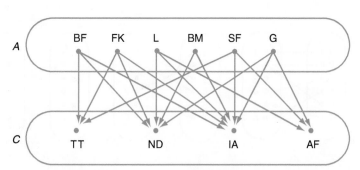

Figure 4-31 Solution to Practice Problem 4.

x	1	2	3
$f(x)$	11	13	10

x	10	11	12	13
$g(x)$	4	5	4	6

Table 4-9 Table for Exercise 3

3. Here are two functions $f : \{1, 2, 3\} \to \{10, 11, 12, 13\}$ and $g : \{10, 11, 12, 13\} \to \{4, 5, 6\}$ whose rules are given in Table 4-9.
   - **(a)** What is $f(1)$? $g(11)$?
   - **(b)** Give arrow diagrams for f and g.
   - **(c)** Which of these compositions can be defined: $g \circ f$, $g \circ g$, $f \circ g$, or $f \circ f$?
   - **(d)** For any of the compositions above that *are* defined, give the domain and codomain, and draw the arrow diagram.

4. Let $c : \mathcal{P}(\{x, y, z\}) \to \mathcal{P}(\{x, y, z\})$ be the function with rule $c(A) = \{x, y, z\} - A$, and let $n : \mathcal{P}(\{x, y, z\}) \to \{0, 1, 2, 3\}$ be the function such that $n(A)$ is the number of elements in the set A. Which composition is defined, $c \circ n$ or $n \circ c$? For the one that is defined, describe the domain and the codomain, and give an arrow diagram for the function.

5. Let $S = \{0, 1, 2, 3, 4, 5, 6, 7, 8, 9, 10\}$. Define $f : S \to S$ by $f(n) = $ the number of letters in the English-language spelling of the number. For example, $f(7) = 5$, since the word "seven" has five letters.
   - **(a)** Give an arrow diagram for f. Is f invertible?
   - **(b)** Give an arrow diagram for $f \circ f$.
   - **(c)** Give an arrow diagram for $f \circ f \circ f$.

6. For each of the functions whose arrow diagrams are shown, provide the missing function's diagram. If it is impossible to do so, or if there is more than one diagram that will work, say so.
   - **(a)** In Figure 4-32, what is $g \circ f$?
   - **(b)** In Figure 4-33, what is g?
   - **(c)** In Figure 4-34, what is f?

7. For each of the functions whose arrow diagrams are shown, provide the missing function's diagram. If it is impossible to do so, or if there is more than one diagram that will work, say so.
   - **(a)** In Figure 4-35, what is f?
   - **(b)** In Figure 4-36, what is g?
   - **(c)** In Figure 4-37, what is f?
   - **(d)** In Figure 4-38, what is g?

8. For each of the functions given below, describe the missing function by giving its domain, codomain, and rule as a set of ordered pairs. If it is impossible to do so, or if there is more than one function that will work, say so.
   - **(a)** If f has domain $\{1, 2, 3, 4\}$, codomain $\{a, b, c, d\}$, and rule $\{(1, b), (2, a), (3, d), (4, c)\}$ and g has domain $\{a, b, c, d\}$, codomain $\{x, y, z\}$, and rule $\{(a, x), (b, y), (c, z), (d, x)\}$, what is $g \circ f$?
   - **(b)** If f has domain $\{1, 2, 3, 4\}$, codomain $\{a, b, c, d\}$, and rule $\{(1, b), (2, a), (3, d), (4, a)\}$ and $g \circ f$ has domain $\{1, 2, 3, 4\}$, codomain $\{A, B, C, D, E\}$, and rule $\{(1, A), (2, C), (3, B), (4, D)\}$, what is g?
   - **(c)** If g has domain $\{a, b, c, d\}$, codomain $\{W, X, Y, Z\}$, and rule $\{(a, X), (b, Y), (c, Z), (d, X)\}$ and $g \circ f$ has domain $\{1, 2, 3, 4\}$, codomain $\{W, X, Y, Z\}$, and rule $\{(1, W), (2, Z), (3, X), (4, Y)\}$, what is f?

9. For each of the functions given below, describe the missing function by giving its domain, codomain, and rule as a set of ordered pairs. If it is impossible to do so, or if there is more than one function that will work, say so.

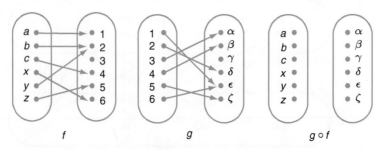

Figure 4-32 Diagrams for Problem 6(a).

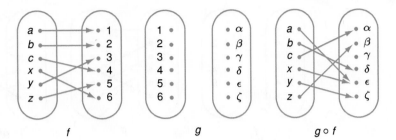

Figure 4-33   Diagrams for Problem 6(b).

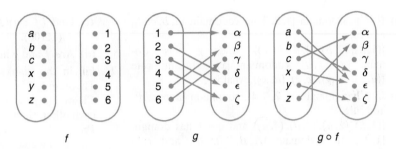

Figure 4-34   Diagrams for Problem 6(c).

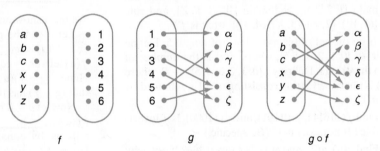

Figure 4-35   Diagrams for Problem 7(a).

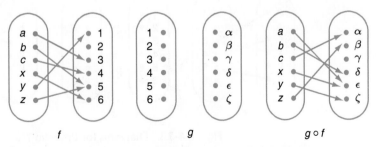

Figure 4-36   Diagrams for Problem 7(b).

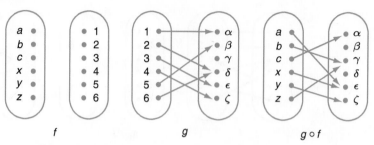

Figure 4-37 Diagrams for Problem 7(c).

(a) If f has domain $\{1, 2, 3, 4\}$, codomain $\{a, b, c, d\}$, and rule
$\{(1, a), (2, b), (3, d), (4, c)\}$ and g has domain $\{a, b, c, d\}$, codomain $\{x, y, z\}$, and rule $\{(a, z), (b, y), (c, z), (d, x)\}$, what is $g \circ f$?

(b) If f has domain $\{1, 2, 3, 4\}$, codomain $\{a, b, c, d\}$, and rule
$\{(1, b), (2, a), (3, d), (4, a)\}$ and $g \circ f$ has domain $\{1, 2, 3, 4\}$, codomain $\{A, B, C, D, E\}$, and rule $\{(1, D), (2, C), (3, E), (4, C)\}$, what is g?

(c) If g has domain $\{a, b, c, d\}$, codomain $\{W, X, Y, Z\}$, and rule
$\{(a, Z), (b, W), (c, X), (d, Y)\}$ and $g \circ f$ has domain $\{1, 2, 3, 4\}$, codomain $\{W, X, Y, Z\}$, and rule $\{(1, W), (2, Z), (3, X), (4, Y)\}$, what is f?

10. Let $\mathcal{B}$ be the set of binary strings* of length 5. Define $f : \mathcal{B} \to \{0, 1, 2, 3, 4, 5\}$, where $f(s)$ is the number of 1's in the string s. Define $g : \{0, 1, 2, 3, 4, 5\} \to \mathcal{B}$, where $g(n)$ is the binary string consisting of n 1's, followed by $5 - n$ 0's.

(a) Find $f(11011)$, $f(01101)$, and $f(11000)$. Is f invertible? If not, why not? (Be specific.)

(b) Find $g(0), g(2)$, and $g(4)$. Is g invertible? If not, why not? (Be specific.)

(c) Find $(f \circ g)(2)$, $(f \circ g)(0)$, $(g \circ f)(11010)$, $(g \circ f)(11100)$.

(d) Are f and g inverses of each other?

11. Fill in the blanks in the first part of the proof of Theorem 1.

Claim If $f : A \to B$ and $g : B \to A$ are inverses of each other, then $g \circ f = \iota_A$ and $f \circ g = \iota_B$.

Proof. Let $f : A \to B$ and $g : B \to A$ be functions that are inverses of each other. From our definition in the previous section, this means that

For all $a \in A$ and $b \in B$, $f(a) = b$ if and only if
$$g(\underline{\hspace{1.5cm}}) = \underline{\hspace{1.5cm}} \qquad (4.1)$$

We must show both $g \circ f = \iota_A$ and $f \circ g = \iota_B$.

Let $a \in A$ be given, and set $b = f(a)$. In this case, $(g \circ f)(a) = g(\underline{\hspace{1.5cm}}) = g(b)$. The fact that $b = f(a)$ tells us, by equation (4.1), that $g(b) = \underline{\hspace{1.5cm}}$. Hence, $(g \circ f)(a) = \underline{\hspace{1.5cm}}$ for all $a \in A$, proving that $(g \circ f)$ has the same rule as ι_A.

Now let $b \in B$ be given, and set $a = g(b)$. In this case, $(f \circ g)(\underline{\hspace{1.5cm}}) = f(a)$. The fact that $a = g(b)$ tells us, by equation (4.1), that $f(\underline{\hspace{1.5cm}}) = b$. Hence, $(f \circ g)(b) = \underline{\hspace{1.5cm}}$ for all $b \in B$, proving that $\underline{\hspace{1.5cm}}$ has the same rule as ι_B. ■

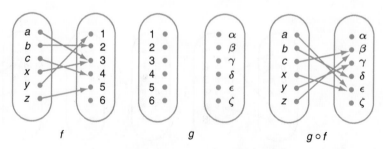

Figure 4-38 Diagrams for Problem 7(d).

---

* A binary string is made up of the characters 0 and 1. The length of the binary string is the number of characters used. For example, the length of the binary string 00110 is 5.

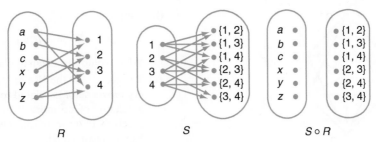

Figure 4-39   Diagrams for Problem 15.

12. Fill in the blanks in the second part of the proof of Theorem 1.

    Claim   If $f : A \to B$ and $g : B \to A$ satisfy $g \circ f = \iota_A$ and $f \circ g = \iota_B$, then f and g are inverses of each other.

    Proof.   Let $f : A \to B$ and $g : B \to A$ be functions satisfying $g \circ f = \iota_A$ and $f \circ g = \iota_B$. To show that f and g are inverses of each other, we must show that

    $$\text{For all } a \in A \text{ and } b \in B, \; f(a) = b$$
    $$\text{if and only if } g(b) = a.$$

    We will prove this biconditional statement using two separate proofs.

    Let $a \in A$ and $b \in B$ be given such that $b = f(a)$. In this case, $g(b) = g(\underline{\hspace{1.5cm}}) = (g \circ f)(a)$. The fact that \underline{\hspace{2cm}} tells us that $(g \circ f)(a) = a$. Hence, $g(b) = \underline{\hspace{2cm}}$, as desired.

    Now let $a \in A$ and $b \in B$ be given such that $a = g(b)$. In this case, $f(\underline{\hspace{1.5cm}}) = f(\underline{\hspace{1.5cm}}) = (f \circ g)(b)$. The fact that \underline{\hspace{2cm}} tells us that $(f \circ g)(b) = \underline{\hspace{2cm}}$. Hence, $f(a) = \underline{\hspace{2cm}}$, as desired. ■

13. For each relation R on the given set A, compare the arrow diagram for R with the arrow diagram for $R \circ R$.

    (a) $A = \{1, 2, 3, 4, 5, 6\}$ and $(x, y) \in R$ means $x < y$.

    (b) $A = \{1, 2, 3, \ldots, 11, 12\}$ and $(x, y) \in R$ means that x is a factor of y.

    (c) $A = \{1, 2, 3, \ldots, 9, 10\}$ and $(x, y) \in R$ means that $x + y$ is divisible by 4.

14. For each relation R described in Exercise 13, describe the relation $R \circ R$ in English.

15. Let R and S be the relations whose arrow diagrams are shown in Figure 4-39. Draw the arrow diagram for the relation $S \circ R$.

16. Let R and S be the relations shown in Table 4-10. In each case, an associated codomain element is listed directly below each domain element. Create a similar table showing the relation $R \circ S$.

    Give an interpretation in English of relations R, S, and $R \circ S$.

17. For the relation R in Example 15 of Section 4.1 (repeated in Figure 4-40), what is the meaning of $R \circ R$? Draw the arrow diagram for this new relation.

    ● **Domain:** The set $S = \{A, B, C, D, E, F\}$ of men's basketball teams in a certain college athletic conference.

    ● **Codomain:** The same set S.

    ● **Rule:** $(x, y) \in R$ (or equivalently, "$x \, R \, y$") means "x beat y this year," where R is given by the arrow diagram below.

18. Let A be the set consisting of the following 12 people: Andy, Angela, Brian, Chris, Clint, Jennifer, Jessica, Julie, Katie, Kristina, Luke, and Paula. Given the relations described below, give English descriptions of the relations *SiblingOf* $\circ$ *MarriedTo* and *MarriedTo* $\circ$ *SiblingOf*, and draw their arrow diagrams.

    Relation *SiblingOf* is defined on A such that $(x, y) \in SiblingOf$ means that x and y are siblings, where

    ● Andy and Katie are siblings.

    ● Chris, Julie, and Brian are siblings.

    ● Kristina and Paula are siblings.

Domain of R	10010	10100	01010	00110	01100	10001	01001	00101	00011	11000
Codomain of R	021	012	111	201	102	030	120	210	300	003
Domain of S	{1, 2}	{1, 3}	{1, 4}	{1, 5}	{2, 3}	{2, 4}	{2, 5}	{3, 4}	{3, 5}	{4, 5}
Codomain of S	11000	10100	10010	10001	01100	01010	01001	00110	00101	00011

Table 4-10   Rules for Relations R (above) and S (below)

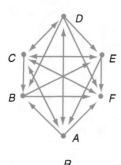

R

Figure 4-40 Diagram for Problem 17.

- Luke and Jennifer are siblings.
- Jessica and Angela are siblings.

Relation *MarriedTo* is defined on A such that $(a, b) \in$ *MarriedTo* means that a is married to b, where

- Andy is married to Jennifer.
- Katie is married to Luke.
- Chris is married to Paula.
- Brian is married to Jessica.
- Kristina is married to Clint, an only child.
- Julie and Angela are single.

19. Let A be the set of all students at your school this semester, B the set of all sections of all courses offered at your school this semester, and C the set of all instructors at your school. If $(x, y) \in R$ means that student x is enrolled in course y, and $(a, b) \in S$ means course a is being taught this semester by instructor b, what is the meaning of the relation $S \circ R$?

20. Let A be the set of all students at your school this semester, B the set of all towns on Earth, and C the set of all instructors at your school. If $(x, y) \in R$ means that student x was born in town y, and $(a, b) \in S \circ R$ means student a was born in the same town as instructor b, what is the meaning of the relation S?

21. Let A be the set of all production facilities, B the set of all machine parts, and C the set of all parts distributors. If $(x, y) \in S$ means that part x is available from distributor y, and $(a, b) \in S \circ R$ means facility a can do business with distributor b, what is the meaning of the relation R with domain A and codomain B?

22. (This exercise continues our look at the "Kevin Bacon Game" in Example 9.) Let A be the set of all people who have ever appeared in a commercially released movie, and let the relation R on A be defined by the fact that $(x, y) \in R$ means actors x and y appeared in a film together.

   (a) Use the UVA "Oracle of Bacon" Web site to find a specific pair of actors a and b such that $(a, b) \in R \circ (R \circ R)$ but $(a, b) \notin (R \circ R)$.

(b) Suppose

$$R \circ R \circ R \circ R = R \circ R \circ R$$

   What would this mean in plain English?

(c) The premise of the "Kevin Bacon Game" is that every movie actor is related to Kevin Bacon through a "chain" of no more than six movies. State this precisely in terms of the relation R, the set A, and Kevin Bacon (KB).

23. Consider the relation R on the set $A = \{0, 1, 2, 3, 4, 5, 6\}$ with the rule that $x R y$ if $x - y$ is divisible by 3. Draw the one-set arrow diagram for R and $R \circ R$, and describe the latter relation with an English sentence.

24. For each relation R whose complete arrow diagrams are shown in Figure 4-41, draw the arrow diagram for the new relations $R \circ R$ and $R \circ (R \circ R)$.

25. For each relation R in Figure 4-41, give an arrow diagram for the relation $R \circ R^{-1}$?

26. Consider once again the relations in Example 8 of Section 4.1.

- Relation R_1

   Domain: The set $\mathbb{S}$ of all students at your college this semester.

   Codomain: The set $\mathbb{C}$ of all classes offered at your college this semester.

   Rule: (x, y) is in R_1 if student x is enrolled in class y this semester.

- Relation R_2

   Domain: The set $A = \{1, 2, 3, 4, 5, 6\}$.

   Codomain: The same set A.

   Rule: (x, y) is in R_2 if $x - y$ is an even integer.

- Relation R_3

   Domain: The set $\mathbb{N}$ of natural numbers.

   Codomain: The set $\mathbb{Z}$ of integers.

   Rule: $R_3 = \{(x, y) \in \mathbb{N} \times \mathbb{Z} : x = y^2\}$, or equivalently, $R_3 = \{(y^2, y) : y \in \mathbb{Z}\}$.

- Relation R_4

   Domain: The set $\mathbb{Z}$ of integers.

   Codomain: The set $\mathbb{N}$ of natural numbers.

   Rule: $R_4 = \{(x, y) \in \mathbb{N} \times \mathbb{Z} : y = x^2\}$, or equivalently, $R_4 = \{(x, x^2) : x \in \mathbb{N}\}$.

(a) What is the meaning of the relation $R_1 \circ R_1^{-1}$? What is the meaning of the relation $R_1^{-1} \circ R_1$?

(b) What is the meaning of the relation $R_2 \circ R_2^{-1}$? What is the meaning of the relation $R_2^{-1} \circ R_2$?

(c) What is the meaning of the relation $R_3 \circ R_3^{-1}$? What is the meaning of the relation $R_3^{-1} \circ R_3$?

(d) What is the meaning of the relation $R_4 \circ R_4^{-1}$? What is the meaning of the relation $R_4^{-1} \circ R_4$?

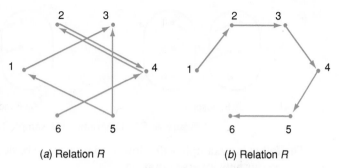

(a) Relation R (b) Relation R

Figure 4-41   Diagrams for Problem 24.

27. For the relation R in Exercise 17, what is the meaning of $R \circ R^{-1}$?

28. Investigate the precise relationship between the relations $R \circ R^{-1}$ and $R^{-1} \circ R$ for an arbitrary relation R with domain A and codomain B.

29. Investigate the precise relationship between the relation $R \circ R^{-1}$ and the identity relations ι_A and ι_B for an arbitrary relation R with domain A and codomain B.

4.3 Properties of Functions and Set Cardinality

So far in this chapter, we have discussed the "invertibility" of functions on more than one occasion. In this section, we will see how this property of a function $f : A \rightarrow B$ tells us about the relative size of the domain and codomain, A and B. In order to take full advantage of this relationship, we will need to write formal proofs about functions. As usual, before we can write proofs, we must decide on the formal definitions that both the READER and AUTHOR will agree on.

Invertibility and Other Properties

We have seen informally that invertiblity is the ability to reverse the rule for a function $f : A \rightarrow B$ and get a new function $g : B \rightarrow A$. In this way, we can think of g as "undoing" the effects of f. In Section 4.1, we called g the inverse of f, but to emphasize the connection to the original function f, we will now use the notation f^{-1}, just as we did with relations.

> **Definition**   The function $f : A \rightarrow B$ is *invertible* if there is a function $f^{-1} : B \rightarrow A$ such that $f(x) = y$ if and only if $f^{-1}(y) = x$. The notation f^{-1} is read as "f inverse," and the symmetry of the definition means that $\left(f^{-1}\right)^{-1} = f$.

Recall that Theorem 1 of Section 4.2 tells us that the above characterization of inverse functions is equivalent to saying that $f^{-1}(f(x)) = x$ and $f(f^{-1}(y)) = y$ for all $x \in A$ and all $y \in B$. This characterization is often easier to use in formal proofs, especially when we are thinking of inverse functions in terms of "opposite" actions.

Figure 4-42   Diagrams for Example 1.

The following example will allow us to review the basic idea of inverse functions while using our new inverse notation.

Example 1

1. Recall that $\mathbb{R}^{\geq 0} = \{x \in \mathbb{R} : x \geq 0\}$ is the set of nonnegative real numbers. Verify that the function $f : \mathbb{R}^{\geq 0} \to \mathbb{R}^{\geq 0}$ defined by $f(x) = x^2$ has as its inverse the function $f^{-1} : \mathbb{R}^{\geq 0} \to \mathbb{R}^{\geq 0}$ defined by $f^{-1}(x) = \sqrt{x}$.

2. If $f : \{a, b, c\} \to \{1, 2, 3\}$ is the function with the rule $f(a) = 2$, $f(b) = 3$, $f(c) = 1$, then $f^{-1} : \{a, b, c\} \to \{1, 2, 3\}$ is the function with the rule $f^{-1}(1) = c$, $f^{-1}(2) = a$, $f^{-1}(3) = b$. Draw the arrow diagrams for f and f^{-1}, and discuss how they are related.

SOLUTION

1. We can verify algebraically that $f(x) = y$ means $x^2 = y$, which since x and y are both nonnegative, implies that $x = \sqrt{y}$, which means $x = g(y)$. The reverse steps are similarly true, so $f(x) = y$ if and only if $x = g(y)$. That is, g is the inverse of f.

   Using the alternate characterization allows us to approach the issue algebraically. Specifically, for every $x \in \mathbb{R}^{\geq 0}$,

   $$g(f(x)) = g(x^2) = \sqrt{x^2} = x \quad \text{(since } x \geq 0\text{)}$$

   and for every $y \in \mathbb{R}^{\geq 0}$,

   $$f(g(y)) = f(\sqrt{y}) = (\sqrt{y})^2 = y$$

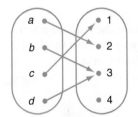

The function g

Figure 4-43   A noninvertible function.

2. In this case, to obtain an arrow diagram for f^{-1}, we simply reverse each of the arrows in the diagram of f, as shown in Figure 4-42.

   $\square$

In addition to making it easy to visualize the inverse of a function, the arrow diagram makes it easy to see why some functions are not invertible. We first saw this at the end of Section 4.1 when we explained the two reasons why the function g in Example 19 (and repeated in Figure 4-43) is not invertible.

The function in this example has two problems that keep it from being invertible. Mathematicians have defined terms that help them describe these situations.

● The first problem with g is that nothing in the domain maps to 4. This causes the reversed arrow diagram (shown in Figure 4-44) to have no arrow coming from 4, which is against the rules for a function. We describe the original function as not being *onto* when this happens.

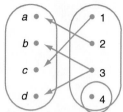

The function g Reversing the arrows

Figure 4-44 The first problem with g.

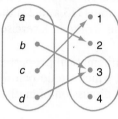 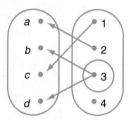

The function g Reversing the arrows

Figure 4-45 The second problem with g.

- The second problem with g is that two elements of the domain map to the same element 3 in the codomain. This causes the reversed arrow diagram (shown in Figure 4-45) to have two arrows coming from 3, which is against the rules for a function. We describe the original function as not being *one-to-one* when this happens.

Example 2 *Which of the following functions are invertible? For any that are not invertible, explain why it is not one-to-one or not onto.*

1. *$f : \mathbb{Z} \to \mathbb{Z}$ given by the rule $f(x) = 2x + 3$ for all $x \in \mathbb{Z}$.*
2. *$g : \mathbb{Z} \to \mathbb{N}$ defined by the rule*

$$g(z) = \begin{cases} -2z & \text{if } z \leq 0 \\ 2z - 1 & \text{if } z > 0 \end{cases}$$

3. *$h : \mathbb{N} \to \mathbb{N}$ defined so that for all $n \in \mathbb{N}$, $h(n)$ is the sum of the digits in the (base ten) numeral n.*

SOLUTION Only the second function is invertible.

1. This function is not *onto* because, for example, there is no value $n \in \mathbb{Z}$ such that $2n + 3 = 0$. Hence, 0 is not an output of the function f. You will prove this function is *one-to-one* in Exercise 11.

2. This function is invertible. The partial arrow diagram in Figure 4-46 suggests that the positive integers are mapping to the odd natural numbers and the remaining integers are mapping to the even natural numbers. This observation is the basis for the proof, which is left for the student in Exercise 13.

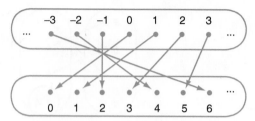

Figure 4-46 An invertible function.

3. This function is not *one-to-one* because, for example, $h(13) = h(22)$. That is, there are two inputs (13 and 22) that have the same output (4). You will prove this function is *onto* in Exercise 12. □

Like any good mathematical properties, the terms *onto* and *one-to-one* can be defined more formally, which we do below. The form of these definitions is important for the subsequent discussion of mathematical proofs about these properties of functions.

Definition

1. A function f is *onto* if everything in the codomain really is an output of f. That is, for every element y in the codomain, there must be (at least one) x in the domain where $f(x) = y$.
2. A function f is *one-to-one* if nothing in the codomain is an output via two different inputs. That is, for every choice of different elements x_1 and x_2 in the domain, $f(x_1)$ and $f(x_2)$ must be different.
3. A function f is a *one-to-one correspondence* if it is both *one-to-one* and *onto*. This is equivalent to saying that f is *invertible*.

These definitions can also be stated specifically in terms of arrow diagrams:

- A function is *onto* if every point in the codomain has an arrow ending at that point.
- A function is *one-to-one* if no point in the codomain has two or more arrows ending at that point.
- A function is a *one-to-one correspondence* (or invertible) if every point in the codomain has *exactly one* arrow ending at that point.

Practice Problem 1   *Which of the following functions are invertible? For each noninvertible function, explain why it is not one-to-one or not onto.*

(a) $c : \mathbb{Z} \to \mathbb{Z}$ *given by* $c(x) = x^3$ *for all* $x \in \mathbb{Z}$.

(b) $s : \mathbb{N} \to \mathbb{N}$ *defined so that* $s(x)$ *is the closest whole number to* $\sqrt{x}$ *for all* $x \in \mathbb{N}$.

(c) $h : \{0, 1, 2, 3, 4\} \to \{1, 2, 4, 6, 8\}$ *given so that* $h(n)$ *is the ones' digit of* 2^n *for all* $n \in \{0, 1, 2, 3, 4\}$.

(d) $g : \{0, 1, 2, 3, 4, 5, 6, 7, 8, 9\} \to \{0, 1, 2, 3, 4, 5, 6, 7, 8, 9\}$ *given so that* $g(n)$ *is the ones digit of* 2^n *for all* $n \in \{0, 1, 2, 3, 4, 5, 6, 7, 8, 9\}$.

Formal Proofs About Functions

Having formal definitions for *one-to-one* and *onto* allows us to discuss formal proofs of these properties. Although these properties are fairly concrete in the context of arrow diagrams, they can quickly seem very abstract when we begin to write proofs. This is very similar to the difference between visualizing a property of sets with Venn diagrams and writing a formal proof of that property. In this book, we will primarily use these proofs as further practice of our mathematical writing skills.

Example 3 *Prove that the function $f : \mathbb{N} \to \mathbb{N}$ with the rule $f(x) = 5x + 7$ is one-to-one.*

SOLUTION To show that $f : \mathbb{N} \to \mathbb{N}$ is *one-to-one*, we must show that "for every choice of different elements x_1 and x_2 in $\mathbb{Z}$, $f(x_1)$ and $f(x_2)$ must be different." If we state this as an "if, then" statement, it reads, "For all $x_1, x_2 \in \mathbb{Z}$, if $x_1 \neq x_2$, then $f(x_1) \neq f(x_2)$." **When we want to prove that a given function is *one-to-one*, we will always prove the contrapositive of this statement.**
Claim Define the function $f : \mathbb{N} \to \mathbb{N}$ with the rule $f(x) = 5x + 7$. For all $x_1, x_2 \in \mathbb{N}$, if $f(x_1) = f(x_2)$, then $x_1 = x_2$.
Proof. Let $x_1, x_2 \in \mathbb{N}$ be given such that $f(x_1) = f(x_2)$. By the definition of f, this means that

$$5x_1 + 7 = 5x_2 + 7$$

By simple algebra (subtract 7 from both sides, then divide by 5 on both sides), we conclude that $x_1 = x_2$, completing the proof. □

Once we made the decision to prove this property using the *contrapositive* of its formal definition, it became so straightforward to prove a function is *one-to-one*, it almost doesn't seem like we are doing anything. To convince you that this in fact, is a proof, try the following problem.

Example 4 *Find the mistake in the following proof that the function $g : \mathbb{R} \to \mathbb{R}$ with $g(x) = x^2$ is one-to-one.*

PROOF *Let $x_1, x_2 \in \mathbb{R}$ be given such that $g(x_1) = g(x_2)$. By the definition of g, this means that*

$$x_1^2 = x_2^2$$

By simple algebra (take the square root of both sides), we conclude that $x_1 = x_2$, completing the proof.

SOLUTION We know the proof must be wrong because the given function is not *one-to-one*, as the partial arrow diagram in Figure 4-47 shows. The easy way to find the mistake in the proof is to play the role of the READER, and when you are asked to give elements x_1 and x_2 with $f(x_1) = f(x_2)$, you pick two domain values whose arrows point to the same codomain value.

● The READER chooses $x_1 = -1$ and $x_2 = 1$ as a response to the first line of the proof.

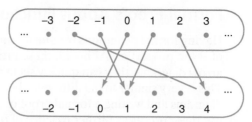

Figure 4-47 Diagram for Example 4.

- The AUTHOR asserts that $x_1^2 = x_2^2$, which we must agree with because of the way the x_1 and x_2 were chosen.

- The AUTHOR then claims that taking the square root of both sides of the equation will make it say, "$x_1 = x_2$," which is **not** true for our choice of x_1 and x_2. Since this is the first false statement that the proof claims is true, we have found the mistake.

The problem is that knowing $x_1^2 = x_2^2$ does not allow you to conclude that $x_1 = x_2$ since you do not know whether the chosen values are positive or negative. □

*Explore more on
the Web.*

Practice Problem 2 *Prove that the function $g := \mathbb{N} \to \mathbb{Q}$ with the rule $g(n) = \frac{5}{n+2}$ is one-to-one.*

The advantage of formal proofs is that they allow you to prove general properties rather than just properties of specific objects. We saw this in Section 2.2 when we proved statements like "If n is odd, then $3n$ is odd," only later to prove the more general rule "The product of two odd integers is odd." It is typical that the more general the property, the more useful it is, but the more abstract the proof. Let's see an example of moving from specific to general with the same proof structure.

In the following proposition, the functions are specifically $f : \mathbb{N} \to \mathbb{N}$ with $f(x) = 5x + 7$ from Example 3 and $g : \mathbb{N} \to \mathbb{Q}$ with $g(n) = \frac{5}{n+2}$ from Practice Problem 2.

Proposition 1 *Since we know that f and g are one-to-one, then we know that the function $h : \mathbb{N} \to \mathbb{Q}$ with the rule $h(x) = g(f(x))$ is also one-to-one. (Recall that the function h defined this way is the composition of g with f, which we usually write as $h = g \circ f$ and say, "h is g of f.")*

PROOF We prove that h is *one-to-one* using the contrapositive of the formal definition of *one-to-one*. Let $x_1, x_2 \in \mathbb{N}$ be given such that $h(x_1) = h(x_2)$. By the definition of h, this means that

$$g(f(x_1)) = g(f(x_2)) \qquad \text{or}$$

$$\frac{5}{f(x_1) + 2} = \frac{5}{f(x_2) + 2}$$

However, when we showed that g is *one-to-one* in Practice Problem 2, we showed that the only way for $\frac{5}{f(x_1)+2} = \frac{5}{f(x_2)+2}$ to be true is if $f(x_1) = f(x_2)$ is true. That is, $5x_1 + 7 = 5x_2 + 7$. Similarly, when we showed that f is *one-to-one* in Example 3, we showed that the only way for $5x_1 + 7 = 5x_2 + 7$ to be true is if $x_1 = x_2$ is true. Since we can conclude that $x_1 = x_2$, the proof is complete. ■

The form of this proof can be generalized to give us a much more useful, universal rule about the composition of functions.

Proposition 2   *If* $f : A \to B$ *is one-to-one and* $g : B \to C$ *is one-to-one, then the new function* $(g \circ f) : A \to C$ *is also one-to-one.*

PROOF   To simplify things we will use the function name h for $(g \circ f)$. That is, $h : A \to C$ is defined by the rule $h(x) = g(f(x))$. We prove that h is *one-to-one* using the contrapositive of the formal definition of *one-to-one*.

Let $x_1, x_2 \in \mathbb{A}$ be given such that $h(x_1) = h(x_2)$. By the definition of h, this means that

$$g(f(x_1)) = g(f(x_2))$$

However, since g is *one-to-one*, the only way for $g(f(x_1)) = g(f(x_2))$ to be true is if $f(x_1) = f(x_2)$ is true. Similarly, since f is *one-to-one*, the only way for $f(x_1) = f(x_2)$ to be true is if $x_1 = x_2$ is true. Since we can conclude that $x_1 = x_2$, the proof is complete. ■

Regardless of whether a function is given in terms of an algebraic formula, an arrow diagram or an abstract description, we always use the same form of argument when we want to prove that the function is *one-to-one*. Remember that this form is based on the **contrapositive** of the definition of *one-to-one*.

> **Claim.** $f : A \to B$ is *one-to-one*.
> *Proof.*   Let $a_1 \in A$ and $a_2 \in A$ be given such that $f(a_1) = f(a_2)$.
>
> $$\vdots$$
>
> **Use known information about** f **to conclude that** $a_1 = a_2$**.**

For the property of being *onto*, it is much easier to connect the formal definition with its informal meaning. Testing whether a function is *onto* can even be viewed as a game played by our old friends, AUTHOR and READER.

Example 5   *Given the function* $f : \mathbb{N} \to \mathbb{Q}^+$ *with the rule* $f(n) = \frac{n^2+1}{n+1}$, *play several turns of the following game:*

- READER *chooses an element* r *from the codomain* $(\mathbb{Q}^+)$.
- AUTHOR *responds with an element* n *from the domain* $(\mathbb{N})$.
- *If* $f(n) = r$, *then* r *really is an output from* f, *so* AUTHOR *gets a point.*
- *Otherwise,* READER *gets a point.*

SOLUTION   We show the moves in Table 4-11. It appears that READER has found a move (choosing 4/3) for which AUTHOR has no response. Hence, it looks like READER will be able to score as many points as she likes, and so she will always win this game. □

READER Chooses (r)	AUTHOR Responds (n)	Is $f(n) = r$?
1	0	Yes (point for A)
5/2	2	No (point for R)
5/2	3	Yes (point for A)
4/3	5	No (point for R)
4/3	6	No (point for R)
4/3	7	No (point for R)

Table 4-11 Solution for Example 5

In terms of an arrow diagram, the value 4/3 chosen by READER seems to be an element of the codomain to which no arrow points. This would imply that this function is *not onto*, but how do we rigorously prove this?

Since the rule for function f is given by an algebraic equation, it makes sense that the explanation should involve algebra. Consider AUTHOR's task as an algebra problem: "Find a natural number n for which $\frac{n^2+1}{n+1} = 4/3$." The tools for solving this equation come from a high school algebra course:

The given equation $\frac{n^2+1}{n+1} = \frac{4}{3}$ can be multiplied by $3(n+1)$ on both sides to yield the equation $3(n^2+1) = 4(n+1)$, which can be simplified to $3n^2 - 4n - 1 = 0$. Applying the quadratic formula to this equation gives us the solutions $n = \frac{4 \pm \sqrt{28}}{12}$.

Since these values for n are not in $\mathbb{N}$, we conclude that it is not possible to solve this equation with a natural number for n. So in this case, even to show the function is *not onto* requires a formal justification.

On the other hand, the game itself suggests the following proof structure for arguing that a function *is* onto. We present this proof structure first as an informal letter from AUTHOR to READER.

Proposition 3 *The function* $f : \mathbb{R}^+ \to (1, \infty)$ *with the rule* $f(x) = \frac{x+1}{x}$ *is onto.* *(Recall** *that the interval notation* $(1, \infty)$ *denotes the set* $\{z \in \mathbb{R} : 1 < z\}$.)

PROOF
Dear READER,

Suppose you choose an element r from the codomain $(1, \infty)$. I will show you an element z of the domain $\mathbb{R}^+$ such that $f(z) = r$.

1. Let $z = \frac{1}{r-1}$.
2. You can check

$$f\left(\frac{1}{r-1}\right) = \frac{\frac{1}{r-1}+1}{\frac{1}{r-1}} = \frac{\frac{r}{r-1}}{\frac{1}{r-1}} = r$$

---

* This notation was introduced in Exercise 21 of Section 3.1.

3. Since $r \in (1, \infty)$, we know that $r > 1$, from which it follows that $r - 1 > 0$, and so $\frac{1}{r-1} \in \mathbb{R}^+$. Hence, the z we have chosen is in the domain regardless of which codomain value for r you use.

Hence, my $z \in \mathbb{R}^+$ will make the equation $f(z) = r$ true, so your choice of r really is an ouput of f.

Sincerely yours,

AUTHOR

Before you try your hand at this sort of proof, let's look at one more sample proof in which the algebra is less messy but the language is much more formal.

Proposition 4   *The function $g : \mathbb{Q} \to \mathbb{Q}$ with the rule $g(x) = 5x - 1$ is onto.*

PROOF   Let $y \in \mathbb{Q}$ (the codomain of g) be given. Take x to be $\frac{y+1}{5}$. Since $y \in \mathbb{Q}$, it follows that $x \in \mathbb{Q}$, the domain of g. Now we can see that

$$g(x) = g\left(\frac{y+1}{5}\right)$$
$$= 5 \cdot \left(\frac{y+1}{5}\right) - 1$$
$$= (y+1) - 1 = y$$

Hence, any chosen y in the codomain is really an output of g, so g is onto.

Notice that much of the "algebra fiddling" is left out of this proof since it is just as logically sound (but much shorter) to demonstrate that a solution works as it is to algebraically derive the solution.

Explore more on the Web.

Practice Problem 3   *Prove that the function $f : \mathbb{Q} \to \mathbb{Q}$ with the rule $f(x) = 3x + 7$ is onto. Prove that the function $g : \mathbb{Q} \to \mathbb{Q}$ with the rule $g(x) = x^2 + 5x - 10$ is not onto.*

Just as with the property of *one-to-one*, the real test for this proof technique is how it holds up when working with abstract functions. The proof of the following proposition is addressed in Exercises 8 and 9 at the end of the section.

Proposition 5   *If $f : A \to B$ and $g : B \to C$ are both onto, then the new function $(g \circ f) : A \to C$ is also onto.*

Just as we are able to summarize our approach for proving functions are *one-to-one*, we can also describe the form of argument when we want to prove that a function is *onto*.

Claim. $f : A \rightarrow B$ is *onto*.
Proof.   Let $b \in B$ be given.

$$\vdots$$

Use known information about f to produce a domain element $a \in A$ for which $f(a) = b$.

We end our discussion of formal proofs with a reminder that a function that is both *one-to-one* and *onto* is *invertible*. Therefore, to formally prove that an abstract function is *invertible*, we can prove that the function is *one-to-one* and *onto* in two separate arguments. We currently have all the components on hand to establish perhaps the most important abstract property of invertible functions.

Theorem 6   *If $f : A \rightarrow B$ is invertible and $g : B \rightarrow C$ is invertible, then the new function $(g \circ f) : A \rightarrow C$ is invertible.*

PROOF   Let invertible functions f and g as described be given. Since the functions are invertible, then it follows they are each *one-to-one*, and this implies, according to Proposition 2, that the composition $(g \circ f)$ is *one-to-one*.

On the other hand, since the functions are invertible, then it follows they are each *onto*, and this implies, according to Proposition 5, that the composition $(g \circ f)$ is *onto*.

Since $(g \circ f)$ is *one-to-one* and *onto*, then this means that $(g \circ f)$ is invertible. ■

Set Cardinality and the Pigeonhole Principle

For finite sets, there is an interesting relationship among the concepts of *one-to-one*, *onto*, *one-to-one correspondence*,* and the sizes of the domain and codomain. We will state the properties as a theorem, and give a very informal explanation in terms of arrow diagrams.

Theorem 7   *Let $f : A \rightarrow B$ be a function, where A and B are finite sets of sizes m and n, respectively.*

1. If f is one-to-one, then $m \leq n$.

2. If f is onto, then $m \geq n$.

3. If f is a one-to-one correspondence, then $m = n$.

PROOF   Envision an arrow diagram for the function f. Since A contains m elements, there are exactly m arrows, one starting at each element of A.

1. If f is one-to-one, then the m arrows all lead to different elements of B. So B has at least these m elements. Hence, $n \geq m$.

---

*A *one-to-one correspondence* is the term for invertible function that we will use in Chapter 5.

2. With only *m* arrows, they can end at no more than *m* different points. In an onto function, every element of *B* is an ending point of an arrow, so there are no more than *m* elements in *B*. Hence, $n \leq m$.

3. This follows from parts (1) and (2), but we can also get a nice visualization. For a one-to-one correspondence, every element of the codomain is pointed to by one and only one of the *m* arrows from the domain elements. Hence, the number *n* of elements in the codomain is exactly *m*.

■

We will make heavy use of the third part of this theorem in Chapter 5. The result contends that if we can find a *one-to-one correspondence* between two sets, then we will know the two sets are the same size. We will use this fact to show that different counting problems have the same answer without being able to answer either problem! However, even the first two parts of the Theorem are useful for conclusions about the relative sizes of sets.

Example 6 *Let A be the set of all three-element subsets of* $\{1, 2, 3, 4, 5\}$*, let B be the set of all two-element subsets of* $\{1, 2, 3, 4\}$*, and let* $f : A \to B$ *be the function so that for all* $X \in A$*,* $f(X)$ *is the set obtained by removing the largest number from the set X. For example,* $f(\{1, 3, 4\}) = \{1, 3\}$*. Prove that f is onto and cite the part of Theorem 7 that allows us to conclude that* $n(A) \geq n(B)$*.*

SOLUTION   *Proof that f is onto.* Let $Y \in B$ be given. That is, *Y* is a two-element subset of $\{1, 2, 3, 4\}$. Let $X = Y \cup \{5\}$. Since $5 \notin Y$ (after all, *Y* is a subset of $\{1, 2, 3, 4\}$), we know that *X* is a three-element subset of $\{1, 2, 3, 4, 5\}$ whose largest element is 5. Hence, $f(X) = Y$, so *Y* is an output of the function *f*.

By the second part of Theorem 7, we can conclude that $n(A) \geq n(B)$.

□

Practice Problem 4   *Let A be the set of all three-element subsets of* $\{1, 2, 3, 4, 5\}$*, let B be the set of all two- or three-element subsets of* $\{1, 2, 3, 4\}$*, and let* $f : A \to B$ *have the rule* $f(S) = S - \{5\}$ *for each* $S \in A$*. Prove that f is onto, and then write the conclusion to the second part of Theorem 7 as an English sentence about subsets.*

The first part of Theorem 7 is closely related to the *basic pigeonhole principle* from Section 2.5. In fact, we can more formally prove the contrapositive of this part of Theorem 7 using that result directly.

Theorem 8 (The Pigeonhole Principle Revisited)   *Let* $f : A \to B$ *be a function, where A and B are finite sets of sizes m and n, respectively. If* $m > n$*, then f is not one-to-one.*

PROOF   Let $f : A \to B$ be a function, where *A* and *B* are finite sets of sizes *m* and *n*, where $m > n$. Define *n* boxes labeled by the elements of *B*. For each element $i \in B$, we put an object *a* from set *A* into box *i* if $f(a) = i$. According to this rule, we have *n* boxes into which we are placing $m > n$ objects. By the basic pigeonhole principle, there must be a box *z* with at least two objects. This means that there are two elements x and y in A for which $f(x) = z$ and $f(y) = z$. This, in turn, means that f is not *one-to-one*. ■

Example 7 *For each of the following statements, specify sets A and B and a function f so that Theorem 8 applies and the fact that f is not one-to-one is equivalent to the conclusion of the statement.*

1. *In a particular card game, a "hand" consists of five cards from a standard deck. Prove that in every hand there is a suit that appears more than once.*
2. *Prove that the last four digits of my Social Security number (SSN) contain two digits whose difference is divisible by 3.*

SOLUTION

1. Let A be the set of all cards in my hand, let B be the set {clubs, hearts, spades, diamonds}, and let $f : A \rightarrow B$ be the function where $f(x)$ is the suit of card x. Since $n(A) = 5$ and $n(B) = 4$, we know from Theorem 8 that f is not one-to-one, and hence there is some element (suit) of the codomain that is mapped to by two elements (cards) in the domain (my hand).

2. Let A be the set of digits in the last four digits of my SSN, let B be the set {0, 1, 2}, and let $f : A \rightarrow B$ be the function where $f(z)$ is the remainder obtained when z is divided by 3. (This definition of f guarantees that $z - f(z)$ is divisible by 3 no matter what z is, a fact proven in Section 2.2.)

   Now you do not necessarily know the size of the set A because you do not know how many *different digits* are in the last four digits of my SSN, so we must proceed in two cases:

   - **Case 1:** Suppose some digit appears twice in the last four digits of my SSN. Then these two digits differ by 0, which is a multiple of 3, and we are done.
   - **Case 2:** Suppose the four digits at the end of my SSN are all different. In this case, since $n(A) = 4$ and $n(B) = 3$, we know from Theorem 8 that f is not one-to-one, and hence there is some element r of the codomain that is mapped to by two elements w and x in the domain. That is, $f(x) - f(w) = 0$. Because $f(w) - w$ and $f(x) - x$ are each divisible by 3, we can choose integers K and L so that $w - f(w) = 3K$ and $x - f(x) = 3L$.

$$w - x = (w - x) + (f(x) - f(w))$$
$$= (w - f(w)) - (x - f(x))$$
$$= 3K - 3L = 3(K - L)$$

   Hence, $w - x$ is divisible by 3.

   □

Practice Problem 5 *Prove that for every set of five positive integers $\{a, b, c, d, e\}$, at least two numbers in the set $\{3^a, 3^b, 3^c, 3^d, 3^e\}$ have the same ones' digit.*

Infinite Sets

So far our discussion about the relative sizes of sets has been directed toward finite sets, but the same ideas can be applied to infinite sets. This raises issues about the foundations of mathematics that, while very interesting, are not really central to this course as a whole. We end this section by addressing some of these issues in order to illustrate some surprising ideas about infinity.

> **Definition**   Sets A and B *have the same cardinality* if there exists an invertible function $f : A \to B$.

Note that we use the word "cardinality" instead of "size" as we start down the path to formal thinking about infinity. We informally think of a *finite set* being one whose cardinality is a natural number, but this leads to a negative definition of infinity (i.e., a set is infinite if it is not finite), and we have seen on more than one occasion that negative definitions are difficult to work with in proofs. Now that we understand the concepts of one-to-one and onto functions, another definition presents itself that is actually much easier to work with.

> **Definition**   A set A is *infinite* if there exists a function $f : A \to A$ that is one-to-one but not onto.

 Example 8 *Show that $\mathbb{N}$ is infinite, using the above definition.*

SOLUTION   Let $f : \mathbb{N} \to \mathbb{N}$ be defined by the rule $f(x) = 2x$. It is straightforward to prove that f is one-to-one, so we leave that detail to the reader. Since the outputs of f are always even integers, there are many examples (like $1 \in \mathbb{N}$) of codomain elements that are not outputs of f. Hence, f is not onto. By our new definition of infinite, this proves that $\mathbb{N}$ is infinite.   □

Practice Problem 6   *Show directly that $\mathbb{Z}$ and $\mathbb{N}$ have the same cardinality.*

This practice problem illustrates one shortcoming of our formal definition of *same cardinality*. Constructing *invertible* functions between sets can be more challenging than finding more natural one-to-one functions. For this reason, the famous German mathematician Georg Cantor (1845–1918) proposed* the following result, published in 1895, that makes it much easier to work formally with infinite sets. This result is typically included in a first course in formal set theory, and even though the idea is fairly intuitive, the proof is well beyond the scope of this book.

Theorem 9 (Cantor–Bernstein Theorem)   *Given sets A and B, if there is a one-to-one function $f : A \to B$ and a one-to-one function $g : B \to A$, then sets A and B have the same cardinality.*

---

*Cantor did not actually prove this. Felix Bernstein (1878–1956) gave the first correct proof.

Example 9 *Use the Cantor–Bernstein theorem to show that $\mathbb{Q}^{\geq 0}$ has the same size as the set $\mathbb{N}$.*

SOLUTION   Since $\mathbb{N} \subseteq \mathbb{Q}^{\geq 0}$, we immediately know that the function $f : \mathbb{N} \to \mathbb{Q}^{\geq 0}$ defined by $f(x) = x$ is one-to-one. On the other hand, the function $g : \mathbb{Q}^{\geq 0} \to \mathbb{N}$ can be defined as follows: Given $r \in \mathbb{Q}^{\geq 0}$, there is a unique choice of relatively prime natural numbers a and b such that $r = \frac{a}{b}$, so let $g(r) = 2^a \times 3^b$. Because the prime factorization of numbers is unique, the function g is also one-to-one. $\square$

All this energy to show that two infinite sets have the same cardinality might seem wasted if you have never thought hard about infinity before. After all, aren't all infinite sets the same size? If you have never given it any thought, you are in good company. Mathematics history suggests that no one had carefully considered questions involving "infinity" until Georg Cantor in the twilight of the nineteenth century. We have already seen one of his important ideas, but his most famous theorem might leave you scratching your head. The basic idea in the proof of this theorem has led to some of the most famous important results about the foundations of mathematics, including Kurt Gödel's famous incompleteness theorems.

Theorem 10 (Cantor's Theorem)   *For every set A, A and $\mathcal{P}(A)$ do not have the same size. (That is, no set is the same size as its power set.)*

PROOF   Let a set A and a function $f : A \to \mathcal{P}(A)$ be given. We will show that f is not invertible by showing that it is not onto. That is, we will produce a member B of the codomain $\mathcal{P}(A)$ that is not equal to $f(x)$ for any $x \in A$. We can define the set B directly as follows:

$$B = \{x \in A : x \notin f(x)\}$$

Since $B \subseteq A$, this means that $B \in \mathcal{P}(A)$, so B is a member of the codomain of f. To see that $f(x) \neq B$ for all $x \in A$, we imagine what would happen if there *is* a $b \in A$ with $f(b) = B$. A contradiction then arises when considering the question "Is $b \in B$?" To see this, we just consider both possible answers:

- **Case 1**: If $b \in B$, then by the definition of B, this means that $b \notin f(b)$. But $f(b)$ *is* B, so this says that $b \notin B$, a contradiction.
- **Case 2**: If $b \notin B$, then this means $b \notin f(b)$, so according to the way the set B is defined, b is one of the elements in B, a contradiction.

Thus, we have the strange contradiction that $b \in B$ if and only if $b \notin B$. Because of this contradiction, we conclude that there is no value of x for which $f(x) = B$, so f is not invertible.   ■

Example 10 *Trace through the proof of Cantor's theorem with a simple example.*

SOLUTION   Suppose $A = \{1, 2, 3, 4\}$ and $f : A \to \mathcal{P}(A)$ is given below:

$$f(1) = \{1, 4\}$$
$$f(2) = \{1, 2, 3\}$$
$$f(3) = \{\}$$
$$f(4) = \{3\}$$

Note that $1 \in f(1), 2 \in f(2)$, but $3 \notin f(3)$ and $4 \notin f(4)$, so the set B defined in the proof is $B = \{3, 4\}$. Clearly in our example, there is no value of x for which $f(x) = B$. $\square$

We saw in Exercise 24 of Section 3.1 that a finite set with n elements has 2^n subsets, so the fact that $n \neq 2^n$ is hardly a news flash. The significant thing about Cantor's proof is that it applies to *all* sets, even infinite ones.

Corollary 11 $\mathbb{N}$ and $\mathcal{P}(\mathbb{N})$ do not have the same size.

The conclusion from this is the part that might make you scratch your head: Cantor showed that there are different sizes of infinity! With this in mind, the set $\mathbb{N}$ takes a special place representing the *smallest* possible infinity.

> *Definition*   An infinite set A is *countable* if it has the same cardinality as $\mathbb{N}$. An infinite set is *uncountable* if it is not countable.

With this terminology, our work in this section can be rephrased as follows:

- $\mathbb{N}$ is countable (from the definition of countable).
- $\mathbb{Z}$ is countable (from Practice Problem 6).
- $\mathbb{Q}^{\geq 0}$ is countable (from Example 9).
- $\mathcal{P}(\mathbb{N})$ is uncountable (from Corollary 11).

Solutions to Practice Problems

1   Only the third function is invertible. Since the rule for the fourth function is the same as the third, obviously the question of invertibility has to do with domain and codomain as well as with the rule.

(a) There is no $a \in \mathbb{Z}$ for which $a^3 = 2$, so 2 is an element of the codomain $\mathbb{Z}$ to which no domain element is mapped. (This function is not onto.)

(b) $s(3) = s(5) = 2$, so more than one domain element maps to the same codomain element. (This function is not one-to-one.)

(c) This is invertible.

(d) g is not invertible since it has both kinds of problems; it is not onto because there is no arrow pointing to 0 or 7 (or any odd number for that matter), and it is not one-to-one because there are multiple arrows pointing to numbers like 2 and 4.

2   (We prove the contrapositive of the definition of one-to-one.) Let $n_1 \in \mathbb{N}$ and $n_2 \in \mathbb{N}$ with $f(n_1) = f(n_2)$ be given. This means that

$$\frac{5}{n_1 + 2} = \frac{5}{n_2 + 2}$$

Multiplying by $(n_1 + 2)(n_2 + 2)$ on both sides of this equation gives us $5 \cdot (n_2 + 2) = 5 \cdot (n_1 + 2)$, which can be algebraically simplified (divide both sides by 5 and then subtract 2 from both sides) to $n_2 = n_1$.

3 We first argue that the function $g : \mathbb{Q} \to \mathbb{Q}$ with the rule $g(x) = x^2 + 5x - 10$ is *not* onto. We will show that there is no input $r \in \mathbb{Q}$ such that $g(r) = -20$, that is, with $r^2 + 5r + 10 = 0$. The quadratic formula tells us that in order to obtain $r^2 + 5r + 10 = 0$, we must have

$$r = \frac{-5 \pm \sqrt{5^2 - 4 \cdot 10}}{2}$$

and since $5^2 - 4 \cdot 10 = -15 < 0$, this is impossible for a rational number r. Hence, no rational number input can make $g(x) = -10$, which means that $-10 \in \mathbb{Q}$ is not an output of g.

Claim The function $f : \mathbb{Q} \to \mathbb{Q}$ with the rule $f(x) = 3x + 7$ is onto.

Proof. Let $r \in \mathbb{Q}$ be given. Create the number $q = \frac{r-7}{3}$. It is easy to see that q is in $\mathbb{Q}$ (since we used only subtraction and division on the rational number r), and we can check:

$$f(q) = 3 \cdot \left(\frac{r-7}{3} \right) + 7$$
$$= (r - 7) + 7$$
$$= r$$

So we have produced an input q that will map to the given rational number r. Hence, every rational number is an output of f. ∎

4 *Proof that f is onto.* Let $Y \in B$ be given. That is, Y is a two-element subset of $\{1, 2, 3, 4\}$ or a three-element subset of $\{1, 2, 3, 4\}$. Since we do not know which, we proceed to argue in two cases:

● Suppose Y is a two-element subset of $\{1, 2, 3, 4\}$. Let $X = Y \cup \{5\}$, and X will be a three-element subset of $\{1, 2, 3, 4, 5\}$ for which $f(X) = X - \{5\} = Y$. Hence, $f(X) = Y$ in this case.

● Suppose Y is a three-element subset of $\{1, 2, 3, 4\}$. Let $Z = Y$, and Z will be a three-element subset of $\{1, 2, 3, 4, 5\}$ for which $f(Z) = Z - \{5\} = Z$. Hence, $f(Z) = Y$ in this case.

In either case, Y is an output of the function f, so we conclude that f is onto. ∎

By the second part of Theorem 7, we can conclude that $n(A) \geq n(B)$. That is, there are at least as many three-element subsets of $\{1, 2, 3, 4, 5\}$ as there are two-element subsets and three-element subsets of $\{1, 2, 3, 4\}$ put together.

5 Let A be the set $\{a, b, c, d, e\}$ alluded to in the problem. With a little investigation, we can determine that the only ones digits that can occur in a number of the form 3^x are digits from the set $\{3, 9, 7, 1\}$, so let this be the set B, and let $f : A \to B$ be the function where $f(x)$ is the ones digit of 3^x. Since $n(A) = 5$ and $n(B) = 4$, we know from Theorem 8 that f is not one-to-one, and hence there is some element (ones digit) of the codomain that is mapped to by two elements (numbers) in the domain (my set of positive integers).

6 We will use a rule that associates positive inputs with even natural numbers and associates negative inputs with odd natural numbers:

$$f(x) = \begin{cases} 2x & \text{if } x \geq 0 \\ -1 - 2x & \text{if } x < 0 \end{cases}$$

Exercises for Section 4.3

1. Each of the following functions is not onto. To demonstrate this, provide an example of an element in the codomain and explain why no element of the domain is associated with it.

   (a) $f : \mathbb{R} \to \mathbb{R}$ with the rule $f(x) = x^2 + 4x + 1$.

   (b) $g : \mathbb{Q}^+ \to \mathbb{Q}^+$ with the rule $g(x) = \frac{x^2 + 9}{x}$.

   (c) $h : [1, \infty) \to [1, \infty)$ with the rule $h(x) = \frac{1}{x+1}$.

2. Each of the following functions is not one-to-one. To demonstrate this, provide an example of two elements of the domain that are associated with the same element of the codomain.

   (a) $f : \mathbb{R} \to \mathbb{R}$ with the rule $f(x) = x^2 + 4x + 1$.

   (b) $g : \mathbb{Q}^+ \to \mathbb{Q}^+$ with the rule $g(x) = \frac{x^2 + 9}{x}$.

   (c) $h : \mathbb{Q} \to \mathbb{Q}^{\geq 0}$ with the rule $h(x) = |3x + 1|$.

3. For each of the following functions, decide if it is one-to-one, onto, invertible, or none of these:

   (a) $f : \mathbb{Q} \to \mathbb{Q}$ with the rule $f(x) = 3x + 1$.

   (b) $g : \mathbb{Q}^+ \to \mathbb{Q}^+$ with the rule $g(x) = 7x + 5$.

   (c) $f : \mathbb{R}^+ \to \mathbb{R}^+$ with the rule $f(x) = \frac{1}{x+1}$.

   (d) $g : (1, \infty) \to (1, \infty)$ with the rule $f(x) = \frac{x}{x-1}$.

4. For each of the following functions, decide if it is one-to-one, onto, invertible, or none of these:

   (a) $f : \mathbb{N} \times \mathbb{N} \to \mathbb{N}$ with the rule $f(a, b) = a + b$.

   (b) $f : \mathbb{N} \times \mathbb{N} \to \mathbb{N}$ with the rule $f(a, b) = 2^a \cdot 3^b$.

   (c) $f : \mathbb{N} \times \mathbb{N} \to \mathbb{N}$ with the rule $f(a, b) = a^2 + b^2$.

5. Let $S = \{a, b, c\}$, and define a function $c : \mathcal{P}(S) \to \mathcal{P}(S)$ by $c(A) = S - A$.

   (a) Is the function c one-to-one?

   (b) Is the function c onto?

   (c) If the function c is invertible, then describe the inverse of c. If c is not invertible, explain why not.

6. Let $S = \{a, b, c\}$ and let c be the function of Exercise 5. Let $n : \mathcal{P}(S) \to \{0, 1, 2, 3\}$ be defined so that $n(X)$ is the

i	0	1	2	3
$s(i)$	$\varnothing$	$\{a\}$	$\{a, b\}$	$\{a, b, c\}$

Table 4-12   Table for Exercise 6

number of elements in X, and let $s : \{0, 1, 2, 3\} \to \mathcal{P}(S)$ be defined by Table 4-12.

(a) Which composition is defined, $c \circ s$ or $s \circ c$? For the one that is defined, describe the domain and the codomain, and give an arrow diagram for the function.

(b) Give an arrow diagram for $s \circ n$. Is $s \circ n$ one-to-one? Is $s \circ n$ onto?

(c) Give an arrow diagram for $n \circ s$. Is $n \circ s$ one-to-one? Is $n \circ s$ onto?

7. Fill in the blanks in the following proof:
   Proposition   If $f : \mathbb{R} \to \mathbb{R}$ is one-to-one, then the function $h : \mathbb{R} \to \mathbb{R}$ with the rule $h(x) = f(2 \cdot x)$ is also one-to-one.
   Proof   We prove that h is one-to-one using the contrapositive of the formal definition of one-to-one. Let $x_1, x_2 \in \mathbb{R}$ be given such that _____. By the definition of h, this means that
   $$f(2 \cdot x_1) = f(2 \cdot x_2)$$
   However, since _____, the only way for $f(2 \cdot x_1) = f(2 \cdot x_2)$ to be true is if _____ is true. Dividing by 2 on both sides of this equation, we conclude that _____, completing the proof.

8. Fill in the blanks in the following proof:
   Proposition   If $f : \mathbb{R} \to \mathbb{R}$ is onto, then the function $h : \mathbb{R} \to \mathbb{R}$ with the rule $h(x) = f(2 \cdot x)$ is also onto.
   Proof.   Let $y \in \mathbb{R}$ be given. Since _____, there must be an $x \in \mathbb{R}$ with $f(x) = y$. If we take $z = $ _____ $\in \mathbb{R}$, it follows that
   $$h(z) = f(2 \cdot z) = f(\underline{\quad\quad})$$
   From this it follows that $h(z) = y$, completing the proof. ∎

9. Generalize the previous exercise to prove that if $f : A \to B$ and $g : B \to C$ are both onto, then the composition $(g \circ f) : A \to C$ is onto.

10. For each of the following statements, specify sets A and B and a function f so that Theorem 8 applies and the fact that f is not one-to-one is equivalent to the conclusion of the statement.

   (a) In a particular card game, an entire standard 52-card deck is divided as evenly as possible among all players at the start of the game. Prove that if there are three players playing, everyone must have at least two cards of the same value at the start of the game.

   (b) The Boatsville Youth Basketball League consists of 10 teams, each playing once a week for a total of 10 weeks. Conclude that every team must play some team twice.

   (c) Prove that for every set of five positive integers $\{a, b, c, d, e\}$, at least two numbers in the set $\{7^a, 7^b, 7^c, 7^d, 7^e\}$ have the same ones' digit.

   (d) Prove that in any set of five integers, there must be at least two whose difference is divisible by 4.

   (e) Prove that in anyone's seven-digit telephone number, there must be two digits whose difference is divisible by 6.

   (f) Each person living in New York City has between 0 and 500,000 hairs on his or her head. According to the 2000 census, there are about 8,010,000 people living in the five boroughs of New York City. Conclude that there are at least two people of the same sex living in the same borough of New York City with exactly the same number of hairs on their heads.

11. Prove that the function $f : \mathbb{Z} \to \mathbb{Z}$ with $f(x) = 2x + 3$ from Example 2 is one-to-one.

12. Prove that the following function from Example 2 is onto: $h : \mathbb{N} \to \mathbb{N}$, where $h(n)$ is the sum of the digits in the (base ten) numeral n.

13. Prove that the function $g : \mathbb{Z} \to \mathbb{N}$ with the rule
   $$g(z) = \begin{cases} -2z & \text{if } z \le 0 \\ 2z - 1 & \text{if } z > 0 \end{cases}$$
   from Example 2 is one-to-one and onto.

14. Let $f : [0, \infty) \to [4, \infty)$ be the function with the rule $f(x) = x^2 + 4$.

   (a) Prove that f is one-to-one.

   (b) Prove that f is onto, and conclude that f is invertible.

   (c) Demonstrate that f is invertible directly by finding an algebraic formula to describe the rule for f^{-1}.

15. Let $g : \mathbb{R} \to \mathbb{R}$ be the function with the rule $g(x) = 5x - 7$.

   (a) Prove that g is one-to-one.

   (b) Prove that g is onto, and conclude that g is invertible.

   (c) Demonstrate that g is invertible directly by finding an algebraic formula to describe the rule for g^{-1}.

16. Let $f : \mathbb{R} \to \mathbb{R}$ be the function with the rule $f(x) = x^3 - 2$.

   (a) Prove that f is one-to-one.

   (b) Prove that f is onto, and conclude that f is invertible.

(c) Demonstrate that f is invertible directly by finding an algebraic formula to describe the rule for f^{-1}.

17. Let $h : [2, \infty) \to (0, 1]$ be the function with the rule $h(x) = \frac{1}{x-1}$.

(a) Prove that h is one-to-one.

(b) Prove that h is onto, and conclude that h is invertible.

(c) Demonstrate that h is invertible directly by finding an algebraic formula to describe the rule for h^{-1}.

18. Let $f : (1, \infty) \to (1, \infty)$ be the function with the rule $f(x) = \frac{x}{x-1}$.

(a) Prove that f is one-to-one.

(b) Prove that f is onto, and conclude that f is invertible.

(c) Demonstrate that f is invertible directly by finding an algebraic formula to describe the rule for f^{-1}.

19. Let $A = \{0, 1, 2\} \times \{0, 1, 2, 3\}$ and $B = \{n : n \text{ is a positive factor of } 500\}$, and let $f : A \to B$ be the function with the rule $f(a, b) = 2^a \times 5^b$.

(a) Prove that f is one-to-one.

(b) Prove that f is onto.

(c) What theorem in this chapter guarantees that A and B have the same cardinality?

(d) Which set (A or B) is it easier to determine the cardinality of? What is the cardinality?

20. Let $A = \{0, 1, 2, 3, 4\} \times \{0, 1, 2\}$ and $B = \{n : n \text{ is a positive factor of } 144\}$, and let $f : A \to B$ be the function with the rule $f(a, b) = 2^a \times 3^b$.

(a) Prove that f is one-to-one.

(b) Prove that f is onto.

(c) What theorem in this chapter guarantees that A and B have the same cardinality?

(d) Which set (A or B) is it easier to determine the cardinality of? What is the cardinality?

For Exercises 21 to 23, we will call a set Σ of characters an *alphabet* and a string using only characters from this set a *word over the alphabet* Σ. For example, if the alphabet is the set $\{a, 1, z\}$, then some possible words are $1a1a$, $1za$, and $za1$.

21. Let A be the set of words over the alphabet $\{a, b, c\}$ with length 5 or less, let B be the set of words over the alphabet $\{a, b\}$ with length 4 or less, and let $f : A \to B$ be the function so that for every word W, $f(W)$ is the word W with every occurrence of the character c deleted. Prove f is one-to-one or demonstrate it is not. Prove f is onto or demonstrate that it is not. Which set is bigger, A or B?

22. With this notation, *a binary string* (from Section 4.1) is simply a word over the alphabet $\{0, 1\}$. Let A be the set of all binary strings of length 6 consisting of exactly two 0's and four 1's, let B be the set of all binary strings of length 5 consisting of two 0's and three 1's, and let $f : A \to B$ be the function so that for all $W \in A$, $f(W)$ is the result of changing the leftmost 1 in W into a 0 and then removing the leftmost character altogether. For example, $f(101101) = 01101$ and $f(011110) = 01110$. Prove f is one-to-one or demonstrate it is not. Prove f is onto or demonstrate that it is not. Which set is bigger, A or B?

23. We will call a word over a given alphabet a *permutation* if the word does not use the same character twice. For example, abc and $bcad$ are permutations over the alphabet $\{a, b, c, d\}$. Let A be the set of all permutations of length 4 over the alphabet $\Sigma = \{a, b, c, d, x, y, z\}$, let B be the set of all subsets of Σ of size 4, and let $f : A \to B$ be the function so that for all $p \in A$, $f(p)$ is the set of characters that occur in p. For example, $f(bcxa) = \{a, b, c, x\}$. Prove f is one-to-one or demonstrate it is not. Prove f is onto or demonstrate that it is not. Which set is bigger, A or B?

24. Consider the function $f : \mathcal{P}(\{1, 2, 3\}) \to \mathcal{P}(\{1, 2\})$ with the rule $f(A) = A - \{3\}$.

(a) Prove that f is onto.

(b) Demonstrate that f is not one-to-one.

(c) Demonstrate that every B in the codomain of f is associated with exactly two elements in the domain of f.

(d) Explain in your own words how the above facts imply that the domain of f is twice the size of the codomain of f.

25. Generalize the argument in the previous exercise to show that $\mathcal{P}(\{1, 2, 3, \ldots, n-1, n\})$ is twice the cardinality of $\mathcal{P}(\{1, 2, 3, \ldots, n-1\})$.

26. Use Exercise 25 as part of a proof by induction that the cardinality of $\mathcal{P}(\{1, 2, 3, \ldots, n\})$ is 2^n for all $n \geq 1$.

27. Prove Part (2) of Theorem 7.

28. Find a one-to-one function from $\mathbb{R}$ to $\mathbb{R}^{\geq 0}$. What result in this chapter allows you to conclude that these two sets have the same size?

29. Suppose $A = \{a, b, c, d\}$ and $f : A \to \mathcal{P}(A)$ is given below:

$$f(a) = \{b\}$$
$$f(b) = \{b, c, d\}$$
$$f(c) = \{a, d\}$$
$$f(d) = \{b, d\}$$
$$f(e) = \{a, b, c, d\}$$

What is the set B constructed by the proof of Cantor's theorem for which $f(x) = B$ is not true for any $x \in A$?

30. Suppose $A = \{a, b, c, d, e\}$ and $f : A \to \mathcal{P}(A)$ is given below:

$$f(a) = \{b, d, e\}$$
$$f(b) = \{a, c, d\}$$
$$f(c) = \{a, d, e\}$$
$$f(d) = \{b, e\}$$
$$f(e) = \{\}$$

What is the set B constructed by the proof of Cantor's theorem for which $f(x) = B$ is not true for any $x \in A$?

31. Suppose $A = \{a, b, c, d, e\}$ and $f : A \to \mathcal{P}(A)$ is given below:

$$f(a) = \{a, b, e\}$$
$$f(b) = \{b, c, d\}$$
$$f(c) = \{c, d, e\}$$
$$f(d) = \{b, d\}$$
$$f(e) = \{e\}$$

What is the set B constructed by the proof of Cantor's theorem for which $f(x) = B$ is not true for any $x \in A$?

4.4 Properties of Relations

One moral of the previous section is the fact that when studying abstract mathematical objects like functions, understanding special properties of the objects (like one-to-one or onto) allows us to apply them to other mathematical problems, just as we did to set cardinality problems at the end of that section. This is a major reason that we study mathematical structures abstractly, as opposed to focusing only on specific functions on familiar domains, for example. We will continue this theme with the more general mathematical structure of binary relations in this section.

The properties of binary relations that will be studied are geared toward two classes of examples of binary relations. We will broadly refer to these as "order relations" and "equivalence relations." The former class should consist of relations that compare objects through phrases like "a is no bigger than b," where the word "bigger" can be interpreted in many ways. Similarly, the latter class consists of relations that equate objects through phrases like "a is the same as b," where the word "same" can have many interpretations.

Of course, not every relation belongs to one of these two classes, but the relevance of these classes to other mathematical topics makes them worthy of our attention. In this section, we will address order relations and discuss proofs of the properties relevant to them. We will discuss equivalence relations in the subsequent section.

Order Relations

We already know several examples of order relations from this course and before. We will start with these to discover the common properties that are always present in notions of "order."

 Example 1 *Draw arrow diagrams for each of the following relations:*

1. *The relation R_1 on the set $\{1, 2, 3, 4\}$ with the rule "$(x, y) \in R_1$ if $x \leq y$."*
2. *The relation R_2 on the set $\mathcal{P}(\{1, 2, 3\})$ with the rule "$(x, y) \in R_2$ if $x \subseteq y$."*
3. *The relation R_3 on the set $\{1, 2, 3, 6\}$ with the rule "$(x, y) \in R_3$ if x divides y."*

SOLUTION The complete arrow diagrams are shown in Figure 4-48. □

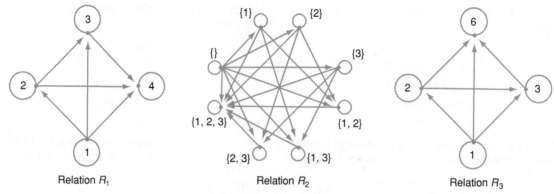

Figure 4-48 Solution to Example 1.

We can simplify the arrow diagram for order relations in several ways, and the complexity of the diagram for R_2 certainly gives us ample motivation to do so. Each of the three examples above satisfies three properties, and these properties allow us to simplify the arrow diagrams significantly.

Definition Let R be a binary relation on a set A.

1. R is said to be *reflexive* if $(a, a) \in R$ for all $a \in A$. In terms of the arrow diagram, this means that every node has a loop.
2. A relation R is called *antisymmetric* if for all $a, b \in A$, if $a \neq b$ and $(a, b) \in R$, then $(b, a) \notin R$. In terms of the arrow diagram, this means that arrows only go in one direction.
3. A relation R is called *transitive* if whenever $(a, b) \in R$ and $(b, c) \in R$, it must also be the case that $(a, c) \in R$. In terms of the arrow diagram, this means that whenever you can follow two arrows to get from node a to node c, you can also get there along a single arrow.

These properties lead to a simplification of the arrow diagram that is named for the German mathematician Helmut Hasse (1898–1979). Of course, when we use a Hasse diagram, we must convey separately that the relation being shown satisfies these three properties, so we will use the following definition to make this easy:

Definition A relation R on a set A is called a *partial order on A* if R is antisymmetric, transitive, and reflexive.

In this case, we will simplify the arrow diagram and refer to it as the *Hasse diagram* of the partial order. A Hasse diagram will simplify the arrow diagram in two ways.

1. We will suppress the loops since these are not necessary when we know in advance that the relation is reflexive.

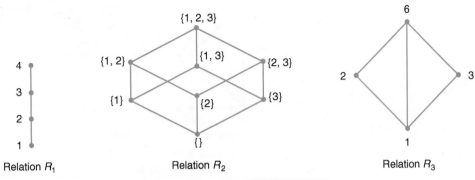

Relation R_1 Relation R_2 Relation R_3

Figure 4-49   Solution to Example 2.

2. We will use line segments instead of arrows where the direction of the relation is always from a node lower on the page to one that is higher on the page. Since we know in advance that the relation is antisymmetric, the arrows would only go in one direction anyway.

3. We only draw a line segment from the (lower) node a to the (higher) node b if $(a, b) \in R$ and there is no $c \in A$ that lies "between" a and b. Since we know in advance that the relation is transitive, any time we see a line from a to b and a line from b to c, we know that the *regular arrow diagram* would have an arrow from a to c—we always suppress this extra arrow in a Hasse diagram.

Example 2   *Draw the Hasse diagrams for the three partial orders in Example 1.*

SOLUTION   The Hasse diagrams shown in Figure 4-49 are significantly simpler than the arrow diagrams we saw in Example 1.   □

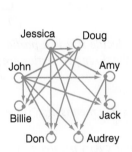

Relation D

Figure 4-50
Diagram for Practice
Problem 1.

Practice Problem 1   *Draw the Hasse diagram for each of the following partial orders:*

(a) *The relation D on the set $E = \{John, Jessica, Doug, Amy, Don, Billie, Audrey, Jack\}$ with the rule $(a, b) \in D$ if a is a descendant of b. Assuming we allow that people are descendants of themselves, the arrow diagram for this relation shown in Figure 4-50 is a partial order.*

(b) *Let R be the relation on the set $A = \{1, 2, 3, 6, 9, 18\}$ given by the rule $(a, b) \in R$ if a divides b.*

To prove properties of binary relations, we must first be sure that we adequately understand the definitions of the properties. Hence, before our discussion turns to proofs, let's consider several examples of relations that may or may not have the three special properties we have defined so far.

Example 3   *Let $A = \{1, 2, 3, 4, 5\}$. Draw an arrow diagram for each of the two relations on A described below, and then give a specific example that justifies each statement that follows.*

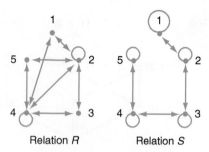

Relation *R*   Relation *S*

Figure 4-51 Diagram for Example 3.

- *Relation R is defined so that x R y means x · y is even.*
- *Relation S is defined so that a S b means |b − a| ≤ 1.*

1. *Relation R is not reflexive.*
2. *Relation S is not transitive.*
3. *The relations R and S are not antisymmetric.*

SOLUTION   We give the two diagrams in Figure 4-51, and then address the statements below. Notice how we use "double arrows" to keep the diagrams less cluttered.

1. Since $1 \in A$ and $(1, 1) \notin R$, R is not reflexive.
2. Since $(1, 2) \in S$, $(2, 3) \in S$, and $(1, 3) \notin S$, we see that S is not transitive.
3. Since both $(1, 2) \in R$ and $(2, 1) \in R$, we see that R is not antisymmetric. The same example shows that S is not antisymmetric either. □

Practice Problem 2   *For each of the following relations, decide if it is a partial ordering. If it is not, give specific examples to show which of the three properties are not satisfied.*

(a) *Let $A = \mathcal{P}(\{1, 2, 3\})$, and define the relation R_1 on A so that $s \, R_1 \, t$ means $n(s \cap t) \neq 0$.*

(b) *Let $B = \{1, 2, 3, 4, 5\}$, and define the relation R_2 on B so that $x \, R_2 \, y$ means $x \cdot y$ is odd.*

(c) *Let $C = \{Rap, Ram, Cram, Map, Arm, Ramp, Camp, Car, Trap, Race, Part\}$, and define the relation R_3 on A so that $x \, R_3 \, y$ means every letter in word x is also in word y.*

Proofs About Properties of Relations

As usual, the form of a proof about an abstract object is directly tied to the formal definitions involved. In the case of the reflexive and transitive properties for relations, the definitions are simple if, then statements, allowing us to incorporate the direct proof structure with which we have become familiar.

The reflexive and transitive properties

 Example 4 Let $A = \{1, 2, 3, 6, 9, 18\}$, and let R be the relation on A given by $R = \{(x, y) \in A \times A : y \text{ is a multiple of } x\}$. Prove that R is reflexive.

SOLUTION   To show that R is reflexive, we must prove to the READER that "for every $a \in A$, $(a, a) \in R$." We do this by inviting the READER to choose any element of A at all and call it a, and then explaining to her how we know (even though we do not know what value of a she picked) that $(a, a) \in R$.

Proof that R is reflexive. Let $a \in A$ be given. We know that $a = 1 \cdot a$, so certainly a is a multiple of a. By the definition of R, this means that $(a, a) \in R$. Hence, R is reflexive. □

In general, a proof that a relation R on a set A is reflexive always has the following form:

Claim. R is reflexive.
Proof. Let $a \in A$ be given.

$$\vdots$$

Use the definition of R to conclude that $(a, a) \in R$.

Explore more on the Web.

Practice Problem 3   Let R be the relation on $\mathbb{Z}$ given by the rule $R = \{(a, b) \in \mathbb{Z} \times \mathbb{Z} : a - b \text{ is even}\}$. Prove that R is reflexive.

 Example 5   Let R be the relation on $\mathbb{Z}$ given by the rule $R = \{(a, b) \in \mathbb{Z} \times \mathbb{Z} : a - b \text{ is even}\}$. Prove that R is transitive.

SOLUTION   The formal definition says that R is transitive means, "If $(a, b) \in R$ and $(b, c) \in R$, then $(a, c) \in R$," so we can once again prove this if, then statement directly.

Proof that R is transitive. Let a, b, and c be given so that $(a, b) \in R$ and $(b, c) \in R$. This means (by the definition of R) that $a - b$ is even and $b - c$ is even. In other words, $a - b = 2 \cdot K$ for some integer K, and $b - c = 2 \cdot L$ for some integer L. But in this case,

$$a - c = (a - b) + (b - c)$$
$$= 2K + 2L$$
$$= 2(K + L)$$

Since $K + L$ is an integer, we can conclude that $a - c$ is even. Therefore, by the definition of R, $(a, c) \in R$. This means that R is transitive. □

In general, a proof that a relation R on a set A is transitive always has the following form:

> **Claim.** R is transitive.
>
> *Proof.*   Let $a, b, c \in A$ be given such that $(a, b) \in R$ and $(b, c) \in R$. **Use the definition of R to state what this means about a, b, and c.**
>
> $$\vdots$$
>
> **Use the definition of R to conclude that $(a, c) \in R$.**

Explore more on the Web.

Practice Problem 4   *Let $A = \{1, 2, 3, 4, 6, 9, 12, 18, 36\}$, and let R be the relation on A with the rule $R = \{(a, b) : b$ is a multiple of $a\}$. Prove that R is transitive.*

The antisymmetric property

Proofs about the antisymmetric property seem somewhat less direct because the definition of *antisymmetric* is a bit more complex. Recall that R is antisymmetric if the following holds:

$$\text{For all } a, b \in A, \text{ if } a \neq b \text{ and } (a, b) \in R, \text{ then } (b, a) \notin R.$$

The additional complexity in this case comes primarily from the fact that the definition involves negative conditions, so we must rewrite it as an equivalent statement that is easier to apply our proof strategies to.

   We discussed earlier that the informal meaning of antisymmetry is that the arrows only go in one direction, so it seems like the statement "For are all $a, b \in A$, if $(a, b) \in R$, then $(b, a) \notin R$" would be just as suitable as the definition of antisymmetry. There is only one problem. If $a = b$ in this statement (remember that using two different variable names does not necessarily mean that the variables cannot have the same value), then the statement would read, "If $(a, a) \in R$, then $(a, a) \notin R$." Hence, no antisymmetric relation could be reflexive, and we do not want these properties to have that kind of interdependence. (See Exercise 14.)

   In light of this distinction, we could rephrase the informal meaning to say, "In the arrow diagram for an antisymmetric relation, the only arrows that go in both directions are loops." The formalization of this statement will be the form on which we will base proofs about antisymmetry.

Proposition 1   *A relation R on a set A is antisymmetric if and only if the following statement is true:*

$$\text{For all } a, b \in A, \text{ if } (a, b) \in R \text{ and } (b, a) \in R, \text{ then } a = b.$$

PROOF   The proof is simply a matter of verifying that this rewording of the definition is logically equivalent to the original statement. (See Exercise 16.) ■

Example 6   *Let $A = \{1, 2, 3, 4, 6, 9, 12, 18, 36\}$, and let R be the relation on A given by $R = \{(x, y) \in A \times A : y$ is a multiple of $x\}$. Prove that R is antisymmetric.*

SOLUTION   *Proof.* Let $a, b \in A$ be given such that $(a, b) \in R$ **and** $(b, a) \in R$. From the definition of R, this means that b is a multiple of a **and** a is a multiple of b. This, in turn, means that $a = K \cdot b$ and $b = L \cdot a$ for some integers K and

L. Combining these equations tells us that $a = K \cdot (L \cdot a)$, from which it follows that $K \cdot L = 1$. Since the elements in *A* are all positive, then *K* and *L* must be *positive* integers, so $K \cdot L = 1$ implies that $K = L = 1$. This, in turn, implies that $a = b$. □

So even though it is not the most intuitive definition of antisymmetry, our proofs that a relation *R* on a set *A* is antisymmetric will always have the following form:

Claim. *R* is antisymmetric.
Proof. Let $a, b \in A$ be given such that $(a, b) \in R$ and $(b, a) \in R$. **Use the definition of *R* to state what this means about *a* and *b*.**

$$\vdots$$

Conclude that $a = b$.

Explore more on the Web.

Practice Problem 5 *Let $A = \mathcal{P}(\{1, 2, 3, 4\})$, and let R be the relation on A with the rule $R = \{(s, t) \in A \times A : s \subseteq t\}$. Prove that R is antisymmetric.*

Other Types of Orders

Transitivity and antisymmetry are the essential properties of "order relations." For example, if *a* is smaller than *b* and *b* is smaller than *c*, then it will follow that *a* is smaller than *c*. Similarly, if *a* is smaller than *b*, then it cannot be the case that *b* is smaller than *a*. This is significant because it will be true whether "smaller" refers to sizes of numbers, areas of triangles, or heights of people.

Example 7 *Each of the following relations conveys a reasonable sense of "order" among the elements of A. Explain why each is not a partial ordering.*

1. *Let $A = \{1, 2, 3, 4, 5\}$ and let $R = \{(a, b) \in A \times A : a < b\}$.*
2. *Let A be the set of classes required for graduation, and let $P = \{(c, d) \in A \times A : Class\ c\ must\ be\ completed\ before\ starting\ class\ d\}$.*

   SOLUTION Relation *R* is not reflexive because, for example, the statement "$1 < 1$" is not true. Similarly, relation *S* is not reflexive because, for example, the statement "Calculus I must be completed before starting Calculus I" is not true. □

In each of these examples, it is the reflexive property that fails. This means that reflexivity is not essential for a relation to seem like an "ordering." However, in each of these examples, reflexivity not only fails but fails spectacularly. The arrow diagrams for each of the previous relations have no loops at all. We will call relations like this *strict partial orders*.

> ***Definition*** A relation R on A is *irreflexive* if for all $a \in A, (a, a) \notin R$. (That is, the arrow diagram of an irreflexive relation R has no loops.) A *strict partial ordering* on the set A is a relation R on A that is transitive, antisymmetric, and irreflexive.

Practice Problem 6 *Let* $A = \{1, 2, 3, 4, 6, 12\}$ *and let*

$$R = \{(m, n) \in A \times A : m \text{ is a proper factor of } n\}$$

Prove that R is irreflexive.

Note that "irreflexive" does *not* merely mean "not reflexive," as the following example shows.

Example 8 *Draw an arrow diagram of a relation on* $A = \{1, 2, 3, 4\}$ *that is transitive and antisymmetric but neither reflexive nor irreflexive.*

SOLUTION The relation in Figure 4-52 is not reflexive since there is no loop at node 2. However, it is not irreflexive since there *is* a loop at node 1. □

Figure 4-52
Solution for
Example 8.

Partial orders (strict or otherwise) are relevant in many applications in the science and business world. Scheduling problems provide one such use of these structures in task management, a topic of common interest to businessmen, engineers, and computer scientists. In its simplest form, we define a relation on a set of tasks so that "a is related to b" means "Task a must be completed before task b can begin." The following particular example shows how this situation is relevant to students at a college.

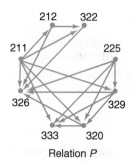

Relation *P*

Figure 4-53 Diagram
for Example 9.

Example 9 *Let P be the relation on the set* $M = \{211, 212, 225, 320, 322, 326, 329, 333\}$ *of math courses defined by the rule* $(a, b) \in P$ *if course a must be completed before beginning course b, where this "prerequisite structure" is given by the arrow diagram in Figure 4-53. Is P a strict partial order? If so, draw the Hasse diagram* for the relation.*

SOLUTION Since there are no loops, the relation is irreflexive. Since there is no pair of nodes with arrows in both directions, the relation is antisymmetric. Transitivity can also be checked by noting that any time there is a "two-arrow" path from a node a to a node b (like $211 \rightarrow 212 \rightarrow 322$) there is an arrow directly from a to b. The Hasse diagram for this strict partial ordering is given in Figure 4-54. □

Another variation on the relations we have studied comes from considering the difference between relations R_1 and R_2 in Example 1:

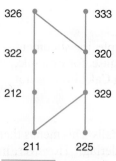

The Hasse diagram for *P*

Figure 4-54 Solution
to Example 9.

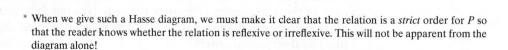

* When we give such a Hasse diagram, we must make it clear that the relation is a *strict* order for P so that the reader knows whether the relation is reflexive or irreflexive. This will not be apparent from the diagram alone!

- The relation R_1 on the set $\{1, 2, 3, 4\}$ with the rule "$(x, y) \in R_1$ if $x \le y$."
- The relation R_2 on the set $\mathcal{P}(\{1, 2, 3\})$ with the rule "$(x, y) \in R_2$ if $x \subseteq y$."

Relation R_1 forces a decision about every pair of numbers, while R_2 allows for the possibility that some pairs of sets are completely unrelated. An unrelated pair in a partial ordering is called an *incomparable pair*, and it is the existence of these pairs that makes the ordering "partial." If we know for a fact that every pair is comparable, then we can say that the ordering is *total* instead of *partial*.

Definition   We say that R is a *total ordering on A* if R is a reflexive, transitive, and antisymmetric relation on A that also satisfies the property

$$\text{For all } a, b \in A, \text{ if } a \ne b, \text{ either } (a, b) \in R \text{ or } (b, a) \in R.$$

As with partial orderings, a *strict total ordering* has the same properties except that it is irreflexive instead of reflexive.

Example 10 *Let $A = \mathcal{P}(\{1, 2, 4, 8\})$. Identify each of the following relations on A as a partial ordering, a total ordering, a strict partial ordering, a strict total ordering, or none of these:*

1. *The relation R_1 is defined so that $(S, T) \in R_1$ means every element of S is less than or equal to every element of T.*
2. *The relation R_2 is defined so that $(S, T) \in R_2$ means $n(S) < n(T)$.*
3. *The relation R_3 is defined so that $(S, T) \in R_3$ means the sum of the elements in S is less than or equal to the sum of the elements in T.*

SOLUTION   Relation R_1 is a partial ordering, R_2 is a strict partial ordering, and R_3 is a total ordering. If any of these surprise you, draw the appropriate Hasse diagram to check!   □

Practice Problem 7   *Let $\mathcal{B}$ be the set of binary strings of length exactly 4. Label each of the following relations on $\mathcal{B}$ as a partial order, a strict partial order, a total order, a strict total order, or none of these.*

(a) $R_1 = \{(\alpha, \beta) \in \mathcal{B} \times \mathcal{B} : \alpha \text{ has fewer 1's than } \beta \text{ has}\}.$

(b) *(For those who are familiar with Section 2.6.) Let $V : \mathcal{B} \to \mathbb{N}$ be the function such that $V(\alpha)$ is the value of the binary numeral α. For example, $V(1100) = 12$ and $V(0011) = 3$. Let $R_2 = \{(\alpha, \beta) \in \mathcal{B} \times \mathcal{B} : V(\alpha) \le V(\beta)\}.$*

Summary

There are two important classes of binary relations: "order relations" and "equivalence relations." We have seen several types of order relation in this section, and discussed how their behavior can be explained in terms of abstract properties such as reflexivity, antisymmetry, and transitivity. We have also continued practicing

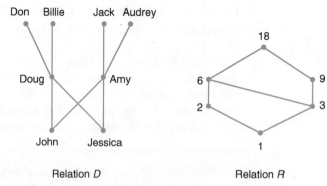

Don   Billie   Jack Audrey

Doug   Amy

John   Jessica

Relation *D*

Relation *R*

Figure 4-55 Solution to Practice Problem 1.

our logic and proof-writing skills by proving these properties do or do not hold for particular relations.

Solutions to Practice Problems

1 The two Hasse diagrams are given in Figure 4-55.

2 None of the relations are partial orders.

(a) R_1 is not antisymmetric since $(\{1\}, \{1, 2\}) \in R_1$ and $(\{1, 2\}, \{1\}) \in R_1$. R_1 is not transitive since $(\{1, 2\}, \{2, 3\}) \in R_1$ and $(\{2, 3\}, \{3\}) \in R_1$ but $(\{1, 2\}, \{3\}) \notin R_1$. R_1 is also not reflexive since $(\emptyset, \emptyset) \notin R_1$.

(b) R_2 is not antisymmetric since $(1, 3) \in R_2$ and $(3, 1) \in R_2$. R_2 is not reflexive since $(2, 2) \notin R_2$.

(c) R_3 is not antisymmetric since (Arm, Ram) $\in R_3$ and (Ram, Arm) $\in R_3$.

3 *Proof.* Let $a \in \mathbb{Z}$ be given. Clearly, $a - a = 0$, and 0 is even since $0 = 2 \cdot 0$. Thus, it follows from the definition of R that $(a, a) \in R$. Hence, R is reflexive. ∎

4 *Proof.* Let $a, b, c \in A$ be given so that $(a, b) \in R$ and $(b, c) \in R$. This means (by the definition of R) that b is a multiple of a and c is a multiple of b. In other words, $b = a \cdot K$ for some integer K, and $c = b \cdot L$ for some in-

teger L. But in this case,

$$c = bL = (aK)L = a(KL)$$

Since KL is an integer, we can conclude that c is a multiple of a. Therefore, by the definition of R, $(a, c) \in R$. ∎

5 *Proof that R is antisymmetric.* Let $S, T \in \mathcal{P}(\{1, 2, 3, 4\})$ be given such that $(S, T) \in R$ **and** $(T, S) \in R$. By the definition of R, this means that $S \subseteq T$ and $T \subseteq S$. But by the definition of equality of sets, this means that $S = T$. ∎

6 *Proof that R is irreflexive.* Let $a \in \{1, 2, 3, 4, 6, 12\}$ be given. According to the definition of "proper factor," a proper factor of a must be smaller than a. Hence, a cannot be a proper factor of a. Therefore, by the definition of R, $(a, a) \notin R$. We can conclude from this that R is irreflexive. ∎

7 R_1 is a strict partial order, and R_2 is a total order.

Exercises for Section 4.4

1. Complete the arrow diagram (Figure 4-56) for each of the following relations on $A = \{1, 2, 3, 4, 5, 6, 7, 8\}$, and decide if it has any of the reflexive, antisymmetric, or transitive properties. For each property a relation does not have, illustrate this failure with a specific example.

(a) $R_1 = \{(1, 1), (1, 2), (1, 4), (1, 8), (2, 2), (2, 4), (2, 8), (3, 3), (3, 6), (4, 4), (4, 8), (5, 5), (6, 6), (7, 7), (8, 8)\}$

(b) $R_2 = \{(2, 2), (2, 4), (2, 6), (2, 8), (3, 1), (4, 4), (4, 6), (4, 8), (5, 1), (5, 3), (6, 6), (6, 8), (7, 1), (7, 3), (7, 5), (8, 8)\}$

(c) $R_3 = \{(1, 1), (1, 3), (1, 5), (1, 7), (2, 2), (2, 4), (2, 8), (3, 3), (3, 5), (3, 7), (4, 2), (4, 4), (4, 8), (5, 3), (5, 7), (6, 6), (6, 8), (8, 2), (8, 4), (8, 8)\}$

(d) $R_4 = \{(1, 3), (1, 5), (1, 7), (2, 2), (2, 4), (2, 6), (2, 8), (3, 5), (3, 7), (4, 4), (4, 6), (4, 8), (5, 7), (6, 6), (6, 8), (8, 8)\}$

2. For each of the following relations on $\mathbb{Z}$, decide if the relation is reflexive or irreflexive. If it does not have one (or both) of these properties, give a specific example to illustrate this.

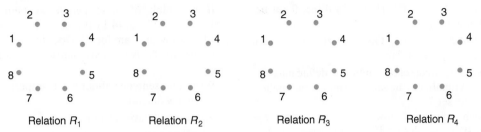

Figure 4-56 Diagrams for Problem 1.

(a) $R_1 = \{(a, b) \in \mathbb{Z} \times \mathbb{Z} : a + b \text{ is even}\}$
(b) $R_2 = \{(a, b) \in \mathbb{Z} \times \mathbb{Z} : a + b \text{ is odd}\}$
(c) $R_3 = \{(a, b) \in \mathbb{Z} \times \mathbb{Z} : a + 2b \text{ is even}\}$

3. For each of the following relations on $\mathbb{Z}$, decide if the relation is antisymmetric. If it is not, give a specific example to illustrate this.

(a) $R_1 = \{(a, b) \in \mathbb{Z} \times \mathbb{Z} : a + b \text{ is even}\}$
(b) $R_2 = \{(a, b) \in \mathbb{Z} \times \mathbb{Z} : a^2 + b \text{ is odd}\}$
(c) $R_3 = \{(a, b) \in \mathbb{Z} \times \mathbb{Z} : ab + b \text{ is odd}\}$

4. For each of the following relations on $\mathbb{Z}$, decide if the relation is transitive. If it is not, give a specific example to illustrate this.

(a) $R_1 = \{(a, b) \in \mathbb{Z} \times \mathbb{Z} : a + b \text{ is even}\}$
(b) $R_2 = \{(a, b) \in \mathbb{Z} \times \mathbb{Z} : a + b \text{ is odd}\}$
(c) $R_3 = \{(a, b) \in \mathbb{Z} \times \mathbb{Z} : a + 2b \text{ is even}\}$

5. Consider the two relations whose diagrams are shown in Figure 4-57.

(a) Describe each of the relations by filling in the blanks:
   - $R_1 = \{(a, b) \in A \times A : \underline{\hspace{1cm}}\}$, where $A = \{\underline{\hspace{1cm}}\}$.
   - $R_2 = \{(a, b) \in B \times B : \underline{\hspace{1cm}}\}$, where $B = \{\underline{\hspace{1cm}}\}$.

(b) Are either of these relations reflexive? If not, add arrows to the diagram to make the relation reflex-

ive. Give a description in words of this new relation in each case.

(c) Are either of these relations transitive? If not, add arrows to the diagram to make the relation transitive. Give a description in words of this new relation in each case.

6. Let A be the set of letters in the English alphabet. For each of the following relations on A, decide if it is reflexive, irreflexive, transitive, or antisymmetric. (Each can satisfy more than one of these properties.)

(a) $R_1 = \{(\alpha, \beta) \in A \times A : \alpha \text{ immediately precedes } \beta \text{ in alphabetical order}\}$
(b) $R_2 = \{(\alpha, \beta) \in A \times A : \alpha \text{ comes before } \beta \text{ in alphabetical order}\}$

7. Let P be the set of people who have ever lived. For each of the following relations on P, decide if it is reflexive, irreflexive, transitive, or antisymmetric. (Each can satisfy more than one of these properties.)

(a) $R_1 = \{(\alpha, \beta) \in P \times P : \alpha \text{ is a child of } \beta\}$
(b) $R_2 = \{(\alpha, \beta) \in P \times P : \alpha \text{ is a descendant of } \beta\}$

8. Let C be the set of airports in the world. For each of the following relations on C, decide if it is reflexive, irreflexive, transitive, or antisymmetric. (Each can satisfy more than one of these properties.)

(a) $R_1 = \{(\alpha, \beta) \in C \times C : \text{There is a direct flight from } \alpha \text{ to } \beta\}$

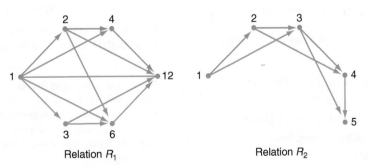

Figure 4-57 Diagrams for Problem 5.

(b) $R_2 = \{(\alpha, \beta) \in C \times C : \text{There is a direct flight between } \alpha \text{ and } \beta\}$

(c) $R_3 = \{(\alpha, \beta) \in C \times C : \text{One can travel between cities } \alpha \text{ and } \beta \text{ by airplane(s)}\}$

9. For a non-empty finite set of numbers S, define $\max S$ to be the largest number in the set S and $\min S$ the smallest number in the set S.

For each of the following relations on $A = \mathcal{P}(\{1, 2, 3, 4\}) - \{\phi\}$, decide if it is reflexive, irreflexive, transitive, or antisymmetric. (Each can satisfy more than one of these properties.)

(a) $R_1 = \{(S, T) \in A \times A : \max S \le \max T\}$

(b) $R_2 = \{(S, T) \in A \times A : \max S \le \min T\}$

(c) $R_3 = \{(S, T) \in A \times A : \max S < \max T\}$

For each property that a relation fails to satisfy, give a specific example to illustrate this.

10. For a finite set of numbers S, define $\sigma(S)$ to be the sum of the numbers in the set S, where $\sigma(\phi) = 0$.

For each of the following relations on $A = \mathcal{P}(\{1, 2, 3, 4\})$, decide if it is reflexive, irreflexive, transitive, or antisymmetric. (Each can satisfy more than one of these properties.)

(a) $R_1 = \{(S, T) \in A \times A : \sigma(S) \le \sigma(T)\}$

(b) $R_2 = \{(S, T) \in A \times A : \sigma(S) < \sigma(T)\}$

(c) $R_3 = \{(S, T) \in A \times A : \sigma(S - T) < \sigma(S \cap T)\}$

For each property that a relation fails to satisfy, give a specific example to illustrate this.

11. Decide whether the relation R on the set $\{1, 2, 3, 4, 5\}$ defined below is reflexive, antisymmetric, or transitive. Predict what this property (or properties) will imply about the diagram of R. Draw the diagram to see if you are correct.

$$R = \{(1, 1), (1, 2), (1, 4), (2, 2), (2, 1), (2, 4), (3, 3),$$
$$(3, 4), (4, 4)\}$$

12. Let the relation R on the set $E = \{\text{David, Mary, Doug, Rob, Don, Billie, John, Jessica, Andrew, Allen, Jacob, Christopher, Saache, Tommy}\}$ be given by

$$R = \{(a, b) : a \text{ is a son of } b\}$$

For those of you who do not know these people, this means

$R = \{$(David, Don), (Doug, Don), (Rob, Don),
  (David, Billie), (Doug, Billie), (Rob, Billie),
  (Andrew, Mary), (Allen, Mary), (Jacob, Mary),
  (John, Doug), (Christopher, Rob),
  (Tommy, David)$\}$

If we think of these people as being numbered $1, 2, 3, \ldots, 12, 13, 14$ in the order they appear above, what is the diagram for R? Does this relation have reflexive, irreflexive, antisymmetric, or transitive properties?

13. Write one sentence about what property shows up in the arrow diagram

(a) Of a reflexive relation.

(b) Of an antisymmetric relation.

(c) Of a transitive relation.

14. Let $A = \{1, 2, 3\}$. Give an example of a relation R on A that is

(a) Transitive and reflexive but not antisymmetric.

(b) Antisymmetric and reflexive but not transitive.

(c) Antisymmetric and transitive but not reflexive.

15. Let $S = \{1, 2, 3\}$. For each of the following relations on $\mathcal{P}(S)$, draw the arrow diagram and decide if the relation is reflexive, antisymmetric, or transitive. If it fails any of these properties, give a specific example to illustrate this. If it has all three properties (i.e., if it is a partial ordering), give the corresponding Hasse diagram.

(a) $R_1 = \{(A, B) \in \mathcal{P}(S) \times \mathcal{P}(S) : A \subseteq B\}$

(b) $R_2 = \{(A, B) \in \mathcal{P}(S) \times \mathcal{P}(S) : B - A = \{3\}\}$

(c) $R_3 = \{(A, B) \in \mathcal{P}(S) \times \mathcal{P}(S) : A \cap B = \emptyset\}$

(d) $R_4 = \{(A, B) \in \mathcal{P}(S) \times \mathcal{P}(S) : A \cap B \ne \emptyset\}$

(e) $R_5 = \{(A, B) \in \mathcal{P}(S) \times \mathcal{P}(S) : n(A) \le n(B)\}$

16. Fill in the missing details in the following proof of Proposition 1.

Proof. Let a relation R on a set A be given, and let $a, b \in A$ be given as well. The equivalence of the two characterizations of antisymmetry is based on the logical equivalence of the statements rather than on context. Consider the following propositional variables:

- Let p be the statement "$a = b$."
- Let q be the statement "$(a, b) \in R$."
- Let r be the statement "$(b, a) \in R$."

In terms of p, q, and r, the compound proposition "If $a \ne b$ and $(a, b) \in R$, then $(b, a) \in R$" is written as

$$((\neg p) \wedge q) \to r \qquad (*)$$

Similarly, the compound proposition "If $(a, b) \in R$ and $(b, a) \in R$, then $a = b$" is written as

Place this statement in the rightmost column of the truth table in Table 4-13, and show that it is equivalent to $(*)$. Since the compound propositions are equivalent regardless of the choice of a and b, it follows that the two characterizations of antisymmetry are equivalent. ∎

p	q	r	$(\neg p) \wedge q$	$((\neg p) \wedge q) \rightarrow r$	
T	T	T			
T	T	F			
T	F	T			
T	F	F			
F	T	T			
F	T	F			
F	F	T			
F	F	F			

Table 4-13 Table for Exercise 16

17. Prove or give a counterexample to each of the following statements:
   (a) If R is a reflexive relation on A, then R^{-1} is a reflexive relation on A.
   (b) If R is an antisymmetric relation on A, then R^{-1} is an antisymmetric relation on A.
   (c) If R is a transitive relation on A, then R^{-1} is a transitive relation on A.

18. Prove or give a counterexample to each of the following statements:
   (a) If R is a reflexive relation on A, then $R \circ R$ is a reflexive relation on A.
   (b) If R is an antisymmetric relation on A, then $R \circ R$ is an antisymmetric relation on A.
   (c) If R is a transitive relation on A, then $R \circ R$ is a transitive relation on A.

19. Give two different examples of a relation R on the set $A = \{a, b, c, d\}$ satisfying $R \circ R = R$.

20. Since the rule for a relation on A is technically a subset of $A \times A$, we can use the usual set operations to combine relations on the same set. Let

$A = \{0, 1, 2, 3, 4\}$ and define relations R_1 and R_2 by $R_1 = \{(0, 1), (1, 2), (0, 3), (1, 4), (0, 0)\}$ and $R_2 = \{(1, 1), (1, 2), (2, 1), (2, 2)\}$.
   (a) What is $R_1 \cup R_2$?
   (b) What is $R_1 \cap R_2$?
   (c) What is $R_1 - R_2$?

21. Prove or give a counterexample to each of the following statements for a given set A:
   (a) If R_1 and R_2 are reflexive relations on A, then the relation $R_1 \cup R_2$ is reflexive.
   (b) If R_1 and R_2 are reflexive relations on A, then the relation $R_1 \cap R_2$ is reflexive.
   (c) If R_1 and R_2 are antisymmetric relations on A, then the relation $R_1 \cup R_2$ is antisymmetric.
   (d) If R_1 and R_2 are antisymmetric relations on A, then the relation $R_1 \cap R_2$ is antisymmetric.

22. Prove that for any relation R on a set A, if R is transitive, then $R \circ R \subseteq R$.

23. Prove that for any relation R on a set A, if $R \circ R \subseteq R$, then R is transitive.

4.5 Equivalence Relations

There are many times when we would like to express a relationship in which objects in a set are thought of as being "the same" for some particular application. For example, in geometry, we might treat two triangles as being "the same" if they are similar triangles. In arithmetic, we treat two fractions $\frac{a}{b}$ and $\frac{c}{d}$ as being "the same" if $ad = bc$. Or perhaps a teacher assigns a group project and gives the same grade to everyone in the same group. In each of these examples, there is an underlying binary relation for which the phrase "a is related to b" loosely means that "a and b are treated the same."

A relation of this type is called an *equivalence relation.** In this section, we give a formal definition of equivalence relation, discuss the properties that characterize these relations, and continue to develop our proof-writing skills with these abstract mathematical objects.

Equivalence Relations and Partitions

We study binary relations on a set because they are able to capture important information within a very simple structure. For example, the two relations below reflect basic properties of positive integers as they apply to the numbers in the set $A = \{1, 2, 3, 4, 5, 6\}$.

 Example 1 *Draw the arrow diagram for each of the following relations:*

1. $R_1 = \{(a, b) \in A \times A : b = a + 1\}$
2. $R_2 = \{(a, b) \in A \times A : a - b \text{ is even}\}$

SOLUTION The diagrams for R_1 and R_2 are shown in Figure 4-58. □

Relation R_2 is our first example of an equivalence relation. In this case, two numbers are treated as being "the same" if they have the same parity—that is, if they are both even or both odd. In this way, the relation R_2 divides the set A into two disjoint (i.e., nonoverlapping) subsets: $\{1, 3, 5\}$, every two elements of which are related by R_2, and $\{2, 4, 6\}$, every two elements of which are related by R_2. We call this collection of subsets a *partition of A*. This is the central feature of our definition of an equivalence relation.

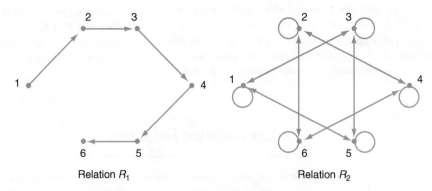

Figure 4-58 Diagrams for Example 1.

---

* Here we think of "equivalence" as being specific to the application at hand, always to be distinct from the notion of "equal" that we will continue using to mean "identical."

> **Definition**   For a set A, a *partition of A* is a set $S = \{S_1, S_2, S_3, \ldots\}$ of subsets of A (each set S_i is called *a part of S*) such that
>
> 1. For all i, $S_i \neq \emptyset$. That is, each part is nonempty.
> 2. For all i and j, if $S_i \neq S_j$, then $S_i \cap S_j = \emptyset$. That is, different parts have nothing in common.
> 3. $S_1 \cup S_2 \cup S_3 \cup \cdots = A$. That is, every element in A is in some part.

> **Definition**   For a set A and a relation R on A, R is an *equivalence relation on A* if there is a partition S of A such that $(x, y) \in R$ if and only if x and y are in the same part of S. In this case, we will refer to S as the *partition of A induced by R*.

Example 2 *Imagine a simplified card game where a player is dealt two face-up cards from a "deck" of five cards numbered $1, 2, 3, 4, 5$. Figure 4-59 shows every possible "hand." If the rules of the game call for the value of your hand to be based on which cards you have and not the order in which they are received, then some of these hands (like hand 1 and hand 5, e.g.) would be considered the same. If the relation R on the set of hands showed above is defined so that $(x, y) \in R$ means hands x and y are considered the same in the game, then what is the partition induced by this relation?*

SOLUTION   If we use the shorthand H_i for "hand i," we can efficiently write this partition as follows:

$$\{\{H_1, H_5\}, \{H_2, H_9\}, \{H_3, H_{13}\}, \{H_4, H_{17}\}, \{H_6, H_{10}\},$$
$$\{H_7, H_{14}\}, \{H_8, H_{18}\}, \{H_{11}, H_{15}\}, \{H_{12}, H_{19}\}, \{H_{16}, H_{20}\}\}$$

☐

Practice Problem 1   *In Example 2, suppose we play a game in which the rules treat two hands as the same if the **sum of the values** on the cards is the same. What is the partition induced by this relation?*

Example 3 *Determine which of the following relations are equivalence relations on the given sets by describing the induced partition:*

1. *$A = \{1, 2, 3, 4, 5, 6, 7, 8, 9\}$, and the relation R on A given by the rule $(a, b) \in R$ means that a and b have a common factor greater than 1.*
2. *The relation R on $\mathbb{Z}$ given by the rule $(x, y) \in R$ means that $x - y$ is divisible by 4.*
3. *$A = \{0, 1, 2, 3, 4, 5, 6\} \times \{1, 2, 3\}$, and the relation R on A given by the rule where $(a, b) \, R \, (c, d)$ means that $ad = bc$.*

SOLUTION   The first relation is not an equivalence relation since in a partition, it would be impossible for 4 and 6 to be in the same part, 6 and 9 to be in the same part, **and** 4 and 9 not to be in the same part. The other two relations *are* equivalence relations with the following induced partitions.

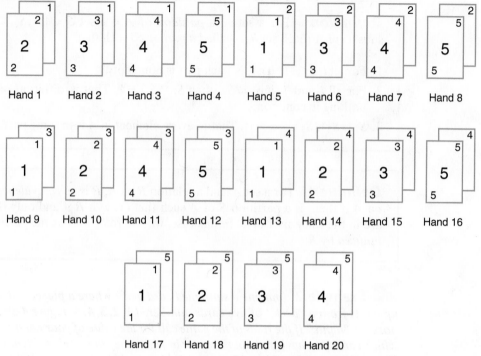

Figure 4-59   Possible card hands for Example 2.

2. The set $\mathbb{Z}$ of integers is partitioned into the four parts by this relation:

$$\{\{\ldots, -8, -4, 0, 4, 8, \ldots\}, \quad \{\ldots, -7, -3, 1, 5, 9, \ldots\},$$
$$\{\ldots, -6, -2, 2, 6, 10, \ldots\}, \quad \{\ldots, -5, -1, 3, 7, 11, \ldots\}\}$$

Another way to describe this partition of the infinite set $\mathbb{Z}$ is as the set $\{P_0, P_1, P_2, P_3\}$, where

- $P_0 = \{a \in \mathbb{Z} : a = 4k \text{ for some } k \in \mathbb{Z}\}$
- $P_1 = \{a \in \mathbb{Z} : a = 4k + 1 \text{ for some } k \in \mathbb{Z}\}$
- $P_2 = \{a \in \mathbb{Z} : a = 4k + 2 \text{ for some } k \in \mathbb{Z}\}$
- $P_3 = \{a \in \mathbb{Z} : a = 4k + 3 \text{ for some } k \in \mathbb{Z}\}$

3. The induced partition of A is the set

$$\{S_0, S_1, S_2, S_3, S_4, S_5, S_6, S_{1/2}, S_{3/2}, S_{5/2}, S_{1/3}, S_{2/3}, S_{4/3}, S_{5/3}\}$$

where

$S_0 = \{(0, 1), (0, 2), (0, 3)\}$	$S_{1/2} = \{(1, 2)\}$
$S_1 = \{(1, 1), (2, 2), (3, 3)\}$	$S_{3/2} = \{(3, 2)\}$
$S_2 = \{(2, 1), (4, 2), (6, 3)\}$	$S_{5/2} = \{(5, 2)\}$
$S_3 = \{(3, 1), (6, 2)\}$	$S_{1/3} = \{(1, 3)\}$
$S_4 = \{(4, 1)\}$	$S_{2/3} = \{(2, 3)\}$
$S_5 = \{(5, 1)\}$	$S_{4/3} = \{(4, 3)\}$
$S_6 = \{(6, 1)\}$	$S_{5/3} = \{(5, 3)\}$

$\square$

Practice Problem 2   *Draw arrow diagrams for the following equivalence relations. In each case, give the partition on A induced by the relation.*

(a) *For $A = \{0, 1, 2, 3, 4, 5, 6\}$, $R = \{(x, y) \in A \times A : (x - 2.5) \cdot (y - 2.5) \geq 0\}$*

(b) *For $A = \{0, 1, 2, 3, 4, 5, 6\}$, $R = \{(x, y) \in A \times A : x^3 - y^3 \text{ is divisible by } 7\}$*

Two properties of equivalence relations should be immediately apparent. For any reasonable notion of "the same," we should know that any element is "the same" as itself, and we should know that the statement "*a* is the same as *b*" is equivalent to the statement "*b* is the same as *a*." The first of these properties should look familiar: An equivalence relation must be reflexive. The latter property is one we have not discussed before, but it is related to the notion of antisymmetry that we studied in our discussion of order relations.

> ***Definition***   A relation R on set A is said to be *symmetric* if for all $a, b \in A$, if $(a, b) \in R$, then $(b, a) \in R$.

In terms of arrow diagrams, a symmetric relation has the property that every pair of nodes connected by an arrow is actually connected by two arrows, one in each direction. So only loops and "double arrows" will be used in the arrow diagrams of symmetric relations.

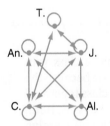

Figure 4-60
Diagram of the symmetric relation in Example 4.

Example 4   *Let $A = \{Tommy, Christopher, Allen, Jacob, Andrew\}$, and define the relation R_3 on A so that $x\,R_3\,y$ means x and y share at least one letter of their names. Draw the arrow diagram for the relation R.*

SOLUTION   The arrow diagram is given in Figure 4-60.   □

As with other properties, we can write proofs about the symmetric property based on the straightforward if, then structure of its definition.

Example 5   *For the relation R on the set $\mathbb{Z}$ given by $R = \{(x, y) \in \mathbb{Z} \times \mathbb{Z} : x - y \text{ is even}\}$, prove that R is symmetric.*

SOLUTION   To show that R is symmetric, we must show that, "If $(a, b) \in R_2$, then $(b, a) \in R$," so we can also prove this if, then statement directly by paying close attention to the definition of R.

Proof. Let $a, b \in \{1, 2, 3, 4, 5, 6\}$ be given so that $(a, b) \in R$. This means (by the definition of R) that $a - b$ is even. In other words, $a - b = 2 \cdot K$ for some integer K. But in this case, $b - a = -(a - b) = 2 \cdot (-K)$, and since $-K$ is an integer too, we can conclude that $b - a$ is even as well. Therefore, by the definition of R, $(b, a) \in R$. We conclude that R is symmetric.   □

In general, a proof that a relation R on a set A is symmetric always has the following form.

Claim. R is symmetric.

Proof.   Let $a, b \in A$ be given such that $(a, b) \in R$. **Use the definition of R to say what this means about a and b.**

$$\vdots$$

Use the definition of R to conclude that $(b, a) \in R$.

Practice Problem 3   *Let $A = \{0, 1, 2, 3, 4, 5\}$ and let R be the relation on A given by the rule $R = \{(a, b) : a^2 - b^2 \text{ is a multiple of } 3\}$.*

(a) Prove that R is reflexive.

(b) Prove that R is symmetric.

It seems reasonable that any equivalence relation should be reflexive and symmetric. The next example shows that to be an equivalence relation, a relation must satisfy more than just these two properties.

Example 6   *Let S be the set of all students at your school this semester, and let the relation R be given so that $(x, y) \in R$ means students x and y have a class together this semester. Explain why R is reflexive and symmetric but is not an equivalence relation on S.*

SOLUTION   Certainly, any student enrolled this semester has a class with him or herself, so R is reflexive. The relation R is clearly symmetric because the statements "x and y have a class together" and "y and x have a class together" express the same thing. However, it is easy to imagine students like Amy, Beth, and Carolyn, where Amy and Beth have a class together, Beth and Carolyn have a different class together, and Amy and Carolyn do not have any classes together. Since the parts of a partition may not overlap, it is impossible for there to be a partition $\mathcal{P}$ where Amy and Beth are in the same part, Beth and Carolyn are in the same part, but Amy and Carolyn are not in the same part. □

In a plot twist worthy of an action thriller, it turns out the property that will alleviate this particular problem is our old friend, transitivity.

Example 7   *Each of the following relations on $\mathbb{Z}$ is reflexive and symmetric. Which of them is also transitive?*

1. $T_1 = \{(a, b) \in \mathbb{Z} \times \mathbb{Z} : b - a \text{ is divisible by } 5\}$

2. $T_2 = \{(a, b) \in \mathbb{Z} \times \mathbb{Z} : a^2 - b^2 \text{ is divisible by } 5\}$

3. $T_3 = \{(a, b) \in \mathbb{Z} \times \mathbb{Z} : |a - b| \leq 2\}$

SOLUTION   The relations T_1 and T_2 are both transitive (see Exercise 9), but relation T_3 is not transitive. To see this, simply notice that $(1, 3) \in T_3$ and $(3, 4) \in T_3$ but $(1, 4) \notin T_3$. □

Practice Problem 4   *Let* $C = \{2, 3, 4, 5, 6, 7, 8, 9, 10\}$, *and define the relation R on C so that* $(a, b) \in R$ *means the greatest common divisor of a and b is* **greater than** 1. *Is this relation transitive? Is it reflexive or symmetric?*

To show that transitivity really is the last piece of the puzzle, we will formally prove that a binary relation that is reflexive, symmetric, and transitive must be an equivalence relation.

Theorem 1   *If the relation R on the set A is reflexive, symmetric, and transitive, then R is an equivalence relation.*

PROOF   For each element $a \in A$, define the set $P_a = \{x \in A : (x, a) \in R\}$. That is, P_a is the set of elements in A that are related to a. To prove that the set $\{P_a : a \in A\}$ is a partition of A, we must verify each of the three properties in the definition of partition from page 315.

1. **Claim:** Each P_a is nonempty.
   Proof.   For each $a \in A$, $(a, a) \in R$ because R is reflexive. From this, it follows that $a \in P_a$; hence, P_a is nonempty.
2. **Claim:** For all $a, b \in A$, if $P_a \cap P_b \neq \emptyset$, then $P_a = P_b$. (Note that this is actually the contrapositive of the statement given in the definition.)
   Proof.   Let $a, b \in A$ be given such that $P_a \cap P_b \neq \emptyset$. This means that there is an element $c \in P_a \cap P_b$. This, in turn, means that $(c, a) \in R$ and $(c, b) \in R$. Now the symmetry of R tells us that $(a, c) \in R$ and $(b, c) \in R$. We can now use the fact that R is transitive. From our knowledge that $(a, c) \in R$ and $(c, b) \in R$, we can conclude (by transitivity) that $(a, b) \in R$. Similarly, from our knowledge that $(b, c) \in R$ and $(c, a) \in R$, we can conclude (by transitivity) that $(b, a) \in R$. See Exercise 5 for details on how to conclude that $P_a = P_b$.
3. **Claim:** The union of all the P's is the set A.
   Proof.   For each $a \in A$, $(a, a) \in R$ because R is reflexive. From this, it follows that for any given $a \in A$, we must have $a \in P_a$, and so every element of A is certainly in the union of all the P's.

In Theorem 1, we established that any relation on a set A that is reflexive, symmetric, and transitive must be an equivalence relation—that is, the relation induces a partition on A. Now that we have the tools for proving that a relation has these properties, we can show that the converse of Theorem 1 is true. We illustrate the idea with an example, and leave the more general proof as an exercise.

Example 8   *Let* $\mathbb{S}$ *be the set of students in a class where a group project is assigned, and define the relation G on* $\mathbb{S}$ *with the rule*

$$(a, b) \in G \text{ if } a \text{ and } b \text{ are in the same group.}$$

Assume that every student is part of exactly one group. Explain why G is reflexive, symmetric, and transitive.

SOLUTION   We have to show that G is reflexive, symmetric, and transitive, so we take one at a time.

● **Claim:** G is reflexive.

Proof. Let $a \in \mathbb{S}$ be given. Certainly, a is in the same group as him or herself. Hence, $(a, a) \in G$.

● **Claim:** G is symmetric.

Proof. Let $(a, b) \in G$ be given. The definition of G tells us that this means a and b are in the same group. But in this case, we can say $(b, a) \in G$ as well.

● **Claim:** G is transitive.

Proof. Let $(a, b) \in G$ and $(b, c) \in G$ be given. The definition of G tells us that this means a and b are in the same group, and b and c are in the same group. Since each student is part of only one group, this means that a and c are in the same group. Hence, we can conclude that $(a, c) \in G$. ☐

Theorem 2   *An equivalence relation R on a set A is reflexive, symmetric, and transitive.*

PROOF   Generalize the solution to Example 8. See Exercise 23. ■

Since we have established this statement and its converse, we have completely characterized equivalence relations in terms of their abstract properties.

Theorem 3   *A relation R on a set A is an equivalence relation on A if and only if R is reflexive, symmetric, and transitive.*

Practice Problem 5   *We have seen that the following relations are reflexive, symmetric, and transitive. Describe the partition of $\mathbb{Z}$ induced by each equivalence relation.*

(a) $T_1 = \{(a, b) \in \mathbb{Z} \times \mathbb{Z} : b - a \text{ is divisible by } 5\}$

(b) $T_2 = \{(a, b) \in \mathbb{Z} \times \mathbb{Z} : a^2 - b^2 \text{ is divisible by } 5\}$

Solutions to Practice Problems

1   The partition of the set of hands (using the notation from Example 2) is

$\{\{H_1, H_5\}, \{H_2, H_9\}, \{H_3, H_6, H_{10}, H_{13}\}, \{H_4, H_7, H_{14}, H_{17}\}, \{H_8, H_{11}, H_{15}, H_{18}\}, \{H_{12}, H_{19}\}, \{H_{16}, H_{20}\}\}$

2   The arrow diagrams are given in Figure 4-61. The partition of A induced by R_1 is $\{\{0, 1, 2\}, \{3, 4, 5, 6\}\}$. The partition of A induced by R_2 is $\{\{0\}, \{1, 2, 4\}, \{3, 5, 6\}\}$.

3   This requires two separate proofs.

(a) *Proof.*   Let $a \in A$ be given. Since $a^2 - a^2 = 0 = 3 \cdot 0$, we know that $a^2 - a^2$ is a multiple of 3, and hence, $(a, a) \in R$.

(b) *Proof.*   Let a and b be given so that $(a, b) \in R$. This means (by the definition of R) that $a^2 - b^2$ is a multiple of 3. In other words, $a^2 - b^2 = 3 \cdot K$ for some integer K. But in this case, $b^2 - a^2 = -(a^2 - b^2) =$ $3 \cdot (-K)$. Since $-K$ is an integer, we can conclude that $b^2 - a^2$ is a multiple of 3 as well. Therefore, by the definition of R, $(b, a) \in R$. ■

4   This relation is reflexive since the greatest common divisor of a and itself is a, and all elements of C are greater than 1. The relation is also symmetric since the greatest common divisor of two numbers does not depend on the order in which the two numbers are given. However, the given relation is *not* transitive since, for example, 4 and 6 are related (their greatest common divisor is 2) and 6 and 9 are related (their greatest common divisor is 3), but 4 and 9 are not related.

5   We learned in Section 2.2 that for integers a and b, $a - b$ is divisible by 5 if and only if a and b leave the same remainder when divided by 5. Hence, there are five parts in the partition of $\mathbb{Z}$ induced by equivalence relation T_1.

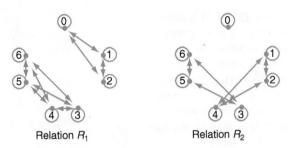

Figure 4-61   Solution for Practice Problem 2.

That is, T_1 induces the partition $\{P_0, P_1, P_2, P_3, P_4\}$ of $\mathbb{Z}$, where

- $P_0 = \{n \in \mathbb{Z} : n \text{ is divisible by } 5\} = \{\ldots, -10, -5, 0, 5, 10, \ldots\}$
- $P_1 = \{n \in \mathbb{Z} : n - 1 \text{ is divisible by } 5\} = \{\ldots, -9, -4, 1, 6, 11, \ldots\}$
- $P_2 = \{n \in \mathbb{Z} : n - 2 \text{ is divisible by } 5\} = \{\ldots, -8, -3, 2, 7, 12, \ldots\}$
- $P_3 = \{n \in \mathbb{Z} : n - 3 \text{ is divisible by } 5\} = \{\ldots, -7, -2, 3, 8, 13, \ldots\}$

- $P_4 = \{n \in \mathbb{Z} : n - 4 \text{ is divisible by } 5\} = \{\ldots, -6, -1, 4, 9, 14, \ldots\}$

Similarly, the partition of $\mathbb{Z}$ induced by equivalence relation T_2 is determined by remainders on division by 5. Specifically, T_2 induces the partition $\{A, B, C\}$ of $\mathbb{Z}$, where

- $A = P_0$ (from above)
- $B = P_1 \cup P_4$ (from above)
- $C = P_2 \cup P_3$ (from above)

Exercises for Section 4.5

1. Complete the arrow diagram in Figure 4-62 for each of the following relations on the set $A = \{1, 2, 3, 4, 5, 6, 7, 8\}$, and decide if it has any of the reflexive, symmetric, or transitive properties. For each property a relation does not have, illustrate this failure with a specific example.

   (a) $R_1 = \{(1, 1), (1, 2), (1, 4), (1, 8), (2, 1), (2, 2), (2, 4), (2, 8), (3, 3), (3, 6), (4, 1), (4, 2), (4, 4), (4, 8), (5, 5), (6, 3), (6, 6), (7, 7), (8,1), (8, 2), (8, 4), (8, 8)\}$

   (b) $R_2 = \{(1, 4), (1, 8), (2, 2), (2, 4), (2, 6), (2, 8), (3, 3), (3, 4), (3, 6), (4, 1), (4, 2), (4, 3), (4, 4), (4, 5), (4, 6), (4, 7), (4, 8), (5, 4), (5, 5), (6, 2), (6, 3), (6, 4), (6, 6), (6, 8), (7, 4), (7, 7), (8, 1), (8, 2), (8, 4), (8, 6), (8, 8)\}$

   (c) $R_3 = \{(1, 1), (1, 3), (1, 5), (1, 7), (2, 2), (2, 4), (2, 6), (2, 8), (3, 5), (3, 7), (4, 4), (4, 6), (4, 8), (5, 7), (6, 6), (6, 8), (8, 8)\}$

2. For each of the following relations on $\mathbb{Z}$, decide if the relation is symmetric. If it is not, give a specific example to illustrate this.

   (a) $R_1 = \{(a, b) \in \mathbb{Z} \times \mathbb{Z} : a + b \text{ is even}\}$

   (b) $R_2 = \{(a, b) \in \mathbb{Z} \times \mathbb{Z} : a + b \text{ is odd}\}$

   (c) $R_3 = \{(a, b) \in \mathbb{Z} \times \mathbb{Z} : a + 2b \text{ is even}\}$

3. Which of the following relations is symmetric, which are antisymmetric, and which are neither? Support any negative conclusions with a specific example.

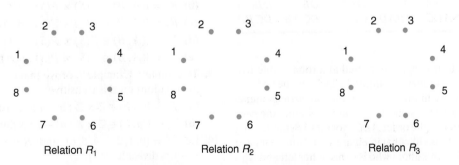

Figure 4-62   Diagrams for Exercise 1.

(a) The relation R on the set $A = \{2, 3, 5, 7\}$ is defined by the rule $R = \{(x, y) \in A \times A : x + y$ is even$\}$.

(b) The relation R on the set $\mathbb{Z}$ is defined by the rule $R = \{(x, y) \in \mathbb{Z} \times \mathbb{Z} : xy + y$ is even$\}$.

(c) The relation R on the set $A = \mathcal{P}(\{1, 2, 3, 4\})$ is defined by the rule $R = \{(s, t) \in A \times A : \sigma(s) = \sigma(t)\}$, where $\sigma(X)$ is the sum of the elements in the set X.

(d) The relation R on the set $\mathbb{Z}$ is defined by the rule $R = \{(x, y) \in \mathbb{Z} \times \mathbb{Z} : 2x + y^2$ is even$\}$.

4. Which of the following are partitions of the set $A = \{1, 2, 3, 4, 5, 6, 7, 8\}$? For each that is not, explain why not.

(a) $\{1, 2, \{3, 4, 5\}, \{6, 7, 8\}\}$

(b) $\{\{1, 5\}, \{6, 7, 2\}, \{4, 3, 5\}, \{8\}\}$

(c) $\{\{1, 4\}, \{6, 8, 2\}, \{3, 5\}, \{7\}\}$

(d) $\{\{1, 8\}, \{4, 3, 5\}, \{7, 2\}\}$

5. In the proof of Theorem 1, we showed that $(a, b) \in R$ and $(b, a) \in R$. Fill in the blanks below to complete the proof that $P_a = P_b$.

Proof of Theorem 1 Continued. To show that $P_a = P_b$, we must show that each set is a subset of the other.

Claim $P_a \subseteq P_b$.

Proof. Let $z \in P_a$ be given. From the definition of P_a, we know that _____ $\in R$, so by the _____ property of R, it follows that _____ $\in R$ as well. But this means that $z \in P_b$, completing the proof of this claim. ∎

Claim $P_b \subseteq P_a$.

Proof. Let $z \in P_b$ be given. From the definition of P_b, we know that _____ $\in R$, so by the _____ property of R, it follows that _____ $\in R$ as well. But this means that $z \in P_a$, completing the proof of this claim. ∎

6. Imagine four people ($A, B, C,$ and D) lined up to enter a restaurant. There are 24 different orders in which they can wait in line:

$ABCD$	$ABDC$	$ACBD$	$ACDB$	$ADBC$	$ADCB$
$BCAD$	$BCDA$	$BDAC$	$BDCA$	$CABD$	$CADB$
$CDAB$	$CDBA$	$DABC$	$DACB$	$DBAC$	$DBCA$
$BACD$	$BADC$	$CBAD$	$CBDA$	$DCAB$	$DCBA$

When the four people are seated at a round table (assuming they are seated counterclockwise in the order in which they are in line), some of the line arrangements end up looking the same. For example, whether the people are seated in the order $ABCD$ or $CDAB$, the seating at the table looks like that in Figure 4-63. Note that each person only cares about who is to his or her left and who is to his or her right, not about which particular chair

Figure 4-63 Example for Problem 6.

he or she sits in. If we think of the relation R on the set S of line arrangements with the rule $(X, Y) \in R$ if arrangements X and Y are the same when seated at a round table, what is the partition of the set S induced by R?

7. Imagine a company that makes charm bracelets, each of which has four charms $A, B, C,$ and D. In the manufacturing process, the charms are lined up to enter a threading machine. There are 24 different orders in which they can enter the machine:

$ABCD$	$ABDC$	$ACBD$	$ACDB$	$ADBC$	$ADCB$
$BCAD$	$BCDA$	$BDAC$	$BDCA$	$CABD$	$CADB$
$CDAB$	$CDBA$	$DABC$	$DACB$	$DBAC$	$DBCA$
$BACD$	$BADC$	$CBAD$	$CBDA$	$DCAB$	$DCBA$

Once the charms are threaded on a bracelet, many of these arrangements look the same. If we think of the relation R on the set $\mathcal{B}$ of line arrangements with the rule $(X, Y) \in R$ if arrangements X and Y look the same when threaded on a bracelet, what is the partition of the set $\mathcal{B}$ induced by R?

8. Let $S = \{1, 2, 3\}$. For each of the following relations on $\mathcal{P}(S)$, draw the arrow diagram and decide if the relation is reflexive, symmetric, or transitive. If it is all three (i.e., an equivalence relation), give the corresponding partition of $\mathcal{P}(S)$.

(a) $R_1 = \{(A, B) \in \mathcal{P}(S) \times \mathcal{P}(S) : A \subseteq B\}$

(b) $R_2 = \{(A, B) \in \mathcal{P}(S) \times \mathcal{P}(S) : A \subset B\}$

(c) $R_3 = \{(A, B) \in \mathcal{P}(S) \times \mathcal{P}(S) : A \cap B = \emptyset\}$

(d) $R_4 = \{(A, B) \in \mathcal{P}(S) \times \mathcal{P}(S) : A \cap B \neq \emptyset\}$

(e) $R_5 = \{(A, B) \in \mathcal{P}(S) \times \mathcal{P}(S) : n(A) = n(B)\}$

9. To complete Example 7, prove that each of the following relations on $\mathbb{Z}$ is transitive:

(a) $T_1 = \{(a, b) \in \mathbb{Z} \times \mathbb{Z} : b - a$ is divisible by 5$\}$

(b) $T_2 = \{(a, b) \in \mathbb{Z} \times \mathbb{Z} : a^2 - b^2$ is divisible by 5$\}$

10. Let $A = \{0, 1, 2, 3, 4, 5, 6\}$ and $R = \{(x, y) \in A \times A : x - y$ is divisible by 3$\}$.

(a) Prove that R is reflexive.

(b) Prove that R is symmetric.

(c) Prove that R is transitive.

(d) What is the partition of A induced by the equivalence relation R?

11. Let $n \geq 2$ be given. Repeat the previous exercise for the relation

$$R = \{(x, y) \in \mathbb{Z} \times \mathbb{Z} : x - y \text{ is divisible by } n\}$$

on the set $A = \mathbb{Z}$.

12. Let the relation R on the set $E = \{$David, Mary, Doug, Rob, Don, Billie, John, Jessica, Andrew, Allen, Jacob, Christopher, Saache, Tommy$\}$ be given by

$$R = \{(a, b) : a \text{ is a son of } b\}$$

For those of you who do not know these people, this means

$R = \{$(David, Don), (Doug, Don), (Rob, Don), (David, Billie), (Doug, Billie), (Rob, Billie), (Andrew, Mary), (Allen, Mary), (Jacob, Mary), (John, Doug), (Christopher, Rob), (Tommy, David)$\}$

What is the meaning of the relation $R \circ R^{-1}$? Does the relation $R \circ R^{-1}$ have reflexive, symmetric, or transitive properties? If it has all three, give the partition of the set E induced by the relation $R \circ R^{-1}$.

13. Write one sentence about what property shows up in the arrow diagram

(a) Of a reflexive relation.

(b) Of a symmetric relation.

(c) Of a transitive relation.

14. Let $A = \{1, 2, 3\}$. Give an example of a relation R on A that is

(a) Transitive and reflexive but not symmetric.

(b) Symmetric and reflexive but not transitive.

(c) Symmetric and transitive but not reflexive.

15. Prove or give a counterexample to each of the following statements about symmetric relations:

(a) If R is a symmetric relation on A, then R^{-1} is a symmetric relation on A.

(b) If R is a symmetric relation on A, then $R \circ R$ is a symmetric relation on A.

(c) If R is any relation on A, then $R \circ R^{-1}$ is a symmetric relation on A.

(d) If R is any relation on A, then $R^{-1} \circ R$ is a symmetric relation on A.

16. Since the rule for a relation on A is technically a subset of $A \times A$, we can use the usual set operations to com-

bine relations on the same set. Let $A = \{0, 1, 2, 3, 4\}$ and define relations R_1 and R_2 by $R_1 = \{(0, 1), (1, 2), (0, 2), (3, 4), (4, 3)\}$ and $R_2 = \{(0, 0), (1, 2), (2, 1), (3, 4), (4, 3), (3, 3), (4, 4)\}$.

(a) What is $R_1 \cup R_2$?

(b) What is $R_1 \cap R_2$?

(c) What is $R_2 - R_1$?

(d) Find the smallest transitive relation R_1' such that $R_1 \subseteq R_1'$. (Such a relation R' is called the *transitive closure* of R.)

(e) Find the transitive closure of relation R_2.

17. Prove or give a counterexample to each of the following statements:

(a) For relations R_1 and R_2 on A, if R_1 is symmetric and $R_2 \subseteq R_1$, then R_2 is symmetric.

(b) For relations R_1 and R_2 on A, if R_1 is reflexive and $R_1 \subseteq R_2$, then R_2 is reflexive.

18. Give a counterexample to each of the following claims about the set $A = \{1, 2, 3, 4, 5\}$.

(a) If R_1 is a reflexive relation on A and R_2 is a symmetric relation on A, then the relation $R_1 \cap R_2$ is both reflexive and symmetric.

(b) If R_1 is a reflexive relation on A and R_2 is a symmetric relation on A, then the relation $R_1 \cup R_2$ is both reflexive and symmetric.

(c) If R_1 is a transitive relation on A and R_2 is a transitive relation on A, then the relation $R_1 \cup R_2$ is also transitive.

19. The relation $R \cup R^{-1}$ can lend insight into the properties of the original relation R.

(a) Let $A = \{0, 1, 2, 3\}$ and $R = \{(0, 1), (0, 0), (1, 2), (2, 1), (0, 3), (2, 2)\}$. What is $R \cup R^{-1}$?

(b) Is $R \cup R^{-1}$ always symmetric?

(c) What is special about $R \cup R^{-1}$ when the relation R is symmetric?

20. Prove each of the following statements for a given set A:

(a) If R_1 and R_2 are symmetric relations on A, then the relation $R_1 \cup R_2$ is symmetric.

(b) If R_1 and R_2 are symmetric relations on A, then the relation $R_1 \cap R_2$ is symmetric.

21. Prove for any relation R on a set A, if R is symmetric, then $R = R^{-1}$. (HINT: Show both $R \subseteq R^{-1}$ and $R^{-1} \subseteq R$ to establish the conclusion.)

22. Prove for any relation R on a set A, if $R = R^{-1}$, then R is symmetric.

23. Prove Theorem 2.

4.6 Numerical Functions in Discrete Math

In this chapter, we have shown that functions are more general than mere manipulations of numbers, but we cannot deny that certain numerical functions are very important in discrete math. We complete the chapter by using some of the more important numerical functions for specific applications.

Exponents and Logarithms

A question that frequently arises in discrete math concerns the size of a finite structure. This often comes up within the context of representing the structure in a computer (which of course has a finite amount of space), but it can be a central issue in mathematics applications as well. The simplest example of this is when the structure is simply a large integer.

 Example 1 *How many digits are in the number* $2^{1,000}$?

If $2^{1,000} = d_m 10^m + d_{m-1} 10^{m-1} + \cdots + d_1 10^1 + d_0$, then we know $10^m \le 2 < 10^{m+1}$. Hence, we need to know what power of 10 is approximately equal to $2^{1,000}$. The logarithm function does exactly this.

> **Definition** The base ten logarithm of a positive real number r is the real number p for which $10^p = r$. In this case, we write $\log_{10} r = p$.

Example 2 *Note that since* $10^0 = 1$, *we write* $\log_{10} 1 = 0$. *Other values require at least a little bit of work.*

1. *Since* $10^3 = 1,000$, *we write* $\log_{10} 1,000 = 3$.
2. *Since* $10^{0.69897} \approx 5$ *(try it on your calculator and see!), we write* $\log_{10}(5) \approx 0.69897$.
3. *Since* $10^{0.30103} \approx 2$ *(try it on your calculator and see!), we write* $\log_{10}(2) \approx 0.30103$.

We now know that we can find the number of digits in $2^{1,000}$ by calculating $\log_{10}(2^{1,000})$, but this is not entirely helpful if we have to calculate $2^{1,000}$ and then find the base ten logarithm of that answer—if we could calculate $2^{1,000}$, we would have already answered the question! However, if we think about the familiar rules for exponents in the right way, we can derive some properties of logarithms that will make finding the answer easy!

Theorem 1 *The following properties hold for all real numbers* x *and* y:

1. $(10^x)^y = 10^{x \cdot y}$
2. $10^x \cdot 10^y = 10^{x+y}$
3. $10^x / 10^y = 10^{x-y}$

We will not prove these properties here (although they are simple to derive for integer values of x and y as you will see in the exercises), but we will use them to derive the corresponding properties of the logarithm function.

Theorem 2 *The following properties hold for all positive real numbers a and b:*

1. $\log_{10}(a^b) = b \cdot \log_{10}(a)$
2. $\log_{10}(a \cdot b) = \log_{10}(a) + \log_{10}(b)$
3. $\log_{10}(a/b) = \log_{10}(a) - \log_{10}(b)$

PROOF We will derive only the first two parts, and we will leave the third part as a practice problem.

1. Let $c = \log_{10}(a)$. This means that $a = 10^c$, and so $a^b = (10^c)^b = 10^{b \cdot c}$ by rule #1 of Theorem 1. This means that $\log_{10}(a^b) = b \cdot c = b \cdot \log_{10}(a)$, as desired.
2. Let $x = \log_{10}(a)$ and $y = \log_{10}(b)$. Then $a = 10^x$ and $b = 10^y$, which means that

$$a \cdot b = 10^x \cdot 10^y$$
$$= 10^{x+y} \text{ by rule \#2 of Theorem 1}$$

But this means that $\log_{10}(a \cdot b) = x + y = \log_{10}(a) + \log_{10}(b)$, as desired.
3. See the practice problem 1 below.

■

We can now solve the problem given in Example 1.

SOLUTION In Example 2 above, we found that $\log_{10}(2) \approx 0.30103$. From rule #1 of Theorem 1,

$$\log_{10}(2^{1,000}) = 1,000 \cdot \log_{10}(2) \approx 1,000 \cdot (0.30103) = 301.03$$

Hence, $10^{301} < 2^{1,000} < 10^{302}$, so we conclude that $2^{1,000}$ is a 302-digit number.

□

Practice Problem 1 *Show how the third part of Theorem 2 can be derived from the third part of Theorem 1.*

The logarithm properties in Theorem 2 allow us to answer a number of different questions about the size of large integers.

 Example 3 *How many digits does* 1,000! *have?*

As before, the real problem here is to compute $\log_{10}(1,000!)$ without computing 1,000! itself first. We would like to take advantage of the product structure of 1,000!, so we investigate a simpler, similar problem. Rule #2 in Theorem 2 gives us a clue about how to find the $\log_{10}$ of a number that is expressed as a product. This is all we need to find the answer to the question in Example 3.

SOLUTION   Since $1{,}000! = 1{,}000 \cdot 999 \cdot 998 \cdots\cdots 4 \cdot 3 \cdot 2 \cdot 1$, property #2 of Theorem 2 can be used to write

$$\log_{10}(1{,}000!) = \log_{10}(1{,}000) + \log_{10}(999) + \log_{10}(998) + \cdots$$
$$+ \log_{10}(2) + \log_{10}(1)$$

The right-hand side of this equation involves summing 1,000 numbers, so it still seems computationally challenging, but this approach alleviates our biggest problem before in that each of the values on the right-hand side is very easy to compute with a calculator. For example, $\log_{10}(1{,}000) = 3$, $\log_{10}(999) \approx 2.9995655$, $\log_{10}(998) \approx 2.99913054$, and so on. To obtain the final answer, it would be easiest to get either a spreadsheet or a calculator involved.[*] The final answer is $\log_{10}(1{,}000!) \approx 2{,}567.6$, so we conclude that

$$10^{2,567} < 1{,}000! < 10^{2,568}$$

which means that 1,000! has 2,568 digits.   □

The only thing special about the base of ten is the connection to the number of digits that is tied to our base ten place value system of writing numerals. Since any positive number b can be used as a base for exponentiation, it follows that any such b can also be the base of a logarithm.

> **Definition**   For any positive real number $b \neq 1$, we write $\log_b r = p$ to mean that $b^p = r$, and we call p *the base b logarithm of r*.

Because the exponent rules in Theorem 1 work for bases other than ten, the same rules that apply to base ten logarithms also apply to logarithms in any base. It is arguable whether the most important base for discrete math is base two or base ten, but base two logarithms are certainly the most common type of logarithm in this particular text. Because computers represent numbers internally as base two numbers, then the number of base two digits in a number is typically more of a concern than base ten digits.

Example 4   *If your calculator displays 10 digits, what is the largest power of 2 that your calculator can display exactly? Explain how you can get the answer just from knowing that $\log_{10}(2) \approx 0.30103$.*

SOLUTION   The first 11-digit number is 10^{10}, so we need to know the largest value of m such that

$$2^m < 10^{10}$$

This is the same as saying that $\log_{10}(2^m) < 10$, which by rule #1 of Theorem 1 means the same thing as saying that $m \cdot \log_{10}(2) < 10$. Since we know

---

[*] On a TI-83 calculator, for example, the command `sum(seq(log₁₀(n),n,1,500))` yields the answer 1134.0864 . . . after a few dramatic seconds.

n	5	6	7	8	9	10
$n\log_2(n)$	11.6	15.5	19.7	24.0	28.5	33.2

Table 4-14 Table for Example 5

that $\log_{10}(2) \approx 0.30103$, this last inequality is roughly the same as saying $m \cdot 0.30103 < 10$, which is true of any $m < 33.22$. We conclude that 2^{33} is the largest power of 2 that is less than 10^{10}. □

Example 5 *Suppose you can represent positive integers up to $2^{31} - 1$. What is the largest value of n for which you can represent n^n exactly?*   (NOTE: $2^{31} - 1$ is the largest number that can be represented by a 32-bit integer in a computer, so your answer tells you the largest value of n for which n^n can be represented as such an integer.)*

SOLUTION We want to find the largest integer n with $n^n < 2^{31}$, so we want $\log_2(n^n) < 31$. But by the properties of logarithms, we know that $\log_2(n^n) = n\log_2(n)$, so we can fill in Table 4-14 to show values for this expression. This means that 9^9 is the largest number that can be represented. □

Practice Problem 2 *Suppose you can represent positive integers up to $2^{63} - 1$. What is the largest value of n for which you can represent n! exactly?*   (NOTE: $2^{63} - 1$ is the largest number that can be represented by a 64-bit integer in a computer, so your answer tells you the largest value of n for which n! can be represented as such an integer.)*

From Reals to Integers

Arguably, the most frequently used numerical functions in discrete math are the functions that convert fractional numbers to integers. We are taught very early on rules for rounding to the nearest whole number, but there are even more basic functions than this.

Definition We define the functions *floor* and *ceiling*, each as functions $\mathbb{R} \to \mathbb{Z}$, as follows:

- $\lfloor x \rfloor =$ the greatest integer less than or equal to x.
- $\lceil x \rceil =$ the least integer greater than or equal to x.

Example 6 *Table 4-15 shows how these rules work on a few different input values. From the definitions of the floor and ceiling functions, explain how these two functions are related to each other.*

SOLUTION We can describe the relationship in two cases as follows:

$$\lceil x \rceil = \begin{cases} \lfloor x \rfloor & \text{if } x \in \mathbb{Z} \\ \lfloor x \rfloor + 1 & \text{if } x \in \mathbb{R} - \mathbb{Z} \end{cases}$$

x	10.1	6	2.9	-1.5
$\lfloor x \rfloor$	10	6	2	-2
$\lceil x \rceil$	11	6	3	-1

Table 4-15 Table for Example 6

A more clever relationship is the following, which one can check is equivalent to the one above:

$$\text{For all } x \in \mathbb{R}, \lfloor x \rfloor = -\lceil -x \rceil$$

One of the common uses of the floor function is to find the number of multiples of a given value in a range of integers. The floor function naturally arises in this context.

Example 7 *How many multiples of 6 are in the set* $\{1, 2, 3, \ldots, 1{,}000\}$?

SOLUTION   The set of multiples of 6 can be written as

$$\{1 \cdot 6, 2 \cdot 6, 3 \cdot 6, 4 \cdot 6, \ldots, \underline{\hspace{1cm}} \cdot 6\}$$

so the question is simply about the largest integer _____ such that _____ $\cdot 6 \leq 1{,}000$. We can find this by dividing $1{,}000/6 = 166.666\ldots$, so we know that $166 \cdot 6 = 996 \leq 1{,}000$ but $167 \cdot 6 = 1{,}002 > 1{,}000$. Hence, there are $\lfloor 1{,}000/6 \rfloor$ mutliples of 6 in the set $\{1, 2, 3, \ldots, 1{,}000\}$.

This example can be easily generalized to give the following basic fact about the floor function:

Proposition 3   *Let k and n be positive integers. The number of multiples of k in the set of integers* $\{1, 2, 3, \ldots, n\}$ *is given by* $\lfloor n/k \rfloor$.

Practice Problem 3   *Use the floor function to express a formula for the number of multiples of k in the set* $\{m, m+1, m+2, \ldots, n\}$. (HINT: *How can you use the result of Proposition 3 directly?*)

Example 8 *Use the floor function to find the number of numbers in the set* $S = \{1, 2, 3, \ldots, 1{,}000\}$ *that are evenly divisible by 3 or 7.*

SOLUTION   Let $A = \{n \in S : n \text{ is divisible by 3}\}$ and $B = \{n \in S : n \text{ is divisible by 7}\}$. According to the inclusion-exclusion principle from Section 3.1,

$$n(A \cup B) = n(A) + n(B) - n(A \cap B)$$

But Proposition 3 tells us that $n(A) = \left\lfloor \frac{1{,}000}{3} \right\rfloor$ is simply the number of multiples of 3 in S, $n(B) = \left\lfloor \frac{1{,}000}{7} \right\rfloor$ is simply the number of multiples of 7 in S, and

$n(A \cap B) = \left\lfloor \frac{1,000}{21} \right\rfloor$ is simply the number of multiples of 21 in S. Hence,

$$n(A \cup B) = n(A) + n(B) - n(A \cap B)$$
$$= \left\lfloor \frac{1,000}{3} \right\rfloor + \left\lfloor \frac{1,000}{7} \right\rfloor - \left\lfloor \frac{1,000}{21} \right\rfloor$$
$$= 333 + 142 - 47 = 428$$

We end this brief discussion with an application of the floor function to another question about large integers.

Example 9 *How many consecutive 0's does the (base ten) number* 72! *have on its right end?*

SOLUTION   If we imagine the large number

$$72 \cdot 71 \cdot 70 \cdots 3 \cdot 2 \cdot 1$$

written as a product of its prime factors,[*] it will look something like this:

$$2^{p_1} \cdot 3^{p_2} \cdot 5^{p_3} \cdot 7^{p_4} \cdots$$

From this point of view, each consecutive 0 on the right end of this large number comes from one of the 2's and one of the 5's in this prime factorization. It seems sensible (and turns out to be true) that there are always more factors of 2 than factors of 5 in $n!$, so the number of 0's at the right end is exactly the same as the value of p_3 in the prime factorization above. Hence the question becomes, "What is the power of 5 in the prime factorization of 72!?"

Consider the numbers being multiplied to form 72!:

$$1 \cdot 2 \cdot 3 \cdot 4 \cdot \underline{5} \cdot 6 \cdot 7 \cdot 8 \cdot 9 \cdot \underline{10} \cdot 11 \cdots 64 \cdot \underline{65} \cdot 66 \cdot 67 \cdot 68 \cdot 69 \cdot \underline{70} \cdot 71 \cdot 72$$

Every fifth number is underlined in the product above, since each of these numbers (and no others) contributes a factor of 5 to the prime factorization of 72!. We cannot write them all in the space above, but we know there would be $\left\lfloor \frac{72}{5} \right\rfloor = 14$ underlined numbers. Those numbers in the list that are multiples of 25 each contribute an *additional* factor of 5 to the prime factorization. There are only two such numbers, 25 and 50, in this particular list. There are no multiples of 125 or higher powers of 5 in the list above, so the total number of factors of 5 in 72! is

$$\left\lfloor \frac{72}{5} \right\rfloor + \left\lfloor \frac{72}{25} \right\rfloor = 14 + 2 = 16$$

Therefore, there are 16 zeroes on the right end of 72!. If you have a computer algebra system, you can compute the 104-digit number

---

[*] The fact that such a representation exists was proven in Section 2.4.

$$72! = 6123445837688608686152 \cdots 689274204160000000000000000$$

to check that this is correct. □

Practice Problem 4   *How many consecutive 0's are on the right end of* 1,000!?

The process we used for the previous example answers a more general question about the prime factorization of factorial numbers, which we state as a proposition below. This is an important idea for some interesting number theoretic results concerning the distribution of prime numbers.

Proposition 4   *For a positive integer n and a prime number p, the power of p in the prime factorization of n! is given by*

$$\left\lfloor \frac{n}{p} \right\rfloor + \left\lfloor \frac{n}{p^2} \right\rfloor + \left\lfloor \frac{n}{p^3} \right\rfloor + \left\lfloor \frac{n}{p^4} \right\rfloor + \cdots$$

PROOF   See Exercise 26 at the end of this section.   ■

Composition of Numerical Functions

There are many reasons for studying the composition of functions. Perhaps the most practical reason comes from the way we solve numerical problems in real life. We always have some fixed set of tools like our knowledge of specific function properties, our calculator's fixed set of function keys, or a programming language with a given set of mathematical functions. In all these cases, we must be able to combine these basic tools as necessary to represent new functions or to carry out complex operations based on the simple ones.

Example 10   *A useful mathematical application of logarithms and integer functions is the calculation of the length of the base b representation of a number. Find an easy function to make this calculation, using the basic functions we have studied in this section.*

SOLUTION   The base b representation of a number n comes from the unique representation

$$n = d_k \cdot b^k + d_{k-1} \cdot b^{k-1} + \cdots + d_1 \cdot b + d_0$$

where each d_i is from $\{0, 1, 2, \ldots, b-1\}$. (These are the base b "digits" of n.) The length of this representation is $k+1$, but how is k related to n? Since $b^k \le n < b^{k+1}$, it follows that $k \le \log_b n < k+1$, and so $k = \lfloor \log_b n \rfloor$. We state our simple formula as the general proposition below.   □

Proposition 5   *For any integer $b \ge 2$, the number of digits in the base b representation of a positive integer n is given by the function*

$$\lfloor \log_b n \rfloor + 1$$

which is built out of the $\log_b$ function, the floor function, and addition.

Example 11 *To determine if a positive integer n is prime, we can simply try dividing the number by prime numbers less than n. However, you never need to check a prime number p greater than $\sqrt{n}$ since if $p > \sqrt{n}$ and $p \cdot m = n$, then $m < \sqrt{n}$. Find a function $f : \mathbb{N} \to \mathbb{N}$ built from the mathematical functions in this section such that $f(n)$ is the largest integer that is less than or equal to $\sqrt{n}$.*

SOLUTION   In general, only those $p < \sqrt{n}$ need to be checked, so the largest value that might ever need to be checked is $\lfloor \sqrt{n} \rfloor$. So we can use the function with the rule $f(n) = \lfloor \sqrt{n} \rfloor$.   □

Arithmetic Operations as Functions

There are many mathematical processes that can be viewed as functions, and this commonality often lends insight into properties shared by many processes. Ordinary arithmetic provides the most common examples.

Example 12 *The function $Sum : \mathbb{Z} \times \mathbb{Z} \to \mathbb{Z}$ defined by the rule*

$$Sum(i, j) = i + j$$

is the usual addition operation on integers. The function $Prod : \mathbb{Z} \times \mathbb{Z} \to \mathbb{Z}$ defined by the rule

$$Prod(i, j) = i \cdot j$$

is the usual multiplication operation on integers.
Write the distributive property of multiplication over addition

$$a \cdot (b + c) = (a \cdot b) + (a \cdot c)$$

in terms of these functions.

SOLUTION   For all $a, b, c \in \mathbb{Z}$,

$$Prod(a, Sum(b, c)) = Sum(Prod(a, b), Prod(a, c))$$

□

Notice that since the inputs are ordered pairs, the technically correct notation for an output of the function $Sum : \mathbb{Z} \times \mathbb{Z} \to \mathbb{Z}$ is $Sum((a, b))$ rather than $Sum(a, b)$, but we will consistently use the latter for simplicity.

Practice Problem 5 *Give a careful definition of the "difference" operation as a function from $\mathbb{Z} \times \mathbb{Z}$ to $\mathbb{Z}$, and express the following property in terms of this function: For all integers a, b, and c,*

$$(a - b) - c = (a - c) - b$$

Solutions to Practice Problems

1 *Proof.* Let $x = \log_{10}(a)$ and $y = \log_{10}(b)$. Then $a = 10^x$ and $b = 10^y$, which means that

$$a/b = 10^x/10^y$$
$$= 10^{x-y} \text{ by rule \#3 of Theorem 1}$$

But this means that $\log_{10}(a/b) = x - y = \log_{10}(a) - \log_{10}(b)$, as desired. ∎

2 We will have to use trial and error to see when $\log_2(n!)$ first reaches 63. Table 4-16 shows values of $\log_2(n!)$ computed (using technology) as the sum

$$\sum_{i=1}^{n} \log_2(i) = \log_2(1) + \log_2(2) + \log_2(3) + \cdots + \log_2(n)$$

From this we see that $2^{61} < 20! < 2^{63} < 21!$, so 20! is the largest factorial that can be represented using a 64-digit binary number.

n	5	10	15	20	21	22
$\log_2(n!)$	6.9	21.8	40.3	61.1	65.5	69.9

Table 4-16   Solution to Practice Problem 2

3 The number of multiples of k in the set $\{m, m+1, \ldots, n\}$ is the number of multiples of k in the set $\{1, 2, \ldots, n\}$ less the number of multiples of k in the set $\{1, 2, \ldots, m-1\}$. According to Proposition 3, this is

$$\lfloor n/k \rfloor - \lfloor (m-1)/k \rfloor$$

4 To account for every factor of 5 in the list of numbers from 1 to 1,000, we add the number of multiples of 5, which is $\lfloor 1{,}000/5 \rfloor = 200$; the number of multiples of 25, which is $\lfloor 1{,}000/25 \rfloor = 40$; the number of multiples of 125, which is $\lfloor 1{,}000/125 \rfloor = 8$; and the number of multiples of 625, which is $\lfloor 1{,}000/625 \rfloor = 1$. Therefore, the exponent of 5 in the prime factorization of 1,000! is

$$\lfloor 1{,}000/5 \rfloor + \lfloor 1{,}000/25 \rfloor + \lfloor 1{,}000/125 \rfloor + \lfloor 1{,}000/625 \rfloor$$
$$= 249$$

Hence, 1,000! ends with 249 zeroes.

5 We can define

$$dif : \mathbb{Z} \times \mathbb{Z} \to \mathbb{Z}$$

with the rule $dif(x, y) = x - y$. The given property then is stated as follows: for all $a, b, c \in \mathbb{Z}$,

$$dif(dif(a, b), c) = dif(dif(a, c), b)$$

Exercises for Section 4.6

1. Fill in Table 4-17 to support the relationship given in Example 6.

2. For each of the following, indicate whether you believe the statement is true for all real numbers x and y, and give a counterexample if not.
   (a) $\lfloor x \rfloor + \lfloor y \rfloor = \lfloor x + y \rfloor$
   (b) $\lceil 2x \rceil = 2 \lceil x \rceil$
   (c) $\lfloor x \rfloor + \lceil x \rceil = 2 \lfloor x \rfloor$
   (d) $-\lfloor x \rfloor = \lceil -x \rceil$
   (e) If $x \geq 0$, then $\lfloor x + 0.5 \rfloor = \lceil x - 0.5 \rceil$.

x	$-x$	$\lfloor -x \rfloor$	$-\lfloor -x \rfloor$	$\lceil x \rceil$
0.5				
10.4				
−5.3				
4				
−17				

Table 4-17   Table for Exercise 1

3. Suppose you know that x is a positive real number, and you would like to build a function that returns the nearest integer to x. Call this function $round(x)$, and show how it is related to the floor function. (If x is equally near two integers, your function should return the higher number.)

4. How many digits does the number 3^{100} have?

5. How many digits does the number $3^{1{,}000}$ have?

6. How many digits does the number $9^{1{,}000}$ have?

7. Fermat numbers are of the form $2^{2^n} + 1$. How many digits (in base ten) does the fifth Fermat number have? The tenth?

8. Let $a_0 = 1$, and $a_n = 2^{a_{n-1}}$ for all $n \geq 1$. How many digits does a_5 have?

9. Explain why the third part of Theorem 2 follows from the first two parts.

10. Investigate the relationship between the number of digits of 2^n and the number of digits of 3^n. Prove the relationship using the properties in Theorem 2.

11. Use the definition of logarithms to explain why the following equation is true:

$$\log_{10} 2 = \frac{1}{\log_2 10}$$

12. Investigate the relationship between the value of $\log_{10} n$ and $\log_n 10$. Prove the relationship using the properties in Theorem 2.

13. How many positive integers less than or equal to 2,000 are divisible by

   (a) 3?

   (b) 5?

   (c) 7?

14. How many positive odd numbers less than 2,000 are divisible by 3?

15. How many positive integers less than or equal to 2,000 are divisible by

   (a) 3 or 5?

   (b) 3 or 7?

   (c) 2 or 5?

   (d) 2 or 7?

   (e) 2, 3, or 5?

16. How many positive odd numbers less than 2,000 are divisible by 3 or 5?

17. Let's call an integer greater than 1 a *faux prime* if it is not divisible by 2, 3, 5, or 7.

   (a) How many positive integers less than or equal to 2,000 are divisible by 2, 3, 5, or 7?

   (b) What percentage of all positive numbers less than or equal to 2,000 are faux primes?

   (c) Use a computer program or a table of primes to find the number of primes less than or equal to 2,000.

   (d) What percentage of the faux primes are really primes?

   (e) Discuss what the previous answer implies about checking numbers less than or equal to 2,000 for being prime.

18. Explain why 100! has two more digits than 99!.

19. How many base ten digits does 100! have?

20. How many base ten digits does 500! have?

21. What is the smallest value of n for which $n!$ exceeds 10,000 digits in length? (Investigate for several values and try to discern a pattern.)

22. How many consecutive 0's occur at the right-hand side of the number 100!?

23. How many consecutive 0's occur at the right-hand side of the number 1,985!?

24. If 3^p divides evenly into 1,992!, what is the largest value that p can have?

25. If 2^p divides evenly into 1,992!, what is the largest value that p can have?

26. Give a proof of Proposition 4 for the case $p = 2$. (You may use the fact that if 2 divides the product $a \cdot b$, then 2 must divide at least one of the numbers a or b.)

27. The following exercises use ideas from Section 2.6.

   (a) If 500! is written using its binary representation, how many consecutive 0's occur on the right-hand side?

   (b) If 500! is written using its hexadecimal representation, how many consecutive 0's occur on the right-hand side?

   (c) If 500! is written using its base six representation, how many consecutive 0's occur on the right-hand side?

28. The following exercises use ideas from Section 2.6

   (a) If 500! is written using its binary representation, how many digits will be needed?

   (b) If 500! is written using its hexadecimal representation, how many digits will be needed?

   (c) If 500! is written using its base six representation, how many digits will be needed?

29. According to the solution for Example 11, to determine if the number 1,361 is prime, we only have to check 1,361 for divisibility by all prime numbers less than $\lfloor\sqrt{1{,}361}\rfloor$. List all the numbers that need to be checked.

30. Repeat the previous exercise using 2,003 instead of 1,361.

31. Let $Pow : \mathbb{R}^+ \times \mathbb{R}^+ \to \mathbb{R}$ be the function with the rule $Pow(x, y) = x^y$. Using this along with the functions defined in Example 12, which of the following are true? For each that is false, provide a specific counterexample.

   (a) For all $x, y, z \in \mathbb{R}^+$, $Pow(Prod(x, y), z)$ $= Prod(Pow(x, z), Pow(y, z))$.

   (b) For all $x, y, z \in \mathbb{R}^+$, $Prod(Pow(x, y), z)$ $= Pow(Prod(x, z), Prod(y, z))$.

   (c) For all $x, y, z \in \mathbb{R}^+$, $Pow(Pow(x, y), z)$ $= Pow(x, Prod(y, z))$.

   (d) For all $x \in \mathbb{R}^+$, $Prod(2, Pow(2, x))$ $= Pow(2, Sum(x, 1))$.

32. Using the functions from the previous exercise and Practice Problem 5, write each of the following facts using the functions $Prod$, Sum, Dif, and/or Pow:

   (a) For all $x, y, z \in \mathbb{R}^+$, $x^{y+z} = x^y \cdot x^z$.

   (b) For all $x, y, z \in \mathbb{R}^+$, $\left(x^2\right)^y = x^{2 \cdot y}$.

   (c) For all $x, y, z \in \mathbb{R}^+$, if $x \leq y$, then

$$z^x + z^y = z^x \cdot \left(1 + z^{y-x}\right)$$

33. Let $Quo : \mathbb{Z} \times \mathbb{Z}^+ \to \mathbb{Z}$ be the function with the rule $Quo(x, y) = \lfloor x/y \rfloor$, and let $Rem : \mathbb{Z} \times \mathbb{Z}^+ \to \mathbb{Z}$ be the function with the rule $Rem(x, y) = \lfloor x \bmod y \rfloor$. Which of the following (which also use the functions from Example 12) are true statements? For each that is false, provide a specific counterexample.

(a) For all $x, y \in \mathbb{Z}^+$, $x = Prod(Quo(x, y), y)$.

(b) For all $x, y, z \in \mathbb{Z}^+$, $Quo(Sum(x, y), z)$
$= Sum(Quo(x, z), Quo(y, z))$.

(c) For all $x, y, z \in \mathbb{Z}^+$, $Rem(Sum(x, y), z)$
$= Sum(Rem(x, z), Rem(y, z))$.

(d) For all $x, y \in \mathbb{Z}^+$, $x = Sum(Prod(Quo(x, y), y), Rem(x, y))$.

4.7 Excursion: Iterated Functions and Chaos

Magic Tricks

Functions play a central role in the relatively new mathematical fields of fractals and chaos theory. Believe it or not, the concept of composition of functions is even present in some magic tricks.

 Example 1 *A magician gives a "volunteer" the following instructions:*

1. *Think of any positive integer.*
2. *Count how many letters it takes to write your number in English. The number of letters is your new number.*
3. *Repeat step 2 until you get the same number twice in a row.*

At this point the magician announces that the volunteer's current number is 4. Why does this trick work?

SOLUTION   Step 2 of the instructions gives rise to the following function definition: $f : \mathbb{N} \to \mathbb{N}$ with the rule

$$f(n) = \text{the number of letters in the English expression for } n$$

If we call the original chosen number a_0, then we are forming a sequence of natural numbers

$$a_0, a_1, a_2, a_3, \ldots$$

with $a_1 = f(a_0)$, $a_2 = f(a_1)$, $a_3 = f(a_2)$, and so on. The entire diagram for the function f cannot be shown since the domain is the infinite set $\mathbb{N}$; however, part of the diagram is shown in Figure 4-64.

The "loop" near the dot for 4 indicates an arrow from 4 to itself. It certainly seems apparent that for values of n greater than 14 (i.e., those not shown), $f(n) < n$. This is, in fact, a true statement, but it is based more on properties of the English language than on mathematics. This along with the diagram is enough to explain why the only value of n for which $f(n) = n$ is $n = 4$, but is this enough to explain why the trick always works? ☐

The next example shows that not every trick like this works out so well.

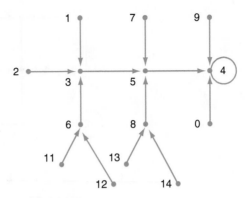

Figure 4-64 Partial arrow diagram for Example 1.

Letters	A, E, I, O, U, N, L, S, T, R	D, G, P	M, C	W, F, H, V, Y	K	B	J, X	Q, Z
Score	1	2	3	4	5	6	8	10

Table 4-18 Table for Example 2

Example 2 *According to the official rules of Scrabble,®* scores are assigned to letters as shown in Table 4-18. Suppose the trick in Example 1 is revised so that instead of getting the new number from the length of the English word, the letters in the words for each number are scored according to the rules in Scrabble® (using the face values of the tiles and without worrying whether there are enough standard tiles to spell each word), and this score is the new number.*

For example, if the original number is 87, then

- *"Eighty-seven" has a score of 21.*
- *"Twenty-one" has a score of 15.*
- *"Fifteen" has a score of 13.*
- *"Thirteen" has a score of 11.*
- *"Eleven" has a score of 9.*
- *"Nine" has a score of 4.*
- *"Four" has a score of 7.*
- *"Seven" has a score of 8.*
- *"Eight" has a score of 9.*

At this point the values will cycle 9, 4, 7, 8, 9, 4, 7, . . . and hence never settle on a fixed value. What does the diagram for the function g : $\mathbb{N} \to \mathbb{N}$ with the rule

$$g(n) = \text{the Scrabble}^® \text{ score for the English expression for } n$$

look like?

---

* Scrabble® ©2005 Hasbro, Inc. Used with permission.

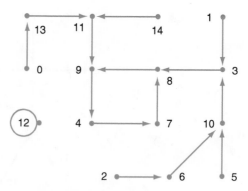

Figure 4-65 Partial arrow diagram for Example 2.

SOLUTION   In the partial diagram shown in Figure 4-65, we can see the pattern $9, 4, 7, 8, 9, 4, 7, \ldots$ reflected in the "cycle" of arrows between these numbers. So although there is a number n (namely 12) for which $g(n) = n$, sometimes we never get to it in our sequence. For this reason, using Scrabble® scores makes for a pretty poor magic trick. $\square$

The idea of repeating the application of a function to form a sequence arises in a variety of otherwise unrelated mathematical problems. We call such a sequence an *iterated function sequence*, and it depends only on the choice of the first term a_0 and the function itself.

Definition   Given a function f on a set A and an initial term $a_0 \in A$, the *iterated function sequence** for f starting at a_0 is

$$a_0, a_1, a_2, \ldots$$

where $a_{n+1} = f(a_n)$. In the event that the sequence starts to repeat—that is, $a_{n+k} = a_n$ for some smallest possible k—we say that $a_n, a_{n+1}, a_{n+2}, \ldots, a_{n+k}$ is a *cycle with period k*.

The preceding examples use functions on numbers, but in a decidedly unmathematical way. We would like to see how we might use mathematics to settle some questions about these iterated function sequences.

 Example 3   *Let $f : \mathbb{N} \to \mathbb{N}$ be defined by*

$$f(n) = \begin{cases} \frac{n}{2} & \text{if } n \text{ is even} \\ n + 3 & \text{if } n \text{ is odd} \end{cases}$$

---

* In the mathematics literature, this is often called the *orbit of a_0 under f*, but we prefer the language here since it emphasizes the role of function iteration in the sequence.

a_0	Iterated Function Sequence
1	$1, 4, 2, 1, 4, 2, \ldots$
2	$2, 1, 4, 2, 1, 4, \ldots$
3	$3, 6, 3, 6, \ldots$
4	$4, 2, 1, 4, 2, 1, \ldots$
5	$5, 8, 4, 2, 1, 4, 2, 1, \ldots$
6	$6, 3, 6, 3, \ldots$
7	$7, 10, 5, 8, 4, 2, 1, 4, 2, 1, \ldots$
8	$8, 4, 2, 1, 4, 2, 1, \ldots$
9	$9, 12, 6, 3, 6, 3, \ldots$

Table 4-19 Solution to Example 3

Show that every iterated function sequence for f with $a_0 \in \mathbb{Z}^+$ either ends with the cycle $3, 6, 3, 6, \ldots$ or the cycle $1, 4, 2, 1, 4, 2, \ldots$.

SOLUTION   Let's first explore the problem a bit, and then we will write a proof. Table 4-19 shows the sequences that arise from various choices of a_0. Notice that as we form each new sequence, it eventually looks like one that we formed earlier. This should remind you of the principle of mathematical induction. Let's see how to prove properties of these sequences with this tool.

Let $P(n)$ be the statement "The iterated function sequence with the first term $a_0 = n$ eventually ends with the cycle $3, 6, 3, 6, \ldots$ or the cycle $1, 4, 2, 1, 4, 2, \ldots$." The table above verifies statements $P(1), P(2), \ldots, P(9)$. Now let $k \geq 9$ be given, assume that statements $P(1), P(2), \ldots, P(k)$ have been verified, and consider statement $P(k + 1)$. That is, we are considering the iterated function sequence with the first term $a_0 = k + 1$. There are two possibilities based on the definition of f:

- If $a_0 = k + 1$ is even, then the iterated function sequence looks like $k + 1, \frac{k+1}{2}, \ldots$. Since $\frac{k+1}{2} < k + 1$, statement $P(\frac{k+1}{2})$ has already been verified, so we know that the iterated function sequence beginning with $\frac{k+1}{2}$ eventually ends with the cycle $3, 6, 3, 6, \ldots$ or the cycle $1, 4, 2, 1, 4, 2, \ldots$. This means that the iterated function sequence beginning with $k + 1$ eventually ends with the cycle $3, 6, 3, 6, \ldots$ or the cycle $1, 4, 2, 1, 4, 2, \ldots$.

- If $a_0 = k + 1$ is odd, then $a_1 = a_0 + 3 = k + 4$ is even, so the iterated function sequence looks like $k + 1, k + 4, \frac{k+4}{2}, \ldots$. Since $k > 3$, it follows that $\frac{k+4}{2} < k + 1$, so statement $P(\frac{k+4}{2})$ has already been verified. Hence, we know that the iterated function sequence beginning with $\frac{k+4}{2}$ eventually ends with the cycle $3, 6, 3, 6, \ldots$ or the cycle $1, 4, 2, 1, 4, 2, \ldots$. This means that the iterated function sequence beginning with $k + 1$ eventually ends with the cycle $3, 6, 3, 6, \ldots$ or the cycle $1, 4, 2, 1, 4, 2, \ldots$.

This completes the induction. $\square$

Practice Problem 1   *Let $f : \mathbb{N} \to \mathbb{N}$ be defined by*

$$f(n) = \begin{cases} \frac{n}{2} & \text{if } n \text{ is even} \\ n + 1 & \text{if } n \text{ is odd} \end{cases}$$

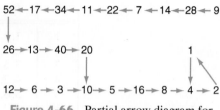

Figure 4-66 Partial arrow diagram for Example 4.

Show that every iterated function sequence for f with $a_0 \in \mathbb{Z}^+$ ends with the cycle $1, 2, 1, 2, 1 \ldots$.

Not every function is as easy to analyze as the two we have seen so far. The following example is called the Collatz problem, the Syracuse problem, or Ulam's problem depending on whom you ask. It is one of the best-known unsolved problems in mathematics.

 Example 4 *Let $g : \mathbb{N} \to \mathbb{N}$ be defined by*

$$g(n) = \begin{cases} \frac{n}{2} & \text{if } n \text{ is even} \\ 3n + 1 & \text{if } n \text{ is odd} \end{cases}$$

Does the iterated function sequence for g always end at the cycle $4, 2, 1, 4, 2, 1, \ldots$?

SOLUTION   Surprisingly, no one knows. A partial diagram for this function is shown in Figure 4-66. Even though many people have extended this picture to huge numbers of terms, the question remains open.   □

Chaos in Dynamical Systems

There are a variety of definitions of the term "chaos" in mathematics and physics, but one thing that they all imply is that chaos lies on the boundary of order and randomness. We will investigate this broad claim in the remainder of this section.

Example 5   *Let $f : \mathbb{R} \to \mathbb{R}$ be defined by $f(x) = x^2 - 0.8$, and consider the iterated function sequence of f with the first term $a_0 = 0.5$ as shown in Table 4-20. Explain this pattern using simple algebra.*

SOLUTION   It looks like the terms a_n with index n even are settling in on one value near -0.7236, while the terms a_n with odd index n are settling in on a value near -0.2764. The sequence is defined by repeatedly applying the function f, so $a_{n+2} = f(a_{n+1}) = f(f(a_n))$. This is the relationship between successive even-indexed terms as well as between successive odd terms. If these really are settling on some value v, then it must be the case that

$$\begin{aligned} v &= f(f(v)) \\ &= f(v^2 - 0.8) \\ &= (v^2 - 0.8)^2 - 0.8 \\ &= v^4 - 1.6v^2 - 0.16 \end{aligned}$$

n	0	1	2	3	4	5	6
a_n	0.5	−0.55	−0.4975	−0.5525	−0.4948	−0.5552	−0.4917

n	7	8	9	10	11	12
a_n	−0.5582	−0.4884	−0.5615	−0.4848	−0.5650	−0.4808

$$\vdots$$

n	494	495	496	497	498	499
a_n	−0.7236	−0.2764	−0.7236	−0.2764	−0.7236	−0.2764

Table 4-20 Table for Example 5

The equation $v = v^4 - 1.6v^2 - 0.16$ is the same (by factoring) as

$$\left(v^2 - v - \frac{4}{5}\right)\left(v^2 + v + \frac{1}{5}\right) = 0$$

We can now use the quadratic formula (twice) to conclude that either $v = \frac{1}{2} - \frac{1}{10}\sqrt{105}$, $v = \frac{1}{2} + \frac{1}{10}\sqrt{105}$, $v = -\frac{1}{2} + \frac{1}{10}\sqrt{5}$, or $v = -\frac{1}{2} - \frac{1}{10}\sqrt{5}$. These four possible values of v are approximately $v \approx -0.5246951$, $v \approx 1.524695$, $v \approx -0.2763932$, and $v \approx -0.7236068$. The last two are those that showed up in our numerical investigation above. This leads to a natural question about why the other values of v did not show up. The answer is discovered by noticing that the above $v^2 - v - \frac{4}{5}$ factor being equal to 0 is precisely the same thing as $v^2 - \frac{4}{5} = v$. In other words, $v = \frac{1}{2} - \frac{1}{10}\sqrt{105}$ and $v = \frac{1}{2} + \frac{1}{10}\sqrt{105}$ are the values that satisfy $f(v) = v$. So if we take as the first term $x_0 = \frac{1}{2} - \frac{1}{10}\sqrt{105}$, the iterated function sequence will be just that constant value forever, and likewise for the first term $x_0 = \frac{1}{2} + \frac{1}{10}\sqrt{105}$.

The three cycles for this function with period 2 or 1 are

$$-0.7236068, -0.2763932, -0.7236068, -0.2763932, \ldots,$$
$$-0.5246951, -0.5246951, -0.5246951, \ldots, \text{ and}$$
$$1.524695, 1.524695, 1.524695, \ldots$$

To see that this function produces no cycles of period 3, we can simply look for the roots of $y = f(f(f(x))) - x$ on its graph shown in Figure 4-67.

We can see that the only two roots are at approximately -0.5246951 and 1.5246951, but these initial terms produced our cycles of period 1. Since there are no other roots, there are no cycles of period 3. ☐

Practice Problem 2 *Find all cycles with period 1 or 2 for the iterated function sequence of* $g(x) = 3x - \frac{1}{x}$.

This seems to use just a little computation and algebra, so the preceding comment about chaotic behavior lying between order and randomness no doubt seems strange. One way this manifests itself in this example lies within the computations themselves. The word "compute" refers to numerical computation devices such as computers or calculators. To see the problem with these devices for chaotic systems, you will need one of them for some experimentation.

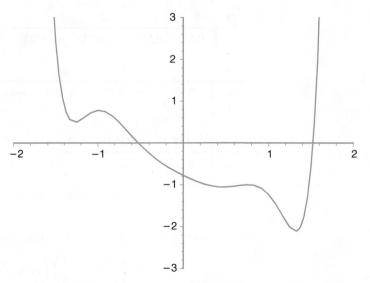

Figure 4-67 Graph of $y = f(f(f(x))) - x$ in Example 5.

Example 6 *Let $f : \mathbb{R} \to \mathbb{R}$ be the same function as in the previous example, and consider the iterated function sequence for f with the first term $a_0 = 1.5246$ (i.e., a little less than the positive value that gave us a cycle of period 1) in Table 4-21. So we see that even though we start the sequence very close to the value where we will get a cycle of period 1, the sequence, is in fact attracted to the cycle of period 2.*

Now consider the iterated function sequence for the same function f but this time with the first term $a_0 = 1.5247$, shown in Table 4-22. This is a little more than the previous initial value, and a very little bit more than the positive value that gave us a cycle of period 1. As you can see, the terms of this sequence quickly exceed the bounds of a calculator. The point is that this is very different from the previous sequence, even though the initial terms for the two sequences were almost identical.

The phenomenon illustrated in the previous example is called "sensitivity to initial conditions," and it is an important property in all studies of chaos theory. It basically means that there can be very different long-term behavior in two systems whose initial terms are almost identical.

The most famous adage about chaos in physical system is attributed to Edward Lorenz (1917–) who wrote, "One meteorologist remarked that if the theory were

n	0	1	2	3	4	5	6
a_n	1.5246	1.5244	1.5238	1.5220	1.5165	1.4997	1.4992

$$\vdots$$

n	94	95	96	97	98	99
a_n	−0.7236	−0.2764	−0.7236	−0.2764	−0.7236	−0.2764

Table 4-21 First Table for Example 6

n	0	1	2	3	4	5	6
a_n	1.5247	1.5247	1.5247	1.5248	1.5251	1.5260	1.5287

n	7	8	9	10	11	12	13
a_n	1.5368	1.5617	1.6390	1.8863	2.7580	6.8063	45.526

n	14	15	16	17	18
a_n	$2,071.8$	4.3×10^6	1.8×10^{13}	3.4×10^{26}	1.1×10^{53}

Table 4-22   Second Table for Example 6

n	0	1	2	3	4	5	6
a_n	0.5	−1.75	1.06	−0.87	−1.24	−0.46	−1.79

n	7	8	9	10	11	12	13
a_n	1.20	−0.56	−1.69	0.85	−1.28	−0.35	−1.88

Table 4-23   Solution to Example 7

correct, one flap of a seagull's wings would be enough to alter the course of the weather forever." Over time Lorenz's seagull became a butterfly, and sensitivity to initial conditions became known as the *butterfly effect*.

The implication of sensitivity to initial conditions is very important in computation. Since digital computers cannot represent most numbers exactly, most values stored in a computer are only approximations to an exact value. For many computations, using an approximation that is within, say, 10^{-12}, of the exact value is good enough, but for iterated function systems, as we saw in the previous example, even this much difference in initial conditions can cause incredibly different results in the long term.

Sensitivity to initial conditions is important in the interpretation of chaos in most disciplines, but it may seem too dramatic to say this property lies "between order and randomness" as we did at the beginning of this section. We will close this section with two examples that perhaps better illustrate the meaning of this puzzling statement.

Example 7 *Examine the iterated function sequences for the function $f : \mathbb{R} \to \mathbb{R}$ defined by $f(x) = x^2 - 2$.*

SOLUTION   Table 4-23 shows values of the iterated function sequence using $a_0 = 0.5$. This sequence not only looks "random" for these small values of n, they do not settle down in any sense even when n is quite large, as Table 4-24 illustrates.   □

n	1,000	1,001	1,002	1,003	1,004
a_n	−1.90	1.62	0.63	−1.60	0.57

Table 4-24   The Sequence for Larger Values of n

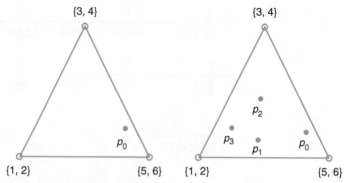

Figure 4-68 The chaos game in Example 8.

This example shows that it is possible to get "random" behavior out of a perfectly determined iterative process. The next example illustrates the position of chaos between order and randomness by iterating a perfectly random process only to find a great deal of order in the result. The "game" is usually attributed to Michael Barnsley.

Example 8 (The Chaos Game) *Begin with a triangle with an initial point p_0 as shown in Figure 4-68 on the left.* * *Roll a fair six-sided die, and mark the point p_1 as follows:*

● *If the outcome is 1 or 2, draw p_1 halfway along an invisible line segment connecting p_0 with the vertex labeled {1, 2}.*

● *If the outcome is 3 or 4, draw p_1 halfway along an invisible line segment connecting p_0 with the vertex labeled {3, 4}.*

● *If the outcome is 5 or 6, draw p_1 halfway along an invisible line segment connecting p_0 with the vertex labeled {5, 6}.*

Repeating this process with each successive point (i.e., forming point p_{i+1} from point p_i with a roll of the die) will give a sequence of points that slowly fill in the triangle. The picture on the right of Figure 4-68 shows the first four points in the simulation of three die tosses resulting in rolls of 2, then 3, and then 1. What will the long-term behavior of this process be, and what will the picture eventually look like?

SOLUTION   The three pictures in Figure 4-69 show the result of this process after 100 die rolls, 1,000 die rolls, and 10,000 die rolls.

This figure is called the Sierpinski triangle, and it is one of the most famous of a class of images called "fractals" that are closely related to chaos. Notice the self-similiarity of parts of the Sierpinski triangle to the whole picture. This is one of the defining properties of fractals.   □

These two examples show that there can be randomness from order and order from randomness. This is the essence of chaos. Current research in chaos has had a profound impact on the way we model real events and the techniques we use

---

* The location of the initial point can be proven to be irrelevant to the impending outcome.

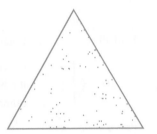 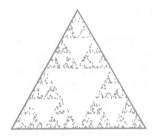 

Figure 4-69   More iterations of the chaos game.

for numerical computation. On a theoretical level, the study of chaos has found applications ranging from astronomy to financial markets. It is surely one of the most important mathematical developments of the last century.

Solutions to Practice Problems

1   After investigating the behavior of the iterated function system $\{a_n\}$ for f, we make the following observation, which we can prove by induction:

Claim   For all $n \in \mathbb{Z}^+$, the iterated function sequence for f with the first term $a_0 = n$ eventually ends with the cycle $1, 2, 1, 2, \ldots$.

Proof by induction on n.   Let $P(n)$ be the statement "The iterated function sequence for f with the first term $a_0 = n$ eventually ends with the cycle $1, 2, 1, 2, \ldots$." First let's look at the sequences that arise from various choices of a_0 in Table 4-25. Let $m \geq 10$ be given such that statements $P(1), P(2), \ldots, P(m-1)$ have all been checked to be true. Now consider the iterated function sequence for f with the first term $a_0 = m$. We know that m is either even or odd, so we argue in two cases.

● **Case 1:** If m is even, then the sequence looks like $m, \frac{m}{2}, \ldots$. Since $\frac{m}{2} < m$, we have already checked

a_0	Iterated Function Sequence
1	$1, 2, 1, 2, 1, 2, \ldots$
2	$2, 1, 2, 1, 2, 1, \ldots$
3	$3, 4, 2, 1, \ldots$
4	$4, 2, 1, \ldots$
5	$5, 6, 3, 4, 2, 1, \ldots$
6	$6, 3, 4, 2, 1, \ldots$
7	$7, 8, 4, 2, 1, \ldots$
8	$8, 4, 2, 1, \ldots$
9	$9, 10, 5, 6, 3, 4, 2, 1, \ldots$

Table 4-25   Solution to Practice Problem 1

statement $P\left(\frac{m}{2}\right)$, so we know the sequence that the iterated function sequence for f begins with $\frac{m}{2}$ eventually ends with the cycle $1, 2, 1, 2, \ldots$. Hence, the iterated function system that starts with $a_0 = m$ must eventually end with the cycle $1, 2, 1, 2, \ldots$.

● **Case 2:** If m is odd, then the sequence looks like $m, m+1, \frac{m+1}{2}, \ldots$. Since $\frac{m+1}{2} < m$ for values of $m \geq 10$, we have already checked statement $P\left(\frac{m+1}{2}\right)$, so we know the sequence that the iterated function sequence for f begins with $\frac{m+1}{2}$ eventually ends with the cycle $1, 2, 1, 2, \ldots$. Hence, the iterated function system that starts with $a_0 = m$ must eventually end with the cycle $1, 2, 1, 2, \ldots$.   ∎

2   To find cycles with period 1, we simply solve the equation $g(x) = x$, which we can do after rewriting the equation $3x - \frac{1}{x} = x$ as $2x^2 - 1 = 0$, which has solutions $x = \pm\frac{\sqrt{2}}{2}$. These two values constitute cycles with period 1.

To find cycles with period 2, we solve the equation $g(g(x)) = x$, which we can do after determining that

$$g(g(x)) = \frac{27x^4 - 19x^2 + 3}{3x^3 - x}$$

and so the equation $g(g(x)) = x$ is equivalent to the equation

$$27x^4 - 19x^2 + 3 = 3x^4 - x^2$$

and we can factor $24x^4 - 18x^2 + 3$ as $3(2x+1)(2x-1)(2x^2-1)$. Hence, $g(g(x)) = x$ when $x = -1/2$, $x = 1/2$, or $x = \pm\frac{\sqrt{2}}{2}$. We have seen that these last two values are actually cycles of g with period 1, so only the values $\pm 1/2$ constitute cycles with period 2.

Exercises for Section 4.7

1. Using the values $A = 1$, $B = 2$, $C = 3$, and so on, we can assign a number to every English word by adding the values of the letters in the word. We will call this the "ordinal value" of a word. With this in mind, we can define the function $f : \mathbb{N} \to \mathbb{N}$ by

$$f(n) = \text{the ordinal value for the English expression for } n$$

   (a) Evaluate $f(1)$ and $f(16)$.

   (b) Give a specific example to show that f is not onto.

   (c) The function f has only one cycle, $216 \to 228 \to 228 \to 255 \to \underline{\quad} \to 216$. What number goes in the blank?

   (d) Explain why this function would not make much of a magic trick.

2. (From [17]) A deck of 32 cards is numbered top to bottom from 1 to 32. The magician slowly deals the cards into three face-up piles, the spectator silently chooses one card and remembers which pile it falls into. After all cards are dealt (into piles numbering 11, 11, and 10), the spectator indicates into which pile his card fell, and the magician sandwiches this pile between the other two, turning all three face-down again in his hand and taking care to put one of the piles of 11 on top of the stack. The deal is repeated, this time with the spectator merely silently noting which pile his card is in and sharing this information after the deal has been completed. The same reassembly of the deck is performed. This is repeated until, after the fourth deal, the magician reveals that the chosen card is in the 17th position of the deck.

   (a) In the magic trick described, let a_0 denote the number (between 1 and 32) of the original card, and a_n the position of the chosen card after n deals. Explain in English why the sequence $a_0, a_1, a_2, \ldots$ is an iterated function sequence for the function $f : \{1, 2, 3, \ldots, 32\} \to \{1, 2, 3, \ldots, 32\}$ given by

$$f(x) = 11 + \left\lceil \frac{x}{3} \right\rceil$$

   (b) Give a complete diagram for the function f, and use your diagram to explain why the trick works.

3. Find all cycles for the function $f : \mathbb{N} \to \mathbb{N}$ defined by

$$g(n) = \begin{cases} \frac{n}{2} & \text{if } n \text{ is even} \\ n + 5 & \text{otherwise} \end{cases}$$

   Prove your answer is correct by mathematical induction.

4. Find all cycles for the function $g : \mathbb{N} \to \mathbb{N}$ defined by

$$g(n) = \begin{cases} \frac{n}{2} & \text{if } n \text{ is even} \\ \frac{n}{3} & \text{if } n \text{ is odd and divisible by 3} \\ n + 1 & \text{otherwise} \end{cases}$$

   Prove your answer is correct by mathematical induction.

5. Lothar Collatz (1910–1990) studied many iterated functions, but there is really no record that he studied the problem that bears his name in Example 4. A similar problem that does appear in his notebooks [38] concerns the function

$$g(x) = \begin{cases} \frac{2n}{3} & \text{if } n \text{ is divisible by 3} \\ \frac{4n-1}{3} & \text{if } n - 1 \text{ is divisible by 3} \\ \frac{4n+1}{3} & \text{if } n + 1 \text{ is divisible by 3} \end{cases}$$

   (a) Find any cycles of g involving only numbers less than 20.

   (b) Collatz's original question concerned the iterated function sequence for g with the first term $a_0 = 8$. Use a calculator or computer to compute the first 20 terms of this sequence.

6. Find all cycles of period 1 for the function $f : \mathbb{R} \to \mathbb{R}$ defined by $f(x) = x^2 - 1$.

7. Find all cycles of period 1 for the function $g : \mathbb{R} \to \mathbb{R}$ defined by $g(x) = 2x - \frac{1}{x}$.

8. Find all cycles of period 2 for the function $f : \mathbb{R} \to \mathbb{R}$ defined by $f(x) = x^2 - 1$.

9. Find all cycles of period 2 for the function $g : \mathbb{R} \to \mathbb{R}$ defined by $g(x) = 2x - \frac{1}{x}$.

10. Find all cycles of period 1 or 2 for the function $f : \mathbb{R} \to \mathbb{R}$ defined by $f(x) = x^2 - 0.5$.

11. Find any cycles of period 1 or 2 for the function $f : \mathbb{R} \to \mathbb{R}$ defined by $f(x) = x^2 - 1.5$.

12. Use a graph to show that there are no cycles of period 3 for the function $f : \mathbb{R} \to \mathbb{R}$ defined by $f(x) = x^2 - 1.5$.

13. A well-known procedure for computing square roots dates back to the Babylonians and is a special case of a procedure from calculus called Newton's method:

    ● Let z be a positive real number whose square root you would like to know.

    ● Let $g : \mathbb{R} \to \mathbb{R}$ be defined by $g(x) = \frac{1}{2}(x + \frac{z}{x})$.

    ● Let a_0 be your best "guess" for $\sqrt{z}$.

    ● The iterated function sequence $a_0, a_1, a_2, \ldots$ forms a sequence of improving approximations for $\sqrt{z}$.

    Find all cycles of period 1 or 2 for the function g.

14. Exercise 13 suggests a general way to approximate the zeroes of functions. For example, if we want to find a

value for x such that $x^3 - 2 = 0$, we can instead find a value of x such that $x = \frac{1}{2}\left(x + \frac{2}{x^2}\right)$ by looking for a cycle of the function $g(x) = \frac{1}{2}\left(x + \frac{2}{x^2}\right)$ with period 1. For each of the equations below, let $g(x)$ be as described. Show that a cycle of g with period 1 will be a solution to the given equation, and use a spreadsheet or calculator* to find an approximation of such a value.

(a) To solve $x^3 - 2 = 0$, let $g(x) = \frac{1}{2}\left(x + \frac{2}{x^2}\right)$.

(b) To solve $e^x - 2 = 0$, let $g(x) = \frac{1}{2}\left(x + \frac{2x}{e^x}\right)$.

(c) To solve $\cos x - x = 0$, let $g(x) = \frac{1}{2}(x + \cos x)$.

15. Find all cycles of period 2 or less for the function $f : [0, 1] \to [0, 1]$ defined by

$$f(x) = \begin{cases} 2x & \text{if } x \leq \frac{1}{2} \\ 2 - 2x & \text{otherwise} \end{cases}$$

(Recall that the interval notation $[0, 1]$ here refers to the set of real numbers x with $0 \leq x \leq 1$.)

16. Using the function f in Exercise 15 and initial values of the form $\frac{2}{n}$, where n is odd, prove that a cycle always occurs. (An interesting research problem is to examine the relationship between the value of $\frac{2}{n}$ and the period of the resulting cycle.)

17. One of the reasons that "sensitivity to initial conditions" is important can be seen when using technology to explore iterated function sequences for the function f given in Exercise 15. Use a spreadsheet to generate 100 terms of several such sequences using different initial values of the form $\frac{2}{n}$, where n is an odd integer. (In an Excel spreadsheet, this can be done by placing $= 2/5$ in cell A1 and then the formula $=$IF(A1<=0.5,2*A1,2-2*A1) in cell A2 and "filled down" through cell A100.) Write a paragraph with supporting data describing what surprising event occurs and how this is a side-effect of sensitivity to initial conditions.

18. Consider the function $g : [0, 1] \to [0, 1]$ with $g(x) = 4x - \lfloor 4x \rfloor$.

(a) Describe in words what g does to an input number x.

(b) Explain why the iterated function sequence for g with the first term $a_0 = \frac{2}{3}$ forms a cycle of period 1.

(c) Use a spreadsheet or calculator to find the first 50 terms of the iterated function sequence for g with the first term $a_0 = \frac{2}{3}$. (In an Excel spreadsheet, this can be done by placing $= 2/3$ in cell A1 and then the formula $=$ 4*A1 - floor (4*A1,1) in cell A2 and "filled down" through cell A50.) Can you explain what is going wrong?

19. The Fibonacci numbers $\{F_n\}$ are defined recursively by the rule

$$F_1 = 1, \quad F_2 = 1, \quad \text{and} \quad F_n = F_{n-1} + F_{n-2} \text{ for all } n \geq 3$$

We can study the ratio of consecutive Fibonacci numbers by letting $r_n = \frac{F_{n+1}}{F_n}$ for each $n \geq 1$.

(a) Prove that $r_n = 1 + \frac{1}{r_{n-1}}$ for all $n \geq 2$.

(b) Find a function $g : \mathbb{R} \to \mathbb{R}$ such that $\{r_n\}$ is an iterated function sequnce for g.

(c) Find all cycles of g with period 1. Write an English sentence relating one of these numbers directly to the Fibonacci numbers.

20. (The following solitaire game is called Bulgarian solitaire and is analyzed in [1].) Take a deck of 45 cards, and divide it into as many face-down piles as you wish. Take one card from each pile, and combine these to form a new pile. Repeat this operation until you start seeing the same configuration more than once. This is a tricky function to make explicit, but hopefully the idea of "cycle" can be understood. Find the unique cycle of period 1.

 # 4.8 Excursion: Growth of Functions

In this section we will see how formal reasoning and mathematical induction can be used in the formal analysis of the complexity of algorithms. While this topic is primarily of interest in the field of computer science, at its heart lies a deeper examination of the relationship between recursive descriptions and closed-form descriptions of sequences. Hence, even those with no interest in computer science will still benefit from studying these ideas.

---

* Issues about when an iterated function sequence "converges" to a value that consitutes a cycle with period 1 are too advanced for this course. It is enough for now that the problems given here all behave nicely.

Complexity of Algorithms

Understanding algorithms is particularly important in the field of computer science. In particular, when presented with more than one way to algorithmically solve a problem, we would like to be able to judge which method is most efficient.

Example 1 *Consider the problem of finding which page in a dictionary a given word is on. There are two obvious algorithmic solutions to the problem:*

1. *Start on page one of the dictionary. If the given word is between (using alphabetical ordering) the first and last words on the page, then find it on the page; otherwise, turn the page and repeat the process.*

2. *Open the dictionary to the middle. If the given word comes before the first word on this page, the word must be in the first half of the dictionary; otherwise, the word must be in the second half of the dictionary. Repeating this process will find first which quarter of the book contains the word, then which eighth of the book contains the word, and so on. Eventually, the fraction of the book you have narrowed the word down to will consist of a single page.*

Which of these is more efficient?

SOLUTION   This is fairly obvious since most people (other than small children) use the latter method. If the dictionary contains 1,000 pages, then the first method involves comparing the given word to as many as 1,000 words, while the latter method will find the correct page after comparing the word to only 10 words.   □

We will measure "efficiency" (or "complexity") by finding a relationship between the size of the input to the algorithm and the amount of time* the algorithm requires to run. Since "run time" is dependent on a particular machine, we measure time complexity by designating a particular time-consuming operation (like multiplying numbers or comparing values), and counting how many times this operation is performed. In this way, we think of the complexity of an algorithm as a function from $\mathbb{N}$, the possible sizes of inputs, to $\mathbb{N}$, the possible number of designated operations required. For example, if $f(n)$ and $g(n)$ represent the maximum number of word comparisons required by the two respective algorithms in Example 1 operating on a dictionary with n pages, then our solution can be written as $f(1,000) = 1,000$ and $g(1,000) = 10$.

In mathematics, to be able to compare the relative merits of different solutions to problems, we often need tools that allow us to see what is important in the midst of the things that are not. The analysis and comparison of algorithms are good examples of this idea. We have already alluded to the idea of measuring efficiency by concentrating on how many times a particular time-consuming operation is performed. For example, suppose you have the choice of two algorithms whose time complexities are measured as $f(n) = \lceil 0.1n^3 \rceil + 10n$ and $g(n) = 10n^2 - n$. Which should you choose? The first

---

* Sometimes, efficiency is measured in terms of memory space required, but we will not address that in this text.

algorithm is better for small values of n—for example, $f(10) = 101$ while $g(10) = 990$. However, if the input size is large, the following example shows that the second algorithm is preferable.

 Example 2 *Prove that for all $n \geq 100$, $f(n) > g(n)$.*

SOLUTION *Let n be given satisfying $n \geq 100$. Using algebraic operations, we obtain:*

$$n \geq 100$$
$$n^3 \geq 100n^2 \quad \text{(multiply by positive number } n^2 \text{)}$$
$$0.1n^3 \geq 10n^2 \quad \text{(multiply by positive number 0.1)}$$

Since $f(n) > 0.1n^3$ and $g(n) < 10n^2$, this establishes the conclusion. □

For large values of n, the presence of n^3 in the first expression causes it to be larger than the second expression, where the highest power of n is n^2. Speaking loosely, we say that the first algorithm has time complexity "roughly n^3," while the time complexity for the second is "roughly n^2." All other details about $f(n)$ and $g(n)$ are insignificant as the size of the input gets larger. These ideas will be developed in more detail as we progress through this section.

A few words about functions and sequences

We have studied number sequences throughout the first two chapters of this book, but now we see that sequences are simply a particular type of function. Specifically, given a function $f : \mathbb{N} \to \mathbb{R}$, we can define a sequence $\{a_n\}$ with the rule $a_k = f(k)$ for all $k \in \mathbb{N}$. Similarly, any sequence with the first term a_0 can be associated with a function from $\mathbb{N}$ to $\mathbb{R}$. In this section, we will interchange these two concepts freely in order to keep notation as straightforward as possible.

Induction and Inequalities

For some algorithms it is reasonably easy to develop a closed formula for the complexity, but for many algorithms this is a difficult if not impossible task. Fortunately, in order to determine when one algorithm is much better than another, we do not necessarily need a closed-form description of each algorithm's complexity.

In Example 3, we imagine an algorithm that requires a_n operations when given an input of size n. Suppose the structure of the algorithm leads us to conclude that a recurrence relation like

$$a_n = 2a_{n-1} + n^3$$

must be satisfied. We have seen in Chapters 1 and 2 that finding a closed formula for such a sequence can be difficult, so it should come as some relief to learn that we do not need to in this section.

 Example 3 *For the recursively defined sequence $a_n = 2a_{n-1} + n^3$ with $a_0 = 0$, prove that $a_n > 2^n$ for every natural number $n \geq 2$.*

n	0	1	2	3	4	5	6	7	8	9	10
a_n	0	1	10	47	158	441	1,098	2,539	5,590	11,909	24,818
2^n	1	2	4	8	16	32	64	128	256	512	1,024

Table 4-26   Table for Example 3

SOLUTION   Perhaps our first instinct is to compare the two expressions for several sample values of n as shown in Table 4-26, and then use mathematical induction to prove the given inequality.

Proof by induction. Let $P(n)$ be the statement "$a_n > 2^n$." From the table, we can see that statements $P(2), P(3), \ldots, P(10)$ are all true. Now let $m \geq 11$ be given such that $P(m)$ is the first statement not yet checked to be true. In particular, we know that statement $P(m-1)$ is true. That is, we know that $a_{m-1} > 2^{m-1}$. Now

$$\begin{aligned}
a_m &= 2 \cdot a_{m-1} + m^3 \\
&> 2 \cdot (2^{m-1}) + m^3 \quad \text{by statement } P(m-1) \\
&= 2^m + m^3 \\
&> 2^m \quad \text{since } m > 1
\end{aligned}$$

This string of relationships implies that $a_m > 2^m$, verifying that statement $P(m)$ is true.   □

Practice Problem 1   *For the recursively defined sequence $b_n = 3b_{n-1} + 2n$ with $b_1 = 1$, prove that $b_n \geq 3^n$ for every natural number $n \geq 3$.*

You might have already guessed that the second "dictionary search" algorithm given in Example 1 really is the best way to find a value in a sorted list. It might not have been as obvious that the number of required comparisons is related to the number of pages in the dictionary using the logarithm and ceiling functions discussed in Section 4.6.

We will talk later about why logarithms are unavoidable when studying certain types of algorithms, but for now, we want to try to understand how they affect the growth rate of recursively defined sequences. In Example 4, c_n represents the maximum number of comparisons required to sort a list of n numbers from smallest to largest using a particular algorithm.

Example 4   *For the recursively defined sequence $c_n = c_{n-1} + \lceil \log_2 (n) \rceil$ with $c_1 = 0$, prove that $c_n \leq n \cdot \lceil \log_2 (n) \rceil$ for every natural number $n \geq 1$.*

SOLUTION   We use Table 4-27 to compare the two expressions for several values of n, and then use mathematical induction to prove the given inequality.

Proof by induction. Let $P(n)$ be the statement "$c_n \leq n \cdot \lceil \log_2 (n) \rceil$." From the table, we can see that statements $P(1), P(2), \ldots, P(6)$ are all true. Now let $m \geq 7$ be given such that $P(m)$ is the first statement not yet checked to be true. In particular, we know that statement $P(m-1)$ is true. That is, we know that

n	1	2	3	4	5	6
c_n	0	1	3	5	8	11
$n \cdot \lceil \log_2(n) \rceil$	0	2	6	8	15	21

Table 4-27   Table for Example 4

$c_{m-1} \leq (m-1) \cdot \lceil \log_2(m-1) \rceil$. Now

$$
\begin{aligned}
c_m &= c_{m-1} + \lceil \log_2(m) \rceil \\
&\leq (m-1) \cdot \lceil \log_2(m-1) \rceil + \lceil \log_2(m) \rceil && \text{by statement } P(m-1) \\
&\leq (m-1) \cdot \lceil \log_2(m) \rceil + \lceil \log_2(m) \rceil && \text{since } \lceil \log_2(m-1) \rceil \leq \lceil \log_2 m \rceil \\
&= m \cdot \lceil \log_2(m) \rceil
\end{aligned}
$$

This string of relationships implies that $c_m \leq m \cdot \lceil \log_2(m) \rceil$, verifying that statement $P(m)$ is true.   □

Note that the preceding proof required knowing a particular fact about logarithms: $\log_2(x-1) \leq \log_2(x)$ for $x \geq 2$. In terms of functions, we say that the $\log_2$ function is *increasing*, a common property among functions that measure complexity. This type of algebraic relationship is typical of the details found in induction proofs of inequalities. Since these details are sometimes tricky, people who deal with growth rates of sequences and functions have developed terms and tools that allow us to work with these ideas at a higher level.

The Language of Function Growth: Θ, O, and Ω

The first step in simplifying our work is to characterize functions based on their "basic growth rates." With practice we will then be able to analyze the growth rate of more complicated recursively defined sequences that use these functions as building blocks. These two ideas together will allow us to understand the complexity of many basic algorithms in computer science.

We begin with some definitions that allow us to classify and compare functions by growth rate.

Definition   For a fixed function $g : \mathbb{N} \to \mathbb{R}^+$, we define the following sets of functions:

1. The set $O(g(n))$ is the set of all functions $f : \mathbb{N} \to \mathbb{R}^+$ such that for all $n \geq N$,

$$f(n) \leq K \cdot g(n)$$

   for some real number K and natural number N. Informally, we can think of $f(n) \in O(g(n))$ as meaning that "$f(n)$ grows no faster than $g(n)$."

2. The set $\Omega(g(n))$ is the set of all functions $f : \mathbb{N} \to \mathbb{R}^+$ such that for all $n \geq N$,

$$f(n) \geq L \cdot g(n)$$

   for some real number L and natural number N. Informally, we can think of $f(n) \in O(g(n))$ as meaning that "$f(n)$ grows at least as fast as $g(n)$."

3. $\Theta(g(n)) = O(g(n)) \cap \Omega(g(n))$. Informally, we can think of $f(n) \in \Theta(g(n))$ as meaning that "$f(n)$ and $g(n)$ grow at the same rate." We often state this by saying, "$f(n)$ is of order $g(n)$."

For example, in Example 3 we showed that $a_n \in \Omega(2^n)$, in Practice Problem 1 we showed that $b_n \in \Omega(3^n)$, and in Example 4 we showed that $c_n \in O(n \log_2(n))$. Before making general observations about these sets of functions, we will look at an example of Θ notation, since that is most significant in our subsequent discussion.

Example 5 SELECTION SORT *is a simple algorithm for sorting a list of numbers, and it can be shown that if a_n is the number of comparisons required to sort a list of n numbers, then a_n can be described recursively by*

$$a_n = a_{n-1} + (n-1) \quad \text{with } a_1 = 0$$

We showed in Section 2.3 that this sequence has the closed form $a_n = \frac{n^2-n}{2}$. Verify that $a_n \in \Theta(n^2)$.

SOLUTION We can equivalently write $a_n = \frac{1}{2}n^2 - \frac{1}{2}n$. For large values of n, the $\frac{1}{2}n$ term is very small compared to the $\frac{1}{2}n^2$ term. For example, when $n = 1,000$, the $\frac{1}{2}n^2$ term is 500,000 and the $\frac{1}{2}n$ term is only 500. For these large values, we reason that a_n is "roughly" $\frac{1}{2}n^2$. To put this conjecture in terms compatible with our Θ notation, we prove the following proposition. □

Proposition 1 *For the sequence given by $a_n = \frac{n^2-n}{2}$, $\frac{1}{4}n^2 \le a_n \le n^2$ for all $n \ge 2$. Hence, $a_n \in \Theta(n^2)$.*

PROOF Let $n \ge 2$ be given. Since n is positive, $\frac{n^2-n}{2} < \frac{n^2}{2}$, and clearly $\frac{n^2}{2} < n^2$. This shows that $a_n \le n^2$. To show that $a_n \ge \frac{1}{4}n^2$, it is much more natural to first rewrite the inequality so that it involves a comparison of a polynomial to 0. In this case the inequality $\frac{n^2-n}{2} \ge \frac{1}{4}n^2$ is the same, after multiplying by 4 and subtracting n^2 from each side, as the inequality $n^2 - 2n \ge 0$. We can now easily argue that if $n \ge 2$, it follows that both n and $n-2$ are positive, in which case we know that the product $n(n-2)$ is positive. That is, $n^2 - 2n \ge 0$, as desired. ■

Through a more general argument, we could show that $\Theta(n^2)$ includes all quadratic polynomials with a positive leading term. Also significant is the fact that other sets like $\Theta(n)$ do *not* include quadratic polynomials. The following example illustrates a proof along these lines.

Example 6 *Show that for the sequence given by $a_n = \frac{n^2-n}{2}$, $a_n \notin \Theta(n)$.*

SOLUTION We show that $a_n \notin O(n)$ (and hence $a_n \notin \Theta(n)$) using a proof by contradiction. Suppose that $a_n \in O(n)$, and let N and K be the values in the definition of O so that $\frac{n^2-n}{2} \le K \cdot n$ for all $n > N$. This is the same, after multiplying through by 2 and adding n to each side, as saying that $n^2 \le (2K + 1)n$ for all $n > N$. Since n is positive, this implies that $n \le 2K + 1$ for all $n > N$. This final statement is absurd regardless of the particular values of N and K, so we have a contradiction. □

In Examples 5 and 6, we see that the growth rate of $a_n = \frac{1}{2}n^2 - \frac{1}{2}n$ is determined by the highest-degree term in the expression. This is true of any polynomial expression, as the following theorem indicates. Although we do not prove this theorem here, the previous examples give some indication of how such a proof might be structured and Exercise 10 outlines the proof of an easy special case.

Theorem 2   *Let $f(n)$ be a polynomial of degree p with positive leading coefficient. That is, $f(n) = c_p n^p + c_{p-1} n^{p-1} + \cdots + c_1 n + c_0$, where $p \in \mathbb{N}$, $c_p > 0$, and each $c_i \in \mathbb{R}$. Then $f(n) \in \Theta(n^p)$. Moreover, for any $q \neq p$, $f(n) \notin \Theta(n^q)$.*

In Example, 5, we had to investigate a bit to decide that the sequence satisfied the property $\frac{1}{4}n^2 \leq a_n \leq n^2$. Since we knew a closed formula for a_n, that was not too hard. For a sequence defined recursively, however, the task may be more difficult. A valuable tool in this sort of investigation is a standard spreadsheet program since it models recursion very nicely. To simplify this process, we give an alternate characterization of O, Ω, and Θ that is easier to use in practice.

Proposition 3   *Let functions $f : \mathbb{N} \to \mathbb{R}^+$ and $g : \mathbb{N} \to \mathbb{R}^+$ be given.*

1. *$f(n) \in O(g(n))$ if and only if there are positive numbers K and N such that $f(n)/g(n) \leq K$ for all $n > N$.*
2. *$f(n) \in \Omega(g(n))$ if and only if there are positive numbers L and N such that $f(n)/g(n) \geq L$ for all $n > N$.*
3. *$f(n) \in \Theta(g(n))$ if and only if there are positive numbers K, L, and N such that $L \leq f(n)/g(n) \leq K$ for all $n > N$.*

Example 7   *For the sequence a_n satisfying $a_n = a_{n-1} + (3n - 1)$, verify that $a_n \in \Theta(n^2)$.*

SOLUTION   We can first examine this sequence empirically using a spreadsheet as follows:

1. Fill Column A with the values of n: 1, 2, 3, ..., 100. (This can be done using "Edit Fill Series.")
2. Put 1 (a value for a_1) in cell B1.
3. Put the formula =B1+3*A2−1 into cell B2.
4. Fill Down the formula to cells B3 through B100.
5. Put the formula =A1^2 into cell C1.
6. Fill Down this formula to cells C2 through C100.
7. Put the formula =B1/C1 into cell D1.
8. Fill Down this formula to cells D2 through D100.

Table 4-28 shows some of the values generated by this spreadsheet.   ☐

n	a_n	n^2	a_n/n^2
20	609	400	1.523
40	2,419	1,600	1.512
60	5,429	3,600	1.508
80	9,639	6,400	1.506
100	15,049	10,000	1.505

Table 4-28 Sequence from Example 7

This evidence leads to the following proposition, part of which is proven below and the rest of which is left for Exercise 16.

Proposition 4 *For the sequence satisfying $a_n = a_{n-1} + (3n - 1)$ and $a_1 = 1$, for all $n \geq 1$, $n^2 \leq a_n \leq 2n^2$. Hence, $a_n \in \Theta(n^2)$.*

PROOF (By induction) We will prove only the statement "$a_n \leq 2n^2$" here. Since $a_1 = 1$, the first statement "$a_1 \leq 2 \cdot 1^2$" is true. Let $m \geq 2$ be given such that the statement "$a_m \leq 2m^2$" is the first one not yet checked to be true. In particular, the statement "$a_{m-1} \leq 2(m-1)^2$" has already been checked to be true. Now

$$a_m = a_{m-1} + 3m - 1 \quad \text{from the recurrence relation}$$
$$\leq 2(m-1)^2 + (3m - 1) \quad \text{by statement } P(m-1)$$
$$= 2m^2 - m + 1$$

Since $m \geq 2$, it follows that $2m^2 - m + 1 \leq 2m^2$. Hence,

$$a_m \leq 2m^2$$

which is precisely the next statement to be checked. ∎

The function growth notation is useful to express the notion that one growth rate is "better" than another. Since we are predominantly concerned with functions that measure the number of operations required by an algorithm, it makes sense that the better function is the one with smaller outputs reflecting a more efficient algorithm. The significance of this can be seen when making a direct connection between the number of operations and the time an algorithm requires to run. Table 4-29 makes this connection for some specific functions assuming the algorithm can perform 10^6 (1 million) operations per second. If we have a choice of implementing an easy algorithm with complexity of order n^2 or doing a bit more work to implement an algorithm with complexity of order $n \log n$, we can see from the table that it can really make a difference, especially if the input is large or if the algorithm will be used repeatedly.

Now using the function growth notation that we have developed, we can state the first of two theorems that tell us the growth rate of a recursively defined sequence based on the form of its recurrence relation. Once again we will not prove this theorem here, but we will opt instead to investigate it further in the exercises.

Theorem 5 *Let p and C be positive real numbers, and suppose that $g(n) \in \Theta(n^p)$. The sequence with recurrence relation $a_n = C \cdot a_{n-1} + g(n)$ has one of three possible growth rates:*

Function f	$n = 100$	$n = 1{,}000$	$n = 10{,}000$	$n = 100{,}000$
$\log_2 n$	0.6×10^{-5} sec	10^{-5} sec	1.3×10^{-5} sec	1.7×10^{-5} sec
n	10^{-4} sec	10^{-3} sec	10^{-2} sec	0.1 sec
$n \log_2 n$	10^{-3} sec	10^{-2} sec	0.133 sec	1.66 sec
n^2	10^{-2} sec	1 sec	100 sec	1 day
n^3	1 sec	17 min	1.5 weeks	30+ years

Table 4-29 Approximate Run Time for Various Algorithm Complexities

1. *If $C < 1$, then $a_n \in \Theta(n^p)$.*
2. *If $C = 1$, then $a_n \in \Theta(n^{p+1})$.*
3. *If $C > 1$, then $a_n \in \Theta(C^n)$.*

Using this theorem, we can say that in Example 7 (where $a_n = a_{n-1} + (3n - 1)$), since $C = 1$ and $(3n - 1) \in \Theta(n)$, it follows that $a_n \in \Theta(n^2)$. Similarly, we can say that in Example 3, the sequence with recurrence relation $a_n = 2a_{n-1} + n^3$ must have $a_n \in \Theta(2^n)$ since in this case $C = 2 > 1$ and $g(n) \in \Theta(n^3)$.

Practice Problem 2 *Use Theorem 5 to predict the growth rate of a sequence a_n satisfying each of the following recurrence relations:*

(a) $a_n = \frac{1}{2}a_{n-1} + (3n - 1)$

(b) $a_n = \frac{3}{2}a_{n-1} + (n^2 + 2n + 3)$

(c) $a_n = a_{n-1} + \sqrt{n}$

Logarithms and Algorithms

Before proceeding to the second theorem that predicts the growth rate of a sequence based on the form of its recursive description, we will pause here to see how logarithms arise naturally in the study of algorithms. We actually need look no further than the "better" algorithm that we used in Example 1 to find the page of a dictionary containing a given word. Following that same process, we can find any given value in any sorted list as follows.

Example 8 *The following algorithm determines whether a sorted list L (sorted lowest to highest) contains a number x. How many comparisons are necessary when a list L with n entries is used as input?*

BINARY SEARCH

[1] Compare[*] x and the middle-most element of L to determine which half of the list x is in.

---

[*] We assume that in a single comparison of two numbers a and b, we can determine if $a = b$, $a < b$, or $a > b$.

[2] Compare x with the middle-most element of that half to determine which quarter of the list x is in.

[3] Compare x with the middle-most element of that quarter to determine which eighth of the list x is in.

[4] And so on.

[5] Once the portion of the list you have narrowed it down to has only one element in it, you can give a "yes" or "no" answer with a single additional comparison.

SOLUTION The longest this process can take is when the given item is not actually in the list, so that is the case we will use to discuss complexity. For example, in a list of 15 elements, the first comparison will split the list into two "halves," each with seven elements. The second comparison will split these into two "quarters," each with three elements. In this way, two more comparisons will rule out the given value from being in the list.

In general, the number of comparisons is the number of times one must repeat the splitting process to be left with a list of length 1, and then one additional comparison to check if the value is in that very small list. The number of times a number n must be halved before the result is no more than 1 can be found algebraically by finding the smallest integer k such that

$$n \cdot \left(\frac{1}{2}\right)^k \leq 1$$

which is the same thing (using logarithms) as

$$k \geq \log_2(n)$$

The exact way in which these values are rounded (using floor or ceiling functions) requires a little more thought, so this is left for Exercise 1 at the end of this section. It turns out that $\lfloor \log_2(n) \rfloor + 1$ comparisons are required to verify that the given number is not in the sorted list of length n. □

Closely related to this "searching algorithm" is a class of puzzles that involves finding a counterfeit coin among many real ones. These are popular in recreational mathematics, and the idea behind them is also fundamentally tied to logarithms.

Example 9 *I have a bag with 16 coins that appear identical, but I know one of them is counterfeit. The real coins weigh 1 oz each and the counterfeit one weighs some different amount. I have a spring scale that tells me the exact weight of whatever I put on it.*

SOLUTION Use the following algorithm:

COUNTERFEIT DETECTOR I

[1] Weigh eight coins. If they do not weigh 8 oz, then the counterfeit coin is among them; otherwise, the counterfeit coin is among the other eight. Keep the eight that include the counterfeit coin.

[2] Weigh four coins. If they do not weigh 4 oz, then the counterfeit coin is among them; otherwise, the counterfeit coin is among the other four. Keep the four that include the counterfeit coin.

[3] Weigh two coins. If they do not weigh 2 oz, then the counterfeit coin is among them; otherwise, the counterfeit coin is among the other two. Keep the two that include the counterfeit coin.

[4] Weigh one of these coins to determine whether it or the other is counterfeit.

This algorithm requires four weighings to find the counterfeit among 16 coins. We can use mathematical induction to verify that it will require k weighings to find a counterfeit among 2^k coins. This can be extended to show that the number of weighings for n coins is $\lceil \log_2 n \rceil$. $\qquad\square$

Proposition 6   *Algorithm* COUNTERFEIT DETECTOR I *requires k weighings to find a counterfeit among 2^k coins.*

PROOF   See Exercise 25. $\qquad\blacksquare$

Logarithms might seem complicated to work with, but from the point of view of functional growth, they can be simplified somewhat by the fact that the base of the logarithm does not affect the growth rate. We will first get a glimpse of why this is true, and then we will discuss further why it is beneficial.

Example 10   *Show that $\log_2 n \in \Theta(\log_3 n)$.*

SOLUTION   Let $x = \log_3 n$. That is, x is the power of 3 that yields n, or $3^x = n$. Since $3 = 2^{\log_2 3}$, it follows that

$$n = 3^x = \left(2^{\log_2 3}\right)^x = 2^{x \cdot \log_2 3}$$

Hence, the power of 2 that yields n is $x \cdot \log_2 3$—that is, $\log_2 n = x \cdot \log_2 3$. From this, it follows that

$$\frac{\log_2 n}{\log_3 n} = \frac{x \cdot \log_2 3}{x} = \log_2 3 \approx 1.5849625$$

This means that $\frac{\log_2 n}{\log_3 n}$ is a constant (i.e., with no dependence on the value of n), so for all $n \geq 1$

$$\log_3 n \leq \log_2 n \leq 2 \log_3 n$$

If you are unfamiliar with logarithms, this might be a bit of a surprise, so you will want to experiment on your calculator to be sure you really believe it. (See Exercise 3.) $\qquad\square$

The same reasoning can be applied to show that any two logarithms with bases greater than one have the same growth rate. Therefore, we can pick a standard base that we understand, and use a logarithm in that base to represent all logarithmic growth rates. In calculus, for example, we use the "natural" logarithmic base e (e is an irrational number approximately equal to 2.718), and we even denote $\log_e x$ as $\ln x$ instead. From our experience, using a standard base of two makes more sense in discrete math.

Proposition 7   *For any base $b > 1$, $\log_b n \in \Theta(\log_2 n)$.*

This proposition justifies why we often just use the phrase "order of $\log n$," with no specific base, to describe logarithmic growth.

The other important thing to know about logarithmic growth is that it is very slow, which for functions measuring complexity is a very good thing. In particular, logarithmic functions grow more slowly than any positive power of n, a fact that can be easily proven using the tools of calculus.

Proposition 8   *If p is any positive real number, then there is a positive number N such that*

$$\log_2 x < x^p$$

for all $x > N$.

PROOF   Ask a calculus teacher and make her day!   ■

The "big-oh" notation gives a natural hierarchy of elementary functions based on their growth rates:

$$\cdots O(\log_2 n) \subseteq \cdots \subseteq O(n^{1/3}) \subseteq O(n^{1/2}) \subseteq O(n) \subseteq O(n^2) \subseteq \cdots \subseteq O(2^n) \cdots$$

We can think of each of these subset relations in its "if, then" form. For example, the relationship $O(n) \subseteq O(n^2)$ can be thought of as "If $f(n) \in O(n)$, then $f(n) \in O(n^2)$." In other words, "If $f(n)$ grows no faster than n, then $f(n)$ grows no faster than n^2."

The hierarchy above says that of all the growth rates we have seen, logarithmic growth rate is the slowest and exponential growth rate is the fastest. Although other growth rates are possible, these (and combinations of them) are sufficient for understanding our main result about the growth of recursively defined sequences.

Divide-and-Conquer Recurrences

A common type of recurrence relation in computer science courses arises from analyzing the complexity of *recursive algorithms*. Although these are among the most important recurrence relations for a computer scientist to understand, they are also among the most difficult to find closed formulas for. Fortunately, our function growth notation can often be used to understand the growth rate of these sequences without needing an exact closed formula.

Definition   A *divide-and-conquer recurrence relation* is one of the form

$$a_n = C \cdot a_{\lfloor n/k \rfloor} + f(n)$$

where C and k are positive constants (with $k > 1$) and f is any function on $\mathbb{N}$.

A typical application is the analysis of the sorting algorithm in Example 12. Before tackling this more difficult real-world example, however, we will first look at a few "abstract" examples to gain some experience with sequences like this.

Example 11 *The first few terms of the recursive description $a_n = 3 \cdot a_{\lfloor n/2 \rfloor} + n$ for $n \geq 2$, with $a_1 = 1$, look like this:*

$$1, 5, 6, 19, 20, 24, 25, \ldots$$

It is difficult to find an exact solution to this kind of recurrence, but we can still analyze the long-term behavior of the sequence. Prove that for all $n \geq 5$, $a_n < n^2 + n$.

SOLUTION   Since $a_5 = 20 < 5^2 + 5$, the first statement is true. Now let $m \geq 6$ be given such that the inequality has been checked for $a_5, \ldots, a_{m-1}$. When we consider the next number in the sequence, we have

$$\begin{aligned}
a_m &= 3 \cdot a_{\lfloor m/2 \rfloor} + m \\
&< 3 \cdot \left((\lfloor m/2 \rfloor)^2 + \lfloor m/2 \rfloor \right) + m \\
&\leq 3 \cdot (m/2)^2 + m/2 + m \\
&= \frac{3}{4}m^2 + \frac{3}{2}m \\
&= (m^2 + m) - \left(\frac{1}{4}m^2 - \frac{1}{2}m \right) \\
&< m^2 + m
\end{aligned}$$

since $\frac{1}{4}m^2 - \frac{1}{2}m > 0$ when $m \geq 5$. This last inequality is somewhat mysterious, but it can be verified by looking at a graph of the parabola $y = \frac{1}{4}x^2 - \frac{1}{2}x$ or proved directly by induction. ☐

As we have seen, there is an important interplay between the components of the recurrence relation. In terms of the definition, the relative sizes of C and k, as well as the growth rate of $f(n)$, play important parts in the overall growth rate of a divide-and-conquer recurrence relation.

Practice Problem 3   *List the first 10 terms of each of the following recurrence relations and compare them to the one in Example 11:*

1. $b_1 = 1$ *and* $b_n = b_{\lfloor n/2 \rfloor} + n$
2. $c_1 = 1$ *and* $a_n = 3 \cdot c_{\lfloor n/2 \rfloor} + 2$
3. $d_1 = 1$, $d_2 = 1$, *and* $d_n = 3 \cdot d_{\lfloor n/3 \rfloor} + n$

Exercise 31 explains how to use a standard spreadsheet program to generate many terms for these sequences. Before studying the growth of these sequences, let's see how this kind of recurrence relation arises from a recursive algorithm.

Example 12   MERGE SORT, *a common recursive algorithm for sorting a list of values, performs the following steps:*

1. *If the given list has only one number, then it already is sorted with no additional required work.*
2. *If the given list has more than one number, then do the following three steps:*
   (a) *Take the first half of the list and ask this program to sort it.*

n	1	2	3	4	5	6	7	8	9	10
s_n	0	1	3	5	8	11	14	17	21	25

Table 4-30 MERGE SORT Complexity for Small Values of n

(b) *Take the second half of the list and ask this program to sort it. (If the list has odd length, take one more number in the first part than in the second part.)*

(c) *Merge the sorted first half with the sorted second half, and the entire list will be sorted.*

Find a recurrence relation for the number of comparisons required by MERGE SORT *to sort a list of length n.*

SOLUTION   Let s_n denote the number of comparisons required to sort a list of n numbers using this algorithm. The first condition above tells us that $s_1 = 0$ since a list with only one number does not require any comparison of numbers to become sorted—it is already sorted. If the list has $n \geq 2$ elements, then it is broken into two parts, the first containing $\left\lceil \frac{n}{2} \right\rceil$ numbers and the second containing $\left\lfloor \frac{n}{2} \right\rfloor$ numbers. By the definition of s, it will take $s_{\lceil n/2 \rceil}$ comparisons to sort the first part and $s_{\lfloor n/2 \rfloor}$ comparisons to sort the second part. With a little thought, we can see that it requires $n - 1$ comparisons to merge two sorted lists (containing n total numbers) together, so we have established the recurrence relation

$$s_n = s_{\lceil n/2 \rceil} + s_{\lfloor n/2 \rfloor} + n - 1$$

Table 4-30 illustrates the first few values of this sequence.   □

The Master Theorem

The final result of this section characterizes the growth rate of a sequence based on the form of its divide-and-conquer recursive description. The complexity of many recursive algorithms can be determined by this theorem.

Theorem 9 (Simplified Master Theorem)   *Let positive constants L and K be given, and let* $q = \log_K(L)$. *The sequence* $\{a_n\}$ *with recurrence relation*

$$a_n = La_{n/K} + g(n)$$

has one of three possible growth rates:

1. *If* $g(n) \in O(n^p)$ *for some positive* $p < q$, *then* $f(n) \in \Theta(n^q)$.
2. *If* $g(n) \in \Theta(n^q)$, *then* $f(n) \in \Theta(n^q \cdot \log n)$.
3. *If* $g(n) \in \Omega(n^r)$ *for some* $r > q$, *then* $f(n) \in \Theta(g(n))$.

We should note that the fraction n/K in the recurrence relation is really shorthand for $\lceil n/K \rceil$ or $\lfloor n/K \rfloor$, since the result is true regardless of which of these is intended. Before we discuss why the theorem is true, let's see how to apply it to the divide-and-conquer recurrence relations we encountered earlier in this section.

Example 13 *Use the master theorem to draw a conclusion about the growth rate of sequences a_n and b_n from Example 11 and Practice Problem 3.*

SOLUTION Since $a_n = 3 \cdot a_{\lfloor n/2 \rfloor} + n$, we can apply the master theorem with $K = 2$, $L = 3$, and $g(n) = n$. In the theorem, $q = \log_2 3 \approx 1.585$. Since $g(n) \in O(n^1)$, the first case of the master theorem holds, so we can conclude that $a_n \in \Theta(n^q)$.

Since $b_n = b_{\lfloor n/2 \rfloor} + n$, we can apply the master theorem with $K = 2$, $L = 1$, and $g(n) = n$. In the theorem, $q = \log_2 1 = 0$. Since $g(n) \in \Omega(n^1)$, the third case of the master theorem holds, and we can conclude that $b_n \in \Theta(n)$. □

Practice Problem 4 *Use the master theorem to draw a conclusion about the growth rate of sequences c_n and d_n from Practice Problem 3.*

We will not prove the master theorem, but we will verify the critical second case with an example. The process that we will use is the same more general process that is used in the proof of the theorem.

Example 14 *Verify that the master theorem is correct for the sequence s_n that measures the complexity of the* MERGE SORT *algorithm in Example 12, for values of n of the form 2^m.*

SOLUTION Without paying attention to floors and ceilings, the recurrence relation for s_n has the form

$$s_n = 2 \cdot s_{n/2} + (n - 1)$$

Using $K = L = 2$ and $g(n) = n - 1$ in the master theorem, we see that $q = \log_2 2 = 1$, so since $g(n) \in \Theta(n^1)$, the theorem says that $s_n \in \Theta(n \log n)$. To verify this independently, we will look at values of s_n, where n has the form 2^m, since in this case, the recurrence is easy to implement.

$$
\begin{aligned}
s_{16} &= 2 \cdot s_8 + g(16) \\
&= 2 \cdot (2 \cdot s_4 + g(8)) + g(16) \\
&= 2 \cdot (2 \cdot (2 \cdot s_2 + g(4)) + g(8)) + g(16) \\
&= 2 \cdot (2 \cdot (2 \cdot (2 \cdot s_1 + g(2)) + g(4)) + g(8)) + g(16) \\
&= 2^4 \cdot s_1 + 2^3 \cdot g(2) + 2^2 \cdot g(4) + 2^1 \cdot g(8) + g(16) \\
&= 2^4 \cdot 0 + 2^3 \cdot g(2^1) + 2^2 \cdot g(2^2) + 2^1 \cdot g(2^3) + g(2^4)
\end{aligned}
$$

Since $g(n) = n - 1$, each $g(2^k)$ is roughly equal to 2^k and so this final sum is roughly 2^4 added to itself four times, or roughly $4 \cdot 2^4$. Notice that this is exactly the value of the formula $n \cdot \log_2 n$ for the value $n = 16$ we started with. □

In the solution to the previous problem, we saw that "unwinding" the divide-and-conquer recurrence relation leads to a sum of roughly $\log_K n$ terms, each of which is the product of a power of K and g evaluated at a power of K. The master theorem simply pits the two terms in each product against each other: If g does not have a high enough growth rate, then the powers of K dominate; if g has too high a growth rate, then the powers of K become irrelevant. The second case of the master

n	1	2	3	4	5	6	7	8	9	10
a_n	1	5	6	19	20	24	25	65	66	70
b_n	1	3	4	7	8	10	11	15	16	18
c_n	1	5	5	17	17	17	17	53	53	53
d_n	1	1	6	7	8	9	10	11	27	28

Table 4-31   Table for Practice Problem 3

theorem shows what happens when neither of the terms dominates the other. Only in this case is the number $(\log n)$ of summands relevant. To understand the details requires only this idea and some familiarity with logarithm properties. Some of these details are addressed further in the exercises.

Solutions to Practice Problems

1   *Proof by induction.* The first statement to be checked is "$b_3 \geq 3^3$." Since $b_1 = 1$, it follows from the recurrence relation that $b_2 = 3 \cdot 1 + 4 = 7$ and $b_3 = 3 \cdot 7 + 6 = 27$. Hence, the first statement is true. Now let $m \geq 4$ be given such that all statements up to and including "$b_{m-1} \geq 3^{m-1}$" have been checked. We now reason as follows about the next statement:

$$\begin{aligned} b_m &= 3 \cdot b_{m-1} + 2m \quad \text{by the recurrence relation} \\ &\geq 3 \cdot 3^{m-1} + 2m \quad \text{by the previously checked} \\ &\qquad\qquad\qquad\quad \text{statement} \\ &= 3^m + 2m \\ &\geq 3^m \quad \text{since } 2m \geq 0 \end{aligned}$$

∎

2   (a) In this case, $C < 1$ and $g(n) = (3n - 1) \in \Theta(n)$, so $a_n \in \Theta(n)$.

(b) In this case, $C = \frac{3}{2} > 1$, so $a_n \in \Theta((3/2)^n)$.

(c) In this case, $C = 1$ and $g(n) = \sqrt{n} \in \Theta(n^{1/2})$, so $a_n \in \Theta(n^{3/2})$.

3   Table 4-31 compares values of each sequence to the corresponding values of the sequence $\{a_n\}$.

4   Since $c_n = 3 \cdot c_{\lfloor n/2 \rfloor} + 2$, we can apply the master theorem with $K = 2$, $L = 3$, and $g(n) = 2$. In the theorem, $q = \log_2 3 \approx 1.585$. Since $g(n) \in \Omega(n^1)$, the first case of the master theorem holds, and we can conclude that $c_n \in \Theta(n^q)$.

Since $d_n = 3 \cdot d_{\lfloor n/3 \rfloor} + n$, we can apply the master theorem with $K = 3$, $L = 3$, and $g(n) = n$. In the theorem, $q = \log_3 3 = 1$. Since $g(n) \in \Theta(n^1)$, the second case of the master theorem holds, and we can conclude that $d_n \in \Theta(n \log n)$.

Exercises for Section 4.8

1. Fill out Table 4-32 to find the number of comparisons required to verify that a given value is **not** among the entries in a sorted list of length n. In your analysis, assume the worst case at each step—namely, at any step in which the list is split unevenly, assume that the value is judged to be in the larger "half." (Be sure to use recursive thinking!)

2. For each of the following rules for the function $f : \mathbb{N} \to \mathbb{N}$, make a table to investigate the ratio $f(2n)/f(n)$ for increasing values of n. For each definition of f, write a sentence of the form "When the input value is doubled, the output is _____."

(a) $f(n) = n$
(b) $f(n) = 5n + 1$
(c) $f(n) = \frac{1}{5}n + 1$
(d) $f(n) = n^2$
(e) $f(n) = 3n^2 + 1$
(f) $f(n) = 2^n$
(g) $f(n) = 2^{n+1}$
(h) $f(n) = \lfloor \sqrt{n} \rfloor$

3. To verify the fact derived in Example 10, use your calculator to evaluate $f(n) = \log_2 n$, $g(n) = \log_3 n$, and $h(n) = f(n)/g(n)$, for values of n from the set $\{2, 3, 4, 9, 10\}$.

In Exercises 4–6, use a computer or calculator spreadsheet as in the following example: To explain why $\frac{1}{2}n^2 - n \in \Theta(n^2)$, we first place the values 1 through 50 in cells A1 through A50; next the formula $= 1/2 * A1^2 - A1$ is placed in cell B1 and filled down through cell

Length of list (n)	1	2	3	4	5	6	7	8	9	10
Number of comparisons	1	2	2	3						
Value of $\lfloor \log_2 n \rfloor$										

Table 4-32 Table for Exercise 1

B50; and finally the formula =B1/A1^2 is placed in cell C1 and filled down through cell C50. The fact that after cell C4, the values in the C column stay between 0.25 and 0.5 supports the claim that $\frac{1}{2}n^2 - n \in \Theta(n^2)$. In the spirit of the exercises below, we would write $\underline{0.25} \cdot n^2 \leq \frac{1}{2}n^2 - n \leq \underline{0.5} \cdot n^2$ for all $n \geq \underline{4}$.

4. Fill in the blanks to make a true statement out of each of the following, using spreadsheet data to support your answers:

   (a) $2n + 1 \in \Theta(n)$ because

   _____ $\cdot n \leq 2n+1 \leq$ _____ $\cdot n$ for all

   $n \geq$ _____

   (b) $\frac{1}{4}n^2 - 10n + 3 \in \Theta(n^2)$ because

   _____ $\cdot n^2 \leq \frac{1}{4}n^2 - 10n + 3 \leq$ _____ $\cdot n^2$

   for all $n \geq$ _____

   (c) $n + \sqrt{n} \in \Theta(n)$ because

   _____ $\cdot n \leq n + \sqrt{n} \leq$ _____ $\cdot n$ for all

   $n \geq$ _____

   (d) $n^2 + \log_2 n \in \Theta(n^2)$ because

   _____ $\cdot n^2 \leq n^2 + \log_2 n \leq$ _____ $\cdot n^2$

   for all $n \geq$ _____

5. Fill in the blanks to make a true statement out of each of the following, using spreadsheet data to support your answers:

   (a) $3n - 1 \in O(n)$ because

   $3n - 1 \leq$ _____ $\cdot n$ for all $n \geq$ _____

   (b) $10n + 7 \in O(n^2)$ because

   $10n + 7 \leq$ _____ $\cdot n^2$ for all $n \geq$ _____

   (c) $\lfloor \log_2 n \rfloor \in O(\sqrt{n})$ because

   $\lfloor \log_2 n \rfloor \leq$ _____ $\cdot \sqrt{n}$ for all $n \geq$ _____

6. Fill in the blanks to make a true statement out of each of the following, using spreadsheet data to support your answers:

   (a) $3n - 7 \in \Omega(n)$ because

   $3n - 7 \geq$ _____ $\cdot n$ for all $n \geq$ _____

   (b) $\frac{1}{5}n^2 \in \Omega(n)$ because

   $\frac{1}{5}n^2 \geq$ _____ $\cdot n$ for all $n \geq$ _____

   (c) $\frac{1}{2}n^2 - 5n \in \Omega(n^2)$ because

   $\frac{1}{2}n^2 - 5n \geq$ _____ $\cdot n^2$ for all $n \geq$ _____

7. In Proposition 1 we used the definition of Θ to show that the sequence given by $a_n = \frac{n^2 - n}{2}$ satisfies $a_n \in \Theta(n^2)$. Do the same thing, but using the alternative characterization given in Proposition 3 instead. Specifically, find positive numbers K and L and prove that $a_n/n^2 = \frac{1}{2} - \frac{1}{2n}$ stays between K and L for all $n \geq 2$.

8. Modify your solution to the previous exercise to prove that $a_n = \frac{n^2 + n}{2}$ satisfies $a_n \in \Theta(n^2)$.

9. Let $a_n = 5n^3 + 4n^2 + 6n + 7$. Show that $a_n \in \Theta(n^3)$ as follows: Taking the ratio a_n/n^3 and simplifying yield $5 + \frac{4}{n} + \frac{6}{n^2} + \frac{7}{n^3}$. Since $n \geq 1$, what is the largest value that $\frac{4}{n} + \frac{6}{n^2} + \frac{7}{n^3}$ can have? How does this (along with Proposition 3) show that $a_n \in \Theta(n^3)$?

10. We can generalize the previous two exercises to prove a special case of Theorem 2: Let $f(n)$ be a polynomial of degree p with positive leading coefficient and *no negative coefficients*. That is, $f(n) = c_p n^p + c_{p-1} n^{p-1} + \cdots + c_1 n + c_0$, where $p \in \mathbb{N}, c_p > 0$, and each $c_i \in \mathbb{R}^{\geq 0}$. Prove that $f(n) \in \Theta(n^p)$.

11. The following three problems give an indication for how Theorem 2 can be proven in general:

    (a) Let $a_n = 5n^3 - 4n^2 - 6n - 7$. Show that $a_n \in \Theta(n^3)$. (HINT: Taking the ratio a_n/n^3 and simplifying yield $5 - \frac{4}{n} - \frac{6}{n^2} - \frac{7}{n^3} = 5 - (\frac{4}{n} + \frac{6}{n^2} + \frac{7}{n^3})$.)

    (b) Let $a_n = 5n^3 + 4n^2 - 6n + 7$. Show that $a_n \in \Theta(n^3)$. (HINT: Observe that $5n^3 - 4n^2 - 6n - 7 \leq a_n \leq 5n^3 + 4n^2 + 6n + 7$ for all $n \geq 0$.)

    (c) Let $a_n = 5n^3 - 4n^2 - 6n + 7$. Show that $a_n \in \Theta(n^3)$.

12. Give a general proof of Theorem 2.

13. To show that $a_n \notin O(n^2)$, we must show that for any value of K, there is an integer n such that $\frac{a_n}{n^2} > K$. For each of the following sequences, find such an integer n for each of the values of K in $\{10, 20, 50, 100\}$:

(a) $a_n = n^3$

(b) $a_n = \frac{1}{2}n^3 + 2n$

(c) $a_n = n^3 - 5n^2 - 30$

(d) $a_n = \frac{1}{2}n^3 - 2n^2 - 3n - 17$

14. To show that $a_n \notin Q(g(n))$, we must show that for any value of K and N, there is an integer $n > N$ such that $\frac{a_n}{g(n)} > K$. Use this fact to prove each of the following statements:

(a) If $a_n = n^2$, show that $a_n \notin O(n)$.

(b) If $a_n = 2n^3 + 3$, show that $a_n \notin O(n^2)$.

(c) If $a_n = 3n$, show that $a_n \notin O(\sqrt{n})$.

15. For each of the following statements, use a spreadsheet to explore the given sequence (as is done in Example 7) and fill in the blanks to create a true statement based on your evidence. You do not need to prove the statements are true.

(a) Given that $a_n = a_{n-1} + (n^2 + 2n)$ and $a_1 = 1$, it follows that

$$\underline{\qquad} \cdot n^3 \le a_n \le \underline{\qquad} \cdot n^3 \text{ for all}$$

$$n \ge \underline{\qquad}$$

(b) Given that $a_n = a_{n-1} + (\frac{1}{2}n + 17)$ and $a_1 = 10$, it follows that

$$\underline{\qquad} \cdot n^2 \le a_n \le \underline{\qquad} \cdot n^2 \text{ for all}$$

$$n \ge \underline{\qquad}$$

(c) Given that $a_n = a_{n-1} + \sqrt{n}$ and $a_1 = 3$, it follows that

$$\underline{\qquad} \cdot n\sqrt{n} \le a_n \le \underline{\qquad} \cdot n\sqrt{n}$$

$$\text{for all } n \ge \underline{\qquad}$$

16. For each of the following statements, use a spreadsheet to explore the given sequence (as is done in Example 7) and fill in the blanks to create a true statement based on your evidence. You do not need to prove the statements are true.

(a) Given that $a_n = 2a_{n-1} + n^2$ and $a_1 = 1$, it follows that

$$\underline{\qquad} \cdot 2^n \le a_n \le \underline{\qquad} \cdot 2^n \text{ for all}$$

$$n \ge \underline{\qquad}$$

(b) Given that $a_n = \frac{2}{3}a_{n-1} + (3n + 17)$ and $a_1 = 12$, it follows that

$$\underline{\qquad} \cdot n \le a_n \le \underline{\qquad} \cdot n \text{ for all}$$

$$n \ge \underline{\qquad}$$

(c) Given that $a_n = 3a_{n-1} - n^3$ and $a_1 = 30$, it follows that

$$\underline{\qquad} \cdot n\sqrt{n} \le a_n \le \underline{\qquad} \cdot n\sqrt{n}$$

$$\text{for all } n \ge \underline{\qquad}$$

17. In the proof of Proposition 4 we proved that a sequence a_n satisfying $a_n = a_{n-1} + (3n - 1)$ must have $a_n \in O(n^2)$. Prove that such a sequence must also have $a_n \in \Omega(n^2)$, thus completing the proof that $a_n \in \Theta(n^2)$.

18. Use induction to prove each of the following inequalities:

(a) Given that $a_n = a_{n-1} + 3n$ and $a_1 = 1$, prove that $a_n \le 3n^2$ for all $n \ge 1$.

(b) Given that $b_n = b_{n-1} + (n - 2)$ and $b_1 = 1$, prove that $b_n \le n^2$ for all $n \ge 1$.

(c) Given that $c_n = c_{n-1} + (3n^2 + n)$ and $c_1 = 1$, prove that $c_n \ge n^3$ for all $n \ge 1$.

(d) Given that $d_n = d_{n-1} + \lceil \sqrt{n} \rceil$ and $d_1 = 1$, prove that $d_n \ge n$ for all $n \ge 1$.

19. Use induction to prove each of the following inequalities:

(a) Given that $a_n = 2a_{n-1} + 1$ and $a_1 = 1$, prove that $a_n \le 2^n - 1$ for all $n \ge 1$.

(b) Given that $b_n = \frac{1}{2}b_{n-1} + n$ and $b_1 = 1$, prove that $b_n \le 2n$ for all $n \ge 1$.

(c) Given that $c_n = 3c_{n-1} + n$ and $c_1 = 1$, prove that $c_n \ge 3^{n-1}$ for all $n \ge 1$.

(d) Given that $d_n = 2d_{n-1} + n^3$ and $d_1 = 1$, prove that $d_n \ge 2^n$ for all $n \ge 2$.

20. Prove each of the following properties for all functions f, g, and h on $\mathbb{N}$:

(a) For every function f on $\mathbb{N}$, $f(n) \in \Theta(f(n))$.

(b) For all functions f and g on $\mathbb{N}$, if $f(n) \in \Theta(g(n))$, then $g(n) \in \Theta(f(n))$.

(c) For all functions f, g, and h on $\mathbb{N}$, if $f(n) \in \Theta(g(n))$ and $g(n) \in \Theta(h(n))$, then $f(n) \in \Theta(h(n))$.

21. Prove each of the following for all functions f, g, and h on $\mathbb{N}$:

(a) If $f(n) \in O(h(n))$ and $C > 1$, then $C \cdot f(n) \in O(h(n))$.

(b) If $f(n) \in O(h(n))$ and $g(n) \in O(h(n))$, then $f(n) + g(n) \in O(h(n))$.

(c) If $f(n) \in O(g(n))$, then $g(n) \in \Omega(f(n))$.

22. Prove each of the following for all functions f and g on $\mathbb{N}$:

(a) If $f(n) \in O(g(n))$ and $g(n) \in O(f(n))$, then $f(n) \in \Theta(g(n))$.

(b) If $f(n) \in O(n^2)$ and $g(n) \in O(n^3)$, then $f(n) + g(n) \in O(n^3)$.

Number of coins (n)	1	2	3	4	5	6	7	8	9	10
Number of weighings										
Value of $\lceil \log_2(n) \rceil$										

Table 4-33   Table for Exercise 26

(c) If $f(n) \in \Theta(n^2)$ and $g(n) \in \Theta(n^3)$, then $f(n) + g(n) \in \Theta(n^3)$.

23. Each of the following statements is false. In each case, provide a specific counterexample that shows the statement is false.

(a) For all $a > 1$, $O(a^n) = O(2^n)$.

(b) For all functions f and g on $\mathbb{N}$, if $f(n) \in O(g(n))$, then $g(n) \in O(f(n))$.

(c) For all functions f and g on $\mathbb{N}$, if $f(n) \in O(n^2)$ and $g(n) \in O(n^2)$, then $f(n) - g(n) = 0$.

(d) If $f(n) \in O(h(n))$ and $g(n) \in O(h(n))$, then $f(n) \cdot g(n) \in O(h(n))$.

24. What property of logarithms explains why $O(\log_b(n^p)) = O(\log_b n)$ for all $p > 0$?

25. Prove by induction that the algorithm COUNTERFEIT DETECTOR I of Example 9 will require k weighings to find a counterfeit among 2^k coins.

26. Describe modifications to the COUNTERFEIT DETECTOR I algorithm that will make it find a counterfeit coin from among n coins given the conditions in Example 9. Complete Table 4-33 to support the claim that the number of weighings required (in the worst case) is $\lceil \log_2(n) \rceil$. (Be sure to use recursive thinking!)

27. I have a bag with 27 coins that appear identical, but I know one of them is counterfeit. The real coins weigh 1 oz each and the counterfeit one weighs more. I have a balance scale that tells me which of two loads is heavier. Using the COUNTERFEIT DETECTOR I algorithm, we can find the counterfeit in five weighings, but the following algorithm does better:

COUNTERFEIT DETECTOR II

(i) Weigh nine coins against nine other coins. If they do not balance, then the counterfeit coin is among the heavier nine; otherwise, the counterfeit coin is among the nine that were not weighed. Keep the nine that include the counterfeit coin.

(ii) Weigh three coins against three other coins. If they do not balance, then the counterfeit coin is among the heavier three; otherwise, the counterfeit coin is among the three that were not

weighed. Keep the three that include the counterfeit coin.

(iii) Weigh one coin against another one. If they do not balance, then the counterfeit coin is the heavier one; otherwise, the counterfeit coin is the one that was not weighed.

This algorithm requires only three weighings to find the counterfeit among 27 coins. Prove by induction that this algorithm will require k weighings to find a counterfeit among 3^k coins.

28. Think of a number in $\{1, 2, 3, \ldots, n\}$, and I will try to guess it. When I guess wrong, you tell me if your number is higher or lower than my guess. What is the smallest number k of guesses I can make to guarantee that I am right on or before the kth guess?

29. Match each of the recursive descriptions on the left with the function on the right that best characterizes the sequence's growth rate.

(a) $s_1 = s_2 = 1, s_n = 2s_{\lfloor n/3 \rfloor} + 1$ (A) $s_n = \frac{7}{6} \cdot n^{\log_2 3}$

(b) $s_1 = 1, s_n = 2s_{\lfloor n/2 \rfloor} + 3$ (B) $s_n = 4n$

(c) $s_1 = 1, s_n = 3s_{\lfloor n/2 \rfloor} + 5$ (C) $s_n = 2 \cdot n^{\log_3 2}$

30. In Exercise 29, the expression given in (B) describes a familiar *linear function*. The other two are called *sublinear* and *superlinear*. Can you tell from their graphs which is which? For positive constants a, b, and c (with $b > 1$), fill in the following blanks regarding the growth rate of the recurrence relation $s_n = a \cdot s_{\lfloor n/b \rfloor} + c$:

(a) If $a < b$, then the closed formula for s_n is a _____ function.

(b) If $a = b$, then the closed formula for s_n is a _____ function.

(c) If $a > b$, then the closed formula for s_n is a _____ function.

31. Prove directly by induction (i.e., do not just cite the master theorem) that the sequence s_n in Example 12, which counts the number of comparisons in the MERGE SORT algorithm, satisfies the inequality $s_n \leq n^2$ for all $n \geq 1$.

k	1	2	3	4	5	6	7	8	9	10
n	2	4	8	16	32	64	128	256	512	1024
c_n	8	28	80	208	512	1,216	2,816	6,400	14,336	31,744
$n \cdot \lceil \log_2(n) \rceil$	2	8	24	64	160	384	896	2,048	4,608	10,240
Ratio	4	3.5	3.3	3.3	3.2	3.2	3.1	3.1	3.1	3.1

Table 4-34   Table for Exercise 32

32. A divide-and-conquer recurrence relation can be examined with a spreadsheet* for special values of n. For example, given the sequence with recursive description $c_n = 2c_{n/2} + 3n$ where $c_1 = 1$, we can easily generate values of this sequence for values of n of the form 2^k. Note that the master theorem predicts that c_n is of order $n \cdot \log(n)$.

- Fill column A with the values of k: 0, 1, 2, 3, …, 20. (This can be done using "Edit Fill Series.")
- Put the formula =2^A1 into cell B1, and Fill Down the formula to cells B2 through B21.
- Put 1 (a value for c_1) in cell C1.
- Put the formula =2*C1+3*B2 into cell C2, and Fill Down the formula to cells C3 through C21.
- Put the formula =B1*log(B1,2) into cell D1, and Fill Down this formula to cells D2 through D21.
- Put the formula =C1/D1 into cell E1, and Fill Down this formula to cells E2 through E21.

The spreadsheet now has select values of n in column B, terms of the given sequence in column C, values of the predicted growth rate $n\log_2(n)$ in column D, and the ratio of the sequence to its predicted growth rate in column E. A sampling of values from this spreadsheet is shown in Table 4-34. Since the ratios in the spreadsheet seem to be settling in around the value of 3, it appears that the values of c_n are trapped between $2n\log_2(n)$ and $4n\log_2(n)$, corroborating the prediction of the master theorem.

Use this same approach to verify the prediction of the master theorem for each of the following sequences:

(a) $c_n = 3c_{n/2} + n$, where $c_1 = 1$.
(b) $c_n = 2c_{n/2} + \frac{1}{2}n^2$, where $c_1 = 10$.
(c) $c_n = 3c_{n/2} + 2n^2$, where $c_1 = 0$.
(d) $c_n = c_{n/2} + 3n$, where $c_1 = 1$.
(e) $c_n = c_{n/2} + 3$, where $c_1 = 3$.

33. Use induction to prove the following inequalities about the sequences from Practice Problem 3:

(a) For all $n \geq 1$, $b_n \leq 2n$.
(b) For all $n \geq 1$, $c_n \leq 2n^2$.
(c) For all $n \geq 1$, $d_n \leq n^2$.

Chapter 4 Summary

4.1 Definitions, Diagrams and Inverses

Terms and concepts

- You should understand the definition of a *function* and the role of *domain*, *codomain* and *rule* in this definition.

- You should understand the definition of a *binary relation* and the role of *domain*, *codomain* and *rule* in this definition.

- You should undertand the definition of the *inverse* of a function or relation.

---

* To see how to use a spreadsheet to examine *all* values of a divide-and-conquer recurrence relation, use the help files in your program. For example, the LOOKUP command in Microsoft Excel® can be used for this purpose.

Working with functions and relations

- You should be able to use and convert between the different representations of a function's rule, including *tables*, *algebraic expressions* and *arrow diagrams*.

- You should be able to use and convert between the different representations of a relation's rule, including *tables*, *algebraic expressions*, *English descriptions*, *one-set arrow diagrams*, and *two-set arrow diagrams*.

- You should be able to determine if a binary relation is a function.

- You should be able to determine the inverse of a binary relation regardless of how the rule is represented.

- You should be able to determine whether or not a given function has an inverse function.

- When a function does have an inverse, you should be able to determine the inverse function regardless of how the function's rule is represented.

- You should be able to prove that two given functions are inverses of each other.

4.2 The Composition Operation

Terms and concepts

- You should know what we mean by the *composition of functions* f and g, written as $f \circ g$, and be able to distinguish between $f \circ g$ and $g \circ f$.

- You should know what the *identity function* is.

- You should know what we mean by the *composition of relations* R and S, written as $R \circ S$, and be able to distinguish between $R \circ S$ and $S \circ R$.

Working with compositions

- You should be able to find the composition of two functions when the rules are given as arrow diagrams, sets of ordered pairs, algebraic equations, or English descriptions. In the latter case, you should understand the English-language meaning of the composite function.

- You should be able to find the composition of two relations when the rules are given as arrow diagrams, sets of ordered pairs, algebraic equations, or English descriptions. In the latter case, you should understand the English-language meaning of the composite relation.

- You should understand the meaning of inverse functions in terms of composition and the identity function, and be able to prove that two given functions are inverse functions using this idea.

4.3 Properties of Functions and Set Cardinality

Terms and concepts

- You should be able to identify when a function is *not invertible*.

- You should be familiar with the terms *one-to-one* and *onto*.

- Given a function $f : A \to B$ on finite sets A to B, you should know how the properties of one-to-one and onto correspond to the relationship between the sizes of the domain and codomain. If you studied the pigeonhole principle in Section 2.5, you should understand how these relationships are equivalent to the basic version of that principle.

- You should be familiar with the term *cardinality* and the formal definition of when two sets have the *same cardinality*.

- You should know the terms *countable* and *uncountable* as they relate to the cardinalities of infinite sets.

Working with function properties and set cardinality

- You should be able to use the terms one-to-one and onto in the context of functions whose rules are given by arrow diagrams, sets of ordered pairs, algebraic equations, or English descriptions.

- You should be able to produce a specific counterexample to illustrate that a given function is *not one-to-one*.

- You should be able to produce a specific counterexample to illustrate that a given function is *not onto*.

- Given any (finite) set A, you should be able to explicitly find the set B Cantor constructed in his famous proof of Theorem 9.

Proofs

- You should be able to write a well-structured mathematical proof that a function is *one-to-one*.

- You should be able to write a well-structured mathematical proof that a function is *onto*.

- You should be able to prove when simple infinite sets like $\mathbb{N}$ or $\mathbb{Z}$ have the same infinite cardinality, and you should be familiar with the proof that two infinite sets like $\mathbb{N}$ and $\mathcal{P}(\mathbb{N})$ do not have the same infinite cardinality.

4.4 Properties of Relations

Terms and concepts

- You should know the meaning of the terms *reflexive*, *antisymmetric*, and *transitive*, and the use of the term *partial order* to describe relations with all three properties.

- You should realize that a *Hasse diagram* is a simplified arrow diagram, and you should be able to draw the Hasse diagram for a relation that is a partial order.
- You should understand the notion of a *strict partial order*, and the corresponding *irreflexive* property.
- You should be able to distinguish between a partial order and a *total order* by looking for *incomparable pairs*.

Investigating and proving properties

- For a given relation, you should be able to determine whether or not it satisfies the various properties we have discussed: reflexive, irreflexive, antisymmetric, transitive.
- You should be able to use the result of this analysis to determine if the relation is a partial order, a strict partial order, or neither.
- You should be able to prove that particular relations satisfy the reflexive property.
- You should be able to prove that particular relations satisfy the transitive property.
- By thinking of the antisymmetric property in a slightly different (but logically equivalent) manner, you should be able to prove that particular relations satisfy the antisymmetric property.

4.5 Equivalence Relations

Terms and concepts

- You should understand what we mean by a *partition* of a set. Partitions were introduced briefly in the preceding chapter.
- You should know the connection between an *equivalence relation* and the corresponding partition. For a given equivalence relation, you should be able to describe the corresponding partition.
- You should know that an equivalence relation is characterized by three properties: *reflexive, symmetric,* and *transitive*. The reflexive and transitive properties are the same properties discussed in the preceding section for partial orders.

Investigating and proving properties

- By exploiting the connection between an equivalence relation and the corresponding partition, you should be able to determine if a given relation is an equivalence relation.
- For a given relation, you should be able to determine whether or not it satisfies the various properties of an equivalence relation: reflexive, symmetric, transitive.
- You should be able to prove that particular relations satisfy the reflexive property, the symmetric property, and the transitive property.

4.6 Numerical Functions in Discrete Math

Terms, concepts, and formulas

- You should understand the *base ten logarithm* of a positive real number r (written $p = \log_{10} r$), and the connection to the expression $10^p = r$.
- You should realize that the same type of function can be defined for other bases. For example, $p = \log_2 r$ means $2^p = r$.
- You should know what we mean by the *floor* and *ceiling* functions, written as $\lfloor x \rfloor$ and $\lceil x \rceil$, respectively.
- You should realize that numerical *operations* (addition, etc.) can be viewed as functions, and be able to interpret and use the corresponding notation $Sum(i, j), Prod(i, j)$, and so on.
- You should be thoroughly familiar with these properties of logarithms, and realize that similar properties hold for other bases.
  - $\log_{10}(a^b) = b \cdot \log_{10}(a)$
  - $\log_{10}(a \cdot b) = \log_{10}(a) + \log_{10}(b)$
  - $\log_{10}(a/b) = \log_{10}(a) - \log_{10}(b)$

Solving problems

- You should be able to use the definition and properties of logarithms to solve various problems. For example: *What is the largest value of n for which you can represent n^n exactly in a 32-bit integer in a computer?*
- You should be able to use the floor and ceiling function to solve various problems. For example: *How many multiples of k are in the set $\{m, m+1, m+2, \ldots, n\}$?*
- You should be able to use compositions of these functions – for example, $\lceil \log_2 n \rceil$ – to solve problems.

4.7 Excursion: Iterated Functions and Chaos

Terms and concepts

- You should know what is meant by the *iterated function sequence for f starting at a_0*, and the associated concept of a *cycle of period k*.
- You should realize that some iterated function sequences demonstrate *sensitivity to initial conditions*, and be able to use this to explain the so-called "*butterfly effect*."
- You should realize that sensitivity to initial conditions is an important property in the mathematical systems studied by *chaos theory*.
- You should be familiar with the *chaos game* and the resulting *Sierpinski triangle*.

Calculations and proofs

- You should be able to calculate values for iterated function sequences for various choices of a_0, giving complete or partial arrow diagrams, and locating cycles when they exist.
- Using algebraic methods, you should be able to locate all cycles of length 1 and length 2 for given functions.
- Using a calculating device such as a spreadsheet, you should be able to demonstrate that certain functions are sensitive to their initial conditions.
- You should be able to prove that a given sequence eventually reaches one of several cycles, using mathematical induction.

4.8 Excursion: Growth of Functions

Terms and concepts

- You should recognize the use of the word *complexity*, or *time complexity*, to indicate the time *efficiency* of an algorithm.
- You should realize that a sequence of real numbers starting with a_0 is equivalent to a function $f : \mathbb{N} \to \mathbb{R}$.
- You should understand the language of function growth. Specifically, you should understand formal statements of this type:
  — $f(n) \in O(g(n))$
  — $f(n) \in \Theta(g(n))$
  — $f(n) \in \Omega(g(n))$
  You should also recognize the informal use of phrases such as "order n^2" to describe function growth.
- You should know, and understand the importance of, the hierarchy of complexities, part of which is given here:

$$...O(\log_2 n) \subseteq ... \subseteq O(n^{1/3}) \subseteq O(n^{1/2}) \subseteq O(n)$$
$$\subseteq O(n^2) \subseteq ... \subseteq O(2^n)...$$

- You should recognize the form of *divide and conquer recurrences*.

Determining growth rates

- You should understand that we use functions to measure the complexity of algorithms, and that the main emphasis of complexity analysis involves comparing the behavior of functions as the size of the input increases.
- You should be able to recognize and make correct statements using O, Θ, and Ω – for example, $\frac{n^2-n}{2} \in O(n^2)$, and $\frac{n^2-n}{2} \notin O(n)$.
- You should be able to examine a sequence empirically using a spreadsheet, in order to investigate its complexity properties.
- You should be able to apply Theorem 5 to establish the growth rate of recursively defined functions.
- You should be able to apply Theorem 9 (the *Simplified Master Theorem*) to establish the growth rate of divide and conquer recurrences.

Proofs

- You should be able to use algebraic methods or mathematical induction to prove that one function is eventually larger than another, in a variety of contexts. For example:
  — The functions may be given in closed form.
  — The functions may be given recursively.
  — The functions may involve the use of the log, *floor*, and *ceiling* functions.
- For particular functions, you should be able to prove statements involving O, Θ, and Ω.

5 | **Combinatorics**

This chapter introduces the basic ideas of combinatorics, one of the major areas of discrete mathematics. Many people refer to combinatorics questions as "counting problems." Although this is an oversimplification, it is true enough of the topics that we will study in this chapter. We have of course seen counting problems in earlier chapters as we studied the sizes of various sets to better understand the nature of operations on sets and as we discussed the connection between properties of a function (like *one-to-one* and *onto*) and the relative sizes of the sets that comprised the function's domain and codomain. Hence, combinatorics is very much part of the other topics we have studied up to this point, and we will soon see that it is essential to the study of probability in the following chapter as well.

We begin the present chapter with some traditional techniques for answering questions of the form "How many...?" Along the way, we will encounter some less traditional techniques such as recursive modeling applied to more difficult problems. These techniques can, in turn, be applied to other topics such as methods for finding closed formulas for recurrence relations. We end the chapter with methods for finding closed formulas for recursively defined sequences of the type that we encountered in the first two chapters of the book.

In all this, an important ingredient in the study of combinatorics is the representation of new objects or situations in terms of simpler objects. A large part of this is the ability to recognize when two problems are actually the same. This will become one of the most valuable skills you can take away from your study of combinatorics.

5.1 Introduction

If you ever browse through a standard pocket dictionary, you will notice that the word "combinatorics" does not appear within its pages. You may conclude that this is a word mathematicians made up on a slow day at the office. The fact is that the field of combinatorics encompasses many subjects of various flavors. Perhaps the best we can do is to examine some common properties that these share to develop a sense of what it is all about.

In combinatorics, we usually deal with finite "structures" (which we discuss below) and the properties of the counting numbers. Combinatorics traditionally addresses three types of questions: *existence*, *enumeration*, and *optimization*. That is, one could ask when a certain prescribed configuration exists, such as a solution to a puzzle. Satisfied that one exists, one can ask how many different solutions there are or the related question of how likely you are to find it. Finally, there may be some measure by which there is a best or worst solution (a solution that takes, e.g., more than 100 years to find would probably be considered "bad"), so one could ask to find a particular one.

For example, a salesperson wishes to visit 10 particular cities, but he has a limited budget. Equipped with an airline schedule, he first sets out to determine if he can find a sequence of flights that will take him to all 10 cities on his budget. Happily, he finds out that he can, so he then greedily tries to determine which airline is the least expensive. Discouraged by the enormous number of ways to schedule his trip and the intractability of the problem of finding the very best deal, he quits his job, becomes a mathematician, and is suddenly more popular at parties.

Enumeration questions especially abound in science and elsewhere: "How many different tickets are there for the Pennsylvania lottery?" "How many samples from a box of spare parts should be tested to reasonably conclude that all parts in the box are acceptable?" "Which of two computer algorithms will run in the shorter amount of time?"

The point is that all three types of combinatorics problems can commonly arise. We will mostly be concerned with enumeration and existence questions in this book. Optimization questions are the subject of broad areas of mathematics such as operations research and calculus of variations, which generate a lot of interest in their own right. We will see some examples of optimization questions, although time will restrain us from discussing the solutions in much detail.

Finite Structures

In combinatorics, we work with the positive integers and the most basic of finite structures, the finite *set* (e.g., $\{1, 2, 3\}$ or $\{\text{John, George}\}$). Note that in a set, the order in which the elements are listed is irrelevant and each element needs to be listed only once—that is, the only thing that distinguishes between two sets is which objects are present and which are not. For example, $\{3, 1, 2, 3\}$ and $\{\text{George, John}\}$ describe the same two sets as above. Enumerative problems in combinatorics are usually stated as questions of the form "How many ways are there to ...?" That is, a finite set of objects or events will be described, and we will be asked how many elements the set contains.

Example 1 *The question "How many U.S. states have a name which begins with the letter 'A'?" is the same as the question "How many elements are in the set {Alabama, Alaska, Arizona, Arkansas}?"*

In just a couple of lines, we have already seen a set of numbers, a set of people, and a set of states. In each case, there is a larger "universe" from which the set is taken. The universes in the examples above might be the set of all living people, the set of all U.S. states, and the set of all natural numbers. In this sense, the sets we described are more properly referred to as *subsets* of the larger sets—that is, sets whose elements all derive from the larger set.

Example 2 *The question "How many ways can two winners be chosen for prizes from a class consisting of just four people {Andrew, Bob, Carly, Diane}?" is the same as the question "How many elements are in the set*

$$\{\{A, B\}, \{A, C\}, \{A, D\}, \{B, C\}, \{B, D\}, \{C, D\}\}?$$

where we use only the first letter of each person's name for brevity.

The previous example made an assumption about the way we should represent "choosing two people for prizes." Using a *set* of two people to represent those two people being chosen implicitly means that (1) there is no difference between the two prizes, and (2) it is impossible for the same person to win both prizes. Without these assumptions, we would have chosen different representations:

Example 3 *Consider once again the question "How many ways can two winners be chosen for prizes from a class consisting of just four people {Andrew, Bob, Carly, Diane}?"*

1. *Assuming that there is a first prize and a second prize, how can this question be written as "How many elements are in the set...?"*
2. *Assuming that these are two different door prizes (so it matters who gets which prize) and that the same person could win both, how can this question be written as "How many elements are in the set...?"*
3. *Assuming that the prizes are identical door prizes (so it doesn't matter who gets which prize) and that the same person could win both, how can this question be written as "How many elements are in the set...?"*

SOLUTION

1. We will simply list the winners so that the first person listed gets first prize and the second person listed gets second prize. Then the question is "How many elements are in the set

$$\{AB, BA, AC, CA, AD, DA, BC, CB, BD, DB, CD, DC\}?"$$

2. We will simply list the winners so that the first person listed gets one door prize and the second person listed gets the other door prize. Then the

question is "How many elements are in the set

$$\{AB, BA, AC, CA, AD, DA, BC, CB, BD, DB, CD, DC, AA, BB, CC, DD\}?"$$

3. Since the prizes are identical, we will simply list the winners in alphabetical order, with a person listed twice if he or she wins both prizes. Then the question is "How many elements are in the set

$$\{AA, AB, AC, AD, BB, BC, BD, CC, CD, DD\}?"$$

$\square$

This means that our original innocent question in Example 2 has four possible different answers, depending on what assumptions are made. We summarize the answers in Table 5-1, showing the number of ways the prizes can be given for each of the four interpretations of the problem. The fact that our very first question had four different answers might understandably cause some concern, but we can avoid these difficulties by being very explicit in the assumptions for our problems. A key ingredient for accomplishing this is to agree on names and definitions for the four different representations used in the problem above. Since all the problems in this chapter can be represented using one of these basic structures in some shape or form, this clarification will go a long way toward avoiding confusion. For this same reason, we should talk a bit more about each of these structures.

In the discussion of the question above, we used two questions about the prizes to determine a precise interpretation of the problem. In terms of the representation of the problem, the first question "Are the prizes different?" really asks the question "Does the order of A and B matter in our *representation*?" If the prizes are different, then we need, for example, both AB and BA on our list—the order in which we list the winners makes a difference in what they win.

An *ordered list* is a finite sequence of objects in which order is important (i.e., there is a first object, a second object, etc.). For example, in $abcb$ we are concerned not only with what objects are there, but also with how many times and in what order they occur. So $bbca$ and abc are ordered lists, each different from $abcb$ and each other even though all three lists just use the letters a, b, and c. Note that these particular examples are ordered lists of letters although the objects in the list can theoretically be anything.

A natural example of an ordered list is a numeral. Every number can be represented as a list of digits, each one from $\{0, \ldots, 9\}$, and clearly the order of the digits is important to the value of the number. For example, 1,132 and 1,321 are each lists of the digits 1, 1, 2, 3, but they represent different numbers because the digits are in different orders.

An *unordered list* is a finite sequence of objects in which we are concerned with how many of each object occur, not with the order in which they are given. For example, 1, 2, 3, 1 and 1, 1, 2, 3 are considered to be the same unordered list of

		Are the prizes different?	
		Yes	**No**
Can a person win both prizes?	**Yes**	16	10
	No	12	6

Table 5-1 Four Interpretations of Example 3

numbers, while 1, 2, 3 is a different one since here the number 1 occurs only once. It might be more natural to think of the numbers as being "types" of things, so 1, 1, 2, 3 means that we are listing two things of type 1 and one thing of types 2 and 3. We will give some natural circumstances in which this is a desirable type of structure below.

A bridge hand[*] is a nice example of an unordered list—certainly the order of the cards in a hand means nothing to a bridge player. The hand is determined only by which cards are present. Another example is a bag of groceries. The fact that a bag contains two oranges and a gallon of milk is enough to determine its contents—the order of the items within the bag is irrelevant. This second example is used by many authors to capture the essence of the unordered list structure to the extent that they use the word *bag* for an unordered list.

We now return our attention to the questions in the table, focusing on the second question "Can a person win both prizes?" If the answer is yes, then entries such as AA are allowed in our list of winners. We can see that the question could be phrased more generally as "Are repetitions allowed?" That is, when we make the list of winners, are we allowed to repeat a person?

The observant reader may have already noticed a redundancy in the fact that a *set* is really an unordered list with the restriction that we can take no more than one of each type of available object. Similarly, an ordered list with this same restriction is usually called a *permutation*. We use the special words "permutation" and "set" only because they refer to commonly encountered objects and we want to be efficient in talking and writing about them. Because we have this special terminology for the situation where repetition is not allowed, we will use the general terms *ordered list* and *unordered list* to indicate the more general situation where repetition is allowed. Table 5-2 summarizes the terminology to be used. Permutations and sets are so important that we have particular notation for these two special cases. The number of r-element subsets (sometimes called r-combinations) of the set $\{1, 2, \ldots, n\}$ is denoted by $C(n, r)$, and the number of permutations of length r using elements of the set $\{1, 2, \ldots, n\}$ is denoted by $P(n, r)$.[†] In the next several sections we will develop tools that will allow us to numerically evaluate $C(n, r)$ and $P(n, r)$, which in turn allows us to answer questions about the more general ordered and unordered lists.

Practice Problem 1   *Decide which of the four structure types (set, unordered list, permutation, or ordered list) best characterizes the objects in each of the following situations. Can any of the following have more than one answer? List the additional assumptions you must make for your representation to be correct.*

		Does order matter?	
		Yes	**No**
Are repetitions allowed?	**Yes**	Ordered list	Unordered list (bag)
	No	Permutation	Set

Table 5-2   Terminology for Four Basic Structures

---

[*] A bridge hand is played in much the same way as a hand in the game of hearts, if that is more familiar.

[†] This may be written differently in other books or on your calculator. The number of permutations of length r can also be expressed as $_nP_r$ or P^n_r. Similarly, the number of subsets of size r can be expressed as $_nC_r$, C^n_r, or $\binom{n}{r}$. We use our notation simply because it is easier to type.

(a) *Dealing a five-card (draw) poker hand*

(b) *Dealing a two-card blackjack hand*

(c) *Creating a game schedule for your favorite sports team*

(d) *Filling your orange plastic jack-o'-lantern for trick or treat.*

Organization in Counting

There is a famous story about the prolific mathematician Carl Friedrich Gauss (1777–1855). As a child, Gauss had a teacher who set his students to doing meaningless sums to occupy their time. The teacher was not very creative even at coming up with meaningless sums, so the students were told to add the numbers from 1 to 100. While the other students started adding 1 to 2, and then this sum to 3, and so on, Gauss noticed that the sum could be written $(1 + 100) + (2 + 99) + (3 + 98) + \cdots + (50 + 51)$—in other words, 101 added to itself 50 times resulting in 5,050. So Gauss finished well before everyone else and was beaten up at recess for being such a geek.

The moral of the story is that some seemingly hard problems can often be done in a simple way if you give some thought to organization before you begin. In our problems, we could try to write down everything we are supposed to be counting and then point to each one and say "one," "two," ..., but we would rather be like Gauss. In fact, one could argue that the main emphasis of this chapter is simply organization. Along this line, we will try throughout this chapter to uncover and understand particularly elegant answers to questions and try not to "get beaten up" in the process.

Example 4 *How many permutations of the letters in MATH are there?*

SOLUTION   If we write them down until we cannot think of any more, we might get **LIST 1** below. The drawback to this is that we might never be sure if the list is complete. Another drawback is that once the list is made, we will have to then count the length of the list, and since it is too long (for most people) to count on their fingers and toes, we may not be sure of the answer we get.

LIST 1: *MATH, AMTH, AMHT, THAM, AHMT, HAMT, HMAT, MHAT, THMA, MHTA, HMTA, HATM, AHTM, MAHT, TMAH, MTHA, HTMA, TMHA, ATMH, TAHM, ATHM, TAMH, MTAH,* and *HTAM*

On the other hand, we might organize our answer as **LIST 2** in Table 5-3, where the permutations are sorted so that each column contains permutations starting with the same letter. (Within each column, the entries are arranged based on the second letter.) It is clear that it is easier to be sure of the size of **LIST 2** than **LIST 1**, but we also notice that it is easier to check for completeness. Thus, by writing the list down in a sensible way, it is much easier to find its length and be sure of it.   ☐

Example 5   *Explain how to organize the outcomes to answer the question "How many ways can two winners be chosen for prizes from a class consisting of just four people {Andrew, Bob, Carly, Diane}?" from the beginning of this section. (In this example, one person is allowed to win both prizes.)*

MATH	AMTH	TMAH	HMAT
MAHT	AMHT	TMHA	HMTA
MTAH	ATMH	TAMH	HAMT
MTHA	ATHM	TAHM	HATM
MHAT	AHMT	THMA	HTMA
MHTA	AHTM	THAM	HTAM

Table 5-3   LIST 2

SOLUTION   As we have seen, the fact that one person is allowed to win both prizes means that repetition is allowed in the representations of the outcomes. Since we are not told if the prizes are different (i.e., if order matters in the representations), we will answer the question under both possible circumstances.

If the prizes are different, then our representations are ordered lists since order matters and repetition is allowed. We will organize the data in Table 5-4 with four columns corresponding to the possible winners of the first prize. Within each column we arrange the rows in alphabetical order by the second prize winner. From this table, it is easy to see that there are 16 possible outcomes.

If the prizes are not different, then our representations are unordered lists since order does not matter and repetition is allowed. This time we can organize the outcomes in Table 5-5 with four columns corresponding to the possible winners of the first prize. To avoid listing the same outcome twice, we now adopt the convention that within each column we only list those outcomes whose second prize winner does not come alphabetically before the first prize winner. This avoids, for example, having both AB and BA (which represent the same outcome) in our table. In this case, the table has columns of varying lengths, but we can easily see that the total is $4 + 3 + 2 + 1 = 10$.   □

Practice Problem 2   *Consider the question "How many ways can two winners be chosen for prizes from a class consisting of just four people {Andrew, Bob, Carly, Diane}?" from the beginning of this section, but assume that we know the same person is not allowed to win both prizes.*

Andrew	Bob	Carly	Diane
AA	BA	CA	DA
AB	BB	CB	DB
AC	BC	CC	DC
AD	BD	CD	DD

Table 5-4   Prizes Are Different in Example 5

Andrew	Bob	Carly	Diane
AA	*BB*	*CC*	*DD*
AB	*BC*	*CD*	
AC	*BD*		
AD			

Table 5-5 Prizes Are the Same in Example 5

(a) *Make a table for the list of outcomes for this question, assuming that the prizes are different. In this case, the outcomes are represented as permutations.*

(b) *Make a table for the list of outcomes for this question, assuming that the prizes are the same. In this case, the outcomes are represented as sets.*

Example 6 *Represent the list of all outcomes of rolling a red six-sided die and a green six-sided die in an organized way.*

SOLUTION   We use Table 5-6 with the result of the green die labeling the columns and the result of the red die labeling the rows. □

Example 7 *Represent the list of all possible results of tossing a penny, a nickel, and a dime together.*

SOLUTION   The list of all possible results of tossing a penny, a nickel, and a dime together can be simplified by using three letters, for example, HTH, to reflect the result where we agree ahead of time to the convention that the first letter describes how the penny lands (in this example, H stands for "heads"), the second letter describes how the nickel lands (in this example, T for "tails"), and the last letter describes how the dime lands. We can illustrate these outcomes using a "tree" structure, where the first branch denotes the result of the penny, the second branch the result of the nickel, and the last branch the result of the dime (see Figure 5-1). One must only imagine walking along the tree from left to right, and think of tossing coins when one chooses a branch. This will turn out to be a useful alternative to imagining tables in many circumstances. □

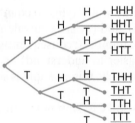

Figure 5-1 Tree of outcomes in Example 7.

	Green 1	Green 2	Green 3	Green 4	Green 5	Green 6
Red 1	(1, 1)	(1, 2)	(1, 3)	(1, 4)	(1, 5)	(1, 6)
Red 2	(2, 1)	(2, 2)	(2, 3)	(2, 4)	(2, 5)	(2, 6)
Red 3	(3, 1)	(3, 2)	(3, 3)	(3, 4)	(3, 5)	(3, 6)
Red 4	(4, 1)	(4, 2)	(4, 3)	(4, 4)	(4, 5)	(4, 6)
Red 5	(5, 1)	(5, 2)	(5, 3)	(5, 4)	(5, 5)	(5, 6)
Red 6	(6, 1)	(6, 2)	(6, 3)	(6, 4)	(6, 5)	(6, 6)

Table 5-6 Solution to Example 6

Practice Problem 3 *Explain an orderly way to organize each of the following lists so that the number in each list is easy to see:*

(a) The list of all permutations of the three letters in the word HAT
(b) The list of all (different-looking) permutations of the four letters in the word BOOK
(c) The list of all three-element sets using letters from the word GAMES

Combinatorial Equivalence

There is a concept in discrete mathematics that seems to be simultaneously the least understood by beginning students and the most ignored by textbooks. On the likely chance that there is a correlation between these two situations, a word on this subject seems appropriate. Formally, it is the idea of a "one-to-one correspondence" between finite sets, but it is perhaps more easily thought of as recognizing when two problems have the same answer. There are several levels of difficulty among applications of this idea. The following examples are meant to illustrate these levels. In them we will be concerned with seeing when two problems are equivalent—we will not actually answer any of the questions now, but you should keep them in mind.

The easiest example of this phenomenon is illustrated by the following pair of questions:

(i) How many even multiples of 3 are there between 1 and 100, inclusive?
(ii) How many multiples of 6 are there between 2 and 96, inclusive?

Notice that if we had explicitly listed the numbers that we were trying to count, in both cases we would have written 6, 12, 18, ..., 96. So (i), and (ii) are equivalent because they are asking you to count exactly the same set of numbers; consequently, you can choose to answer (i) or (ii) (whichever you find easier to understand) and you know that the other one has the same answer. Unfortunately, this example is so simple as to be of essentially no use to us later on.

Perhaps more relevant is the next application, which is still easy, but not as silly as the previous example.

Example 8 *Without answering either question, explain why (i) and (ii) have the same answer.*

(i) How many multiples of 3 are there between 100 and 300, inclusive?
(ii) How many integers are there between 34 and 100, inclusive?

SOLUTION For each number x in the list counted by (ii), the number $3x$ is in the list counted by (i). This can be illustrated as follows:

List (i)	102	105	108	⋯	297	300
	↕	↕	↕		↕	↕
List (ii)	34	35	36	⋯	99	100

Since each item in list (i) corresponds to one item in list (ii) and vice versa, these two lists have the same length. □

Most relevant to our purposes are examples where we can create a correspondence between questions about different structures.

Example 9 *Without answering either question, explain why (i) and (ii) have the same answer.*

(i) *How many ways are there to distribute three balls—one red, one blue, and one green—to 10 people? (Someone getting more than one ball is allowed.)*

(ii) *How many integers are there from 0 to 999, inclusive?*

SOLUTION   Since in (i) we are considering a set of distributions of balls among people and in (ii) we are looking at a set of numbers, we are obviously not counting the same set, but we claim the two sets that are being counted do have the same size. Certainly, this could be the case "accidentally," but we are interested in finding a reason for it. And if we can find a reason for it, then as before, there's no need to answer both questions. In this example, we can imagine the people in (ii) being numbered $0, 1, \ldots, 9$, and we will record three numbers from left to right as "who gets the red ball," "who gets the blue ball," and "who gets the green ball."

For example,

Distribution	Person 1 gets red Person 0 gets blue Person 1 gets green	Person 2 gets red Person 2 gets blue Person 9 gets green	Person 3 gets red Person 7 gets blue Person 5 gets green
	↕	↕	↕
Integer	101	229	375

So for each way of distributing the balls, there is a corresponding integer between 0 and 999. Observing that every such integer corresponds to a unique distribution of balls, we can conclude that (i) and (ii) are counting sets of the same size. □

This sort of argument always requires two components: a rule establishing a correspondence between the problems, and an informal explanation as to why every element in the set associated with each problem corresponds to a unique element in the set associated with the other problem. The previous example illustrates the usefulness of this technique as well—we have reduced the complicated-sounding problem (i) to a trivial one (ii). This is why the recognition of equivalence between two problems is fundamentally important. It allows us to "transform" an unfamiliar problem into a familiar one.

We saw above that sometimes establishing that two problems will have the same answer can be useful. In no subject is this more true than in combinatorics. The next example shows the power of this idea even in the absence of any good way to deal with either problem.

Example 10 *Without answering either question, explain why (i) and (ii) have the same answer.*

(i) *How many sets of size 2 can be made using elements from $\{1, 2, 3, \ldots, 9\}$?*

(ii) *How many sets of size 7 can be made using elements from $\{1, 2, 3, \ldots, 9\}$?*

SOLUTION We must describe how to link each entry in one list with a unique entry in the other list, so let's try it by considering a generic entry in the list described by (i). Let S denote the set $\{1, 2, 3, \ldots, 9\}$. With each two-element subset $T = \{a, b\}$ of S, associate the seven-element set $S - T$, consisting of the seven elements in $\{1, 2, 3, \ldots, 9\}$ *other than* $\{a, b\}$. For example,

$$\{3, 5\} \qquad \{1, 9\} \qquad \{2, 3\} \qquad \{6, 7\} \qquad \cdots$$
$$\updownarrow \qquad\qquad \updownarrow \qquad\qquad \updownarrow \qquad\qquad \updownarrow$$
$$\{1, 2, 4, 6, 7, 8, 9\} \quad \{2, 3, 4, 5, 6, 7, 8\} \quad \{1, 4, 5, 6, 7, 8, 9\} \quad \{1, 2, 3, 4, 5, 8, 9\} \cdots$$

This rule is fully and uniquely reversible, so this links the entries in the two lists with a one-to-one correspondence. ☐

Example 11 *Without answering either question, explain why (i) and (ii) have the same answer.*

(i) *How many different outcomes are there in flipping a coin five times in a row?*

(ii) *How many sets can be made using elements from $\{1, 2, 3, 4, 5\}$?*

SOLUTION Picture five numbered blank spaces into which you will enter "H" or "T" as shown below:

_____	_____	_____	_____	_____
1	2	3	4	5

We will describe how to build an entry in the list for (ii) given a generic entry in the list for (i). A generic entry for (i) can be written as an ordered list of length 5 using H's and T's to stand for "heads" and "tails." For example, THHTH is such an entry. We place the entry into the five numbered blank spaces, then simply write down (inside the {} brackets) the positions of the H's. In our example of THHTH, the H's occur in the second, third, and fifth positions, so we write down the set $\{2, 3, 5\}$. ☐

Explore more on the Web.

Practice Problem 4 *Illustrate the rule of the preceding example by filling in the missing entries in Table 5-7.*

We conclude this section with one final example that describes a complicated correspondence that we will see again in this chapter.

Result of coin tosses	THHTT	HTTTT			HHHHH	
Subset of {1, 2, 3, 4, 5}			{2, 5}	∅		{1, 4, 5}

Table 5-7 Table for Practice Problem 4

 Example 12 *The following are equivalent:*

(i) *How many positive integer solutions are there to* $x + y + z = 21$?

(ii) *How many two-element subsets of* $\{1, 2, \ldots, 20\}$ *are there?*

SOLUTION   We must describe how to link each entry in one list with a unique entry in the other list, so let's try it by considering a generic object described by (i), an ordered list of three positive integers x, y, z for which $x + y + z = 21$. For example, $1 + 1 + 19$, $2 + 3 + 16$, and $3 + 2 + 16$ are all different solutions counted in (i). Notice that once x and y are chosen, there is no choice about the value of z since the three numbers must sum to 21.

Our correspondence will be to associate with each solution x, y, z the subset $\{x, x + y\}$. Since y cannot be zero, we can be sure that x and $x + y$ are two different values, and since z is positive, the largest $x + y$ can be is 20. Hence, we can be sure that the set $\{x, x + y\}$ is a two-element subset of $\{1, 2, \ldots, 19, 20\}$, so this set will certainly be listed in (ii) above.

Table 5-8 illustrates some examples of this correspondence. To see that this rule is reversible, we consider an item in (ii) and show that there is an item in (i) corresponding to it. For example, consider the subset $\{14, 17\}$. By the way the rule was given, it is clear that x must be 14, the smaller element of the set. Then since $x + y$ must equal 17, we see that y is 3. Finally, $x + y + z = 21$ shows that z is 4. It is easy to check that the set linked to this solution $14 + 3 + 4 = 21$ using the above correspondence is none other than the set $\{14, 17\}$. We can describe this pattern in general terms as follows. Given a subset $\{a, b\}$ of $\{1, 2, \ldots, 20\}$ with $a < b$, let $x = a$ and $y = b - a$, and then let $z = 21 - b$.

We can consequently be sure that our rule matches the answers to questions (i) and (ii) in a one-to-one manner. Hence, we can be sure that these two questions have the same answer **even though we do not yet know what this answer is!**   □

Practice Problem 5   *In Example 12, we checked the reversibility of the original rule using the following reasoning: Given a set $\{a, b\}$ with $a < b$, let $x = a$, $y = b - a$,*

Solution to $x + y + z = 21$		Subset $\{a, b\}$ of $\{1, 2, \ldots, 20\}$
$1 + 1 + 19$	→	{1, 2}
$3 + 2 + 16$	→	{3, 5}
$2 + 3 + 16$	→	{2, 5}
$16 + 1 + 4$	→	{16, 17}
$18 + 2 + 1$	→	{18, 20}

Table 5-8 Table for Example 12

Set $\{a, b\}$	$\{1, 2\}$	$\{19, 20\}$	$\{1, 20\}$	$\{1, 10\}$	$\{10, 20\}$	$\{11, 12\}$
Solution x, y, z						
Set $\{x, x + y\}$						

Table 5-9   Table for Practice Problem 5

and $z = 21 - b$, and then the solution x, y, z will correspond to $\{a, b\}$. For each of the two-element subsets in Table 5-9, give the corresponding solution to $x + y + z = 21$, and check that the set $\{x, x + y\}$ is, in fact, $\{a, b\}$.

Solutions to Practice Problems

1 (a) A five-card poker hand has no repetition and the order in which the cards are received is irrelevant, so this is a set.

   (b) A two-card blackjack hand has no repetition, but since the first card is "down" and the second is "up" for all to see, the order does matter for the game, so this is a permutation.

   (c) This depends on the sport—the order of games certainly matters but repetition in a schedule is allowed in some sports (like baseball) and not others (like the schedule of Venus Williams's tennis matches at Wimbledon, a single-elimination tournament). Hence, this can be either a permutation or simply an ordered list.

   (d) Anyone who has dumped his or her candy on the floor and sorted it knows that the only thing that matters is how many of each piece of candy you get. Hence, this is an unordered list (of candy types) in which repetition is allowed.

2 (a) In Table 5-10, each column contains those permutations for which the person whose name is at the top of the column receives the first prize.

   (b) In Table 5-11, each column contains those sets for whom the person which name is at the top of the column is the first of the winners in alphabetical order.

Andrew	Bob	Carly	Diane
AB	BA	CA	DA
AC	BC	CB	DB
AD	BD	CD	DC

Table 5-10   Solution for Practice Problem 2(a)

Andrew	Bob	Carly	Diane
$\{A, B\}$	$\{B, C\}$	$\{C, D\}$	
$\{A, C\}$	$\{B, D\}$		
$\{A, D\}$			

Table 5-11   Solution for Practice Problem 2(b)

3 In each case we can make a table.

   (a) The list of all permutations of the three letters in the word HAT can be arranged as in Table 5-12 so that entries in the same column start with the same letter. This gives us three columns, each of which will have two rows, for a total of six objects.

   (b) The list of all different-looking permutations of the four letters in the word BOOK can be arranged as in Table 5-13 so that entries in the same column start with the same letter. This gives us three columns, but not every column has the same number of rows. Altogether this give us $3 + 3 + 6 = 12$ entries.

   (c) The list of all three-element sets using letters from the word GAMES can be arranged so that entries in the same column have the same first letter in alphabetical order. For example, in the E column goes every three-element set of letters for which E is the first letter in the set in alphabetical order. In this case, that means those sets that contain an E but

H	A	T
HAT	AHT	THA
HTA	ATH	TAH

Table 5-12
Solution for Practice Problem 3(a)

B	*O*	*K*
BOOK	OBOK	KBOO
BOKO	OBKO	KOBO
BKOO	OOBK	KOOB
	OOKB	
	OKOB	
	OKBO	

Table 5-13   Solution for
Practice Problem 3(b)

A	*E*	*G*
{A, G, M}	{E, G, M}	{G, S, M}
{A, G, S}	{E, S, G}	
{A, S, M}	{E, M, S}	
{A, E, G}		
{A, E, M}		
{A, E, S}		

Table 5-14   Solution for Practice
Problem 3(c)

not an *A*. With this organization, there will be three columns (do you see why there is no *M* column and no *S* column?), each of which has a different number of rows, so the best we can do is organize our list as in Table 5-14. This table does not have a nice shape, but we can tell that it is complete. There are $6 + 3 + 1 = 10$ such three-element sets.

4   The completed table is shown in Table 5-15.

5   The completed table is shown in Table 5-16.

Exercises for Section 5.1

1. As in Examples 2 and 3, convert each question into a question of the form "How many elements are in the set...?" In case of ambiguity, clearly state any assumptions you make about order and repetition. If the corresponding set is large, just describe how the elements are listed, and list five or six sample elements of the set.

   (a) As a reward for a job well done, a child is allowed to reach into a bag and grab any two candy bars. The bag contains five each of the following: KrazyKat Bar, MilkyMorning Bar, LusciousLemon Bar. How many different ways are there for the child to make her choice?

   (b) If the child in the previous problem plans to give the first bar she takes out of the bag to her brother and eat the second one herself, how many different ways are there for her to make her choice?

   (c) When packing for a short trip, a traveler chooses three shirts from a closet that contains nine shirts. How many ways are there to do this?

   (d) A shopper is buying three shirts from a store that stocks nine different types of shirts. How many ways are there to do this, assuming the shopper is willing to buy more than one of the same shirt?

Result of coin tosses	THHTT	HTTTT	THTTH	TTTTT	HHHHH	HTTHH
Subset of {1, 2, 3, 4, 5}	{2, 3}	{1}	{2, 5}	∅	{1, 2, 3, 4, 5}	{1, 4, 5}

Table 5-15   Table for Practice Problem 4

Set {a, b}	{1, 2}	{19, 20}	{1, 20}	{1, 10}	{10, 20}	{11, 12}
Solution x, y, z	1 + 1 + 19	19 + 1 + 1	1 + 19 + 1	1 + 9 + 11	10 + 10 + 1	11 + 1 + 9
Set {x, x + y}	{1, 2}	{19, 20}	{1, 20}	{1, 10}	{10, 20}	{11, 12}

Table 5-16   Solution for Practice Problem 5

(e) An ice cream shop offers eight flavors of ice cream. How many different two-scoop dishes are possible if the two scoops are placed side by side in the dish?

(f) An ice cream shop offers eight flavors of ice cream. How many different two-scoop dishes are possible if the two scoops are placed one on top of the other?

2. Follow the directions for Exercise 1.

(a) A state lottery game consists of drawing a ping pong ball from each of three machines containing ten balls labeled from 0 to 9. How many possible outcomes are there for the game? (A winning ticket must match the numbers drawn in the same order they were drawn.)

(b) A security box guarding a sensitive area has three buttons colored red, green, and blue. To enter the area, you must enter a security code, pressing four buttons in succession. How many different security codes are there?

(c) A store sells apples, bananas, and oranges. I want to buy two pieces of fruit to have for lunch. How many possible ways are there to do this?

(d) In the previous problem, if I'm buying one piece for myself and one for my best friend, how many possible ways are there for me to buy two pieces of fruit?

(e) Near the end of a clearance sale, a store has only three bicycles left for sale—a black one, a green one, and a red one. I want to buy two bicycles. How many possible ways are there to do this?

(f) In the previous problem, if I'm buying one bike for myself and one for my best friend, how many possible ways are there for me to buy two bikes?

3. For the situations described in Exercises 1 and 2, decide which of the four structure types (set, unordered list, permutation, or ordered list) best characterizes the situation.

4. Decide which of the four structure types (set, unordered list, permutation, or ordered list) best characterizes the objects in each of the following situations. Can any of the following have more than one answer?

(a) Dealing a 13-card hand for the card games bridge or hearts

(b) Selecting three officers—president, vice president, and secretary—for a club

(c) Rolling a pair of dice

(d) Grabbing 12 marbles from a jar containing red, green, and blue marbles

(e) Ordering pizza toppings from a menu of choices

(f) Setting a batting order for a baseball team

5. Example 6 presents an organized list of all outcomes of rolling a red six-sided die and a green six-sided die. Use that list to answer these questions.

(a) How many possible outcomes are there?

(b) Of these, for how many is the sum of the values on the two dice equal to 7? For how many is the sum of the values equal to 11? If you roll a pair of dice, are you more likely to roll a 7 or to roll an 11?

(c) For how many of the outcomes is the sum of the values 5 or less? What percentage of the possible outcomes does this represent?

6. Example 6 presents an organized list of all outcomes of rolling a red six-sided die and a green six-sided die. Use that list to answer these questions.

(a) Of the listed outcomes, how many are "doubles" (i.e., both dice have the same value)? What percentage of the possible outcomes does this represent?

(b) How many are doubles **and** have the sum of the two dice values less than 4?

(c) How many have a 5 on exactly one of the dice?

(d) How many have a 5 on at least one of the dice?

7. In Example 7, we presented a tree as a handy representation of the possible results of tossing a penny, a nickel, and a dime. Use that tree to answer these questions.

(a) How many possible outcomes are there?

(b) Of these, how many have exactly one head?

(c) How many have at least one head?

(d) How many have exactly two heads?

8. Use the tree from Example 7 to answer these questions.

(a) Of the outcomes shown, for how many does the result on the penny match the result on the dime?

(b) Which is more likely, that the result on the penny matches the result on the dime, or that they do not match?

(c) For how many of the outcomes do all three coins match?

(d) For how many of the outcomes do exactly two of the coins match?

(e) For how many outcomes is the number of tails less than the number of heads?

9. The tree in Figure 5-2 represents a "best of 3" tennis match between two players labeled A and B. In the ordered list representation, the letters A and B indicate a set won by Player A or B, respectively, where the first letter refers to the first set, the second letter to the second set, and the third letter to the third set when necessary. Use this tree to answer the questions that follow.

(a) How many different results are there?

(b) Of these, for how many does the match end after two games?

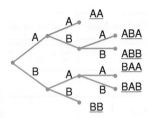

Figure 5-2 Game tree for Exercise 9.

(c) For how many does Player A win in exactly three games?

(d) Explain why the answer to part (c) is the same as the answer to this question: "How many ordered lists of length 2 taken from {*A*, *B*} have exactly one A?" (HINT: To win in exactly three games, how many of the first two games must Player A have won?)

10. The idea in Exercise 9 can be easily extended to a "best of 5" match in which the first player to win three sets wins the match. Answer the following questions about this type of match:

(a) Give a tree to represent a "best of 5" match between Players A and B.

(b) How many ways are there for the match to end after three games?

(c) How many ways are there for Player A to win in exactly four games?

(d) Explain why the answer for part (c) is the same as the answer to the question "How many ordered lists of length 3 taken from {*A*, *B*} have exactly two A's?"

(e) How many ways are there for Player A to win in exactly five games?

(f) Fill in the blanks. The answer to part (e) is the same as the answer to the question "How many ordered lists of length _____ taken from {*A*, *B*} have exactly _____ A's?"

11. In Example 4, we presented an organized list of all permutations of the letters in the word MATH. Use that list to answer these questions.

(a) How many permutations are there? Can you answer this without actually counting them all?

(b) Make a copy of the table in Example 4, and change each *H* to a *T*. Now some of the entries in the table are the same. For example, both *MATH* and *MAHT* have become *MATT*. Divide the *original* table entries into groups using the rule "Two original entries go in the same group if they now look the same on your copy." How many groups are there? How big is each group?

(c) Use part (b) to answer the question "How many distinct arrangements are there for the letters in the word *MATT*?" as a division problem.

12. Answer these questions by organizing the items to be counted in any way you like.

(a) How many different outfits can John form if he has three shirts (red, green, and yellow) to choose from, and two pairs of pants (black, white)?

(b) A new board game has a standard six-sided die, and a spinner with three colors, red, white, and blue. A player takes a turn by tossing the die and spinning the spinner. How many different possible results are there?

(c) A security box guarding a sensitive area has three buttons colored red, green, and blue. To enter the area, you must enter a security code, pressing four buttons in succession. How many different security codes are there?

(d) A shopper wants to buy a total of three pieces of fruit, and he only likes apples, bananas, and oranges. How many different ways are there for him to select the three pieces of fruit?

(e) How many arrangements are there of the letters in the word HEAR?

(f) Using your answer to part (e) as a starting point, how many arrangements are there of the letters in the word HEARD?

13. Answer these questions by organizing the items to be counted in any way you like.

(a) Sara does an online search for flights to visit her grandmother. She is given a choice of four flights to her grandmother's city, and a choice of three flights for the return trip. How many different ways are there for her to schedule her trip?

(b) How many different-looking arrangements are there of the letters in the word DEED?

(c) How many arrangements are there of the letters in the word GAMES?

(d) How many different-looking arrangements are there of the letters in the word PUZZLE?

(e) If we did a national survey asking people for their favorite two days of the week, how many different responses could we possibly get?

14. A *binary sequence* is simply an ordered list using only digits chosen from {0, 1}, where we usually suppress the commas that ordinarily separate list items. For example, 010, 110, and 011 are different binary sequences of length 3.

(a) How many binary sequences are there of length 3?

(b) How many binary sequences are there of length 4? Of these, how many end with a 1?

(c) How many binary sequences of length 4 use exactly two 1's and two 0's?

(d) How many binary sequences of length 5 do not have two adjacent 1's?

(e) How many binary sequences of length 5 look the same reading them forward as backward? (For example, 11011 counts since its reverse is also 11011, but 10110 does not count since it is not identical with its reverse, 01101.)

15. Answer these questions about the list of positive integers less than 100,000.

(a) How many multiples of 7 are there?

(b) How many times does the digit 9 appear?

(c) How many perfect squares are there?

16. Each of the following problems is about counting the factors of numbers in an organized way:

(a) How many positive factors does 300 have? (HINT: Make a table with three columns: Those divisible by 4, those divisible by 2 but not 4, and those not divisible by 2. Use the fact that $300 = 2^2 \cdot 3 \cdot 5^2$ to help you organize your table.)

(b) How many positive factors does 600 have?

(c) How many positive factors does 9,576 have? (HINT: $9,576 = 2^3 \cdot 3^2 \cdot 7 \cdot 19$.)

(d) A whole number is square-free if it is evenly divisible by no perfect squares (other than 1).

   i. How many of the positive factors of 300 are square-free?

   ii. How many of the positive factors of 9,576 are square-free?

17. This exercise explores a method that can be used to attack problems of the type given in Exercise 16.

(a) How many ordered lists of length 2 with entries taken from the set {0, 1, 2} are there?

(b) The answer to part (a) is the same as the answer to the question "How many positive factors does 36 have?" To see this, link each ordered list a, b to the positive factor $2^a 3^b$. Fill in Table 5-17 with examples showing how this correspondence works.

(c) Explain in your own words why this correspondence matches each ordered list entry to exactly one positive factor of 36, and is reversible. (HINT: Write $36 = 4 \cdot 9 = 2^2 \cdot 3^2$.)

18. These questions have the same answer.

   i. How many distinguishable arrangements of the letters in *OIIOO* are there?

   ii. How many two-element subsets of {1, 2, 3, 4, 5} are there?

A function that demonstrates this fact takes a given arrangement of the letters in *OIIOO*, and returns the list of positions containing the *I*'s.

Ordered list	Factor of 36
0, 0	$2^0 3^0 = 1$
0, 1	
1, 2	
2, 2	
	4
	6
	36

Table 5-17 Table for Exercise 17

(a) Describe in words the reverse function.

(b) Fill in Table 5-18 to illustrate the correspondence.

19. These questions have the same answer.

   i. How many eight-digit binary sequences have three 1's, no two of which are adjacent?

   ii. How many three-element subsets of {1, 2, 3, 4, 5, 6} are there?

Here is a function that demonstrates this fact: Given an eight-digit binary sequence containing three 1's, no two of which are adjacent, write down the three positions containing the 1's. Decrease the second number you wrote by 1, and decrease the third number by 2.

(a) Describe in words the reverse function.

(b) Fill in Table 5-19 to illustrate the correspondence.

OIIOO	{2, 3}
OOOII	
OOIOI	
IOOOI	
	{1, 4}
	{3, 4}
	{1, 3}
	{2, 5}

Table 5-18 Table for Exercise 18

10101000	$\{1, 2, 3\}$
01010100	
00100101	
00010101	
	$\{1, 3, 5\}$
	$\{1, 3, 6\}$
	$\{3, 4, 5\}$
	$\{2, 4, 6\}$

Table 5-19 Table for Exercise 19

20. Explain why (i) and (ii) have the same answer in each of the pairs of questions below. You do not need to actually answer either question.

   (a) i. How many distinguishable arrangements of the letters in *ABABA* are there?

   ii. How many five-digit binary sequences are there with exactly three 1's?

   (b) i. How many two-element subsets of the set $\{1, 2, 3, 4, 5\}$ are there?

   ii. How many three-element subsets of the set $\{1, 2, 3, 4, 5\}$ are there?

   (c) i. How many permutations of all five of the objects in $\{a, 1, x, 3, 9\}$ are there?

   ii. How many permutations of length 4 of the objects in $\{a, 1, x, 3, 9\}$ are there?

   (d) i. How many ways are there to flip three heads in five tosses of a coin?

   ii. How many three-element subsets of $\{1, 2, 3, 4, 5\}$ are there?

21. Explain why (i) and (ii) have the same answer in each of the pairs of questions below. You do not need to actually answer either question.

   (a) Rooks attack each other if they are on the same row or column of a chessboard.

   i. How many ways can eight rooks (which all look the same) be placed on a chessboard so that no two rooks can attack each other?

   ii. How many arrangements are there of the letters *ABCDEFGH*?

   (b) A positive factor of an integer n is a positive integer by which n can be evenly divided.

   i. How many positive factors does 537,824 have? (HINT: This number is $2^5 \cdot 7^5$.)

   ii. How many different outcomes result from rolling a pair of distinguishable six-sided dice?

   (c) i. How many different paths can a rook take to move from the lower-left corner to the upper-right corner of a chessboard if it never moves downward or leftward?

   ii. How many different arrangements are there of the following characters?

   $$AAAAAAABBBBBBB$$

22. The following problems look at the factors of a special kind of number that occurs frequently in discrete mathematics. The general explanation of the pattern is quite difficult to formalize at this level, but students who like a good challenge should enjoy working on it.

   (a) What is the largest power of 2 that evenly divides 11! ? NOTE: We write $n!$ (read "n factorial") for the product

   $$(n)(n-1)(n-2)\cdots(2)(1).$$

   (b) What is the largest power of 2 that evenly divides 234!?

   (c) Notice that if you write 11 in base two, it is 1011 that has three 1's in it, and $11 - 3 = 8$, and that 2^8 is the answer to the question in part (a). Now notice that 234 in base two is 11101010 that has five 1's in it, and $234 - 5 = 229$, and that 2^{229} is the answer to part (b). Show that this is no coincidence—establish that for any positive integer n, the above process will produce the exponent of the highest power of 2 that evenly divides $n!$.

5.2 Basic Rules for Counting

In this section we apply our ideas about organization in counting to answer some standard kinds of problems. In particular, we will derive "formulas" for two of the finite structures from the previous section.

Example 1 *In a small Southern state, a license plate consists of one or two letters from $\{A, L, B, M\}$ followed by four or three (respectively) digits. How many license plates are possible?*

SOLUTION Let's call a license plate "type I" if it has one initial letter and "type II" if it has two initial letters. The list of the type I license plates can be organized as shown in Table 5-20. This is a table with 10,000 rows and four columns, so we can say it has 40,000 entries.

Table 5-21 shows a similar organization of the type II license plates. It will only have 1,000 rows, but it will have more columns. In particular, the first row of the table will consist of the letter combinations shown in Table 5-21 along with the digits 000. Thus, the table of type II license plates will have 1,000 rows and 16 columns (one for each of the letter combinations). We can conclude that there are 16,000 type II plates. Combining this with the number of type I plates, we determine that the answer to the original question is $40,000 + 16,000 = 56,000$. □

The above example used an organized way of counting the possible license plates, but we had to make some decisions along the way as to how to break up the problem. Also, when we counted the type II license plates, we did not even write down the list. Instead, we just thought about what the list would look like in tabular form. This was enough to get a visual image of the list, allowing us to subsequently determine the size of the list. The basic organizational techniques in this problem come up so often that it is convenient to give them names so that we can talk about them more easily.

The Rule of Products

Formally, the rule of products states that the number of ordered pairs with a first coordinate from set A and a second coordinate from set B is equal to the product of the number of elements in each set. Informally, the rule says that if we can organize

$A0000$	$L0000$	$B0000$	$M0000$
$A0001$	$L0001$	$B0001$	$M0001$
⋮	⋮	⋮	⋮
$A9999$	$L9999$	$B9999$	$M9999$

Table 5-20 Organizing the Type I Plates in Example 1

AA	LA	BA	MA
AL	LL	BL	ML
AB	LB	BB	MB
AM	LM	BM	MM

Table 5-21   Organizing the Type II
Plates in Example 1

our list into a table with m rows and n columns, then we know that the list has $(m)(n)$ entries total. We saw this in Example 1 when we counted each type of license plate.

Example 2 *How many ways are there to order a meal consisting of one sandwich and one beverage at a restaurant that serves five different sandwiches and six different beverages?*

SOLUTION   Call the sandwiches $1, 2, 3, 4, 5$ and the beverages A, B, C, D, E, F. Then the orders could be written as in Table 5-22. Since this is a table with six rows and five columns, the list has a total of 30 entries. □

There is a more practical way to state the rule of products that does not involve writing down the actual table:

Rule of products: If each entry in a list can be created by first selecting one of x objects and then one of y objects, then the list has a total of $(x)(y)$ entries. In terms of sets, this means that $n(A \times B) = n(A) \cdot n(B)$ for all finite sets A and B.

In the example above, the entries are made by first selecting one of the six objects in $\{A, B, C, D, E, F\}$ and then one of the five objects in $\{1, 2, 3, 4, 5\}$, so the list has length $(5)(6) = 30$.

The advantage to this latter way of thinking of the product rule is that it is easier to picture adding "steps" to a selection process than it is to add dimensions to a table.

A1	A2	A3	A4	A5
B1	B2	B3	B4	B5
C1	C2	C3	C4	C5
D1	D2	D3	D4	D5
E1	E2	E3	E4	E5
F1	F2	F3	F4	F5

Table 5-22   Organizing Orders in
Example 2

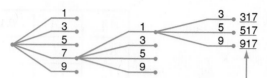

2. choose a different tens' digit.   Resulting number chosen

1. Choose a ones' digit.   3. choose a different hundreds' digit.

Figure 5-3   Part of the selection tree.

Example 3   *How many numbers between 100 and 1,000 have three distinct odd digits? (For example, 153 should be counted but not 133 or 123.)*

SOLUTION   Each of these numbers can be constructed in a three-step process. First choose a number from the five digits in $\{1, 3, 5, 7, 9\}$ and make this the ones digit. Next choose the tens digit from the four digits left after our ones digit is removed from $\{1, 3, 5, 7, 9\}$. Third, choose a hundreds digit from the three digits left after our first two choices have been removed from $\{1, 3, 5, 7, 9\}$. Applying the products rule gives us

$$5 \cdot 4 \cdot 3 = 60$$

such numbers. □

A *three-step* process corresponds to a *three-dimensional* table if we hold onto our original analogy. A more versatile visual image for these multistep processes is a *selection tree* as shown in Figure 5-3 for the example above.

The branches shown indicate that 7 was chosen as the ones digit and 1 was chosen as the tens digit. At the right side of the tree we see all final outcomes that correspond to these choices. If you can visualize the entire tree, you will see that each of the five original branches has four branches, giving a total of 20 branches after the second step. Each of these 20 branches, in turn, has three branches in the last step, so altogether the final list at the right will consist of $(5)(4)(3) = 60$ numbers, just as our solution predicted.

Practice Problem 1   *Use the product rule to answer these questions.*

(a) *How many binary sequences of length 3 are there?*

(b) *In how many ways can one be dealt the first two cards in blackjack, first card face-down and second card face-up? Assume that you are using a standard deck of 52 cards.*

The Rule of Sums

Formally, the rule of sums states that the number of elements in the union of two disjoint sets is equal to the sum of the number of elements in each set. Informally, this means that if the list being enumerated can be broken into two disjoint parts, then the number of entries in the whole list is the sum of the number of entries in the two parts. We applied this rule in Example 1 when we added the number of type 1 plates to the number of type 2 plates to get the total number of license plates.

Rule of sums: If the list to count can be split into two disjoint pieces of size x and y, then the original list has $x + y$ entries. In terms of sets, we can write this as $n(A \cup B) = n(A) + n(B)$, provided A and B are disjoint.

Recall that the product rule, stated for a two-step process, had an obvious generalization to a multiple-step process. Similarly, we stated the rule of sums for two disjoint pieces, but it generalizes to more than two pieces, **provided that every two pieces are disjoint.**

Example 4 *How many positive integers less than 1,000 consist of distinct digits from* $\{1, 3, 7, 9\}$?

SOLUTION We split our problem into three disjoint pieces: those with one digit, those with two digits, and those with three digits. So the solution is the sum of (i) the number of one-digit numbers using distinct digits from $\{1, 3, 7, 9\}$, (ii) the number of two-digit numbers using distinct digits from $\{1, 3, 7, 9\}$, and (iii) the number of three-digit numbers using distinct digits from $\{1, 3, 7, 9\}$, and each of (i), (ii), and (iii) can be calculated using the product rule as above. (Try it now!) So the answer is $4 + (4)(3) + (4)(3)(2) = 40$. □

We now continue with more complex counting problems that combine the sum and product rules.

Example 5 *In how many ways can one win a dice game played with three distinguishable dice in which a winning roll is one that has at least two values that are the same?*

SOLUTION If we think of the dice as being colored red, green, and white, then we can represent an outcome as an ordered list of length 3 with elements taken from $\{1,2,3,4,5,6\}$ in which the first number is the roll of the red die, the second number is the roll of the green die, and the third number is the roll of the white die.

In this manner every winning roll looks like one of the following: XXY, XYX, YXX, or XXX, where X and Y are different numbers from $\{1,2,3,4,5,6\}$. In each of the first three of these cases, there are six ways to choose X and five ways to choose Y, while in the fourth case there are only six outcomes altogether (namely 111, 222, 333, 444, 555, and 666). There are a total of $(3)(6)(5) + 6 = 96$ winning rolls. □

Practice Problem 2 *If we roll a six-sided die three times and record the result as an ordered list of length 3, how many of the possible outcomes contain exactly one 1. (*Hint: *Divide the list into three disjoint sets: those where the 1 is the first roll, those where it is the second, and those where it is the third.)*

Example 6 *How many three-letter sequences of Greek letters do not consist of three of the same letter? (There are 24 letters in the Greek alphabet.)*

SOLUTION Each three-letter sequence of Greek letters is either (i) a string of three of the same letter or (ii) a string that we are supposed to count. So by the sum rule, the total number of three-letter sequences of Greek letters is the number of sequences of type (i) plus the number of sequences of type (ii).

By the product rule, the total number of three-letter sequences of Greek letters is (24)(24)(24), and clearly the number of sequences of type (i) is 24 since there is one such sequence for each letter. Therefore, the number of sequences of type (ii) is $24^3 - 24 = 13,800$. □

Example 6 is a slight twist of the sum rule that is fairly common. If a problem asks for the number of a certain type of objects that have a specific property, then by knowing the **total** number of this type of object and the number of these objects that do **not** have the property, we can find the answer to the original problem by simple subtraction. In fact, any time the total number of objects is obvious, instead of answering a given counting question, we can answer the *complementary problem* that asks how many of the objects are **not** counted by the original question. Since every problem has an associated complementary problem, one always has the freedom to choose whether the original problem or its complement seems more approachable. This is sometimes just a matter of personal taste, but often it makes the difference between an elegant solution and a mess.

Rule of complements: If there are x objects, and y of those objects have a particular property, then the number of those objects that do **not** have that particular property is $x - y$. In terms of sets, using U for the universal set, we can write this as $n(A') = n(U) - n(A)$ for all sets A with elements from U.

Example 7 *How many five-digit numbers use distinct digits from $\{0, \ldots, 6\}$? How many of these are odd? How many are even?*

SOLUTION   Place the digits left to right since the leftmost (ten thousands' place) digit is the only one with a special restriction (i.e., it cannot be 0). The product rule applied to this five-step process will give us $(6)(6)(5)(4)(3) = 2,160$.

For the second part, do the same thing, but this time place the ones' digit first (it now has the greatest restriction since it must be odd). The leftmost digit should be placed next (since it is the second most restricted), and then the final digits can be placed in any order. These five steps can be completed in $(3)(5)(5)(4)(3) = 900$ ways.

For the third part, since every five-digit number using the digits $\{0, \ldots, 6\}$ is either odd or even, the answer to the third part can be found by subtracting the previous answer from the total: $2,160 - 900 = 1,260$. □

Practice Problem 3   *How many rolls of three distinguishable dice result in the largest number showing being either a 5 or a 6? (For example, the roll 554 counts since the largest number showing is 5, but the roll 334 does not count since the largest number showing is 4.)*

One difficulty in using the sum rule comes from the fact that sometimes the "natural" way to break a list into pieces does not create disjoint pieces. Since we saw that the sum rule can be stated in terms of the size of the union of sets, you might recall from Section 3.1 that a more general fact is true. This more general statement makes the rule of sums easier to implement in practice.

Example 8 *If we roll a six-sided die three times and record the result as an ordered list of length 3, in how many of the $6^3 = 216$ possible outcomes are there exactly one 1 or exactly one 6?*

SOLUTION   We will use the set notation from Section 3.1. Let A be the set of all rolls in which there is exactly one 6, and let B be the set of all rolls in which there is exactly one 1. It is easy (see Practice Problem 2) to count $n(A) = 3 \cdot 5^2 = 75$ and $n(B) = 3 \cdot 5^2 = 75$, but we cannot apply the sum rule to answer the question because A and B are not disjoint. However, it is possible to account for the overlap between A and B. Namely, $A \cap B$ is the set of all rolls in which there is exactly one 6 **and** exactly one 1, so $n(A \cap B) = 3 \cdot 2 \cdot 4 = 24$ (see Exercise 5). Hence to find $n(A \cup B)$, we can add $n(A) + n(B)$ and then subtract the overlap, $n(A \cap B)$. That is,

$$n(A \cup B) = n(A) + n(B) - n(A \cap B)$$
$$= 75 + 75 - 24 = 126$$

$\square$

In Section 3.1 we called this the principle of inclusion-exclusion. It is very useful for problems where there is a natural way to break down the problem into cases that are not necessarily disjoint. Note that the following formal statement is still true even when the pieces are disjoint, so this is truly a more general rule of sums.

Rule of sums with overlap: If the list to count can be split into two pieces of size z and y, and the pieces have z objects in common, then the original list has $x + y - z$ entries. In terms of sets, we can write this as $n(A \cup B) = n(A) + n(B) - n(A \cap B)$ for all sets A and B.

Although this rule does generalize to more than two pieces, the generalization is not as easy to apply. In general, if you divide a problem into more than two pieces, make sure there is no overlap among the pieces.

Algorithms for Counting

As we progressed through the examples in this section, we informally developed a way of counting by imagining ourselves building each entry in a list through a sequence of steps. The idea of writing an informal *algorithm* for generating objects is of fundamental importance in discrete math. For those familiar with computer programming, what we are doing is imagining writing a program to print the entire list that we want. For example, the following "pseudo-code" would print the list associated with the problem, "How many odd numbers having distinct digits are between 100 and 1,000?"

```
For each U from {1,3,5,7,9} do
  For each H from {1,2,3,4,5,6,7,8,9} with H ≠ U do
    For each T from {0,1,2,3,4,5,6,7,8,9} with T ≠ H and T ≠ U do
      Print H,T,U
```

We definitely do not care to write code for all our problems, so we will usually say something like "Choose a ones' digit from {1, 3, 5, 7, 9}, choose a different hundreds' digit from {1, 2, ..., 9}, and choose a tens' digit from {0, 1, ..., 9} different from the first two choices." This describes the same algorithm in English. Now we can look at either description and say that there are five ways to do the first step, eight ways to do the second step (regardless of what choice was made for the first step), and eight ways to do the third step (again regardless of what choices were made in the first two steps). By the product rule, this algorithm generates $(5)(8)(8) = 320$ samples.

The parenthetical conditions above are very important. It must be the case that the number of choices at each stage is independent of previous choices.

Example 9 *If we asked, "How many numbers ≥200 can be formed using distinct digits from {0, ..., 6}?" we could write the following three-step algorithm:*

[1] Choose a ones' digit.

[2] Choose a different tens' digit.

[3] Choose a hundreds' digit different from the other two.

Analyze this algorithm using the product rule.

SOLUTION When we analyze the algorithm, we see that there are seven choices in completing the first step, six choices in the second step, leaving us only to count the choices in the last step. But here is the problem: If we choose, say, 3 as our ones' digit and 5 as our tens' digit, then the hundreds' digit must be chosen from the three objects in {2, 4, 6}. On the other hand, if we chose 2 as our ones' digit and 0 as our tens' digit, then the hundreds' digit must be chosen from the four objects in {3, 4, 5, 6}. So the number of choices for the hundreds' digit **depends** on the previous choices.

All we can conclude from this attempted solution is that this particular *algorithm* escapes analysis with the products rule, so either the algorithm or the analysis should be changed for this problem. See Exercise 28 at the end of this section to fix the solution. □

Practice Problem 4 *Explain why the following algorithm **does not** correctly generate the list of all results of rolling a sum of 10 on three distinguishable six-sided dice. (We will represent the outcomes of the dice rolls as an ordered list of length 3 with elements taken from {1, 2, 3, 4, 5, 6}.)*

[a] Choose any element of {1, 2, 3, 4, 5, 6} for the first roll.

[b] Choose any element of {1, 2, 3, 4, 5, 6} for the second roll.

[c] Fill in the third number by subtracting the sum of the previous two numbers from 10.

*Applying the product rule to this algorithm tells us that there will be $(6)(6)(1) = 36$ ways to roll three dice to get a sum of 10. **This is wrong!***

General Formulas to Count Ordered Lists and Permutations

We are now ready to apply the concepts in this section to derive "counting formulas" for some of the basic finite structures we introduced in Section 5.1.

Theorem 1 *The number of ordered lists from* $\{1, \ldots, n\}$ *of length* r *is* n^r.

PROOF Think of the r places for entries in the ordered list. Choose any number from $\{1, \ldots, n\}$ for the first space, any number from $\{1, \ldots, n\}$ for the second space, and so on for all r spaces. By the product rule, there will be

$$\underbrace{(n)(n) \cdots (n)}_{r \text{ times}} = n^r$$

of these ordered lists. ■

Example 10 *How many* r*-digit binary sequences (i.e., ordered lists from* $\{0, 1\}$ *of length* r*) are there?*

SOLUTION By Theorem 1, there are 2^r of them. □

Definition

1. Remember that a permutation is just an ordered list in which no element is repeated. The number of permutations* from $\{1, \ldots, n\}$ of length r is denoted by $P(n, r)$.
2. We write $n!$ for $P(n, n) = (n)(n-1) \cdots (3)(2)(1)$ with the stipulation that $0! = 1$. We read the notation $n!$ as "n factorial."

Theorem 2 *For any nonnegative integers* n *and* r, $P(n, r) = n \cdot (n - 1) \cdot (n - 2) \cdots (n - r + 1)$. *Note that if* $n < r$, *then this product will be 0, consistent with our definition; and that if* $n \geq r$, *this product can be written as* $\frac{n!}{(n-r)!}$.

PROOF Thinking of it in the same way as in the proof of Theorem 1 except with the constraint that at each step we may only choose from objects not already selected, we have n choices for the first space, $n - 1$ choices for the second space, and so on, giving us the desired formula (by the product rule) after all r spaces have been filled. ■

Example 11 *How many batting orders for a baseball team are possible from a roster of 20 players?*

SOLUTION A batting order is an ordered list from the set of available players of length 9 without repetition—a permutation. So by Theorem 2, there are $P(20, 9) = \frac{20!}{11!} = (20)(19) \cdots (12) \approx 6 \times 10^{10}$ of them. □

Example 12 *How many ways are there to arrange seven people in a line?*

SOLUTION Call the people 1, 2, 3, 4, 5, 6, and 7. An arrangement in the line of these people corresponds to a permutation from $\{1, \ldots, 7\}$ of length 7, and by Theorem 2, there are $P(7,7) = 7! = 5,040$ of these. □

---

* Other notation that is sometimes used includes $_nP_r$ and P_r^n.

Example 13 *How many ways are there for three married couples to arrange themselves in a movie theater line, assuming that spouses always stand together?*

SOLUTION Call the couples Smith, Jones, and Williams. Form the movie lines in a four-step process as follows. Choose a permutation of the three names in $3! = 6$ ways, choose an order for the couple first in this permutation to stand in $2! = 2$ ways, choose an order for the couple second in the permutation to stand in $2! = 2$ ways, and choose an order for the couple last in the permutation to stand in $2! = 2$ ways. By the product rule, there are $(6)(2)(2)(2) = 48$ ways. □

Practice Problem 5 *You have a geography test consisting of a column of 10 countries and a column of 20 export products—your task is to match each country to its top export, and you are instructed that exactly 10 of the export products will be correct answers (each for only one listed country) and the other 10 are red herrings.* * *If you have not studied at all, how many different ways are there to complete the test by just guessing?*

Solutions to Practice Problems

1 (a) We can form the binary sequence in three steps, from left to right: (i) Select the first digit, (ii) select the second digit, (iii) select the third digit. There are two ways to do each step, and thus $2 \cdot 2 \cdot 2 = 8$ binary sequences. This is illustrated by the tree shown in Figure 5-4. For example, choosing a 0 in the first step, a 1 in the second step, and a 1 in the third step will result in the sequence 011 shown in blue in Figure 5-4.

(b) This problem is implicitly a two-step process: deal the face-down card, then deal the face-up card. There are clearly 52 ways to do the first step. Then, no matter which card has been dealt, there are 51 cards left, so there are 51 ways to do the second step. Hence, the number of different two-card blackjack hands is $52 \times 51 = 2,652$.

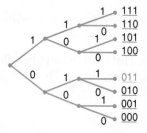

Figure 5-4 Tree for Practice Problem 1(a).

2 There are $1 \cdot 5 \cdot 5$, where the only 1 is the first roll (one choice for the first roll, five for the second, and five for the third). Similarly, there are $5 \cdot 1 \cdot 5$, where the only 1 is the second roll, and $5 \cdot 5 \cdot 1$ where the only 1 is the third roll. The total is $3 \cdot 5^2 = 75$.

3 As in Example 5, we will represent an outcome as an ordered list of length 3 with elements taken from $\{1, 2, 3, 4, 5, 6\}$. The largest number rolled must be either 5 or 6, so in order to not be counted, a roll would have to be represented as an ordered list of length 3 from the set $\{1, 2, 3, 4\}$. There are $4^3 = 64$ such ordered lists, leaving $6^3 - 4^3 = 152$ rolls that must have a 5 or 6.

4 Follow the algorithm's instructions to see what it is doing. For example, if we choose 3 in the first step and 4 in the second step, we end up with the "roll" 343, but if we choose 1 in the first step and 1 in the second step, we end up with 118, which is not a possible dice roll. Similarly, if we choose 6 in the first step and 5 in the second step, we cannot have a negative result on the third die! The problem with this algorithm is that it generates several things that we do not want. Hence, the answer of 36 it gives us is too large. The correct answer is 27, which we will see how to get in the next section.

5 For each of the 10 countries, you are choosing one of the 20 products, so we can represent your answer sheet as an ordered list of length 10 where the elements come from the 20 products without repetition. The total number of these answer sheets is then $P(20, 10) = 670,442,572,800$, a big enough number that studying is definitely better than guessing!

---

*Unless the top export of a country *is* red herring, in which case this item is a wild goose.

Exercises for Section 5.2

Be clear on how you are organizing your thoughts and applying the basic counting rules in answering the following questions.

1. **(a)** How many ordered lists with entries from $\{1, \ldots, 6\}$ of length 10 are there?

   (b) How many of the above never have a 6 appear?

2. **(a)** A true-false test contains 10 questions. In how many ways can a student answer the questions if every question is answered?

   (b) A true-false test contains 10 questions. In how many ways can a student answer the questions if some questions may be left unanswered?

   (c) A matching test contains 10 questions, and each answer must be used exactly once. In how many ways can a student answer the questions if every question is answered?

3. A certain mid-Atlantic state has a simple rule for its license plates: Use three letters followed by four digits.

   (a) How many possible plates are there?

   (b) Suppose the state court has ruled that 97 of the possible three-letter combinations are offensive and cannot be used. Now how many possible plates are there?

4. A certain large Northeastern state has a strange rule for the two-letter combinations that are legal for its license plates. To keep people's plates from accidentally spelling possibly offensive words, they use the rule that any license plate that has a vowel as its first letter must also have a vowel as its second letter. For example, *CD* and *OE* are legal combinations, but *ON* is not. How many two-letter prefixes are possible on license plates in this state? (They consider only the five letters *A, E, I, O,* and *U* to be vowels.)

5. If we roll a six-sided die three times and record the result as an ordered list of length 3, in how many of the $6^3 = 216$ possible outcomes are there exactly one 1 and exactly one 6? (HINT: Form the entries in the list using three steps: Decide which position will contain the 1, decide which position will contain the 6, and fill in the remaining position.)

6. In how many ways can a club with 17 members elect a president, vice president, and secretary (assuming no person can fill more than one office) for each situation described?

   (a) There are no restrictions.

   (b) Susan has removed herself from consideration for president due to a busy schedule, but she is willing to serve in either of the other offices.

   (c) Sam has indicated that he will serve as president only if Mary is named as the vice president.

   (d) The club's bylaws require that last year's vice president becomes this year's president.

7. An organization has 10 male and 7 female members. In how many ways can the organization elect a president, vice president, and secretary for each situation described?

   (a) The president must be female.

   (b) The president can be either sex, but the vice president must be female and the secretary must be male.

   (c) The president and vice president must be of the same sex.

   (d) All three officers may not be of the same sex.

8. An organization has 8 math majors, 12 computer science majors, and 6 science majors. In how many ways can the organization elect a president, vice president, and secretary for each situation described?

   (a) The three officers are all the same major.

   (b) At least one of the officers must be a science major.

   (c) Either the president or the vice president must be a math major.

9. We will represent the results of three tosses of a six-sided die as an ordered list of length 3 with entries from $\{1, \ldots, 6\}$.

   (a) How many different results are possible?

   (b) In how many of these are all three tosses different?

10. We will represent the results of four tosses of a coin as an ordered list of length 4 with entries from $\{H, T\}$.

    (a) How many different results are possible?

    (b) In how many of these are the first two tosses the same?

11. We will represent the results of five cards being dealt from a standard deck of 52 cards as a permutation of length 5 with entries from the set of cards.

    (a) How many different results are possible?

    (b) In how many of these does the first card have the same value as the last card?

    (c) In how many do all five cards have the same suit?

12. **(a)** How many arrangements are there of the letters in the word *MATCH*?

    (b) Of these, how many have the letters *M* and *A* side by side in that order? (HINT: Think of *MA* as a single object, so you are counting arrangements of the four objects *MA, T, C,* and *H*.)

    (c) How many have the letters *M* and *A* side by side but in the order *AM*?

13. **(a)** How many arrangements are there of the letters in the word *EMPHATIC*?

(b) Of these, how many have the letters *E* and *M* side by side in either *EM* or *ME* order?

14. A blackjack hand consists of two cards, the first of which is dealt face-down and the second face-up.

(a) In how many blackjack hands is the face-up card an ace and the face-down card a 10?

(b) In how many blackjack hands does the face-up card have a value of jack, queen, or king, while the face-down card has a value between 2 and 9, inclusive?

(c) In how many blackjack hands is the face-up card an ace and the face-down card a club?

15. A combination lock has three numbers in the combination, each in the range 1 to 50.

(a) How many different combinations are there?

(b) How many of the combinations have no duplicate numbers?

(c) How many of the combinations have the first and second number matching?

(d) How many of the combinations have exactly two of the three numbers matching?

16. There are 16 marbles numbered 1 to 16 in a box. Marbles 1 to 5 are red, marbles 6 to 8 are green, and marbles 9 to 16 are blue. I draw out four marbles, one at a time without replacing them, and record the result as an ordered list of four color/number combinations. For example, I write R4, G6, R1, B10 if the marbles drawn are, in order, #4, #6, #1, #10.

(a) How many possible results are there?

(b) Of these, for how many are both the first and last marbles red?

(c) For how many are the first and second marbles different colors?

(d) For how many are all four marbles the same color?

17. In Exercise 16, how do your answers to the questions change if each drawn marble is replaced in the box before the next marble is drawn?

18. On a TV game show, nine scarves numbered 1 to 9 are placed in a basket and thoroughly mixed. Then they are drawn out one at a time.

(a) How many possible results are there?

(b) Of these, how many have all five odd numbers first, followed by four even numbers?

19. How many of the results counted in Exercise 18 follow the pattern "even, odd, even, odd, even, odd, even, odd, odd?"

20. A guest speaker in a department seminar chooses six audience members. One after another he asks them their birth month, writing their responses as an ordered list of length 6.

(a) How many different results are possible?

(b) Of these, for how many are there no duplicate months listed?

(c) What percent of the possible results have a duplicate month? If the speaker always bets each audience a dollar that there will be a duplicate, is this a good bet for the speaker?

21. The professor in your class asks each of the 30 students in the class his or her birthday, writing their responses as an ordered list of length 30. Assume no one is born on February 29.

(a) How many different results are possible?

(b) Of these, for how many are there no duplicate birthdays listed?

(c) What percent of the possible results have a duplicate birthday? Is this percent higher than you expected, lower than you expected, or about what you expected?

22. If we toss a coin 10 times in succession and record the outcomes as an ordered list of length 10 using entries from $\{H, T\}$, how many of the 2^{10} possible outcomes satisfy each of these conditions?

(a) Begin with three "heads" in a row.

(b) End with three "tails" in a row.

(c) Begin with three "heads" in a row and end with three "tails" in a row.

(d) Begin with three "heads" in a row or end with three "tails" in a row.

23. A certain club is forming a recruitment committee consisting of five of its members. They have calculated that there are 8,568 different ways to form this committee. Of these, 700 have exactly one woman, 2,520 have exactly two women, 3,360 have exactly three women, 1,680 have exactly four women, and 252 have exactly five women.

(a) How many committees have at least three women?

(b) How many committees have no women?

(c) How many committees have at most two women?

24. The club of the preceding exercise has two members named Jack and Jill. They have calculated that 2,380 of the committees have Jack on them, 2,380 have Jill, 1,820 have Jack but not Jill, 1,820 have Jill but not Jack, and 560 have both Jack and Jill.

(a) How many committees have either Jack or Jill?

(b) How many committees have neither Jack nor Jill?

(c) Jack and Jill are carpooling, so they insist that if either one is on the committee, the other person must also be on the committee. How many committees meet this condition?

(d) Jack and Jill have had a fight. Jack says, "If Jill is on the committee, I won't be." Jill says, "If Jack is on the committee, I won't be." How many of the committees meet this condition?

25. The following questions follow up on the principle of inclusion–exclusion, discussed in Section 3.1 and Example 8 of this section.

(a) Of the $6^3 = 216$ results of rolling a six-sided die three times discussed in Example 8, how many of them include getting exactly one 1, exactly one 2, or exactly one 6?

(b) There are $6^4 = 1,296$ results of rolling a six-sided die four times in succession. How many of these include getting exactly one 1, exactly one 2, or exactly one 6?

26. Write each of the following as a product (e.g., $P(6, 4) = 9 \cdot 8 \cdot 7 \cdot 6$), and then evaluate the product:

(a) $P(19, 5)$

(b) $P(6, 6)$

(c) $P(120, 17)$

27. Fill in the blanks to write each of the following products in the form $P(n, r)$:

(a) $9 \cdot 8 \cdot 7 \cdot 6 = P(_____, _____)$

(b) $10 \cdot 9 \cdot 8 \cdot 7 \cdot 6 \cdot 5 \cdot 4 \cdot 3 \cdot 2 \cdot 1 = P(_____, _____)$

(c) $365 \cdot 364 \cdot 363 \cdot \cdots \cdot 338 = P(_____, _____)$

(d) $(m+1) \cdot (m) \cdot (m-1) \cdot (m-2) \cdot (m-3) = P(_____, _____)$

(e) $k \cdot (k-1) \cdot (k-2) \cdot \cdots \cdot (k-m) = P(_____, _____)$

28. Solve the problem in Example 9 by changing the algorithm so that the product rule *can be* applied.

29. Here is an algorithm for Exercise 7(b), which counts the ways to elect officers in an organization of 10 males and 7 females, if the president can be either sex, but the vice president must be female and the secretary must be male: "There are 17 choices for the president, 7 for the vice president, and 10 for the secretary, giving $17 \cdot 7 \cdot 10 = 1,190$ different results." This is incorrect. Explain why, and give a correct algorithm.

30. Here is an algorithm for Exercise 11(b), which counts the five-card hands where the first card dealt has the same value as the last card dealt: "There are 52 choices for the first card, 51 for the second, 50 for the third, 49 for the fourth, and 3 for the last, giving $52 \cdot 51 \cdot 50 \cdot 49 \cdot 3 = 19,492,200$ possible hands." This is incorrect. Explain why, and give a correct algorithm.

31. Here is an algorithm for Exercise 8(c), which counts the ways to elect officers where either the president or vice president must be a math major (The organization has 8 math majors, 12 computer science majors, and 6 science majors.) "If the president is a math major, there are 8 ways to choose the president, 25 for the vice president, and 24 for the secretary. There are the same number when the vice president is a math major, so the total is $2 \cdot 8 \cdot 25 \cdot 24 = 9,600$." This is incorrect. Explain why, and give a correct algorithm.

32. Here is an algorithm for Exercise 24(a): "There are 2,380 committees that have Jack, and 2,380 that have Jill. By the sum rule, there are $2,380 + 2,380 = 4,760$ that have either Jack or Jill." Explain the error, and correct the algorithm.

33. For part (d) of Exercise 24, a student reasons correctly that we want to count committees that do not include both Jack and Jill, and that this is logically the same as "don't have Jack, or don't have Jill." However, his counting algorithm is incorrect. Explain why, and correct it. The algorithm is: "By the complement rule, there are $8,568 - 2,380 = 6,188$ committees that don't have Jack, and $8,568 - 2,380 = 6,188$ committees that don't have Jill. By the sum rule, there are $6,188 + 6,188 = 12,376$ that either don't have Jack, or don't have Jill."

34. How many five-digit numbers have all distinct digits and are odd?

35. How many numbers from the set $\{1, 2, \ldots, 9,999\}$ use exactly two different digits? (For example, 1,121 should be counted, but 3,333 and 1,231 should not.)

36. How many four-digit numbers use the digit 7?

37. How many positive integers have distinct digits, all of which are odd?

38. How many positive integers have distinct digits?

39. (a) How many four-digit odd numbers have digits only from the set $\{1, \ldots, 5\}$?

(b) How many of these have at least one digit repeated?

(c) How many of these have only one digit repeated?

40. (a) Using only the digits 1, 2, 3, 4, 5, and 6, how many five-digit numbers can be formed?

(b) How many of these have at least one 5?

(c) How many of them have either no 5 or no 6?

41. In how many ways can five families of four stand in a movie theater line if each family must stay together?

42. How many subsets of $\{1, 2, 3, 4, 5\}$ have two elements? How does this compare with the number of permutations of length 2 with entries from $\{1, 2, 3, 4, 5\}$?

43. Prove by induction that for all $n \geq 2$, the number of two-element subsets of $\{1, \ldots, n\}$ is $\frac{n \cdot (n-1)}{2}$.

5.3 Combinations and the Binomial Theorem

We have seen that the rules of the sums and products are remarkably simple building blocks that can be used to solve many enumeration problems. In this section, we will extend these rules to do even more.

Counting with Equivalence Classes

 Example 1 *How many two-element subsets of* $\{1, 2, 3, 4\}$ *are there?*

The difference between permutations and subsets is simply a matter of whether we care about the order of the entries—in a permutation we do and in a subset we do not. Specifically, we consider $\underline{2}\ \underline{3}$ and $\underline{3}\ \underline{2}$ to be different permutations while we consider $\{2, 3\}$ and $\{3, 2\}$ to be the same set.

We saw in Section 4.5 that when we have a notion of "equivalence" on a set of objects, we call this an *equivalence relation*. More important, we call a set of objects that are equivalent to one another an *equivalence class*.[*] In this example, there is an equivalence relation on the set of all permutations of length two with entries from $\{1, 2, 3, 4\}$. This equivalence relation can be described as "would look the same if they were sets." For example, we would say that

$$\{\underline{2}\ \underline{3}, \underline{3}\ \underline{2}\}$$

is an equivalence class since the permutations therein would look the same if they were sets.

Here are some other equivalence classes for this equivalence relation:

- $\{\underline{1}\ \underline{3}, \underline{3}\ \underline{1}\}$
- $\{\underline{4}\ \underline{2}, \underline{2}\ \underline{4}\}$
- and others

This provides the key idea for answering the question in Example 1.

SOLUTION   (To Example 1) We know that the number of permutations of length 2 with entries taken from $\{1, 2, 3, 4\}$ is $P(4, 2) = 4 \cdot 3 = 12$. If we use the equivalence relation "would be the same if they were sets," each equivalence class contains two permutations, so we will have $\frac{12}{2} = 6$ equivalence classes. Table 5-23 shows the 12 permutations in the 6 equivalence classes. Each equivalence class (by its very definition) identifies a single two-element subset of $\{1, 2, 3, 4\}$. In this problem, there is a small enough number of these to simply list:

$$\{1, 2\}, \quad \{1, 3\}, \quad \{1, 4\}, \quad \{2, 3\}, \quad \{2, 4\}, \quad \{3, 4\}$$

So the answer to the question is 6, but more important, we have developed an approach to a more general question.   □

---

[*] In the language of Section 4.5, equivalence classes are simply the parts of the partition induced by an equivalence relation. The important thing is that equivalence classes form a collection of nonempty subsets of S, for which every element of S is in exactly one subset, or *class*.

$\frac{1\ 2}{2\ 1}$	$\frac{1\ 3}{3\ 1}$	$\frac{1\ 4}{4\ 1}$	$\frac{2\ 3}{3\ 2}$	$\frac{2\ 4}{4\ 2}$	$\frac{3\ 4}{4\ 3}$

Table 5-23 Equivalence Classes in Example 1

Practice Problem 1 *Use the idea of equivalence classes to explain why the number of three-element subsets of* $\{1, 2, 3, 4, 5\}$ *is* $\frac{P(5,3)}{6}$.

Definition

1. We will write $C(n, r)$ (usually read "*n* choose *r*") for the number* of sets from $\{1, \ldots, n\}$ of size r.

2. Sets from $\{1, \ldots, n\}$ of size r are also called *r-combinations* from $\{1, 2, \ldots, n\}$.

Theorem 1 $C(n, r) = \frac{P(n,r)}{r!}$. *Note that if* $n \geq r$, *this can also be written as* $\frac{n!}{r!(n-r)!}$.

PROOF We imagine stretched out before us the permutations of length r with entries taken from $\{1, 2, \ldots, n\}$, and we concentrate on the equivalence classes using the same equivalence relation as in Example 1. A permutation of length r from $\{1, 2, \ldots, n\}$ consists of an ordered list with r different objects; hence, we know that any of the $r!$ ways to arrange these r objects will look the same as the original permutation if they were sets (i.e., if the order of them did not matter).

Therefore, each of the equivalence classes on the $P(n, r)$ total permutations is made up of $r!$ permutations, and each of these equivalence classes corresponds to a single subset of $\{1, 2, \ldots, n\}$. Hence, the number of r-element subsets of $\{1, 2, \ldots, n\}$ is

$$\frac{P(n, r)}{r!}$$

■

Example 2 *How many five-person committees can be formed from the 100-member U.S. Senate?*

SOLUTION This is the definition of $C(100, 5)$—choose five members from the 100-element set, where order does not matter and repetitions are not allowed. By Theorem 1, this is $P(100, 5)/5! = \frac{100!}{5!95!} \approx 7.5 \times 10^7$. □

Practice Problem 2 *In draw poker, a player receives a hand of five cards from a standard deck. A flush is a hand that has all five of its cards of the same suit. (There*

---

* Other notation sometimes used includes $_nC_r$, C_r^n, and $\binom{n}{r}$.

are four suits in a standard deck, with 13 cards in each suit.) How many draw poker hands are flushes?

Before we begin to solve problems using our nice formula for $C(n, r)$, let's look at one more example of the idea of counting with equivalence classes.

Example 3 *How many ways are there to arrange six children holding hands in a circle?*

SOLUTION   Note that given two such arrangements, as long as all the children are holding the same hands on the same sides in the two configurations, the arrangements are considered the same. For example, the two arrangements in Figure 5-5 are considered the same.

Now if the children stood in a line instead of a circle, we know they can be arranged in $6! = 720$ ways. Let's say that two of these linear arrangements are "equivalent" if they amount to the same thing when you wrap them in a circle. (Imagine the children holding hands in their line and then the two children on the ends holding hands to complete the circle so that "left-to-right" in the line becomes "clockwise" in the circle.) For example, Table 5-24 shows the arrangements in a line that all become the same circular arrangement shown in Figure 5-5. In fact, each equivalence class will consist of six arrangements in a line, and each equivalence class corresponds to one circular arrangement. Therefore, the number of circular arrangements is $\frac{720}{6} = 120$.   □

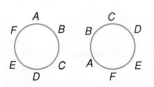

Figure 5-5   Equivalent circular arrangements.

Applications of Combinations

There are many counting problems where the solution can be viewed as choosing r-element subsets of a given set of size n. We can also combine this idea with the rules of sums, products, and complements developed in the preceding section.

Example 4   *A club of ten women and eight men is forming a five-person steering committee. Clearly, there are $C(18, 5)$ possible committees. Of these, how many are possible for each of these situations?*

1. The committee contains exactly three women.

2. The committee contains at least three women.

3. Jack and Jill refuse to work together, so the committee does not contain both of them.

SOLUTION

1. We fill the committee in a two-step process—select three women, and then select two men. There are $C(10, 3)$ ways to do the first step, then $C(8, 2)$ ways

| ABCDEF | BCDEFA | CDEFAB | DEFABC | EFABCD | FABCDE |

Table 5-24   An Equivalence Class of Linear Arrangements

to do the second step. By the products rule, the total number of committees is

$$C(10, 3) \cdot C(8, 2) = 120 \cdot 28$$
$$= 3{,}360$$

2. We divide our solution into disjoint sets: those with exactly three women, those with exactly four women, and those with exactly five women. For each part, we use reasoning similar to part 1. The total is

$$C(10, 3) \cdot C(8, 2) + C(10, 4) \cdot C(8, 1) + C(10, 5) \cdot C(8, 0)$$
$$= 120 \cdot 28 + 210 \cdot 8 + 252 \cdot 1$$
$$= 5{,}292$$

3. There are at least three ways to attack this problem. The details are considered in Exercise 21.

   (a) Divide the list into three disjoint parts: those that have Jack but not Jill, those that have Jill but not Jack, and those that have neither.
   (b) Solve the complementary problem.
   (c) Divide the list into two overlapping parts: (i) those that don't have Jack, and (ii) those that don't have Jill. Then use the "sums rule with overlap." ☐

Explore more on the Web.

Practice Problem 3   *How many of the committees described in Example 4 contain at least one man?*

This type of reasoning easily extends to situations where there are more than two types of objects to place in the set.

Example 5 *If we select a committee of size 10 from a group that contains 25 Democrats, 28 Republicans, and 14 Independents, how many of the possible committees will have exactly 5 Democrats, 4 Republicans, and 1 Independent?*

SOLUTION   Apply the product rule to the three-step process, (i) choose the Democrats, (ii) then the Republicans, and (iii) then the Independents. There are $C(25, 5) \cdot C(28, 4) \cdot C(14, 1)$ such committees.   ☐

We can even apply reasoning about choosing subsets to problems that do not appear to be about sets. In the second part of the next example, we count ordered lists using an algorithm where one step involves combinations. This indicates an important correspondence between sets and ordered lists that will be used a great deal in the remainder of this chapter.

Example 6 *A coin is tossed five times, and the results are recorded as an ordered list from {H, T} of length 5.*

1. *How many possible outcomes are there?*
2. *Of these, how many contain exactly three heads?*

3. *Generalize the previous result to explain why* $C(5,0)+C(5,1)+C(5,2)+C(5,3)+C(5,4)+C(5,5)=2^5$.

SOLUTION

1. By the product rule, there are $2\cdot2\cdot2\cdot2\cdot2=2^5=32$ possible outcomes.

2. We number the positions in the list with the numbers 1 to 5, and record the positions that contain heads. For example, HHTHT corresponds to the selection $\{1,2,4\}$. The question can thus be translated to "How many ways are there to choose the three positions to be occupied by the heads?" This is the same question as "How many three-element subsets of $\{1,2,3,4,5\}$ are there?" The answer is $C(5,3)=10$.

3. By part 1, there are 2^5 possible outcomes. We count those outcomes a different way, by dividing them into six disjoint sets: those with no heads, those with one head, and so on. By reasoning similar to part 2, there are $C(5,0)$ with no heads, $C(5,1)$ with one head, and so on. By the sum rule, the total is $C(5,0)+C(5,1)+C(5,2)+C(5,3)+C(5,4)+C(5,5)$. Since the two answers must be the same, this establishes the result. □

The Binomial Theorem and Arithmetic Triangle

Combinations are also important outside the setting of a discrete math course. The most familiar use of them is probably their role in the so-called binomial theorem. This simple result explains algebraically how to multiply out the polynomial

$$(1+x)^n = \underbrace{(1+x)\cdot(1+x)\cdots\cdots(1+x)}_{n\text{ terms}}$$

Believe it or not, this fact was of central importance to the mathematicians who laid down the foundations for the development of calculus[*] in the seventeenth century and earlier.

In algebra, a *binomial* is simply a polynomial with two terms. Thanks to the distributive property, the product of binomials $(1+x)\cdot(1+x)\cdot(1+x)$ expands algebraically to become

$$(1+x)\cdot(1+x)\cdot(1+x) = 1\cdot1\cdot1+x\cdot1\cdot1+1\cdot x\cdot1+x\cdot x\cdot1$$
$$+1\cdot1\cdot x+x\cdot1\cdot x+1\cdot x\cdot x+x\cdot x\cdot x$$

This expansion can be simplified by combining terms. For example, the terms $x\cdot1\cdot1$, $1\cdot x\cdot1$, and $1\cdot1\cdot x$ each simplify to x, so when they are added together, the result simplifies to $3x$. Notice that each term in the resulting sum is formed by choosing either the 1 or the x from each of the three binomials being multiplied. Each of the terms in the example above comes from choosing one x from among the

---

[*] In fact, an early great discovery of Sir Isaac Newton (1643–1727) himself was the generalization of this fact for values of n other than natural numbers.

three available x's, so we could have guessed ahead of time that there were going to be $C(3, 1)$ of these terms to be combined.

There is nothing special about this example, of course. This happens for any number of $1 + x$ terms being expanded. For example, when the product

$$(1 + x) \cdot (1 + x) \cdot (1 + x) \cdot (1 + x)$$

is expanded to

$$(1 + x)^4 = 1 + 4x + 6x^2 + 4x^3 + x^4$$
$$= C(4, 0) + C(4, 1)x + C(4, 2)x^2 + C(4, 3)x^3 + C(4, 4)x^4$$

we see that the coefficient of x^k is $C(4, k)$ for each x^k on the right-hand side.

This general connection between the numbers $C(n, k)$ and the expansion of $(1 + x)^n$ is the binomial theorem.

Theorem 2 (The Binomial Theorem)   *The coefficient of the x^k term in the expansion of $(1 + x)^n$ is $C(n, k)$. More formally,*

$$(1 + x)^n = C(n, 0) + C(n, 1)x + C(n, 2)x^2 + \cdots + C(n, n)x^n$$
$$= \sum_{k=0}^{n} C(n, k)x^k$$

Explore more on the Web.

Practice Problem 4   *What is the coefficient of x^8 in*

(a) *The expansion of $(1 + x)^{10}$?*

(b) *The expansion of $(1 + x^2)^{10}$?*

```
            1
          1   1
        1   2   1
      1   3   3   1
    1   4   6   4   1
  1   5  10  10   5   1
1   6  15  20  15   6   1
```

Figure 5-6 The first seven rows of the AT.

The binomial theorem is usually accompanied by the more visually appealing arithmetic triangle (AT), also called Pascal's triangle* or Tartaglia's triangle. The triangle is simply a table of the values of $C(n, k)$ in the shape of a triangle, so mathematically it is nothing new for us. However there are many patterns within the triangle that might not be apparent without this visual representation. The first few rows are shown in Figure 5-6.

In order to have the proper correspondence, we refer to the row and entry numbers starting with 0. That is, the top row is row 0 and the leftmost entry in each row is entry 0. Using this convention, we see that entry k in row n is the number $C(n, k)$.

Practice Problem 5   *What does the binomial theorem state if we substitute the value 1 for the variable x? How can this be interpreted as a fact about the arithmetic triangle?*

---

* It might seem ironic, but we use the term arithmetic triangle out of respect for Blaise Pascal (1623–1662). After all, this is the term he himself used when he wrote about it in 1653.

When we form the triangle, we are struck by a pattern: When two successive entries in a row are added, one gets the entry immediately below them. Formally, it seems that

$$C(n, k) + C(n, k+1) = C(n+1, k+1) \qquad (*)$$

For example, we expect that entry 2 in row 5 will be $C(4, 2) + C(4, 1) = 6 + 4$, and if we check using our counting formula for $C(5, 2)$, we get $C(5, 2) = \frac{5 \cdot 4}{2!} = 10$ that is the same thing.

To account for this pattern, we need look no further than the binomial theorem itself. We know that $(1 + x)^5 = (1 + x) \cdot (1 + x)^4$, so the binomial theorem tells us that

$$\begin{aligned}
\sum_{k=0}^{5} C(5, k)x^k &= (1 + x) \cdot \left(1 + C(4, 1)x^1 + C(4, 2)x^2 + C(4, 3)x^3 + x^4\right) \\
&= (1 + x) \cdot \left(1 + 4x + 6x^2 + 4x^3 + x^4\right) \\
&= 1 + (4 + 1)x + (6 + 4)x^2 + (4 + 6)x^3 + (1 + 4)x^4 + x^5
\end{aligned}$$

By comparing the coefficients on the left-hand side of the long equation above with those on the right-hand side, we see the pattern clearly. For example, the coefficient of x^2 on the left is $C(5, 2)$, while the coefficient of x^2 on the right is $C(4, 2) + C(4, 1)$. This means that the additive property of forming the arithmetic triangle can be accounted for by the way we multiply polynomials.

Practice Problem 6 *Multiply* $(1 + x)^5 = 1 + 5x + 10x^2 + 10x^3 + 5x^4 + x^5$ *by* $(1 + x)$, *and compare the resulting coefficients to row 6 of the arithmetic triangle.*

Another way to account for patterns in the arithmetic triangle is with our old friend, mathematical induction.

 Example 7 *Show that the following is true for all $n \geq 2$:*

$$C(2, 2) + C(3, 2) + C(4, 2) + \cdots + C(n, 2) = C(n+1, 3)$$

SOLUTION   Let $P(n)$ be the statement "$C(2, 2) + C(3, 2) + C(4, 2) + \cdots + C(n, 2) = C(n+1, 3)$." The first statement $P(2)$ simply states, "$C(2, 2) = C(3, 3)$," which is certainly true. Now let $m \geq 2$ be given such that $P(2), P(3), \ldots, P(m-1)$ have all been checked to be true, and we are ready to consider statement $P(m)$.

$$\begin{aligned}
&(C(2, 2) + C(3, 2) + \cdots + C(m-1, 2)) + C(m, 2) \\
&= (C(m, 3)) + C(m, 2) \text{ by } P(m) \\
&= C(m+1, 3) \text{ by } (*)
\end{aligned}$$

This shows that statement $P(m)$ is true, completing the induction.   □

Practice Problem 7 *Use mathematical induction to prove that the following is true for all $n \geq 4$:*

$$C(4,4) + C(5,4) + \cdots + C(n,4) = C(n+1,5)$$

We conclude this introduction to the arithmetic triangle with a magic trick attributed to Harry Lorayne in [25].

Example 8 *A spectator removes five cards from a deck (which has no face cards) and lays them in a face-up row. The magician then removes a card and places it face-down well above the other cards. The spectators are given the following instructions:*

- *Form a row of four face-up cards above those that are there by adding the values of two consecutive cards in the row of five and subtracting 10 whenever the sum exceeds 10. Find a card that has this value and place it above and between the two cards.*
- *Use the same rule to form a row of three face-up cards and then a row of two face-up cards.*
- *When the rule is applied one last time, the performer turns over his selected card to reveal that it is the correct value for the top of the triangle.*

  How does the magician determine which card to place at the top?

SOLUTION If the cards are $A, B, C, D,$ and E, the prediction is determined by computing

$$A + E + 4 \times (B + D - C)$$

with appropriate subtraction of the 10's when this expression exceeds 10.
For example, if the initial cards are $1, 4, 3, 5,$ and 7, the performer computes

$$1 + 7 + 4 \times (4 + 5 - 3) = 32$$

which means his prediction is 2. The spectator obediently forms the row $5, 7, 8, 2$ according to the given rules, followed by the row $2, 5, 10$, and then the row $7, 5$. At this point it is clear that the top card should be a 2, which the performer reveals to be the case. But why does this work? □

Solutions to Practice Problems

1 Instead of giving a table that includes all 60 of the permutations counted by $P(5,3)$, we will focus on the equivalence classes. For example, the permutations in

$$\{\underline{1}\ \underline{3}\ \underline{5},\ \underline{1}\ \underline{5}\ \underline{3},\ \underline{3}\ \underline{1}\ \underline{5},\ \underline{3}\ \underline{5}\ \underline{1},\ \underline{5}\ \underline{1}\ \underline{3}, \underline{5}\ \underline{3}\ \underline{1}\}$$

would all be the same if they were sets (and to no other permutation would this apply), so this is an equivalence class. The fact that there are six items in this class is no accident since we know there are 3! arrangements of any three distinct objects. Hence, every equivalence

class consists of six permutations, and each equivalence class corresponds to only one set ($\{1,3,5\}$ in the example). Therefore, the number of three-element subsets of $\{1,2,3,4,5\}$ is $\frac{P(5,3)}{6}$.

2 If we were creating a flush on purpose, we could do it with the following two-step algorithm:

- Choose a suit.
- Choose five cards from this suit for your hand.

Since there are four ways to complete the first step and $C(13, 5) = \frac{(13)(12)(11)(10)(9)}{5!} = 1{,}287$ ways to complete the second step, by the products rule there are a total of $(4)(1{,}287) = 5{,}148$ flushes possible.

3 Although we could find the sum of the number with one man, the number with two men, ..., and the number with five men, it is easier to solve the complementary problem. The solution then is the difference between the total number of committees and the number with no men. This is $C(18, 5) - C(10, 5) = 8{,}568 - 252 = 8{,}316$.

4 (a) $(1+x)^{10} = C(10, 0) + C(10, 1)x + C(10, 2)x^2 + C(10, 3)x^3 + \cdots$, so the coefficient of x^8 is $C(10, 8) = 45$.

(b) $(1+x^2)^{10} = C(10, 0) + C(10, 1)(x^2) + C(10, 2)(x^2)^2 + C(10, 3)(x^2)^3 + \cdots$. Since $(x^2)^4 = x^8$, the coefficient of x^8 in this expansion is $C(10, 4) = 210$.

5 If $x = 1$, the binomial theorem states

$$(1+1)^n = C(n, 0) + C(n, 1) + C(n, 2) + \cdots + C(n, n)$$

which shows up in the arithmetic triangle as the fact that the sum of the entries in row n is 2^n.

6 Multiplying $1 + 5x + 10x^2 + 10x^3 + 5x^4 + x^5$ by $(1 + x)$ results in

$$1 + (1 + 5)x + (5 + 10)x^2 + (10 + 10)x^3 + (10 + 5)x^4 + (5 + 1)x^5 + x^6$$

which has coefficients $1, 6, 15, 20, 15, 6, 1$, identical to row 6 of the arithmetic triangle.

7 Let $P(n)$ be the statement "$C(4, 4) + C(5, 4) + C(6, 4) + \cdots + C(n, 4) = C(n+1, 5)$." The first statement $P(4)$ simply states, "$C(4, 4) = C(5, 5)$," which is certainly true. Now let $m \geq 4$ be given such that $P(4), P(5), \ldots, P(m-1)$ have all been checked to be true, and consider the next statement $P(m)$.

$$(C(4, 4) + C(5, 4) + \cdots + C(m-1, 4)) + C(m, 4)$$
$$= (C(m, 5)) + C(m, 4) \quad \text{by } P(m-1)$$
$$= C(m+1, 5) \quad \text{by the Pascal triangle pattern}$$

This shows that $P(m)$ is true, completing the induction.

Exercises for Section 5.3

1. (a) List all the permutations of length 2 from the set $\{a, b, c, d, e\}$.

(b) Organize the permutations into equivalence classes for the equivalence relation "would be the same if they were sets."

(c) How big is each equivalence class? How many are there?

(d) How many subsets of $\{a, b, c, d, e\}$ have two elements?

2. (a) How many permutations of length 3 from the set $\{a, b, c, d, e\}$ are there?

(b) For the permutation acd of length 3, list all the elements of its equivalence class for the equivalence relation "would be the same if they were sets."

(c) Repeat part (b) for the permutation dae.

(d) How big is each equivalence class formed as in parts (b) and (c)?

(e) How many subsets of $\{a, b, c, d, e\}$ have three elements?

3. Refer to Example 3.

(a) List all elements in the equivalence class for the following order of the children in a line: $ACDBFE$.

(b) List all elements in the equivalence class for the following order of the children in a line: $ADFBCE$.

4. Four people (Al, Betty, Cindy, and Dan) are to be seated at a round table.

(a) List all the possible orderings of the four names, organized into equivalence classes by the relation "the same if placed in a circle."

(b) How many ways are there to accomplish the seating?

5. In how many ways can four married couples be seated at a round table so that spouses sit together? (HINT: Do this in two steps. Place the couples' surnames in a circle, then for each couple decide whether the husband sits to the left or the right of his wife.)

6. How many ways are there for five married couples to stand in a circle with the condition that all spouses are next to each other?

7. In how many ways can four boys and four girls stand in a circle if they must alternate boy-girl-boy-girl?

8. In how many ways can four boys and four girls stand in a circle if all the boys stand together and all the girls stand together?

9. How many subsets of $\{1, 3, 5, 7, 9\}$ have two elements?

10. How many subsets of $\{1, \ldots, 6\}$ have an odd number of elements?

11. In how many ways can six shuttle vans line up at the airport?

12. How many variations in first-, second-, and third-place finishes are possible in a 100-yard dash with six runners?

13. How many committees of three people can be formed from a club with 17 members?

14. On a TV show, there are eight contestants. One of them (call her Jill) has won a prize dinner, and is allowed to select three of the others to join her. How many ways are there for her to make her selection?

15. The 9 men and 12 women in the Math Club need to form a fundraising committee.

   (a) How many possible committees of four people can be formed? $C(21,4)$

   (b) How many of these four-person committees have the same number of women as men? $C(9,2)\cdot C(12,2)$

   (c) How many four-person committees have more women than men? $C(9,1)\cdot C(12,3)+C(12,4)$

16. This problem refers to the Math Club from the previous problem. Jill (a woman) is the president, and Jack (a man) is the vice president. Answer each of the questions above with the added condition that Jack or Jill but not both must serve on the committee.

17. A bag contains a dozen oranges, two of which are rotten. A sample of three oranges is taken from the bag.

   (a) In how many ways can the sample be taken (how many different samples are there)?

   (b) Of these, how many contain exactly one rotten orange?

   (c) How many contain exactly two rotten oranges?

   (d) How many contain no rotten oranges?

18. Suppose a shipment of 100 computers contains four defective computers, and we choose a sample of six computers.

   (a) How many different samples are there? $C(100,6)$

   (b) Of these, how many samples contain all four defective computers? What percent of the total does this represent? $C(4,4)\cdot C(96,2)$

   (c) How many samples contain one or more defective computers? What percent of the total does this represent? $C(100,6)-C(96,6)$

19. You conduct an experiment in which you interview a large number of families, each of which has eight children. For each family, you write down the gender (M for male, F for female) for the eight children, in order from oldest to youngest.

   (a) How many possible results are there?

   (b) Of these, how many consist of either all M's or all F's?

   (c) How many have exactly three M's?

20. You are tracking your favorite baseball player by writing down his performance in 10 successive plate appearances using an ordered list of length 10.

   (a) If you use H for any time he gets a hit and N when he doesn't, how many possible results are there? 2^{10}

   (b) Of the total, how many include exactly six hits? $C(10,6)$

   (c) Of the total, how many of them include fewer than three hits? $C(10,0)+C(10,1)+C(10,2)$

21. Finish part 3 of Example 4. That is, use each of the three suggested strategies, and verify that all give the same numerical answer.

22. Three members (Mary, Sue, and Tom) of a 20-person office are carpooling, so they insist on never working separately. That is, whenever one of them is on a committee, all three must be. How many committees of size 7 meet this requirement?

23. How many committees of five men and four women can be formed from an organization with 43 women and 47 men?

24. Suppose in the previous scenario that among the members there are 20 married couples and there is a rule prohibiting spouses from serving on the same committee. How many five-man, four-woman committees can be formed under this new restriction?

25. In a standard deck of 52 cards, how many five-card hands have at least one king?

26. There are five red, three green, and eight blue marbles in a box. $C(16,4)$

   (a) In how many ways can a sample of four be selected, without replacement and without regard to order? (That is, reach in and grab four marbles at once).

   (b) Of these, how many have all four selected the same color? $C(5,4)+C(8,4)$

   (c) How many have two colors, with two of each color? $C(5,2)\cdot C(3,2)+$ all other combos

27. A committee of size 6 is being formed from a group of 10 Republicans, 8 Democrats, and 4 Independents.

   (a) How many different committees are possible?

   (b) Of these, how many have exactly two of each political persuasion?

28. Your friend spins a spinner with three colors (red, green, blue), and you record the results on six consecutive spins.

   (a) How many possible results are there?

   (b) Of these, how many have exactly two of each color?

29. One game in the Pennsylvania Lottery requires the purchase of a ticket that has 11 different numbers taken from {1, 2, 3, ..., 79, 80}. The order of the numbers does not matter.

   (a) How many different lottery tickets are there?

   (b) Once a week, the lottery draws seven different numbers from a rolling barrel of ping pong balls numbered 1 through 80. How many different lottery tickets can contain all seven winning numbers?

(c) Calculate the ratio of possible tickets to possible winning tickets.

30. How many odd five-digit numbers have distinct digits and no occurrence of the digit 9? How many of these have their digits in decreasing order?

31. What is the coefficient of z^5 in the expansion of $(1+z)^8$?

32. What is the coefficient of t^6 in the expansion of $\left(1+t^2\right)^7$? What is the coefficient of t^5 in this same expansion?

33. What is the sum of every other entry in row n of the arithmetic triangle? What does the binomial theorem say when $x = -1$? How are these questions related?

34. Investigate the following problem for some small values of n, and then explain the general statement based on the binomial theorem.

What is the coefficient of h (i.e., h^1) in the expansion of

$$\frac{(1+h)^n - 1}{h} ?$$

35. Prove by mathematical induction that for all $n \geq 0$,

$$C(n, 0) + C(n, 1) + C(n, 2) + \cdots + C(n, n) = 2^n$$

36. Prove by mathematical induction that for all $n \geq 0$,

$$C(n, 1) + 2 \cdot C(n, 2) + 3 \cdot C(n, 3) + \cdots +$$
$$n \cdot C(n, n) = n \cdot 2^{n-1}$$

37. Prove by mathematical induction that for all $n \geq 1$,

$$C(n, 1)x + 2 \cdot C(n, 2)x^2 + 3 \cdot C(n, 3)x^3 + \cdots +$$
$$n \cdot C(n, n)x^n = nx(1+x)^{n-1}$$

38. If you know something about differential calculus, show how the identity in Exercise 37 follows from the binomial theorem using the idea of derivative.

39. What entry in row 7 is the same as $C(5, 2) \cdot C(2, 0) + C(5, 1) \cdot C(2, 1) + C(5, 0) \cdot C(2, 2)$? (Think about the binomial theorem applied to all parts of the true equation $(1 + x)^2 \cdot (1 + x)^5 = (1 + x)^7$ to see why this is true.)

40. Generalize the previous problem to give a general description of the "multiplication" of two rows of the triangle.

41. Another true statement about polynomials is that

$$x \cdot (1 + (1 + x) + (1 + x)^2 + \cdots + (1 + x)^n)$$
$$= (1 + x)^{n+1} - 1$$

(a) Explain why the above equation is true of polynomials.

(b) Compare the coefficients of x^2 on either side of this equation, and explain why this proves

$$C(1, 1) + C(2, 1) + C(3, 1) + \cdots + C(n, 1)$$
$$= C(n + 1, 2)$$

(c) What fact about the arithmetic triangle is explained by comparing coefficients of x^3 in this same polynomial identity?

(d) State the general rule at work in the last two exercises as a pattern in the arithmetic triangle. (Draw a picture.)

42. Let k be a fixed positive integer. Prove the following statement by induction on $n \geq k$:

$$C(k, k) + C(k + 1, k) + C(k + 2, k) + \cdots + C(n, k)$$
$$= C(n + 1, k + 1)$$

Draw a picture to illustrate this statement in the arithmetic triangle.

43. Use the mod 10 arithmetic system from Section 2.7 to explain the secret of the card trick in Example 8.

5.4 Binary Sequences

One of the fundamental building blocks in combinatorics is the binary sequence. We have already seen that the number of binary sequences of length n is 2^n, and we have enough tools at hand to count the number of binary sequences with any prescribed number of 0's and 1's.

 Example 1 *How many binary sequences are there with five 1's and three 0's?*

SOLUTION   We are trying to count ordered lists, but it will not work to simply place the eight symbols left-to-right. Instead, think of eight blank spaces as shown below:

1	2	3	4	5	6	7	8

Now the algorithm is only two steps long: (i) Choose five of the eight spaces and put the five 1's there, and then (ii) put the three 0's in the remaining three spaces. There are therefore $C(8, 5) \times C(3, 3) = C(8, 5)$ such binary sequences.

Alternatively, you could (i) choose three of the eight spaces and put the three 0's there, and (ii) put the five 1's in the remaining five spaces. This algorithm then produces $C(8, 3) \times C(5, 5) = C(8, 3)$ binary sequences. Notice that we have implicitly shown that $C(8, 5) = C(8, 3)$ since they are both answers to the same problem! □

Theorem 1 *The number of binary sequences with r 1's and n − r 0's is $C(n, r)$ or $C(n, n − r)$.*

PROOF This is a generalization of Example 1. ■

Practice Problem 1 *Use Theorem 1 along with the sums rule to find the number of binary sequences of length 5 that use an odd number of 1's.*

The same type of algorithm that generates binary sequences can be used to solve other types of counting problems.

 Example 2

1. *How many ordered lists of 10 letters chosen from {m, a, t} have exactly three m's?*
2. *Of these, how many also have exactly four a's?*

SOLUTION

1. Again, we are trying to count ordered lists with repetitions allowed, but we have the same peculiar restriction—we do not have an unlimited number of each letter. We must use the letter *m* exactly three times, so it it will not do to place a first letter, a second letter, and so on. Instead, imagine 10 blank spaces and perform the following two-step algorithm:

1	2	3	4	5	6	7	8	9	10

Step 1. Choose three of the 10 spaces and put the three *m*'s in these spaces.
Step 2. Fill the remaining seven spaces from left to right with letters chosen from {a, t}.

For example, if we choose spaces {2, 4, 9} in the first step, we have

	m		m					m	
1	2	3	4	5	6	7	8	9	10

There are now seven empty slots to fill with a's and t's, and this would be true no matter which spaces were chosen in the first step. Reasoning as in Example 1, we see that there are $C(10, 3)$ ways to do the first step of the algorithm. The second step corresponds to selecting an ordered list of length 7 from a set of two elements, for which the count is $2 \cdot 2 \cdot 2 \cdot 2 \cdot 2 \cdot 2 \cdot 2 = 2^7$. By the rule of products, there are $C(10, 3) \cdot 2^7 = 15,360$ ways to complete the two-step process.

2. The algorithm is very similar.

Step 1. Choose three of the 10 spaces and put the three m's in these spaces.

Step 2. Choose four of the remaining seven spaces and place the four a's in those spaces.

Step 3. Finally, place t's in the remaining three spaces.

For example, if we choose $\{2, 4, 9\}$ in the first step, and $\{1, 5, 6, 10\}$ in the second step, we have

a	m		m	a	a			m	a
1	2	3	4	5	6	7	8	9	10

After placing the t's, we have the ordered list $amtmaattma$.

Again, there are $C(10, 3)$ ways to do the first step. In the second step we choose four of the seven remaining spaces, and there are $C(7, 4)$ ways to do this. There is only $1 \cdot 1 \cdot 1 = 1^3 = 1$ way to do the third step. The total count is $C(10, 3) \cdot C(7, 4) \cdot 1 = 4,200$. □

Example 3 *How many distinguishable arrangements of the letters in the word MISSISSIPPI are there?*

SOLUTION An arrangement of the letters in this word will be an ordered list of length 11, chosen from the letters $\{M, I, S, P\}$ that make up the word, with the restriction that there must be one M, four I's, four S's, and two P's. We imagine 11 empty spaces, and perform a four-step algorithm.

Step 1. Choose one of the 11 spaces, and put the M in this space.

Step 2. Choose four of the remaining 10 spaces, and put the four I's in these spaces.

Step 3. Choose four of the remaining six spaces, and place the four S's in these spaces.

Step 4. Finally, place the 2 P's in the remaining two spaces.

For example, if we choose space $\{4\}$ in the first step, and then spaces $\{2, 3, 8, 10\}$ in the second step and $\{1, 6, 7, 11\}$ in the third step, we will have

S	I	I	M		S	S	I		I	S
1	2	3	4	5	6	7	8	9	10	11

With only the P's left to place, we will have the arrangement $SIIMPSSIPIS$ after all four steps.

The analysis of this algorithm is simple. There are $C(11, 1)$ ways to do the first step, $C(10, 4)$ ways to do the second step, $C(6, 4)$ ways to do the third step, and $C(2, 2)$ ways to do the fourth step. So the final answer to this question is the product

$$C(11, 1)\, C(10, 4)\, C(6, 4)\, C(2, 2) = (11)(210)(15)(1) = 34{,}650$$

Explore more on the Web.

Practice Problem 2

(a) *Show that the answer to the preceding example does not depend on the particular order in which the letters are placed by using an algorithm that places first the P's, then the S's, then the M, and finally the I's.*

(b) *How many ways are there to arrange the letters in the word ALABAMA?*

Unordered Lists with Repetition Allowed

We can also apply binary sequences to find a counting formula for the last of the four types of finite structures described at the beginning of this chapter, unordered lists in which repetition is allowed. The prototypical example of this structure is a bag of grocery items, where there may be several of the same type of object (like apples) and the order in which the objects are placed in the bag is irrelevant. Because of this analogy, unordered lists with repetition allowed are also known as *bags*.

We will count these structures by representing them with binary sequences and then applying the knowledge about binary sequences we have developed in this section. It is best to look first at a simple example, so we begin by revisiting an idea we first encountered in Section 5.1.

Example 4 *Without answering either question, explain why these two questions have the same answer.*

(i) *How many ways can one fill a bag with 10 pieces of fruit at a store that sells only apples, bananas, and peaches? (We are assuming that the store has at least 10 of each type of fruit available.)*

(ii) *How many solutions in* nonnegative *integers are there to the following equation?*

$$a + b + c = 10$$

SOLUTION   To show that question (i) and (ii) have the same answer, we can represent a bag of fruit (counted by (i)) as a list of three numbers a, b, c in which a represents the number of apples, b the number of bananas, and c the number of peaches. In every such list, a, b, and c will be nonnegative integers that sum to 10. This sets up a one-to-one correspondence between bags of fruit and solutions to the equation. Here are a couple of specific examples of this correspondence.

● A bag with four apples, one banana, and five peaches corresponds to the equation $4 + 1 + 5 = 10$.

● The solution $8 + 0 + 2 = 10$ corresponds to a bag with eight apples, no bananas, and two peaches.

Example 5 *Without answering either question, explain why these two questions have the same answer.*

(i) *How many solutions in* nonnegative *integers are there to the following equation?*

$$a + b + c = 10$$

(ii) *How many binary sequences of length 12 have exactly two 1's and ten 0's?*

SOLUTION   Set up a correspondence by linking a solution (a, b, c) to the equation in (ii) to the binary sequence made of a 0's, then a 1, then b more 0's, then another 1, and finally c 0's.

(2,1,7) $\longleftrightarrow$ 001010000000

(1,7,2) $\longleftrightarrow$ 010000000100

(5,0,5) $\longleftrightarrow$ 000001100000

(3,7,0) $\longleftrightarrow$ 000100000001

(0,10,0) $\longleftrightarrow$ 100000000001

Figure 5-7 The correspondence in Example 5.

$$(a, b, c) \longleftrightarrow \underbrace{0\ldots0}_{a}1\underbrace{0\ldots0}_{b}1\underbrace{0\ldots0}_{c}$$

Since $a + b + c = 10$, the resulting sequence will be of length 12, and will contain exactly two 1's and ten 0's. For example, part of this correspondence is shown in Figure 5-7. This "linking" is a one-to-one correspondence.   □

The significance of these correspondences should now be clear. By Theorem 1, we know that the answer to question (iii) is $C(12, 2)$. Therefore, this is also the number of solutions described in question (ii), and hence it is also the number of bags described by question (i).

We summarize the more general connection between these three counting problems in the following lemma.

Lemma 2   *The following three questions have the same answer for all natural number values of n and r:*

● *How many unordered lists with repetitions allowed (bags) of size r with entries from a set of size n are there?*

● *How many nonnegative integer solutions to the equation $x_1 + x_2 + \cdots + x_n = r$ are there?*

● *How many binary sequences of length $r + n - 1$ with exactly r 0's are there?*

Practice Problem 3   *For the particular questions in Examples 4 and 5, illustrate the correspondences between the bags of fruit, the equation solutions, and the binary sequences by filling in the missing entries in Table 5-25.*

This allows us to see the power of the one-to-one correspondence in combinatorics. Because we know how to answer one of the three questions in Lemma 2, we know how to answer all three.

Bag of Fruit	Equation	Binary Sequence
1 apple, 3 bananas, 6 peaches		
	$4 + 2 + 4 = 10$	
		100100000000
		000000000011

Table 5-25 Table for Practice Problem 3

Theorem 3 *Let natural numbers n and r be given.*

1. *The number of solutions to the equation $x_1 + \cdots + x_n = r$ using nonnegative integers is $C(r + n - 1, r)$.*

2. *The number of unordered lists of length r taken from a set of size n, with repetitions allowed, is $C(r + n - 1, r)$.*

3. *The number of bags of r pieces of fruit that can be bought at a store with n types of fruit available is $C(r + n - 1, r)$.*

    PROOF This simply combines Lemma 2 with Theorem 1. ■

Applications of Theorem 3

Many people find the counting formula derived in Theorem 3 difficult to use in specific situations, so we will give some practical advice for a variety of problems. Note that it is perfectly correct to simply use Theorem 3 as it is written, but if you have trouble remembering it, you should consider our recommendations for answering these types of questions.

Example 6 *How many bags of 20 pieces of candy can one buy from a store that sells four types of candy?*

    SOLUTION While it is possible to match this up directly with Theorem 3, it is easier to relate it to more fundamental facts as follows:

Step 1. To what nonnegative integer equation do these bags of candy correspond?

        Solution. If we let w, x, y, and z represent the number of pieces of the four types of candy, the equation would be $w + x + y + z = 20$.

Step 2. Choose a particular solution to this equation, and draw a picture of a binary sequence that corresponds to this solution. To remember how to do this, imagine a string of twenty 0's that must be broken into four parts with sizes w, x, y, and z, respectively. To partition the string of twenty 0's into four parts, insert three **1**'s as dividers within the string so that w 0's occur before the first **1**, x 0's occur between the first and second **1**'s, and so on.

        Solution. Using this idea, we see that the solution $w = 3, x = 0$, $y = 10, z = 7$ corresponds to the binary sequence

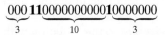

Step 3. What are the length and the number of 0's in this binary sequence? How many binary sequences are there with these attributes?

        Solution. This sequence has length 23 with twenty 0's and three 1's. There are a total of $C(23, 20)$ or, equivalently, $C(23, 3)$ of these. □

    So the correct first step in a problem of this type is to find a representation of the objects being counted as solutions to a nonnegative integer equation. Sometimes this objective requires some creativity, as the remaining examples illustrate.

Example 7 *How many bags of 10 pieces of fruit can be purchased at a store that carries apples, bananas, peaches, and pears if we insist on getting **at least one of each type of fruit**?*

SOLUTION We can form such a bag of fruit in two steps: (i) Put one of each type of fruit into the bag, and (ii) put in six more pieces of any type. There's only one way to do the first step. For the second step, we can let a, b, c, and d represent the number of additional apples, bananas, peaches, and pears, respectively. Clearly, $a + b + c + d = 6$ and the variables can be any nonnegative integer, so these solutions can be represented as strings of six 0's, into which three **1**'s have been placed as dividers. There are a total of $C(6 + 4 - 1, 6)$ such binary sequences. Hence, there are $C(9, 6) = 84$ bags of this type. □

Note that because of the original analogy between bags and solutions to equations, this same problem could be presented in the following different way.

Example 8 *How many positive integer solutions are there to the equation*

$$w + x + y + z = 10?$$ $6 + 4 - 1 \quad (9)$

SOLUTION *The problem has the same answer, $C(9, 6)$, as Example 7.* □

Practice Problem 4 *Use the idea in Example 7 to answer the following question: "How many nonnegative integer solutions are there to the equation $w + x + y + z = 10$ if w is required to be at least 3 and x must be positive?"*

Explore more on the Web.

Our final application comes from games that involve rolling dice. We consider these partly because the next chapter focuses on probabilities involved in games, and many games use dice to generate their moves. This kind of problem also illustrates that some very natural questions have some surprisingly difficult answers.

Example 9 *How many different outcomes are possible if I toss a regular six-sided die four times? In how many of these do the four rolls sum to 14?*

SOLUTION We can represent an outcome as an ordered list from $\{1, \ldots, 6\}$ of length 4—that is, as x_1, x_2, x_3, x_4, where x_i is the result of the ith toss of the die. There are $6^4 = 1{,}296$ of these.

To answer the second question, we think of the above representation of the four rolls of the die; then the condition that the sum is 14 corresponds to having $x_1 + x_2 + x_3 + x_4 = 14$. So if we can say how many solutions (x_1, x_2, x_3, x_4) there are to this equation with the condition that each x_i is from $\{1, \ldots, 6\}$, then this will answer the original question.

Let's first think about how many solutions to $x_1 + x_2 + x_3 + x_4 = 14$ there are with only the condition that each $x_i \geq 1$. Example 8 essentially showed that this is the same thing as the number of nonnegative integer solutions to the equation $x_1' + x_2' + x_3' + x_4' = 10$, which according to Theorem 3 is $C(10 + 4 - 1, 10) = 286$.

Now back to the original problem—that is, we wish to have each $x_i \leq 6$ too. In the above analysis, we counted many solutions that could not correspond to dice rolls. For example, we counted $(1, 1, 3, 9)$ since it corresponds to the following binary sequence with fourteen 0's and three 1's:

$$0\ 1\ 0\ 1\ 0\ 0\ 0\ 1\ 0\ 0\ 0\ 0\ 0\ 0\ 0\ 0\ 0$$

The strategy we will use will be to subtract from the total number 286 the number of solutions that **do not** correspond to die rolls. So we wish to count the number of solutions to $x_1 + x_2 + x_3 + x_4 = 14$ (with each $x_i \geq 1$) that have *either $x_1 \geq 7$ or $x_2 \geq 7$ or $x_3 \geq 7$ or $x_4 \geq 7$*. Note that it is impossible to have, say, $x_1 \geq 7$ *and* $x_2 \geq 7$ at the same time in a solution to $x_1 + x_2 + x_3 + x_4 = 14$, so the "bad" solutions (i.e., those not corresponding to die rolls) can be separated into four types, depending on which of the four quantities is too big.

Now how many solutions to $x_1 + x_2 + x_3 + x_4 = 14$ (with each $x_i \geq 1$) are there in which $x_1 \geq 7$? To answer this, it is easiest to think about this as a problem about bags of fruit. This is equivalent to asking "How many bags of 14 pieces of fruit can be bought from a store that carries apples, bananas, cantalopes, and peaches, if we insist on getting at least one of each type of fruit **and** we must get at least seven apples?" Example 7 suggests that we form these bags in two steps: First, place seven apples, one banana, one cantalope, and one peach into the bag; second, place four more pieces freely into the bag. There is only one way to do the first step, and by Theorem 3 there are $C(4 + 4 - 1, 4)$ ways to do the second step, so there are $C(7, 4) = 35$ total bags of this type.

This means that there are 35 solutions to $x_1 + x_2 + x_3 + x_4 = 14$ (with each $x_i \geq 1$) in which $x_1 \geq 7$. Similarly, there are also 35 with $x_2 \geq 7$, 35 with $x_3 \geq 7$, and 35 with $x_4 \geq 7$, giving us a total of $4 \times 35 = 140$ "bad" solutions.

Therefore, there are $286 - 140 = 146$ solutions with each x_i from $\{1, \ldots, 6\}$, and hence 146 ways that the die rolls can sum to 14. $\square$

In this example, we were lucky that it was impossible to have two or more variables that were "bad" at the same time in the previous problem. If this had happened (e.g., if the sum of the four dice had to be 16 instead of 14), we would have had to incorporate the *principle of inclusion-exclusion* first discussed in Section 3.1.

Practice Problem 5 *In how many ways can one roll a sum of 17 on six distinguishable six-sided dice?*

Summary of Counting Basic Structures

In the first four sections of this chapter, we have introduced four types of finite structures and provided "formulas" that count them. Table 5-26 summarizes what we have determined so far. In each case, the entries in the structure come from the set $\{1, 2, \ldots, n\}$.

Solutions to Practice Problems

1 The binary sequences described must have either one 1, three 1's, or five 1's. Applying Theorem 1 for each of these cases, we have $C(5, 1)$ binary sequences of length

5 with one 1, $C(5, 3)$ binary sequences of length 5 with three 1's, and $C(5, 5)$ binary sequences of length 5 with five 1's. The answer then is $C(5, 1) + C(5, 3) + C(5, 5)$.

What?	How Many?
Ordered lists of length r	n^r
Permutations of length r	$P(n, r)$
Unordered lists of size r	$C(r + n - 1, r)$
Sets of size r	$C(n, r)$

Table 5-26 A Summary of Counting Formulas

2 (a) This algorithm gives $C(11, 2) \cdot C(9, 4) \cdot C(5, 1) \cdot C(4, 4) = 34{,}650$, which is the same count as in the example.

(b) We imagine seven blanks into which we place the letters in ALABAMA according to the following algorithm:

- Choose four of the seven blanks, and place A's in them.
- Choose one of the remaining three blanks, and place the L in it.
- Choose one of the remaining two blanks, and place the B in it.
- Choose the one remaining blank, and place the M in it.

There are $C(7, 4)$ ways to do the first step, three ways to do the second step, two ways to do the third step, and one way to do the last step, so by the products rule there are

$$C(7, 4) \cdot 3 \cdot 2 \cdot 1 = 210 \text{ distinguishable arrangements}$$

3 Table 5-27 shows the completed table.

4 First put the required three apples and one banana into the bag, then purchase six additional pieces of fruit. There is one way to do the first step. The second step corresponds to binary sequences with six 0's (the fruit) and three 1's (the dividers separating the groups of 0's). There are $C(9, 6) = C(9, 3) = 84$ different bags of fruit meeting the conditions.

5 As in Example 9, we first convert the equation

$$x_1 + x_2 + x_3 + x_4 + x_5 + x_6 = 17$$

with positive integer solutions to the corresponding equation

$$x_1 + x_2 + x_3 + x_4 + x_5 + x_6 = 11$$

with nonnegative solutions. Now following Example 9, the number of nonnegative solutions to this equaiton, with each $x_i \le 5$, is

$$C(16, 5) - 6 \cdot C(10, 5) = 2{,}856$$

Exercises for Section 5.4

1. How many binary sequences of length 8 satisfy each of the following conditions?
  (a) Exactly six 1's
  (b) Exactly two 0's
  (c) At least two 1's

2. How many binary sequences of length 10 satisfy each of the following conditions?
  (a) The same number of 0's and 1's $C(10, 5)$
  (b) At most two 0's $C(10, 0) + C(10, 1) + C(10, 2)$
  (c) At least eight 1's $same \ as \ b$

Bag of Fruit	Equation	Binary Sequence
1 apple, 3 bananas, 6 peaches	$1 + 3 + 6 = 10$	010001000000
4 apples, 2 bananas, 4 peaches	$4 + 2 + 4 = 10$	000010010000
2 bananas, 8 peaches	$0 + 2 + 8 = 10$	100100000000
10 apples	$10 + 0 + 0 = 10$	000000000011

Table 5-27 Solution to Practice Problem 3

3. A ternary sequence is a sequence of digits chosen from $\{0, 1, 2\}$. How many ternary sequences of length 12 satisfy each listed condition?

(a) Exactly eleven 1's

(b) Exactly three 1's and two 0's

(c) No 1's

4. You are tracking your favorite baseball player by writing down his performance in 10 successive plate appearances using an ordered list of length 10.

(a) If you use H for a hit, S for a strikeout, B for a base-on-balls, and O for anything else, how many results are possible?

(b) Of the total, how many have exactly three H's?

(c) Of the total, how many have exactly four H's, exactly one S, exactly two B's, and exactly three O's?

5. How many ordered lists, of length 10 with repetitions allowed, are there satisfying each condition?

(a) Using the set $\{a, b, c, d\}$

(b) Using the set $\{x, y, z\}$ and having exactly three x's

(c) Using the set $\{0, 1, 2, 3, 4\}$, having exactly two 0's and three 2's, and not beginning with 0

6. (Compute exact numerical answers for (a) and (b), and then answer (c) using complete sentences.)

(a) How many binary sequences of a length at most 5 have exactly three 1's?

(b) How many binary sequences of length 6 have exactly four 1's?

(c) Why do these two questions have the same answer?

7. How many 10-digit numbers are there that use exactly three 1's, three 2's, and four 3's?

8. How many properly written numbers (i.e., leading 0's are not allowed) use each digit $(0, 1, 2, \ldots 9)$ exactly once?

9. How many properly written numbers use each digit $(0, 1, 2, \ldots 9)$ exactly twice?

10. How many arrangements are there of the letters in the word MATHEMATICS?

11. How many arrangements are there of the letters from your last name?

12. How many arrangements are there of the letters in the word MISSISSIPPI that begin with either an I or an S?

13. In Example 3 and Practice Problem 2, we extended the idea for counting binary sequences to count any kind of sequence made with a fixed set of symbols. In this exercise, we will develop an alternative way to answer these questions using our more fundamental notion of equivalence classes.

(a) There are $7! = 5{,}040$ arrangements of the letters in ALABAMA, given that the seven letters all look different. On the other hand, if we did not use the goofy fonts, some arrangements, like ALAAMAB and ALAAMAB for example, would look the same. How many of the arrangements of the seven goofy letters look like ALABAMA when the goofiness is removed?

(b) Use the answer to part (a) along with the fact that there are 5,040 different arrangements of the seven distinct letters in ALABAMA, to determine the number of distinguishable arrangements of the letters in ALABAMA.

14. For Examples 4 and 5, illustrate the correspondence between the original bag of fruit, the equation solution, and the final binary sequence by filling in the missing entries in Table 5-28.

15. I purchase a bag of 12 pieces of fruit from a store that sells apples, bananas, peaches, oranges, pears, and pineapples. Using the strategy outlined in the section:

(a) Describe the equation and binary sequence that would represent these bags: (i) two of each kind of fruit; (ii) six apples and six pears; (iii) 12 oranges.

(b) Describe the bags and equations corresponding to these binary sequences: (i) 00110011001000000; (ii) 10000001010010010; (iii) 00010001000100011.

16. How many ways are there to fill a bag with 20 pieces of fruit at a store that sells apples, bananas, and peaches?

Bag of fruit	Equation	Binary sequence
5 apples, 5 bananas		
	$5 + 0 + 5 = 10$	
		011000000000
		100000000001

Table 5-28 Table for Exercise 14

17. How many ways are there to fill a bag with 20 pieces of fruit at a store that sells apples, bananas, peaches, and oranges?

18. How many ways are there to fill a bag with 20 pieces of fruit at FruitMart, a store that sells 30 types of fruit?

19. How many solutions using nonnegative integers are there to the equation $x + y + z = 11$?

20. How many solutions using nonnegative integers are there to the equation $w + x + y + z = 10$?

21. How many solutions using nonnegative integers are there to the equation $w + x + y + 2z = 10$?

22. How many different bags of produce can I bring back from the store, assuming that there are apples, bananas, oranges, and peaches available; I buy at least one of each; and I purchase exactly 15 pieces of fruit altogether? $C(14, 11)$ $4 + 11 - 1$

23. Repeat the preceding exercise, but this time assuming you buy at least two of each.   16 total choices

24. How many solutions using positive integers are there to the equation $x + y + z = 13$? $13 - 3 = 10$ $10 + 3 - 1 = 12$ Choose 10

25. How many positive integer solutions are there to the equation

$$a + b + 2c = 10?$$

(HINT: Break the problem into four cases depending on whether $c = 1$, $c = 2$, $c = 3$, or $c = 4$.)

26. Translate this problem from Section 5.1 into a "bag" problem, and solve it: "How many ways can two winners be chosen for prizes from a class consisting of just four people {Andrew, Bob, Carly, Diane}, assuming that the prizes are identical door prizes (so it doesn't matter who gets which prize) and that the same person could win both?" (HINT: First rewrite the problem in the form "How many unordered lists of length _____ can be made from the set {_____}, with repetitions allowed?")

27. Translate the following into a "bag" problem, and solve. The Computer Club has 15 members, and plans to give away 35 identical refrigerator magnets to the members. How many ways are there to accomplish this if each member must receive at least one magnet?

28. How many ways are there to distribute 200 apples among 43 women and 47 men if each woman must get at least two apples and each man must get at least one apple? $C(90, 67)$ $90 + 67 - 1$ $C(156, 67)$

29. How many ways are there to distribute 200 apples among 42 Democrats, 47 Republicans, and 5 Independents if each Democrat and Republican must get at least one apple and at least one Independent gets less than four apples?

30. Generalize Example 8 to find a general formula for the number of positive integer solutions to the equation

$$x_1 + x_2 + \cdots + x_n = r$$

31. There are $6^4 = 1,296$ different outcomes from rolling four distinguishable ordinary (six-sided) dice. In how many of these is the sum of the dice 12?

5.5 Recursive Counting

At the beginning of Section 5.1, we related the ingenuity displayed by Gauss as a child in computing the sum of the first 100 positive integers. This sort of clever organization has been exploited throughout this chapter to derive all sorts of slick devices for counting complicated sets of objects. In this section, we will examine a form of reasoning that is valuable in many aspects of discrete mathematics but especially in the study of computer science.

Examples

Despite contradictory rumors, recursion is not a fabricated tool to make beginning students dizzy. The human brain naturally uses recursive reasoning in many situations. For example, when Gauss announced that the sum $1 + 2 + 3 + \cdots + 100$ was 5,050, the rest of the class was probably suitably impressed. However, suppose the teacher had then responded with "Alright Mr. Smarty-Pants, Prince of Mathematics, now do *this* sum!"

$$1 + 2 + 3 + \cdots + 99 + 100 + 101$$

No student in the class would have started over again. They would have simply added 101 to the 5,050 that Gauss already computed and arrived at 5,151 all at the same time.

We will model other problems with this same point of view. That is, we will imagine that the answer to the proposed question is known when $n = 100$, and we will think of a way to use this information to arrive at the answer when $n = 101$. Of course, there is nothing sacred about the number 100 so our explanation ought to be completely general. If at this point this seems closely related to mathematical induction, you are on the right path.

Example 1 *Find a recursive model for the number of games that will occur in the first round of a round-robin tournament with n teams. (In the first round of a round-robin tournament, each team plays all other teams once.)*

SOLUTION Let a_n denote the number of first-round games in a tournament of n teams. For example, $a_1 = 0$ and $a_2 = 1$. Let's imagine that we know everything there is to know about tournaments with $n - 1$ teams, but we are now faced with n teams to play in the first round. Use the letter Z to denote the last team to enter. Without Z there would have been $n - 1$ teams in the tournament so there would have been a_{n-1} first-round games, but Z is there and must play each of those $n - 1$ teams as well—this accounts for an additional $n - 1$ games. There is a total then of $a_{n-1} + (n - 1)$ first-round games in a tournament with n teams. More simply stated,

$$a_n = a_{n-1} + (n - 1)$$

This along with the knowledge that $a_1 = 0$ is enough to find the value of a_n for any n. □

Example 2 *Find a recursive model for $P(n, r)$, the number of r-permutations from $\{1, \ldots, n\}$.*

SOLUTION The presence of two variables makes this a little more elusive, but it's not too hard. Imagine we already are experts on permutations of all lengths on sets of size $n - 1$, and we wish to enumerate the r-permutations of $\{1, \ldots, n\}$. We will construct these with a two-step algorithm:

● Choose an element from $\{1, \ldots, n\}$ for the first position.

● Place an $(r - 1)$-permutation from the remaining $(n - 1)$-element set in the remaining $r - 1$ positions.

As this algorithm will form all r-permutations from $\{1, \ldots, n\}$, the product rule tells us that

$$P(n, r) = n \cdot P(n - 1, r - 1)$$

This along with the fact that $P(n, 0) = 1$ for all n is enough to generate any value of $P(n, r)$, so this is an adequate recursive model. □

Practice Problem 1 *Let $d_{n,k}$ denote the number of ways to roll a sum of n using k distinguishable regular (six-sided) dice. Explain why the following relationship is true:*

$$d_{n,k} = d_{n-1,k-1} + d_{n-2,k-1} + d_{n-3,k-1} + d_{n-4,k-1} + d_{n-5,k-1} + d_{n-6,k-1}$$

Example 3 *Let b_n be the number of subsets of $\{1, 2, \ldots, n\}$ that do not contain consecutive numbers as elements. For example, if $n = 4$, the eight sets*

$$\{1, 3\}, \{2, 4\}, \{1, 4\}, \{1\}, \{2\}, \{3\}, \{4\}, \emptyset$$

all have this property so we would say $b_4 = 8$. Find a recursive model for b_n.

SOLUTION   It is easy to verify that $b_1 = 2$ (by listing the sets $\{1\}$ and $\emptyset$) and $b_2 = 3$ (by listing the sets $\{1\}$, $\{2\}$, and $\emptyset$). Now for a specific $n > 2$, imagine that we know everything about these kinds of subsets of $\{1, 2, \ldots, k\}$ for all $k < n$, and we are trying to understand subsets of $\{1, 2, \ldots, n\}$. Any subset of $\{1, 2, \ldots, n\}$ either contains the element n or it does not.

● If a subset contains n, then in order to include no consecutive numbers, the remaining elements of the subset must come from $\{1, 2, \ldots, n - 2\}$ and include no consecutive numbers. Because of our (imagined) expertise, we know there are b_{n-2} ways to choose these remaining elements.

● If a subset does not contain n, then it is actually a subset of $\{1, 2, \ldots, n - 1\}$ with no consecutive numbers, and our (imagined) expertise tells us there are b_{n-1} such subsets.

These two cases account for all subsets of $\{1, 2, \ldots, n\}$ with no consecutive numbers, so $b_n = b_{n-1} + b_{n-2}$. □

Practice Problem 2   *Let c_n denote the number of binary sequences of length n. Use a recursive model to explain why*

$$c_n = 2c_{n-1}$$

Example 4 *The classic puzzle "The Towers of Hanoi" consists of n disks of different diameters and three pegs. (Figure 5-8 illustrates $n = 6$.) The only rule is that no larger disk can be stacked on a smaller disk. The objective is to move all the disks from the leftmost peg to the rightmost peg one disk at a time. Let H_n denote the number of required moves to accomplish the objective, and find a recursive model for H_n.*

SOLUTION   It is easy to see that $H_1 = 1$. Assume that we are masters of the Tower of Hanoi with $n - 1$ disks, and that someone has given us a Tower of Hanoi puzzle with n disks stacked (legally) on the leftmost peg. The puzzle can be solved in the following steps:

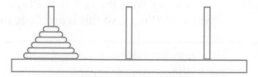

Figure 5-8   Towers of Hanoi.

n	1	2	3	4	5	6
$H(n)$	1	3	7	15	31	63

Table 5-29 Number of Moves for the n-disk Tower of Hanoi

- Move the topmost $n-1$ disks from the left to the center peg.
- Move the bottommost (i.e., largest) disk from the left peg to the right peg.
- Move the $n-1$ disks from the center peg to the right peg.

Since the first step of this operation is equivalent to solving an $(n-1)$-disk Tower of Hanoi puzzle, we know from our assumed expertise that there are H_{n-1} required moves to accomplish the first step. Clearly, only one move is required to accomplish the second step, and again the last step requires the "solution" of an $n-1$-disk puzzle and consequently requires H_{n-1} additional moves. These three steps then solve the puzzle with n disks so we have $H_n = H_{n-1} + 1 + H_{n-1}$, or

$$H_n = 2H_{n-1} + 1$$

□

Example 4 illustrates an interesting phenomenon. The recursive description of H_n was easy to justify, but if we had written out a few of the H_n numbers, we would see the strong pattern shown in Table 5-29. We leave the proof of this pattern for more practice with induction.

Practice Problem 3 *Use induction to show that the H_n of Example 4 satisfies $H_n = 2^n - 1$.*

Example 5 *The end of a tennis game can sometimes be strange since a player must essentially win the game by two points. Specifically, if the game is tied late in the game (this is called being "tied at deuce"), then the game continues until one player has won two points more than the other player. In how many different ways can a tennis game tied at deuce still be tied at deuce 10 points later?*

SOLUTION Let T_n be the number of ways this can happen after n points. If we look at some examples, we might see a pattern among the values of T_n.

- If the game is tied at deuce, then the game cannot be tied again after only one point; therefore, we would say that $T_1 = 0$. Extending this reasoning, we can see that in fact after any odd number of points have been played, the game cannot be tied again; hence, $T_n = 0$ whenever n is odd.
- Let's consider T_2, the number of ways that a game tied at deuce is tied at deuce again two points later. We will represent the play of the games as a list of letters from $\{A, B\}$, where A represents a point won by one player and B represents a point won by the other player. The only sequences of two points that will leave the game tied again are AB and BA; hence, $T_2 = 2$.

- We can deal with T_4 in a similar way by simply listing all possibilities for four points: $ABAB$, $ABBA$, $BAAB$, and $BABA$. Therefore, $T_4 = 4$.

Now suppose that we have mastered counting the possibilities for games lasting *less than n* points, and we are confronted with having to count how many games that start at deuce are at deuce again after n points (assuming n is an even number). These games look like one of the following:

- AB_____, where the blank contains a sequence representing a game that starts tied at deuce and is again tied at deuce $n - 2$ points later. We know that there are T_{n-2} possibilities to go in the blank.
- BA_____, where the blank contains a sequence representing a game that starts tied at deuce and is again tied at deuce $n - 2$ points later. We know once again that there are T_{n-2} possibilities to go in the blank.

Since in each of these two cases there are T_{n-2} possibilities, the total number of ways this can happen is $2 \cdot T_{n-2}$. In other words,

$$T_n = 2 \cdot T_{n-2}$$

This along with the facts that $T_1 = 0$ and $T_2 = 2$ is enough to determine the value of T_n for any n. ☐

Solutions to Practice Problems

1 Imagine one die is a different color than the others— let's say it's red. A roll of all k dice that sum to n falls into one of six cases, depending on the outcome of the red die.
- If the red die is a 1, then the other $k - 1$ dice must sum to $n - 1$; there are $d_{n-1,k-1}$ ways for this to happen.
- If the red die is a 2, then the other $k - 1$ dice must sum to $n - 2$; there are $d_{n-2,k-1}$ ways for this to happen.
- If the red die is a 3, then the other $k - 1$ dice must sum to $n - 3$; there are $d_{n-3,k-1}$ ways for this to happen.
- If the red die is a 4, then the other $k - 1$ dice must sum to $n - 4$; there are $d_{n-4,k-1}$ ways for this to happen.
- If the red die is a 5, then the other $k - 1$ dice must sum to $n - 5$; there are $d_{n-5,k-1}$ ways for this to happen.
- If the red die is a 6, then the other $k - 1$ dice must sum to $n - 6$; there are $d_{n-6,k-1}$ ways for this to happen.

Altogether this means that

$$d_{n,k} = d_{n-1,k-1} + d_{n-2,k-1} + d_{n-3,k-1}$$
$$+ d_{n-4,k-1} + d_{n-5,k-1} + d_{n-6,k-1}$$

2 Each binary sequence of length n consists of either 0 or 1 as its leftmost entry followed by a binary sequence of length $n - 1$. Hence,

$$c_n = 2c_{n-1}$$

3 From Example 4, we know that $H_1 = 1$ and $H_n = 2H_{n-1} + 1$. Let $P(n)$ be the statement "$H_n = 2^n - 1$." Since $2^1 - 1 = 1$, we know that $P(1)$ is true. Now suppose $P(1), P(2), \dots, P(m - 1)$ have all been checked to be true for some given integer $m \geq 2$. Then

$$H_m = 2H_{m-1} + 1 \quad \text{from Example 4}$$
$$= 2(2^{m-1} - 1) + 1 \text{ by } P(m - 1)$$
$$= 2^m - 1$$

Exercises for Section 5.5

1. Find a recursive model for a_n, the number of n-digit numbers that do not use the digit 0.

2. For Practice Problem 1, give enough information about "initial values" so that all values of $d_{n,k}$ can be determined.

3. For Example 5, find a formula in terms of n for T_n, and prove your formula is correct using mathematical induction.

4. Use the recurrence relation $P(n,r) = n \cdot P(n-1, r-1)$ (with $P(n,0) = 1$ for all $n \geq 0$) from Example 2 to prove by induction that $P(n,n) = n!$ for all $n \geq 0$.

5. Use the recurrence relation $P(n,r) = n \cdot P(n-1, r-1)$ (with $P(n,0) = 1$ for all $n \geq 0$) from Example 2 to prove each of the following:

   (a) For all $n \geq 1$, $P(n,1) = n$

   (b) For all $n \geq 2$, $P(n,2) = n \cdot (n-1)$ (You will need to use (a).)

   (c) For all $n \geq 3$, $P(n,3) = n \cdot (n-1) \cdot (n-2)$ (You will need to use (b).)

6. Recall that $C(n,k)$ is defined to be the number of k-element subsets of the set $\{1, 2, \ldots, n\}$. These can be generated using the following recursive algorithm: (i) Decide whether to use the largest possible number n or not; (ii) fill in the remaining elements chosen from $\{1, 2, \ldots, n-1\}$ to form a k-element set. Explain how this accounts for the fact that

$$C(n,k) = C(n-1, k) + C(n-1, k-1)$$

7. Find a recursive model for c_n, the number of ways to cover a $2 \times n$ "chessboard" with "dominoes" that each cover a 1×2 area.

8. Let d_n denote the number of ways that n letters can be put into n envelopes so that no letter goes in the correct envelope. The numbers d_n are traditionally called *derangement* numbers.

   The great Swiss mathematician Leonhard Euler (1707–1783) showed that these numbers satisfy the recurrence relation

$$d_n = (n-1) \cdot (d_{n-1} + d_{n-2}) \quad \text{for } n \geq 3$$

Find the values of d_1 and d_2, and then use this recursive description to compute the value of d_5.

9. Prove by induction on $n \geq 1$ that the derangement numbers from the previous exercise satisfy

$$d_n = n! \left(\frac{1}{0!} - \frac{1}{1!} + \frac{1}{2!} - \frac{1}{3!} + \cdots + (-1)^n \frac{1}{n!} \right)$$
$$= n! \left(\sum_{k=0}^{n} (-1)^k \frac{1}{k!} \right)$$

10. Find a recursive model for t_n, the number of ways in which $2n$ tennis players can be partitioned into n first-round matches in a tournament.

11. Find a recursive model W_n, the number of ways that a tennis game tied at deuce is over after exactly n more points are played. (This problem uses terminology from Example 5.)

12. Find a recursive model (in terms of n and k) for $f_{n,k}$, the number of ways to fill a bag with n pieces of fruit at a store that sells k types of fruit.

13. Find a recursive model for s_n, the number of binary sequences of length n that have no consecutive 1's.

14. Find a recursive model for w_n, the number of properly written n-digit numbers that do not have consecutive 1's as digits.

15. A *partition* of the set $X = \{1, 2, \ldots, n\}$ is a collection $S_1, S_2, \ldots$ of nonempty subsets of X that do not overlap (i.e., when $i \neq j$, $S_i \cap S_j \neq \emptyset$) but that together account for the entire set (i.e., $S_1 \cup S_2 \cup \cdots = X$). Find a recursive model for counting B_n, the number of different partitions of $\{1, 2, \ldots, n\}$. (These are called the *Bell numbers* in combinatorics.)

5.6 Excursion: Solving Recurrence Relations

This section is focused on the issue of finding closed formulas for recursively defined sequences. This problem is a central fixture in Section 1.2 of this book, although we were not brave enough at the time to attempt any formal procedures for finding closed formulas. In that section and the subsequent sections on mathematical induction in Chapter 2, we contented ourselves with building an understanding of, and being able to prove, the relationships between the two types of sequence definitions. In this section, we will develop some methods for moving from recursive descriptions to closed formulas, and we will see some applications of these methods.

Difference Tables

When we played "guess the next number" in Section 1.2, we often looked at the differences between successive terms to look for a pattern. This sometimes led to a recursive description, but it rarely led immediately to a closed formula. The method of *difference tables* will allow us to find a closed formula without ever producing a recursive description under special circumstances. We will use the sequences in the following example to develop this method over the next few pages.

Example 1 *Find the next three terms in each of the following sequences, and express the general pattern as a recurrence relation:*

1. 2, 5, 8, 11, 14, _____, _____, _____, . . .
2. 6, 11, 19, 30, 44, _____, _____, _____, . . .
3. 1, 5, 14, 30, 55, 91, _____, _____, _____, . . .

For the sequence 2, 5, 8, 11, 14, . . . , it is easy to form the sequence of differences between successive terms as follows:

$$2 \underset{3}{\to} 5 \underset{3}{\to} 8 \underset{3}{\to} 11 \underset{3}{\to} 14 \to \cdots$$

In this discussion, it is important that we index our original sequence beginning with 0. Hence, in the sequence above, we have $s_0 = 2$, $s_1 = 5$, and so on. Formally, we define the sequence of differences for a given sequence $\{s_n\}$ as follows.

Definition   For a given sequence of numbers $\{s_n\}$ beginning with s_0, we define the sequence $\{\Delta_n\}$ of first differences by the rule

$$\Delta_n = s_{n+1} - s_n$$

for all $n \geq 0$. Hence, the first term in the sequence of first differences is $\Delta_0 = s_1 - s_0$.

Table 5-30 shows the original sequence and the sequence of first differences in convenient table form. The examination of differences is of central importance to mathematics. Indeed, the entire field of differential calculus is a close examination of certain differences. The connection between sums and differences in "continuous" mathematics is reflected by the fundamental theorem of calculus itself. In the context of "finite differences," this relationship is slightly more transparent. Consider the sum of the numbers in the Δ_n row of Table 5-30:

n	0	1	2	3	4	5	6	7
s_n	2	5	8	11	14	17	20	23
Δ_n	3	3	3	3	3	3	3	. . .

Table 5-30   First Differences of the Sequence s_n

$$\Delta_{m-1} + \Delta_{m-2} + \cdots + \Delta_2 + \Delta_1 + \Delta_0$$
$$= (s_m - s_{m-1}) + (s_{m-1} - s_{m-2}) + \cdots + (s_2 - s_1) + (s_1 - s_0)$$
$$= s_m - s_0$$

Since the first differences are all three in the example, we know that $\Delta_{m-1} + \Delta_{m-2} + \cdots + \Delta_2 + \Delta_1 + \Delta_0 = 3m$, and so

$$3m = s_m - s_0$$

which means that $s_m = 3m + 2$.

This is certainly more complicated than it needs to be for a sequence as simple as this, but we have laid a foundation of tools that can be applied to the more interesting sequences. The primary one (which you are asked to prove formally in the exercises) is the relationship between sums and differences.

Theorem 1   (*Fundamental Theorem of Sums and Differences*) *For any sequence* $\{s_n\}$ *with first differences* $\Delta_k = s_{k+1} - s_k$, *and any* $n \geq 1$,

$$s_n - s_0 = \sum_{k=0}^{n-1} \Delta_k$$

Example 2 *Apply Theorem 1 and the procedure above to find a closed formula for the sequence*

$$6, 11, 19, 30, 44, \ldots$$

When we form the difference table as we did before, the sequence of first differences is not constant, so we need to extend our definition of differences so that it can be repeated. We define the sequence of *second differences* by forming the differences of the sequence of first differences. (Likewise, we can define the sequence of *third differences* by forming the differences of the sequence of second differences, and so on, but that will not be necessary for this example.) The general definition is recursive in nature.

> **Definition**   For a given sequence of numbers $\{s_n\}$ beginning with s_0 and an integer $k \geq 2$, we define the sequence $\{\Delta_n^k\}$ of kth differences by the rule
>
> $$\Delta_n^k = \Delta_{n+1}^{k-1} - \Delta_n^{k-1}$$
>
> for all $n \geq 0$. Hence, the first term in the sequence of kth differences is $\Delta_0^k = \Delta_1^{k-1} - \Delta_0^{k-1}$. For consistency, we refer to the sequence of first differences as $\{\Delta_n^1\}$ instead of $\{\Delta_n\}$ in this context.

Now the *complete* difference table for the given sequence $6, 11, 19, 30, 44, \ldots$ from Example 2 is given in Table 5-31. We can easily find a closed formula for Δ_n, but let's do it as an application of Theorem 1 just for fun. According to the theorem,

n	0	1	2	3	4
s_n	6	11	19	30	44
Δ_n	5	8	11	14	$\cdots$
Δ_n^2	3	3	3	$\cdots$	$\cdots$

Table 5-31 Difference Table for Example 2

we should have

$$\Delta_m - \Delta_0 = \sum_{k=0}^{m-1} \Delta_k^2$$
$$= \sum_{k=0}^{m-1} 3$$
$$= 3m$$

Since $\Delta_0 = 5$, we conclude that $\Delta_m = 3m + 5$, and this formula clearly matches the terms shown. Now applying Theorem 1 again, we have

$$s_m - s_0 = \sum_{k=0}^{m-1} \Delta_k$$
$$= \sum_{k=0}^{m-1}(3k + 5)$$
$$= 3 \cdot \sum_{k=0}^{m-1} k + \sum_{k=0}^{m-1} 5$$
$$= 3 \cdot \frac{(m-1) \cdot m}{2} + 5 \cdot m \quad \text{by Proposition 1 from Section 2.3}$$
$$= \frac{3}{2}m^2 + \frac{7}{2}m.$$

Therefore, since $s_0 = 6$, we can conclude that $s_m = \frac{3}{2}m^2 + \frac{7}{2}m + 6$.

This process is straightforward but rather tedious. There is a remarkable simplification of this process whenever the difference table eventually has a constant row, as it does in these previous examples.

Theorem 2 *For a sequence $\{s_n\}$ whose k^{th} differences are constant, for all $n \geq 0$,*

$$s_n = s_0 + \sum_{i=1}^{k} \Delta_0^i \cdot C(n, i)$$

Example 3 *Use Theorem 2 to find a closed formula for the sequence*

$$1, 5, 14, 30, 55, 91, \ldots$$

n	0	1	2	3	4	5
s_n	**1**	5	14	30	55	91
Δ_n^1	**4**	9	16	25	36	$\cdots$
Δ_n^2	**5**	7	9	11	$\cdots$	$\cdots$
Δ_n^3	**2**	2	2	$\cdots$	$\cdots$	$\cdots$

Table 5-32   Difference Table for Example 3

SOLUTION   Table 5-32 shows the complete difference table for this sequence. The numbers in bold are s_0, Δ_0^1, Δ_0^2, and Δ_0^3, so according to Theorem 2,

$$s_n = s_0 + \Delta_0^1 \cdot C(n,1) + \Delta_0^2 \cdot C(n,2) + \Delta_0^3 \cdot C(n,3)$$

$$= \mathbf{1} + \mathbf{4} \cdot n + \mathbf{5} \cdot \frac{n(n-1)}{2} + \mathbf{2} \cdot \frac{n(n-1)(n-2)}{6}$$

$$= 1 + \frac{13}{6}n + \frac{3}{2}n^2 + \frac{1}{3}n^3 \qquad \square$$

Practice Problem 1   *Find a formula for the sum of the first n fourth powers*

$$\sum_{k=1}^{n} k^4$$

The previous section showed how sometimes we can easily find a recursive description of the answer to a counting problem. Difference tables can also be used to find closed formulas for some of these descriptions.

Example 4   *Let r_n be the number of regions of the plane created by n lines, no two of which are parallel and no three of which intersect in any single point. Find a recursive description of r_n and use this to find a closed formula utilizing Theorem 2.*

SOLUTION   In Figure 5-9, we draw some pictures to see the first few values of the sequence $\{r_n\}$.

Because of the rules for placing the lines, when each new line is drawn, it must cross each of the previously drawn $n-1$ lines. This will create n new regions. Hence, the recurrence relation $r_n = r_{n-1} + n$ holds for this sequence. We can use this pattern to extend the pattern in the picture as we create the difference table for this sequence in Table 5-33. According to Theorem 2, the closed formula for this sequence is

$$r_n = 1 + C(n,1) + C(n,2)$$

which can be simplified to

$$r_n = 1 + \frac{n(n+1)}{2} \qquad \square$$

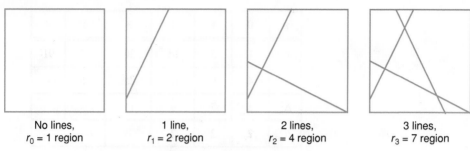

No lines,
$r_0 = 1$ region

1 line,
$r_1 = 2$ region

2 lines,
$r_2 = 4$ region

3 lines,
$r_3 = 7$ region

Figure 5-9 First few cases for Example 4.

Proving Theorem 2

The key to proving Theorem 2 is the following fact about Pascal's arithmetic triangle. The proof of this fact is addressed in Exercise 12 at the end of this section. The format of the proof of Theorem 2 is induction on $k \geq 1$, although it is one of the more difficult induction proofs in this book.

Figure 5-10 The hockey stick identity with $m = 2$ and $n = 3$.

Theorem 3   (*Hockey Stick Identity*) *For all $m \geq 0$ and $n \geq 0$,*

$$C(m, m) + C(m + 1, m) + \cdots + C(m + n, m) = C(m + n + 1, m + 1) \qquad (5.3)$$

This fact obtains its name from the picture that you get when you illustrate it in Pascal's arithmetic triangle as shown in Figure 5-10.

PROOF   (of Theorem 2) Let $P(k)$ be the statement "For any sequence $\{s_n\}$ whose k^{th} differences are constant, $s_n = s_0 + \sum_{i=1}^{k} \Delta_0^i \cdot C(n, i)$ for all $n \geq 0$." Statement $P(1)$ simply says, "For any sequence $\{s_n\}$ whose first differences are constant, $s_n = s_0 + \Delta_0^1 \cdot C(n, 1)$ for all $n \geq 0$." This is established by Exercise 13 at the end of this section.

Now let $m \geq 2$ be given such that statements $P(1), \ldots, P(m - 1)$ have all been checked to be true. In particular, $P(m - 1)$ tells us about sequences with constant $(m - 1)^{\text{th}}$ differences. To prove the next statement, we let any sequence $\{s_n\}$ be given whose m^{th} differences are constant. Consider the sequence $\{\Delta_n\}$ of first differences of $\{s_n\}$. This new sequence $\{\Delta_n\}$ has constant $(m - 1)^{\text{th}}$ differences. More important, the ith differences for the sequence $\{\Delta_n\}$ are precisely the $(i + 1)^{\text{th}}$ differences for the original sequence $\{s_n\}$. So by statement $P(m - 1)$, we know that

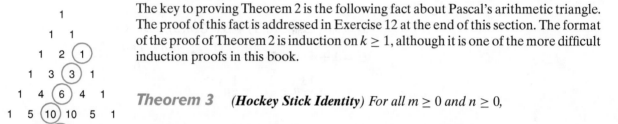

n	0	1	2	3	4
r_n	1	2	4	7	11
Δ_n	1	2	3	4	$\cdots$
Δ_n^2	1	1	1	$\cdots$	$\cdots$

Table 5-33   Difference Table for Example 4

for all $n \geq 0$,

$$\Delta_n = \Delta_0 + \sum_{i=1}^{m-1} \Delta_0^{i+1} \cdot C(n, i)$$

We know that

$$s_n = s_0 + \sum_{j=0}^{n-1} \Delta_j \quad \text{by Theorem 1}$$

$$= s_0 + \sum_{j=0}^{n-1} \left(\Delta_0 + \sum_{i=1}^{m-1} \Delta_0^{i+1} \cdot C(j, i) \right) \quad \text{by statement } P(m-1)$$

$$= s_0 + \sum_{j=0}^{n-1} \Delta_0 + \sum_{j=0}^{n-1} \sum_{i=1}^{m-1} \Delta_0^{i+1} \cdot C(j, i)$$

$$= s_0 + n \cdot \Delta_0 + \sum_{i=1}^{m-1} \left(\sum_{j=0}^{n-1} C(j, i) \right) \cdot \Delta_0^{i+1} \quad \text{from rearranging the sum}$$

$$= s_0 + \Delta_0 \cdot C(n, 1) + \sum_{i=1}^{m-1} C(n, i+1) \cdot \Delta_0^{i+1} \quad \text{by Theorem 3}$$

$$= s_0 + \sum_{i=1}^{m} \Delta_0^i \cdot C(n, i)$$

The induction proof is complete at this point, but it is highly recommended that you trace through it with a particular sequence in mind to fully understand the steps. ∎

Other Types of Recurrence Relations

When we study solutions to recurrence relations, we look at different types of recurrences and try to find a general solution for each type. There are some common techniques that can be used to find these general solutions, but these would be appropriate through further study in combinatorics. In the remainder of this section, we will look at two particular types of recurrence relations, and use induction to prove what we can about them. We will learn more about some of the recurrence relations we saw earlier in the book, and we will see some common applications of recurrence relations. Let's first see why the method of difference tables is not enough for understanding recurrence relations in general.

 Example 5 *Complete the difference table for the sequence*

$$5, 13, 37, 109, 325, 973, \ldots$$

SOLUTION   Table 5-34 shows the first two rows of the difference table. The first differences change by a factor of 3 as you look left to right, as do the second differences. Hence, the pattern in the differences will **not** eventually be constant, but they have settled down into the same simple pattern.   □

Because the differences are never constant, we cannot find a closed formula for the sequence in Example 5 using our method of difference tables. However, the

n	0	1	2	3	4	5
s_n	5	13	37	109	325	973
Δ_n	8	24	72	216	648	$\cdots$
Δ_n^2	16	48	144	432	$\cdots$	$\cdots$

Table 5-34   Partial Difference Table for Example 5

difference table above is still useful in determining that each term in the original sequence is roughly three times the previous term. A sequence based on this kind of pattern is the first of two special types of recurrence relation that we will study in the remainder of this section.

Linear First-Order Recurrences

> **Definition**   A *linear first-order recurrence relation* is one of the form
> $$a_n = C \cdot a_{n-1} + f(n)$$
> where C is any constant and f is any (nonrecursive) function.

Sequences with this sort of recursive pattern are easy to understand when the function f in the definition is simple. We will be able to recognize these cases when we see a row in the difference table in which the values increase by a constant factor, as they do in Example 5 above. To illustrate how this is helpful, we will find a closed formula for the sequence in Example 5.

Example 6 *Find a closed formula for the sequence* $5, 13, 37, 109, 325, 973, \ldots$ *whose difference table is given in Example 5.*

SOLUTION   We break down the solution into a sequence of observations:

1. $\Delta_n = 8 \cdot 3^n$ for all $n \geq 0$.
2. By Theorem 1, $s_n - s_0 = \sum_{i=0}^{n-1} \Delta_i = 8 \cdot \sum_{i=0}^{n-1} 3^i$.
3. From Exercise 13 of Section 2.3 (concerning the sum of a geometric sequence), we know that

$$\sum_{i=0}^{m-1} 3^i = \frac{3^m - 1}{2}$$

4. Therefore,

$$s_n = 5 + 8 \cdot \sum_{i=0}^{n-1} 3^i$$
$$= 5 + 8 \cdot \left(\frac{3^n - 1}{2} \right)$$
$$= 1 + 4 \cdot 3^n$$

$\square$

While it is possible to find a closed formula for many linear first-order recurrence relations in this way, it is more efficient to prove a general result that can be used for a large class of sequences.

Theorem 4   *Given the recurrence relation $a_n = b \cdot a_{n-1} + c$, with $b \neq 1$, the closed formula $a_n = L \cdot b^n + \frac{c}{1-b}$ satisfies the recurrence relation for any value of L.*

PROOF   We show that any closed formula of the given form satisfies the given recurrence. If $a_n = L \cdot b^n + \frac{c}{1-b}$, then $a_{n-1} = L \cdot b^{n-1} + \frac{c}{1-b}$, and so

$$b \cdot a_{n-1} + c = L \cdot b^n + \frac{c}{1-b} \cdot b + c$$
$$= L \cdot b^n + \frac{c}{1-b}$$
$$= a_n$$

which means the recurrence relation is satisfied by this closed formula. ∎

In practice, we will typically know a starting point for the sequence in addition to a recurrence relation, so we can use the starting value to determine an appropriate value of L in the general closed formula given above.

Example 7 *Find a recursive description for the following sequence, and use Theorem 4 to find a closed formula:*

$$1, 3, 11, 43, 171, \ldots$$

SOLUTION   The recurrence relation for this sequence is $a_n = 4a_{n-1} - 1$, so $b = 4$ and $c = -1$ above, giving us a solution of the form $a_n = L \cdot 4^n + \frac{1}{3}$. In order for $a_1 = 1$ to be true, we solve $L \cdot 4^1 + \frac{1}{3} = 1$, and get $L = \frac{1}{6}$. Therefore,

$$a_n = \frac{1}{6} \cdot 4^n + \frac{1}{3}$$

is the closed formula for the given sequence. □

Example 8 *Suppose you save money at a bank that pays 5% per year in simple interest. For example, if you have $1,000 in the bank all year, then at the end of the year, the bank adds $50 to your account. If you open the account with $100, and in addition to the interest that the bank pays, on the last day of each year, you deposit $100 more into the account, how much money will you have at the end of 10 years? n years?*

SOLUTION   Let M_n be the amount of money you have at the **beginning** of the nth year. We can figure out the value of your account for the first few years:

● $M_1 = 100$
● $M_2 = 100 + 5 + 100 = 205$
● $M_3 = 205 + 0.05 \cdot (205) + 100 = 315.25$

In general, we see the recursive pattern

$$M_n = M_{n-1} + 0.5 \cdot M_{n-1} + 100$$
$$= 1.05 \cdot M_{n-1} + 100$$

We can use Theorem 4 with $b = 1.05$ and $c = 100$ to conclude that the closed formula

$$M_n = L \cdot (1.05)^n + \frac{100}{1 - 1.05}$$
$$= L \cdot (1.05)^n - 2{,}000$$

will describe the same sequence, for an appropriate value of L. To find this value of L, we simply take a known value like $M_1 = 100$ and solve the equation

$$100 = L \cdot (1.05)^1 - 2{,}000$$

which tells us $L = 2{,}000$. Therefore, the closed formula for this sequence is

$$M_n = 2{,}000 \cdot (1.05)^n - 2{,}000$$
$$= 2{,}000 \cdot ((1.05)^n - 1)$$

We can check a few values to make sure this agrees with our work above:

- $M_1 = 2{,}000 \cdot \left((1.05)^1 - 1\right) = 100$
- $M_2 = 2{,}000 \cdot \left((1.05)^2 - 1\right) = 205$
- $M_3 = 2{,}000 \cdot \left((1.05)^3 - 1\right) = 315.25$

The given question about the end of the tenth year is answered by the value of $M_{11} = 2{,}000 \cdot \left((1.05)^{11} - 1\right) \approx 1{,}420.68$, which is the amount of money in the bank at the beginning of the eleventh year. Similarly the amount of money in the bank at the end of n years is given by $M_{n+1} = 2{,}000 \cdot \left((1.05)^{n+1} - 1\right)$. □

Practice Problem 2   *Every spring, a nearby lake is stocked with 5,000 catfish. Due to fishing and environmental conditions, the population of catfish n weeks after the stock date is given by*

$$p_n = p_{n-1} + 0.05 \cdot p_{n-1} - 500$$

starting with the first week and continuing until there are no catfish left. Find a closed formula for the number of catfish n weeks after the stock date, and use your answer to project the number of catfish there will be in the lake.

Second-Order Recurrence Relations

A second-order recurrence relation is one where each term in a sequence of numbers can be related to the two preceding terms in the sequence. The most famous example of this is the sequence of Fibonacci numbers, defined by $F_1 = F_2 = 1$ with $F_n =$

$F_{n-1} + F_{n-2}$ for $n \geq 2$. The remainder of this section explores the general appearance of closed formulas for this type of recursive description.

Example 9

1. *Show that no matter what numbers are used for C and K, the closed formula $a_n = C \cdot 2^n + K \cdot 3^n$ satisfies the recurrence relation*

$$a_n = 5a_{n-1} - 6a_{n-2}$$

2. *Find appropriate values of C and K in the previous part to determine the closed formula for the following sequence with recurrence relation $a_n = 5a_{n-1} - 6a_{n-2}$:*

$$3, 7, 17, 43, 113, 307, \ldots$$

SOLUTION

1. If $a_n = C \cdot 2^n + K \cdot 3^n$, then it follows that

$$a_{n-1} = C \cdot 2^{n-1} + K \cdot 3^{n-1} \qquad \text{and} \qquad a_{n-2} = C \cdot 2^{n-2} + K \cdot 3^{n-2}$$

so

$$\begin{aligned} 5a_{n-1} - 6a_{n-2} &= 5\left(C \cdot 2^{n-1} + K \cdot 3^{n-1}\right) - 6\left(C \cdot 2^{n-2} + K \cdot 3^{n-2}\right) \\ &= (10C - 6C)2^{n-2} + (15K - 6K)3^{n-2} \\ &= C \cdot 2^n + K \cdot 3^n \\ &= a_n \end{aligned}$$

2. Since $a_1 = 3$ and $a_1 = C \cdot 2 + K \cdot 3$, we know that $2C + 3K = 3$. Since $a_1 = 7$ and $a_1 = C \cdot 4 + K \cdot 9$, we know that $4C + 9K = 7$. Combining these two equations tells us that $C = 1$ and $K = \frac{1}{3}$. Hence, the closed formula for this sequence is

$$a_n = 2^n + \frac{1}{3}3^n = 2^n + 3^{n-1}$$

□

Theorem 5 *If r and s are two different solutions to the equation $x^2 = cx + d$, then no matter what numbers are used for C and K, the closed formula $a_n = C \cdot r^n + K \cdot s^n$ satisfies the recurrence relation*

$$a_n = c \cdot a_{n-1} + d \cdot a_{n-2}$$

PROOF   See Exercise 21 at the end of this section.   ■

Example 10 *Find the closed formula for the Fibonacci numbers, which satisfy*

$$F_n = F_{n-1} + F_{n-2} \quad \text{with } F_1 = F_2 = 1$$

SOLUTION   Since the quadratic equation $x^2 = x + 1$ has solutions (derived from the quadratic formula)

$$x = \frac{1 \pm \sqrt{5}}{2}$$

we know from Theorem 5 that a sequence with closed formula

$$F_n = C \cdot \left(\frac{1+\sqrt{5}}{2}\right)^n + K \cdot \left(\frac{1-\sqrt{5}}{2}\right)^n$$

will satisfy the Fibonacci numbers' recurrence relation. In addition, the specific values $F_1 = 1$ and $F_2 = 1$ give us the following two equations in terms of the constants C and K:

$$1 = C \cdot \left(\frac{1+\sqrt{5}}{2}\right)^1 + K \cdot \left(\frac{1-\sqrt{5}}{2}\right)^1$$

$$1 = C \cdot \left(\frac{1+\sqrt{5}}{2}\right)^2 + K \cdot \left(\frac{1-\sqrt{5}}{2}\right)^2$$

The solution to this system of equations is $C = \frac{1}{\sqrt{5}}$ and $K = \frac{-1}{\sqrt{5}}$, so the Fibonacci numbers have the closed formula

$$F_n = \frac{1}{\sqrt{5}}\left(\frac{1+\sqrt{5}}{2}\right)^n - \frac{1}{\sqrt{5}}\left(\frac{1-\sqrt{5}}{2}\right)^n$$

$\square$

Practice Problem 3   *Find the closed formula for the sequence* $1, 5, 13, 41, 121, \ldots$ *satisfying the recurrence relation* $a_n = 2a_{n-1} + 3a_{n-2}$.

Solutions to Practice Problems

1 Table 5-35 shows the table of differences for the sequence $0, 1, 17, 98, 354, \ldots$ So we have

$$\sum_{k=1}^{n} k^4 = 0 + 1 \cdot C(n, 1) + 15 \cdot C(n, 2) + 50 \cdot C(n, 3)$$
$$+ 60 \cdot C(n, 4) + 24 \cdot C(n, 5)$$

With some work (or some help from a computer algebra system), this can be simplified to

$$\sum_{k=1}^{n} k^4 = \frac{n(2n+1)(n+1)(3n^2+3n-1)}{30}$$

2 We can restate the recurrence relation for this sequence as

$$p_n = 1.05 \cdot p_{n-1} - 500$$

The first few values of this sequence are given in Table 5-36. According to Theorem 4 with $b = 1.05$ and $c = -500$, the closed formula for this sequence is

$$p_n = L \cdot (1.05)^n + \frac{-500}{-0.05}$$
$$= L \cdot (1.05)^n + 10,000$$

In order for $p_0 = 5,000$, we must have $L = -5,000$, so we have the closed formula

$$p_n = -5,000 \cdot (1.05)^n + 10,000$$

If we solve when this is equal to 0 (using logarithms), we find that $p_{14} \approx 100$ and $p_{15} \approx -395$, so the last time there are catfish in the lake is after 14 weeks of the season.

3 Since the quadratic equation $x^2 = 2x + 3$ has solutions $x = 3$ and $x = -1$, we know from Theorem 5 that a sequence with closed formula

$$a_n = C \cdot 3^n + K \cdot (-1)^n$$

will satisfy the given recurrence relation. In addition, the specific values $a_1 = 1$ and $a_2 = 5$ give us the

n	0	1	2	3	4	5	6
c_n	**0**	1	17	98	354	979	2275
Δ_n	**1**	16	81	256	625	1296	$\cdots$
Δ_n^2	**15**	65	175	369	671	$\cdots$	$\cdots$
Δ_n^3	**50**	110	194	302	$\cdots$	$\cdots$	$\cdots$
Δ_n^4	**60**	84	108	$\cdots$	$\cdots$	$\cdots$	$\cdots$
Δ_n^5	**24**	24	$\cdots$	$\cdots$	$\cdots$	$\cdots$	$\cdots$

Table 5-35 Difference Table for Practice Problem 1

n	0	1	2	3	4	5	6
p_n	5,000	5,150	5,208	5,168	5,026	4,778	4,416

Table 5-36 Difference Table for Practice Problem 2

following two equations in terms of the constants C and K:

$$1 = C \cdot (3)^1 + K \cdot (-1)^1$$
$$5 = C \cdot (3)^2 + K \cdot (-1)^2$$

The solution to this system of equations is $C = \frac{1}{2}$ and $K = \frac{1}{2}$, so this sequence has the closed formula

$$a_n = \frac{1}{2}(3^n + (-1)^n)$$

Exercises for Section 5.6

1. Find closed formulas for each of the following sequences:
   (a) $3, 10, 17, 24, 31, \ldots$
   (b) $7, 13, 25, 43, 67, \ldots$
   (c) $1, 3, 8, 16, 27, 41, \ldots$
   (d) $1, 10, 35, 84, 165, 286, \ldots$

2. Find closed formulas for each of the following sequences:
   (a) $s_0 = 4, s_n = s_{n-1} + 3$ for all $n \geq 1$
   (b) $s_0 = 0, s_n = s_{n-1} + 2n$ for all $n \geq 1$
   (c) $s_0 = 7, s_n = s_{n-1} + 3n - 5$ for all $n \geq 1$
   (d) $s_0 = 7, s_n = s_{n-1} + n^2 - 1$ for all $n \geq 1$

3. Find a closed formula for each of the following sums:
   (a) $\sum_{k=1}^{n}(4k - 3)$
   (b) $\sum_{k=1}^{n}(3k + 1)$
   (c) $\sum_{k=1}^{n}(4k^2 + 3k - 1)$
   (d) $\sum_{k=1}^{n} k^3$

4. In a certain round robin basketball tournament, every team must play every other team exactly twice. Let g_n be the number of games that must be played if n teams participate in the tournament.

   (a) By imagining what happens to a tournament with $n - 1$ teams if it is joined by one more team, find a recursive description for the sequence $\{g_n\}$.
   (b) Use Theorem 2 to determine a closed formula for g_n.

5. There is a slightly shorter way to state Theorem 2 for a sequence $\{s_n\}$ with constant k^{th} differences. For all $n \geq 0$,

$$s_n = \sum_{i=0}^{k} \Delta_0^i \cdot C(n, i)$$

How must we interpret the notation Δ_0^0 for this to be consistent with the theorem?

6. Find closed formulas for each of the sequences whose recursive description is given.
   (a) $a_1 = 2, a_n = 2a_{n-1} + 3$ for all $n \geq 1$
   (b) $a_1 = 1, a_n = 3a_{n-1} + 2$ for all $n \geq 1$
   (c) $a_1 = 0, a_n = 5 - 2a_{n-1}$ for all $n \geq 1$
   (d) $a_1 = 3, a_n = 3 - a_{n-1}$ for all $n \geq 1$

7. Suppose your house mortgage (the amount you owe on your house) grows at a rate of 0.5% per month.

You originally borrow $100,000 for the house, and every month you make a payment of $1,000. How many months will it take before you owe nothing on the house?

8. A bacteria culture grows at a rate of 10% per day.

   (a) If this morning the culture has 1,000,000 bacteria, how many days will it take for this number to double?

   (b) Suppose at the beginning of each day (excluding this morning), you remove a sample of 50,000 bacteria for testing. Now how many days will it take for the original 1,000,000 bacteria to double?

   (c) Suppose at the beginning of each day (excluding this morning), you remove a sample of 200,000 bacteria for testing. What happens under this scenario?

9. Find a recursive description for the sequence in Example 5, and use Theorem 4 to derive the closed formula for the sequence.

10. We saw in Practice Problem 3 of Section 5.5 that the recurrence $a_n = 2 \cdot a_{n-1} + 1$ (with $a_0 = 0$) is satisfied by the closed formula $a_n = 2^n - 1$. If instead we are given the condition $a_0 = k$, what is the closed formula for a_n?

11. If $a_n = b \cdot a_{n-1} + c$, with $b \neq 1$ and we are given $a_0 = 1$, what is the value of the constant L in Theorem 4 in terms of b and c?

12. Prove Theorem 1 by induction on n.

13. For any sequence $\{s_n\}$ whose first differences are the constant K, show by induction on n that $s_n = s_0 + Kn$ for all $n \geq 0$.

14. Prove by induction on $n \geq 1$ that if $p(x)$ is an nth degree polynomial, then the nth differences for the sequence $p(0), p(1), p(2), \ldots$ are constant.

15. Use the procedure in Example 6 to find a closed formula for each of the following sequences:

    (a) $a_0 = 1, a_n = 2a_{n-1} + n$ for all $n \geq 1$

    (b) $a_0 = 1, a_n = 3a_{n-1} - 2n + 1$ for all $n \geq 1$

    (c) $a_0 = 1, a_n = 2a_{n-1} + n^2$ for all $n \geq 1$

16. Prove Theorem 4 by applying the method of difference equations to the generic recurrence relation of the form

$$a_n = b \cdot a_{n-1} + c$$

    for constants $b \neq 1$ and c.

17. For each of the following second-order recurrence relations, find a general closed formula (using constants C and K as in Theorem 5) for a sequence that satisfies it:

    (a) $a_n = 7a_{n-1} - 12a_{n-2}$

    (b) $a_n = 6a_{n-1} - 8a_{n-2}$

    (c) $a_n = a_{n-1} + 12a_{n-2}$

    (d) $a_n = 6a_{n-2} - a_{n-1}$

18. For each of the recurrence relations in Exercise 17, find the closed formula for the sequence satisfying that recurrence along with the initial conditions $a_0 = 0$ and $a_1 = 1$.

19. For each of the following second-order recurrence relations, find a general closed formula (using constants C and K as in Theorem 5) for a sequence that satisfies it:

    (a) $a_n = a_{n-1} + 3a_{n-2}$

    (b) $a_n = 2a_{n-1} + 2a_{n-2}$

    (c) $a_n = 2a_{n-1} - 5a_{n-2}$ (NOTE: This problem requires the use of complex numbers.)

    (d) $a_n = 2(a_{n-1} - a_{n-2})$ (NOTE: This problem requires the use of complex numbers.)

20. For each of the recurrence relations above, find the closed formula for the sequence satisfying that recurrence along with the initial conditions $a_0 = 1$ and $a_1 = 1$.

21. Prove Theorem 5.

22. In Theorem 5, there is an assumption that the quadratic equation $x^2 = cx + d$ has **two different** roots. If this is not the case, then the closed formulas will have a different form. The following proposition is to be proved in Exercise 24.

    Proposition. *If $x^2 = cx + d$ has only one real solution r, then no matter what numbers are used for C and K, the closed formula $a_n = (C + K \cdot n)r^n$ satisfies the recurrence relation*

$$a_n = c \cdot a_{n-1} + d \cdot a_{n-2}$$

    Use this proposition to find a general closed formula for a sequence satisfying each of the following recurrence relations:

    (a) $a_n = 4a_{n-1} - 4a_{n-2}$

    (b) $a_n = -6a_{n-1} - 9a_{n-2}$

    (c) $a_n = 8a_{n-2} - 16a_{n-1}$

23. For each of the recurrence relations above, find the closed formula for the sequence satisfying that recurrence along with the initial conditions $a_0 = 1$ and $a_1 = 1$.

24. Prove the proposition in Exercise 22.

Chapter 5 Summary

5.1 Introduction

Terms and concepts

- You should recognize these terms and their relationship to the questions concerning ordering and repetition:
  - *Ordered list*—order matters, repetitions are allowed
  - *Unordered list* or *bag*—order does not matter, repetitions are allowed
  - *Permutation*—order matters, repetitions are not allowed
  - *Set*—order does not matter, repetitions are not allowed
- You should recognize that:
  - $P(n, r)$ counts the r-element permutations drawn from a set of size n.
  - $C(n, r)$ counts the r-element subsets of a set of size n (also referred to r-combinations from the set)
- You should realize that some textbooks, calculators, and instructors use alternative notation for $P(n, r)$ and for $C(n, r)$.
- You should know that *combinatorial equivalence* refers to the use of a 1-1 correspondence to demonstrate that two counting problems have the same solution.

Strategies for counting

- You should understand the utility of skillful organization as an aid in counting.
- You should be able to apply a variety of organizational schemes in various applications, including at least the following:
  - Tables where each row/column is completely filled
  - Tables where some rows/columns have more entries than others
  - Trees
- As specific examples of this organization, you should thoroughly understand the table used to count outcomes of rolling two dice, and the tree used to count outcomes of multiple coin tosses.
- You should be able to demonstrate that two sets are the same size, without knowing the size of either set, by describing a 1-1 correspondence between the two sets (that is, by using combinatorial equivalence).

5.2 Basic Rules for Counting

Terms and concepts

- You should be able to state and apply these rules:
  - The *rule of products* (or *product rule*)—in set symbols, $n(A \times B) = n(A) \cdot n(B)$
  - The *rule of sums* (or *sum rule*)—in set symbols, $n(A \cup B) = n(A) + n(B)$, provided $A \cap B$ is empty
  - The *rule of sums with overlap*—in set symbols, $n(A \cup B) = n(A) + n(B) - n(A \cap B)$
  - The *rule of complements* (or *complement rule*)—in set symbols, $n(A') = n(U) - n(A)$ where U is the universal set
- You should recall that a *permutation* is a list in which order matters and repetitions are not allowed, and that $P(n, r)$ counts the r-element permutations drawn from a set of size n.
- You should recognize that $n!$ ("*n factorial*") is the same as $P(n, n)$.

Formulas

- $n! = (n)(n - 1) \cdots (3)(2)(1)$ and $0! = 1$
- $P(n, r) = n \cdot (n - 1) \cdot (n - 2) \cdots (n - r + 1)$
- $P(n, r) = \frac{n!}{(n-r)!}$, provided that $n \geq r$
- The number of ordered lists of length r (repetitions allowed), drawn from a set of size n, is n^r

Using the basic counting rules

- Concerning the rule of products:
  - You should be able to apply the rule to situations where the list to be counted can be considered to be constructed in multiple steps. (We described *algorithms* that carried out these constructions.)
  - It is necessary that the number of ways to complete each step is independent of how the earlier steps were completed.
- Concerning the rule of sums:
  - You should be able to apply the rule to situations where the list to be counted can be split into pieces.
  - It is necessary that every pair of pieces must be disjoint.
  - In the case of two pieces, the inclusion-exclusion principle allows us to apply the rule by subtracting the overlap of the two pieces. (There are corresponding

methods for three or more pieces, but they are more complex.)

- Concerning the rule of complements:
  - You should recognize that it is frequently easier to count the complement of a set.
  - Once you have counted the complement, you should be able to apply the complement rule to solve the original problem.
- You should be able to solve problems involving combinations of these three rules.
- You should realize that the formulas for counting ordered lists and permutations come from applying the product rule. Thus, many problems have alternative solutions:
  - as an application of the product rule; or
  - as an application of ordered lists or permutations.

5.3 Combinations and the Binomial Theorem

Terms and concepts

- You should recall that $C(n, r)$ counts the r-element subsets drawn from a set of size n, and that an r-element subset is also called an *r-combination*.
- You should know that the *binomial theorem* tells how to expand $(1 + x)^n$, and that the result can be phrased in terms of $C(n, r)$.
- You should know that the *arithmetic triangle* (frequently called *Pascal's triangle*) provides a visual tool for constructing the coefficients in the binomial theorem.

Formulas

- $C(n, r) = \frac{P(n,r)}{r!}$
- $C(n, r) = \frac{n!}{r!(n-r)!}$, provided that $n \geq r$
- The binomial theorem:

$$(1 + x)^n = C(n, 0) + C(n, 1)x + C(n, 2)x^2 + \cdots$$

$$+ C(n, n)x^n = \sum_{k=0}^{n} C(n, k)x^k$$

Counting with equivalence classes

- You should understand the use of equivalence classes to derive the formula for $C(n, r)$.
- You should be able to solve problems by taking a known set and splitting it into equivalence classes each of the same size—for example, counting circular arrangements.

Using $C(n, r)$ to count

- You should recognize that many different types of problems can be interpreted as counting r-element subsets of a set with n elements.
- You should be able to combine the rule for counting subsets with the sum, product, and complement rules from the previous section.

The binomial theorem

- You should be able to use the theorem to determine the coefficient of x^k in various expansions.
- You should be able to create the first several rows of the arithmetic triangle, and relate it to the binomial theorem.
- You should be able to explore various patterns in the arithmetic triangle, using tools such as induction and substituting particular values for x in the binomial theorem.

5.4 Binary Sequences

Terms and concepts

- You should recall that a *binary sequence* is an ordered list of 1s and 0s.
- You should recall that an unordered list with repetitions allowed is also called a *bag*.

Counting binary sequences

- You should be able to use r–combinations to count binary sequences of length n with r zeros.
- You should be able to generalize this algorithm to count sequences drawn from sets other than $\{0, 1\}$—for example, ordered lists of length 10 drawn from $\{m, a, t\}$ containing exactly 3 m's and 2 a's.
- You should be able to use this method to count rearrangements for sets containing repetitions (for example, the distinct arrangements of the letters in the word *MISSISSIPPI*).

Unordered lists with repetitions (bags)

- You should be able to relate these problem types, using 1-1 correspondences:
  - Counting unordered lists with repetitions
  - Counting non–negative integer solutions to equations such as $a + b + c = 10$.
- You should be able to use binary sequences to solve both these types of problems.

- You should be able to generalize the algorithm to other situations (for example, adding the restriction that a and b must be positive and c must be at least 3).

Formulas

- You now have formulas for each of the four types of list:

What?	How many?
Ordered lists of length r	n^r
Permutations of length r	$P(n, r)$
Unordered lists of size r	$C(r + n - 1, r)$
Sets of size r	$C(n, r)$

However, more important than memorizing these formulas is understanding, and being able to apply, the reasoning that leads to the formulas.

5.5 Recursive Counting

Terms, concepts, and skills

- You should recall that a *recursive formula* defines each term of a sequence using an expression involving one or more earlier terms of the sequence. You should also recall that one or more initial terms of the sequence must be given as numbers.
- When we use recursive formulas to solve counting problems, we refer to the result as a *recursive model* of the situation.
- You should be able to create a recursive model for a variety of counting problems. Each such recursive model involves two components:
  - Determine the count for one or more small versions of the problem.
  - For the general version of the problem, imagine that all smaller versions of the problem have been solved.

Use the solutions for the smaller versions to develop a recursive formula for the general situation.

5.6 Excursion: Solving Recurrence Relations

Terms and concepts

- You should know that the terminology *recurrence relation* refers to recursively defined sequences, and that *solving* a recurrence relation refers to finding a closed formula for the sequence.
- You should be able to build a *difference table* for a recursively defined sequence, or for a sequence whose first few terms are given explicitly.
- You should understand the related concepts: *first differences, second difference, third differences*, and so on.
- You should know that a *first order* recurrence relation defines each item of the sequence in terms of the previous item. Similarly, for a *second order* recurrence relation, each term in the sequence can be related to the two preceding terms in the sequence.

Solving recurrence relations

- You should be able to use one or more of these techniques to solve particular recurrence relations:
  - Apply Theorem 1 when the first differences are constant.
  - Apply Theorem 2 whenever the difference table eventually has a constant row.
  - Apply Theorem 1 when the first differences have the form $\Delta_n = c \cdot d^n$.
  - Apply Theorem 4 to first order recurrence relations of the form $a_n = b \cdot a_{n-1} + c$, with $b \neq 1$.
  - Apply Theorem 5 to second order recurrence relations of the form $a_n = c \cdot a_{n-1} + d \cdot a_{n-2}$ when $x^2 = cx + d$ has two different solutions.

6 | **Probability**

Probability is an important topic in the study of games. Most games people play involve some element of chance. Many times the element of chance is entirely based on some simple random device like a pair of dice or a deck of cards, while the game itself consists of countless uses of the simple device along with many decisions and hidden strategies. The issues of decisions and strategies are the subject of what it is formally known as game theory in mathematics. In this chapter, we strive to understand how probabilities connected with some large event (like a game) are related to those connected with the simple events (like a dice roll) that drive them.

We will start off by simply seeing how the ideas from combinatorics are related to questions of probability. We will see that the standard rules in counting and even the recursive model of thinking readily spill over into the study of probability. In some sense, this chapter will reinforce those skills of organized algorithmic and recursive thinking that we had built in the previous chapter. When things begin to get complicated, we will find that matrices are helpful for keeping track of an abundance of information, so we introduce matrix arithmetic and see what kind of problems matrices can help us solve.

6.1 Introduction

In *An Essay on Probability*, Augustus De Morgan (1806–1871) wrote, "I consider probability as meaning the state of mind with respect to an assertion, a coming event, or any other matter on which absolute knowledge does not exist." This will be our starting point as we seek the origins of the mathematics of probability.

In the year 1654, 12 years after the birth of Isaac Newton and the death of Galileo Galilei, a French gambler named Chevalier De Méré (1607–1684) had a problem. It concerned a fair method of splitting a "pot" in a game of chance that is interrupted partway through. The heart of the matter can be illustrated by comparing the following two questions:

1. What is the probability that four rolls of a single die result in at least one roll of 6?

2. What is the probability that 24 rolls of a pair of dice result in at least one roll of "double 6"?

De Méré would have reasoned that in the first problem there are six possible outcomes for each roll and we are performing $4 = (2/3)(6)$ rolls, while in the second problem there are 36 possible outcomes for each roll and we are performing $24 = (2/3)(36)$ rolls. Since in both cases we perform 2/3 as many rolls as there are total outcomes per roll, the answer should be the same. The problem facing De Méré was that he had observed in practice that the first probability was greater than $\frac{1}{2}$, while the second was less than $\frac{1}{2}$. His conclusion was that there must be a basic inconsistency in mathematics. He communicated the problem to his friend, the prominent mathematician and infamous neurotic, Blaise Pascal (1623–1662). Pascal wrote to the great amateur mathematician Pierre Fermat (1601–1665) about the question, initiating a series of letters between Pascal and Fermat that effectively resulted in the birth of the new mathematical field of probability.*

You have surely already guessed that De Méré had discovered no basic inconsistency in mathematics. As we shall see in this chapter, the answer to the first question is about 0.52, while the answer to the second one is roughly 0.49. The fact that De Méré was able to "observe" this difference in practice explains why he made his living gambling. The fact that he did not understand it perhaps explains why he did not make his living doing mathematics.

The first thing to do is to establish the language of probability and some of the basic facts that we will use. In probability, we are interested in the outcomes of certain experiments. By *experiment* we can mean anything from rolling dice to buying a lottery ticket to, well, performing a laboratory experiment. We will assume at first that we only perform experiments under which every possible outcome is *equally likely*. We will see later some easy ways to deal with situations where the outcomes are not equally likely, but in general this is one of the complexities that makes the study of probability interesting. We will refer to the set of outcomes of a given experiment as its *sample space*.

Defining what is meant by "equally likely" is a rather sticky point primarily appealing to one's intuition. We might say that two outcomes are equally likely if a rational person would not favor one outcome over the other in many repetitions of the experiment. For example, if the experiment is the toss of a fair coin, then there

---

* Actually, there was an Italian mathematician named Gerulamo Cardano (1501–1576) who published a gambler's handbook, the *Liber de Ludo Aleae*, in which many of the basic problems of probability were discussed. For some reason, though, he is only mentioned in footnotes when this subject is introduced.

is no reason to bet that heads will come up more often than tails in arbitrarily many tosses, so we will consider the outcomes of heads and tails to be equally likely in this experiment. On the other hand, if the experiment is the roll of a fair (six-sided) die in a game, then one might argue that there are only two outcomes: either I get a 6 or I do not get a 6. No one who has played games with dice would believe that these two outcomes are equally likely, however.

Example 1 *Suppose our experiment is to roll a pair of (six-sided) dice and record the sum of the face-up numbers we see. How should we represent rolls of a pair of dice so that the outcomes are equally likely?*

SOLUTION   There are three reasonable ways to represent the result of rolling two dice.

● We could describe all outcomes by giving the set of all possible sums:

$$\{2, 3, 4, 5, 6, 7, 8, 9, 10, 11, 12\}$$

In this case, there are 11 possible outcomes, but anyone who has played dice games will tell you that getting a sum of 2 is not as likely as getting a sum of 7.

● We could describe all outcomes by giving the set of all possible rolls as follows:

double 1, double 2, double 3, double 4, double 5, double 6,
$$\{1, 2\}, \{1, 3\}, \{1, 4\}, \{1, 5\}, \{1, 6\}, \{2, 3\}, \{2, 4\}, \{2, 5\}, \{2, 6\},$$
$$\{3, 4\}, \{3, 5\}, \{3, 6\}, \{4, 5\}, \{4, 6\}, \{5, 6\}$$

It is more subtle, but a fact, that in this list, the outcome $\{1, 2\}$ is twice as likely to occur as the outcome "double 1." To see this, consider the dice as being colored, one green and one red. In this case, getting "red 1, green 2" and "red 2, green 1" would both be considered the outcome $\{1, 2\}$, while only the roll "red 1, green 1" would be considered the outcome "double 1."

● Using the above analogy, we could describe all outcomes by describing the rolls as ordered pairs, with the first number representing the red die and the second number representing the green die. Only in this representation are the outcomes equally likely. The outcomes are shown in Table 6-1.   □

	Green 1	Green 2	Green 3	Green 4	Green 5	Green 6
Red 1	(1, 1)	(1, 2)	(1, 3)	(1, 4)	(1, 5)	(1, 6)
Red 2	(2, 1)	(2, 2)	(2, 3)	(2, 4)	(2, 5)	(2, 6)
Red 3	(3, 1)	(3, 2)	(3, 3)	(3, 4)	(3, 5)	(3, 6)
Red 4	(4, 1)	(4, 2)	(4, 3)	(4, 4)	(4, 5)	(4, 6)
Red 5	(5, 1)	(5, 2)	(5, 3)	(5, 4)	(5, 5)	(5, 6)
Red 6	(6, 1)	(6, 2)	(6, 3)	(6, 4)	(6, 5)	(6, 6)

Table 6-1   Equally Likely Outcomes for Example 1

With any particular experiment we will associate a set S of outcomes that will be called the *sample space*. For an experiment, it will interest us to know the proportion of outcomes in which a certain *event* takes place—we will call such outcomes *successful*. Technically, an event is defined to be any set of an experiment's outcomes. For example, having a winning lottery ticket is an event in the experiment of buying a lottery ticket. Of course, having a losing ticket is an event, too, even though it might feel weird to call this outcome "successful." We can now define what we mean by probability.

Definition   Given an experiment with a sample space S of equally likely outcomes and an event E, the *probability of the event* (denoted by $Prob(E)$) is the ratio of the number of successful outcomes to the total number of outcomes. That is,

$$Prob(E) = \frac{n(E)}{n(S)}$$

Practice Problem 1   *How many of the 36 equally likely outcomes listed in Example 1 have a sum of 10? Using the definition above, what is the probability of getting a sum of 10 when a pair of dice is rolled?*

Example 2   *Consider the experiment of drawing two cards from the top of a standard deck of 52 cards, and the event E of the two cards having the same value.*

1. *Describe the set S of all outcomes, represented so they are equally likely.*
2. *Describe the event E in terms of your representation.*
3. *Compute $Prob(E) = n(E)/n(S)$.*

SOLUTION

1. We can describe the sample space S as the set of all permutations of length 2 with entries from the deck of 52 cards. There are $(52)(51)$ of these.
2. The outcomes in E can be formed with a two-step process: Choose any card for the first card and then choose any card of the same value for the second card. There are $(52)(3)$ of these.
3. Now the probability calculation is easy:

$$Prob(E) = \frac{n(E)}{n(S)} = \frac{(52)(3)}{(52)(51)} = \frac{1}{17}$$

□

Practice Problem 2   *Consider the experiment of tossing a coin five successive times, and the event E that the last two tosses have the same result.*

1. *Describe the set S of all outcomes, represented so they are equally likely.*
2. *Describe the event E in terms of your representation.*
3. *Compute $Prob(E) = n(E)/n(S)$.*

Sometimes in the study of probability, it is easier to find the probability that a given event does *not* happen. For this reason, we define the *complement* of an event E as those outcomes in the sample space that are *not* in E. This definition is consistent with the Chapter 3 notion of a set's complement relative to an understood universe. In the context of probability, it is traditional to use the notation $\overline{E}$ for the complement of E.

Example 3 *What is the probability that for a six-sided die rolled three times the same result comes up more than once?*

SOLUTION   We will describe the outcomes in the sample space S as lists of length 3 with entries from $\{1, 2, 3, 4, 5, 6\}$. If E denotes the set of outcomes that use the same number more than once, then the complement $\overline{E}$ is the set of those outcomes that use all different numbers. That is, $\overline{E}$ is the set of permutations of length 3 with entries from $\{1, 2, 3, 4, 5, 6\}$, so in this case, it is easy to compute $n(\overline{E}) = P(6, 3) = 120$. Since $n(S) = 6^3 = 216$, we have

$$Prob(E) = \frac{n(E)}{n(S)} = \frac{n(S) - n(\overline{E})}{n(S)} = 1 - \frac{n(\overline{E})}{n(S)} = 1 - \frac{120}{216} \approx 0.44$$

This solution gives us the general relationship between the probabilities of an event and its complement.

Proposition 1   *Given an event E,*

$$Prob(E) + Prob(\overline{E}) = 1$$

This means that once we know one of these quantities, we can easily find the other.

Example 4 *What is the probability that in a group of six people, two will have birthdays in the same month, assuming that all months are equally likely?*

SOLUTION   We will let an outcome be represented as an ordered list of length 6 from $\{1, \ldots, 12\}$ (for the 12 months). For example, if the first person has a birthday in January, the second person in February, and everyone else in August, we will represent this as $1, 2, 8, 8, 8, 8$. The question asks for the probability of the event E, which is the set of all outcomes in which some number occurs more than once. To illustrate Proposition 1, we will instead find the probability of the complementary event $\overline{E}$, which is the set of all outcomes in which no number occurs twice.

There are a total of $12^6 = 2{,}985{,}984$ outcomes, which we assume are equally likely. An outcome is in $\overline{E}$ if it *does not* contain a repeated number, which actually means that the outcome is a *permutation* of length 6 from $\{1, \ldots, 12\}$. Hence, there are $P(12, 6) = 665{,}280$ outcomes in $\overline{E}$, so we have

$$Prob(\overline{E}) = \frac{P(12,6)}{12^6} = \frac{665{,}280}{2{,}985{,}984} \approx 0.22$$

We conclude that $Prob(E) = 1 - Prob(\overline{E}) \approx 0.78$.

Practice Problem 3 *If everyone in your class writes down his or her favorite two-digit number, what is the probability that two people will have written down the same number?*

We conclude this section with some experiments involving cards that we will investigate further in the exercises.

Example 5 *We have a packet of three cards in which one is blue on both sides, one is red on both sides, and one is red on one side and blue on the other. One of three cards is chosen and placed on the table in full view of both of us. We now wager on the color of the face-down side of the card.*

You will give me $5 if the face-down side is the same color as the (visible) face-up side, and I will give you $6 otherwise. For example, if the chosen card has a red side up, then the face-down side is either red or blue, and you win 50% of the time. So you should definitely play this game, right?

SOLUTION Label the faces of the cards (first card) B_1 and B_2, (second card) R_1 and R_2, and (third card) R_3 and B_3. The six possible (equally likely!) outcomes of the "experiment" of drawing a card and placing it on the table are $(B_1, B_2), (B_2, B_1), (R_1, R_2), (R_2, R_1), (B_3, R_3)$, and (R_3, B_3), where in each pair, the first entry denotes the face that is up and the second entry denotes the face that is down. In only the last two of these would you win, giving you an actual probability of $\frac{1}{3}$ of winning. □

Example 6 *Try playing the following game of solitaire with a normal deck of 52 cards. Shuffle the deck and then go through it removing any two adjacent cards that are the same color (red or black). Do this until there are no instances of adjacent cards of the same color in the deck. If you exhaust the deck, you are a winner. What is the probability of winning?*

SOLUTION Since the values of the cards do not matter, we will think of a deck as having just 26 indistinguishable red cards and 26 indistinguishable black cards. So the total number of arrangements of the deck is $C(52, 26)$ (like a binary sequence with twenty-six 1's and twenty-six 0's). An arrangement that results in a win seems very hard to analyze, but there is actually a simple argument. It is left to you to verify that the following procedure always results in an arrangement that will win.

Take any arrangement of 13 red and 13 black cards (there are $C(26, 13)$ of these) and any arrangement of the other 13 reds and 13 blacks. Now weave together these two arrangements alternating taking a card from the first arrangement and a card from the second arrangement. Believe it or not, this will always result in a winning arrangement and there are no other winning arrangements except for those you get this way!

Therefore, there are $C(26, 13)C(26, 13)$ winning arrangements, so the probability of winning this game is

$$C(26, 13)^2/C(52, 26) \approx 22\%$$

□

Practice Problem 4 *What is the probability of winning the solitaire game in the previous example if we use* **two** *standard decks of cards shuffled together?*

Solutions to Practice Problems

1 From the table in Example 1, we can simply list the outcomes $(4, 6)$, $(6, 4)$, and $(5, 5)$ that have a sum of 10. Since there are three of these, the probability that two rolled dice will sum to 10 is $\frac{3}{36} = \frac{1}{12}$.

2 **(a)** The outcomes can be represented as ordered lists of length 5 with entries from $\{H, T\}$, so there are $2^5 = 32$ outcomes in the sample space S.

  (b) The event E consists of those outcomes that look like ___ HH or ___ TT. In either of the two cases, there are $2^3 = 8$ outcomes, so $n(E) = 16$.

  (c) It is now easy to compute

$$Prob(E) = \frac{n(E)}{n(S)} = \frac{16}{32} = 0.5$$

  So there is a 50% chance that the last two tosses will be the same.

3 We will assume there are 32 people in the class—your numbers might be different. We will represent the outcomes as ordered lists of length 32 with entries from $\{10, 11, \ldots, 98, 99\}$, so there are 90^{32} outcomes in our sample space S. If E is the event that two people have the same number, then the complement $\overline{E}$ is the event that everyone has different numbers. So $\overline{E}$ is the set of permutations of length 32 with entries from $\{10, 11, \ldots, 98, 99\}$, of which there are $P(90, 32)$. Hence,

$$Prob(\overline{E}) = \frac{n(\overline{E})}{n(S)} = \frac{P(90, 32)}{90^{32}} \approx 0.00184$$

and so

$$Prob(E) = 1 - Prob(\overline{E}) \approx 1 - 0.00184 = 0.99816$$

That is, there is a better than 99.8% chance of two people picking the same number.

4 The argument is the same, but now there are 52 red cards and 52 black cards in the entire pack of 104 cards, so the answer changes to

$$C(52, 26)^2 / C(104, 52) \approx 15.5\%$$

Exercises for Section 6.1

1. If one card is to be drawn from a standard deck of cards, find the probability that
   (a) The card is a spade.
   (b) The card is a red ace.
   (c) The card is a 3 or an 8.
   (d) The card is either red or an ace.

2. A gumball machine contains 20 yellow gumballs, 40 green gumballs, 25 white gumballs, and 15 black gumballs. If one gumball is purchased, find the probability that
   (a) The gumball is yellow.
   (b) The gumball is green or white.
   (c) The gumball is not black.

3. If two dice, one red and one green, are rolled, find the probability that
   (a) Exactly one of the dice is a 6.
   (b) The sum of the dice is 5.
   (c) The sum of the dice is at least 10.
   (d) The value on the red dice is less than the value on the green die.
   (e) The values on the two dice are different.

4. If a coin is tossed five times, find the probability that
   (a) The first toss matches the second toss.
   (b) We get five heads.
   (c) We get at least one result of tails.

5. For each of the following experiments, describe the sample space S so that the outcomes as described are equally likely:
   (a) Three (six-sided) dice are rolled.
   (b) A club with 20 members elects a president and a vice president.
   (c) A coin is flipped six times.
   (d) Three people choose cards from a deck.

6. The following probability questions refer to the experiments in the previous exercise:
   (a) What is the probability that if three dice are rolled, all three dice have the same value?
   (b) If a club's members consist of 8 men and 12 women, and everyone is equally likely to be chosen an officer (either president or vice president), what is the probability that among these two officers are one man and one woman?

(c) What is the probability that a coin tossed six times results in three heads and three tails (in any order)?

(d) What is the probability that three people choose cards of the same suit from a standard deck of 52 cards?

7. If a blackjack hand (consisting of two cards, one face-down and one face-up) is dealt from a standard deck of cards, find the probability that

(a) The face-down card is an ace.

(b) The face-down card is a spade and the face-up card is a heart.

(c) The face-down card is an ace and the face-up card is a spade.

(d) The two-cards have the same suit.

(e) The face-up card has a value of a 10, jack, queen, or king.

(f) One card is an ace and the other has a value of a 10, jack, queen, or king.

8. A club with 12 female members and 11 male members is electing a president and vice president. How many election outcomes are possible? Assuming each outcome is equally likely, find the probability that

(a) Both officers are male.

(b) The president is female.

(c) The two officers are of the same sex.

9. The club in the previous exercise is forming a budget committee consisting of three club members. How many committees can be formed? Assuming each committee is equally likely to be appointed, find that probability that

(a) The committee includes neither the club's president nor vice president.

(b) The committee has at least one female.

(c) The committee is all-female.

(d) David and Jenny (a married couple) are not both on the committee.

10. In a box of lightbulbs there are 12 good bulbs and 4 defective bulbs. If five bulbs are to be removed, find the probability that

(a) All five bulbs are good.

(b) At most one bulb is bad.

11. A packet of five cards consists of the ace of spades, the ace of hearts, the 2 of spades, the 3 of diamonds, and the 4 of clubs. If the packet is thoroughly shuffled, what is the probability that the two aces end up side by side?

12. In one version of poker, each player gets a five-card hand of cards from a standard deck.

(a) A *flush* is a hand consisting of all cards from the same suit. What is the probability that a five-card hand is a flush?

(b) A *straight* is a hand consisting of card values in consecutive order, regardless of suits, where an ace can be counted as either the highest value in the deck or the lowest value in the deck (but not both). What is the probability that a five-card hand is a straight?

(c) A *straight flush* is a hand that is both a straight and a flush. What is the probability that a five-card hand is a straight flush?

13. The combination to a lock consists of an ordered list of three digits from $\{0, 1, 2, \ldots, 9\}$. Assume that every such list is equally likely to be guessed by a thief trying to guess the correct combination. Find the probability that

(a) The combination guessed uses the digit 9.

(b) The combination guessed uses no repeated digits.

(c) The combination guessed has the first two digits correct.

14. What is the probability that in a class of 16 students two will have birthdays on the same day (assuming that each day is equally likely)? You may either ignore February 29 or try the more difficult problem of accounting for it.

15. How does the answer in Exercise 14 change when the number of students is doubled?

16. On a TI-83 calculator, the command line Prod(seq((366-I)/365,I,1,20)) will give the probability that everyone in a group of 20 people has a different birthday. Use this (or other technology you have at your disposal) to answer the following questions about the birthday problem:

(a) In a group of 40 people, what is the probability that at least two people share a birthday?

(b) How many people must be in a group for there to be a 0.95 probability that at least two people will share a birthday?

(c) How many people must be in a group for there to be a 0.99 probability that at least two people will share a birthday?

17. Some institutions use the last four digits of a student's Social Security number as a "student number." Assuming that all four-digit strings of digits are equally likely to occur, what is the probability that there will be no duplication of student numbers in an incoming freshman class consisting of 500 new students. (NOTE: You will need a technology tool like the one discussed in the previous problem.)

18. In a particular lottery game, a purchased ticket consists of seven distinct numbers from $\{1, 2, \ldots, 80\}$. On Wednesday nights, someone's grandmother draws 11 numbered ping pong balls from a set of balls numbered from $\{1, \ldots, 80\}$. Anyone whose seven ticket numbers all appear among the 11 drawn numbers is a winner. If I always buy a ticket with numbers

11, 22, 33, 44, 55, 66, 77, what is my probability of winning in a given week?

19. The following questions refer to a simple dice game:

(a) Suppose Jessica and John each roll a six-sided die. What is the probability that John's roll is higher than Jessica's?

(b) Suppose Jessica rolls an eight-sided die and John rolls a six-sided die. What is the probability that John's roll is higher than Jessica's?

(c) Generalize your answer to the previous problem for an n-sided die versus an m-sided die. (Assume that $n \geq m$.)

20. Suppose that a nonnegative integer solution to the equation $w + x + y + z = 10$ is chosen at random (each one being equally likely to be chosen). What is the probability that in this particular solution w is less than or equal to 2?

21. A man goes to the store to buy seven candy bars. The store sells five different types of candy. Assuming each choice is equally likely to occur, what is the probability that he gets at least one bar of each type?

22. If 10 apples are distributed among five children (so that every possible distribution is equally likely), what is the probability that every student gets at least one apple?

23. If n apples are distributed among 32 children (so that every possible distribution is equally likely), what is the probability (in terms of n) that every student gets at least one apple?

24. Use the answer to Exercise 23 to decide how many apples should be distributed so that there is a 0.99 probability every student gets at least one apple.

25. Use index cards to make the set of cards described in Example 5. Conduct the experiment 10 times recording how many times (out of 10) the face-down color was different from the face-up color. Combine these results with those of others in the class. Does the $\frac{1}{3}$ probability given in the example seem reasonable?

26. Play the game of solitaire described in Example 6 ten times, and record the number of times the game is "won."

27. Verify the claim about winning configurations of the deck of cards for the solitaire game described in Example 6.

6.2 Sum and Product Rules for Probability

Mathematics is replete with systems in which formal rules are used to build answers to complicated questions using an understanding of simpler problems. In the study of probability, when a complex event is made up of simpler events, we fully expect to be able to compute the probability of the complex event in terms of the probabilities of the simpler parts. To create such a system, we need only think about the ways simple events can be combined to form complex events.

The Sum Rule

In Chapter 5, we found that counting the elements in the union of two sets is particularly simple when the sets have no elements in common. As we shall see, the same principle applies in the calculation of probability for complex events.

> **Definition** Two events are said to be *disjoint* (or *mutually exclusive*) if they cannot occur simultaneously.

The term "disjoint" is compatible with the meaning of disjoint as it applies to sets when you remember that technically an event is just a named set of outcomes, and sets are said to be disjoint if they share no common member.

Example 1 *For a toss of a die, the events "getting a 3" and "getting a 4" are disjoint. However, for two tosses of a die, the events of "getting a 3 on the first roll" and "getting a 4 on the second roll" are not disjoint since a roll of 3 followed by a roll of 4 is an outcome in both events.*

Practice Problem 1 *For each of the experiments given below, decide if the events described are disjoint:*

1. *When tossing a coin four times, let E_1 be the event that there are exactly three heads and E_2 be the event that there are exactly two heads.*
2. *When choosing four cards, let E_1 be the event that the cards have the same value and E_2 be the event that the cards have the same suit.*
3. *When choosing a committee of three people from a club with 8 men and 12 women, let E_1 be the event that the committee has a woman and let E_2 be the event that the committee has a man.*

□

Theorem 1 (The Sum Rule) *If E_1 and E_2 are disjoint events in a given experiment, then the probability that E_1 or E_2 occurs is the sum of Prob(E_1) and Prob(E_2). That is,*

$$Prob(E_1 \text{ or } E_2) = Prob(E_1) + Prob(E_2)$$

for disjoint events.

Example 2 *In the game of backgammon, one uses rolls of two dice to generate moves, and there are usually combinations of dice outcomes that are important. For example, one might hope for either a 5 on at least one of the dice or a sum of 5 on the two dice. Verify that these are two disjoint events, and find the probability of this happening.*

SOLUTION   We will use E_1 for the event of "getting at least one 5" and E_2 for the event of "getting a sum of 5." It should be clear that E_1 and E_2 are disjoint since it is impossible for a single roll of the two dice to be counted in both events. Using the list of 36 equally likely events in Example 1, we can easily find the number of outcomes in E_1 and E_2. In Table 6-2, we list the 11 outcomes in E_1 in **bold** and the 4 outcomes in E_2 as underlined.

Since E_1 and E_2 are disjoint,

$$Prob(E_1 \text{ or } E_2) = Prob(E_1) + Prob(E_2) = \frac{11}{36} + \frac{4}{36} = \frac{5}{12}$$

□

It is not necessary for two events to be disjoint to find the probability that either happens, but the sum rule shows us that when two events *are* disjoint, the probability of either event happening is related to the probability of the two events in a very simple way.

$$
\begin{array}{cccccc}
(1,1) & (1,2) & (1,3) & \underline{(1,4)} & \mathbf{(1,5)} & (1,6) \\
(2,1) & (2,2) & \underline{(2,3)} & \overline{(2,4)} & \mathbf{(2,5)} & (2,6) \\
(3,1) & \underline{(3,2)} & \overline{(3,3)} & (3,4) & \mathbf{(3,5)} & (3,6) \\
\underline{(4,1)} & \overline{(4,2)} & (4,3) & (4,4) & \mathbf{(4,5)} & (4,6) \\
\overline{\mathbf{(5,1)}} & \mathbf{(5,2)} & \mathbf{(5,3)} & \mathbf{(5,4)} & \mathbf{(5,5)} & \mathbf{(5,6)} \\
(6,1) & (6,2) & (6,3) & (6,4) & \mathbf{(6,5)} & (6,6)
\end{array}
$$

Table 6-2 Outcomes for Example 2

Theorem 2 (The General Sum Rule)   *If E_1 and E_2 are any events in a given experiment, then the probability that E_1 or E_2 occurs is given by*

$$
Prob(E_1 \ or \ E_2) = Prob(E_1) + Prob(E_2) - Prob(E_1 \ and \ E_2)
$$

If E_1 and E_2 are disjoint, then $E_1 \cap E_2 = \emptyset$, so $Prob(E_1 \ and \ E_2) = 0$.

This general rule is more versatile but requires that we calculate the probability of the two events occurring simultaneously. This can sometimes be rather subtle, so we will discuss this at length after some examples and practice using simple instances of the general sum rule.

Example 3 *What is the probability that three cards chosen from a standard deck of cards consist of either three face cards (i.e., cards having a value of jack, queen, or king) or three cards of the same suit?*

SOLUTION   We will represent the cards as subsets of size 3 taken from the deck of 52 cards; hence, $n(S) = C(52, 3)$. Let E_1 denote those subsets consisting of three face cards, and let E_2 denote those subsets consisting of three cards of the same suit. Since there are 12 face cards in a standard deck, $n(E_1) = C(12, 3)$, and since there are four suits, each with 13 cards, $n(E_2) = 4 \cdot C(13, 3)$. Now $E_1 \cap E_2$ is the event that the subset consists of three face cards of the same suit. Since there are only three face cards in each suit, this means that $n(E_1 \cap E_2) = 4$. The general sum rule says that

$$
\begin{aligned}
Prob(E_1 \ or \ E_2) &= Prob(E_1) + \text{Prob}(E_2) - Prob(E_1 \ and \ E_2) \\
&= \frac{C(12,3)}{C(52,3)} + \frac{4 \cdot C(13,3)}{C(52,3)} - \frac{4}{C(52,3)} \\
&= \frac{1,360}{22,100} \approx 0.0615
\end{aligned}
$$

$\square$

Practice Problem 2   *What is the probability that when tossing a fair coin five times, either the first two tosses have the same result, or the last two tosses have the same result?*

The Product Rule

In the preceding problems, the intersection of E_1 and E_2 was easy to count because it was small, but we will need to have better tools available to deal with more complicated problems. In some cases, the tool is very easy to understand.

Example 4 *Suppose a die is rolled and a card is drawn from a standard deck. What is the probability that the die and the card both show an even number value?*

SOLUTION We represent the outcomes as the set of ordered pairs (a, b), where $a \in \{1, 2, 3, 4, 5, 6\}$ and b is a standard playing card. Hence, the sample space has 6×52 outcomes. The successful outcomes are the ordered pairs (a, b), where $a \in \{2, 4, 6\}$ and $b \in \{2C, 4C, 6C, 8C, 10C, 2H, 6H, \ldots\}$. Hence, the event E consists of 3×20 outcomes.

Therefore,

$$Prob(E) = \frac{3 \times 20}{6 \times 52} = \frac{3}{6} \times \frac{20}{52}$$

This is simply the probability that the die is even *times* the probability that the card is even.

□

It seems like the probability of two events happening is simply the product of the probabilities that each event happens. This is the nicest possible rule, but unfortunately, it does not always work. Just as we need the assumption that two events are disjoint in order to apply the nicest version of the sum rule, we must acknowledge that an assumption on the events is also necessary in order for this nice version of the product rule to be correct.

Definition Two events are said to be *independent* if the occurrence of one event is not influenced by the occurrence (or nonoccurrence) of the other event.

Example 5 *For two tosses of a coin, the events "getting a result of heads on the first toss" and "getting a result of heads on the second toss" are independent. Explain why in the experiment of dealing two cards from a standard deck, however, the events "first card chosen is a black ace" and "second card chosen is a black ace" are not independent.*

SOLUTION The probability that the second card chosen is a black ace depends on whether the first card is, or is not, a black ace. Specifically, if the first event fails, then there are two successful outcomes for the second card, but if the first event succeeds, then there is only one successful outcome for the second card.

□

Practice Problem 3 *For each of the experiments given below, decide if the events described are independent:*

1. *When rolling a six-sided die four times, let E_1 be the event that the first two rolls sum to 7 and let E_2 be the event that the last two rolls sum to 10.*

2. *When choosing a committee of three people from a club with 8 men and 12 women, let E_1 be the event that the committee has a woman and let E_2 be the event that the committee has a man.*

3. *In a household with four children, let E_1 be the event that the first child is male and let E_2 be the event that at least half of the children are female.*

Theorem 3 (The Product Rule) *If E_1 and E_2 are independent events in a given experiment, then the probability that both E_1 and E_2 occur is the product of $Prob(E_1)$ and $Prob(E_2)$. That is,*

$$Prob(E_1 \ and \ E_2) = Prob(E_1) \cdot Prob(E_2)$$

for independent events.

Example 6 *Suppose I have a "loaded" die for which the probability of a 6 appearing is $\frac{1}{2}$, while the probability of each of the other faces appearing is $\frac{1}{10}$. What is the probability of getting a 5 and then a 6 on two tosses of the loaded die?*

SOLUTION   The events E_1, "getting a 5 on the first toss," and E_2, "getting a 6 on the second toss," are independent, so by the product rule,

$$Prob(E_1 \ and \ E_2) = Prob(E_1) \cdot Prob(E_2) = \frac{1}{10} \cdot \frac{1}{2} = \frac{1}{20}$$

$\square$

Practice Problem 4 *For each experiment in Practice Problem 3, compute $Prob(E_1 \ and \ E_2)$ along with $Prob(E_1) \cdot Prob(E_2)$. Check that these values are equal only in the cases where the events are independent.*

Just as we were able to drop the assumption of "disjoint events" to create a general sum rule, we can likewise drop the assumption of "independent events" to create a general product rule. If two events are not independent, then the occurrence or nonoccurrence of one of them affects the likelihood of the other one happening. To generalize the product rule, we must first make this more precise.

> **Definition**   Given events E_1 and E_2 for some experiment, we define the *probability of E_1 given E_2*, denoted by $Prob(E_1|E_2)$, as the probability that E_1 happens given that E_2 occurs. Note that if E_1 and E_2 are independent, then $Prob(E_1|E_2) = Prob(E_1)$.

This sort of probability is called *conditional probability* since it expresses the probability of one event happening *if* something else occurs as well.

Example 7 *In Practice Problem 3, we found that when choosing a committee of three people from a club with 8 men and 12 women, the event E_1 of the committee having a woman on it and the event E_2 of the committee having a man on it are not independent. Compute $Prob(E_1|E_2)$.*

SOLUTION   To make the discussion easier to follow, we will refer to event E_1 as W and to event E_2 as M. So we are calculating $Prob(W|M)$, the probability that a committee contains a woman, given that it contains a man. The "given that it contains a man" condition implies that our sample space of equally likely outcomes consists of those committees that contain at least one man. We consider the complement (those committees with no men), and calculate that

$$n(\overline{M}) = C(12, 3)$$

so there are $C(20, 3) - C(12, 3) = 920$ committees in our sample space. Now, within this sample space, which committees contain a woman? These are the ones containing one man and two women (there are $C(8, 1) \cdot C(12, 2)$ of these) or two men and one woman (there are $C(8, 2) \cdot C(12, 1)$ of these). Hence,

$$Prob(E_1|E_2) = Prob(W|M) = \frac{C(8, 1) \cdot C(12, 2) + C(8, 2) \cdot C(12, 1)}{920}$$

$$= \frac{864}{920} = \frac{108}{115}$$

□

Practice Problem 5   *For the preceding example, calculate Prob(M|W), the probability that the committee contains a man given that it must contain a woman.*

In the solutions to the example and practice problem, we essentially argued that

$$Prob(E_1|E_2) = \frac{n(E_1 \cap E_2)}{n(E_2)}$$

and that

$$Prob(E_2|E_1) = \frac{n(E_1 \cap E_2)}{n(E_1)}$$

These ideas lead to our general product rule.

Theorem 4 (The General Product Rule)   *If E_1 and E_2 are any events in a given experiment, then the probability that both E_1 and E_2 occur is given by*

$$Prob(E_1 \text{ and } E_2) = Prob(E_2) \cdot Prob(E_1|E_2)$$
$$= Prob(E_1) \cdot Prob(E_2|E_1)$$

Note that if E_1 and E_2 are independent, then this says the same thing as Theorem 3.

In words, this theorem says that we can choose either of the two events and calculate its probability, then multiply that by the *conditional* probability of the other event, *given that the chosen event occurs*. In many situations, it is helpful to picture the events as taking place one after the other, but this is not necessary for the theorem to be applied.

Example 8   *Two marbles are chosen from a bag containing three red, five white, and eight green marbles.*

1. What is the probability that both are red?

2. What is the probability that one is white and one is green?

SOLUTION

1. We can solve this using the methods of the preceding section, but it will be instructive to see how the product rule applies. We picture the marbles being removed one at a time, and define the event R_1 as "the first marble is red," and the event R_2 as "the second marble is red." The general product rule states

$$Prob(R_1 \text{ and } R_2) = Prob(R_1) \cdot Prob(R_2|R_1)$$

Clearly, $Prob(R_1)$ is $\frac{3}{16}$. To calculate $Prob(R_2|R_1)$, we imagine that the event "first marble is red" occurs, and see that of the remaining 15 marbles there are two red ones, giving $\frac{2}{15}$. The final answer is $\frac{3}{16} \cdot \frac{2}{15} = \frac{1}{40}$.

2. To apply the product rule, it is again convenient to imagine the marbles being drawn out one at a time. Given this viewpoint, there are two distinct ways we could get the desired result: white for the first marble and green for the second, or green for the first and white for the second. Again, we use meaningful names for the events (e.g., G_2 denotes the event "the second marble is green.") In this problem, we apply the sum rule and *then* the product rule, obtaining

$$Prob(W_1 \text{ and } G_2) + Prob(G_1 \text{ and } W_2) = Prob(W_1) \cdot Prob(G_2|W_1)$$
$$+ Prob(G_1) \cdot Prob(W_2|G_1)$$
$$= \frac{5}{16} \cdot \frac{8}{15} + \frac{8}{16} \cdot \frac{5}{15} = \frac{1}{3}$$

Practice Problem 6   *When you deal a blackjack hand (first card face-down and second card face-up), let T indicate the event that the face-down card has a value of 10 (i.e., a 10, jack, queen, or king), and let A stand for the event that the face-up card is an ace.*

1. *Calculate Prob(T|A), the probability that the face-down card has a value of 10 given that the face-up card is an ace. Use this to calculate Prob(T and A).*
2. *Calculate Prob(T and A) by using Prob(A|T) instead. Is the answer the same?*

Conditional probability is extremely important in understanding the results of statistical surveys, medical procedures, or quality control experiments. In these applications, there are no theoretical probabilities—everything is determined by data collection. The mathematical field of *statistics* addresses the connection between data sets and the inferences one can draw from them.

Example 9   *Drug testing of athletes can be a tricky business because medical tests are not always accurate for many different reasons. Suppose that a test for steroids has been determined (by extensive experimentation and data analysis) to have the following characteristics:*

● *When the athlete has used steroids, the test result is correct with probability 0.995 and incorrect with probability 0.005*

• *When the athlete has not used steroids, the test result is correct with probability* 0.98 *and incorrect with probability* 0.02.

An athlete at a certain event is randomly chosen for drug testing. Let P be the event that the steroid test is positive, and let S be the event that the athlete has used steroids. Assuming that 3% of all athletes at that event are using illegal steroids (i.e., $Prob(S) = 0.03$), what is $Prob(P \text{ and } S)$? What does this mean?

SOLUTION According to our general product rule,

$$Prob(P \text{ and } S) = Prob(P|S) \cdot Prob(S)$$

$Prob(P|S)$ is the probability that the steroid test is positive given that the athlete really has used steroids, and we are given that this probability is 0.995. We are also given that $Prob(S) = 0.03$; hence,

$$Prob(P \text{ and } S) = 0.995 \cdot 0.03 = 0.02985$$

This means that there is a probability of 0.02985 that a tested athlete chosen at random will be a steroid user *and* be caught by the test.

□

Bernoulli Trials

We conclude this section by showing how to use the sum rule and product rule together to solve a type of problem that is very common in the study of games. In this kind of problem, there is an experiment that is based on the independent repetition of a simpler experiment. In this case, we refer to each instance of the simpler experiment as a *trial** to distinguish it from the larger experiment.

We have already seen experiments like this. For example, rolling a single die once is a simple experiment that is repeated four successive times to form the experiment in part of Practice Problem 1.

Example 10 *Suppose a baseball player gets a hit with probability $\frac{1}{3}$ every time he steps to the plate. What is the probability that the player gets exactly one hit in four tries?*

The simple experiment here is the batter appearing at the plate to try to get a hit, and this trial is repeated four times. We will represent outcomes for this larger experiment as an ordered list of length 4 with entries taken from $\{H, N\}$, where H represents getting a hit and N represents not getting a hit. For example, $\underline{H}\,\underline{N}\,\underline{N}\,\underline{H}$ represents the player getting hits on his first and fourth try, but not on his second or third try.

It is important to notice that **the outcomes for this larger experiment are not equally likely!** For example, since it is more likely to not get a hit than to get a hit each time at the plate, we expect that the outcome $\underline{N}\,\underline{N}\,\underline{N}\,\underline{N}$ is more likely than the outcome $\underline{H}\,\underline{H}\,\underline{H}\,\underline{H}$. This is precisely why this problem differs from those using our basic approach to probability problems in the previous section.

---

* In the more general study of probability, these are called *Bernoulli trials* in honor of the Swiss mathematician Jacob Bernoulli (1654–1705) who wrote one of the earliest books dedicated to probability.

Since we are given that the batter's probability of getting a hit is always $\frac{1}{3}$, the successive trials are independent, and so we can use the product rule. In the example above,

$$Prob(\underline{H}\,\underline{N}\,\underline{N}\,\underline{H}) = Prob(H) \cdot Prob(N) \cdot Prob(N) \cdot Prob(H)$$

$$= \left(\frac{1}{3}\right) \cdot \left(\frac{2}{3}\right) \cdot \left(\frac{2}{3}\right) \cdot \left(\frac{1}{3}\right)$$

$$= \left(\frac{1}{3}\right)^2 \left(\frac{2}{3}\right)^2$$

Notice that the probability of the batter *not* getting a hit is $\frac{2}{3}$. We are now ready to solve the example problem.

SOLUTION

- Let E_1 be the event of getting the sequence $\underline{H}\,\underline{N}\,\underline{N}\,\underline{N}$.
- Let E_2 be the event of getting the sequence $\underline{N}\,\underline{H}\,\underline{N}\,\underline{N}$.
- Let E_3 be the event of getting the sequence $\underline{N}\,\underline{N}\,\underline{H}\,\underline{N}$.
- Let E_4 be the event of getting the sequence $\underline{N}\,\underline{N}\,\underline{N}\,\underline{H}$.

These events are certainly disjoint, so by the sum rule the probability that one of them happens is simply

$$Prob(E_1 \text{ or } E_2 \text{ or } E_3 \text{ or } E_4) = Prob(E_1) + Prob(E_2) + Prob(E_3) + Prob(E_4)$$

$$= Prob(\underline{H}\,\underline{N}\,\underline{N}\,\underline{N}) + Prob(\underline{N}\,\underline{H}\,\underline{N}\,\underline{N})$$

$$+ Prob(\underline{N}\,\underline{N}\,\underline{H}\,\underline{N}) + Prob(\underline{N}\,\underline{N}\,\underline{N}\,\underline{H})$$

$$= \left(\frac{1}{3}\right)\left(\frac{2}{3}\right)^3 + \left(\frac{1}{3}\right)\left(\frac{2}{3}\right)^3$$

$$+ \left(\frac{1}{3}\right)\left(\frac{2}{3}\right)^3 + \left(\frac{1}{3}\right)\left(\frac{2}{3}\right)^3$$

$$= 4\left(\frac{1}{3}\right)\left(\frac{2}{3}\right)^3$$

$\square$

It is possible to visualize all the outcomes of four successive attempts at hitting with the same kind of tree structure that we encountered in Chapter 5. At each branch in the tree shown in Figure 6-1, the simple experiment is performed. If the baseball player gets a hit, the upper branch is followed (labeled H), and if the player does not get a hit, the lower branch (labeled N) is followed. Thus, all 16 possible results of the four plate appearances are presented in a visual way.

This tree structure is already fairly large, so drawing it is not a particularly effective way to solve problems, but visualizing it internally can be very useful. Notice that the 16 final outcomes are precisely the four-digit binary sequences using symbols H and N. It is this fact that we will take advantage of in the next section to solve general problems of this type.

Practice Problem 7   *What is the probability that the player from Example 10 will get at least one hit in four plate appearances?*

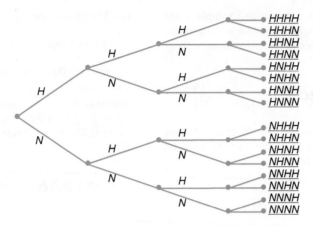

Figure 6-1 Decision tree for four tries at hitting.

Solutions to Practice Problems

1 (a) When tossing a coin four times, no outcome can consist of "exactly two heads" and also "exactly three heads"; hence, these two events are disjoint.

(b) When choosing four cards, if the four cards all have the same value, then they cannot all have the same suit; hence, these two events are disjoint.

(c) When choosing a committee of three people from a club with 8 men and 12 women, there are many ways in which the committee can include a woman and a man, so these events are not disjoint.

2 Let E_1 be the event of getting the same result on the first two tosses, and let E_2 be the event of getting the same result on the last two tosses. It is straightforward to count

- $n(S) = 2^5 = 32$
- $n(E_1) = 2^4 = 16$
- $n(E_2) = 2^4 = 16$
- $n(E_1 \cap E_2) = 2^3 = 8$

Therefore,

$$Prob(E_1 \text{ or } E_2) = Prob(E_1) + Prob(E_2)$$
$$-Prob(E_1 \text{ and } E_2)$$
$$= \frac{16}{32} + \frac{16}{32} - \frac{8}{32}$$
$$= \frac{3}{4}$$

3 (a) The events are independent. The first two rolls have no influence on the last two rolls.

(b) The events are not independent. The probability of the committee having a man (E_2) is different when E_1 occurs than it is when E_1 does not occur. Specifi-

cally, if E_1 does not occur, then E_2 happens for sure (i.e., its probability is 1), and if E_1 does occur, then E_2 is not guaranteed to happen (i.e., its probability is less than 1).

(c) The events are not independent. Four-children households where the oldest child is a girl are more likely to have two or more girls than four-children households where the oldest child is a boy.

4 The two quantites are equal only in the first part, which is the only part that describes independent events.

(a) We have $Prob(E_1 \text{ and } E_2) = \frac{6 \times 3}{6^4}$, while $Prob(E_1) \cdot Prob(E_2) = \frac{6}{36} \times \frac{3}{36}$.

(b) We have

$$Prob(E_1 \text{ and } E_2) = 1 - Prob(\text{no men or no women})$$
$$= 1 - \left(\frac{C(8,3)}{C(20,3)} + \frac{C(12,3)}{C(20,3)} \right)$$
$$\approx 0.7579$$

while

$$Prob(E_1) \cdot Prob(E_2) = \left(1 - \frac{C(8,3)}{C(20,3)} \right)$$
$$\times \left(1 - \frac{C(12,3)}{C(20,3)} \right) \approx 0.7674$$

(c) Assuming there is a $\frac{1}{2}$ probability that any given child will be male, we have $Prob(E_1 \text{ and } E_2) = \frac{4}{16} = 0.25$, while $Prob(E_1) \cdot Prob(E_2) = \frac{1}{2} \times \frac{11}{16} = 0.34375$.

5 The sample space this time is the set of all committees with at least one woman, of which there are (by using the complement) $C(20,3) - C(8,3) = 1,084$. Of these,

the ones that contain a man are, as in Example 7, the committees containing one man and two women or two men and one woman. As calculated in the example, there are 864 of these, so

$$Prob(E_2|E_1) = Prob(M|W) = \frac{864}{1,084} = \frac{216}{271}$$

6 (a) If the face-up card is an ace, then there are 51 other cards in the deck, and 16 of these are either a 10, jack, queen, or king. Hence,

$$Prob(T|A) = \frac{16}{51}$$

By the product rule,

$$Prob(T \text{ and } A) = Prob(T|A) \cdot Prob(A) = \frac{16}{51} \cdot \frac{4}{52}$$

Exercises for Section 6.2

1. For each of the following experiments, decide whether the given events are disjoint. For each that is not, describe something that is in both events.
    (a) **Experiment**: Drawing two cards. **Event 1**: Getting at least one ace. **Event 2**: Getting at least one club.
    (b) **Experiment**: Rolling two dice. **Event 1**: Getting at least one 5. **Event 2**: Getting a sum of 8.
    (c) **Experiment**: Drawing two cards. **Event 1**: Getting two aces. **Event 2**: Getting two clubs.

2. A card is drawn from an ordinary deck of 52 cards. Show how to use the basic sum rule to find the probability that the card is
    (a) An ace or a jack.
    (b) A diamond or a black face card (jack, queen, or king).
    (c) An even number value or a red face card.

3. A card is drawn from an ordinary deck of 52 cards. Show how to use the general sum rule to find the probability that the card is
    (a) An ace or a heart.
    (b) An ace or a black card.
    (c) A diamond, a club, or a king.

4. For each of the following experiments, decide whether the given events are independent. For each that is not, demonstrate that $Prob(E_1 \text{ and } E_2) \neq Prob(E_1) \cdot Prob(E_2)$.
    (a) **Experiment**: Dealing two cards from a standard deck. **Event 1**: The first card is an ace. **Event 2**: The second card is a 10, jack, queen, or king.

(b) The product rule gives

$$Prob(T \text{ and } A) = Prob(T) \cdot Prob(A|T) = \frac{16}{52} \cdot \frac{4}{51}$$

which is the same result.

7 Let E denote the event of getting at least one hit. The only way he does *not* get at least one hit is if he ends up at $\underline{NNNN}$ in the tree, so this is the complementary event $\overline{E}$. The product rule tells us that the probability of $\overline{E}$ is

$$Prob(\underline{N}\ \underline{N}\ \underline{N}\ \underline{N}) = Prob(N) \cdot Prob(N) \cdot Prob(N)$$
$$\cdot Prob(N)$$
$$= \left(\frac{2}{3}\right)^4$$

Hence, we know that $Prob(E) = 1 - Prob(\overline{E}) = 1 - (2/3)^4 \approx 0.8025$.

(b) **Experiment**: Rolling a red die and a green die. **Event 1**: Getting a red 5. **Event 2**: Getting a sum of 8.
    (c) **Experiment**: Rolling a red die and a green die. **Event 1**: Getting a red 5. **Event 2**: Getting a green 6.

5. Two cards are drawn from a deck, with replacement. (This means that one person chooses a card, looks at it, and returns it, and then another person chooses a card, looks at it, and returns it.) Show how to use the product rule to find the probability that
    (a) The first card is an ace and the second card is black.
    (b) Both cards are spades.
    (c) Neither card has a value from {2, 3, 4, 5}.
    (d) At least one card is an ace.   (HINT: Consider the complementary problem.)

6. A fair six-sided die is cast, a fair coin is tossed, and a card is drawn from a standard deck.
    (a) What is the probability that you tossed a heads or drew an ace?
    (b) What is the probability that you rolled a 4 or drew an ace?
    (c) What is the probability that you rolled a 4, tossed a heads, and drew an ace?

7. What is the probability that when a pair of dice are rolled, either (at least) one die shows a 5 or the dice sum to 8?

8. What is the probability that when a pair of dice are rolled, either (at least) one die shows a 4 or the dice sum to 10?

9. What is the probability of rolling three dice and obtaining a sum between 5 and 10 (inclusive)?

10. John tosses a penny four times and Jessica tosses a nickel four times. What is the probability that at least one of them gets four results of heads?

11. John and Jessica each toss a coin four times. What is the probability that at least one of them gets four results that are the same (i.e., either four heads or four tails)?

12. John, Jessica, and Anna each toss a coin four times. What is the probability that at least one of them gets four results that are the same?

13. Suppose that we each roll a six-sided dice. You win if the sum of our rolls is even, and I win otherwise. What is the probability that I win? Now suppose I win if the sum is a multiple of 3 and you win otherwise—what is the probability that I win now?

14. Show that it is more likely to get at least one 6 on four rolls of a single die than at least one "double 6" on 24 rolls of a pair of dice, by computing the probabilities exactly. (These are the questions from the beginning of Chapter 6.)

15. What is the probability of rolling a six-sided die six times and having all the numbers 1 through 6 result (in any order)?

16. What is the probability of rolling a six-sided die six times and having only odd numbers result?

17. What is the probability of getting a straight or a flush in a five-card poker hand? (See Exercise 12 in Section 6.1 for more on the rules of poker.)

18. What is the probability of having nothing good in a five-card poker hand? (That is, no two cards have the same value, and the hand is neither a straight nor a flush.)

19. In blackjack, your opponent receives a face-down card and a face-up card. In particular, you can see your opponent's face-up card. This situation is a natural application of conditional probability. For each question below, specify events E_1 and E_2 so that the given question is about $Prob(E_2|E_1)$, and then answer the probability question.

(a) What is the probability that his face-down card is a 10, jack, queen, or king, given that you see his face-up card is an ace?

(b) What is the probability that his face-down card is a 10, jack, queen, or king, given that you see his face-up card is an ace and you know your own hand consists of two kings?

(c) What is the probability that his face-down card is a 10, jack, queen, or king, given that you see his face-up card is an ace and you know your own hand consists of a 4 and a 5?

20. In certain poker games, five-card hands consist of three face-down cards and two face-up cards. In particular, you can see two of the cards in your opponent's hand. For each question below, specify events E_1 and E_2 so that the given question is about $Prob(E_2|E_1)$, and then answer the probability question.

(a) What is the probability that his hand is a flush, given that you can see he has the ace and 5 of clubs?

(b) What is the probability that his hand is a flush, given that you can see he has the ace and 5 of clubs, and you know your own hand consists of {4H, 6D, 6S, 10S, QD}?

(c) What is the probability that his hand is a flush, given that you can see he has the ace and 5 of clubs, and you know your own hand consists of all clubs?

21. In the poker games described in Exercise 20, five-card hands consist of three face-down cards and two face-up cards. In particular, you can see two of the cards in your opponent's hand. For each question below, specify events E_1 and E_2 so that the given question is about $Prob(E_2|E_1)$, and then answer the probability question.

(a) What is the probability that his hand is a straight, given that you can see he has the 2 of diamonds and the 6 of clubs?

(b) What is the probability that his hand is a straight, given that you can see he has the 3 of diamonds and the 6 of clubs?

(c) What is the probability that his hand is a straight, given that you can see he has the 5 of diamonds and the 6 of clubs?

22. Consider the experiment and events of Example 7.

(a) Find $Prob(E_1|\overline{E_2})$ (also written $Prob(W|\overline{M})$ in the example).

(b) Find $Prob(E_2|\overline{E_1})$ (also written $Prob(M|\overline{W})$ in the example).

(c) In this example, verify that

$$Prob(E_2|E_1)Prob(E_1) + Prob(E_2|\overline{E_1})Prob(\overline{E_1})$$
$$= Prob(E_2)$$

and that

$$Prob(E_1|E_2)Prob(E_2) + Prob(E_1|\overline{E_2})Prob(\overline{E_2})$$
$$= Prob(E_1)$$

(d) Explain in words why the previous statement should be true for any events E_1 and E_2.

23. Using the information and notation from Example 9,

(a) What is $Prob(P \text{ and } \overline{S})$, and what does it mean?

(b) Using the solution in Example 9 and the previous answer, determine $Prob(P)$.

(c) What is $Prob(S|P)$, and what does it mean?

(d) What is $Prob(S|\overline{P})$, and what does it mean?

24. What is the probability that the baseball player in Example 10 gets exactly two hits in four tries?

25. What is the probability that a fair six-sided die rolled five times comes up 6 exactly once?

26. What is the probability that in six tosses of a fair coin an even number of heads' arise? How does this answer change if we sneak in a coin that comes up heads 75% of the time?

6.3 Probability in Games of Chance

The last example in Section 6.2 illustrates a common problem when analyzing probabilities in games, so it is worthwhile to address it here in more generality.

Theorem 1   *Given a simple experiment, called a Bernoulli trial*, and an event that occurs with probability p, if the trial is repeated independently n times, then the probability of having exactly k successes is*

$$C(n, k) \cdot p^k \cdot (1 - p)^{n-k}$$

PROOF   We represent the outcomes of the experiment as ordered lists of length n with entries from $\{S, F\}$, where S denotes success and F failure of the trial. The number of different such lists that use exactly k S's is $C(n, k)$ since these are merely binary sequences. Since $Prob(S) = p$ and hence $Prob(F) = 1 - p$, the probability of any particular outcome with k S's and $n - k$ F's is $p^k \cdot (1 - p)^{n-k}$. Hence, the probability of having k successes is this probability times the number of outcomes of this sort, $C(n, k)$. ∎

Example 1   *What is the probability that in 10 successive rolls of a fair, six-sided die, we get exactly five results of 6?*

SOLUTION   This is a direct application of Theorem 1, with $n = 10$, $k = 5$, and $p = 1/6$, but we will go through this problem to be sure that the above proof is clear.

We will represent the outcomes as ordered lists of length 10 using the entries S and F, where S denotes successfully rolling a 6 and F denotes the failure to roll a 6. Getting exactly five results of 6 corresponds to those lists that use exactly five S's and five F's. Since these are effectively binary sequences, the number of these is $C(10, 5) = 252$, and the probability of each one happening can be found using the product rule to be

$$Prob(S) \cdot Prob(S) \cdot Prob(S) \cdot Prob(S) \cdot Prob(S) \cdot$$
$$Prob(F) \cdot Prob(F) \cdot Prob(F) \cdot Prob(F) \cdot Prob(F)$$
$$= \left(\frac{1}{6}\right)^5 \left(\frac{5}{6}\right)^5$$

---

* Jacob Bernoulli (1654–1705) was a pioneer in the study of probability.

Therefore, the probability that one of these occurs is

$$C(10, 5)\left(\frac{1}{6}\right)^5\left(\frac{5}{6}\right)^5 \approx 0.013$$

Explore more on the Web.

Practice Problem 1 *Suppose we have a fake coin that comes up heads with probability $\frac{4}{7}$. What is the probability that in 10 successive tosses of this fake coin, we get at least eight results of heads?*

Many sporting events are based on a series of games that are played until one team wins a specified number. For example, in many playoffs, teams play a "best-of-seven" series, meaning that they play games until one team has won four games (the majority of the seven possible games). This is somewhat confusing since the series stops immediately once one team has won four games, so a best-of-seven series might consist of a total of four games, five games, six games, or seven games. Analyzing these series is related to Theorem 1, but we can already see there will be some significant differences. One way to see the different issues is to look at a simple version of a series from the point of view of a tree diagram.

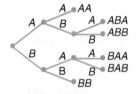

Figure 6-2
Best-of-three series.

Example 2 *If team A wins every game it plays with probability $\frac{3}{4}$, then what is the probability that team A wins a best-of-three series over team B?*

SOLUTION We can visualize all possible series using the tree structure in Figure 6-2. Here we denote a series as an ordered list with entries from $\{A, B\}$, where A denotes a game that team A wins and B denotes a game that team B wins. Note that the number of games in the series is either two or three, so we will have to apply Theorem 1 with care. The three series in which team A is the winner are AA, ABA, and BAA. These series are disjoint events, so we can simply add the probabilities of each:

$$Prob(A \text{ wins the series}) = Prob(AA) + Prob(ABA) + Prob(BAA)$$
$$= \left(\frac{3}{4}\right)^2 + \left(\frac{3}{4}\right)^2 \cdot \left(\frac{1}{4}\right) + \left(\frac{3}{4}\right)^2 \cdot \left(\frac{1}{4}\right)$$
$$= \frac{27}{32}$$

Example 3 *Suppose the local Barons team wins every game they play with a probability of 0.75. What is the probability that they would win a best-of-seven playoff series (i.e., they play games until one team has won four games) by a margin of four games to two?*

SOLUTION The series could be represented as a list of B's and O's (for Barons and Opponents) that uses two O's and four B's and ends with a B. For example, the list BOBBOB represents a series in which the Barons win games 1, 3, 4, and 6. Since these series must end with a B, the only choices to be made involve how to distribute the two O's and three B's among the first five games; there are $C(5, 2) = 10$ of these altogether. By the product rule, the series BOBBOB above would happen with probability

$$\left(\frac{3}{4}\right) \cdot \left(\frac{1}{4}\right) \cdot \left(\frac{3}{4}\right) \cdot \left(\frac{3}{4}\right) \cdot \left(\frac{1}{4}\right) \cdot \left(\frac{3}{4}\right) = \left(\frac{3}{4}\right)^4 \left(\frac{1}{4}\right)^2$$

Since any of the other nine possible 4-2 (Barons) series have the same probability of occurring, the final answer (by the sum rule) is the above probability added to itself 10 times. That is, the probability of a 4-2 Barons win is

$$10 \cdot \left(\frac{3}{4}\right)^4 \cdot \left(\frac{1}{4}\right)^2 \approx 0.20$$

$\square$

Practice Problem 2 *Using the same assumptions as in the previous example, what is the probability that the Barons win a best-of-seven series by a margin of four games to three?*

Explore more on the Web.

Example 4 *After game 1 of the World Series (a best-of-seven series) in 1992, the announcers took great pleasure in announcing that over the previous 20 years, it had happened 12 times that the team which won the first game went on to win the series. They seemed to be suggesting that 60% is surprisingly high. Is it?*

SOLUTION Let the teams be called A and B, so that we can denote a World Series as we did for the series in Example 3. We will assume for the moment that the probability of A winning any particular game is $\frac{1}{2}$ and that the games are independent events. Our sample space will be the set of all World Series in which team A wins the first game. The problem will be to determine the probability that something in our space results in A winning the series. Note, however, that each outcome in the sample space is not equally likely since we do not have a fixed series length among items in the sample space.

Therefore, we will break the problem into cases based on the total number of games in a series, and then we will apply the rule of sums. Assume team A wins the first game. Then

- In order to have a four-game series, team A must win the next three games. There is a $(\frac{1}{2})^3 = \frac{1}{8}$ probability that this happens.
- In order to have a five-game series, it must look like $A\_\_\_A$, where the blanks are filled by any ordered list of two A's and one B. There are $C(3, 2) = 3$ of these lists, and the probability of each of them happening is $(\frac{1}{2})^4 = \frac{1}{16}$. Thus, the probability is $\frac{3}{16}$.
- In order to have a six-game series, it must look like $A\_\_\_\_A$, where the blanks are filled by any ordered list of two A's and two B's, of which there are $C(4, 2) = 6$. The probability of each is $\frac{1}{32}$, so the probability for this case is $\frac{6}{32}$.
- In order to have a seven-game series, it must look like $A\_\_\_\_\_A$, where the blanks are filled by any ordered list of two A's and three B's, of which there are $C(5, 2) = 10$. The probability of each is $\frac{1}{64}$, so the probability for this case is $10/64$.

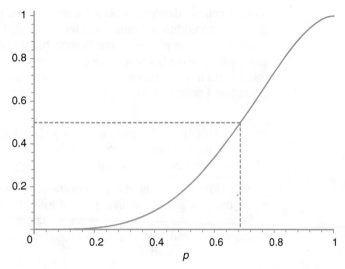

Figure 6-3   Probability of series win versus probability of game win.

By the sum rule, the probability that team A wins the series after winning the first game is $\frac{1}{8} + \frac{3}{16} + \frac{6}{32} + \frac{10}{64} = \frac{21}{32} \approx 65.6\%$.

So in fact, 12 out of the last 20 is slightly lower than one would expect, with each team having a $\frac{1}{2}$ probability of winning each game.   □

Example 5   *Very few teams in sports history have come back to win a best-of-seven series after being down two games to none. What would the probability p of team A winning any given game over team B have to be so that team A will have at least a 50% chance of coming back from a 2-0 deficit to claim the series?*

SOLUTION   If A starts off down two games to none, A can only come back and win the series by winning four of the next five games, making the remainder of the series look like one of the following:

- A wins the next four games in a row with probability p^4.
- The remainder of the series looks like _ _ _ _ _A, where the four blanks consist of three A's and one B in any order, with probability $4 \cdot (1 - p) \cdot p^4$.

Hence, the probability that team A comes back to win the series is

$$4(1 - p)p^4 + p^4 = p^4 (5 - 4p)$$

The graph of this expression as a function of p in Figure 6-3 shows us that the probability team A wins after losing the first two games exceeds 0.5 when p is at least 0.69.   □

Tennis is an example of a game that does not have a finite set of outcomes. Any number of points can make up a tennis game, since to win the game a player must "win by two." The scoring in tennis is archaic. Individual point totals take on values of 0, 15, 30, 40, or "game" in which the convention is that once tied at 40 (such a tie score is called "deuce"), a player must win two points in a row to win the game, but there are no additional numerical scores used. These seem, like strange conventions if you have never played tennis before, but you can always think of tennis as being a game where you play to a score of four but you have to win by two. For this reason, there is no fixed number of points in a tennis game. It is possible to have a tennis game last 4 points or 40.

Example 6 *Suppose a tennis player has a 3/5 probability of winning each point she plays. What is the probability that she wins a game by a score of "game"-30. (This would be a score of 4-2 if it was any other kind of game.)*

SOLUTION   Using A for our preferred player and B for her opponent, a game of this type must look like a list of four A's and two B's, where the rightmost letter (i.e., the final point) is an A. There are only $C(5, 2) = 10$ such binary sequences, and each one has a probability of

$$\left(\frac{3}{5}\right)^4 \left(\frac{2}{5}\right)^2$$

of occurring, so the probability that a game ends with this score is

$$10 \left(\frac{3}{5}\right)^4 \left(\frac{2}{5}\right)^2$$

□

The more interesting problem in tennis is how to analyze games that at some point are tied at deuce. We will see how to deal with this in Section 6.5.

Solutions to Practice Problems

1   There are three disjoint cases to consider: Either we get exactly eight heads, we get exactly nine heads, or we get all 10 heads. According to Theorem 1, the probabilities for these cases are $C(10, 8) \left(\frac{4}{7}\right)^8 \left(\frac{3}{7}\right)^2$, $C(10, 9) \left(\frac{4}{7}\right)^9 \left(\frac{3}{7}\right)^1$, and $C(10, 10) \left(\frac{4}{7}\right)^{10} \left(\frac{3}{7}\right)^0$, respectively. Hence, the probability that one of these cases occurs is the sum

$$C(10, 8) \left(\frac{4}{7}\right)^8 \left(\frac{3}{7}\right)^2 + C(10, 9) \left(\frac{4}{7}\right)^9 \left(\frac{3}{7}\right)^1$$
$$+ C(10, 10) \left(\frac{4}{7}\right)^{10} \left(\frac{3}{7}\right)^0 \approx 0.1255$$

2   As in Example 3, such a series (i.e., a 4-3 Barons win) must end with a B, so it must look like a sequence of

three O's and three B's (in some order) followed by a B. For example, BOBBOOB is the sequence BOBBOO followed by the letter B. There are $C(6, 3) = 20$ such sequences, and each represents a series that has probability

$$\left(\frac{3}{4}\right)^4 \cdot \left(\frac{1}{4}\right)^3$$

Therefore, the probability that the Barons will win 4-3 is

$$20 \cdot \left(\frac{3}{4}\right)^4 \cdot \left(\frac{1}{4}\right)^3 \approx 0.0989$$

Exercises for Section 6.3

1. What is the probability that in seven rolls of a six-sided die, the result of 1 appears exactly five times?

2. What is the probability that in seven rolls of a six-sided die, the result of 1 appears *at least* five times?

3. What is the probability of getting exactly 3 heads on 10 tosses of a fair coin?

4. What is the probability that in 10 tosses of a fair coin, the result of heads appears at least eight times? Compare this to your answer in Practice Problem 1.

5. What is the probability of getting exactly one 6 on 10 tosses of a fair six-sided die?

6. What is the probability of getting at least three heads on five tosses of a fair coin?

7. What is the probability of getting at least two 3's on five tosses of a fair six-sided die?

8. What is the probability of getting more heads than tails on four tosses of a fair coin?

9. What is the probability of getting more 6's than 5's on four tosses of a fair six-sided die?

10. What is the probability when tossing a pair of dice of getting either a 5 on one die, a sum of 5, or a sum of 8? (This is the kind of problem that arises in backgammon.)

11. What is the probability that in five dealt cards, exactly two cards will have the same value? (This is called "a pair" in poker.)

12. In military clubs, a dice game is played with rules similar to those of draw poker. A player rolls five dice and these five values comprise his or her hand. The player then has the option to re-roll any number of the dice to improve the result. Suppose I roll a pair of 3's and re-roll the other three dice.

    (a) What is the probability that my new hand contains three of a kind?

    (b) What is the probability that my new hand contains two pair?

    (c) What is the probability that my new hand contains three of one value and two of some different value? (This hand is called a full house in draw poker.)

13. In the dice game described in Exercise 12, suppose I leave two dice with values 3 and 4 on the table and re-roll the other three dice. Answer questions (a–c) in Exercise 12 for this new situation.

14. The Bears and the Cougars play a best-of-three series in which the probability of the Bears winning each individual game is constant. The Bears have a 0.9 probability of winning the series. What is the probability of the Bears winning each individual game?

15. Which is larger, the probability that a fair coin will come up heads exactly three times in six tosses, or the probability that a fair (six-sided) die will come up as a 5 or 6 exactly twice in four tosses?

16. For each value of k from 0 to 8, find the probability of getting exactly k heads on eight tosses of a fair coin. Create a bar graph with the values 0 through 8 along the horizontal axis, with the height of each bar determined by the answers to the probability questions. This graph illustrates what is called the *binomial probability distribution*.

17. As a group, repeat the experiment "Toss a fair coin eight times and record the number of heads" a total of 100 times. Create a bar graph with the values 0 through 8 along the horizontal axis, with the height of the bar labeled k determined by the proportion of all outcomes of the experiment that had exactly k heads. Compare this bar graph to the one drawn in the previous exercise.

18. Christopher's favorite numbers are 1 and 4, Jessica likes 2, 3, and 5, and John's favorite number is 6. What is the probability that on nine rolls of a fair die, each child sees his or her favorite numbers three times?

19. Assume that Kenny always gets a hit with probability 1/3 and a base-on-balls with probability 1/6.

    (a) Determine the probability that in five plate appearances, Kenny gets at least two hits.

    (b) Determine the probability that in five plate appearances, Kenny gets a base-on-balls at least twice.

    (c) Determine the probability that in five plate appearances, Kenny gets on base with a hit more times than he gets on with a base-on-balls.

20. We assumed in Example 3 that the Barons had a probability of $\frac{3}{4}$ of winning every game they played. What is the probability that the Barons would win a best-of-seven-game series? (Use the results from Example 3 and Practice Problem 2.)

21. John wins each set of tennis he plays with probability $\frac{5}{9}$. Read the following two questions, and make an intuitive guess about which answer will be bigger. Then answer both questions to see if you are right.

    (a) If he is behind one set to none, what it is the probability that he can come back to win a best-of-five-set tennis match?

    (b) If he is behind two sets to one, what is the probability that he can come back to win a best-of-five-set tennis match?

22. Suppose that team A has a 60% chance of winning any given game against team B.

    (a) What is the probability that team A wins a best-of-three series against team B?

    (b) What is the probability that team A wins a best-of-five series against team B?

    (c) What is the probability that team A wins a best-of-seven series against team B?

23. **Home Field Advantage.** Many best-of-seven series are scheduled so that the first two games are played at the home field of team A, the next three are played at the home field of team B, and the last two are played back at the home field of team A. (Of course, if the series ends early, some games will not be played.) Assume that both teams win on their own home field with probability $\frac{3}{5}$. What is the probability that team A (the first home team) wins the series?

6.4 Expected Value in Games of Chance

Gambling games are based on chance occurrences like getting cards of all the same suit or rolling a sum of 8 on two dice three times in a row. We do not think of a game like baseball or tennis as being *based* on chance occurrences, but we often make implicit assumptions along these lines. For example, we think that a .320 hitter is more likely to get a hit today than a .220 hitter. In Section 6.3 we made assumptions about the probability that one team beats another, even though the contest is surely not random. In this section we will analyze some other aspects of games related to the explicit or implicit probabilities involved in them.

We have already seen examples of how to compute probabilities for "large" events given information about "smaller" events. For example, we analyzed the probability of winning a best-of-seven series in terms of the probability of winning an individual game. Another aspect of larger events that is also of interest is the average number of the smaller events that make up the larger one. For example, what is the average number of games that a best-of-seven series lasts? How many points are needed to resolve a tennis game that is tied at deuce? To answer these kinds of questions, we will need to investigate the mathematical idea of average values.

The Definition of Average Value

Example 1 *Suppose that among 100 households there are 30 with no children, 25 with one child, 30 with two children, 10 with three children, and 5 with four children. What is the average number of children per household?*

SOLUTION   There are a total of $(0)(30) + (1)(25) + (2)(30) + (3)(10) + (4)(5) = 135$ children among the 100 households, for an average of 1.35 children per household.

□

We can arrive at the same answer by combining the idea of a "weighted average" with our knowledge of probability. The number of children in each household is a number from $\{0, 1, 2, 3, 4\}$, and the probability that each of these numbers occurs is simply the proportion of households with that number of children. That is, if we think of our probability experiment as choosing a household, we will have the following:

- *Prob*(there are 0 children) $= \frac{30}{100}$
- *Prob*(there is 1 child) $= \frac{25}{100}$
- *Prob*(there are 2 children) $= \frac{30}{100}$
- *Prob*(there are 3 children) $= \frac{10}{100}$
- *Prob*(there are 4 children) $= \frac{5}{100}$

A more efficient way to represent this is through the notion of a *random variable*. This is simply a measurement of some numerical value associated with each possible outcome of a probability experiment. In this case, the random variable X is the number of children in the household chosen by the experiment. This allows us to write $Prob(X = 0)$, for example, instead of $Prob$(there are 0 children) as we did above.

Definition   For a given probability experiment, let X be a random variable whose possible values come from the set of numbers $\{x_1, \ldots, x_n\}$. Then the *expected value of X*, denoted by $E[X]$, is the sum

$$(x_1) \cdot Prob(X = x_1) + (x_2) \cdot Prob(X = x_2) + \cdots + (x_n) \cdot Prob(X = x_n)$$

This is sometimes called the *average value* of the random variable, thinking of the average of the values X takes on over many repetitions of the experiment.

We can use this definition to find an alternative solution to Example 1:

$$E[X] = (0) \cdot Prob(X = 0) + (1) \cdot Prob(X = 1) + (2) \cdot Prob(X = 2)$$
$$+ (3) \cdot Prob(X = 3) + (4) \cdot Prob(X = 4)$$
$$= (0) \cdot \frac{30}{100} + (1) \cdot \frac{25}{100} + (2) \cdot \frac{30}{100} + (3) \cdot \frac{10}{100} + (4) \cdot \frac{5}{100}$$
$$= \frac{(0)(30) + (1)(25) + (2)(30) + (3)(10) + (4)(5)}{100}$$
$$= \frac{25 + 60 + 30 + 20}{100} = 1.35$$

Practice Problem 1   *Let X represent the number of siblings of a person in your class. Collect all the necessary data about X, and use the definition to compute $E[X]$.*

Often, the term *average value* is more appropriate than *expected value* for our intuition about the meaning of $E[X]$. Consider the following example.

Example 2 *On the loaded die from Example 6 in Section 6.2, what is the expected value on one roll? Compare your answer to the expected value on one roll of a fair die.*

SOLUTION   For that die, the probability of a 6 is $\frac{1}{2}$, with a probability of $\frac{1}{10}$ for each of the other faces. Let X represent the outcome of one roll—so for either die X has possible values from $\{1, 2, 3, 4, 5, 6\}$. For the loaded die, we can compute

$$E[X] = (1)\left(\frac{1}{10}\right) + (2)\left(\frac{1}{10}\right) + (3)\left(\frac{1}{10}\right) + (4)\left(\frac{1}{10}\right) + (5)\left(\frac{1}{10}\right) + (6)\left(\frac{1}{2}\right)$$
$$= \frac{1}{10}(15) + \frac{1}{2}(6) = 4.5$$

and similarly, for the fair die,

$$E[X] = (1)\left(\frac{1}{6}\right) + (2)\left(\frac{1}{6}\right) + (3)\left(\frac{1}{6}\right) + (4)\left(\frac{1}{6}\right) + (5)\left(\frac{1}{6}\right) + (6)\left(\frac{1}{6}\right)$$
$$= \frac{1}{6}(21) = 3.5$$

□

Value of X	1	2	3	4	5	6
Number of rolls (approx.)	5,000	5,00	5,000	5,000	5,000	5,000

Table 6-3   Approximate Distribution of 30,000 Die Rolls

In each case, we claim that the expected value on one roll of a die is not a whole number. Of course, we never really expect to roll a 3.5 on a die, so the word "expectation" is a poor choice in this context. The quantity 3.5 is actually our prediction for the *average* of all the values of X that we will see in many, many trials of the experiment.

In this case, imagine rolling a fair die 30,000 times. Because the die is fair, we expect the distributions of rolls to roughly break down as shown in Table 6-3. So if we find the average of all 30,000 recorded values of X, we will have

$$\frac{\text{Sum of } X \text{ values}}{\text{Number of trials}}$$

$$\approx \frac{(1)(5,000) + (2)(5,000) + (3)(5,000) + (4)(5,000) + (5)(5,000) + (6)(5,000)}{30,000}$$

$$= (1)\frac{5,000}{30,000} + (2)\frac{5,000}{30,000} + (3)\frac{5,000}{30,000} + (4)\frac{5,000}{30,000} + (5)\frac{5,000}{30,000} + (6)\frac{5,000}{30,000}$$

$$= (1)\frac{1}{6} + (2)\frac{1}{6} + (3)\frac{1}{6} + (4)\frac{1}{6} + (5)\frac{1}{6} + (6)\frac{1}{6} = 3.5$$

This interpretation of expected value as "long-term average" is particularly relevant in games that are to be played many, many times such as gambling and lottery games.

Example 3   *Suppose you pay $2 each time to play the following game: Two dice are rolled, and you win $5 for each 6 that comes up. Do you expect to win more than you pay if you play many, many times?*

SOLUTION   Let X represent the amount of money you win in one play of the game. Hence, X can take values from $\{0, 5, 10\}$, and

- $Prob(X = 0) = Prob(\text{getting no 6's}) = \frac{5}{6} \cdot \frac{5}{6} = \frac{25}{36}$
- $Prob(X = 5) = Prob(\text{getting one 6}) = C(2, 1)\frac{1}{6} \cdot \frac{5}{6} = \frac{10}{36}$
- $Prob(X = 10) = Prob(\text{getting two 6's}) = \frac{1}{6} \cdot \frac{1}{6} = \frac{1}{36}$

Hence,

$$E[X] = 0 \cdot Prob(X = 0) + 5 \cdot Prob(X = 5) + 10 \cdot Prob(X = 10)$$

$$= 0 \cdot \frac{25}{36} + 5 \cdot \frac{10}{36} + 10 \cdot \frac{1}{36}$$

$$= \frac{60}{36} \approx \$1.67$$

Since this is less than the $2 it costs to play the game each time, we can see that you will actually *lose* an average of about 33 cents for each time you play over the long term.

□

Practice Problem 2   *Suppose in Example 3 the payoff is $5 for rolling one 6, and $25 for rolling two 6's. Now is it worth $2 to play?*

 Example 4   *In Example 5 of Section 6.1, the following card game is described:*

We have a packet of three cards in which one is blue on both sides, one is red on both sides, and one is red on one side and blue on the other. One of three cards is chosen and placed on the table in full view of both of us. We now wager on the color of the face-down side of the card. You will give me $5 if the face-down side is the same color as the (visible) face-up side, and I will give you $6 otherwise.
    What is the expected value of my winnings each time the game is played?

SOLUTION   Let X represent the number of dollars I win in each play of the game. Then X can only take one of the two values from $\{-6, +5\}$, and from our solution to Example 5, we know that

- $Prob(X = -6) = Prob(\text{the red/blue card is chosen}) = \frac{1}{3}$, and
- $Prob(X = 5) = Prob(\text{the red/blue card is not chosen}) = \frac{2}{3}$

So $E[X] = (-6) \cdot \frac{1}{3} + (5) \cdot \frac{2}{3} = \frac{4}{3} \approx \$1.33$. This means that, on average, I will win $1.33 for each time we play this game, so in the long term, it is a bad idea for you to play.

□

Expectation in Bernoulli Trials

A fundamental idea for relating small trials to larger experiments is the structure of Bernoulli trials developed in Section 6.3. The concept is that an experiment is made up of a sequence of independent trials whose individual probabilities are well understood. In this case, there is an easy general statement we can make about the expected number of successful trials in an experiment.

Theorem 1   *Suppose an experiment consists of the independent repetition of a trial n times, and the probability of that trial's individual success is p each time it is performed. If X denotes the number of successful trials in this experiment, then $E[X] = n \cdot p$.*

This theorem follows easily from the basic rule for expected value:

$$E[X + Y] = E[X] + E[Y]$$

but we will prove it by induction instead. This not only allows us to see an induction proof in a probability context, but it also illustrates a conceptual framework that is essential in the next section.

PROOF   (By induction on n.) Let $P(n)$ denote the statement "The average number of successes in n Bernoulli trials with probability p is $n \cdot p$." Statement $P(1)$ is "The average number of successes in one Bernoulli trial with probability p is p." Letting X denote the number of successes in one trial, we know that X takes on values from $\{0, 1\}$, and so by the definition of expected value

$$E[X] = (0) \cdot Prob(X = 0) + (1) \cdot Prob(X = 1) = p$$

Now let $m \geq 1$ be given such that statements $P(1)$, $P(2)$, ..., $P(m-1)$ have all been checked to be true and we are ready to consider the next statement $P(m)$. Let X denote the number of successes in an experiment consisting of m Bernoulli trials. Just as we did in Section 6.3, we can represent outcomes of this experiment as ordered lists of length m whose elements come from $\{S, F\}$, where we think of S and F as representing "success" and "failure."

Every outcome of the experiment looks like one of the following, where each blank is to be filled by any ordered list of length $m - 1$ with entries from $\{S, F\}$.

Case 1. S_____. In this case, the total number of successes is one more than the number of S's in the blank.

Case 2. F_____. In this case, the total number of successes is exactly the number of S's in the blank.

By the definition of expected value,

$$
\begin{aligned}
E[X] &= (\text{Number of successes in case 1}) \cdot (\text{Probability of case 1 happening}) \\
&\quad + (\text{Number of successes in case 2}) \cdot (\text{Probability of case 2 happening}) \\
&= (1 + (m-1) \cdot p) \cdot (p) + ((m-1) \cdot p) \cdot (1-p) \\
&= (p + m \cdot p^2 - p^2) + (m \cdot p - p - m \cdot p^2 + p^2) \\
&= m \cdot p
\end{aligned}
$$

That is, $E[X] = m \cdot p$, completing the proof of statement $P(m)$. ■

Example 5 *A softball player gets a hit with probability $\frac{3}{5}$ each time she steps to the plate. In any given game, this player has exactly five plate appearances. Use the definition of expected value to show that the expected number of hits per game is 3. Compare this with the prediction made by Theorem 1.*

SOLUTION   Let X denote the number of hits in a game. Then X takes on values from the set $\{0, 1, 2, 3, 4, 5\}$. By the definition of expected value,

$$
\begin{aligned}
E[X] &= (0)Prob(X = 0) + (1)Prob(X = 1) + \cdots + (5)Prob(X = 5) \\
&= (0)C(5,0)\left(\frac{2}{5}\right)^5 + (1)C(5,1)\left(\frac{3}{5}\right)^1\left(\frac{2}{5}\right)^4 + (2)C(5,2)\left(\frac{3}{5}\right)^2\left(\frac{2}{5}\right)^3 \\
&\quad + (3)C(5,3)\left(\frac{3}{5}\right)^3\left(\frac{2}{5}\right)^2 + (4)C(5,4)\left(\frac{3}{5}\right)^4\left(\frac{2}{5}\right)^1 + (5)C(5,5)\left(\frac{3}{5}\right)^5 \\
&= 3
\end{aligned}
$$

This is the same as the value $(5) \cdot \left(\frac{3}{5}\right)$ predicted by the theorem. □

Practice Problem 3   *Use the definition of expected value to show that the average number of results of heads in an experiment consisting of tossing a coin three times is 1.5.*

Average Values in Sports

We conclude this section by solving some problems that come up in sports when two opponents play a best-of-five or a best-of-seven series. We once again model a series as a sequence of independent games. Since the total number of games in a series is not fixed, two issues arise. First, we cannot directly use the result in Theorem 1. Second, the average number of games in a series becomes an interesting quantity to study.

Example 6 *If two teams A and B play a best-of-five series,* and if team A has a $\frac{1}{4}$ probability of winning any given game, then what is the average number of games in the series?*

SOLUTION   The variable X representing the number of games in a series has values from the set $\{3, 4, 5\}$. In order to compute $E[X]$, we need to know $Prob(X = 3)$, $Prob(X = 4)$, and $Prob(X = 5)$.

As in the previous section, we will represent a series as an ordered list of A's and B's in which the winner appears three times, the loser appears two or fewer times, and the winner appears last. For example, $AABA$ denotes a series in which A wins games $1, 2$, and 4 and B wins game 3.

- For X to be 3, we must have either AAA or BBB, so

$$Prob(X = 3) = \left(\frac{1}{4}\right)^3 + \left(\frac{3}{4}\right)^3$$
$$= 0.4375$$

- For X to be 4, we must have either $\_\_\_A$ or $\_\_\_B$, where in the first case the blanks are filled by two A's and one B ($C(3, 1) =$ three ways) and in the second case the blanks are filled by two B's and one A (three ways). Each of the three series in the first case occurs with probability $\left(\frac{1}{4}\right)^3 \cdot \left(\frac{3}{4}\right)$, and each of the three in the second case occurs with probability $\left(\frac{1}{4}\right) \cdot \left(\frac{3}{4}\right)^3$. Therefore,

$$Prob(X = 4) = 3 \cdot \left(\frac{1}{4}\right)^3 \cdot \left(\frac{3}{4}\right) + 3 \cdot \left(\frac{1}{4}\right) \cdot \left(\frac{3}{4}\right)^3$$
$$= 0.3515625$$

- Since $X = 5$ is the only other possible case, we know that $Prob(X = 3) + Prob(X = 4) + Prob(X = 5) = 1$, so

$$Prob(X = 5) = 1 - (Prob(X = 3) + Prob(X = 4))$$
$$= 1 - 0.4375 - 0.3515625 = 0.2109375$$

Therefore, the average number of games is

$$E[X] = 3 \cdot Prob(X = 3) + 4 \cdot Prob(X = 4) + 5 \cdot Prob(X = 5)$$
$$= 3 \cdot (0.4375) + 4 \cdot (0.3515625) + 5 \cdot (0.2109375)$$
$$\approx 3.77 \text{ games}$$

---

* Recall that this means they play games until one team has won three games.

Explore more on the Web.

Practice Problem 4 *If two teams A and B play a best-of-three series, and if team A has a $\frac{2}{3}$ probability of winning any given game, then what is the average number of games in the series?*

Since this analysis is similar for any problem involving series, it is very tempting to answer a more general question once so that we can think about specific questions on a higher level.

Example 7 *If two teams A and B play a best-of-seven series, then assuming that A wins a game with probability p (so B wins with probability $1 - p$) and that the games are independent events, what is the average number of games needed to settle the series in terms of the variable p? For what value of p will the series take the longest?*

SOLUTION The variable X representing the number of games a series takes has values from among $\{4, 5, 6, 7\}$. In order to compute $E[X]$, we need to know $Prob(X = n)$ for each of $n \in \{4, 5, 6, 7\}$.

Denote a series as an ordered list of A's and B's in which the winner appears four times, the loser appears three or fewer times, and the winner appears last. For example, $AABBAA$ denotes a series in which A wins games 1, 2, 5, and 6 and B wins games 3 and 4. Note that $ABAAAB$, for example, does not denote a possible series since the winner, A, does not appear last.

- For X to be 4, we must have either $AAAA$ or $BBBB$, so

$$Prob(X = 4) = p^4 + (1 - p)^4$$

- For X to be 5, we must have either $\_\_\_\_A$ or $\_\_\_\_B$, where in the first case the blanks are filled by three A's and one B (four ways) and in the second case the blanks are filled by three B's and one A (four ways). Each of the four series in the first case occurs with probability $p^4(1 - p)$, and each of the four in the second case occurs with probability $p(1 - p)^4$. Therefore,

$$Prob(X = 5) = 4p^4(1 - p) + 4p(1 - p)^4$$

- Similarly for X to be 6, we must have either $\_\_\_\_\_A$ or $\_\_\_\_\_B$, where the first case has three A's and two B's in the blanks ($C(5, 3) = 10$ ways) and the second case has three B's and two A's in the blanks (also 10 ways). Each series in the first case occurs with probability $p^4(1 - p)^2$, while each series in the second case occurs with probability $p^2(1 - p)^4$, so

$$Prob(X = 6) = 10p^4(1 - p)^2 + 10p^2(1 - p)^4$$

- Using the same analysis for a seven-game series, one can determine that

$$Prob(X = 7) = 20p^4(1 - p)^3 + 20p^3(1 - p)^4$$

Therefore,

$$
\begin{aligned}
E[X] &= 4[p^4 + (1 - p)^4] + 20[p^4(1 - p) + p(1 - p)^4] \\
&\quad + 60[p^4(1 - p)^2 + p^2(1 - p)^4] + 140[p^4(1 - p)^3 + p^3(1 - p)^4] \\
&= 4 + 4p + 4p^2 + 4p^3 - 52p^4 + 60p^5 - 20p^6
\end{aligned}
$$

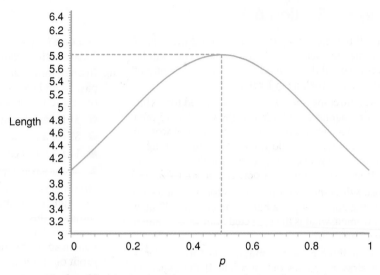

Figure 6-4 Series length versus probability of game win.

In the case where $p = \frac{1}{2}$ (when we would expect that $E[X]$ should be at its largest), this means that $E[X] = 93/16 = 5.8125$. The graph that shows the expected value for all values of p is given in Figure 6-4.

Solutions to Practice Problems

1   Compare your answer with that of someone else in your class.

2   We make a simple change in Example 3:

$$E[X] = 0 \cdot Prob(X = 0) + 5 \cdot Prob(X = 5)$$
$$+ 25 \cdot Prob(X = 25)$$
$$= 0 \cdot \frac{25}{36} + 5 \cdot \frac{10}{36} + 25 \cdot \frac{1}{36}$$
$$= \frac{75}{36} \approx \$2.08$$

Since this is more than the \$2 it costs to play the game each time, you will win an average of about 8 cents for each time you play over the long term, so this is a game you should play, although no one can blame you for not getting too excited about it.

3   Let X denote the number of heads tossed. Since the coin is tossed three times, X takes on values from the set $\{0, 1, 2, 3\}$. By the definition of expected value,

$$E[X] = (0) \cdot Prob(X = 0) + (1) \cdot Prob(X = 1)$$
$$+ (2) \cdot Prob(X = 2) + (3) \cdot Prob(X = 3)$$
$$= (0) \cdot C(3, 0) \left(\frac{1}{2}\right)^3 + (1) \cdot C(3, 1) \left(\frac{1}{2}\right)^3$$

$$+ (2) \cdot C(3, 2) \left(\frac{1}{2}\right)^3 + (3) \cdot C(3, 3) \left(\frac{1}{2}\right)^3$$
$$= (0 + 3 + 6 + 3) \cdot \left(\frac{1}{8}\right) = 1.5$$

4   The variable X representing the number of games in a series has values 2 or 3.

●   For X to be 2, we must have either AA or BB, so

$$Prob(X = 2) = \left(\frac{2}{3}\right)^2 + \left(\frac{1}{3}\right)^2$$
$$= \frac{5}{9}$$

●   Since $X = 3$ is the only other case, we know that

$$Prob(X = 3) = 1 - Prob(X = 2) = \frac{4}{9}$$

Therefore, the expected number of games is

$$E[X] = 2 \cdot Prob(X = 2) + 3 \cdot Prob(X = 3)$$
$$= 2 \cdot (5/9) + 3 \cdot (4/9)$$
$$= \frac{22}{9} \approx 2.44 \text{ games}$$

Exercises for Section 6.4

1. If you roll two dice many times and record the sum of the two dice in each case, what do you expect will be the average of all these sums? What if you use a pair of loaded dice like the one in Example 2?

2. Suppose a three-person committee is formed for a club by drawing names out of a hat. If the club has 15 men and 10 women, what is the expected number of women on the committee? How do you explain the meaning of this to someone who cringes at the thought of "expecting" a fractional number of women?

3. Suppose a three-person committee is formed for a club by drawing names out of a hat. If the club has 30 men and 20 women, what is the expected number of women on the committee?

4. Suppose a three-person committee is formed for a club by drawing names out of a hat. If the expected number of women on the committee is 2.0 and there are 10 men in the club, how many women are in the club?

5. Suppose five cards are drawn from a standard deck. What is the expected number of aces among the five cards?

6. If two cards are drawn from a standard deck of cards, what is the expected sum of the values on the cards? (Use 10 as the value of a jack, queen, or king, 11 as the value of an ace, and the numerical value for all other cards.)

7. In any given course, Amy has a $\frac{1}{3}$ probability of getting an A, a $\frac{1}{4}$ probability of getting a B, a $\frac{1}{5}$ probability of getting a C, a $\frac{1}{6}$ probability of getting a D, and a $\frac{1}{20}$ probability of getting an F. Amy's Dad is a firm believer in carrots over sticks, so he agrees to pay her $10 for each A, $5 for each B, and $1 for each C she receives. How much does Amy expect to get each semester that she takes five courses?

8. In Exercise 19 of Section 6.3, we assumed Kenny has a $\frac{1}{3}$ probability of getting a hit and a $\frac{1}{6}$ probability of drawing a base-on-balls every time he appears at the plate. Suppose he negotiates in his contract for an end-of-the-season bonus of $1,000 for each hit and $100 for each base-on-balls he gets. If he can reasonably expect 600 plate appearances this year, how much does he expect his bonus will be?

9. Here is a real gambling game that we can analyze for fairness. A player pays $10 for the privilege to play. The player tosses a die three times with the following payoff: $10 for the first die that shows a 6, (an additional) $15 for the second die that shows a 6, and (an additional) $20 for the third die to show a 6. What is the expected amount of money the player will win playing this game

one time? How much can the casino expect to make if 10,000 of these games are played each day?

10. In a new lottery game that has been proposed, the player pays $2 and receives five fair dice. These dice are rolled once, and the player receives
- $1 for a roll containing two of the same value;
- $2 for a roll containing three of the same value;
- $4 for a roll containing four-of-a-kind; and
- $8 for a roll containing five-of-a-kind.

How much can the player expect to win on average for each time this game is played? If 100,000 people in the state play this game each day, how much money will the state make in one year from this game? How much can you make the payoff for five-of-a-kind and still make a profit on this game?

11. Use the identity in Exercise 36 of Section 5.3 and the definition of expected value to find the expected number of heads if six fair coins are tossed.

12. Use the identity in Exercise 36 of Section 5.3 and the definition of expected value to find the expected number of heads if N fair coins are tossed. Check your answer with the prediction in Theorem 1.

13. Use the identity in Exercise 37 of Section 5.3 and the definition of expected value to find the expected number of times a 1 is seen in 10 rolls of a fair die.

14. Use the identity of Exercise 37 of Section 5.3 and the definition of expected value to find the expected number of times a 1 is seen in N rolls of a fair die. Check your answer with the prediction in Theorem 1.

15. What is the expected number of boys in a two-children household in which at least one child is known to be a boy? Compare this answer to the expected number of boys in a two-children household in which the *older* child is known to be a boy. Why are these not the same?

16. If the Bisons have a $\frac{3}{5}$ probability of winning any given game against the Mustangs, what will be the average length of a best-of-five series between these two teams?

17. If the Sharks have a $\frac{2}{3}$ probability of winning any given game against the Minnows, what will be the average length of a best-of-five series between these two teams?

18. Repeat Example 7 for a best-of-three series.

19. Repeat Example 7 for a best-of-five series.

20. If two teams A and B play a best-of-seven series, and if team A has a $\frac{3}{5}$ probability of winning any given game, then what is the expected number of games in the series?

21. In a best-of-seven series between two evenly matched teams (i.e., each team has a $\frac{1}{2}$ probability of winning any given game), what is the average margin of victory of

the series winner over the series loser? For example, a team winning a series four games to one has a margin of victory of $4 - 1 = 3$.

22. Find the real average of lengths of best-of-seven series in some professional sport. Compare to Example 7.

23. Use a graph of the expression in Example 7 to determine what p must be in order for the expected length of the series to be less than 5.

24. Flip a coin until it comes up heads recording the number of required tosses. Do this 20 times and find the average number of required tosses. Now compute the theoretical expected number of tosses to compare with your result. (You may need to use the following fact about infinite sums: $\frac{1}{2} + \frac{1}{4} + \frac{1}{8} + \frac{1}{16} + \cdots = 1$.)

6.5 Recursion Revisited

One of the themes in Section 5.5 was the art of counting the number of sequences of a particular type using a recursive description of the sequences. The basic principle involved is that when it comes time to construct sequences of length n, you may take the point of view (essentially by mathematical induction) that you already know how to construct sequences of length $n - 1$ or less.

For example, to construct a binary sequence of length n, we can use the two-step algorithm: (i) Choose a first digit; (ii) choose a binary sequence of length $n - 1$ to follow the first digit.* This same approach can be applied to questions about probability experiments.

Example 1 *This problem revisits Exercise 24 of Section 6.4. Use recursive reasoning to find the average number of tosses of a fair coin that it takes to get a result of heads for the first time.*

SOLUTION   The set of all possible games is infinite, so we cannot simply count the total number of outcomes as part of answering the probability question. However, we can use recursive reasoning instead. The set S of all possible sequences of tosses for this game looks like this:

$$S = \{H, TH, TTH, TTTH, TTTTH, TTTTTH, TTTTTTH, \ldots\}$$

Let x be the average number of flips to get the first result of heads. When we start a new game, there is a 1/2 probability that the game will be over immediately (i.e., the game will last one flip) and a 1/2 probability that it will last one flip longer than average. Formally, this means that

$$x = \frac{1}{2} \cdot (1) + \frac{1}{2} \cdot (1 + x)$$

and the solution to this equation is clearly $x = 2$. □

We can visualize the recursive nature of the set S in Example 1 using a decision tree structure. The important feature in Figure 6-5 is that all the ovals are identical, and each occurs with probability 0.5 within the next larger oval.

---

* It follows that the *number* of binary sequences of length n is simply twice the number of binary sequences of length $n - 1$.

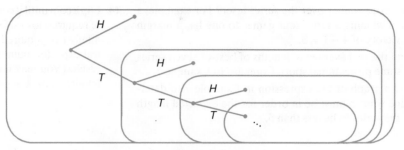

Figure 6-5 The recursive nature of S.

Practice Problem 1 *How many flips on average will it take for a coin to get its second result of heads?*

Remember that in the game of tennis you must win by at least two points, so if there is a tie late in the game (this situation is called "deuce"), the game could conceivably go on for a long time until someone has a two-point lead. A tennis game at deuce is equivalent to a game that starts with a score of 0-0 and ends when either player has a two-point lead, if that is easier to visualize.

Example 2 *Suppose two tennis players are evenly matched (so each has a probability of $\frac{1}{2}$ of winning any given point). If the two players are tied at deuce, what is the average number of additional points that must be played to resolve the game?*

SOLUTION   We will write a list of A's and B's as before to signify a list of points in the order in which they are won. The set of all possible outcomes of the experiment looks like

$$S = \{AA, BB, ABAA, BAAA, ABBB, BABB, ABBAAA, ABBABB, \ldots\}$$

so the expected length is the (infinite) sum of the length of each game times the probability of that game occurring. To avoid the infinite sum, we use a recursive model:

Case 1. The game looks like AA.

Case 2. The game looks like BB.

Case 3. The game looks like AB_____, where the blank can be filled with any game from S above.

Case 4. The game looks like BA_____, where the blank can be filled with any game from S above.

The average length of the games in S then is

(Length of games in case 1)(Prob. of having a game in case 1)

+ (Length of games in case 2)(Prob. of having a game in case 2)

+ (Length of games in case 3)(Prob. of having a game in case 3)

+ (Length of games in case 4)(Prob. of having a game in case 4)

The key here is to notice that the average length of games in cases 3 and 4 is **two more** than the overall average length of games in S. Using their observation, we can write the average length of games in S as

$$(2) \cdot \left(\frac{1}{2}\right)^2 + (2) \cdot \left(\frac{1}{2}\right)^2 + (2 + \textit{avg. length in } S) \cdot \left(\frac{1}{2}\right)^2$$
$$+ (2 + \textit{avg. length in } S) \cdot \left(\frac{1}{2}\right)^2$$

If we let the variable a represent the average length of games in S, this expression becomes

$$a = 2 \cdot (1/4) + 2 \cdot (1/4) + (2 + a) \cdot (1/4) + (2 + a) \cdot (1/4)$$
$$= 1 + \frac{2 + a}{2}$$

from which it follows that $a = 4$. □

Practice Problem 2 *Fill in the details of the following solution to the problem "If A and B are playing tennis and A wins each point with probability $\frac{2}{3}$, what is the probability that A eventually wins a game that is currently tied at deuce?"*

Representing games as we did in the previous example, we have four different cases for what a tennis game can look like:

Case 1. *The game looks like AA.*

Case 2. *The game looks like BB.*

Case 3. *The game looks like AB_____, where the blank can be filled with any game from S.*

Case 4. *The game looks like BA_____, where the blank can be filled with any game from S.*

Because the points played are independent, the probability that A wins such a game is

(Prob. of having a game in case 1) · (Prob. of A winning a game in case 1)
+(Prob. of having a game in case 2) · (Prob. of A winning a game in case 2)
+(Prob. of having a game in case 3) · (Prob. of A winning a game in case 3)
+(Prob. of having a game in case 4) · (Prob. of A winning a game in case 4)

Let w denote the probability that A wins this kind of game. The analysis above can be translated into the equation

$$w = \left(\frac{2}{3}\right)^2 \cdot (\underline{\hspace{1.5cm}}) + \left(\frac{1}{3}\right)^2 \cdot (\underline{\hspace{1.5cm}}) + \left(\frac{2}{3}\right)\left(\frac{1}{3}\right) \cdot (\underline{\hspace{1.5cm}})$$
$$+ \left(\frac{1}{3}\right)\left(\frac{2}{3}\right) \cdot (\underline{\hspace{1.5cm}})$$

which can be solved algebraically to yield $w = \underline{\hspace{1.5cm}}$. □

A tennis game is just a particular instance of a more general type of game studied in what is known as the *gambler's ruin problem*. Imagine that when two tennis players are tied at deuce, each player is given two "markers." When either player wins a point, she takes a marker from her opponent, and the game ends when one player is out of markers. This leads to exactly the same rules for resolving a deuce game in tennis, and it is easy to see how to generalize it to what we will call the *Hank and Ted game*.

In this game, two players, whom we will always call Hank and Ted, each begin with a pile of markers. A turn in the game consists simply of a coin being flipped. If the result of the toss is heads, then Hank wins a marker from Ted, and otherwise Ted wins a marker from Hank. The game is over when either player is out of markers. In a gambling context, the end of the game occurs when one player is ruined. Get it?

Clearly, the person who starts with more markers is more likely to win all the markers. Also, if both players begin with the same number of markers, it seems clear that they should each have a $\frac{1}{2}$ probability of winning the game. But this is perhaps where our intuition abandons us. There are a couple of interesting questions in the careful analysis of this game. Because there are infinitely many different ways the game can be resolved, we will have to use recursive thinking to answer them.

 Example 3 *In the Hank and Ted game played with a fair coin,*

1. *Is it better to have 6 markers to your opponent's 4 markers or 7 markers to his 5, or does it not matter?*

2. *Will a game that starts with 8 markers to 1 last longer or shorter on average than a game that starts with 2 markers to 4?*

SOLUTION

1. Suppose the game is being played with a total of M markers. Let p_n be the probability that Hank wins if he starts with n markers (and so Ted starts with $M - n$ markers). Clearly, $p_0 = 0$ and $p_M = 1$. Also, if Hank starts with n markers and if the first coin toss is heads (probability $= \frac{1}{2}$), then the rest of the game is played (independent of the first toss) with Hank starting with $n + 1$ markers. Similarly, if the first toss is tails, then the rest of the game is played with Hank starting with $n - 1$ markers. By the rules of sums and products, this means that

$$p_n = \frac{1}{2}p_{n+1} + \frac{1}{2}p_{n-1} \quad \text{for every } n \geq 1 \tag{6.1}$$

To solve this, note that equation (6.1) can be rewritten as

$$p_{n+1} - p_n = p_n - p_{n-1} \quad \text{for every } n \geq 1$$

This means that the difference between any two consecutive values of p_i is a constant—let's call it k. We know that $p_M - p_0 = 1 - 0 = 1$ and it is easy to see that

$$p_M - p_0 = (p_M - p_{M-1}) + (p_{M-1} - p_{M-2}) + \cdots + (p_1 - p_0) = (M)(k)$$

Therefore, $k = \frac{1}{M}$. Moreover, using the same argument as above, we have that $p_n = p_n - p_0 = (n)(k)$, so we determine that $p_n = \frac{n}{M}$ for all n. This means that Hank has a $\frac{6}{10}$ probability of winning if he starts with 6 markers to Ted's 4, and he has a $\frac{7}{12}$ probability of winning with 7 markers to Ted's 5.

2. Suppose again that the game is being played with M total markers, and let d_n represent the expected duration (number of coin tosses) of a game in which Hank starts with n markers. Clearly, $d_0 = d_M = 0$. Also, as before there is a $\frac{1}{2}$ probability that the game's first coin toss is heads and this move is followed by a game in which Hank has $n + 1$ markers, which has an expected duration of d_{n+1}. Similarly, there is a $\frac{1}{2}$ probability that the game's first toss is tails, which must be followed by a game of expected duration d_{n-1}. So the expected duration of the game in which Hank starts with n markers (where $n \geq 1$) is

$$d_n = \frac{1}{2}(d_{n+1} + 1) + \frac{1}{2}(d_{n-1} + 1) \qquad (6.2)$$

We will not explicitly solve this, but the method would be similar to that above. We can easily check that $d_n = n \times (M - n)$ satisfies the relation (6.2) with the correct initial values. (See Exercise 23 below.) This means that the expected duration of a game that starts with 8 markers to 1 is $(8)(1) = 8$ coin tosses, and the game that starts with 2 markers to 4 also has an expected duration of $8 = (2)(4)$ coin tosses. □

Practice Problem 3 *Suppose the Hank and Ted game is played with an unfair coin that comes up heads with probability $\frac{2}{3}$. How should you modify recurrence relations (6.1) and (6.2) to model this new game?*

Solutions to Practice Problems

1 Let x denote the average length of such a sequence. Each sequence of coin tosses looks like one of the following:

- $T\underline{\hspace{1cm}}$, where the blank contains a sequence of H's and T's that stops when the second heads is tossed. This means that the average length of the sequences that go in the blank is x; hence, the average length of all sequences in this case is $x + 1$.

- $H\underline{\hspace{1cm}}$, where the blank contains a sequence of H's and T's that stops when the first heads is tossed. We know from Example 1 that the average length of the sequence in the blank is 2; hence, the average length of all sequences in this case is 3.

Since each case above occurs with probability $\frac{1}{2}$, we can put this together to conclude that

$$x = \frac{1}{2} \cdot (x + 1) + \frac{1}{2} \cdot 3$$

which has solution $x = 4$.

2 The blanks occur at the end of the solution as follows:

$$w = \left(\frac{2}{3}\right)^2 \cdot \underline{(1)} + \left(\frac{1}{3}\right)^2 \cdot \underline{(0)} + \left(\frac{2}{3}\right)\left(\frac{1}{3}\right) \cdot \underline{(w)}$$
$$+ \left(\frac{1}{3}\right)\left(\frac{2}{3}\right) \cdot \underline{(w)}$$

which can be solved algebraically to yield $w = \underline{4/5}$.

3 Following the reasoning that gave us (6.1) and (6.2), we will have

$$p_n = \frac{2}{3}p_{n+1} + \frac{1}{3}p_{n-1} \quad \text{for every } n \geq 1$$

and

$$d_n = \frac{2}{3}(1 + d_{n+1}) + \frac{1}{3}(1 + d_{n-1}) \quad \text{for every } n \geq 1$$

Exercises for Section 6.5

1. What is the expected number of rolls of a six-sided die that is rolled until a 1 appears?

2. How many times on average will three dice have to be tossed for the sum to exceed 12 for the first time?

3. A pair of dice are thrown until at least one of the dice comes up 1 for the first time. How many tosses, on average, are required?

4. Generalize the previous three exercises by proving the statement "If a trial is successful with probability p and an experiment consists of performing the trial

repeatedly until it is successful for the first time, the expected number of trials in the experiment is $\frac{1}{p}$."

5. The following problems are *not* examples of recursive problems and the previous exercise cannot be applied to them:

   (a) If a deck of cards is shuffled well and cards are dealt face-up until the first ace appears, how many cards would you expect on average to deal?

   (b) If a deck of cards is shuffled well and cards are dealt face-up until the first spade appears, how many cards would you expect on average to deal?

   (c) Explain why Exercise 4 does not help with these problems.

6. How many tosses of a coin on average will it take before you have seen at least one head and at least one tail?

7. Let h_n be the expected number of tosses required to obtain the nth head on repeatedly tossing a fair coin. Generalize the solution to Practice Problem 1 to find a recursive description for h_n.

8. Which are you more likely to see first if you repeatedly toss a coin, a head followed immediately by a head or a tail followed immediately by a head? (Be careful. This problem is not like the others.)

9. If the server in a tennis game has a $\frac{9}{10}$ probability of winning each point she plays, what is the probability she will eventually win a game that is now tied at deuce?

10. If the server in a tennis game has a $\frac{9}{10}$ probability of winning each point she plays, on average how many more points will a game last that is now tied at deuce?

11. If the server in a tennis game has a $\frac{3}{4}$ probability of winning each point she plays, on average how many more points will a game last that is now tied at deuce?

12. If the server in a tennis game has a $\frac{3}{4}$ probability of winning each point she plays, what is the probability she will eventually win a game that is now tied at deuce?

13. Generalize the previous four exercises. Specifically, if the server in a tennis game has probability p of winning each point she plays, what is the probability she will eventually win a game that is now tied at deuce? What is the expected number of points of such a game?

14. The following questions refer to Example 2 and tie in with the coin-tossing tennis experiments in the exercises of the previous section. A and B are playing tennis starting with a score of 0-0. Remember that the first player to win at least four points wins and being ahead by at least two points wins the game. Assume each player has a $\frac{1}{2}$ probability of winning any given point.

   (a) What is the probability that Player A wins after exactly four points are played? Exactly five points? Exactly six points? Exactly seven points? Exactly eight points?

   (b) What is the probability that the game will be stuck at "deuce" after exactly n points? (Think recursively.)

   (c) What is the probability that Player A wins after exactly n points are played? (Use the previous fact.)

15. If the server wins a tennis point with probability $\frac{2}{3}$, what is the probability that the server wins a tennis game that starts at 0-0?

16. If the server wins a tennis point with probability $\frac{2}{3}$, what is the average length of a tennis game that starts 0-0?

17. When Pete and André play table tennis, they like to play to seven with the rule that one has to win a game by **three** points. This exercise suggests a way to analyze the length of this kind of game once it is tied at the end of the game. Assume that Pete and André each have a probability of $\frac{1}{2}$ of winning each point they play.

   (a) Let t represent the average length of the game once it is tied (at the end of the game when the "win-by-three" rule applies), let a represent the average length of a game in which André is two points ahead of Pete, and let p represent the average length of a game in which Pete is two points ahead of André. Using Example 2 as a guide, explain why

   $$t = \frac{1}{2} \cdot (2 + t) + \frac{1}{4} \cdot (2 + a) + \frac{1}{4} \cdot (2 + p)$$

   (b) In the same spirit, explain why

   $$a = \frac{1}{2} \cdot 1 + \frac{1}{4} \cdot (2 + a) + \frac{1}{4} \cdot (2 + t) \quad \text{and}$$
   $$p = \frac{1}{2} \cdot 1 + \frac{1}{4} \cdot (2 + p) + \frac{1}{4} \cdot (2 + t)$$

   (c) Use algebra to determine the values of the unknowns t, a, and p from the three equations we have found.

18. What is the probability that Hank wins the Hank and Ted game in which Hank starts with 6 markers to Ted's 8 markers? Play the game a number of times with a coin or online to support this theoretical result.

19. What is the expected length of the Hank and Ted game in which Hank starts with 6 markers to Ted's 8 markers? Play the game a number of times with a coin or online to support this theoretical result.

20. Without using the "formulas" derived in Example 3, explain why it makes sense that the Hank and Ted game in which Hank and Ted each start with 7 markers should last exactly one move longer than the game described in Exercise 19.

21. Based on the analysis in this section, fill in the blanks in the sentence below:

    In the Hank and Ted game played with a fair coin, if Hank starts with X markers and Ted starts with Y markers, then the probability that Hank wins is _____,

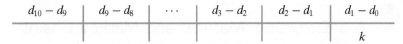

$d_{10} - d_9$	$d_9 - d_8$	$\cdots$	$d_3 - d_2$	$d_2 - d_1$	$d_1 - d_0$
					k

Table 6-4   Table for Exercise 22(b)

the probability that Ted wins is _____, and the expected length of the game is _____ moves.

22. These problems involve a careful analysis of the numbers d_n in Example 3. Assume for these problems that we are playing the Hank and Ted game with a fair coin, where Hank starts the game with 6 markers to Ted's 4 markers. (Hence, $d_0 = d_{10} = 0$.)

   (a) Use equation (6.2) to show that

   $$d_{n+1} - d_n = d_n - d_{n-1} - 2$$

   (b) Letting the letter k represent $d_1 - d_0$, use your previous answer to fill in Table 6-4.

   (c) Use the values in the table above with the fact that

   $$(d_{10} - d_9) + (d_9 - d_8) + (d_8 - d_7) + \cdots + (d_1 - d_0)$$
   $$= d_{10} - d_0 = 0$$

   to find a value of k.

   (d) Use this value of k to fill in Table 6-5.

23. Algebraically verify that the function $d_n = (n)(M - n)$ satisfies the recurrence relation (6.2) and the conditions $d_0 = d_M = 0$.

24. In Practice Problem 2, we addressed a Hank and Ted game in which Hank has snuck an unfair coin into the game that comes up heads $\frac{2}{3}$ of the time. In the solution to that problem, we showed that the probability p_n of Hank winning when he has n markers satisfies the recurrence relation

   $$p_n = \frac{2}{3} p_{n+1} + \frac{1}{3} p_{n-1}$$

   Fill in the details below to find the probability that Hank wins this game if each player starts with 10 markers.

   (a) Show the algebra that allows the given recurrence relation to be rewritten as

   $$p_{n+1} - p_n = \frac{1}{2}(p_n - p_{n-1}) \qquad (6.3)$$

   (b) Leting $k = p_1 - p_0$, explain why (6.3) implies that

   $$p_{n+1} - p_n = \frac{k}{2^n}$$

for all $n \geq 0$.

   (c) Use the fact that

   $$p_{20} - p_0 = (p_{20} - p_{19}) + (p_{19} - p_{18}) + \cdots$$
   $$+ (p_2 - p_1) + (p_1 - p_0)$$

   to explain why $p_{20} - p_0 = k\left(\frac{2^{20}-1}{2^{19}}\right)$.

   (d) Show that for $1 \leq n \leq 20$,

   $$p_n = \left(\frac{2^{19}}{2^{20} - 1}\right)\left(\frac{2^n - 1}{2^{n-1}}\right)$$

   and use this result to answer the given question.

25. In Practice Problem 2, we addressed a Hank and Ted game in which Hank has snuck an unfair coin into the game that comes up heads $\frac{2}{3}$ of the time. In the solution to that problem, we showed that the expected length d_n of a game when Hank has n markers satisfies the recurrence relation

   $$d_n = \frac{2}{3}(d_{n+1} + 1) + \frac{1}{3}(d_{n-1} + 1)$$

   For the game in which each player starts with 10 markers,

   (a) Show algebraically that any closed formula with the form

   $$d_n = c\left(1 - \frac{1}{2^n}\right) - 3n$$

   satisfies the recurrence relation.

   (b) Find the value of c so that $d_{20} = 0$.

   (c) Compute the value of d_{10}.

   (d) Find the initial distribution of markers for which the game is expected to last the longest.

26. In Exercise 24, we examined the Hank and Ted game played with a coin that comes up heads with probability $\frac{2}{3}$. If Hank starts such a game with 10 markers, how many markers should Ted start with in order for the game to be fair? (That is, how can we assure that the probability of each player winning is as close to $\frac{1}{2}$ as possible?)

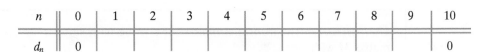

n	0	1	2	3	4	5	6	7	8	9	10
d_n	0										0

Table 6-5   Table for Exercise 22(d)

6.6 Excursion: Matrices and Markov Chains

The recurrence relations from the previous section all came from the same idea—each game is modeled as some initial outcome followed by the remainder of the game. In addition, the games involved had well-defined states determined by a score or a distribution of markers. Problems involving states and rules for moving between the states can be more easily modeled through the mathematical operation of matrix multiplication.

In this section we will develop the concept of matrix multiplication in the context of the probability rules we have studied so far. Students already familiar with adding and multiplying matrices will learn a new way to think about these operations, so the beginning of the section should not be skipped. In order to investigate real games using these tools, some form of technology capable of matrix arithmetic should be used, since the matrices involved can be large and the number of operations performed with them can be even larger.

We will introduce the idea of matrix multiplication using the Hank and Ted game played with three total markers. The "matrix" involved is simply an array with rows and columns labeled with the four possible states of the game as follows:

State 1 is when Hank has 0 markers to Ted's 3.

State 2 is when Hank has 1 markers to Ted's 2.

State 3 is when Hank has 2 markers to Ted's 1.

State 4 is when Hank has 3 markers to Ted's 0.

The entries in the array will be the probabilities of the game moving from one state to another in one move of the game. For example, the entry in row 2, column 1 is the probability of the game changing *in one move* from state 2 (Hank 1, Ted 2) to state 1 (Hank 0, Ted 3), which is $\frac{1}{2}$ since this transition happens if the coin comes up tails. On the other hand, the entry in row 2, column 2 is the probability of the game changing *in one move* from state 2 (Hank 1, Ted 2) to state 2—such a transition is impossible, so the probability of it happening is 0.

If we continue doing this for all pairs of states of the game, we will arrive at the 4×4 array:

$$M = \begin{array}{l} \text{Row 1 (state 1)} \to \\ \text{Row 2 (state 2)} \to \\ \text{Row 3 (state 3)} \to \\ \text{Row 4 (state 4)} \to \end{array} \begin{bmatrix} 1 & 0 & 0 & 0 \\ 1/2 & 0 & 1/2 & 0 \\ 0 & 1/2 & 0 & 1/2 \\ 0 & 0 & 0 & 1 \end{bmatrix}$$

We call this the *transition matrix* for this game since it shows the probabilities of all one-move transitions between states in the game. In the formal definition below and in the remainder of this section, we use the notation $M_{i,j}$ to refer to the entry in row i and column j of a matrix M.

> **Definition**   The *transition matrix* for a game with states $1, 2, \ldots, n$ is the matrix M with
>
> $$M_{i,j} = Prob(\text{the game changes from state } i \text{ to state } j \text{ in one move})$$

Notes:

1. We are making an implicit assumption that the moves of the game are independent events, so the entries in the transition matrix are constant probabilities and the product rule can be correctly used.

2. Since state 1 and state 4 represent the end of the game, the transition matrix shows a probability of 1 that the game stays in whichever of these two states the game ends up in. This is the way in which we deal with the indefinite length of the game. We essentially let all games continue forever, eventually becoming trapped in one of the "winning" states. These states are called *absorbing states*.

3. The defining characteristics of this type of matrix are that all the entries are probabilities and every row has entries that sum to 1. That is, any matrix with these two characteristics is the transition matrix for *some* game.

Matrix Multiplication

Now that we have captured all the probabilities for how the Hank and Ted game can change in a single move, a natural next question is "What is the probability of the game moving from state i to state j in *two moves*?" We will use the notation $Prob(i \rightarrow * \rightarrow j)$ to represent this probability. Answering this type of question is simply an application of the sum and product rules for probability. For example, to compute the probability of the game moving from state 3 to state 3 in *two moves* (i.e., $Prob(3 \rightarrow * \rightarrow 3)$), we recognize that this may be accomplished only by moving from state 3 to state 2 and then state 2 to state 3. Extending our notation a bit and using the product rule, we can compute

$$Prob(3 \rightarrow 2 \rightarrow 3) = \left(\frac{1}{2}\right)\left(\frac{1}{2}\right)$$

In general, the product rule tells us that

$$Prob(3 \rightarrow x \rightarrow 3) = Prob(3 \rightarrow x) \cdot Prob(x \rightarrow 3)$$

Of course, moving from state 3 to state 4 to state 3 is impossible since state 4 is an absorbing state, but this fact is correctly reflected in the above rule:

$$Prob(3 \rightarrow 4 \rightarrow 3) = Prob(3 \rightarrow 4) \cdot Prob(4 \rightarrow 3)$$
$$= (M_{3,4}) \cdot (M_{4,3})$$
$$= \left(\frac{1}{2}\right)(0) = 0$$

In some games, there might be more ways to get from state 3 to state 3 in two moves, but we can account for all possible intermediate stages by simply invoking the sum rule. For example, to fully consider all the possible ways to go from state 3 to state 3, we should write

$$Prob(3 \rightarrow * \rightarrow 3) = Prob(3 \rightarrow 1 \rightarrow 3) + Prob(3 \rightarrow 2 \rightarrow 3)$$
$$+ Prob(3 \rightarrow 3 \rightarrow 3) + Prob(3 \rightarrow 4 \rightarrow 3)$$

which in this case just gives us the same answer as before:

$$Prob(3 \rightarrow * \rightarrow 3) = 0 + \frac{1}{4} + 0 + 0 = \frac{1}{4}$$

This computation can be made directly from the transition matrix

$$M = \begin{bmatrix} 1 & 0 & 0 & 0 \\ 1/2 & 0 & 1/2 & 0 \\ 0 & 1/2 & 0 & 1/2 \\ 0 & 0 & 0 & 1 \end{bmatrix}$$

by looking at row 3, which reflects the probabilities of where the game can go **from** state 3, and column 3, which reflects the probabilities of how the game can get **to** state 3. There we see

$$Prob(3 \to 1 \to 3) = Prob(3 \to 1) \cdot Prob(1 \to 3) = M_{3,1} \cdot M_{1,3} = (0)(0)$$

$$Prob(3 \to 2 \to 3) = Prob(3 \to 2) \cdot Prob(2 \to 3) = M_{3,2} \cdot M_{2,3} = \left(\frac{1}{2}\right)\left(\frac{1}{2}\right)$$

$$Prob(3 \to 3 \to 3) = Prob(3 \to 3) \cdot Prob(3 \to 3) = M_{3,3} \cdot M_{3,3} = (0)(0)$$

$$Prob(3 \to 4 \to 3) = Prob(3 \to 4) \cdot Prob(4 \to 3) = M_{3,4} \cdot M_{4,3} = \left(\frac{1}{2}\right)(0)$$

In other words, the probability of going from state 3 to state 3 in two steps is

$$M_{3,1} \cdot M_{1,3} + M_{3,2} \cdot M_{2,3} + M_{3,3} \cdot M_{3,3} + M_{3,4} \cdot M_{4,3}$$

That is, row 3 of M is multiplied by column 3 of M term for term and the results added. We will refer to this operation as *row-column multiplication*, and it is always possible to perform as long as the "row" and the "column" in question contain the same number of entries.

The operation of *matrix multiplication* is just the extension of this simple idea to an entire matrix.

Definition   Given matrices M and N where the number of entries in the rows of M is the same as the number of entries in the columns of N, we define the *product $M \cdot N$* to mean the new matrix P so that the entry row i, column j of P is the row-column product of row i from M and column j from N. Formally, we write $P = M \cdot N$ to mean that

$$P_{i,j} = M_{i,1} \cdot N_{1,j} + M_{i,2} \cdot N_{2,j} + M_{i,3} \cdot N_{3,j} + \cdots$$

Note that the matrix products $M \cdot N$ and $N \cdot M$ are not defined to be the same, and we shall see below that they usually are not. Also note that it is absolutely essential that M have rows the same length as the columns of N for the definition of $M \cdot N$ to make sense.

It is advisable to stop here and practice matrix multiplication, especially if it is new to you.

Practice Problem 1   *Let* $A = \begin{bmatrix} 1 & 0 & 0 & 0 \\ 1/2 & 0 & 1/2 & 0 \\ 0 & 1/2 & 0 & 1/2 \end{bmatrix}$, $B = \begin{bmatrix} -1 & 0 \\ \frac{1}{3} & \frac{2}{3} \\ 0 & -1/2 \\ 0 & 0 \end{bmatrix}$,

$C = \begin{bmatrix} -1 & 2 \\ 0 & -1 \end{bmatrix}$, *and* $I = \begin{bmatrix} 1 & 0 \\ 0 & 1 \end{bmatrix}$.

Compute each of the following or explain why it cannot be done:

1. $A \cdot B$
2. $B \cdot A$
3. $C \cdot B$
4. $B \cdot C$
5. C^2
6. $C \cdot I$
7. $I \cdot C$
8. $B \cdot I$

Based on your answers above, why is I known as an identity matrix?

Returning to the transition matrix M for the 3-marker Hank and Ted game, we can now form the entire matrix product $M \cdot M$, which can be written M^2.

$$M^2 = \begin{bmatrix} 1 & 0 & 0 & 0 \\ 1/2 & 0 & 1/2 & 0 \\ 0 & 1/2 & 0 & 1/2 \\ 0 & 0 & 0 & 1 \end{bmatrix} \cdot \begin{bmatrix} 1 & 0 & 0 & 0 \\ 1/2 & 0 & 1/2 & 0 \\ 0 & 1/2 & 0 & 1/2 \\ 0 & 0 & 0 & 1 \end{bmatrix} = \begin{bmatrix} 1 & 0 & 0 & 0 \\ 1/2 & 1/4 & 0 & 1/4 \\ 1/4 & 0 & 1/4 & 1/2 \\ 0 & 0 & 0 & 1 \end{bmatrix}$$

The notion of matrix multiplication has been based on the use of the sum and product rules for probabilities to find that the probability of a game moving from state 3 to state 3 in **two** moves is the entry in row 3, column 3 of the product $M \cdot M = M^2$. In general, the entry in row i, column j of M^2 is the probability that the game progresses from state i to state j in two moves. This fact, in turn, can be further generalized to the following theorem.

Theorem 1   *If M is an n × n transition matrix reflecting the one-move transition probabilities for states 1 through n of a game, then for any integer $k \geq 1$, the entry in row i, column j of the matrix M^k is the probability of the game moving from state i to state j in k moves.*

PROOF   The proof is done by induction on k and is left for Exercise 23.   ■

A side effect of this is the fact that M^k itself is another transition matrix, so the entries must all be probabilities with rows summing to 1. This is significant because it means that when we raise a transition matrix to a large power (as we will do next), we will not get huge numbers. It also meshes with our understanding of the games themselves. For example, if we use a matrix M to model a children's game where a single die is used to generate moves, then the matrix M^2 accurately models the same game with each turn consisting of rolling the die twice.

Exploiting the Matrix Model

There are two points to doing all of this. First of all, there are standard computer applications and advanced calculators that are very good at doing matrix arithmetic. Consequently, by modeling our problem in these terms, we can bring the computational power of our machines to work on the problem without having to write any specialized code. The second point to using matrix models comes from the wealth of mathematical knowledge about systems that have matrix representations. This includes ways to give elegant solutions to our problems and more advanced analysis than we will see in this course. You will most likely see these so-called Markov chain techniques in a first course in linear algebra.

We now return to our example and the analysis of the simple Hank and Ted game with 3 markers. We saw that from the transition matrix

$$M = \begin{bmatrix} 1 & 0 & 0 & 0 \\ 1/2 & 0 & 1/2 & 0 \\ 0 & 1/2 & 0 & 1/2 \\ 0 & 0 & 0 & 1 \end{bmatrix}$$

for the game (which describes how the game could progress from each state to each other state in one move of the game) we could find the matrix

$$M^2 = \begin{bmatrix} 1 & 0 & 0 & 0 \\ 1/2 & 1/4 & 0 & 1/4 \\ 1/4 & 0 & 1/4 & 1/2 \\ 0 & 0 & 0 & 1 \end{bmatrix}$$

which describes the probabilities of getting from each state to each other state in exactly **two moves** of the game. Theorem 1 tells us that M^3 similarly describes the probabilities of getting from each state to each other state in exactly **three moves** of the game, and so on.

 Example 1 *Given that*

$$M^3 = \begin{bmatrix} 1 & 0 & 0 & 0 \\ 5/8 & 0 & 1/8 & 1/4 \\ 1/4 & 1/8 & 0 & 5/8 \\ 0 & 0 & 0 & 1 \end{bmatrix}$$

1. What is the probability that a game in state 2 is in state 1 three moves later?
2. What is the probability that a game starting with Hank 2, Ted 1 has Hank 1, Ted 2 after three tosses of the coin?

SOLUTION

1. According to Theorem 1, this is the row 2, column 1 entry in M^3, which is $\frac{5}{8}$.
2. Since Hank 2, Ted 1 is state 3 and Hank 1, Ted 2 is state 2, this question is answered by the row 3, column 2 entry in M^3, which is $\frac{1}{8}$. □

Practice Problem 2 *Given that*

$$M^4 = \begin{bmatrix} 1 & 0 & 0 & 0 \\ 5/8 & 1/16 & 0 & 5/16 \\ 5/16 & 0 & 1/16 & 5/8 \\ 0 & 0 & 0 & 1 \end{bmatrix}$$

1. What is the probability that a game in state 2 is in state 4 four moves later?
2. What is the probability that a game starting with Hank 2, Ted 1 has Hank 1, Ted 2 after four tosses of the coin?

Computing Large Powers with Technology

If we have a computer or calculator available, we can just as easily compute large powers like

$$M^{1,000} \approx \begin{bmatrix} 1 & 0 & 0 & 0 \\ 0.666\ldots & 10^{-301} & 0 & 0.333\ldots \\ 0.333\ldots & 0 & 10^{-301} & 0.666\ldots \\ 0 & 0 & 0 & 1 \end{bmatrix}$$

This allows us to collect information about the game after an enormous number of moves. Remember that the original matrix was set up so that instead of the game actually ending, it just gets stuck eventually in either state 1 or 4. So, for example, the fact that the probability of a game progressing from state 2 to state 4 and being stuck there in 1,000 moves is $0.333\ldots$ (the entry in row 2, column 4 entry of $M^{1,000}$) means that a game starting with Hank 1, Ted 2 ends in a win for Hank within 1,000 moves with probability $0.333\ldots$.

The following facts follow similarly from the matrix $M^{1,000}$:

- A game starting with Hank 1, Ted 2 ends up as a win for Ted within 1,000 moves with probability $0.666\ldots$.
- A game starting with Hank 1, Ted 2 has Hank 1, Ted 2 after 1,000 moves with probability 10^{-301}, a pitifully small number.
- A game starting with Hank 1, Ted 2 has Hank 2, Ted 1 after 1,000 moves with probability 0.

In summary, after 1,000 moves, the probability that the Hank 1, Ted 2 game ends in a win for Hank is about $\frac{1}{3}$, the probability that the Hank 1, Ted 2 game ends in a win for Ted is about $\frac{2}{3}$, and the probability that the game is still going on is very, very small. This is consistent with the solution to the general Hank and Ted problem that we formalized in Exercise 21 of Section 6.5.

These observations can be generalized as the following corollary to Theorem 1.

Corollary 2   *Given a transition matrix M with absorbing state j, the probability that a game beginning in state i will eventually be absorbed in state j is given by the row i, column j entry of the matrix*

$$M_\infty = \lim_{n \to \infty} M^n$$

provided this matrix exists.

For those that have not seen the limit notation used before, it merely means that if the matrix powers "settle down" on a single matrix as n grows without bound, then that single matrix is the limit of the powers of M. We will not address the mathematical formalities of this statement, but rather, we will rely on experimentation and intuition about the nature of the games to which we apply this idea. For example, by looking at the $M^{1,000}$ above as well as

$$M^{10,000} = \begin{bmatrix} 1 & 0 & 0 & 0 \\ 0.666\ldots & 10^{-3,010} & 0 & 0.333\ldots \\ 0.333\ldots & 0 & 10^{-3,010} & 0.666\ldots \\ 0 & 0 & 0 & 1 \end{bmatrix}$$

it appears that the powers are settling on the matrix

$$M_{\infty} = \begin{bmatrix} 1 & 0 & 0 & 0 \\ 2/3 & 0 & 0 & 1/3 \\ 1/3 & 0 & 0 & 2/3 \\ 0 & 0 & 0 & 1 \end{bmatrix}$$

Hence, the entries in column 1 and column 4 give all the probabilities of a game ending up in state 1 (a win for Ted) and state 4 (a win for Ted), respectively.

Let's put together all this information to observe the complete analysis of a game.

Example 2 *Suppose Hank and Ted play their game starting with 5 total markers. There are six possible states in the play of this game:*

- **State 1** *is $H = 0$, $T = 5$.*
- **State 2** *is $H = 1$, $T = 4$.*
- **State 3** *is $H = 2$, $T = 3$.*
- **State 4** *is $H = 3$, $T = 2$.*
- **State 5** *is $H = 4$, $T = 1$.*
- **State 6** *is $H = 5$, $T = 0$.*

Find the transition matrix for this game, and find and interpret M_{∞}.

SOLUTION   For the transition matrix

$$M = \begin{bmatrix} 1 & 0 & 0 & 0 & 0 & 0 \\ 1/2 & 0 & 1/2 & 0 & 0 & 0 \\ 0 & 1/2 & 0 & 1/2 & 0 & 0 \\ 0 & 0 & 1/2 & 0 & 1/2 & 0 \\ 0 & 0 & 0 & 1/2 & 0 & 1/2 \\ 0 & 0 & 0 & 0 & 0 & 1 \end{bmatrix}$$

Starting state	*Prob*(Ted wins)	*Prob*(Hank wins)
Hank 1, Ted 4	4/5	1/5
Hank 2, Ted 3	3/5	2/5
Hank 3, Ted 2	2/5	3/5
Hank 4, Ted 1	1/5	4/5

Table 6-6   Probabilities for the Hank and Ted Game with 5 Markers

we can compute the following:

$$M^{500} \approx \begin{bmatrix} 1 & 0 & 0 & 0 & 0 & 0 \\ 0.8 & 10^{-46} & 0 & 10^{-46} & 0 & 0.2 \\ 0.6 & 0 & 10^{-46} & 0 & 10^{-46} & 0.4 \\ 0.4 & 10^{-46} & 0 & 10^{-46} & 0 & 0.6 \\ 0.2 & 0 & 10^{-46} & 0 & 10^{-46} & 0.8 \\ 0 & 0 & 0 & 0 & 0 & 1 \end{bmatrix}$$

$$M^{5,000} \approx \begin{bmatrix} 1 & 0 & 0 & 0 & 0 & 0 \\ 0.8 & 10^{-460} & 0 & 10^{-460} & 0 & 0.2 \\ 0.6 & 0 & 10^{-460} & 0 & 10^{-460} & 0.4 \\ 0.4 & 10^{-460} & 0 & 10^{-460} & 0 & 0.6 \\ 0.2 & 0 & 10^{-460} & 0 & 10^{-460} & 0.8 \\ 0 & 0 & 0 & 0 & 0 & 1 \end{bmatrix}$$

From this it appears that the powers are settling on the matrix

$$M_\infty = \begin{bmatrix} 1 & 0 & 0 & 0 & 0 & 0 \\ 4/5 & 0 & 0 & 0 & 0 & 1/5 \\ 3/5 & 0 & 0 & 0 & 0 & 2/5 \\ 2/5 & 0 & 0 & 0 & 0 & 3/5 \\ 1/5 & 0 & 0 & 0 & 0 & 4/5 \\ 0 & 0 & 0 & 0 & 0 & 1 \end{bmatrix}$$

From this information we can record in Table 6-6 the probability of either player winning from any given starting position.   □

Even though we have so far only addressed questions that have already been answered in Section 6.5, there is a benefit to the additional computational efforts required by the matrix model: The new method is much, much more versatile. Exercise 24 of Section 6.5 showed the difficulties with the Hank and Ted game when an unfair coin is used to play. In contrast, the matrix model requires a minimal amount of change to reflect this kind of game.

Practice Problem 3   *For the Hank and Ted game played with 4 markers using a coin that comes up heads $\frac{3}{5}$ of the time, find the transition matrix M, and then find and interpret M_∞.*

Example 3   *In a tennis game tied at deuce, the game will end when one player is two points ahead of the other. Assuming that Player A has a $\frac{3}{4}$ probability of winning* *each point, use a matrix model to find the probability that Player A wins a game that starts tied at deuce.*

SOLUTION   This game has five states:

● **State 1** is "Player A wins."
● **State 2** is "Player A up one point."
● **State 3** is "The game is tied."
● **State 4** is "Player A down one point."
● **State 5** is "Player A loses."

The transition matrix for this game is

$$M = \begin{bmatrix} 1 & 0 & 0 & 0 & 0 \\ 3/4 & 0 & 1/4 & 0 & 0 \\ 0 & 3/4 & 0 & 1/4 & 0 \\ 0 & 0 & 3/4 & 0 & 1/4 \\ 0 & 0 & 0 & 0 & 1 \end{bmatrix}$$

so we can compute

$$M^{100} \approx \begin{bmatrix} 1 & 0 & 0 & 0 & 0 \\ 0.975 & 10^{-22} & 0 & 10^{-22} & 0.025 \\ 0.900 & 0 & 10^{-22} & 0 & 0.100 \\ 0.675 & 10^{-22} & 0 & 10^{-22} & 0.325 \\ 0 & 0 & 0 & 0 & 1 \end{bmatrix}$$

$$M^{1,000} \approx \begin{bmatrix} 1 & 0 & 0 & 0 & 0 \\ 0.975 & 10^{-212} & 0 & 10^{-212} & 0.025 \\ 0.900 & 0 & 10^{-212} & 0 & 0.100 \\ 0.675 & 10^{-212} & 0 & 10^{-212} & 0.325 \\ 0 & 0 & 0 & 0 & 1 \end{bmatrix}$$

From this it appears that the powers are settling on the matrix

$$M_{\infty} = \begin{bmatrix} 1 & 0 & 0 & 0 & 0 \\ 0.975 & 0 & 0 & 0 & 0.025 \\ 0.900 & 0 & 0 & 0 & 0.100 \\ 0.675 & 0 & 0 & 0 & 0.325 \\ 0 & 0 & 0 & 0 & 1 \end{bmatrix} = \begin{bmatrix} 1 & 0 & 0 & 0 & 0 \\ 39/40 & 0 & 0 & 0 & 1/40 \\ 9/10 & 0 & 0 & 0 & 1/10 \\ 27/40 & 0 & 0 & 0 & 13/40 \\ 0 & 0 & 0 & 0 & 1 \end{bmatrix}$$

Hence, the probability that a game starting tied (in state 3) ends up as a win for Player A (in state 1) is $\frac{9}{10}$, the entry in row 3, column 1 of this matrix. This is the same answer as in Exercise 12 of Section 6.5. We can further tell from the matrix model that even if Player A is down one point in this game, he still has a $\frac{27}{40}$ probability of eventually winning the game.   □

Exercise 17 of Section 6.5 investigated a game having a "win-by-three" rule. We saw that this was considerably more complicated than the win-by-two rule of tennis or volleyball. Once again, the matrix model is versatile enough to handle this variation with only minor changes.

Practice Problem 4 *Assume that a game of table tennis starts tied, and the winner is the first player to win* three *points more than the other player. If Player A has a $\frac{2}{3}$ probability of winning each point, use a matrix model to find the probability that Player A eventually wins a game that starts off tied.* (HINT: *This game has seven states.*)

Expected Values with the Matrix Model

Now we will see what else we can discover from our matrix model. Notice that we can sum the entries of row 3, column 2 of the matrices M, M^2, M^3, and M^4 to combine the information they contain about going from state 3 to state 2 in four moves or less. In fact, if we wish to have this information about all states of the game simultaneously, we should perform this summation for each position in the matrix. This is exactly how we will define matrix addition:

> **Definition** Given matrices A and B, we denote by $A + B$ the new matrix whose row i, column j entry comes from adding $A_{i,j}$ to $B_{i,j}$ for each i and j. Note that this definition only makes sense if matrices A and B are the same size, both in number of rows and number of columns.

Example 4 *The Hank and Ted game played with a total of 3 markers has transition matrix*

$$M = \begin{bmatrix} 1 & 0 & 0 & 0 \\ 1/2 & 0 & 1/2 & 0 \\ 0 & 1/2 & 0 & 1/2 \\ 0 & 0 & 0 & 1 \end{bmatrix}$$

where state i corresponds to Hank having $i - 1$ markers and Ted having $4 - i$ markers. Discuss the meaning of the sum $M + M^2 + M^3 + M^4$ in terms of the game.

SOLUTION   We calculated each of the matrix powers M^2, M^3, and M^4 earlier in this section, so

$$M + M^2 + M^3 + M^4 = \begin{bmatrix} 4 & 0 & 0 & 0 \\ 9/4 & 5/16 & 5/8 & 13/16 \\ 13/16 & 5/8 & 5/16 & 9/4 \\ 0 & 0 & 0 & 4 \end{bmatrix} \quad (6.4)$$

has $\frac{5}{8}$ as its entry in row 3, column 2. We might at first glance take this to mean that the probability of going from state 3 to state 2 in four moves or less is $\frac{5}{8}$. However, it should be troubling that some of the other entries in this matrix (like the 9/4 in row 2, column 1) are greater than 1. This should at least indicate to us that the entries are *not* probabilities.

In order to understand what these entries *do* mean, we need some new notation. We will represesent an outcome in four moves as

$$\_ \rightarrow \_ \rightarrow \_ \rightarrow \_$$

where we will use in the blanks either the state numbers or an $*$ if we don't care about the state. For example, the event of starting in state 3 and being in state

2 three moves later is denoted by

$$3 \rightarrow * \rightarrow * \rightarrow 2 \rightarrow *$$

The fact that the row 3, column 2 entry of M^3 is $\frac{1}{8}$ means that

$$Prob(3 \rightarrow * \rightarrow * \rightarrow 2 \rightarrow *) = \frac{1}{8}$$

On the other hand the event of going from state 3 to state 2 in one move, denoted by

$$3 \rightarrow 2 \rightarrow * \rightarrow * \rightarrow *$$

has probability $\frac{1}{2}$ since the row 3, column 2 entry in M is $\frac{1}{2}$.

The problem is there are some outcomes like $3 \rightarrow 2 \rightarrow 3 \rightarrow 2 \rightarrow 3$ that are in both events (hence, these events are not disjoint), and so the sum $\frac{1}{8} + \frac{1}{2} = \frac{5}{8}$ is **not** the probability of going from state 3 to state 2 in one or three moves.

We should not give up on this sum too quickly, however—it does turn out to have some significance. Notice that in the above example the *only* outcomes that are counted twice are those that entail the game which starts at state 3 being in state 2 twice in the four moves. Of course beyond four moves, it is true in general that

The sum of the row 3, column 2 entries in M, M^2, M^3, M^4, ... adds k times the probability that a game starting in state 3 is in state 2 exactly k times.

This means that the sum of these entries is the same as

$(1) \times Prob$(game starting in state 3 is in state 2 one time)
$+(2) \times Prob$(game starting in state 3 is in state 2 two times)
$+(3) \times Prob$(game starting in state 3 is in state 2 three times)
$+(4) \times Prob$(game starting in state 3 is in state 2 four times) $+ \cdots$

This is exactly our formula for expected values! □

It follows from this type of reasoning that in the sum $M + M^2 + M^3 + \cdots$, the entry in row i, column j is the expected number of times that a game which starts at state i will be in state j. This is not an accurate statement when $i = j$ because nowhere in the analysis have we accounted for the fact that a game starting in state i is in state i one time without moving at all. So we are careful with our wording in the following result, which we state without proof.

Theorem 3 *Given a transition matrix M for a game, in the matrix sum*

$$Q = M + M^2 + M^3 + M^4 + \cdots + M^n$$

the entry in row i, column j is the expected number of times that a game which begins in state i will enter state j within the first n turns of the game.

Once again, for this to be useful, we need n to be fairly large. In fact, the theoretical result we are after involves understanding the sum in Theorem 3 as a limit process in which $n \rightarrow \infty$. There is a hidden difficulty with this strategy as it stands now. With the use of technology, it is easy to see that sums like

$$M + M^2 + \cdots + M^{100} \approx \begin{bmatrix} 100 & 0 & 0 & 0 \\ 66.2 & 0.33 & 0.67 & 32.8 \\ 32.8 & 0.67 & 0.33 & 66.2 \\ 0 & 0 & 0 & 100 \end{bmatrix} \tag{6.5}$$

or

$$M + M^2 + \cdots + M^{500} \approx \begin{bmatrix} 500 & 0 & 0 & 0 \\ 332.9 & 0.33 & 0.67 & 166.1 \\ 161.1 & 0.67 & 0.33 & 332.9 \\ 0 & 0 & 0 & 500 \end{bmatrix} \tag{6.6}$$

have some large entries. In fact, these examples indicate that as n gets large, so do some of the entries in the matrix sum, and so we will never see these sums "settling down" on some fixed matrix.

If we consider our Hank and Ted example, the large entries that we saw in the evaluation of $M + M^2 + \cdots + M^{100}$ can be understood by recalling an observation we made about the individual powers of M. Each power M^k is a transition matrix in its own right, and so it has probabilities for entries and rows that sum to 1. Hence, when we form the sum $M + M^2 + \cdots + M^{100}$, we will have rows that sum to 100, and when we sum 500 powers of M, we will have rows that sum to 500, so it is clear that some entries are going to get large without bound as we add more and more powers. This is a real problem for thinking of the powers as "going to infinity."

However, one of the nice properties of our matrix model is that if we simply ignore the rows and columns of the transition matrix corresponding to the absorbing states, then the remaining matrix still models the other states perfectly well. For example, if we let N be the result of ignoring states 1 and 4 in the matrix M for the Hank and Ted game with 3 markers from the start of this section:

$$N = \begin{bmatrix} 0 & 1/2 \\ 1/2 & 0 \end{bmatrix}$$

then computing

$$Q = N + N^2 + \cdots + N^{500} \approx \begin{bmatrix} 0.33 & 0.67 \\ 0.67 & 0.33 \end{bmatrix} \tag{6.7}$$

distills an accurate picture of states 2 and 3 from the mess that was shaping up in (6.6).

We can now finally see how to use this to analyze the expected length of the original Hank and Ted game with 3 markers that has driven our discussion so far.

Example 5 *Use the matrix results above to analyze the expected length of the Hank and Ted game played with 3 total markers.*

SOLUTION This is a fairly simple game since it only has two possible starting points, $H = 1, T = 2$ or $H = 2, T = 1$. These correspond to states 2 and 3, respectively. We remarked before that it seems like

$$M^n \to \begin{bmatrix} 1 & 0 & 0 & 0 \\ 2/3 & 0 & 0 & 1/3 \\ 1/3 & 0 & 0 & 2/3 \\ 0 & 0 & 0 & 1 \end{bmatrix} \quad \text{as } n \to \infty$$

From this we were able to give the probabilities of reaching either ending state from any beginning state.

In the matrix sum (6.7), we saw

$$Q = N + N^2 + \cdots + N^{500} \approx \begin{bmatrix} 0.33 & 0.67 \\ 0.67 & 0.33 \end{bmatrix}$$

Since the sum of the entries in first row above seems to be 1, we can state that

The expected number of times that a game which begins in state 2 (which corresponds to the first row in N) enters a nonabsorbing state is 1.

Summing the entries in the second row of N gives us a similar statement about games that begin in state 3.

Since every game must consist of some number of moves among nonabsorbing states followed by a single move into an absorbing state, the expected number of moves in a game that starts in state 2 is $1 + 1 = 2$ moves. The same number of moves is expected of a game that starts in state 3, the game starting with $H = 2, T = 1$.

$\square$

These results can be summarized in the following corollary to Theorem 3.

Corollary 4   *Suppose N is the transition matrix for the nonabsorbing states (also called* transient states*) of a game, and the infinite matrix sum*

$$Q = M + M^2 + M^3 + \cdots$$

exists. Then the expected length of the game that starts in the state corresponding to row i is 1 more than the sum of the entries in row i of Q.

Practice Problem 5   *If the Hank and Ted game is played with 4 total markers and a coin that comes up heads $\frac{3}{5}$ of the time, how many moves on average will the game last if each player starts with 2 markers?*

In Section 6.5, we also addressed problems where a recursive process was involved but where there were no winning probabilities to worry about. In these problems, the *only* issue is the expected duration of the process. In terms of matrices, these problems are typically modeled with a transition matrix having only one absorbing state reflecting the single way that the process can end.

Example 6   *Use matrices to analyze the game where a coin is tossed until it first comes up heads. In particular, use the matrix model to explain why the average number of tosses for this to happen is 2.*

SOLUTION   If we think of state 1 as heads and state 2 as tails, then this game must continue in state 2 until it first goes to state 1, at which time it is over. The following 2×2 matrix describes the transition probabilities:

$$M = \begin{bmatrix} 1 & 0 \\ 1/2 & 1/2 \end{bmatrix}$$

Since state 1 is the absorbing state, we work instead with $N = \begin{bmatrix} 1/2 \end{bmatrix}$, and estimate the sum

$$Q = N + N^2 + N^3 + \cdots$$

by calculating the first 100 terms $N + N^2 + N^3 + \cdots + N^{100} = \begin{bmatrix} 1 - 1/2^{100} \end{bmatrix} \approx [1]$. From this estimate we see that the game which begins in state 2 will enter state 2 on average one time. The game also requires one flip to enter the absorbing state 1, so the expected number of tosses in this game is $1 + 1 = 2$. $\square$

Our final example shows how matrix models can be used to analyze children's board games. These games usually consist of a small number of states, fixed probabilities for moving between states, and no strategy or decisions in playing the games.

Example 7 *Consider the following scaled-down version of a children's game. Each of two children starts with a cherry tree with four cherries, and they take turns spinning a spinner with four equally likely outcomes. Three of these outcomes instruct the child to remove one, two, or three cherries, respectively, from her tree, and the fourth position tells the child to put all his cherries back on the tree. What is the expected number of spins for a child to get all his cherries off of the tree?*

SOLUTION   The following 5×5 matrix represents the transitions for the game:

$$M = \begin{array}{c} 0\text{ cherries} \\ 1\text{ cherries} \\ 2\text{ cherries} \\ 3\text{ cherries} \\ 4\text{ cherries} \end{array} \begin{bmatrix} 1 & 0 & 0 & 0 & 0 \\ 3/4 & 0 & 0 & 0 & 1/4 \\ 2/4 & 1/4 & 0 & 0 & 1/4 \\ 1/4 & 1/4 & 1/4 & 0 & 1/4 \\ 0 & 1/4 & 1/4 & 1/4 & 1/4 \end{bmatrix}$$

From this we can compute the following matrix Q as

$$Q = N + N^2 + N^3 + \cdots$$

where N is the submatrix of M corresponding to the transient states of the game:

$$N = \begin{array}{c} \text{State 1} \\ \text{State 2} \\ \text{State 3} \\ \text{State 4} \end{array} \begin{bmatrix} 0 & 0 & 0 & 1/4 \\ 1/4 & 0 & 0 & 1/4 \\ 1/4 & 1/4 & 0 & 1/4 \\ 1/4 & 1/4 & 1/4 & 1/4 \end{bmatrix}$$

If we estimate Q using the first 500 terms of this sum, we get

$$Q = N + N^2 + N^3 + \cdots + N^{500} \approx \begin{bmatrix} 0.19 & 0.15 & 0.12 & 0.49 \\ 0.49 & 0.19 & 0.15 & 0.61 \\ 0.61 & 0.49 & 0.19 & 0.76 \\ 0.76 & 0.61 & 0.49 & 0.95 \end{bmatrix}$$

From this matrix Q, we can extract the information that, for example, when starting in state 4, the game will land an average of 0.76 moves in state 1, 0.61 moves in state 2, 0.49 moves in state 3, and 0.95 moves in state 4. Therefore, the game lands in some transient state on average

$$0.76 + 0.61 + 0.49 + 0.95 \approx 2.8 \text{ moves}$$

which means the game lasts on average a total of approximately $2.8 + 1 = 3.8$ moves. □

Summary of Markov Chain Techniques

Suppose we are given a process with a finite number of states (each is either transient or absorbing) and a transition matrix M that models the probabilities of changing states. Then

- The entry in row i, column j of M^n is the probability that a process which begins in state i will end in state j exactly n transitions later.
- The entry in row i, column j of the matrix limit

$$M_\infty = \lim_{n \to \infty} M^n$$

(if the limit exists) is the probability that a process which begins in state i eventually settles into state j.

- The entry in row i, column j of

$$Q = N + N^2 + N^3 + \cdots$$

(where N is formed by deleting from M the rows or columns of the absorbing states) is the expected number of times that a process which begins in state i will enter state j.

- Adding 1 to the sum of the entries in row i of Q (from above) gives the average number of transitions needed for a process that begins in state i to be absorbed.

Obtaining Exact Answers

In this short section, we will try to get a feel for the mathematics involved in obtaining exact answers to some of these problems. This is important since we can be in situations where it is not obvious that a process is almost certainly in an absorbing state after 100 or 1,000 transitions the way it is with our small dice and coin games. The proofs of some of these facts are not too hard—they are within the scope of a typical first linear algebra course in college—but they would take us too far afield to pursue in this book.

We have seen that the advantage to only paying attention to the nonwinning states (*transient states*) is that in a game guaranteed to end, the entries in N^n all tend to 0 as n gets large without bound, and this fact turns out to be enough to guarantee that the entries in the infinite sum

$$N + N^2 + N^3 + \cdots$$

do not grow without bound.

Once we are confident that this infinite sum exists, it turns out to be fairly easy to find even without doing huge technology-intensive calculations. Instead of computing huge powers of matrices, we simply apply the following polynomial fact from Exercise 14 in Section 2.3:

$$(1 - x)(x + x^2 + x^3 + \cdots) = x$$

In terms of matrices, this means that

$$(I - N)(N + N^2 + N^3 + \cdots) = N$$

where I is the identity matrix* having 0's everywhere except down the main diagonal where it has 1's.

Example 8 *Show how the above fact can be applied to the Hank and Ted game played with a total of 3 markers.*

SOLUTION We have already established that this problem uses the following transition matrices:

$$M = \begin{bmatrix} 1 & 0 & 0 & 0 \\ 1/2 & 0 & 1/2 & 0 \\ 0 & 1/2 & 0 & 1/2 \\ 0 & 0 & 0 & 1 \end{bmatrix} \quad \text{and} \quad N = \begin{bmatrix} 0 & 1/2 \\ 1/2 & 0 \end{bmatrix}$$

which means that

$$I - N = \begin{bmatrix} 1 & -1/2 \\ -1/2 & 1 \end{bmatrix}$$

Our next goal is to find a matrix that we can multiply by $I - N$ to get I as the result. Without a little background we cannot give a simple method for doing this here, so we will just leave it to a calculator or computer algebra system at this point. The *inverse* of $I - N$ is

$$\begin{bmatrix} 4/3 & 2/3 \\ 2/3 & 4/3 \end{bmatrix}$$

a result that can be much more easily checked than found. Observe that

$$\begin{bmatrix} 4/3 & 2/3 \\ 2/3 & 4/3 \end{bmatrix} \cdot \begin{bmatrix} 1 & -1/2 \\ -1/2 & 1 \end{bmatrix} = \begin{bmatrix} 1 & 0 \\ 0 & 1 \end{bmatrix}$$

This means that we have a much easier way to compute the infinite sum:

$$N + N^2 + N^3 + \cdots = (I - N)^{-1} \cdot N = \begin{bmatrix} 4/3 & 2/3 \\ 2/3 & 4/3 \end{bmatrix} \cdot \begin{bmatrix} 0 & 1/2 \\ 1/2 & 0 \end{bmatrix}$$

$$= \begin{bmatrix} 1/3 & 2/3 \\ 2/3 & 1/3 \end{bmatrix}$$

Note that this is the same result we inferred from looking at the first 500 terms of the sum in Example 5. Now we know for sure that this is correct. □

Example 9 *(Example 6 revisited.) Use the exact matrix model to explain why the average number of tosses for the first heads to come up is 2.*

SOLUTION Recall that state 1 is heads and state 2 is tails, and the following is the transition matrix for the problem:

$$M = \begin{bmatrix} 1 & 0 \\ 1/2 & 1/2 \end{bmatrix}$$

---

* The 2×2 identity matrix was used in Practice Problem 1.

Since state 1 is the absorbing state, we work instead with $N = [\,1/2\,]$, and compute the sum

$$N + N^2 + N^3 + \cdots$$

by calculating instead $(I - N)^{-1} \cdot N = [1/2]^{-1} \cdot [1/2] = [1]$. That is, the game that begins in state 2 will enter state 2 on average one time. The game also requires one flip to enter the absorbing state 1, so the expected number of tosses in this game is $1 + 1 = 2$ as we saw before. $\qquad\square$

Example 10 *(Example 7 revisited.) Use the exact matrix methods to find the expected number of spins for a child to get all her cherries off of the tree in the children's game in Example 7*

SOLUTION   Recall that the following 5×5 matrix represents the transitions for the game:

$$M = \begin{bmatrix} 1 & 0 & 0 & 0 & 0 \\ 3/4 & 0 & 0 & 0 & 1/4 \\ 2/4 & 1/4 & 0 & 0 & 1/4 \\ 1/4 & 1/4 & 1/4 & 0 & 1/4 \\ 0 & 1/4 & 1/4 & 1/4 & 1/4 \end{bmatrix}$$

From this we can compute the matrix

$$Q = N + N^2 + N^3 + \cdots$$

where N is the submatrix of M corresponding to the transient states of the game:

$$N = \begin{bmatrix} 0 & 0 & 0 & 1/4 \\ 1/4 & 0 & 0 & 1/4 \\ 1/4 & 1/4 & 0 & 1/4 \\ 1/4 & 1/4 & 1/4 & 1/4 \end{bmatrix}$$

We saw that to find Q we should compute

$$Q = (I - N)^{-1} N \approx \begin{bmatrix} 0.19 & 0.15 & 0.12 & 0.49 \\ 0.49 & 0.19 & 0.15 & 0.61 \\ 0.61 & 0.49 & 0.19 & 0.76 \\ 0.76 & 0.61 & 0.49 & 0.95 \end{bmatrix}$$

This agrees with our computation of Q in Example 7, so all the conclusions there are still valid here. $\qquad\square$

In this section, there have been several unproven theorems and some perhaps mysterious statements, but we hope that seeing these tools used will serve to motivate you to learn more about linear algebra. We have perhaps raised as many questions as we have answered. Does the matrix $(I - N)^{-1}$ always exist? Is there a quick way to find M^n for large values of n? These questions are central to the foundation of the subject of Markov chains. They are some great questions for your linear algebra instructor when you take that course later in your academic career.

Solutions to Practice Problems

1 (a) $A \cdot B = \begin{bmatrix} -1 & 0 \\ -1/2 & -1/4 \\ 1/6 & 1/3 \end{bmatrix}$

(b) $B \cdot A$ is undefined.

(c) $C \cdot B$ is undefined.

(d) $B \cdot C = \begin{bmatrix} 1 & -2 \\ -1/3 & 0 \\ 0 & 1/2 \\ 0 & 0 \end{bmatrix}$

(e) $C^2 = \begin{bmatrix} 1 & -4 \\ 0 & 1 \end{bmatrix}$

(f) $C \cdot I = C$

(g) $I \cdot C = C$

(h) $B \cdot I = B$

I is called an identity matrix because it does not change any (compatible) matrix multiplied by it.

2 (a) According to Theorem 1, this is the row 2, column 4 entry in M^4, which is $\frac{5}{16}$.

(b) Since Hank 2, Ted 1 is state 3 and Hank 1, Ted 2 is state 2, this question is answered by the row 3, column 2 entry in M^4, which is 0. This makes sense if you think about it.

3 For the Hank and Ted game played with 4 markers, there are five possible states:

● **State 1** is $H = 0$, $T = 4$.
● **State 2** is $H = 1$, $T = 3$.
● **State 3** is $H = 2$, $T = 2$.
● **State 4** is $H = 3$, $T = 1$.
● **State 5** is $H = 4$, $T = 0$.

The unfair coin makes the transition matrix M somewhat asymmetric:

$$M = \begin{bmatrix} 1 & 0 & 0 & 0 & 0 \\ 2/5 & 0 & 3/5 & 0 & 0 \\ 0 & 2/5 & 0 & 3/5 & 0 \\ 0 & 0 & 2/5 & 0 & 3/5 \\ 0 & 0 & 0 & 0 & 1 \end{bmatrix}$$

We can compute

$$M^{100} \approx \begin{bmatrix} 1 & 0 & 0 & 0 & 0 \\ 0.585 & 10^{-16} & 0 & 10^{-16} & 0.415 \\ 0.308 & 0 & 10^{-16} & 0 & 0.692 \\ 0.123 & 10^{-16} & 0 & 10^{-16} & 0.877 \\ 0 & 0 & 0 & 0 & 1 \end{bmatrix} \text{ and}$$

$$M^{1,000} \approx \begin{bmatrix} 1 & 0 & 0 & 0 & 0 \\ 0.585 & 10^{-160} & 0 & 10^{-160} & 0.415 \\ 0.308 & 0 & 10^{-160} & 0 & 0.692 \\ 0.123 & 10^{-160} & 0 & 10^{-160} & 0.877 \\ 0 & 0 & 0 & 0 & 1 \end{bmatrix}$$

From this it appears that the powers are settling on the matrix

$$M_\infty \approx \begin{bmatrix} 1 & 0 & 0 & 0 & 0 \\ 0.585 & 0 & 0 & 0 & 0.415 \\ 0.308 & 0 & 0 & 0 & 0.692 \\ 0.123 & 0 & 0 & 0 & 0.877 \\ 0 & 0 & 0 & 0 & 1 \end{bmatrix}$$

Since state 5 is the winning state for Hank, this means that

● In the game that starts Hank 1, Ted 3, Hank has probability ≈ 0.415 of winning.
● In the game that starts Hank 2, Ted 2, Hank has probability ≈ 0.692 of winning.
● In the game that starts Hank 3, Ted 1, Hank has probability ≈ 0.877 of winning.

4 For a "win-by-three" game, we have seven states:

● **State 1** is "Player A wins."
● **State 2** is "Player A up two points."
● **State 3** is "Player A up one point."
● **State 4** is "Game tied."
● **State 5** is "Player A down one point."
● **State 6** is "Player A down two points."
● **State 7** is "Player A loses."

The transition matrix M for this game is

$$M = \begin{bmatrix} 1 & 0 & 0 & 0 & 0 & 0 & 0 \\ 2/3 & 0 & 1/3 & 0 & 0 & 0 & 0 \\ 0 & 2/3 & 0 & 1/3 & 0 & 0 & 0 \\ 0 & 0 & 2/3 & 0 & 1/3 & 0 & 0 \\ 0 & 0 & 0 & 2/3 & 0 & 1/3 & 0 \\ 0 & 0 & 0 & 0 & 2/3 & 0 & 1/3 \\ 0 & 0 & 0 & 0 & 0 & 0 & 1 \end{bmatrix}$$

With some experimentation, it appears that the powers of M are settling on the matrix

$$M_\infty \approx \begin{bmatrix} 1 & 0 & 0 & 0 & 0 & 0 & 0 \\ 0.984 & 0 & 0 & 0 & 0 & 0 & 0.016 \\ 0.952 & 0 & 0 & 0 & 0 & 0 & 0.048 \\ 0.889 & 0 & 0 & 0 & 0 & 0 & 0.111 \\ 0.762 & 0 & 0 & 0 & 0 & 0 & 0.238 \\ 0.508 & 0 & 0 & 0 & 0 & 0 & 0.492 \\ 0 & 0 & 0 & 0 & 0 & 0 & 1 \end{bmatrix}$$

Hence, a game that starts tied (i.e., in state 4) has a probability of approximately 0.889 of being a win for Player A.

5 This problem uses the same transition matrix M as Practice Problem 3, and the matrix N of transient states for

this game looks like this:

$$N = \begin{bmatrix} 0 & 3/5 & 0 \\ 2/5 & 0 & 3/5 \\ 0 & 2/5 & 0 \end{bmatrix}$$

Summing powers of N allows us to form

$$Q = N + N^2 + \cdots + N^{500} \approx \begin{bmatrix} 0.462 & 1.154 & 0.692 \\ 0.769 & 0.923 & 1.154 \\ 0.308 & 0.769 & 0.462 \end{bmatrix}$$

Exercises for Section 6.6

1. Hank and Ted start a game with Hank having 4 markers to Ted's 2 markers, but they roll a die for each move and Ted wins a marker from Hank with probability $\frac{2}{3}$. So Hank starts with twice as many markers, but Ted has twice the probability of winning each time. Give a transition matrix for this game, being clear about what the states are.

2. John and Jessica start a game with 4 markers each. On each move, John rolls a six-sided die and Jessica rolls an eight-sided die with the rule that whoever rolls the higher number wins a marker and no markers change hands when they roll the same value. Give a transition matrix for this game, being clear about what the states are.

3. Pete and André decide to play a game of table tennis where they start tied, and play until one player has won *four* points more than the other. Assuming that André has a $\frac{2}{3}$ probability of winning each point, give a transition matrix for this game, being clear about what the states are.

4. A certain mathematical board game consists of 16 squares, each one labeled with a subset of $\{1, 2, 3, 4\}$. On each move, a player rolls a four-sided die labeled 1,2,3,4—if they are on square S and they roll the value k, they move to square $S \cup \{k\}$. Note that this might constitute a move from a square to itself. Give a transition matrix for this game, being clear about what the states are.

5. Here is a miniature version of a children's board game for your analyzing pleasure. The game board is shown in Figure 6-6, and a four-sided die (with sides labeled 1, 2, 3, and 4) is used to generate the moves. A game piece starts at square A and moves left-to-right across each row, working its way up to square F in the top row. Give a transition matrix for this game, being clear about what the states are.

6. A game is played on the board in Figure 6-7 using a six-sided die to generate the moves. A game piece starts on square A and moves left-to-right, finishing the game when it makes it to square B. Give a transition matrix

The middle row of this result tells us about the game beginning in state 3 (with Hank and Ted each with 2 markers). Since the sum of the entries in the middle row is approximately $0.769 + 0.923 + 1.154 = 2.846$, the average length of a game that starts in state 3 is approximately $2.846 + 1 = 3.846$ moves.

for this game, being clear about what the states are. To what problem that was addressed earlier in this chapter does this game correspond?

7. Play the game in Exercise 1 ten times to get a feeling for the probability each player wins and the average length of this game.

8. Play the game in Exercise 2 ten times to get a feeling for the probability each player wins and the average length of this game.

9. Play the game in Exercise 3 ten times to get a feeling for the probability each player wins and the average length of this game.

10. Play the game in Exercise 4 ten times to get a feeling for the average length of this game.

11. Play the game in Exercise 5 ten times to get a feeling for the average length of this game.

Here is a quick way to compute a fairly large power of a matrix: Find M^2 by multiplying $M \times M$, then find M^4 by multiplying $M^2 \times M^2$. In this same way, we can find M^8 by multiplying $M^4 \times M^4$ and then M^{16} by multiplying $M^8 \times M^8$. Apply this strategy to answer Exercises 12 to 14.

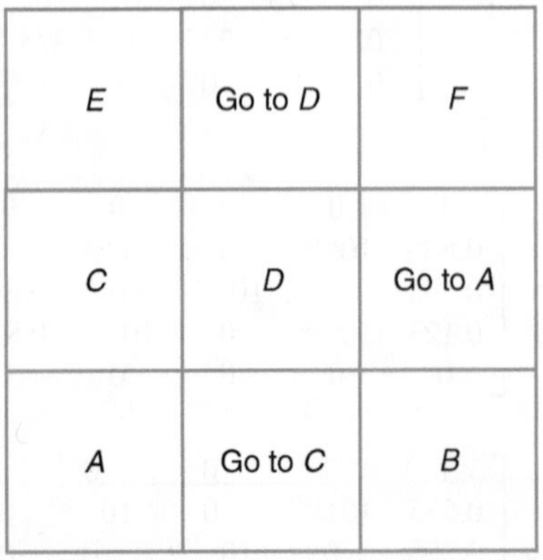

Figure 6-6 Game board for Exercise 5.

A	Go to *A*	Go to *A*	Go to *A*	Go to *A*	Go to *A*	B

Figure 6-7 Game board for Exercise 6.

12. In the game described in Exercise 1, what is the probability that Hank wins within 16 moves?

13. In the game described in Exercise 2, what the probability that John wins within 16 moves?

14. In the game described in Exercise 3, what is the probability that André wins within 16 moves?

15. Give the transition matrix for a best-of-five series between evenly matched opponents, and use this matrix to find the average length of such a series. (NOTE: Such a series must be over after five games, so you do not need to calculate large matrix powers.)

16. Give the transition matrix for a best-of-five series in which team *A* has a $\frac{3}{5}$ probability of winning each game, and use this matrix to find the average length of such a series. (This revisits Exercise 16 from Section 6.4.)

17. Give the transition matrix for a best-of-five series in which team *A* has a $\frac{2}{3}$ probability of winning each game, and use this matrix to find the average length of such a series. (This revisits Exercise 17 from Section 6.4.)

18. Suppose that team *A* has a $\frac{3}{5}$ probability of winning each game in a best-of-five series against team *B*. Use the transition matrix from your solution to Exercise 16 to find the probability that team *A* wins the series. (This revisits Exercise 22(b) from Section 6.3.)

19. Let $M = \begin{bmatrix} 1/2 & 0 \\ 0 & 1/2 \end{bmatrix}$, and prove by induction that for all $n \geq 1$,

$$M^n = \begin{bmatrix} (1/2)^n & 0 \\ 0 & (1/2)^n \end{bmatrix}$$

20. Let $M = \begin{bmatrix} 1 & 0 \\ 1/2 & 1/2 \end{bmatrix}$, and prove by induction that for all $n \geq 1$,

$$M^n = \begin{bmatrix} 1 & 0 \\ 1 - (1/2)^n & (1/2)^n \end{bmatrix}$$

21. Let $M = \begin{bmatrix} 0 & 1/2 \\ 1/2 & 0 \end{bmatrix}$, and prove by induction that for all $n \geq 1$,

$$M^{2n} = \begin{bmatrix} (1/4)^n & 0 \\ 0 & (1/4)^n \end{bmatrix}$$

and

$$M^{2n-1} = \begin{bmatrix} 0 & 2(1/4)^n \\ 2(1/4)^n & 0 \end{bmatrix}$$

22. Let $M = \begin{bmatrix} 0 & 1/2 \\ 1/2 & 0 \end{bmatrix}$, and prove by induction that for all $n \geq 1$,

$$M + M^2 + M^3 + \cdots + M^{2n} =$$
$$\begin{bmatrix} \frac{1}{3}(1 - (1/4)^n) & \frac{2}{3}(1 - (1/4)^n) \\ \frac{2}{3}(1 - (1/4)^n) & \frac{1}{3}(1 - (1/4)^n) \end{bmatrix}$$

Explain how this proves that the infinite sum

$$M + M^2 + M^3 + \cdots = \begin{bmatrix} 1/3 & 2/3 \\ 2/3 & 1/3 \end{bmatrix}$$

23. Prove Theorem 1 by induction on $k \geq 1$.

The remaining problems require the use of technology to perform the appropriate matrix operations.

24. What is the expected length of the game described in Exercise 1?

25. What is the expected length of the game described in Exercise 2?

26. What is the expected length of the game described in Exercise 3?

27. What is the expected length of the game described in Exercise 4?

28. What is the expected length of the game described in Exercise 5?

29. What is the expected length of the game described in Exercise 6?

30. Give the transition matrix for a tennis game between equally matched opponents, and use this matrix to find the average length of such a series.

31. If the server wins a tennis point with probability $\frac{2}{3}$, what is the probability that the server wins a tennis game that starts at 0-0? (This revisits Exercise 15 from Section 6.5.)

32. If the server wins a tennis point with probability $\frac{2}{3}$, what is the average length of a tennis game that starts at 0-0? (This revisits Exercise 16 from Section 6.5.)

Chapter 6 Summary

6.1 Introduction

Terms and concepts

- You should understand how the terms *experiment*, *outcome*, *equally likely*, and *sample space* are used in the study of probability.
- You should know what we mean by an *event*, and by *successful* outcomes of an experiment.
- You should recognize the notation $Prob(E)$ used to denote "the probability of event E."

Formulas

- $Prob(E) = \frac{n(E)}{n(S)}$ (provided the outcomes in S are equally likely).
- $Prob(E) + Prob(\overline{E}) = 1$.

Sample spaces and probability calculations

- You should be able to give sample spaces for a variety of situations and to determine whether the given sample space has equally likely outcomes. In particular, you should be able to give an equally likely sample space for rolling a pair of dice.
- Using the counting techniques of the preceding chapter, you should be able to calculate probabilites using the formula $Prob(E) = \frac{n(E)}{n(S)}$.
- You should realize that sometimes it is easier to calculate the probability that an event does *not* occur. In this case, you should be able to use the formula $Prob(E) + Prob(\overline{E}) = 1$ to calculate $Prob(E)$.

6.2 Sum and Product Rules for Probability

Terms and concepts

- You should know what it means for two events to be *disjoint* (or *mutually exclusive*).
- You should know what it means for two events to be *independent*, and you should realize that this is **not** the same as disjoint.
- You should understand the idea of *conditional probability*, along with its notation: $Prob(E_1|E_2)$, the *probability of E_1 given E_2*
- You should be able to distinguish a *trial* (or *Bernoulli trial*) from the larger experiment it is a part of.

Formulas

- Sum rule:
  - For disjoint events: $Prob(E_1 \text{ or } E_2) = Prob(E_1) + Prob(E_2)$
  - General form: $Prob(E_1 \text{ or } E_2) = Prob(E_1) + Prob(E_2) - Prob(E_1 \text{ and } E_2)$
- Product rule:
  - For independent events: $Prob(E_1 \text{ and } E_2) = Prob(E_1) \cdot Prob(E_2)$
  - General form: $Prob(E_1 \text{ and } E_2) = Prob(E_2) \cdot Prob(E_1|E_2) = Prob(E_1) \cdot Prob(E_2|E_1)$
- The sum rule generalizes for three or more disjoint events. Applying the general sum rule to events that are not disjoint requires a careful use of the Inclusion-Exclusion principle, and is not generally recommended for beginners.
- The product rule generalizes for three or more independent events. With some care, the generalized product rule can be applied to three or more events even if they are not independent.

Calculating probabilities

- You should be able to combine the use of the product and sum rules (and their generalizations), along with the methods of the preceding section, to calculate probabilities for a wide variety of applications.
- You should understand when it is permissible to use the simpler versions of the product and sum rules.

6.3 Probability in Games of Chance

Terms, concepts, and skills

- You should be able to distinguish a *trial* (or *Bernoulli trial*) from the larger experiment it is a part of.
- You should be able to solve "Bernoulli trials" problems of the form: "If the probability of success for each trial is p, what is the probability of obtaining exactly k successes in n trials?" The formula is given as $C(n, k) \cdot p^k \cdot (1 - p)^{n-k}$, but knowing the method that leads to that formula is preferable to memorizing the formula.
- You should be able to apply this method to analyze probabilities for "*best of*" series.

6.4 Expected Value in Games of Chance

Terms, concepts, and formulas

● You should understand that a *random variable* is simply a measurement of some numerical value associated with each possible outcome of a probability experiment.

● For a random variable X, you should know the meaning of the *expected value of X* (sometimes called the *average value* of X).

● You should recognize the notation $E[X]$ for the expected value, and be able to apply the formula $E[X] = (x_1) \cdot Prob(X = x_1) + (x_2) \cdot Prob(X = x_2) + \cdots + (x_n) \cdot Prob(X = x_n)$.

Calculating expected values

● You should be able to calculate expected values for a wide variety of situations. For example, the random variable X might represent:

  – The winnings for a game of chance.

  – The value showing on a die, spinner, etc.

  – The number of siblings a person has.

  – The number of times per year a person sees the doctor.

  – The number of hits a player gets in a baseball game.

  – The number of games in a "best of seven" series.

  – and so on

The essential ingredients are that the variable can take on a finite number of numerical values and that we can calculate the associated probability for each of those values.

● You should know the formula $n \cdot p$ for the average number of successes in n Bernoulli trials with probability p, but you should also be able to perform the necessary calculations to verify that formula for specific situations.

6.5 Recursion Revisited

Calculating expected values

● You should be able to extend the notion of expected value to random variables with an infinite number of possible outcomes, by using recursive reasoning. For example:

  – If you roll a single die repeatedly, how long on average does it take to roll a 6?

  – If you start with $50 and bet $1 repeatedly on a game where you have probability 1/2 of winning, how long can you expect to be able to play before going broke?

● Similarly, you should be able to use recursive reasoning to answer other probability questions, such as, "If a game of tennis is tied at deuce and A has probability 2/3 of winning each point, what is the probability that A will win the game?"

● Specifically, you should be able to apply these methods to the *Hank and Ted game* and other similar games.

6.6 Excursion: Matrices and Markov Chains

Terms and concepts

● You should understand the use of the word *state* to describe situations that can occur in a game, and the idea of a *transition matrix* representing probabilities of moving from state to state in the game.

● You should be able to distinguish *absorbing states* from *transient states*.

● You should recognize the standard use of algebraic operations applied to matrices: for example, $A + B$, $A \cdot B$, A^2.

● You should recognize the use of the symbol I to denote the *identity matrix*, and the use of the notation A^{-1} to indicate the *inverse* of the matrix A.

● You should recognize M_∞, informally, as the matrix the large powers of M are "settling on." Formally, $M_\infty = \lim_{n \to \infty} M^n$ provided the limit exists.

● The section consistently uses M for the transition matrix, N for the same matrix with the rows and columns for absorbing states removed, and Q for the matrix $N + N^2 + N^3 + \cdots$

Matrix operations

● For given matrices A and B, you should know how to calculate $A + B$ and $A \cdot B$, and you should know when these operations are possible.

● You should know the special properties possessed by the identity matrix I.

● You may have learned how to use computer software to calculate large powers of a matrix; or perhaps your instructor will supply the answers to those calculations for you.

● A similar comment applies to the calculation of matrix inverses.

Modeling games with matrices

● You should know how to build the transition matrix M for a game with a finite number of states.

- You should be able to compute probabilities for moving from state to state in 2 moves, in 3 moves, and so on, using matrix multiplication.
- You should be able to interpret the meaning of large powers of the transition matrix, for example M^{1000}, correctly using the notation M_∞ as part of your interpretation.

- You should be able to calculate expected values by interpreting the matrix $Q = N + N^2 + N^3 + \ldots$ (where N is the transition matrix with the rows and columns for absorbing states removed).
- You should be able to calculate Q exactly as $(I - N)^{-1} N$, with an assist from computer software or your instructor for calculating $(I - N)^{-1}$.

7

Graphs and Trees

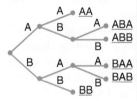

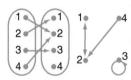

Figure 7-1 Game tree for best-of-three series.

Figure 7-2 Arrow diagrams for functions and relations.

The story of graph theory is a fascinating testimony to the study of recreational mathematics. Many early results in the area apply to nothing more than puzzles on maps or chessboards, and yet today graph theory comprises some of the fastest growing branches of applied mathematics. Graph theoretical concepts are used to design computer circuits, create production schedules, optimize communication networks, and countless other modern tasks. In the realm of abstract mathematics, graphs are often used to add visualization or simplification to difficult concepts. In fact, we have already seen examples of graphs and trees in earlier sections of this book:

- In determining all possible outcomes of a "best-of-three" match between Players A and B, we visualized all possible matches using a "game tree" as in Figure 7-1.

- We have also used "arrow diagrams" to help us visualize properties of functions and relations as in Figure 7-2.

All these diagrams are examples of graphs. The first is a special type of graph called a *tree*. The second and third are called *directed graphs* to emphasize the role of the arrows in the diagrams. In this chapter, not only will we see what these various diagrams have in common, but we will also study a number of additional applications of graphs and trees.

7.1 Graph Theory

Origins and Euler

The origin of graph theory is usually traced to a paper written by the great Swiss mathematician Leonhard Euler (1707–1783). In his paper Euler writes,

> The branch of geometry that deals with magnitudes has been zealously studied throughout the past, but there is another branch that has been almost unknown up until now; Leibniz spoke of it first, calling it the "geometry of position." (*geometria situs*). This branch of geometry deals with the relations dependent on position alone; it does not take magnitudes into consideration, nor does it involve calculation of quantities.

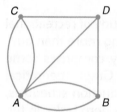

Figure 7-3 The seven bridges of Königsberg.

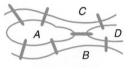

Figure 7-4 Abstract map of Königsberg.

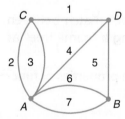

Figure 7-5 Labeled picture of Königsberg.

In this respect, graph theory represents one of the great turning points in the history of mathematics, involving none of the "classical" mathematics of the time. The inspiration for Euler's paper was a popular problem of the day—namely to determine if in the city of Königsberg (whose crude map is shown in Figure 7-3), it was possible to take a walk through the city using each of the seven bridges exactly once. Euler's simple solution to the question introduced the idea of modeling a problem with what we now call a graph. His insight was the observation that the most important factors in the problem were the four regions of the city (labeled A, B, C, and D in the figure) and how they were connected by the seven bridges between those regions. Figure 7-4 shows a graph that emphasizes just these important factors.

In this graph, the *vertices* (i.e., the points in the picture) represent the regions of the city. The *edges* (i.e., the line segments or arcs connecting the points) represent the bridges. Observe that there are two edges between vertices A and B, representing the two bridges joining regions A and B. Because there is no bridge directly joining regions B and C, the graph has no edge between vertices B and C.

Let us imagine Euler taking a walk about the city, using the integers 1 to 7 as labels for the seven bridges as shown in Figure 7-5. We can completely describe Euler's *walk* by listing, in turn, the regions he visits and the bridges he uses to get from region to region. (This is the style Euler uses to describe walks in his original paper.) For example, starting in region D, he might cross bridge 1 to region C, then use bridge 3 to go to region A, and so on. We can represent this information very efficiently as $D, 1, C, 3, A, \ldots$. Here are some other possible walks:

Walk #1: $A, 6, B, 5, D$

Walk #2: $A, 2, C, 3, A$

Walk #3: $D, 5, B, 7, A, 3, C, 1, D$

Walk #4: $D, 1, C, 2, A, 3, C, 2, A, 4, D, 5, B, 6, A, 7, B, 5, D$

Walk #5: $A, 2, C, 1, D, 4, A, 6, B, 7, A, 3, C$

The walk sought by the citizens of Königsberg has a special property—it must use each edge exactly once. Such a walk has come to be called an *Eulerian trail*. If it also begins and ends at the same vertex (i.e., in the same region of the city) it is called an *Eulerian circuit*. Some of these sample walks (#2, #3, and #4) begin and end at the same vertex, but none of the walks are Eulerian trails. Walks #1, #2, #3, and #5 do not use all the edges, and walk #4 uses edges 2 and 5 twice each.

So how did Euler prove that there is no Eulerian trail through the city of Königsberg? He simply considered the number of times such a walk would have to pass through each region of the city. For example, walk #5 above is a trail that begins at vertex *A* and ends at vertex *C*, so vertices *B* and *D* only occur within the walk. Hence, for each edge used to enter vertices *B* or *C*, there is a different edge used to leave it—that is, the edges used to traverse these "internal" vertices occur in pairs. Hence, we know that walk #5 uses an even number of edges connected to vertex *B* and an even number of edges connected to vertex *D*. Since in the entire graph of the city, there are an odd number of edges connected to vertex *B* and an odd number connected to vertex *D*, the direct conclusion of this analysis is that vertices *B* and *D* cannot be "interior" points in any walk that uses every edge in the Königsberg graph exactly once. More significantly, this argument can be generalized to give us the following.

Proposition 1 *In any graph, if there are an odd number of edges connected to a vertex x, then x cannot be an interior vertex (i.e., a vertex other than the starting or stopping point) in an Eulerian trail.*

We will not prove this proposition formally here for lack of formal definitions of many of the terms, but the preceding paragraph outlines the logic involved. This proposition completely resolves the Königsberg bridge problem since all four vertices in the Königsberg graph are connected to an odd number of edges, making all four of them incapable of being in the interior of an Eulerian trail. The only possible conclusion is that the Königsberg graph does not have an Eulerian trail!

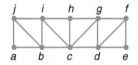

Figure 7-6 Graph for Practice Problem 1.

Practice Problem 1 *Explain why the graph shown in Figure 7-6 does not have an Eulerian trail.*

There are two comments we can make about Euler's solution to this problem. First, viewing the problem *abstractly* (i.e., with only the crucial factors present) was the key to solving the concrete problem about the city. Second, Euler did not simply solve this one problem. He essentially answered this type of question for any city whose map is given. Of course, to this point we have only seen a condition that allows us to say that a graph has no Eulerian trail. In his original paper, Euler also outlined a way to give a positive answer even for complicated cities, but we should develop proper definitions and proof tools before telling the rest of the story.

Terminology and Notation

To avoid getting ahead of ourselves, we need to develop some terminology. We begin with some terms that are suggested by the bridge problem, and then we will look at some ideas that are motivated by other applications. As you will see, most terms used in graph theory are chosen to elicit images of the particular property they name. Unfortunately, the terminology is not entirely standardized. If you read another book on graph theory, you may find different names for the same concept, or even occasionally the same name used for a different concept. The lesson here is to always pay attention to the definitions.

Definition

1. A *graph G* consists of two sets V and E. The elements of V are called *vertices* (or *nodes*), and the elements of E are called *edges*. Each edge is associated with one or two vertices, called its *endpoints*. In the diagram we draw the edge as a line segment or curved arc joining the endpoints.

2. If an edge has only one endpoint, then the edge joins the vertex to itself. This is called a *loop*.

3. If two edges have the same endpoints, they are called *multiple edges* or *parallel edges*.

4. Two nodes that are joined by an edge are said to be *adjacent* nodes.

5. A *walk* in a graph is a sequence $v_1 e_1 v_2 e_2 \ldots v_n e_n v_{n+1}$ with $n \geq 0$ of alternating vertices and edges, which begins and ends with a vertex and where each edge in the list lies between its endpoints. If the beginning vertex is the same as the ending vertex, we say the walk is *closed*. The *length* of a walk is the number of edges in the walk. A walk of length 0 is called a *trivial walk*.

6. A *trail* is a walk with no repeated edges, and a *path* is a walk with no repeated vertices. A *circuit* is a closed trail, and a *trivial circuit* is a circuit with one vertex and no edges. A trail or circuit is called *Eulerian* if it uses every edge in the graph.

7. A *cycle* is a nontrivial circuit in which the only repeated node is the first/last one.

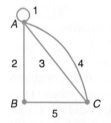

Figure 7-7
Graph for
Example 1.

 Example 1 *Let's try out this terminology on the graph in Figure 7-7.*

1. *The graph has three nodes and five edges.*

2. *Edge 1 is a loop, since it has only one endpoint—that is, A is joined to itself by edge 1. Edges 3 and 4 are parallel.*

3. *Here are some walks in the graph:*

   (a) *$B, 5, C, 5, B, 2, A, 2, B$. This closed walk repeats edge 5 so it is not a trail.*

   (b) *$A, 2, B, 5, C$. This walk is a path since it does not repeat any vertices.*

   (c) *B. This is considered a trivial walk and a trivial circuit.*

   (d) *$A, 1, A, 2, B, 5, C, 3, A$. This walk is a circuit since it is a closed trail that starts and ends at vertex A.*

   (e) *$A, 2, B, 5, C, 3, A, 1, A, 4, C$. This walk is an Eulerian trail since it uses every edge exactly once. The first part of this walk $A, 2, B, 5, C, 3, A$ (obtained by taking everything from the walk between two occurrences of vertex A) is a cycle.*

Practice Problem 2 *By adding one more edge to the graph in Figure 7-7, create a graph with an Eulerian circuit.*

Graphs in Applications

As we have seen, graphs are a fairly natural representation of real-world problems, and as such they are extremely versatile, perhaps to a fault. There are so many variations on these structures that it makes organizing them for study somewhat difficult. Every introductory textbook must make some concessions in the variety of graphs to study in order to build a coherent picture for students to learn the basic tools of the trade. We pause here briefly to illustrate some of this variety *before* making these inevitable concessions.

In any application of graph theory, there are many decisions that will have to be made at the outset. For example, a graph might be used to show a small airline's flights among a set of four cities. In Figure 7-8 we show two different graphs that describe these flights.

In the first graph, the existence of an edge simply indicates that there is a connection between the two cities (i.e., at least one flight in each direction). There is no reason to label the edges since referring to the endpoints (e.g., {Rome, Madrid}) unambiguously describes an edge. Since the first graph has no loops and no multiple edges (this is what we will call a *simple* graph), its edges can always be referenced in this way. Moreover, when describing a walk in a simple graph, we can simply list the vertices of the walk in the order they are traversed.

In the second graph, we explicitly show all the flights labeled by their flight numbers, and we use arrows to indicate the direction of the flight. When arrows are put on the edges to indicate direction, we call the graph a *directed* graph.

Which version of the graph is more appropriate? That depends on the intended application. The first example is the type you are likely to find in an ad for the airline, showing the cities it serves and the connections between them. The second captures more detailed information. Either graph can be used to determine that there is no direct flight from Madrid to London. To answer the question, "How can I get from Madrid to London on this airline?" we can also use either graph to get the solution

Go from Madrid to Rome to Paris to London.

This walk can be unambiguously described as "Madrid, Rome, Paris, London" in the first graph, or, if we wish to give more detailed instructions, we can describe the following walk in the second graph:

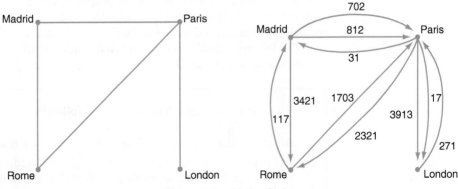

Figure 7-8 Airline flights.

Go from Madrid to Rome by flight 3421, from Rome to Paris by flight 1703, and from Paris to London by flight 3913.

So the amount of information that is conveyed in a graph depends entirely on the application at hand. We will focus on the simplest possible structure of vertices and edges in order to master the essential concepts of this important subject.

More Terms and Notation

The following definitions formally state some of the preceding ideas, along with some others that will be used in subsequent discussions. We postpone the formal definition of a directed graph until Section 7.4.

Definition

1. A *simple graph* is a graph with no loops and no multiple (parallel) edges.
2. The unordered list notation $[a, b]$ indicates an edge with endpoints a and b. This notation is ambiguous if the graph has multiple edges, but we will still use it if doing so will not cause confusion. In a simple graph, an edge with endpoints a and b can also be represented as the two-element set $\{a, b\}$.
3. If the graph is directed, we use (a, b) rather than $[a, b]$ to indicate a directed edge from a to b. This use of the usual *ordered pair* notation emphasizes that in a directed graph the order of the vertices connected by the edges is significant.

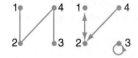

Figure 7-9 Graphs for Example 2.

Example 2 *For the graphs shown in Figure 7-9,*

1. *The picture on the left shows a simple graph with the three edges* $[1, 2]$, $[2, 4]$, *and* $[4, 3]$.
2. *The picture on the right is an arrow diagram for a function. As a graph, it is a directed graph with the four edges* $(1, 2)$, $(2, 1)$, $(3, 3)$, *and* $(4, 2)$.

It is worth noting that even though directed graphs will receive less emphasis in this chapter, much of what we say for graphs has an analog for the case of directed graphs. For example, if the bridges in Königsberg had been one-way streets, we would use a directed graph and we would be discussing Eulerian circuits in a directed graph.

The following definition contains a few additional terms we will use in the continuing discussion of the Königsberg bridge problem.

Definition In a graph G, we use the following terminology:

1. An edge e is said to be *incident with* a node v if v is an endpoint of e.
2. The *degree* of a node v, denoted by $deg(v)$, is the number of times v appears as an endpoint of an edge. That is, $deg(v)$ is the number of edges that are incident with v, except that loops are counted twice.

3. A graph G is *connected* if there is a walk between any pair of distinct nodes.
4. A graph H is a *subgraph* of a graph G if all nodes and edges in H are also nodes and edges in G.
5. A *connected component* of a graph G is a connected subgraph H of G such that no other connected subgraph of G containing H exists.

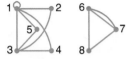

Figure 7-10 Graph for Example 3.

Example 3 *Refer to the graph consisting of 8 nodes and 12 edges shown in Figure 7-10.*

1. *The degree of node 3 is 4—this node appears in the four edges* $[1, 3]$, $[3, 2]$, $[5, 3]$, *and* $[3, 4]$. *The degree of node 6 is 3.*
2. *There is no walk that begins at node 4 and ends at node 6, so the graph is not connected.*
3. *The graph has two connected components.*

Practice Problem 3   Answer the following questions for the graph given above in Example 3:

(a) *Find the degree of node 1.*
(b) *There are two connected components, which we will call H_1 and H_2. Subgraphs H_1 and H_2 are graphs in their own right, each consisting of nodes and edges. Give a complete list of the nodes and edges that make up each of these subgraphs.*
(c) *Find an Eulerian trail in the connected component that includes node 6.*
(d) *Find a subgraph that is connected and contains just the nodes 2, 3, 4, and 5. Explain why this subgraph is not a connected component of the graph.*

Eulerian Graphs

We are now prepared to finish discussing Euler's achievement. So far we have only seen a condition we can use in order to get a "no" answer to the question of finding an Eulerian trail in a given graph. To complete the story, we need to see what conditions will guarantee a "yes" answer. For the other half of the picture, we will focus on the question of a city having an Eulerian circuit instead of an Eulerian trail.

Definition   A graph G is *Eulerian* if there is a circuit in G that involves every edge exactly once. Recall that such a circuit is called an *Eulerian circuit*.

Euler found a simple way to decide if any given connected graph is Eulerian.

Theorem 2   *Let G be a connected graph. The graph G is Eulerian if and only if every node in G has even degree.*

The proof, which is very elegant and uses the recursive structure of graphs, will be given in the next section where we discuss the proof techniques for graphs

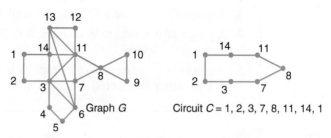

Graph G

Circuit $C = 1, 2, 3, 7, 8, 11, 14, 1$

Figure 7-11 Finding a circuit in G.

in more detail. The following example illustrates the process for constructing an Eulerian circuit in a connected graph in which every vertex has even degree. Notice where the construction process repeats itself recursively—this will (eventually) be the induction step in our proof.

 Example 4 *Let G be the graph given on the left in Figure 7-11.*

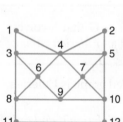

$H_1 \rightarrow$

H_2

Figure 7-12 The induction step.

1. *First we find any circuit C, as shown in Figure 7-11 on the right. (We can do this formally by starting at node 1, and walking along unused edges until we get back to node 1.) Notice that we simply list vertices in our description of C since this unambiguously determines the intervening edges in this simple graph.*

2. *We next form the graph G' by removing from G all the edges in the circuit C. To simplify this example, we also remove those vertices (1 and 2) that have all of their edges removed by this process. The resulting graph G' is not connected, so we will call its components H_1 and H_2 in Figure 7-12 and continue. Since these are smaller connected graphs with every vertex of even degree, this same process will determine that both H_1 and H_2 have Eulerian circuits, respectively, $C_1 = 3, 4, 5, 6, 7, 11, 12, 13, 6, 3, 11, 13, 14, 3$ and $C_2 = 8, 9, 10, 8$.*

3. *We can now piece together C, C_1, and C_2. Specifically, in $C = 1, 2, 3, 7, 8, 11, 14, 1$, we replace the 3 with C_1 and the 8 with C_2 to get the Eulerian circuit:*

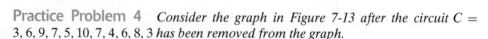

$$1, 2, \underbrace{3, 4, 5, 6, 7, 11, 12, 13, 6, 3, 11, 13, 14, 3}_{C_1}, 7, \underbrace{8, 9, 10, 8}_{C_2}, 11, 14, 1$$

Practice Problem 4 *Consider the graph in Figure 7-13 after the circuit $C = 3, 6, 9, 7, 5, 10, 7, 4, 6, 8, 3$ has been removed from the graph.*

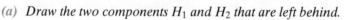

Figure 7-13 Graph for Practice Problem 4.

(a) *Draw the two components H_1 and H_2 that are left behind.*

(b) *Find Eulerian circuits C_1 and C_2, respectively, for the two components.*

(c) *Paste together C, C_1, and C_2 to make an Eulerian circuit for the original graph G.*

Graphs with Eulerian Trails

If you have been really paying attention in this section or else were never paying attention in elementary school, you might have noticed the similarity of the previous problem to a puzzle that usually makes the third grade circuit (so to speak). The

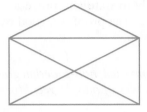 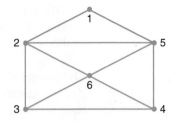

Figure 7-14 The envelope and its graph.

problem is to draw the picture in Figure 7-14 on the left (an "envelope") without lifting pencil from paper and without retracing any line. This problem was originally introduced in Chapter 1. We now have the mathematical sophistication to allow a thorough analysis of the problem.

This is the same problem as finding a trail in the graph on the right in Figure 7-14, using every edge exactly once. (Remember that we call this an *Eulerian trail*.) It might sound like we are asking that the graph be Eulerian, but we are not. The difference is that we are not looking for a circuit—that is, we don't care in this puzzle whether the pencil ends up where it starts! Take a moment to trace the figure as described if you have never done it before.

How closely related is this problem to the one about the bridges of Königsberg? Can we find a general result similar to Theorem 2 about this kind of graph? These questions have an easy and elegant answer. However, before we can give a complete characterization of graphs with Eulerian trails, we need to establish a couple of simple facts about graphs in general.

Theorem 3   *In any graph, the sum of the degrees of the vertices is equal to twice the number of edges. In symbols,*

$$\sum_{i=1}^{n} \deg(v_i) = 2m$$

where $v_1, v_2, \ldots, v_n$ are the vertices of the graph and m is the number of edges in the graph.

PROOF   Each edge has two (not necessarily distinct) endpoints, and the degree of a node counts the number of times it appears as an endpoint of an edge. So, in summing the degrees of the nodes, we count each edge exactly twice.   ■

Corollary 4   *In any graph G, the number of nodes with odd degree is even.*

PROOF   In Chapter 2 you proved that the sum of two even numbers is even, that the sum of two odd numbers is even, and that the sum of an odd and an even number is odd. It follows that, in calculating the sum of all the node degrees, the even numbers certainly sum to an even number, and from Theorem 3, the entire sum is even. Therefore, the odd numbers must sum to an even number too. And if the sum of a bunch of odd numbers is even, then there must be an even number of them.   ■

We are now ready to state and prove a general solution to all puzzles of the envelope-drawing type. The proof is a good example of letting previous results do all the hard work for you.

Theorem 5   *A connected, non-Eulerian graph G has an Eulerian trail if and only if G has exactly two nodes of odd degree. Moreover, the trail must begin and end at these two nodes.*

PROOF   First, suppose that G is a connected graph that is not Eulerian but does have an Eulerian trail, say, starting with node v and ending with node w. By Proposition 1, we know that every node except possibly v and w has even degree. Since the graph is not Eulerian, at least one node has odd degree (by Theorem 2). So, by the previous corollary, at least two nodes have odd degree. The only possibility is that v and w are the only two nodes of odd degree in G.

Now suppose that G is a connected graph with exactly two nodes of odd degree, say, x and y. Form the graph G' by adding to G the edge $[x, y]$. Now G' has all nodes of even degree, so it must have an Eulerian circuit by Theorem 2. Deleting the edge $[x, y]$ from this Eulerian circuit gives us an Eulerian trail in G from x to y. ∎

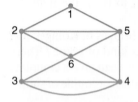

Figure 7-15 Adding an edge makes the envelope Eulerian.

Example 5 *The graph of the "envelope" we saw earlier can thus be modified by adding an edge between its two nodes (3 and 4) of odd degree to get the graph shown in Figure 7-15. The solution method of Example 4 can then be used to find the Eulerian circuit 3, 2, 1, 5, 4, 6, 5, 2, 6, 3, 4, 3 in this new graph. Dropping the last edge [4, 3] from this circuit gives us the Eulerian trail 3, 2, 1, 5, 4, 6, 5, 2, 6, 3, 4 in the original graph.*

Practice Problem 5   *Suppose the Eulerian circuit you found had been*

$$1, 5, 4, 6, 5, 2, 6, 3, 4, 3, 2, 1$$

Explain how to use this circuit to find the trail in the original envelope.

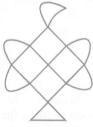

Explore more on the Web

Practice Problem 6   *How can one draw the picture in Figure 7-16 without lifting the pencil or retracing any part of the figure? (This picture of a bird comes from the sona sand drawing tradition of the Chokwe people of South Central Africa.*)*

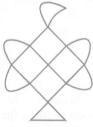

Figure 7-16 Picture for Practice Problem 6.

Example 6 *Here is a magic trick based on Eulerian graphs. The explanation is left as an exercise for the reader. You are the magician. Take a standard set of dominoes and discard the double 6, double 5, double 4, double 3, double 2, double 1, and double blank pieces. Have one spectator choose a single domino and give it to you. Turn your back and have a second spectator line up the remaining dominoes using the conventional rules that the faces touching must share the same number. The spectator*

---

* To learn more about this interesting story-telling tradition, see *Geometry from Africa* by Paulus Gerdes (MAA, 1999), Washington, DC, pp. 156–205.

should only have to make one such line—allow other spectators to help if he or she is having difficulty. Now you reveal (dramatically) the numbers on the ends of the line, and show that they are the same two numbers on the domino originally chosen.

Solutions to Practice Problems

1 Vertices c, f, h, and j are each connected to an odd number of edges. By Proposition 1, none of these can be "interior" vertices of an Eulerian trail, and certainly all four of them cannot make up the starting and ending points of a trail. Hence, this graph has no Eulerian trail.

2 The Eulerian trail in the example starts at A and ends at C. If we add another edge joining A and C, we can use that edge to get back to A and complete the Eulerian circuit.

3 (a) The degree is 6. Don't forget that for a loop the node is counted as an endpoint twice, once for each "end" of the edge.

(b) H_1 has nodes 1, 2, 3, 4, 5 and edges [1, 1], [1, 2], [1, 3], [1, 4], [1, 5], [2, 3], [3, 4], [3, 5], and H_2 has nodes 6, 7, 8 and edges [6, 7], [6, 8], [7, 8], and a second [6, 7] edge.

(c) 6, 7, 8, 6, 7 is one solution.

(d) Along with nodes 2, 3, 4, 5 include edges [3, 5], [3, 2], [3, 4]. This is not a connected component because it is contained within the connected subgraph H_1.

4 $C_1 = 4, 2, 5, 4, 3, 1, 4$ and $C_2 = 8, 9, 10, 12, 11, 8$ are Eulerian circuits for H_1 and H_2, respectively, shown in Figure 7-17. Pasting together these with $C = 3, 6, 9, 7, 5, 10, 7, 4, 6, 8, 3$ gives us the Eulerian circuit

$$3, 6, 9, 7, 5, 10, 7, \underbrace{4, 2, 5, 4, 3, 1, 4}_{C_1}, 6, \underbrace{8, 9, 10, 12, 11, 8}_{C_2}, 3$$

for the original graph G.

5 A circuit can be started at any node in the circuit. So we could rewrite the circuit we found so it either begins

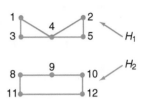

Figure 7-17 Solution to Practice Problem 4.

or ends with the edge we want to drop. One possibility is 4, 3, 2, 1, 5, 4, 6, 5, 2, 6, 3, 4. If we drop the first edge, we get the trail given in the example. Dropping the last edge gives 4, 3, 2, 1, 5, 4, 6, 5, 2, 6, 3.

6 If we label the nodes as shown in Figure 7-18, the picture can be drawn by visiting the nodes in order

$$1, 2, 3, 4, 6, 9, 7, 5, 8, 10, 11, 12, 11, 9, 6, 4, 7, 10, 8, 5, 3, 1$$

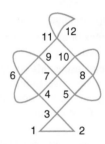

Figure 7-18 Solution to Practice Problem 6.

Exercises for Section 7.1

1. Refer to the graph G in Figure 7-19.

(a) Find the number of nodes and the number of edges.

(b) Find the degree of each node.

(c) Compare the sum of the degrees with the number of edges. How are these two numbers related?

(d) List all cycles in the graph.

2. Refer to the graph G_1 in Figure 7-19.

(a) How many nodes are there?

(b) List the degrees of the nodes in G_1.

(c) What is the sum of the degrees of the nodes in G_1?

(d) How many edges in G_1 are there?

3. Refer to the graph G_2 in Figure 7-19.

(a) How many nodes are there?

(b) How many edges are there?

(c) List the degrees of the nodes in G_2.

(d) Construct a second graph H with the same number of nodes and edges, but having a node with a higher degree than any node in G_2.

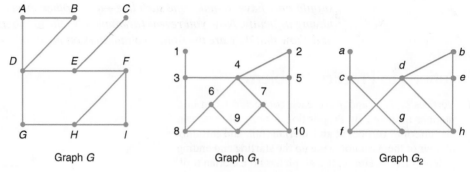

Figure 7-19 Graphs for Exercises 1, 2, and 3.

4. Refer to the graph in Figure 7-20.

(a) Find two different circuits with a length of exactly 7 that begin and end at 1.

(b) Find a walk from 1 to 8 that uses five edges and is not a trail.

(c) Find a trail from 3 to 5 that uses all the edges incident with 4.

(d) Find a circuit starting and ending at 8 that is not a cycle.

(e) Find a cycle of length 4.

5. Consider the graph in Figure 7-20.

(a) List the degrees of all the nodes. Note that all the degrees are even.

(b) Create a circuit using this strategy: Start at node 1, and choose any edge to move to another node. At that node, choose any unused edge to move to another node. Continue until you get back to node 1.

(c) Repeat the previous exercise. Make at least one different choice so that you get a different circuit.

(d) Do you think you will always eventually get back to the original node no matter which edge is chosen at each step? Will this happen for any connected graph you start with?

6. For each of these walks, find a shorter walk that begins and ends with the same nodes as the given walk, and which does not visit any node more than once.

(a) 1, 2, 3, 2, 4

(b) 4, 4, 5, 6, 2

(c) 9, 3, 2, 7, 8, 2, 4

(d) 1, 2, 3, 4, 5, 6, 7, 4, 8, 3, 9

(e) 3, 5, 6, 5, 4, 8, 9, 3, 2, 8, 7

(f) 8, 9, 2, 4, 6, 8, 11, 4, 2, 5, 7, 3, 2, 10

7. The *degree sequence* of a graph is the list of degrees of the nodes of the graph, listed from largest degree to smallest. For example, the degree sequence of the graph in Exercise 4 is 6, 6, 4, 4, 4, 4, 2, 2, 2, 2.

(a) Give an example of a simple graph with degree sequence 3, 3, 2, 2, 2.

(b) Give an example of a simple graph with degree sequence 3, 2, 2, 2, 1.

(c) Give an example of a simple graph with degree sequence 3, 3, 2, 2, 2, 2.

(d) Explain why there is no simple graph with degree sequence 7, 6, 5, 4, 3, 2, 1.

(e) Explain why there is no simple graph with degree sequence 3, 3, 3, 2, 2, 1, 1.

(f) Explain why there is no simple graph with degree sequence 5, 5, 3, 3, 1, 1.

8. Each of the graphs in Figure 7-21 is not Eulerian. In each case, add the fewest edges possible to create a graph that is Eulerian.

9. For each of the graphs in Figure 7-22, find an Eulerian circuit or an Eulerian trail, or else explain why neither is possible.

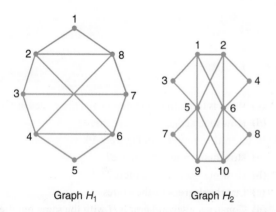

Figure 7-20 Graphs for Exercises 4 and 5.

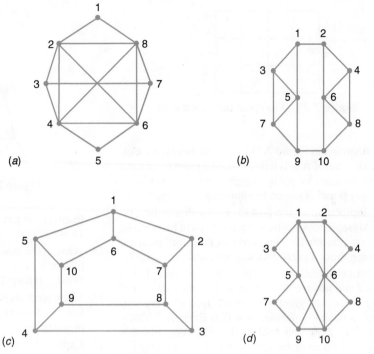

Figure 7-21   Graphs for Exercise 8.

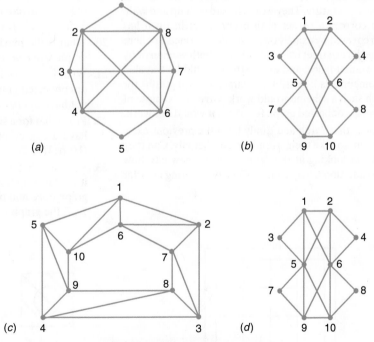

Figure 7-22   Graphs for Exercise 9.

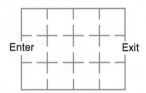

Figure 7-23   Floorplan for Exercise 10.

10. The floorplan in Figure 7-23 shows the Boatsville College Museum of Art. Is it possible for a guest to visit the entire museum by going though every doorway once and only once? (Convert the floorplan to a graph.)

11. The floorplan in Figure 7-24 shows the Boatsville College Museum of Art after renovations have been made. Is it now possible for a guest to visit the entire museum by going though every doorway once and only once?

12. The formal induction proof (in the next section) of Theorem 2 will involve the statement $P(n)$ that says, "For every connected graph G with n edges, if every vertex of G has even degree, then G is Eulerian." Verify $P(0)$, $P(1)$, $P(2)$, and $P(3)$ by drawing every possible connected graph of the appropriate size satisfying the hypothesis, and verifying the conclusion for each graph.

13. Sarah and Emily decided to go to a part of town (see the map in Figure 7-25—a node represents a corner and an edge represents a street) to distribute their radical political literature. They would like to split up and walk from corner to corner until every street in town has been covered by exactly one of them only once. So there should be no street traveled either by both women or by one woman twice, and every street should be traveled by someone. Is it possible? Characterize all graphs for which such a scheme would work. Give an example of a simple, connected graph for which it would not work.

14. Suppose that Sarah and Emily from the previous exercise want to take their crusade to another city. Can they tell from looking at the "graph" of the new city how many additional helpers they will have to bring in order

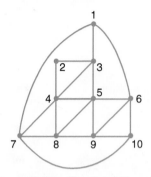

Figure 7-25   Graph for Exercise 13.

to cover the city in the sense of the previous exercise? How?

15. What is the minimum number of times that one must lift one's pencil to draw each picture shown in Figure 7-26?   (HINT: This is related to Exercise 8.)

16. From your investigation in Exercise 15, make a conjecture about the number of times one must lift one's pencil to draw *any* given picture.

17. Explain the magic trick in Example 6.

18. What is the minimum number of edges that a simple, connected graph with n vertices can have? (You do not have to prove your answer is correct.)

19. In light of the previous exercise, explain why there is no connected graph with degree sequence $3, 3, 2, 2, 1, 1, 1, 1, 1, 1$ (as defined in Exercise 7).

20. What is the maximum number of edges that a simple graph with n vertices can have? Justify your answer!

21. What is the maximum number of edges that a simple, disconnected graph with n vertices can have? (You do not have to prove your answer is correct.)

22. Prove that for a simple graph G in which all the n nodes have a degree of least 2, there are at least n edges. (HINT: Use Theorem 3.)

23. Let's say that a connected graph is *almost Eulerian* if it contains a closed walk that uses every edge in the graph once and one edge in the graph twice. In Figure 7-27, the graph shown on the left is "almost Eulerian"

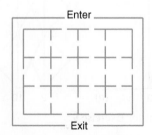

Figure 7-24   Floorplan for Exercise 11.

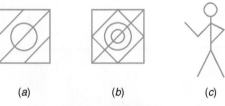

(a)   (b)   (c)

Figure 7-26   Graphs for Exercise 15.

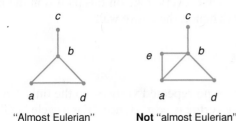

"Almost Eulerian"   **Not** "almost Eulerian"

Figure 7-27   Graphs for Exercise 23.

since the closed walk $W = a, b, c, b, d, a$ uses the edge $[b, c]$ twice and every other edge once. On the other hand, the graph on the right is **not** "almost Eulerian." Find a condition like the one in Theorem 2 that completely characterizes connected graphs that are almost Eulerian.

7.2 Proofs About Graphs and Trees

Our goal in this section is to discuss the proof techniques most commonly used in the study of graphs. In many cases, we will illustrate an idea first with simple graphs in order to keep the notation and definitions as streamlined as possible. Remember that in a simple graph, we can identify an edge by specifying its two endpoints, and we can represent a walk as a list of vertices only.

Direct Proofs of Graph Properties

We begin by noticing that except for a higher degree of notational complexity, our basic ideas for direct proofs still hold. Specifically, we try to write the statement to be proven in a simple "if, then" form so that we can invite the READER to give us any example satisfying the hypothesis, and we strive to convince her that the conclusion must also hold true for her example.

In our first example of a direct proof, we will need to use the following simple fact about connected graphs.

Lemma 1   *A graph is connected if and only if for every pair of different vertices a and b in G, there is a path in G starting at a and ending at b.*

The proof of this lemma follows from Proposition 6 that is itself proved a bit later in this section. That proof will use mathematical induction, so we prefer to postpone it for the moment.

Proposition 2   *Let G be a simple, connected graph, and let a and b be vertices in G such that there is no edge between a and b.* **If** *G′ is the graph formed by adding the edge {a, b} to G,* **then** *G′ has a cycle that contains the edge {a, b}.*

PROOF   Let a simple, connected graph G be given, let a and b be vertices in G such that there is no edge between a and b, and form the graph G' by adding the edge $\{a, b\}$ to G.

Since G is connected, we know by Lemma 1 there is a path

$$P = v_0, v_1, \ldots, v_k$$

where $v_0 = a$ and $v_k = b$ in G. Of course, every edge on this path is in the larger graph G' and the edge $\{a, b\}$ is in G'; hence, the entire walk

$$W = v_0, v_1, \ldots, v_k, v_0$$

is in G'. Since P is a path, it contains no repeated vertices, so the only repeated vertex in W is the beginning/ending vertex $v_0 = a$. Hence, W is a cycle in G'. ■

This property of simple graphs and its converse are very important for other proofs in this section, so we have included the converse as a "fill-in-the-blank" proof in Exercise 2 at the end of this section. You might want to try it before reading on.

Proposition 3   *When an edge is removed from a cycle in a connected graph, the result is a graph that is still connected.*

PROOF   See Exercise 2 for the outline of this direct proof. ■

Just as with any direct proof, it is our prerogative to prove the contrapositive of the statement instead of the original if, then statement. The following proof shows how this might work.

Proposition 4   *For every connected graph G with at least one edge, if G has no cycles, then G has at least one vertex of degree 1.*

It is impossible for a connected graph with at least one edge to have a vertex of degree 0, so the contrapositive statement is "For every connected graph G with at least one edge, if G has every vertex of degree at least 2, then G has a cycle." This is the form of the statement we will prove.

PROOF   (Of the contrapositive statement.) Let G be a connected graph with at least one edge and having every vertex of degree at least 2. Let n be the number of vertices in G, and choose v_0 to be any vertex in G. Now choose vertices to build a walk as follows:

- Since v_0 has a degree of at least 2, we can choose vertex v_1 to be the other endpoint of an edge incident with v_0.
- Since v_1 has a degree of at least 2, and we have already used one edge $[v_0, v_1]$ incident with v_1, we can choose vertex v_2 to be the second endpoint of a different edge incident with v_1.
- Since v_2 has a degree of at least 2, and we have already used one edge $[v_1, v_2]$ incident with v_2, we can choose vertex v_3 to be the second endpoint of a different edge incident with v_3.
- And so on until vertex v_n has been chosen.

This walk uses $n + 1$ vertices, but the entire graph G only has n vertices, so (by the pigeonhole principle, Practice Problem 3 of Section 2.5, to be precise) there must be a first value j such that $v_i = v_j$ for some $i < j$. This choice of i and j will guarantee that

$$v_i, v_{i+1}, \ldots, v_j$$

is a cycle in G. ∎

The type of graph under scrutiny in the previous proposition is an important one from several points of view.

> **Definition**   A *tree* is a connected, simple graph that has no cycles. Vertices of degree 1 in a tree are called *leaves* of the tree.

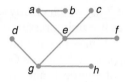

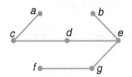

Figure 7-28 Graphs for Example 1.

Example 1   *Answer each of these questions for each of the trees shown in Figure 7-28:*

1. *How many nodes does the graph have? How many edges does the graph have?*
2. *How many leaves does the tree have?*
3. *Choose any edge and remove it. Is the resulting graph a tree?*
4. *Choose any two nodes and add an edge joining them. Is the resulting graph a tree?*

SOLUTION

1. The tree on the top has eight nodes and seven edges. The tree on the bottom has seven nodes and six edges.
2. In the tree on the top, vertices b, c, d, f, and h are leaves. In the tree on the bottom, vertices a, b, and f are leaves.
3. In each case, the result is not connected, so it is not a tree.
4. In each case, the result has a cycle, so it is not a tree.   □

The graphs in the previous example do not look much like the game trees and decision trees we have seen in earlier sections of this book. Those examples have even more structure than a typical graph, so we will defer studying them in detail until the excursion (Section 7.6) at the end of this chapter.

Explore more on the Web.

Practice Problem 1   *Prove that, for every tree G, removing any edge from G will result in a disconnected graph.* (HINT: *Write this statement in if, then form, and then consider the contrapositive.*)

Trees are often described in two different ways, providing us with two of the reasons why they are important. Some people think of a tree as a graph with the most edges possible without containing cycles—we can call this being "a maximal acyclic graph." This is essentially what we proved in Proposition 2. On the other hand, some people think of a tree as a graph with the least edges possible while remaining connected, which we could call being "a minimal connected graph." This is the essence of Practice Problem 1. Like many other ideas in discrete math, the most useful thing is to be able to think of a tree in *either* way, depending on the situation.

Induction Proofs of Graph Properties

We have seen that our usual direct proof style can be used with the graph theory definitions. However, the main proof technique for studying properties of graphs is mathematical induction. If you have only ever proved numerical facts using induction, this technique might seem strange when applied to the decidedly nonnumerical world of graph theory.

Proposition 5 *For every tree G with at least one edge, G has at least **two** leaves (i.e., vertices with degree 1).*

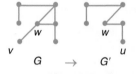

Figure 7-29
The base case for the proof of Proposition 5.

PROOF   Let $P(n)$ be the statement "Every tree with n edges has at least two leaves." Since the only tree with one edge is shown in Figure 7-29, which clearly has both vertices with degree 1, it follows that $P(1)$ is as true as it is uninteresting.

Let $m \geq 2$ be given such that statements $P(1), P(2), \ldots, P(m-1)$ have all been checked to be true. In considering the next statement $P(m)$, we let a tree G with m edges be given. By Proposition 4, there is a vertex v of degree 1, and so there is only one edge $\{v, w\}$ in G that has v as an endpoint. Form the graph G' by removing vertex v and edge $\{v, w\}$ from G.

Now G' is a tree with $m - 1$ edges, so since we have already checked statement $P(m-1)$, we can conclude that G' has at least two leaves. But w is the only vertex in G' that has a different degree in G' than it does in G. Choose one of the leaves in G' (other than w, if w is even a choice) and call it u. The degree of u in G is the same as the degree of u in G'. Hence, u and v are leaves in G. ∎

Since the structure of the induction proofs in this section is quite a bit more complicated than other proofs in this book, we will follow many of them with an example that traces the steps of the proof with a concrete example.

Example 2 *(illustrating Proposition 5.)* *Assuming that we have checked every possible tree with five or fewer edges, suppose we are given the tree G with six edges in Figure 7-30. The steps in the proof call for us to choose vertices v and w as shown in G. The graph G' on the right shows the leaf u that is described in the proof.*

Figure 7-30
Illustrating Proposition 5.

The same type of proof can be used on structures *within* a graph (like subgraphs, walks, circuits, etc.) as well.

Proposition 6 *In any simple graph G with distinct vertices a and b, if there is a walk from vertex a to vertex b, then there is a path from a to b.*

PROOF   (By induction.) Let the simple graph G be given, and $P(n)$ be the statement "For all vertices $a \neq b$ in G, if there is a walk of length n from a to b in G, then there is a path from a to b in G."

Statement $P(1)$ is true since a walk from a to b of length 1 must look like a, b, which is a path from a to b.

Let $m \geq 2$ be given such that statements $P(1), \ldots, P(m-1)$ have already been checked, and we are now considering statement $P(m)$. Let vertices a and b in G be given, and let the walk $W = v_0, v_1, v_2, \ldots, v_{m-1}, v_m$ be a walk of length m with $v_0 = a$ and $v_m = b$.

Now it might be the case that W is actually a path from a to b, which would mean that we have our desired conclusion. So let's address the other possible case, where W is not itself a path. By the definition of path, this means that some vertex is repeated in W. That is, in the list of vertices that describes W, there are values $i < j$ such that $v_i = v_j$.

Form the walk W' as follows. If $i = 0$, $W' = v_j, v_{j+1}, \cdots, v_m$; otherwise, if $j = m$, $W' = v_0, v_1, \cdots, v_i$; otherwise, $W' = v_0, \cdots, v_i, v_{j+1}, \cdots, v_m$. This walk still connects a to b, but the length of W' is less than the length of W. In particular, if we let k be the length of W', then we know that statement $P(k)$ has already been checked, and so it has already been verified that there is a path from a to b. This is our desired conclusion for statement $P(m)$, and so this completes the induction. ∎

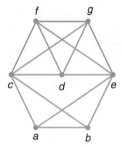

Figure 7-31
Illustrating
Proposition 6.

*Explore more on
the Web*

Example 3 (*illustrating Proposition 6.*) *In the graph shown in Figure 7-31, consider the walk $W = a, c, d, e, f, d, g, e, b$ of length 8 that is not itself a path. If we match this with the notation of the proof above, we would have this:*

$$W = \begin{matrix} a & c & d & e & f & d & g & e & b \\ v_0 & v_1 & v_2 & v_3 & v_4 & v_5 & v_6 & v_7 & v_8 \end{matrix}$$

Since W is not a path, we know it must have repeated vertices. If we look at the example, we see that $v_2 = v_5$, so when we form W', we simply remove everything in between v_2 and v_5 like this:

$$W' = \begin{matrix} a & c & d & g & e & b \\ v_0 & v_1 & v_2 & v_6 & v_7 & v_8 \end{matrix}$$

Practice Problem 2 *By adapting the proof of Proposition 6, prove that in any simple graph G, if there is a circuit containing the edge $\{v, w\}$, then there is a cycle containing the edge $\{v, w\}$.*

Many of our induction proofs about graphs will have this same basic form. It looks considerably more complex than some of our other forms, but it is really no different from the form of induction we have been using in other settings.

Claim. Every graph G with property p also has property q.

Proof by induction. Let $P(n)$ be a statement tantamount to "Every graph on n edges with property p also has property q." Verify the first few statements $P(1), P(2), \ldots$ by listing all possible graphs with property p with the given number of edges.

Let m be given such that statements $P(1), P(2), \ldots, P(m-1)$ have all been checked. To prove statement $P(m)$, let a graph G with property p and having m edges be given. Designate an edge e (either READER chooses or AUTHOR dictates) and form the graph G' by removing e from G. Graph G' or its connected components will be graphs with fewer than m edges and so the fact that the statements $P(1), \ldots, P(m-1)$ have already been checked will lead us to conclude that G' (or its components) has property q.

Use the fact that G' (or its components) has property q to conclude that G itself has property q.

We illustrate this basic proof form on the following important result about trees. Note that there are many other ways to prove this same result, one of which is given in the exercises at the end of this section.

Theorem 7 *If T is a tree with n edges, then T has n + 1 vertices.*

PROOF   Let $P(n)$ be the statement "Every tree with n edges has $n + 1$ vertices." Since the only tree with 0 edges consists of one vertex and no edges, it follows that $P(0)$ is an uninteresting but completely true statement.

Let $m \geq 1$ be given such that statements $P(0), P(1), \ldots, P(m-1)$ have all been checked to be true. In considering the next statement $P(m)$, we let a tree T with m edges be given. Choose any edge $\{v, w\}$ in T, and form the graph T' by removing this edge from T.

Now T' has two components H_1 and H_2, each of which is a tree. Let k_1 and k_2 be the number of edges in H_1 and H_2, respectively. Since $k_1 + k_2 = m - 1$, it follows that $k_1 < m$ and $k_2 < m$, so we know that statements $P(k_1)$ and $P(k_2)$ have each already been checked to be true. Hence, there are $k_1 + 1$ vertices in H_1 and $k_2 + 1$ vertices in H_2. Every vertex in T is in one of H_1 or H_2, so the number of vertices in T is

$$(k_1 + 1) + (k_2 + 1) = (k_1 + k_2) + 2 = (m - 1) + 2 = m + 1$$

Hence, statement $P(m)$ has been checked to be true, completing the induction.   ∎

Tree *T*

Figure 7-32
Illustrating the
proof of
Theorem 7.

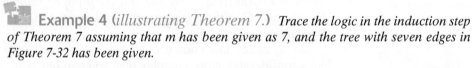

Example 4 (*illustrating Theorem 7.*) *Trace the logic in the induction step of Theorem 7 assuming that m has been given as 7, and the tree with seven edges in Figure 7-32 has been given.*

SOLUTION   We are assuming that statements $P(0), P(1), \ldots, P(6)$ have already been checked. In the given tree with seven edges, since edge $\{v, w\}$ is identified as in the first picture shown in Figure 7-33, we delete it to get the trees H_1 and H_2 shown in the second picture. The tree H_1 has four edges, so statement $P(4)$ says that H_1 has five vertices. Similarly, tree H_2 has two edges,

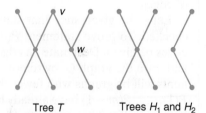

Tree *T*   Trees H_1 and H_2

Figure 7-33   Illustrating the proof
of Theorem 7.

and so by statement $P(2)$, we know that H_2 has three vertices. Hence, T has a total of $5 + 3 = 8$ vertices, as predicted by statement $P(7)$.   □

We can also see how the induction proof form allows us to give a careful proof of Theorem 2 of Section 7.1.

Theorem 2 revisited   Let G be a connected graph. The graph G is Eulerian if and only if every node in G has even degree.

PROOF   Before we begin the proof, we note that it is trivially true (and uninteresting) if G consists of a single vertex and no edges. The remainder of the proof assumes that G contains at least one edge.

First we show that if G is Eulerian, then every node has even degree. Suppose that G has an Eulerian circuit, and let x stand for any node of the graph. We will show that x has even degree. We write the Eulerian circuit as a walk $v_1 e_1 v_2 e_2 \ldots v_n e_n v_1$, observing that because it is a circuit the last node is the same as the first. Since G is connected, x is the same as one or more of the vertices v_1 through v_n. Each time (if any) it appears in the *middle* of the walk (as one of the vertices $v_2 \ldots v_n$), it is the endpoint for two different edges, the one just before it, and the one just after it. (In general, the node labeled v_i is the endpoint for e_{i-1} and e_i.) If it appears at the *beginning* (as vertex v_1), again it is the endpoint for two different edges, e_1 and e_n. So the degree of node x can be computed as twice the number of occurrences of x in the list of vertices v_1 through v_n. That is, the degree of x is even.

Next we show that if every node in G has even degree, then G is Eulerian. We do this by induction on the number of edges. Let $P(n)$ be the statement "Every connected graph with n edges in which every node has even degree has an Eulerian circuit." It is easy to check that the statement $P(1)$ is true since there are very few graphs with one edge. Now let $m \geq 2$ be given such that we know how to find an Eulerian circuit in any connected graph where each node has even degree and in which there are fewer than m edges. (That is, we have checked $P(1)$, $P(2)$, ..., $P(m-1)$.)

Consider a connected graph G that has m edges in which every node has even degree. In particular, every node has a degree greater than 1, so by the contrapositive of Proposition 4, there must be a cycle C in G. Form the graph G' by removing the edges involved in C from G. By the inductive assumption, each component $H_1, \ldots, H_k$ of G' has an Eulerian circuit $C_1, \ldots, C_k$. Each of these must meet C, say, at $v_1, \ldots, v_k$, so we can "insert" each one into C (i.e., arrange C_i to start and end with v_i, and then write C_i in place of v_i in C) to form the desired circuit.   ■

Graph Algorithms and Spanning Trees

We will address our last graph proof technique in the context of a specific example. In developing algorithms to solve problems, we will often construct graphs (or subgraphs) by an iterative process. To show that the end result of the algorithm is a success, we will prove that each step in the process maintains some desired property. Ultimately, this is another proof by induction, but treating it separately will allow us to concentrate on these first examples of graph algorithms.

Chapter 7 / Graphs and Trees

> **Definition** Let G be a simple, connected graph. The subgraph T is a *spanning tree of G* if T is a tree and every node in G is a node in T.

Does every connected graph necessarily have a spanning tree? The answer is yes. In Exercise 24 you are asked to prove this directly using induction. However, it is also possible to give an algorithm for *finding* a spanning tree. One simply removes edges that are part of a cycle until there are no longer any cycles left. More formally,

> **Spanning Tree Algorithm**
>
> - Begin with a simple, connected graph G_0.
> - For each $i \geq 1$, as long as there is a cycle in G_{i-1},
>   choose an edge e in any cycle of G_{i-1}, and form the subgraph G_i of G_{i-1} by deleting e from G_{i-1}.
> - The final result G_k will be a spanning tree of G_0. This is a spanning tree.

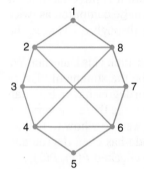

Figure 7-34 Graph for Practice Problem 3.

Practice Problem 3 *Refer to the graph in Figure 7-34 for the following problems:*

(a) *Apply the preceding algorithm to the graph. In each step, list the cycle you choose, and the edge you choose from that cycle.*

(b) *How many edges are there in the resulting spanning tree?*

(c) *Do you think everyone in the class got the same answer for the first part? Do you think everyone got the same answer for the second part?*

Proposition 8 *For any simple, connected graph G, the algorithm above produces a spanning tree of G.*

PROOF Suppose the algorithm is executed starting with $G_0 = G$, where G is the given simple, connected graph. We will use induction to prove the statement $P(n)$: "The graph G_n is a connected subgraph of G that uses all the vertices of G," for values of n from 0 to k. The first statement $P(0)$ says, "The graph G_0 is a subgraph of G that uses all the vertices of G." Since $G_0 = G$, this statement is silly but quite true.

Let the integer m (with $1 \leq m \leq k$) be given such that $P(m)$ is the first statement not yet checked. That is, we have already checked statements $P(0), P(1), \ldots, P(m-1)$, and we are ready to consider statement $P(m)$. Since graph G_m is obtained by removing an edge from a cycle in the connected (by $P(m-1)$) graph G_{m-1}, we know from Proposition 3 that graph G_m is connected. Since (by $P(m-1)$) graph G_{m-1} uses all the vertices of G and we do not remove any vertices in forming G_m, we know that G_m uses all the vertices in G. Finally, since (by $P(m-1)$) graph G_{m-1} is a subgraph of G and G_m is a subgraph of G_{m-1}, we can infer that G_m is a subgraph of G. Hence, statement $P(m)$ is true, completing the induction.

How does this prove that the algorithm works? We need to say one more thing. The algorithm only stops when it produces a graph G_k having no cycles. The induction proof above ensures that G_k is a connected subgraph of G using all the vertices of G. Hence, by the definition of "spanning tree" we can conclude that G_k is a spanning tree of G. ∎

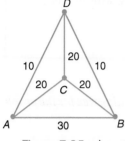

Figure 7-35 A weighted graph.

The idea of a spanning tree as a graph's "skeleton" is important for understanding many abstract properties of connected graphs, but there is a practical implication when the graph represents costs in a connected network. For example, suppose the graph shown in Figure 7-35 represents the potential connections of phone lines among towns A, B, C and D, where the numbers on the edges represent the cost of establishing that connection. Constructing lines that make a spanning tree would provide basic communication between all the towns, but finding a spanning tree with the *smallest possible total cost* would be of additional interest to the local phone company.

For such a "weighted" graph, it is not enough simply to find any spanning tree—we would like to find one whose total weight is as low as possible. When faced with a problem of this type, it is typical to try to design an algorithm that efficiently constructs a solution to the problem. We will see one example of such an algorithm along with a proof that it really does work. The exercises consider two other algorithms, including one that generalizes our simple algorithm for finding a spanning tree. Since we intend to write proofs about these algorithms, the first order of business is to establish the formal definitions.

> **Definition**   A *weighted graph* is a graph $G = (V, E)$ along with a function $w : E \to \mathbb{R}$ that associates a numerical weight to each edge. If G is a weighted graph, then T is a *minimal spanning tree of G* if it is a spanning tree and no other spanning tree of G has smaller total weight.

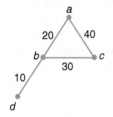

Figure 7-36 Weighted graph for Example 5.

Example 5   *The weighted graph G shown in Figure 7-36 has three spanning trees: $T_1 = \{[a, b], [a, c], [b, d]\}$ with total weight 70; $T_2 = \{[a, b], [b, c], [b, d]\}$ with total weight 60; and $T_3 = \{[a, c], [b, c], [b, d]\}$ with total weight 80. T_2 is a minimal spanning tree since each of the other two trees has a larger total weight.*

Practice Problem 4   *Find a simple, connected weighted graph that has more than one minimal spanning tree.*

Here is one possible algorithm for finding a minimal spanning tree. First, we present the algorithm, along with an example illustrating its use. We then prove that it does indeed yield a minimal spanning tree.

> **Prim's Minimal Spanning Tree Algorithm**
>
> - Given a connected, simple graph G with $n + 1$ nodes.
> - Let v_0 be any node in G, and let $T_0 = \{v_0\}$ be a tree with one node and no edges.

● For each k from $\{1, 2, \ldots, n\}$,

    Let $E_k = \{e$ an edge in $G : e$ has one endpoint in T_{k-1} and the other endpoint not in $T_{k-1}\}$.

    Let e_k be the edge in E_k with the smallest weight. (In case of a tie, choose any edge of the smallest weight.)

    Let T_k be the tree obtained by adding edge e_k (along with its node not already in T_{k-1}) to T_{k-1}.

● T_n is the tree returned by the algorithm.

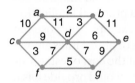

Figure 7-37
Illustrating Prim's
algorithm.

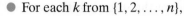

Example 6   *Use Prim's algorithm to find a minimal spanning tree for the graph in Figure 7-37.*

SOLUTION   Let T_0 be the tree with one node a and no edges. The remainder of the algorithm can be traced as shown in Table 7-1. The total weight of the final tree T_6, shown in Figure 7-38, is $2 + 3 + 6 + 7 + 3 + 5 = 26$.   □

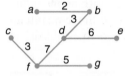

Figure 7-38   Solution
to Example 6.

Practice Problem 5   *In step 4 there was another choice for e_4. What minimal spanning tree do you get if you use that other choice?*

We conclude our discussion by proving that this algorithm does, indeed, yield a minimal spanning tree.

Lemma 9   *For each $k \geq 0$, T_k is included in a minimal spanning tree of G.*

PROOF   We prove by induction on k the statement "T_k is included in a minimal spanning tree of G," which we denote by $P(k)$. Since T_0 consists of only one node and no edges, it is contained in every spanning tree of G, and hence the statement $P(0)$ is true. Now let $m \geq 1$ be given such that $P(0), P(1), \ldots, P(m-1)$ have all been checked to be true, and consider the tree T_m. According to the algorithm, T_m was formed by adding the edge e_m to the tree T_{m-1}. By the inductive hypothesis $P(m-1)$, we know that the tree T_{m-1} is included in a minimal spanning tree of G. Let's call this minimal spanning tree **T**. The edge e_m either is in **T** or it is not, so we argue by cases.

Case 1. If it so happens that the edge e_m is actually in **T**, then it must be the case that T_m is included in **T**, which means that statement $P(m)$ is true.

k	E_k	e_k	**Nodes of T_k**
1	$\{[a, b], [a, c], [a, d]\}$	$[a, b]$	$\{a, b\}$
2	$\{[a, c], [a, d], [b, e], [b, d]\}$	$[b, d]$	$\{a, b, d\}$
3	$\{[a, c], [b, e], [c, d], [d, f], [d, e], [d, g]\}$	$[d, e]$	$\{a, b, d, e\}$
4	$\{[a, c], [c, d], [d, f], [d, g], [e, g]\}$	$[d, f]$	$\{a, b, d, e, f\}$
5	$\{[a, c], [c, d], [c, f], [d, g], [e, g], [f, g]\}$	$[c, f]$	$\{a, b, c, d, e, f\}$
6	$\{[d, g], [e, g], [f, g]\}$	$[f, g]$	$\{a, b, d, e, f, g, c\}$

Table 7-1   Tracing Prim's Algorithm

Case 2. If e_m is *not* an edge in **T**, then we can form the new graph H by adding e_m to **T**. By Exercise 12, this new graph will have exactly one cycle. Let f be the edge on this cycle with the smallest possible weight, and let **T'** be the graph formed by removing f from H. By Exercise 13, **T'** is a spanning tree of G. Since the weight of f is no more than the weight of e_m, it follows that the weight of **T'** is no more than the weight of **T**. Hence, **T'** is a minimal spanning tree of G in which T_m is included. This establishes statement $P(m)$ in this case as well, completing the induction. ∎

Theorem 10   *The tree T_n returned by Prim's algorithm is a minimal spanning tree for the weighted simple, connected graph G on $n + 1$ nodes.*

PROOF   Since the algorithm adds one node at each step, it follows that T_k has $k + 1$ nodes for all k. This means that T_n has $n + 1$ nodes, so T_n is a spanning tree of G. By Lemma 9, T_n is included in a minimal spanning tree of G, so T_n must *be* that minimal spanning tree of G. ∎

Solutions to Practice Problems

1 *Proof.* The contrapositive can be written, "For every simple, connected graph G, if there is an edge we can remove from G and still have a connected graph, then G has a cycle."

Let a simple, connected graph G be given such that there is an edge $\{a, b\}$ in G whose removal from G creates a connected graph G^*. Since G^* is connected, there is a path P in G^* from a to b by definition of "connected." This path is also in the graph G as is the edge $\{a, b\}$, so adding the edge $\{a, b\}$ to P forms a cycle in G. ∎

2 *Proof by induction.* Let the simple graph G with edge $\{v, w\}$ be given, and $P(n)$ be the statement, "If there is a circuit of length n containing $\{v, w\}$ in G, then there is a cycle containing $\{v, w\}$ in G."

A circuit with length 3 containing $\{v, w\}$ in a *simple* graph must look like v, w, x, v, where x is different from v and w. Any such circuit must be a cycle, so statement $P(3)$ is true. Now let $m \geq 4$ be given such that statements $P(1), \ldots, P(m-1)$ have already been checked, and we are now considering statement $P(m)$. Let $C = u_0, u_1, u_2, \ldots, u_{m-1}, u_m$ with $u_m = u_0$ be a circuit of length m that contains the edge $\{v, w\}$. We can choose values i and j with $j > i$ such that $u_j = u_i$ and j is as small as possible. If $i = 0$ and $j = m$, then there are no other repeated vertices in C, which means that C is a cycle. Since C contains $\{v, w\}$, this establishes statement $P(m)$ for this case.

If $i > 0$ or $j < m$, then we can break C into the following two smaller circuits:

$$C' = u_i, \ldots, u_j$$
$$C'' = u_j, \ldots, u_m, u_1, \ldots, u_i$$

(Note that in C'' one of the walks $u_j, \ldots, u_m$ or $u_1, \ldots, u_i$, but not both, can be "empty." See Exercise 4 for more on this.) Since every edge of the circuit C appears in one of these new circuits, we know that edge $\{v, w\}$ is contained in one of them. The new circuit that contains $\{v, w\}$ has length (call it k) less than m, and we know that statement $P(k)$ has already been checked. Hence, we conclude that there is a cycle in G that contains edge $\{v, w\}$. This is our desired conclusion for statement $P(m)$, and so this completes the induction. ∎

3 (a) One possible solution is to (1) remove [8, 1] from the cycle 1, 2, 3, 4, 5, 6, 7, 8, 1, (2) remove [6, 7] from the cycle 3, 4, 6, 7, 3, (3) remove [3, 4] from the cycle 3, 4, 8, 2, 3, (4) remove [5, 6] from the cycle 4, 5, 6, 4, and (5) remove [2, 3] from the cycle 2, 3, 7, 8, 2.

(b) There are seven edges.

(c) There are many ways to do the first part, but they all leave seven edges in the second part.

4 There are many possible answers. One is to change the weight of [a, c] in the example to 30.

5 Figure 7-39 shows a minimal spanning tree.

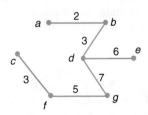

Figure 7-39   Solution to Practice Problem 5.

Exercises for Section 7.2

1. Each of the following statements is false. In each case, demonstrate this by supplying a counterexample.

   (a) In any simple graph G, if W is a closed walk, then W contains a cycle.

   (b) In any connected graph G, if every vertex in G has degree 3, then G must contain a cycle of length 3.

   (c) In any graph G, some pair of vertices must have the same degree.

2. Fill in the blanks to complete the following proof of Proposition 3:

   Proposition When an edge is removed from a cycle in a connected graph, the result is a graph that is still connected.

   Proof. Let G be a connected graph with the cycle

   $$C = v_0, v_1, v_2, \ldots, v_n, v_0$$

   and let G' be the graph obtained by removing the edge $[v_0, v_1]$ from G.

   To show that G' is connected, we have to show that every pair of nodes has a walk between them in G'. Let nodes x and y be given. Since _____, there is a walk

   $$W = x, x_1, x_2, \ldots, x_m, y$$

   where consecutive nodes form edges in G. Either edge $[v_0, v_1]$ is used in walk W or it is not, so we argue in two cases.

   Case 1. If edge $[v_0, v_1]$ is not used in walk W, then

   _____

   is a walk from x to y with edges in G'.

   Case 2. If edge $[v_0, v_1]$ *is* used in this walk, then

   _____

   is a walk from x to y with edges in G'. Since there is a walk from x to y in G' in either case, we conclude that G' is connected. ∎

3. (This is identical to Exercise 6 in Section 7.1.) For each of these walks, find a shorter walk that begins and ends with the same nodes as the given walk, and that does not visit any node more than once.

   (a) $1, 2, 3, 2, 4$

   (b) $4, 4, 5, 6, 2$

   (c) $9, 3, 2, 7, 8, 2, 4$

   (d) $1, 2, 3, 4, 5, 6, 7, 4, 8, 3, 9$

   (e) $3, 5, 6, 5, 4, 8, 9, 3, 2, 8, 7$

   (f) $8, 9, 2, 4, 6, 8, 11, 4, 2, 5, 7, 3, 2, 10$

4. Within each of the given circuits C, identify the circuits C' and C'' described in the solution to Practice Problem 2. In each case, decide which of the two new circuits contain the edge $\{a, b\}$.

   (a) $C = a, b, c, d, b, e, a$

   (b) $C = b, c, d, e, b, a, f, b$

   (c) $C = c, b, d, e, f, d, g, b, a, c$

   (d) $C = a, e, b, f, e, d, a, b, c, a$

5. For each circuit C in Exercise 4, find a cycle that contains the edge $\{a, b\}$.

6. Use Exercise 2 and Theorem 7 to prove that for any simple, connected graph G, if G has exactly one cycle, then G has the same number of nodes and edges.

7. Use the pigeonhole principle (Practice Problem 3 of Section 2.5) to prove that every simple graph with at least two vertices must have two vertices with the same degree. (HINT: First explain why a simple graph with n vertices cannot have both a vertex of degree 0 and a vertex of degree $n - 1$.)

8. Prove the following statement using an argument similar to the proof of Proposition 4:

   Let $m \geq 0$ be given. In a simple, connected graph G, if every vertex in G has degree at least m, then there is a path in G of length at least m.

9. Prove that in a graph G on n nodes, the longest path in G uses no more than $n - 1$ edges.

10. Prove that for every connected graph G, if G has no cycles, then for every pair of vertices a, b in G, there is only one path from a to b in G. (HINT: Consider the contrapositive statement.)

11. Prove that if G is a tree with at least one edge, then deleting any edge from G will result in a graph with exactly two connected components.

12. Prove that if G is a tree with (two different) nodes u and v and $\{u, v\}$ is not an edge in G, then adding edge $\{u, v\}$ to G will form a graph with exactly one cycle. (HINT: Proposition 2 tells us that there is at least one cycle in this new graph.)

13. Prove that if G is a simple, connected graph with exactly one cycle and e is an edge on that cycle, then the graph formed by removing e from G is a tree.

14. Fill in the blanks to complete an alternative proof to Proposition 7.

   Proposition For every $n \geq 1$, any tree with n edges must have exactly $n + 1$ nodes.

   Proof by induction on $n \geq 1$. Let $P(n)$ represent the statement "Any tree G with n edges has $n + 1$ nodes."

The first statement to be checked is $P(0)$, which states,

"_____."

The only graph satisfying the hypothesis of this statement is the one shown below, and this graph does have one vertex.

Hence, $P(0)$ is true. Now let $m \geq 1$ be given such that $P(m)$ is the first statement that has not yet been checked. Let a tree G with m edges be given. By Proposition _____, there is a vertex v in G with degree 1. Let G' be the graph resulting from removing this vertex and its single edge from G.

Now G' has _____ edges, so by statement $P(\underline{\qquad})$ (which has already been checked to be true), we know that G' has _____ nodes. Since _____, G has _____ nodes. This verifies statement $P(m)$, completing the induction. ∎

15. The induction proof of Proposition 5 involves the statement $P(n)$ that says, "For every tree T, if T has n edges, then T has at least two vertices of degree 1." Verify $P(1)$, $P(2)$, $P(3)$, and $P(4)$ by drawing every graph of the appropriate size satisfying the hypothesis, and verifying the conclusion in each case.

16. Let G be a tree, and let k be the number of vertices in G whose degree is at least 3. Prove that G has at least $k + 2$ leaves. (HINT: Use Theorem 3 from Section 7.1.)

17. The induction proof of Theorem 7 involves the statement $P(n)$ that says, "If G is a tree with n edges, then G has $n + 1$ vertices." Verify $P(0)$, $P(1)$, $P(2)$, and $P(3)$ by drawing every possible graph of the appropriate size satisfying the hypothesis, and verifying the conclusion in each case.

18. Fill in the blanks to complete the proof that goes with Exercise 22 in Section 7.1. Notice that here the induction is on the number of *vertices*, not on the number of edges in the graph, so the proof looks a bit different.
    Proposition A simple, connected graph with $n \geq 1$ vertices must have at least $n - 1$ edges.
    Proof by induction. Let $P(n)$ represent the statement "Any simple, connected graph G with n vertices has at least $n - 1$ edges." The first statement to be checked is $P(1)$, which states,

"_____."

The only graph satisfying the hypothesis of this statement consists of one vertex and 0 edges. Hence, $P(1)$ is true. Now let $m \geq 2$ be given such that $P(m)$ is the first statement that has not yet been checked. Let a connected graph G with m vertices be given. Now either G has a vertex of degree 1 or else all its vertices have degree at least 2, so we argue in two cases.

Case 1. If there is a vertex v in G with degree 1, then let G' be the graph resulting from removing this vertex and its single edge. Now G' has _____ vertices, so by statement $P(\underline{\qquad})$ (which has already been checked to be true), we know that G' has at least _____ edges. Since _____, G has at least _____ edges. This verifies statement $P(m)$.

Case 2. If every vertex in G has degree at least 2, then by Exercise _____ in Section 7.1, G has at least m edges. This verifies statement $P(m)$ without even needing to use the induction hypotheses.

In either case, we have verified statement $P(m)$, completing the induction. ∎

19. The induction proof in Exercise 18 involves the statement $P(n)$ that says, "For any simple, connected graph G, if G has n vertices, then G has at least $n - 1$ edges." Verify $P(1)$, $P(2)$, $P(3)$, and $P(4)$ by drawing every possible simple, connected graph of the appropriate size satisfying the hypothesis, and verifying the conclusion for each graph.

20. Prove that if G is a simple graph with n vertices, k connected components, and no cycles, then G has $n - k$ edges. (HINT: Use mathematical induction.)

21. Fill in the missing steps in the following proof of the correct answer to Exercise 20 in Section 7.1.
    Claim For every simple graph G with n vertices, if G is not connected, then G has no more than $\frac{(n-1)(n-2)}{2}$ edges.
    Proof. Let G be a simple, disconnected graph with n vertices, and let H_1 and H_2 be subgraphs of G such that every vertex in G is in exactly one of H_1 or H_2, and there are no edges connecting a vertex in H_1 with a vertex in H_2. Let k be the number of vertices in H_1, so there are $n - k$ vertices in H_2. ∎
    Claim By Exercise _____ in Section 7.1, the maximum number of edges in H_1 is $\frac{k(k-1)}{2}$, and the maximum number of edges in H_2 is $\frac{(n-k)(n-k-1)}{2}$. Hence, the maximum number of edges in G is

    $$f(k) = \frac{k(k-1)}{2} + \frac{(n-k)(n-k-1)}{2},$$
    where _____ $\leq k \leq$ _____

    If we treat n as a constant, this function of k is the parabola

    $$f(k) = \underline{\qquad} k^2 + \underline{\qquad} k + \underline{\qquad}$$

    with a minimum value occurring at $k =$ _____ and maximum value occurring at $k =$ _____ and $k =$ _____.

    Hence, the maximum number of edges G can have is _____.

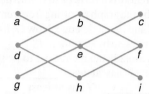

Figure 7-40 Graph for Exercise 22.

22. This exercise uses terminology from Section 4.5. For a graph G with vertex set V, define the relation R on V such that

$$a\,R\,b \text{ means there is a walk from } a \text{ to } b \text{ in } G.$$

(a) Prove that the relation R is reflexive, symmetric, and transitive, and hence R is an equivalence relation.

(b) Give the partition of V induced by R for the graph shown in Figure 7-40.

(c) Describe in general the partition of V induced by R using the terminology of this chapter.

23. Find all minimal spanning trees in each graph shown in Figure 7-41.

24. Using Proposition 3, write an induction proof of the statement "For all $n \geq 0$, every connected graph with n edges has a spanning tree."

25. Consider the following alternative algorithm (due to Kruskal) for finding a minimal spanning tree. Use it to find a minimum spanning tree in each of the graphs in Exercise 23.

● Given a simple, connected weighted graph G with at least one edge, let T_0 begin as a graph with the same set of nodes as G but with no edges.

● List the edges of G from lightest to heaviest: $e_1, e_2, e_3, \ldots, e_m$.

● For each i from 1 to m, let

$$T_i = \begin{cases} T_{i-1} + e_i & \text{if } e_i \text{ can be added to } T_{i-1} \text{ without} \\ & \quad \text{creating a cycle} \\ T_{i-1} & \text{otherwise} \end{cases}$$

26. Prove that for any simple, connected weighted graph G, for each graph T_i produced by Kruskal's algorithm (in Exercise 25), there is a minimal spanning tree T of G such that T_i is a subgraph of T.

27. Fill in the details of the following alternate proof of Proposition 4:

Claim Every tree has at least one leaf.

Proof. Let T be a tree with n edges. We know from Proposition 7 that T has _____ nodes. If each of these nodes has degree greater than 1, then the sum of the degrees of the nodes is at least _____. But according to Theorem 3 of Section 7.1, the sum of the degrees of the nodes in T is _____. Since _____, it is therefore impossible for every node in T to have degree greater than 1. That is, T has at least one node of degree 1. This node is a leaf by definition. ∎

28. Modify the proof in Exercise 27 to show that every tree has at least two nodes of degree 1.

29. The map in Figure 7-42 shows the main roads in the Borough of Boatsville, along with the cost of plowing each street during a heavy snowstorm. Find the cheapest set of roads that must be plowed in order for everyone to be able to travel from any point to any other point in town.

30. Consider the following algorithm on a simple, connected graph G:

● Let $G_0 = G$.

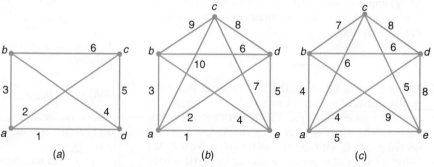

Figure 7-41 Graphs for Exercise 23.

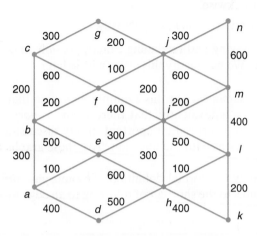

Figure 7-42   Graph for Exercise 29.

- Repeat the following as long as possible: If G_i has a cycle, let e be the most expensive edge on that cycle, and let G_{i+1} be the graph obtained by removing e from G_i.
- The resulting graph has no cycles. Return this as the result T.

Find a minimal spanning tree of the graph in Exercise 29 using this algorithm.

31. For each of the graphs in Exercise 23, find a minimal spanning tree using the algorithm in Exercise 30.

32. Prove by induction that in the algorithm in Exercise 30, each G_i is connected. Conclude that the resulting graph T is a spanning tree of G.

33. Prove by induction that in the algorithm described in Exercise 30, each G_i contains a minimal spanning tree of G. Conclude that this algorithm produces a minimal spanning tree of G.

7.3 Isomorphism and Planarity

In this section, we get back to studying more concepts of graph theory that can be applied directly to problems both practical and recreational. Because the proofs of some of these concepts are better left to a specialized course in graph theory, "proof" will play a less prominent role in this section than it did in the last. Before we get to the applications, however, we need to address one more abstract property of great practical importance.

Isomorphic Graphs

Example 1   *Exercise 6 of Section 7.1 asked for an example of a simple graph with six vertices having degrees 3, 3, 2, 2, 1, and 1, respectively. The solutions in Figure 7-43 are all correct. Which graph is not like the other two?*

SOLUTION   If we imagine the graphs being constructed of tacks (for nodes) and elastic string (for edges), then we can see that graphs (*a*) and (*c*) would look exactly the same if we did the following to graph (*a*):

- Move tacks 1 and 6 so they fall below tacks 2 and 5, respectively.

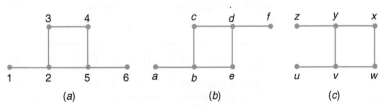

Figure 7-43   Graphs for Example 1.

- Rotate the entire picture 90° clockwise.
- Rename tacks 1, 2, 3, 4, 5, 6 as z, y, x, w, v, u, respectively.

On the other hand, this type of moving and relabeling cannot possibly make graph (b) look like the other two. We will justify this claim a little later.   □

We will express the notion of "sameness" in this example by saying that the graphs (a) and (c) are *isomorphic*, or equivalently, that there is an *isomorphism* between the two graphs. This concept is crucially important in any discussion of graphs, because it makes us come to terms with the question "Are these graphs the same?"

To get the idea, we will give only the formal definition of isomorphic *simple* graphs since it is easiest to state, thanks to the shortcut of referring to an edge by the set of its endpoints.

Definition   Simple graphs G and H are called *isomorphic* if there is a one-to-one and onto function f from the nodes of G to the nodes of H such that $\{v, w\}$ is an edge of G if and only if $\{f(v), f(w)\}$ is an edge of H. The function f is called an *isomorphism*. Hence, an isomorphism is simply a rule associating nodes that preserves the edges joining the nodes.

So formally graphs (a) and (c) in Example 1 are isomorphic using the function f described as

$$1 \mapsto z, 2 \mapsto y, 3 \mapsto x, 4 \mapsto w, 5 \mapsto v, 6 \mapsto u$$

This is the correspondence we determined by our thought experiment with tacks and strings. Table 7-2 gives an explicit list of the edges $\{a, b\}$ in the first graph and the corresponding edges $\{f(a), f(b)\}$ in the second graph.

Explore more on the Web.

Practice Problem 1   *The two graphs shown in Figure 7-44 are isomorphic, as can be seen by the rule*

$$1 \mapsto b, 2 \mapsto d, 3 \mapsto a, 4 \mapsto e, 5 \mapsto c, 6 \mapsto f$$

(a) *Describe a thought experiment with tacks and string that would give this particular isomorphism.*

(b) *In the style of Table 7-2, list the edges in the graph on the left and the corresponding edges in the graph on the right for this particular isomorphism.*

It is difficult, especially with large examples, to construct an explicit function between two graphs to demonstrate they are isomorphic. It is easier to look for

Edges in graph (a)	{1, 2}	{2, 3}	{3, 4}	{4, 5}	{2, 5}	{5, 6}
Edges in graph (c)	{z, y}	{y, x}	{x, w}	{w, v}	{y, v}	{v, u}

Table 7-2   Corresponding Edges for Graphs (a) and (c) in Example 1

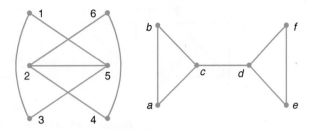

Isomorphic graphs

Figure 7-44 Graphs for Practice Problem 1.

evidence that the graphs are *not* isomorphic. If we cannot find such evidence, this suggests (but does not prove) that the two graphs are the same. Here is a partial list of things to look for when deciding if two graphs are really the same or not. The proof of this proposition is based solely on the formal definition of isomorphism, but it will not be presented here.

Proposition 1 *Two graphs that are isomorphic to one another must have*

1. *The same number of nodes.*
2. *The same number of edges.*
3. *The same number of nodes of any given degree.*
4. *The same number of cycles.*
5. *The same number of cycles of any given size.*

Example 2 *The two graphs shown in Figure 7-45 have the same number of nodes and the same number of edges, and all the nodes in each are degree 3. However, the graphs are not isomorphic. Prove it.*

SOLUTION   Graph *H* has cycles of length 3 while graph *G* does not. Informally, this means that *H* cannot be redrawn to look like *G*. Formally, we can cite the last part of Proposition 1 to conclude that graphs *G* and *H* are not isomorphic.   □

Practice Problem 2   *Find all isomorphic pairs among the graphs shown in Figure 7-46. If you decide a pair is not isomorphic, say how you know this. If you decide a pair is isomorphic, give the isomorphism function.*

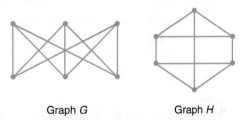

Graph *G*   Graph *H*

Figure 7-45 Graphs for Example 2.

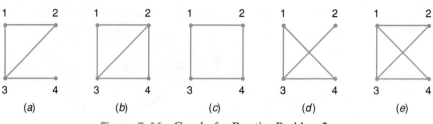

Figure 7-46 Graphs for Practice Problem 2.

Planar Graphs

We have seen on more than one occasion that an appealing feature of graphs is the fact that their structure is based on a simple relationship between vertices and edges and not on precisely how the graph is drawn. However, there are times when the way a graph is drawn *is* extremely important.

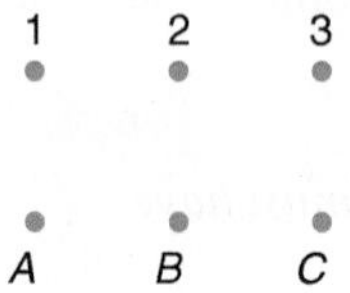

Example 3 (*the utilities puzzle.*) *Given three houses, is there any way to connect them to their electricity, cable TV, and telephone companies without one service's wires crossing over or under the others? You may assume that the houses are arranged any way you like. It is easier to state as "Can one connect each of nodes 1, 2, 3 to each of nodes A, B, C in Figure 7-47 with (not necessarily straight) edges so that no two edges cross?"*

Figure 7-47
Connect letters
to numbers
without crossing.

Of course, this new interest in the way a graph can be drawn comes with the realization that we do not yet have the tools to answer a question like this. So using the lessons of history as a guide, we will try to put the problem into as general a framework as possible.

> ### Definition
>
> 1. A simple, connected graph is called *planar* if there is a way to draw it (on a plane) so that no edges cross (i.e., the only place two edges can meet is at a node). We will call such a "drawing" of a graph on a plane surface with no edge-crossings an *embedding* of the graph in the plane.
> 2. A graph is called *bipartite* if its set of nodes can be partitioned into two disjoint sets S_1 and S_2 so that every edge in the graph has one endpoint in S_1 and one endpoint in S_2.
> 3. The *complete graph* on n nodes, denoted by K_n, is the simple graph with nodes $\{1, \ldots, n\}$ and an edge between every pair of distinct nodes.
> 4. The *complete bipartite graph* on n, m nodes, denoted by $K_{n,m}$, is the simple bipartite graph with nodes $S_1 = \{a_1, a_2, \ldots, a_n\}$ and $S_2 = \{b_1, b_2, \ldots, b_m\}$ and with edges connecting each node in S_1 to every node in S_2.

Example 4 *K_4 and $K_{3,2}$ are shown in Figure 7-48. Show that each is planar by redrawing it with no edge-crossings.*

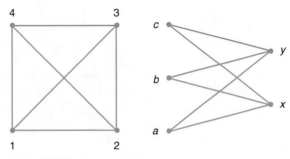

Figure 7-48 K_4 and $K_{3,2}$.

SOLUTION   To show a graph is planar, we need only draw its embedding in the plane. Figure 7-49 shows embeddings of K_4 and $K_{3,2}$, respectively.   □

Using this terminology, the utilities puzzle is asking the question "Is $K_{3,3}$ planar?" This question illustrates in a simplistic way another significant application of graphs. When designing an electronic circuit with logic gates as we did in Section 3.5, there is a set of objects that must be interconnected by "wires" in a specific way. It is easy to see that this corresponds to a graph. There is a cost attached to creating a "bridge" of one wire over another, so the number of edge-crossings should be minimized when designing a circuit. The same sort of issue arises in the design of the integrated circuits that make your computer work. We will not address the problem of "minimizing edge-crossings" in this course, but to get a flavor of this area of mathematics, we will discuss some applications of graphs with *no* edge-crossings, the planar graphs.

The next proposition gives two particular examples of small nonplanar graphs. We will devote the rest of this section to building a framework that will allow us to show that these two graphs, along with many others, are not planar.

Proposition 2   *The graphs K_5 and $K_{3,3}$ shown in Figure 7-50 are not planar.*

Before getting to the proof of this proposition, we should point out that Kazimierz Kuratowski (1896–1980) proved in 1930 that **these are essentially the only two nonplanar graphs**! Of course, this is not strictly true since a larger graph containing K_5 as a subgraph—for example K_6 or K_7—cannot be planar since K_5 is not. Kuratowski actually proved the following.

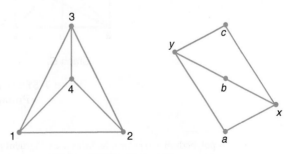

Figure 7-49 K_4 and $K_{3,2}$ are planar.

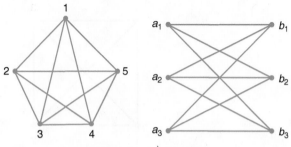

Figure 7-50 K_5 and $K_{3,3}$.

Theorem 3 (Kuratowski's Theorem)   *A graph G is planar if and only if it contains no "copies" of $K_{3,3}$ or K_5 as subgraphs.*

We will not go into what a "copy" of $K_{3,3}$ or K_5 is since it has no bearing on the rest of the course, but it is not too hard to imagine. We will not prove this theorem here. The proof involves some facts about the "topology" of the plane and is typically studied in upper-level courses devoted to graph theory.

Practice Problem 3   *One of the graphs shown in Figure 7-51 is planar and the other is nonplanar. Which is which, and how do you know? For the planar graph, show how to draw it in the plane with no edges crossing.*

Euler's Formula for Planar Graphs

In order to find an easy proof of Proposition 2, we need to understand the structure of planar graphs in general. Once again, we find ourselves face to face with the work of Leonhard Euler, who made a discovery about solid geometry that turns out to be related to our study of planar graphs.

The ancient Greeks knew that there were only five types of regular* polyhedra, but their argument to prove this was based on the measure of the angles at each vertex. Even though these and other solids had been studied in depth from the time of the ancient Greeks, in the mid-eighteenth century Euler made a surprising

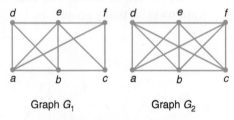

Graph G_1 Graph G_2

Figure 7-51   Graphs for Practice
Problem 3.

---

* A polyhedron is *regular* if its faces are congruent polygons and every vertex is incident with the same number of edges. These regular polyhedra are often called the *Platonic solids*.

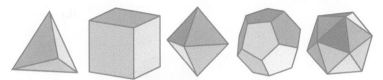

Figure 7-52   The five Platonic solids.

discovery about a relationship between the number of vertices, edges, and faces of polyhedra.

Example 5 *Record the number of vertices, edges, and faces of each of the five regular polyhedra shown in Figure 7-52.*

SOLUTION   Table 7-3 shows these values for each of the five Platonic solids. □

Euler observed that for each of these solids, the total number of faces and vertices is 2 more than the number of edges. According to [5], Euler wrote about this and other observations in a letter to Christian Goldbach in 1750, which concludes,

> I find it surprising that these general results in solid geometry have not previously been noticed by anyone, so far as I am aware; and furthermore, that the important ones ... are so difficult that I have not yet been able to prove them in a satisfactory way.

For many years, Euler was unable to find a satisfactory proof of his observation because he kept thinking of it as a geometry problem. In 1813, Augustin Cauchy (1789–1857) published a paper in which he represented the polyhedra as what we now call a planar graph. In this setting, Cauchy was able to give an elegant proof. Once again, a mathematician looked beyond layers of irrelevant complexity to see the essential ingredients of a problem. And once again, the underlying structure can be studied using graph theory.

Solid	Faces	Vertices	Edges
Tetrahedron	4	4	6
Cube	6	8	12
Octahedron	8	6	12
Dodecahedron	12	20	30
Icosahedron	20	12	30

Table 7-3   Vertices, Edges, and Faces of the Platonic Solids

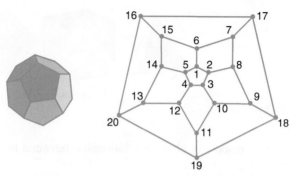

Figure 7-53   The planar graph of the dodecahedron.

Example 6   *Create a graph for the dodecahedron where the vertices and edges for the graph are the vertices and edges (respectively) of the solid, but so that the edges do not cross.*

SOLUTION   Once again, imagine that the vertices and edges are made of tacks and elastic string. In Figure 7-53, if you pick one face of the dodecahedron on the left and stretch it out, all the other vertices and edges will lie in the interior as shown in the figure on the right. ☐

Explore more on the Web.

Practice Problem 4   *Create a graph for the octahedron where the vertices and edges for the graph are the vertices and edges (respectively) of the solid, but so that the edges do not cross.*

Once this transformation is made, we have a planar graph to work with. Cauchy proved that Euler's formula works on the transformed pictures in the plane, and so it must work for the original solids. In order to give a proof for any planar graph, we must make sure that the notion of a "face" is carefully defined, since not every planar graph arises from transforming a polyhedron.

> **Definition**   For a planar graph *G* embedded in the plane, a *face* of the graph is a region of the plane created by the drawing. Since the plane is an unbounded surface, every embedding of a finite planar graph will have exactly one unbounded face.

It is easy to forget about the unbounded face in a planar graph. In Example 6, it might have bothered you if you counted the faces in the planar graph representation and only came up with 11 faces. The cycle 16, 17, 18, 19, 20, 16 is the boundary for the unbounded face in this embedding of the graph.

Example 7   *The picture in Figure 7-54 shows a planar graph G drawn in two different ways. For each drawing, identify the faces by giving the cycle that creates each face, and highlight the unbounded face. Note that the number of faces is the same in each case, but the cycles that border the faces are different.*

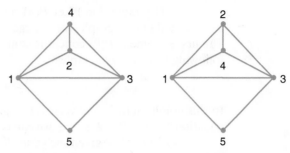

Figure 7-54   Graphs for Example 7.

SOLUTION   We describe each face by giving the cycle that "borders" it in Table 7-4.   □

It is not at all obvious that the number of faces of a planar graph does not depend on a particular embedding of the graph in the plane. Euler's formula establishes that the number of faces is determined by the number of edges and vertices, which are obviously not dependent on how the graph is drawn.

To prove Euler's theorem, we need a couple of facts about embeddings of graphs in the planes. These are the results that depend on the structure of the plane, which we will not prove here. See Exercise 18 to see the role of a "plane surface" in these facts that seem so obvious.

Lemma 4   *Given an embedding of a graph in the plane,*

1. *Removing any edge from a cycle in the graph decreases by 1 the number of faces in the graph; and*
2. *Every edge on a cycle in the graph borders exactly two faces.*

Theorem 5 (Euler's Formula for Planar Graphs)   *For any connected planar graph G embedded in the plane with V vertices, E edges, and F faces, it must be the case that*

$$V + F = E + 2$$

	Left drawing	Right drawing
	1, 2, 4, 1	1, 2, 4, 1
	1, 3, 5, 1	1, 3, 5, 1
	2, 3, 4, 2	2, 3, 4, 2
	1, 2, 3, 1	1, 3, 4, 1
Unbounded →	1, 4, 3, 5, 1	1, 2, 3, 5, 1

Table 7-4   Cycles Bordering Faces in Example 7

PROOF (By induction.) Let $P(n)$ be the statement "For every embedding of a connected planar graph with n edges, V, vertices, and F faces, $V + F = n + 2$." The first statement $P(0)$ concerns connected graphs with no edges, of which the following is the only example:

•

In this graph there is one vertex, 0 edges, and one (unbounded) face. Since the equation $1 + 1 = 0 + 2$ is true, we can conclude that statement $P(0)$ is true.

Now let $m \geq 1$ be given such that $P(m)$ is the first statement yet to be checked. Let G be a connected planar graph with m edges and V vertices with F faces as embedded in the plane. Either G has a cycle or it does not, so we will argue in two cases.

Case 1. If G has no cycles, then G is a tree. Since there are no cycles, G has only the one (unbounded) face. By Theorem 7 of Section 7.2, $V = m + 1$. Hence, we have $V + F - 2 = (m + 1) + 1 - 2 = m$. Thus, statement $P(m)$ is true in this case.

Case 2. If G has a cycle, then form the graph G' by removing an edge from this cycle. By Proposition 3 of Section 7.2, G' is still connected. Clearly, the number of edges in G' is $m - 1$, and by Lemma 4, the number of faces in G' is $F - 1$. Using the fact that statement $P(m - 1)$ has already been checked to be true, we know that the formula in statement $P(m - 1)$ can be applied to graph G'. Specifically,

$$V + (F - 1) = (m - 1) + 2$$

Adding 1 to both sides makes this equation say, "$V + F = m + 2$," which is statement $P(m)$.

■

This result implies that the number of faces of a planar graph does not depend on the particular embedding of that graph in a plane! Hence, we can unambiguously talk about the "number of faces of a planar graph" without mentioning the way the graph is drawn.

The Utilities Puzzle Resolved

After a considerable digression, we now have enough tools to prove that $K_{3,3}$ is not a planar graph and thereby establish that the utilities puzzle has no solution. Because Euler's formula provides a relationship between the number of faces and the number of edges in a planar graph, we can use it to show that some graphs have too many edges to be planar. This is an effective tool for graphs, like $K_{3,3}$, for which every edge is on a cycle.

> **Definition** Let G be a planar graph in which every edge is on a cycle, and let f be a face created by an embedding of G in the plane. The *size of* f is defined to be the number of edges on the cycle in G that makes up the boundary of f.

Lemma 6   *In any planar graph G in which every edge is on a cycle, the sum of the face sizes is twice the number of edges in the graph.*

PROOF   This follows directly from the second part of Lemma 4 in a manner very similar to the proof of Theorem 3 of Section 7.1.   ∎

Theorem 7   *In any planar graph G in which every edge is on a cycle, if k is the size of the smallest cycle, then the number of edges in G is at least $\frac{k}{2}$ times the number of faces in G.*

PROOF   If the smallest cycle has length k, then when we sum the face sizes, we get a result that is at least $k \cdot F$, where F is the number of faces in G. By Lemma 6, the sum of the face sizes is $2m$, where m is the number of edges in G. Putting together these two facts yields the inequality

$$2m \geq k \cdot F$$

from which the desired result follows.   ∎

Figure 7-55 $K_{3,3}$ is not planar.

Proposition 2   *(revisited)* The graphs $K_{3,3}$ and K_5 are not planar.

PROOF   We will show that $K_{3,3}$ (see Figure 7-55) is not planar, and leave K_5 for the practice problem below. We will argue by contradiction, assuming that $K_{3,3}$ *is* planar and deriving a contradictory statement from this assumption. To use the tools available to us, we need to take a quick inventory of the structure of $K_{3,3}$:

- $K_{3,3}$ has six vertices and nine edges.
- Every edge in $K_{3,3}$ is on a cycle.
- The shortest cycle in $K_{3,3}$ has length 4.

If $K_{3,3}$ is planar, then by Euler's formula, it must have exactly five faces. Since the smallest length of a cycle in $K_{3,3}$ is 4, then by Theorem 7, the number of edges is at least 2 times the number of faces. That is, $9 \geq 2 \cdot 5$. Since this false statement logically follows from the assumption that $K_{3,3}$ is planar, we can conclude that $K_{3,3}$ is not planar.   ∎

Practice Problem 5   *Complete the proof of Proposition 2 by explaining why K_5 is not planar.*

Solutions to Practice Problems

1   (a) Exchange tack 2 with tack 5.

   (b) $\{1,3\}$ $\{1,5\}$ $\{2,4\}$ $\{2,5\}$ $\{2,6\}$ $\{3,5\}$ $\{4,6\}$
         $\{b,a\}$ $\{b,c\}$ $\{d,e\}$ $\{d,c\}$ $\{d,f\}$ $\{a,c\}$ $\{e,f\}$

2   Nonmatching edge count shows these pairs are not isomorphic: (*a*) and (*b*), (*a*) and (*e*), (*b*) and (*c*), (*c*) and (*d*), (*c*) and (*e*). Other isomorphisms are ruled out using degrees. Graph (*a*) has a node of degree 3 and (*c*) does not. (*d*) has a node of degree 1 and (*b*) does not. (*d*) has a node of degree 1 and (*e*) does not. The only possibilities left are (*a*) to (*d*) and (*b*) to (*e*). We find that (*a*) and (*d*) are isomorphic using the correspondence $1 \mapsto 1, 2 \mapsto 4, 3 \mapsto 3, 4 \mapsto 2$. Graphs (*b*) and (*e*) are isomorphic using the correspondence $1 \mapsto 2, 2 \mapsto 1, 3 \mapsto 3, 4 \mapsto 4$.

3   In Figure 7-56, the second graph is nonplanar since it contains a copy of $K_{3,3}$. The first graph is planar, and can be redrawn without edge-crossings as shown in Figure 7-56.

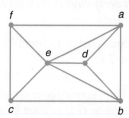

Figure 7-56 Solution to Practice Problem 3.

4 The graph is shown in Figure 7-57.

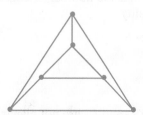

Figure 7-57 Solution to Practice Problem 4.

Exercises for Section 7.3

1. See Figure 7-58. Explain why graphs G_1 and G_2 are not isomorphic. Explain why graphs H_1 and H_2 are not isomorphic.

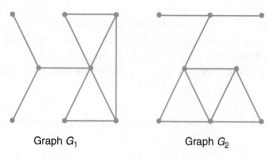

(a)   Graph G_1 Graph G_2

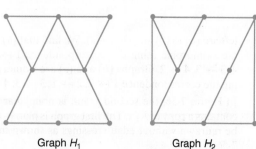

(b)   Graph H_1 Graph H_2

Figure 7-58 Graphs for Exercise 1.

5 *Proof by contradiction.* Suppose K_5 is planar. Since K_5 has 5 vertices and 10 edges, Euler's formula tells us that K_5 will have seven faces when embedded in the plane. However, every edge in K_5 is on a cycle and the shortest cycle in K_5 has length 3, so by Theorem 7, the number of edges in K_5 will be at least $\frac{3}{2}$ times the number of faces. This means that

$$10 \geq \frac{3}{2} \cdot 7$$

which is clearly false. Hence, the original assumption must be wrong, so K_5 is not planar. ∎

2. Explain why graphs (*a*) and (*b*) in Example 1 are not isomorphic.

3. Two of the graphs shown in Figure 7-59 are isomorphic to each other. Which ones?

4. The *degree sequence* of a graph is the list of degrees of the nodes of the graph, listed from largest degree to smallest. For example, the degree sequence of the graph G_2 shown above is 5, 4, 3, 3, 3, 3, 2, 1.

   (a) Construct two connected, simple graphs with degree sequence 3, 3, 2, 2, 2, and explain why your two graphs are not isomorphic.

   (b) Construct two connected, simple graphs with degree sequence 3, 2, 2, 2, 1, and explain why your two graphs are not isomorphic.

   (c) Construct two connected, simple graphs with degree sequence 3, 3, 2, 2, 2, 2, and explain why your two graphs are not isomorphic.

5. Find two simple graphs, each with six nodes, with every node of degree 3, but not isomorphic to each other. (HINT: Use $K_{3,3}$ as one of your graphs, and note that $K_{3,3}$ has no cycles of length 3.)

6. Give two examples of a simple, connected graph with eight nodes, each node having degree 4, that are not isomorphic to each other (and explain why your graphs are not isomorphic). (HINT: Make one graph planar and the other nonplanar.)

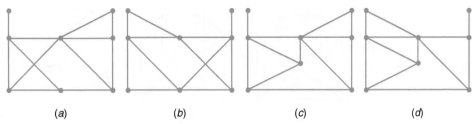

Figure 7-59 Graphs for Problem 3.

7. Give three examples of a simple, connected graph with 10 nodes, each node having degree 5. Explain why no two are isomorphic.

8. Show that each of five houses can be connected to two utilities without lines crossing. How many houses would be too many to connect each to two utilities without lines crossing? Explain your answer completely.

9. In the induction proof of Theorem 5, we used the statement $P(n)$: "For every embedding of a connected planar graph with n edges, V vertices, and F faces, $V + F = n + 2$." Verify statements $P(1), P(2), P(3)$, and $P(4)$ by drawing every possible graph satisfying the hypothesis and showing that the conclusion is true in each case.

10. Show that the two graphs shown in Figure 7-60 are planar by redrawing each in the plane with no edge-crossings.

11. Use Theorem 7 and Euler's formula to prove that each of the graphs shown in Figure 7-61 is not planar. (NOTE: The first graph is a cube with every face diagonal added, the second is a cube with every internal diagonal added, and the last graph is called the *Petersen graph*, which arises in many different contexts in graph theory.)

12. Decide if each of the graphs in Figure 7-62 is planar or nonplanar. If it is planar, draw an embedding of the graph in the plane. If it is nonplanar, explain how you know.

13. For each of the following families of nonregular polyhedra, find the number of vertices, edges, and faces, and verify that Euler's formula holds for each of them. (Examples of each are given in Figure 7-63.)

(a) A prism whose base is a regular k-sided polygon and whose sides are squares. (The first example in the figure shows this prism with $k = 6$.)

(b) An "antiprism" whose top and bottom are regular k-sided polygons and whose sides are alternating isosceles triangles. (The middle example in the figure shows this antiprism with $k = 6$.)

(c) A pyramid whose base is a regular k-sided polygon. (The third example in the figure shows this pyramid with $k = 4$.)

14. For each of the three sample polyhedra in the previous problem, draw the planar graph representation.

15. Let G be a simple, connected planar graph in which every edge is on a cycle. Prove that if G has n vertices and m edges, then

$$m \leq 3n - 6$$

(HINT: The smallest cycle a simple graph can have has length 3.)

16. Let G be a simple, connected planar graph in which every edge is on a cycle. Prove that if every vertex of G has degree d and every face is bound by b edges, then

$$b \cdot F = 2 \cdot E = d \cdot V$$

and so Euler's formula tells us that

$$E \left(\frac{2}{d} + \frac{2}{b} - 1 \right) = 2$$

Show that there are only five possible choices of $d \geq 3$ and $b \geq 3$ for which $\frac{2}{d} + \frac{2}{b} - 1$ has a positive value. Decide which choice of (d, b) is associated with each of the five Platonic solids.

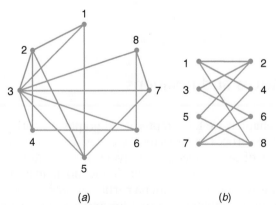

Figure 7-60 Graphs for Problem 10.

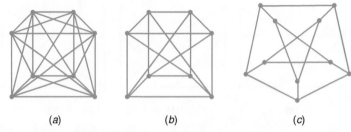

Figure 7-61 Graphs for Problem 11.

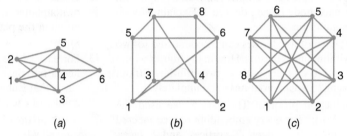

Figure 7-62 Graphs for Problem 12.

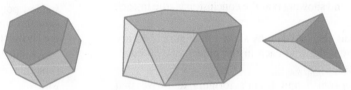

Figure 7-63 Examples for Problem 13.

17. Demonstrate that the utilities puzzle can be solved if you lived on the surface of a doughnut-shaped planet.

18. Demonstrate that each fact in Lemma 4 is false for graphs drawn on the surface of a doughnut.

7.4 Connections to Matrices and Relations

In the first three sections of this chapter, we have represented graphs primarily as pictures, using points to indicate the vertices and arcs for the edges. This is certainly appropriate considering the origins of the subject and the types of applications we have studied. One big drawback to this representation is that it does not lend itself to manipulation by a computer. In this section, we consider a particular representation that is more amenable to the use of a computer as a tool for answering questions about graphs. In the process, we shall discover some interesting connections with other topics in discrete mathematics.

Adjacency Matrices

For any graph, we can store information about the number of edges connecting each pair of vertices in a rectangular grid by using the rule "In row i, column j of the grid, write the number of edges connecting node i and node j."

Consider the example shown in Figure 7-64. In row 1 of the grid, we place a **1** in column 2 since there is **one** edge connecting nodes 1 and 2, and we place a **2** in column 5 since there are **two** edges connecting nodes 1 and 5. Of course, we place zeros in columns 1, 3, and 4, because there are no edges joining node 1 to nodes 1, 3, or 4. The following definition formalizes this idea and introduces some notation.

Definition   Given a graph G with vertex set $V = \{v_1, v_2, \ldots, v_n\}$ and edge set E, we define the *adjacency matrix* of G as follows. The matrix M is an $n \times n$ array of natural numbers, which we imagine having rows and columns labeled as follows:

$$
\begin{array}{c}
\text{Columns } 1, 2, \ldots, n \\
\begin{matrix}
\text{Row } 1 \to \\
\text{Row } 2 \to \\
\text{Row } 3 \to \\
\vdots \\
\text{Row } n \to
\end{matrix}
\left[\phantom{\begin{matrix} x \\ x \\ x \\ x \\ x \end{matrix}} \right]
\end{array}
$$

The entry in row i, column j (referred to as the (i, j) − entry of M or, more concisely, M_{ij}) is defined as

$$M_{ij} = \text{ the number of edges connecting } v_i \text{ and } v_j \text{ in } G$$

The name "adjacency matrix" comes from the fact that a pair of vertices connected by an edge are said to be *adjacent*. A subtle feature of the definition is that the adjacency matrix for a graph depends on how the nodes have been labeled, so it is a good practice to specify the labeling of vertices in addition to the corresponding matrix, especially as you try to compare answers with others.

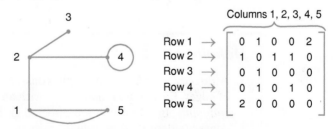

Figure 7-64   The adjacency matrix of a graph.

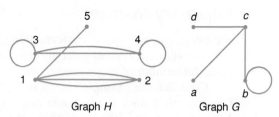

Graph H   Graph G

Figure 7-66 The graph for Practice Problem 1.

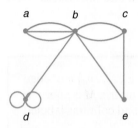

Figure 7-65 The graph for Example 1.

Example 1 *Figure 7-65 shows a graph with vertex set $\{a, b, c, d, e\}$. Use the "obvious" labeling strategy (a as v_1, b as v_2, etc.) to form the adjacency matrix of this graph.*

SOLUTION   We simply let M_{ij} represent the number of edges joining vertices i and j, resulting in the matrix shown below. Note that the labels on the rows and columns help us keep track of which row and column correspond to which vertex—they are not part of the adjacency matrix itself.

$$
\begin{array}{c}
 \\
a \\ b \\ c \\ d \\ e
\end{array}
\begin{array}{c}
\begin{array}{ccccc} a & b & c & d & e \end{array} \\
\left[\begin{array}{ccccc}
0 & 3 & 0 & 0 & 0 \\
3 & 0 & 2 & 1 & 1 \\
0 & 2 & 0 & 0 & 1 \\
0 & 1 & 0 & 2 & 0 \\
0 & 1 & 1 & 0 & 0
\end{array}\right]
\end{array}
$$

☐

Practice Problem 1   *Give the adjacency matrices for the two graphs shown in Figure 7-66. For each matrix, identify the values of M_{13}, M_{31}, M_{25}, and M_{44}.*

Directed Graphs and Matrix Multiplication

The most important generalization for the remainder of this section concerns *directed graphs*. We saw enough in Section 7.1 to know that when drawing a picture of a directed graph, we use arrows rather than line segments to connect the edges. Even in this case, we can still use an adjacency matrix to store information about those edges. We think of M_{ij} as indicating the number of arrows from vertex i to vertex j. Of course, if there are no parallel edges, this count will be either 0 or 1.

We begin by giving a careful definition of a directed graph and its adjacency matrix.

Definition

1. A *directed graph*, like a graph, consists of a set V of vertices and a set E of edges. Each edge is associated with an ordered pair of vertices called its *endpoints*. In other words, a directed graph is the same as a graph, but the edges are described as *ordered* pairs rather than *unordered* pairs.

2. If the endpoints for edge *e* are *a* and *b* in that order, we say *e* is an edge *from a to b*, and in the diagram we draw the edge as a straight or curved arrow from *a* to *b*.

3. For a directed graph, we use (a, b) rather than $[a, b]$ to indicate an edge from *a* to *b*. This emphasizes that the edge is an *ordered* pair, by utilizing the usual notation for ordered pairs.

4. A *walk* in a directed graph is a sequence $v_1 e_1 v_2 e_2 \ldots v_n e_n v_{n+1}$ of alternating vertices and edges that begins and ends with a vertex, and where each edge in the list lies between its endpoints in the proper order. (That is, e_1 is an edge from v_1 to v_2, e_2 is an edge from v_2 to v_3, and so on.) If there is no chance of confusion, we omit the edges when we describe a walk.

5. The *adjacency matrix* for a directed graph with vertices $\{v_1, v_2, \ldots, v_n\}$ is the $n \times n$ matrix where M_{ij} (the entry in row *i*, column *j*) is the number of edges from vertex v_i to vertex v_j.

Example 2 *In Section 7.5 we will see that graphs can be used to represent games. As a simple example, consider a two-player game where there is a single pile of 10 stones and each player may remove one or two stones at a time on his or her turn. The graph given in Figure 7-67 models this game. The labels on the nodes represent the number of stones left at the end of a turn, and each edge represents a legal move. For example, the existence of the edge from node 6 to node 4 indicates there is a legal move that takes the pile from six stones to four stones.*

Write the set of edges in set-builder notation. Give a walk of length 1 in the graph. Give a walk of length 7 from 10 to 0 in the graph.

SOLUTION   The set of vertices is $V = \{0, 1, 2, 3, \ldots, 10\}$ and the set of edges is

$$E = \{(x, y) \in V \times V : x = y + 1 \text{ or } x = y + 2\}$$

That is, the set of edges is

$$E = \{(10, 9), (10, 8), (9, 8), (9, 7), \ldots, (3, 2), (3, 1), (2, 1), (2, 0), (1, 0)\}$$

A walk of length 1 just corresponds to an edge like, for example, 9, 7. We can describe one possible walk of length 7, listing only the vertices, as 10, 8, 7, 5, 4, 3, 1, 0. This walk corresponds to a sequence of moves making up a complete game. □

Practice Problem 2   *Consider the graph of Example 2.*

(a) *Explain why using a directed graph is important for modeling the game.*

(b) *If you label the vertices in the order* 1, 2, 3, 4, 5, 6, 7, 8, 9, 10, 0 *(so that row 1 and column 1 correspond to vertex 1, row and column 2 to vertex 2, ..., row and*

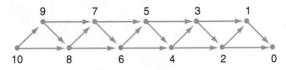

Figure 7-67 The graph for Example 2.

column 11 to vertex 0), give the value of these entries of the adjacency matrix: M_{87},
$M_{53}, M_{82}, M_{28}, M_{34}, M_{43}, M_{66}$.

(c) Explain why $M_{34} \neq M_{43}$ for the adjacency matrix. Could that happen for an ordinary graph?

The representation of directed graphs with matrices is not just for bookkeeping purposes. We will see that there are many properties of directed graphs that can be formally checked using mathematical operations on the corresponding matrices. To do this, we need to first figure out the meaning of specific matrix operations for the underlying graph. For example, given a matrix M for a graph G, what does M^2 tell us about G? We will take the investigation of this question as our starting point for the discussion. The student who wishes to review some basics of matrix operations should consult Appendix B of the text before reading on.

Example 3 *Consider the directed graph G in Figure 7-68 with vertices $\{1, 2, 3, 4, 5, 6\}$, and its adjacency matrix M. Look carefully at the way the product $M \cdot M = M^2$ is defined, and see what it tells us about G.*

SOLUTION   We know, for example, that the entry in row 2, column 3 of $M \cdot M$, comes from multiplying row 2 of M by column 3 of M. That is,

$$\begin{bmatrix} 1 & 0 & 0 & 0 & 1 & 0 \end{bmatrix} \cdot \begin{bmatrix} 1 \\ 0 \\ 0 \\ 0 \\ 1 \\ 1 \end{bmatrix} = (1)(1) + (0)(0) + (0)(0) + (0)(0) + (1)(1) + (0)(1)$$

$$= 2$$

So we get the answer 2 as the row 2, column 3 entry of M^2 specifically because of the two $(1)(1)$ terms in the above sum. We analyze each in turn:

- The first $(1)(1)$ term comes from the product $M_{21} \cdot M_{13}$, so it corresponds to the fact that in M, both the row 2, column 1 entry (M_{21}) and the row 1, column 3 entry (M_{13}) are 1's. This means there is an edge from 2 to 1, and also an edge from 1 to 3. So there is the length-2 walk 2, 1, 3 from node 2 to node 3.

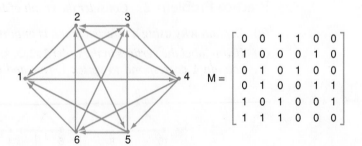

Figure 7-68   The graph for Example 3.

- Similarly, the second (1)(1) term comes from the product $M_{25} \cdot M_{53}$, so it corresponds to the fact that both M_{25} and M_{53} are 1. This means there is the length-2 walk 2, 5, 3 from node 2 to node 3.

It is easy to check by inspecting the graph that, in fact, these are the only length-2 walks from 2 to 3. Hence, we conjecture that the row i, column j entry in M^2 actually counts the number of walks from node i to node j that are exactly of length 2 (i.e., which use two edges). $\qquad\square$

Practice Problem 3 *There were no parallel edges in the previous example. The directed graph whose adjacency matrix appears below extends the graph in Example 3. Draw in the additional edges, and use this picture to decide if the entry in row 2, column 3 of $M \cdot M$ still counts the number of length-2 walks from node 2 to node 3:*

$$M = \begin{bmatrix} 0 & 0 & 4 & 1 & 0 & 0 \\ 2 & 0 & 0 & 0 & 3 & 0 \\ 0 & 1 & 0 & 1 & 0 & 0 \\ 0 & 1 & 0 & 0 & 1 & 1 \\ 1 & 0 & 1 & 0 & 0 & 1 \\ 1 & 1 & 5 & 0 & 0 & 0 \end{bmatrix}$$

Careful attention to the interpretation of matrix multiplication in terms of the adjacency matrix for a directed graph G allows us to prove the following theorem. The proof uses induction on the length of the walks. The details are left as an exercise at the end of this section.

Theorem 1 *Let M be the adjacency matrix of a directed graph G with vertex set $\{1, 2, 3, \ldots, n\}$. The row i, column j entry of M^k counts the number of k-step walks from node i to node j in the graph G. Consequently, the row i, column j entry of*

$$M^1 + M^2 + M^3 + \cdots + M^k$$

counts the number of walks from node i to node j of length 1 or 2 or 3 or ... or k—that is, nontrivial walks of length k or less.

PROOF   See Exercise 13 at the end of this section. $\qquad\blacksquare$

Example 4 *This problem revisits the directed graph of Example 3 repeated in Figure 7-69. Compute matrices M, M^2, M^3, and $M + M^2 + M^3$, and use them to answer the following questions about the graph G:*

1. *How many walks of length 3 are there from node 5 to node 3?*
2. *How many walks are there from node 5 to node 3 of length less than or equal to 3?*
3. *Give a complete list of those walks.*

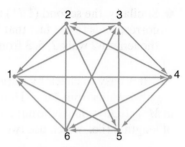

Figure 7-69 The graph for Example 4.

SOLUTION

$$M = \begin{bmatrix} 0 & 0 & 1 & 1 & 0 & 0 \\ 1 & 0 & 0 & 0 & 1 & 0 \\ 0 & 1 & 0 & 1 & 0 & 0 \\ 0 & 1 & 0 & 0 & 1 & 1 \\ 1 & 0 & 1 & 0 & 0 & 1 \\ 1 & 1 & 1 & 0 & 0 & 0 \end{bmatrix}, \quad M^2 = \begin{bmatrix} 0 & 2 & 0 & 1 & 1 & 1 \\ 1 & 0 & 2 & 1 & 0 & 1 \\ 1 & 1 & 0 & 0 & 2 & 1 \\ 3 & 1 & 2 & 0 & 1 & 1 \\ 1 & 2 & 2 & 2 & 0 & 0 \\ 1 & 1 & 1 & 2 & 1 & 0 \end{bmatrix}, \quad M^3 = \begin{bmatrix} 4 & 2 & 2 & 0 & 3 & 2 \\ 1 & 4 & 2 & 3 & 1 & 1 \\ 4 & 1 & 4 & 1 & 1 & 2 \\ 3 & 3 & 5 & 5 & 1 & 1 \\ 2 & 4 & 1 & 3 & 4 & 2 \\ 2 & 3 & 2 & 2 & 3 & 3 \end{bmatrix}$$

$$M + M^2 + M^3 = \begin{bmatrix} 4 & 4 & 3 & 2 & 4 & 3 \\ 3 & 4 & 4 & 4 & 2 & 2 \\ 5 & 3 & 4 & 2 & 3 & 3 \\ 6 & 5 & 7 & 5 & 3 & 3 \\ 4 & 6 & 4 & 5 & 4 & 3 \\ 4 & 5 & 4 & 4 & 4 & 3 \end{bmatrix}$$

1. Since $M^3_{53} = 1$, there is only one walk of length 3 from node 5 to node 3.
2. Since the row 5, column 3 entry in $M + M^2 + M^3$ is 4, there are four walks from node 5 to node 3 of length 3 or less in G.
3. The walk of length 1 is 5, 3, the length 2 walks are 5, 6, 3 and 5, 1, 3, and the length 3 walk is 5, 6, 1, 3. □

Practice Problem 4 *Answer the three questions in Example 4 for walks from node 6 to node 2. Answer the same three questions for walks from node 1 to node 1.*

Connections to Binary Relations

One of the most fascinating things about mathematics is that ideas that seem to be unrelated turn out to have connections between them, and that those connections sometimes make hard problems easier to solve. In this brief subsection, we examine the connections among directed graphs, adjacency matrices, and the binary relations we studied in Chapter 4.

Recall from Section 4.1 that a binary relation on a set A is simply a subset of $A \times A$. Compare this to the notion of a directed graph, consisting of a set V of vertices and a set E of edges, where each edge is associated with an ordered pair of vertices.

Provided there are no parallel edges in the graph, each edge in E can be thought of as the ordered pair (v_1, v_2), where v_1 and v_2 are the endpoints of the edge. In this case, *the edge set of a directed graph with no parallel edges is a binary relation on the vertex set!*

Because of this connection between directed graphs and binary relations, we can extend some of our terminology.

Definition

- Given a relation R on a set A, we can form the directed graph (A, R). We will refer to this as *the graph of the relation R*. Note that the picture of such a graph is exactly what we called the arrow diagram of the relation in Chapter 4.
- The *adjacency matrix for a relation* is simply the adjacency matrix for the corresponding directed graph. Note that the adjacency matrix for a relation has only 0's and 1's as entries.

This correspondence gives us three ways to think of relations. First, a relation is a set of ordered pairs. Second, we can obtain a useful visualization by thinking of the relation as a directed graph. Third, we can obtain a form useful for calculations by considering the adjacency matrix. These connections can be formalized as follows:

Binary Relations, Directed Graphs, and Adjacency Matrices

For a relation R on the set $A = \{1, 2, 3, \ldots, n\}$, the following statements are equivalent for any $a, b \in A$:

1. $(a, b) \in R$ (which we write sometimes as aRb).
2. There is a directed edge from node a to node b in the graph of R.
3. There is a 1 in the row a, column b entry of the adjacency matrix for R.

 Example 5

1. *Let $A = \{1, 2, 3, 4\}$, and let R be the relation $\{(1, 2), (1, 4), (3, 3), (2, 1), (4, 2)\}$. Determine the corresponding graph and adjacency matrix.*

2. *For the adjacency matrix*

$$\begin{bmatrix} 0 & 1 & 0 & 1 & 0 \\ 0 & 0 & 1 & 0 & 0 \\ 1 & 0 & 0 & 1 & 1 \\ 0 & 0 & 0 & 0 & 1 \\ 0 & 0 & 0 & 0 & 0 \end{bmatrix}$$

write the corresponding relation S on the set $\{1, 2, 3, 4, 5\}$ and give its graph.

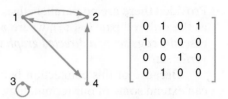

Figure 7-70 The graph and adjacency
matrix for R.

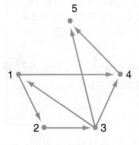

Figure 7-71 The graph
for relation S.

SOLUTION

1. The relation has an ordered pair $(1, 2)$, indicating an edge from node 1 to node 2 in the graph and a 1 in the $(1, 2)$ entry of the matrix. The complete graph and the adjacency matrix for the relation R are given in Figure 7-70.

2. The fact that $M_{31} = 1$ indicates that $(3, 1)$ should be in the relation, and that there is an edge from vertex 3 to vertex 1 in the graph. Reasoning in this way for each entry of 1 in the matrix, we obtain the relation $\{(1, 2), (1, 4), (2, 3), (3, 1), (3, 4), (3, 5), (4, 5)\}$. The graph is shown in Figure 7-71.

☐

These connections will allow us to apply our understanding of any one of these structures to better understand the other two. Let's look at an example of this process.

Boolean Operations and Composition of Relations

One of the benefits of having these alternative ways to look at relations comes from a connection between relation composition and matrix multiplication. To develop this idea, we recall the definition of relation composition, applied to the situation of forming the composition of a relation with itself, and then we will translate this definition into the language of directed graphs and adjacency matrices.

> **Definition**   Let R be a relation on the set A. The composition $R \circ R$ is the relation on A given by the rule "$(a, c) \in R \circ R$ if there is an element $b \in A$ such that $(a, b) \in R$ and $(b, c) \in R$."

To see the connection between this definition and graphs and adjacency matrices, consider what this statement says for a relation R on the set $A = \{1, 2, \ldots, n\}$.

- If G is the graph of R and H is the graph of $R \circ R$, then there is an edge from a to c in H if there is an element $b \in A$ such that there are edges in G from a to b and from b to c.

- If M is the adjacency matrix for the relation R and N is the adjacency matrix for $R \circ R$, then "$N_{ac} = 1$ if for some b, $M_{ab} = 1$ and $M_{bc} = 1$."

a	b	$a \cdot b$	$a + b$
0	0	0	0
0	1	0	1
1	0	0	1
1	1	1	1

Table 7-5   Boolean Arithmetic

The first correspondence means that $R \circ R$ captures information about the existence of walks of length 2 in the graph of R, but what does the second correspondence mean? The answer comes from realizing that in an adjacency matrix, an (i, j) entry of 1 or 0 can be interpreted as values "true" or "false," respectively, for the statement "There is an edge from i to j in the graph of R."

Since we have already seen that matrix multiplication counts the *number* of paths in a graph, and all we want is a "true or false" answer, the clever idea here is to use matrix multiplication to compute the adjacency matrix for $R \circ R$, but to modify the operation to only provide true or false answers. Fortunately, we already know of a system of arithmetic that applies in situations where only true or false answers are desired. We saw these operations in Section 3.4 when we studied Boolean algebra. Table 7-5 shows the relevant interpretation of "plus" and "times" that need to carry over from that section. For an adjacency matrix A, the idea here is to compute the matrix product $A \cdot A$ using the Boolean arithmetic operations on the entries. We will call this the *Boolean product* of A with itself, and we will denote the result $A^{(2)}$ to distinguish it from the ordinary matrix product A^2. Let's compare these two operations in an example, and see how the Boolean operation calculates exactly what we need to know about the relation $R \circ R$.

Example 6   *Consider the relation R on the set $\{1, 2, 3, 4\}$ whose graph G and adjacency matrix A are given in Figure 7-72. Using $B = A^2$ and $C = A^{(2)}$, compare b_{31} to c_{31} and b_{23} to c_{23}.*

SOLUTION

1. In the matrix $B = A^2$, row 3 $\times$ column 1 looks like

$$b_{31} = 1 \cdot 1 + 0 \cdot 0 + 0 \cdot 1 + 1 \cdot 1 = 1 + 0 + 0 + 1 = 2$$

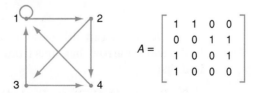

Figure 7-72   The graph and adjacency matrix for Example 6.

We interpret this as "There is an edge from 3 to 1 and an edge from 1 to 1, and there is an edge from 3 to 4 and from 4 to 1; hence 3, 1, 1 is a walk in G of length 2 from node 3 to node 1, and 3, 4, 1 is another walk in G of length 2 from node 3 to node 1, so there is a total of two walks in G of length 2 from node 3 to node 1."

In the matrix $C = A^{(2)}$, row 3 × column 1 looks the same, except we use Boolean arithmetic to evaluate the final answer:

$$c_{31} = 1 \cdot 1 + 0 \cdot 0 + 0 \cdot 1 + 1 \cdot 1 = 1 + 0 + 0 + 1 = 1$$

We can think of this as "There is a walk in G of length 2 from 3 to 1 to 1, and there is a walk in G of length 2 from 3 to 4 to 1, so there is certainly at least one walk in G of length 2 from node 3 to node 1."

2. Similar calculations for b_{23} and c_{23} yield

$$b_{23} = 0 \cdot 0 + 0 \cdot 1 + 1 \cdot 0 + 1 \cdot 0 = 0 + 0 + 0 + 0 = 0$$
$$c_{23} = 0 \cdot 0 + 0 \cdot 1 + 1 \cdot 0 + 1 \cdot 0 = 0 + 0 + 0 + 0 = 0$$

In this case, the answers are the same because an answer of 0 to the question "How many walks in G of length 2 are there from node 2 to node 3?" is the same as a 0 (meaning "false") answer to the question "Is there a walk in G of length 2 from node 2 to node 3?" □

We can now look at the entire matrix $A^{(2)}$ and see its connection to the composite relation $R \circ R$.

Example 7 Let $R = \{(1, 1), (1, 2), (2, 3), (2, 4), (3, 1), (3, 4), (4, 1)\}$ be the relation from Example 6 with adjacency matrix

$$A = \begin{bmatrix} 1 & 1 & 0 & 0 \\ 0 & 0 & 1 & 1 \\ 1 & 0 & 0 & 1 \\ 1 & 0 & 0 & 0 \end{bmatrix}$$

Find the relation $R \circ R$, and compute the matrix $A^{(2)}$.

SOLUTION Using the Boolean product operation, we have

$$A^{(2)} = \begin{bmatrix} 1 & 1 & 1 & 1 \\ 1 & 0 & 0 & 1 \\ 1 & 1 & 0 & 0 \\ 1 & 1 & 0 & 0 \end{bmatrix}$$

To find $R \circ R$, we can either use the definition of composition or we can resort to the two-set arrow diagrams from Section 4.2 as illustrated in Figure 7-73 to find that

$$R \circ R = \{(1, 1), (1, 2), (1, 3), (1, 4), (2, 1), (2, 4), (3, 1), (3, 2), (4, 1), (4, 2)\}$$

Notice that $A^{(2)}$ is the adjacency matrix for $R \circ R$, just as we expected. □

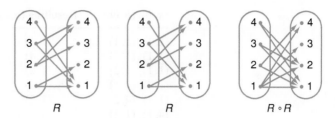

Figure 7-73 The solution to Example 7.

This example illustrates the direct connection between $R \circ R$ and $M^{(2)}$, where M is the adjacency matrix for R. We can formally state this connection as the following theorem. The proof of this theorem is based directly on the ideas from this section, but we will not include it here.

Theorem 2   *If R is a binary relation on a set A with adjacency matrix M, then the matrix $M^{(2)}$ is the adjacency matrix for the relation $R \circ R$ on the set A.*

Practice Problem 5   *For the relation $R = \{(1, 4), (2, 3), (1, 5), (2, 2), (3, 2), (5, 3)\}$ on the set $\{1, 2, 3, 4, 5, 6\}$, draw the directed graph of R, and use it to find $R \circ R$ as a set of ordered pairs. Find the adjacency matrix M of R, and use it to compute $M^{(2)}$. Verify that $M^{(2)}$ is the adjacency matrix for $R \circ R$.*

Application to Transitivity

One application of composition explored in Exercises 21 and 22 of Section 4.4 gave a simple characterization of a transitive relation. This will give us, in turn, a simple test for transitivity of a binary relation using the adjacency matrix of the relation. We will learn about this test after a brief review of the concept of transitivity.

> **Definition**   Let R be a relation on the set A, that is, R is a subset of $A \times A$. R is *transitive* if whenever $(a, b) \in R$ and $(b, c) \in R$, it is also true that $(a, c) \in R$.

Example 8   *Here are three relations on $A = \{1, 2, 3, 4, 5\}$. Which one is not transitive?*

1. *Define R_1 by saying that $a R_1 b$ if a and b have the same number of letters in their English-language spelling. For example, $4 R_1 5$ since the words four and five each have four letters.*
2. *$R_2 = \{(a, b) : a^2 - b^2 \leq 5\}$.*
3. *$R_3 = \{(1, 1), (1, 4), (4, 1), (1, 2), (4, 2), (5, 5)\}$.*

SOLUTION

1. Not only is R_1 transitive, but it is also reflexive and symmetric, so this is an equivalence relation and it partitions the set A into the subsets $\{1, 2\}$, $\{3\}$, and $\{4, 5\}$.

$(a, b), (b, c)$	Resulting (a, c)	Is $(a, c) \in R_3$?
$(1, 1), (1, 1)$	$(1, 1)$	Yes
$(1, 1), (1, 4)$	$(1, 4)$	Yes
$(1, 1), (1, 2)$	$(1, 2)$	Yes
$(1, 4), (4, 1)$	$(1, 1)$	Yes
$(1, 4), (4, 2)$	$(1, 2)$	Yes
$(4, 1), (1, 1)$	$(4, 1)$	Yes
$(4, 1), (1, 2)$	$(4, 2)$	Yes

Table 7-6   Transitive Relation in Example 8

2. Relation R_2 is not transitive since $(3, 2) \in R_2$ and $(2, 1) \in R_2$ but $(3, 1) \notin R_2$.
3. Relation R_3 is transitive, as shown by the complete list of pairs (a, b) and $(b, c) \in R_3$ shown in Table 7-6.

□

Practice Problem 6   *For $A = \{1, 2, 3\}$ decide which of the following relations on A is transitive:*

(a) $R_1 = \{(1, 1), (2, 2), (3, 3), (1, 2), (2, 3)\}$
(b) $R_2 = \{(1, 3), (3, 1), (1, 1), (3, 3)\}$
(c) $R_3 = \{(1, 3), (3, 1)\}$

We have seen that checking a relation for transitivity can be fairly tedious, but if we write the definition of transitivity in terms of relations, graphs, and matrices, we will be able to take advantage of seeing the situation from several points of view.

Proposition 3   *Let a transtive relation R on a set A be given along with its associated directed graph G and adjacency matrix M. Then for all $a, b, c \in A$, the following are true:*

- *If aRb and bRc, then aRc.*
- *In the graph G, if there is an edge from a to b and an edge from b to c, then there is an edge from a to c. We can rephrase this as "In the graph G, if there is a length-2 walk from a to c, then there is an edge from a to c."*
- *If $M_{ab} = 1$ and $M_{bc} = 1$ then $M_{ac} = 1$.*

PROOF   This is just a translation of the transitive property of binary relations into the language of the other structures.   ■

Example 9   *Give the graph and adjacency matrix for relation R_3 of Example 8, and use them to discuss the transitivity of R_3.*

SOLUTION   The matrix and graph are given in Figure 7-74. The easiest way to check transitivity is to use

$$M = \begin{bmatrix} 1 & 1 & 0 & 1 & 0 \\ 0 & 0 & 0 & 0 & 0 \\ 0 & 0 & 0 & 0 & 0 \\ 1 & 1 & 0 & 0 & 0 \\ 0 & 0 & 0 & 0 & 1 \end{bmatrix}$$

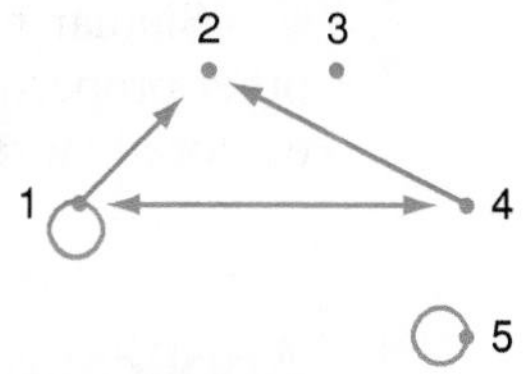

Figure 7-74 The solution to Example 9.

$$M^{(2)} = \begin{bmatrix} 1 & 1 & 0 & 1 & 0 \\ 0 & 0 & 0 & 0 & 0 \\ 0 & 0 & 0 & 0 & 0 \\ 1 & 1 & 0 & 0 & 0 \\ 0 & 0 & 0 & 0 & 0 \end{bmatrix}$$

which locates length-2 walks. When we compare this to M, we see that whenever there is a length-2 walk, there is an an edge, so the relation is transitive. □

This idea can be turned into a nice computational test for transitivity that is useful if you have some form of technology aid* for the matrix calculations. To make this test easy to state, we need one more piece of notation.

Definition Suppose M and N are two adjacency matrices of the same size. We write $M \leq N$ to mean that, for every pair of indices i and j, $M_{ij} \leq N_{ij}$. (That is, $N_{ij} = 1$ whenever $M_{ij} = 1$.)

Proposition 4 *If M is the adjacency matrix for a binary relation R, then R is transitive if and only if $M^{(2)} \leq M$.*

PROOF This statement combines the statement of transitivity in terms of direct graphs in Proposition 3 with the interpretation of $M^{(2)}$ given in Theorem 5. It also follows directly from Exercises 22 and 23 of Section 4.4. ■

Practice Problem 7 *Verify that Proposition 4 is correct for each of these relations on $A = \{1, 2, 3\}$ from Practice Problem 6:*

(a) $R_1 = \{(1, 1), (2, 2), (3, 3), (1, 2), (2, 3)\}$
(b) $R_2 = \{(1, 3), (3, 1), (1, 1), (3, 3)\}$
(c) $R_3 = \{(1, 3), (3, 1)\}$

---

* Appendix B explains how to use a TI-83 calculator for matrix arithmetic.

Similar matrix conditions can be found for checking that a relation satisfies other properties (reflexive, symmetric, etc.) of interest from Chapter 4. These will be explored more in the exercises.

Application to Connectivity

This section began with an application of matrices to the problem of counting the number of walks in a graph. There is a variation of this idea that can answer a related question. Although it might be *interesting* to know how many walks there are in a graph, the existence of walks is often all you really need to know about. An example of this would be a test to see if a graph is connected. This would certainly be a nice thing to have a computer check for us, so we should try to understand this property of graphs in terms of operations on the corresponding adjacency matrices. At this point, it should come as no surprise that the answer to the yes or no question "Is there a walk from node i to node j?" is related to the Boolean product of the adjacency matrix. To make this precise, we will need slightly more general notation than we have used before.

Definition For an adjacency matrix M, we will write
$$M^{(k)} = \underbrace{M \cdot M \cdots \cdot M}_{k \text{ times}}$$
where the $\cdot$ operation on the right is the Boolean product.

This notation allows us to restate Theorem 1 in terms of the existence of walks.

Theorem 5 *Let M be the adjacency matrix of a graph G with nodes $\{1, 2, 3, \ldots, n\}$. The row i, column j entry of $M^{(k)}$ is 1 if and only if there exists a k-step walk in G from node i to node j.*

PROOF See Exercise 14. ∎

Example 10 *Consider the graph G with its adjacency matrix A and its Boolean adjacency matrix B shown in Figure 7-75. Compare A^3 and $B^{(3)}$, and give an interpetation of each.*

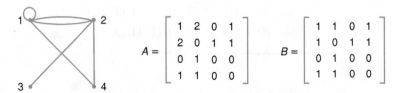

$$A = \begin{bmatrix} 1 & 2 & 0 & 1 \\ 2 & 0 & 1 & 1 \\ 0 & 1 & 0 & 0 \\ 1 & 1 & 0 & 0 \end{bmatrix} \qquad B = \begin{bmatrix} 1 & 1 & 0 & 1 \\ 1 & 0 & 1 & 1 \\ 0 & 1 & 0 & 0 \\ 1 & 1 & 0 & 0 \end{bmatrix}$$

Figure 7-75 The Boolean adjacency matrix for Example 10.

SOLUTION   We compute

$$A^3 = \begin{bmatrix} 15 & 17 & 3 & 9 \\ 17 & 8 & 6 & 9 \\ 3 & 6 & 0 & 2 \\ 9 & 9 & 2 & 5 \end{bmatrix}, \qquad B^{(3)} = \begin{bmatrix} 1 & 1 & 1 & 1 \\ 1 & 1 & 1 & 1 \\ 1 & 1 & 0 & 1 \\ 1 & 1 & 1 & 1 \end{bmatrix}$$

As we have seen before, the entries in A^3 count walks of length 3 between various pairs of nodes. We can interpret $B^{(3)}$ as providing "existence" information (instead of a count) about walks of length 3. Hence, we can tell that node 3 is not connected to itself by a length-3 walk, but every other pair of nodes is. Note that this information is also readily available in A^3. □

To check for connectivity in a graph, we do not care as much about the length of a walk in a graph as about its existence. For a given pair of nodes i and j, we really want to know if there is a walk from i to j of length 1, one of length 2, or one of length 3, and so on. Since we are already using Boolean arithmetic with the entries of the matrices, the use of the logical "or" in this statement can easily be incorporated into those calculations with one more piece of special notation.

> **Definition**   If M and N are adjacency matrices of the same size, the matrix $M \vee N$ is the adjacency matrix whose (i, j) entry is $M_{ij} + N_{ij}$, using the "plus" operation from Boolean arithmetic.

Using this operation, it is easy to build a matrix that tells us about the existence of walks of length k or less.

Corollary 6   *Let M be the adjacency matrix of a graph G with nodes $\{1, 2, 3, \ldots, n\}$. The row i, column j entry of*

$$M^{(1)} \vee M^{(2)} \vee M^{(3)} \vee \cdots \vee M^{(k)}$$

is 1 if and only if there exists a walk from node i to node j of length 1, 2, 3, . . . or k—that is, a nontrivial walk of length k or less.

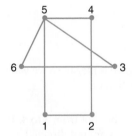

Figure 7-76 The graph for Example 11.

Example 11   *Form the Boolean adjacency matrix of the graph G shown in Figure 7-76, calculate $M^{(1)}$, $M^{(2)}$, $M^{(3)}$, and explain the meaning of entries in $M^{(1)} \vee M^{(2)} \vee M^{(3)}$ in an English sentence.*

SOLUTION

$$M^{(1)} = \begin{bmatrix} 0 & 1 & 0 & 0 & 1 & 0 \\ 1 & 0 & 0 & 1 & 0 & 0 \\ 0 & 0 & 0 & 0 & 1 & 1 \\ 0 & 1 & 0 & 0 & 1 & 0 \\ 1 & 0 & 1 & 1 & 0 & 1 \\ 0 & 0 & 1 & 0 & 1 & 0 \end{bmatrix}, \quad M^{(2)} = \begin{bmatrix} 1 & 0 & 1 & 1 & 0 & 1 \\ 0 & 1 & 0 & 0 & 1 & 0 \\ 1 & 0 & 1 & 1 & 1 & 1 \\ 1 & 0 & 1 & 1 & 0 & 1 \\ 0 & 1 & 1 & 0 & 1 & 1 \\ 1 & 0 & 1 & 1 & 1 & 1 \end{bmatrix}, \quad M^{(3)} = \begin{bmatrix} 0 & 1 & 1 & 0 & 1 & 1 \\ 1 & 0 & 1 & 1 & 0 & 1 \\ 1 & 1 & 1 & 1 & 1 & 1 \\ 0 & 1 & 1 & 0 & 1 & 1 \\ 1 & 0 & 1 & 1 & 1 & 1 \\ 1 & 1 & 1 & 1 & 1 & 1 \end{bmatrix}$$

$$M^{(1)} \vee M^{(2)} \vee M^{(3)} = \begin{bmatrix} 1 & 1 & 1 & 1 & 1 & 1 \\ 1 & 1 & 1 & 1 & 1 & 1 \\ 1 & 1 & 1 & 1 & 1 & 1 \\ 1 & 1 & 1 & 1 & 1 & 1 \\ 1 & 1 & 1 & 1 & 1 & 1 \\ 1 & 1 & 1 & 1 & 1 & 1 \end{bmatrix}$$

The final result is a matrix where every entry is 1. This tells us that every vertex in the graph with adjacency matrix M can be reached from any vertex (including itself) in one, two, or three steps. $\square$

Practice Problem 8 *In the graph of Example 11, list all the pairs of vertices that do not have walks of length 1 or 2 from one to the other, by considering just $M^{(1)} \vee M^{(2)}$.*

Following the example and practice problem above and realizing that the longest path in a graph on n vertices has a length of at most $n-1$ (as proved in Exercise 7 of Section 7.2), we have the following corollary to Theorem 5.

Corollary 7 *If G is a graph on n vertices with adjacency matrix M, then G is connected if and only if the matrix*

$$M^{(1)} \vee M^{(2)} \vee M^{(3)} \vee \cdots \vee M^{(n-1)}$$

consists of all 1's.

We should note that this is not the most efficient way to check if a graph is connected, but it is a very easy test to implement in any computer language in which basic matrix operations are available.

Solutions to Practice Problems

1 The adjacency matrix for G is

$$M = \begin{bmatrix} 0 & 3 & 0 & 0 & 1 \\ 3 & 0 & 0 & 0 & 0 \\ 0 & 0 & 1 & 2 & 0 \\ 0 & 0 & 2 & 1 & 0 \\ 1 & 0 & 0 & 0 & 0 \end{bmatrix}$$

for which $M_{13} = 0$, $M_{31} = 0$, $M_{25} = 0$, $M_{44} = 1$. The adjacency matrix for H is

$$M = \begin{matrix} a \\ b \\ c \\ d \end{matrix} \begin{bmatrix} 0 & 0 & 1 & 0 \\ 0 & 1 & 1 & 0 \\ 1 & 1 & 0 & 1 \\ 0 & 0 & 1 & 0 \end{bmatrix}$$

for which $M_{13} = 1$, $M_{31} = 1$, $M_{44} = 0$, and M_{25} does not exist since there is no fifth node.

2 (a) In this game you are not allowed to put stones back in the pile. For example, the move from 10 to 8 is le-

gal, but the move from 8 to 10 is illegal, so we want an edge in the former direction only.

(b) 1, 1, 0, 0, 0, 1, 0

(c) There is an edge from 4 to 3 but not from 3 to 4. In an ordinary graph, since edges are not directed, we will always have $M_{ij} = M_{ji}$.

3 The multiplication gives $(2)(4) + (0)(0) + (0)(0) + (0)(0) + (3)(1) + (0)(5) = 8 + 3 = 11$. The $(2)(4)$ is $M_{21} \cdot M_{13}$, so it corresponds to the fact that in G there are four edges from node 2 to node 1, and also two edges from node 1 to node 3. If we apply the counting techniques of Chapter 5, there are $2 \cdot 4 = 8$ length-2 walks from 2 to 1 to 3. Likewise, there are $3 \cdot 1 = 3$ such walks from 2 to 5 to 3. Since you can get from node 2 to node 3 by way of either node 5 *or* node 1, there are $8 + 3 = 11$ total walks of length 2. The matrix product does still count the walks.

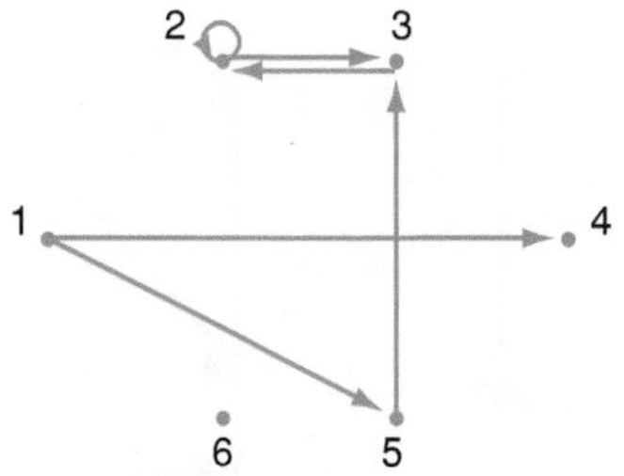

$$M = \begin{bmatrix} 0 & 0 & 0 & 1 & 1 & 0 \\ 0 & 1 & 1 & 0 & 0 & 0 \\ 0 & 1 & 0 & 0 & 0 & 0 \\ 0 & 0 & 0 & 0 & 0 & 0 \\ 0 & 0 & 1 & 0 & 0 & 0 \\ 0 & 0 & 0 & 0 & 0 & 0 \end{bmatrix}$$

Figure 7-77 Solution to Practice Problem 5.

4 $M_{62} = 1$, indicating the walk 6, 2. $M_{62}^2 = 1$, indicating the length-2 walk 6, 3, 2. $M_{62}^3 = 3$, indicating the three walks 6, 1, 3, 2, 6, 1, 4, 2, and 6, 3, 4, 2. Similarly, $M_{11} = 0$ and $M_{11}^2 = 0$, indicating no walks of length 1 or 2, but $M_{11}^3 = 4$, indicating the walks 1, 4, 2, 1, 1, 4, 6, 1, 1, 4, 5, 1, and 1, 3, 2, 1.

5 The graph and adjacency matrix M for the relation R are given in Figure 7-77.

$$M^{(2)} = \begin{bmatrix} 0 & 0 & 0 & 1 & 1 & 0 \\ 0 & 1 & 1 & 0 & 0 & 0 \\ 0 & 1 & 0 & 0 & 0 & 0 \\ 0 & 0 & 0 & 0 & 0 & 0 \\ 0 & 0 & 1 & 0 & 0 & 0 \\ 0 & 0 & 0 & 0 & 0 & 0 \end{bmatrix} \cdot \begin{bmatrix} 0 & 0 & 0 & 1 & 1 & 0 \\ 0 & 1 & 1 & 0 & 0 & 0 \\ 0 & 1 & 0 & 0 & 0 & 0 \\ 0 & 0 & 0 & 0 & 0 & 0 \\ 0 & 0 & 1 & 0 & 0 & 0 \\ 0 & 0 & 0 & 0 & 0 & 0 \end{bmatrix}$$

$$= \begin{bmatrix} 0 & 0 & 1 & 0 & 0 & 0 \\ 0 & 1 & 1 & 0 & 0 & 0 \\ 0 & 1 & 1 & 0 & 0 & 0 \\ 0 & 0 & 0 & 0 & 0 & 0 \\ 0 & 1 & 0 & 0 & 0 & 0 \\ 0 & 0 & 0 & 0 & 0 & 0 \end{bmatrix}$$

Using the graph of R, we can find $R \circ R = \{(1, 3), (2, 2), (2, 3), (3, 2), (3, 3), (5, 2)\}$. This is indeed the relation for which $M^{(2)}$ is the adjacency matrix.

6 Only relation R_2 is transitive, as Table 7-7 shows. As for the other relations, $(1, 2) \in R_1$ and $(2, 3) \in R_1$ but $(1, 3) \notin R_1$, and $(1, 3) \in R_3$ and $(3, 1) \in R_3$ but $(1, 1) \notin R_3$.

$(a, b), (b, c)$	Resulting (a, c)	Is $(a, c) \in R_2$?
(1, 3), (3, 1)	(1, 1)	Yes
(1, 3), (3, 3)	(1, 3)	Yes
(3, 1), (1, 1)	(3, 1)	Yes
(3, 1), (1, 3)	(3, 3)	Yes
(1, 1), (1, 1)	(1, 1)	Yes
(1, 1), (1, 3)	(1, 3)	Yes
(3, 3), (3, 1)	(3, 1)	Yes
(3, 3), (3, 3)	(3, 3)	Yes

Table 7-7 Table for Practice Problem 6.

7 Letting M_i be the adjacency matrix for R_i, we have

$$M_1 = \begin{bmatrix} 1 & 1 & 0 \\ 0 & 1 & 1 \\ 0 & 0 & 1 \end{bmatrix}, \quad M_2 = \begin{bmatrix} 1 & 0 & 1 \\ 0 & 0 & 0 \\ 1 & 0 & 1 \end{bmatrix},$$

$$M_3 = \begin{bmatrix} 0 & 0 & 1 \\ 0 & 0 & 0 \\ 1 & 0 & 0 \end{bmatrix}$$

and so

$$M_1^{(2)} = \begin{bmatrix} 1 & 1 & 1 \\ 0 & 1 & 1 \\ 0 & 0 & 1 \end{bmatrix}, \quad M_2^{(2)} = \begin{bmatrix} 1 & 0 & 1 \\ 0 & 0 & 0 \\ 1 & 0 & 1 \end{bmatrix},$$

$$M_3^{(2)} = \begin{bmatrix} 1 & 0 & 0 \\ 0 & 0 & 0 \\ 0 & 0 & 1 \end{bmatrix}$$

To test the relations for transitivity, we can see that $M_1^{(2)} \nleq M_1$ since the (1, 3) entry in $M_1^{(2)}$ is 1, while the (1, 3) entry in M_1 is 0, so R_1 is not transitive. A similar remark applies to the (1, 1) entry to demonstrate that $M_3^{(2)} \nleq M_3$, showing relation R_3 is not transitive. However, it is the case that $M_2^{(2)} \leq M_2$, so R_2 *is* a transitive relation.

8 Since

$$M^{(1)} \vee M^{(2)} = \begin{bmatrix} 1 & 1 & 1 & 1 & 1 & 1 \\ 1 & 1 & 0 & 1 & 1 & 0 \\ 1 & 0 & 1 & 1 & 1 & 1 \\ 1 & 1 & 1 & 1 & 1 & 1 \\ 1 & 1 & 1 & 1 & 1 & 1 \\ 1 & 0 & 1 & 1 & 1 & 1 \end{bmatrix}$$

there are no length 1 or length 2 walks from node 2 to node 3 and there are no length 1 or length 2 walks from node 2 to node 6.

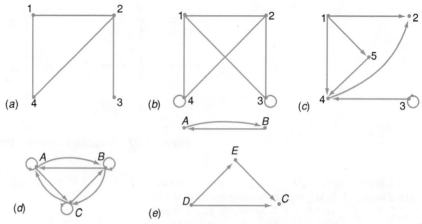

Figure 7-78   Graphs for Problem 1.

Exercises for Section 7.4

1. Give the adjacency matrix for each graph in Figure 7-78.
2. Give the adjacency matrix for these graphs:
   (a) The bridges of Königsberg (page 506)
   (b) The envelope of Figure 7-14 (page 513)
   (c) K_5, the complete (simple) graph with vertex set $\{1, 2, 3, 4, 5\}$
3. Explain how to use the adjacency matrix to calculate the degree of a node for these situations:
   (a) A simple graph (no parallel edges, no loops).
   (b) Loops are allowed but not parallel edges.
   (c) Parallel edges are allowed but not loops.
   (d) Both loops and parallel edges are allowed.
4. For a directed graph, the *out-degree* of a node is the number of edges beginning at the node, and the *in-degree* is the number of edges ending at the node. Explain how to use the adjacency matrix of a directed graph to calculate the in-degree and out-degree of a node for these situations:
   (a) No parallel edges, no loops.
   (b) Loops are allowed but not parallel edges.
   (c) Parallel edges are allowed but not loops.
   (d) Both loops and parallel edges are allowed.
5. Use Theorem 1 to count the number of walks of length 3 or less from node 1 to node 6 in each graph shown in Figure 7-79. Check each answer by directly counting the walks in the graph.
6. For each graph in Problem 5, give the corresponding relation on the set $\{1, 2, 3, 4, 5, 6\}$ as a set of ordered pairs.
7. In solving Exercise 5, you calculated M^3 for the adjacency matrices of the graphs.

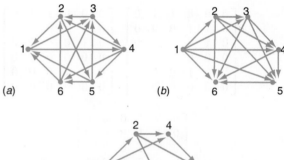

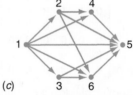

Figure 7-79   Graphs for Problem 5.

(a) Use this to determine $M^{(3)}$, the matrix that would result using Boolean operations rather than ordinary arithmetic operations.
(b) Use the result of part (a) to determine $R \circ R \circ R$, where R is the relation corresponding to the graph.
8. For each given relation on the set $A = \{1, 2, 3, 4, 5, 6\}$, draw the graph and give the adjacency matrices for relations R and $R \circ R$.
   (a) $R = \{(1, 2), (2, 3), (3, 4), (4, 5), (5, 6)\}$
   (b) $R = \{(1, 3), (3, 1), (1, 5), (5, 1), (3, 5), (5, 3), (1, 1), (3, 3), (5, 5)\}$
9. Suppose G is a graph.

Node	1	2	3	4	5	6	7	8	9
Number	2	3	1	4	0	2	0	1	0

Node	Number
1	5
2	2
3	0
4	3
5	2
6	0
7	4
8	2
9	7

Table 7-8   Tables for Exercise 10

Node	1	2	3	4	5	6	7	8	9
Number	a_{61}	a_{62}	a_{63}	a_{64}	a_{65}	a_{66}	a_{67}	a_{68}	a_{69}

Node	Number
1	b_{13}
2	b_{23}
3	b_{33}
4	b_{43}
5	b_{53}
6	b_{63}
7	b_{73}
8	b_{83}
9	b_{93}

Table 7-9   Table for Exercise 11

(a) Using Theorem 5, we see that the Boolean expression $M^{(1)} \vee M^{(2)} \vee M^{(3)}$ determines all pairs of nodes between which there is a walk of length 3 or less. What does $M^{(1)} \vee M^{(2)} \vee M^{(3)} \vee M^{(4)} \vee M^{(5)}$ determine?

(b) If we want to determine the existence of nontrivial walks of length 7 or less, what calculation could we do?

(c) If we want to determine the existence of nontrivial walks of length 9 or less, what calculation could we do?

(d) Suppose G has 10 nodes. Using Theorem 5, what calculation could we do to determine the existence of nontrivial walks of all possible lengths?

(e) Generalize your answer for part (d) to a graph with n nodes.

10. This problem develops some ideas necessary to prove Theorem 1. The formal proof can be found in Exercise 13.

  Suppose a graph has nine nodes labeled 1, 2, 3, ..., 9. In Table 7.8, the left-hand table gives the number of one-step walks from node 6 to each of the nodes in the graph, and the right-hand table gives the number of four-step walks from various nodes to node 3.

(a) Notice that there are two one-step walks from node 6 to node 1, and five four-step walks from node 1 to node 3. How many five-step walks are there from node 6 to node 3 that begin by going to node 1?

(b) How many five-step walks are there from node 6 to node 3 that begin by going to node 2?

(c) How many five-step walks are there from node 6 to node 3 that begin by going to node 8?

(d) How many five-step walks are there from node 6 to node 3 that begin by going to node 9?

(e) How many five-step walks are there from node 6 to node 3?

11. This problem continues the development of ideas for Theorem 1.

(a) Repeat what you did in the previous problem, but using Table 7-9 where symbols have replaced the numbers.

(b) The notation in part (a) uses a_{6t} to indicate the number of one-step walks from node 6 to node t, and b_{t3} for the number of four-step walks from node t to node 3. Using the similar notation of a_{8t} to indicate the number of one-step walks from node 8 to node t, and b_{t2} for the number of four-step walks from node t to node 2, write a formula for the number of five-step walks from node 8 to node 2.

(c) Generalize to obtain a formula for the number of five-step walks from node i to node j.

(d) Fill in the blanks. If M is the adjacency matrix for the graph, the first table in part (a) gives row _____ of M, and the second table gives column _____ of M^4.

12. This problem continues the development of ideas for Theorem 1.

    (a) We will use induction on k, where k is the power to which we are raising the matrix M. Let $P(k)$ be the statement "For every pair of integers i and j, the row i, column j entry of M^k counts the number of k-step walks from node i to node j." Write $P(1)$ carefully.

    (b) Explain why $P(1)$ is true. (HINT: A one-step walk is the same as what?)

    (c) Write $P(k-1)$ carefully.

13. Prove Theorem 1 by induction on $k \geq 1$. In the induction step you should connect the fact that $M^k = M \cdot M^{k-1}$ with a recursive counting model for the number of walks of length k between a fixed pair of nodes. The previous two problems contain related ideas.

14. Explain how Theorem 5 follows from Theorem 1.

15. For each of the following sets A and relations R on A, give the graph and the adjacency matrix for the relation:

    (a) $A = \{1, 2, 3, 4\}$, $R = \{(1, 1), (2, 2), (3, 3), (4, 4)\}$

    (b) $A = \{1, 2, 3, 4\}$, $R = \{(1, 1), (2, 3), (3, 2), (4, 1),$ $(4, 4)\}$

    (c) $A = \{1, 2, 3, 4\}$, $R = \{(1, 2), (2, 3), (1, 3), (3, 1),$ $(3, 2), (2, 1)\}$

    (d) $A = \{1, 2, 3, 4\}$, $R = \{(1, 1), (2, 2), (3, 3), (4, 4),$ $(1, 3), (3, 1), (1, 4), (2, 4)\}$

    (e) $A = \{1, 2, 3, 4\}$, $R = \{(1, 2), (2, 1), (1, 3), (3, 1),$ $(1, 1), (2, 2), (3, 3)\}$

    (f) $A = \{1, 2, 3, 4, 5\}$, $R = \{(1, 3), (2, 4), (3, 5), (4, 1),$ $(5, 4), (3, 1), (2, 2), (1, 5)\}$

    (g) $A = \{1, 2, 3, 4, 5\}$, $R = \{(1, 2), (2, 3), (3, 4), (4, 5),$ $(5, 4), (4, 3), (3, 2), (2, 1)\}$

16. Using the notation of Proposition 4, we can test a relation with adjacency matrix M for being reflexive by checking if $I \leq M$, where I is the matrix* of the same size as M with 1's on the main diagonal and 0's elsewhere. For each relation in Exercise 15, use this test to determine if the relation is reflexive.

17. Prove that for any relation R with adjacency matrix M, the matrix $I \vee M$ is the adjacency matrix of a reflexive relation. This new relation is called the *reflexive closure* of R. It is the smallest reflexive relation extending R.

18. Find the adjacency matrix, graph, and set of ordered pairs descriptions of the reflexive closure of each relation in Exercise 15.

19. We can test a relation with adjacency matrix M for being symmetric by checking if $M = M^T$, where M^T is the matrix† formed by transposing the rows and columns of M. For each relation in Exercise 15, use this test to determine if the relation is symmetric.

20. Prove that for any relation R with adjacency matrix M, the matrix $M \vee M^T$ is the adjacency matrix of a symmetric relation. This new relation is called the *symmetric closure* of R. It is the smallest symmetric relation extending R.

21. Find the adjacency matrix, graph, and set of ordered pairs descriptions of the symmetric closure of each relation in Exercise 15.

22. Use the test given in Proposition 4 to determine if each relation in Exercise 15 is transitive. For those that are not, fill in the blanks in this sentence: "____ R ____ and ____ R ____, but not ____ R ____."

23. The smallest transitive relation that extends a given relation R is called the *transitive closure of R*. If M is the adjacency matrix for a relation R, then the adjacency matrix for the transitive closure of R is

    $$M^{(1)} \vee M^{(2)} \vee M^{(3)} \vee \cdots \vee M^{(k)}$$

    where k is the length of the longest path in the graph of R. We will see how this procedure works by finding the transitive closure of the following relation on the set $A = \{1, 2, 3, 4, 5, 6\}$ with the arrow diagram shown in Figure 7-80.

    $$R = \{(x, y) \in A \times A : x + 1 = y\}$$

    (a) Use the graph to make a complete list of pairs of nodes (a, b) for which there is a walk from node a to b.

    (b) Use the result of part (a) to write the transitive closure of the corresponding relation.

    (c) Write the adjacency matrix M for the graph.

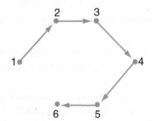

Figure 7-80 Diagram for Problem 23.

---

* This matrix is called the *identity matrix*. See Appendix B for more on matrices.

† This matrix is called the *transpose of M*. See Appendix B for more on matrices.

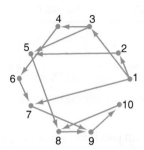

Figure 7-81 Diagram for Problem 24.

(d) Calculate $M^{(1)} \vee M^{(2)} \vee M^{(3)} \vee M^{(4)} \vee M^{(5)}$ to determine the transitive closure of the relation. Compare your result to what you found in part (b).

24. Let the relation R on the set $C = \{1, 2, 3, 4, 5, 6, 7, 8, 9, 10\}$ be given by

$$R = \{(a, b) : \text{There is a direct flight from city } a \text{ to city } b\}$$

where the flights are described by the directed graph in Figure 7-81. Find the matrix for the transitive closure of R. Write an English sentence explaining the meaning of the transitive closure in terms of flights and cities.

25. Find the adjacency matrix, graph, and set of ordered pairs descriptions of the transitive closure of each relation in Exercise 15.

26. For a transitive relation, we know that $M^{(2)} \leq M$—if there is a 1 in row a, column b of $M^{(2)}$, then there must be a 1 in row a, column b of M. A friend has conjectured that, in fact, $M^{(2)} = M$ for transitive relations. Either prove this statement is true, or find a counterexample. (HINT: What does this conjecture say in terms of walks in a graph?)

7.5 Graphs in Puzzles and Games

Throughout this chapter, we have seen important ideas in graph theory motivated by problems about puzzles. We will turn the tables in this section, taking some of the basic graph structures we have developed and applying them to puzzles and games we have not seen before.

Wolves, Goats, and Cabbages

The following version of a truly ancient puzzle is generally attributed to Alcuin of York (735–804), a friend of Charlemagne.

Example 1 *A traveler has three possessions, a wolf, a goat, and a cabbage, which he must transport across a river. The catch is that, if left alone, the wolf will eat the goat or the goat will eat the cabbage, and naturally the boat can hold only the traveler with one possession at a time.*

The first thing that comes to mind when considering this puzzle is probably "That must be a pretty big cabbage." Hopefully, the second thing that comes to mind is how this puzzle might be modeled using a graph.

SOLUTION We will represent a "legal" state of this puzzle—that is, one in which nothing is being eaten—as a pair in which the first entry tells who or what is on the departing shore and the second entry tells who or what is on the arriving shore. We use $\emptyset$ to stand for "no one." For example, the pair (WC, TG) means that the wolf and the cabbage are on the departing shore and the traveler and the goat are on the arriving shore. The pair $(WCTG, \emptyset)$ means that everything is on the departing shore.

We can make a complete list of the legal states, since it turns out there are only 10 of them. We organize them by listing first those states with three items

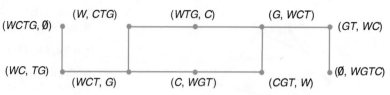

Figure 7-82   A wolf, a goat, and a (really big) cabbage.

on the departing shore, then those states with two items on the departing shore, and so on.

$$(WCTG, \emptyset) \quad (WCT,G) \quad (WTG,C) \quad (CGT,W) \quad (WC,TG)$$
$$(GT,WC) \quad (W,CTG) \quad (C,WGT) \quad (G,WTC) \quad (\emptyset, WGTC)$$

We then draw a graph in which the legal states form the nodes, and we draw an edge between two states if it is possible to get from one to the other in one crossing. Thus, we have the graph in Figure 7-82.

Notice that a solution to the puzzle is simply a path from $(WCTG, \emptyset)$ to $(\emptyset, WGTC)$. ☐

From this point of view, we can see that there are two solutions to the puzzle, and they take the same number of crossings to get across the river. The graph model not only helps us find a solution to the puzzle, but also it allows us to look at *all* solutions—a valuable attribute of graph theoretic solutions.

Practice Problem 1   *Suppose the traveler is actually traveling by ferry boat and is charged an extra $5 fee each time he crosses the river with a live animal. How should he get all his items across the river for as little money as possible?*

Instant Insanity

The previous problem illustrates a fundamental idea in the modeling of puzzles and games using graphs—each node represents a "state" of the puzzle and edges connect states that can be reached one from the other in one "move." When this happens, the crucial structure in the graph model is a *path* from the beginning state to end state. This type of model will come up again in the context of two player games at the end of this section, but first we will look at a different way to model a puzzle using a graph.

The Instant Insanity™* puzzle consists of four cubes with faces colored green, red, blue, or white. The goal is to stack the cubes into a tower so that each of the four colors is represented on each of the four sides of the tower. Figure 7-83 shows how the four cubes are colored. On cube 1, for example, there is a white face in front, a white face to the right, and a red face on top, while there is a blue face to the left, a green face on back, and a green face on the bottom.

---

* INSTANT INSANITY is a trademark of Winning Moves, Inc.,© 2004 Winning Move, Inc. However, variations of the game have been around for over a century. It appears to have been patented in the U.K. in various forms in the early twentieth century, and it was marketed in the U.S. in the 1960s in the form we consider here, by Parker Brothers. It has recently been re-released by Winning Moves, Inc.

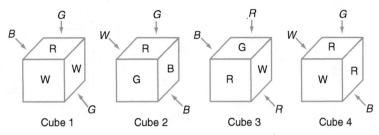

Figure 7-83 Instant Insanity cubes.

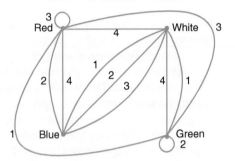

Figure 7-84 The graph of the puzzle.

The puzzle can be modeled with a graph in the following way. Label four nodes of a graph Red, Blue, Green, and White. Now connect two nodes with an edge labeled *i* if cube *i* has a pair of opposite faces with the colors of the nodes. For example, since cube 4 has a white face opposite a red face, there is an edge from White to Red labeled with a 4. This process forms the graph with four nodes and twelve labeled edges shown in Figure 7-84.

Now that we have a simple representation of the puzzle, we need to form a strategy for solving it. The strategy requires a little bit of ingenuity, but it is easy to understand once you try it. We will first solve the easier problem of getting all four colors on just the front and back columns while ignoring the other pair of columns. We will call this a *partial solution* to the puzzle. In order to get the other columns into position without disturbing our "solved" column, each cube can be held by its front and back faces and rotated to change the exposed faces on the left and right columns until the left and right sides are solved as well.

To formalize this strategy, we need to know how to use the graph to find a partial solution to the puzzle.

> *Definition* In the graph *G* of an Instant Insanity™ puzzle, a *good subgraph* of *G* is one using all four nodes and one edge, with each label such that each node has degree 2.

For any good subgraph *H*, we can create a partial solution to the puzzle by stacking the cubes in numerical order so cube *i* is placed in such a way that the forward/backward-facing colors match the endpoints of the edge labeled *i* in *H*. Since every node in *H* has degree 2, every color will be represented on exactly two

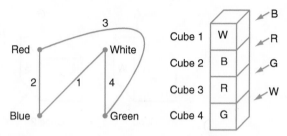

Figure 7-85   A good subgraph yields a partial solution.

cubes, so we can easily arrange them in order that each color appears on the front column once and on the rear column once. Figure 7-85 shows a good subgraph and its accompanying partial solution of the puzzle.

At this point, we might start rotating each cube (fixing its front and back faces) to try all possibilities. This would not take an entirely unreasonable amount of time, but we can do much better by simply using the graph again. If we can find a second good subgraph *that does not use any of the same edges that our first one did*, then we can rotate each cube as we mentioned before until the partial solution corresponding to the second subgraph occurs in the left and right faces of the stack. This will completely solve the puzzle.

If we cannot find a second good subgraph in this way, however, we can only conclude that our first good subgraph was a poor choice, not necessarily that there is no solution. The first subgraph we already found for the puzzle at hand illustrates this shortcoming very nicely.

Example 2   *In order to find a second good subgraph that does not use any of the edges from the first, we can simply delete the edges used by the first and look for a good subgraph in the graph, shown in Figure 7-86, that remains. Explain why this new graph has no good subgraph.*

SOLUTION   To get a good subgraph, we must choose exactly one of the edges labeled 1. But once we do that, it is impossible for the Green node to have an even degree. (The degree will be 1 if we use the edge labeled 2 from Blue to White, and the degree will be 3 if we use the loop labeled 2 at Green.) We conclude that there is no good subgraph.   □

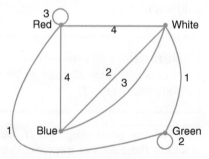

Figure 7-86   This solution cannot be completed.

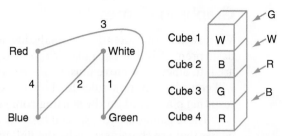

Figure 7-87 Another good subgraph and partial solution.

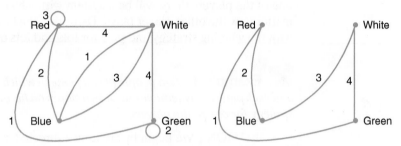

Figure 7-88 Finding a second good subgraph.

Unfortunately, not just any choice of a first good subgraph will necessarily lead to a complete solution. However, if we look at the original graph again, we can find *another* candidate for the "first good subgraph," given in Figure 7-87 with its corresponding partial solution. As we will see, this one works out better for us in the end.

Now applying our strategy again, we see that the leftmost graph in Figure 7-88 shows the original graph with the edges from the new good subgraph removed, and the rightmost graph in the same figure shows the second good subgraph we were hoping to find to complete the solution.

To solve the puzzle now, stack the cubes in the partial solution shown in Figure 7-87, and rotate cube 1 (keeping the front and back faces fixed) until the red and green faces are to the left and right, respectively. Repeat this for each of the other cubes using the good subgraph in Figure 7-88 to determine which faces to put to the left and right. This gives the full solution to the puzzle shown in Figure 7-89.

In hindsight, there was no need to even distinguish the two subgraphs. A complete solution corresponds simply to two partial solutions that can be fit together. In the language of graphs, we can state the following.

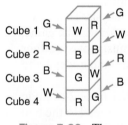

Figure 7-89 The solved puzzle.

Theorem 1 *An Instant Insanity puzzle has a solution if and only if the graph of the puzzle contains two good subgraphs that do not share any edges.*

Since the graph of such a puzzle has only 4 nodes and 12 edges in the first place, this condition has a huge advantage over "trial and error" for solving the puzzle. It is also much easier to give an explanation in terms of graphs when an Instant Insanity puzzle has no solution. This additional point is investigated in the exercises at the end of this section.

Graphs in Games

We now turn our attention to analyzing strategies for some two-player games. The basic premise is that each player removes stones from the playing board until none are left. Since each move must remove something, the game will eventually end. We will consider different rules for how the stones are to be removed and analyze the resulting play based on the starting configuration.

These games are just special cases of a class of games called *finitely progressive games* that are characterized by the finiteness of moves at any time and a guarantee that the game must end in a win for one of the players. It is a fact (that we will not prove) that all such games have a winning strategy for one of the players. That is, for one of the players there will be a system with which that player will always win, no matter how the other player plays. Therefore, the best thing to do is to be the player with the winning strategy and play for lots and lots of money.

Example 3   *Two people play a game in which there is a single pile of 10 stones and each player may remove one or two stones on his or her move. Determine a winning strategy for one of the players.*

SOLUTION   We begin by creating a directed graph to represent the play of this game. In the graph of Figure 7-90, node i represents the game board with i stones on it, and an edge from i to j (with $i > j$) means that with one move the game board can change from i stones to j stones. For this game, we have an edge from i to j if $i - j = 1$ or 2.

We can picture the game as being played on the graph itself with two players taking turns moving to adjacent nodes in order to build a walk from node 10 to node 0. We will call a node "good" if moving to that node results in an inevitable win for the player, and "bad" if it results in a loss.

It is useful in this type of game to do an "end-game" analysis. That is, we begin by focusing on what will happen in the last few moves. To win the game, Sue must move to node 0. For that to happen, her opponent must move to either node 2 or node 1. Hence, the winning node 0 is a "good" node, and nodes 1 and 2 are "bad" nodes.

Now, how can Sue force her opponent to move to either node 2 or node 1? She must move to node 3, since from node 3, the only moves lead to the bad nodes 1 and 2. Hence, node 3 is also a good node. And since Sue can guarantee that she can reach node 3 by forcing her opponent to first move to either node 4 or node 5, we conclude that nodes 4 and 5 are bad nodes.

Repeating this argument, we can determine that nodes 6 and 9 are good nodes, whereas nodes 7, 8, and 10 are bad. Since the game starts at the bad node 10, Sue wants to be the player to move first, and she begins the game by removing one stone, thus moving to the good node 9. From there, no matter

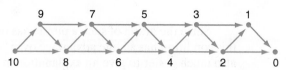

Figure 7-90   The graph of the simple stone removal game.

what her opponent does he will be at a bad node, and Sue can move to the good node 6. On her next move she can reach the good node 3 and then finally the winning node 0.

Note that if the game begins with only nine stones, then it would be the player to go second with the winning strategy for precisely the same reason! □

For any two-player game, we can repeat this process on its underlying graph structure.

> **Definition**   Given any game with a finite number of states, the *graph of the game* is a directed graph with nodes representing each possible state of the game board and with an edge from node i to node j if it is possible for the board to change from state i to state j as the result of one move by either player. In the graph of a finitely progressive game, there are no cycles and every ending state (node) represents a win for one of the players.

In terms of this graph, we may always think of the play of a game as simply constructing a walk from the starting configuration to the winning configuration where the players alternate choosing the "next" node to visit. In the previous example, note that the key to either player's winning is to move to one of the "good" nodes in the set $K = \{9, 6, 3, 0\}$ as shown in Figure 7-91. Once a player moves to one of these nodes, he or she can apply the outlined strategy to eventually win the game. Finding a set of nodes like this will be the key to winning any game. We will call such a set of nodes the *kernel* of the game.

> **Definition**   In the directed graph of any game, a set K of nodes with the following properties is called the *kernel* of the graph:
>
> (i) The winning node (in this case node 0) is in K.
> (ii) From any node not in K, there is an edge to a node in K.
> (iii) There are no edges from nodes in K to other nodes in K.
>
> We will often refer to the kernel as the set of *good* nodes in the graph.

The three properties of a kernel actually dictate an algorithm for finding the kernel from the graph. The following process will result in all the good nodes circled and all the bad nodes crossed out.

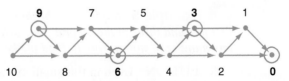

Figure 7-91   Good positions in the simple stone removal game.

> 1. Start with the winning node circled.
> 2. Repeat the following two steps as long as there are nodes that have not been crossed out or circled:
>   (a) Cross out every node that points to a circled node.
>   (b) Circle every node that only points to crossed-out nodes.

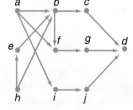

Figure 7-92 Kernel of the game in Example 4.

Example 4   *Figure 7-92 shows the graph of a game in which the winner is the first to reach node d. Find the kernel of this game using the steps above. Which player has a winning strategy?*

SOLUTION   We obtain good and bad states in this order: (1) d is good; (2) c, g, and j are bad since they have edges to d; (3) f and i are good since their only edges lead to bad states; (4) a, b, and e are bad since they have edges to f or i; (5) h is good since its only edges lead to bad states. The kernel is $\{d, f, h, i\}$.   □

Practice Problem 2   *Suppose two players play a game with a pile of 20 stones, and use the rule that on each move, either one, two, or three stones may be removed. Find the kernel of the graph for this game. Which player has a winning strategy?*

The following theorem can be proved using an induction argument that is similar to the algorithm above. It is proven in precisely this setting in Alan Tucker's book on applied combinatorics (see [47]) for those who are interested in reading more about it.

Theorem 2   *The graph of any finitely progressive game has a unique kernel.*

Combining this fact with the strategy we outlined above gives us the following solution to who has the winning strategy in any finitely progressive game.

Corollary 3   *Given a game with kernel K, if the starting configuration of the game is in K, then Player 2 has a winning strategy. If the starting configuration is not in K, then Player 1 has a winning strategy.*

More Stone Removal Games

We will continue with the theme of stone removal games because they are among the easiest games to learn how to play. In each of these games, there are two players, and the person to remove the last stone wins the game.

Example 5   *Suppose we begin with **two** piles of stones with four stones in the first pile and three in the second. On each turn, a player chooses a pile and removes any number of stones from it. Find the kernel of this game.*

SOLUTION   Even in this small example, the graph of this game is too complicated to draw easily. In Figure 7-93, we label a node (a, b) for the state of the game when there are a stones in the first pile and b stones in the second

(4, 3)	(3, 3)	(2, 3)	(1, 3)	(0, 3)
(4, 2)	(3, 2)	(2, 2)	(1, 2)	(0, 2)
(4, 1)	(3, 1)	(2, 1)	(1, 1)	(0, 1)
(4, 0)	(3, 0)	(2, 0)	(1, 0)	(0, 0)

Figure 7-93 Kernel for
Example 5.

pile, but we do not draw any of the edges. Instead, we will just have to mentally visualize that the edges lead from each node to all the other nodes to the right of it on the same row, and to all the nodes below it in the same column. Since the winning node is labeled $(0, 0)$, we can use the algorithm we developed to separate the good nodes from the bad ones in the following order:

1. The winning node $(0, 0)$ is circled.
2. All nodes with edges leading to $(0, 0)$ are crossed out. Specifically, this means $(0, 3), (0, 2), (0, 1), (4, 0), (3, 0), (2, 0)$, and $(1, 0)$ are bad positions.
3. There is only one node that has all its edges leading *only* to crossed-out nodes. We circle the node $(1, 1)$, which represents two piles of one stone each.
4. All nodes with edges leading to $(1, 1)$ are crossed out. Specifically, positions $(1, 3), (1, 2), (4, 1), (3, 1)$, and $(2, 1)$ are all bad.
5. Continuing this process, we will end up with positions $(2, 2)$ and $(3, 3)$ also circled and all other positions crossed out.

Therefore, the kernel for this small game is $K = \{(0, 0), (1, 1), (2, 2), (3, 3)\}$ □

This game might seem too small to be of interest, but it is not hard to generalize the kernel to be correct for a game with the same rules but more stones to start with. In making this generalization, we will also take the opportunity to give a proof that a set of positions satisfies the three defining properties of a kernel.

Proposition 4 *In the stone removal game of Example 5, if the game starts with piles of size $m \geq n \geq 0$, then the kernel is*

$$K = \{(a, b) \in \{1, 2, \ldots, m\} \times \{1, 2, \ldots, n\} : a = b\}$$

That is, a good state is one in which the two piles are of equal size.

PROOF   We will check that K satisfiies properties (i), (ii), and (iii) in the definition of "kernel." We will do this by thinking of the game itself.

(i) The winning position is when there are no stones in either pile. This node is labeled $(0, 0)$ in the graph, and this certainly satisfies the definition of K.

(ii) Let a position in the game be given that is *not* in K. By the definition of K, this position must consist of two piles of unequal sizes, say, x and y with

$x > y \geq 0$. A single move of removing $x - y$ stones from the larger pile will result in game position (y, y), which is in K.

(iii) Suppose we have a game position that *is* in K which is not the winning configuration. That is, the game position is labeled (k, k) for some $k > 0$. Any move from this position has to change the number of stones in one and only one of the piles, resulting in a game position not in K.

Therefore, K is the kernel of this game. ■

This means that the player who can leave two equal piles of stones at the end of his or her turn will win the game. Therefore, if the game starts with two piles of equal size, then Player 2 can apply the winning strategy, and if the game starts with two piles of unequal size, then Player 1 can apply the winning strategy.

Practice Problem 3 *Suppose a game starts with two piles of stones, one with four stones and the other with three stones, and each player can remove one or two stones from a single pile on each turn. Use the algorithm on the picture in Figure 7-94 to find the kernel of this game.*

The Game of Nim

Example 5 is the simplest version of a classic game called *Nim*, which made its mathematical debut in a 1902 article (see [7]) by Charles Bouton. In Nim, there are many piles of stones of various sizes, and players alternate removing stones from the board. The only rule for removal is that on a single move, a player cannot remove stones from more than one pile. There is no limit on the number—indeed, a whole pile could be removed in one move. As before, the winner is the player to remove the last stone.

The problem with Nim is that the graph of the game can be quite large. For example, if the game starts with piles of size 5, 6, and 8, there are $6 \cdot 7 \cdot 9 = 378$ game board states possible, so the graph would have 378 nodes. Although we will not draw these large graphs, we will think about them as we try to uncover the strategy that we know must exist from the previous discussion.

For the general game of Nim, we will not attempt to "discover" the kernel as we did in the previous examples. We will just describe the kernel first, and then check to make sure it satisfies the three defining properties of a kernel. We will also see that it

(4, 3)	(3, 3)	(2, 3)	(1, 3)	(0, 3)
(4, 2)	(3, 2)	(2, 2)	(1, 2)	(0, 2)
(4, 1)	(3, 1)	(2, 1)	(1, 1)	(0, 1)
(4, 0)	(3, 0)	(2, 0)	(1, 0)	(0, 0)

Figure 7-94 Kernel for Practice Problem 3.

is a natural generalization of the kernel of the two-pile Nim game in Example 5. In order to describe the kernel of the general game, we will first need some facts about binary notation from Section 2.6.

Recall that every nonnegative integer has a unique representation in base two. For example, $23 = 2^4 + 2^2 + 2^1 + 2^0$, so the decimal numeral 23 can be represented as 10111 in base two. Note that if we have two base two numerals, say, 10111 and 101, we can consider them both to be the same length by rewriting the second one as 00101. In general, if we have any number of base two numbers, we can consider them all having the same number of digits by adding leading 0's to any that are too short.

Now given a set of base two numbers all of the same length, we can construct a table using one base two number for each row. For example, if the numbers are $\{5, 6, 8\}$, they are represented in base two by $\{0101, 0110, 1000\}$ and put into the following tabular form:

$$
\begin{array}{rcccc}
5 = & 0 & 1 & 0 & 1 \\
6 = & 0 & 1 & 1 & 0 \\
8 = & 1 & 0 & 0 & 0
\end{array}
$$

We complete the table by adding one more row consisting of a four-digit base two number constructed in the following way: The first digit is a 1 if there are an odd number of 1's above it (i.e., among the first digits in the other three rows) and a 0 otherwise; the second, third, and fourth digits are constructed using this same rule applied to the number of 1's above each of them, respectively. In the above example, we would add the four digits 1 0 1 1. We will call the last row constructed in this way the *direct sum* of the previous rows. In fact, we will call this row the direct sum of the original numbers themselves. In our example, we have the binary number 1011 as the direct sum of 5, 6, and 8, as shown in Table 7-10.

$$
\begin{array}{rcccc}
5 = & 0 & 1 & 0 & 1 \\
6 = & 0 & 1 & 1 & 0 \\
8 = & 1 & 0 & 0 & 0 \\
\hline
\text{Direct sum} & 1 & 0 & 1 & 1
\end{array}
$$

Table 7-10 Direct Sum of 5, 6, and 8

Practice Problem 4 *What is the direct sum of the numbers 3, 12, and 13?*

What does all this have to do with the game? We can use the idea of direct sum to describe what the kernel of the general Nim game looks like.

Definition Given a game position with k piles with $s_1, s_2, \ldots, s_k$ stones, respectively, find the direct sum of $s_1, s_2, \ldots, s_k$. Call this the *binary number of the game position*.

Theorem 5 *For the game of Nim with k piles of stones, the set*

$$K = \{\text{game positions that have a binary number of } 000\cdots 0\}$$

is the kernel of the game.

PROOF We will check that K satisfies the defining properties (i), (ii), and (iii) of a kernel.

(i) The winning position in the game consists of all piles of size 0, and the direct sum of any number of 0's is $000\cdots0$.

(ii) Let a game position *not* in K be given. That is, the binary number of the game position contains at least one 1. We must find a move in the game that results in a position which *is* in K. To do this, locate the first 1 (from left to right) in the binary number of the configuration—let's say it is in column i. Now there must be a row, let's say it's row j, in which a 1 also occurs in column i. Remove the appropriate number of stones from pile i so that every binary digit in row i that is above a 1 in the bottom row is changed. (So a 1 is changed to a 0 and a 0 is changed to a 1.) Now the bottom row will be $000\cdots0$, meaning this new game position is in K.

(iii) Let a game position in K other than the winning position be given. By the definition of K, the binary number of this position is $000\cdots0$, so any change in a single row will have to introduce at least one 1 in the bottom row (in the same positions as the digit changes in the changed row), leading to a game position not in K.

It is rather difficult to put the winning strategy into words, so let us look at a particular play of the game instead to see the strategy in action.

Example 6 *Analyze the game that starts with three piles with five, four, and two stones, respectively.*

SOLUTION   The corresponding table of binary numbers with direct sum is shown in Table 7-11. So the initial position of the game is not in the kernel. This means that Player 1 will have a winning strategy. Player 1 looks for the first 1 in the bottom row and finds it in the second column. He looks above to find a 1 in the second column of one of the rows and finds it in row 3. By removing one stone from the third pile, the 010 in the third row becomes a 001 (i.e., the second and third digits are changed), and so the direct sum will now look like it does in Table 7-12. This configuration is in the kernel, so Player 1 is well on his way to winning the game.   □

Explore more on the Web.

Practice Problem 5   *What is the correct first move for Player 1 in a Nim game that starts with piles of size 3, 12, and 13?*

1st pile (5)	1	0	1
2nd pile (4)	1	0	0
3rd pile (2)	0	1	0
Direct sum	0	1	1

Table 7-11   Direct Sum in Example 6

1st pile (5)	1	0	1
2nd pile (4)	1	0	0
3rd pile (1)	0	0	1
Direct sum	0	0	0

Table 7-12 Direct Sum After the Move

Solutions to Practice Problems

1 Referring to the graph in Figure 7-83,
- The solution corresponding to the walk

$$(WCTG, \emptyset) \to (WC, TG) \to (WCT, G) \to (W, CTG)$$
$$\to (WTG, C) \to (G, WCT) \to (TG, WC)$$
$$\to (\emptyset, WCTG)$$

requires an additional $25 animal-crossing fee.
- the solution corresponding to the walk

$$(WCTG, \emptyset) \to (WC, TG) \to (WCT, G) \to (C, WTG)$$
$$\to (CTG, W) \to (G, WCT) \to (TG, WC)$$
$$\to (\emptyset, WCTG)$$

requires an additional $20 animal-crossing fee.
Hence, the second solution is cheaper.

2 The kernel states are determined by this analysis. The winning state 0 is in the kernel. States 1, 2, and 3 have edges leading to 0, so they are bad. From state 4 the only edges lead to bad states, so it is good and goes into the kernel. If we continue in this fashion, the kernel is

$$\{0, 4, 8, 12, 16, 20\}$$

Since the game begins in a kernel state (with 20 stones), Player 2 will be able to successfully move to each kernel state and eventually win the game.

3 Referring to the picture in the problem:

(a) The winning node $(0, 0)$ is circled.
(b) Nodes $(1, 0), (2, 0), (0, 1), (0, 2)$ are crossed out since they all would have edges pointing to $(0, 0)$.
(c) Nodes $(3, 0), (0, 3)$, and $(1, 1)$ are circled since they only lead to crossed-out positions.
(d) Nodes $(4, 0), (3, 1), (3, 2), (2, 3), (2, 1), (1, 3), (1, 2)$ are crossed out since each leads to some circled node.
(e) Nodes $(4, 1), (3, 3)$, and $(2, 2)$ are circled since each leads only to crossed-out nodes.
(f) Nodes $(4, 2)$ and $(4, 3)$ are crossed out.

Thus, $K = \{(0, 0), (3, 0), (0, 3), (1, 1), (4, 1), (2, 2), (3, 3)\}$ is the kernel.

4 In binary, $3 = 11, 12 = 1100$, and $13 = 1101$, so the direct sum computation is shown in Table 7-13.

5 Since the direct sum is 0010, the correct first move for Player 1 is to remove two stones from the first pile, taking it from three stones to one stone.

3 =	0	0	1	1
12 =	1	1	0	0
13 =	1	1	0	1
Direct sum	0	0	1	0

Table 7-13 Direct Sum for Practice Problem 4

Exercises for Section 7.5

1. Two friends have 2 gallons (8 quarts) of water in a pail. They also have two (empty) jars, one holding 5 quarts and the other 3. Using just these measuring devices, how can they split the water so that 4 quarts are in the larger jar and 4 quarts remain in the pail? Produce a complete graph model for this puzzle similar to the one in the wolf, goat, and cabbage puzzle, and find all solutions to the water puzzle in terms of the properties of the graph.

2. Suppose the pail in the previous problem holds 12 quarts of water, the two jars hold 7 quarts and 5 quarts, and the goal is still to split the water evenly with 6 quarts in both the pail and the larger jar. Produce a complete graph model for this puzzle, and find all solutions to the water puzzle in terms of the properties of the graph.

3. Suppose the pail in the previous problem holds 10 quarts of water, the two jars hold 6 quarts and 4 quarts, and the goal is still to split the water evenly with 5 quarts in both the pail and the larger jar. Produce a complete graph model for this puzzle, and explain in terms of the properties of the graph why it has no solution.

4. Puzzles like the wolf, goat, and cabbage puzzle exist in many cultures (see [2]). A slightly different version comes from the Kabjlie region of Algeria. This puzzle involves a man traveling with a jackal, a goat, and a

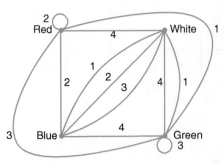

Figure 7-95 Puzzle for Problem 5.

bundle of hay, but in this version, the man can take up to two of these things with him on each crossing. Describe the complete graph model so that you can show *all* the solutions to this puzzle in terms of your graph. Is there a solution that seems to be better than others? Explain why.

5. Draw the (unfolded) cubes for an Instant Insanity™ puzzle with the graph in Figure 7-95.

6. Give an example of an Instant Insanity™ puzzle that does not even have one partial solution. You can draw the cubes or just the graph model.

7. Solve each of the Instant Insanity™ puzzles shown in Figure 7-96 or explain in terms of the graph model why no solution is possible.

8. Suppose two players play a game with two piles of stones (initially with seven stones in each pile) and the rule that on a given turn it is permissible to (a) choose a pile and remove any number of stones from it, or (b) remove exactly one stone from each pile. As usual, the person to take the last stone wins. Find the kernel of this game,

and decide which player has a winning strategy in this game.

9. Suppose two players play a game with two piles of stones (initially with five stones in one pile and seven stones in the other) and the rule that on a given turn it is permissible to (a) choose a pile and remove any number of stones from it, or (b) remove an equal number of stones from each pile. As usual, the person to take the last stone wins. Find the kernel of this game, and decide which player has a winning strategy in this game.

10. Suppose two players play a game with two piles of stones (initially with seven stones in each pile) and the rule that on a given turn it is permissible to remove up to three stones total on each turn, but at least one stone must be removed from each pile if possible. Find the kernel, and decide which player has a winning strategy in this game.

11. In [27], Martin Gardner presents a game played on a chessboard with a queen. The queen is placed on the chessboard anywhere in the topmost row or leftmost column. Two players alternate moving the queen legally (i.e., horizontally, vertically, or diagonally) but never to the left or up, until someone puts the queen on the bottom right square. The player who does this is the winner.

   (a) Which player has a winning strategy? (Be specific. It might depend on the initial placement of the queen.) Find a "stone removal" game that is equivalent to this game.

   (b) What if the same game is played but with a piece that has the combined moving abilities of a queen and a knight, again restricted so that leftward or upward play is illegal?

12. This game is played on a chessboard with a single rook (which can move any number of spaces horizontally or vertically). The rook starts on the top row of the chessboard, and players alternate legally moving the rook only down or to the right. The winner is the player who moves the rook to the bottom rightmost square. Find the kernel of this game, and decide which player has a winning strategy for each possible starting position (in the top row) for the rook.

13. You are playing Nim and the current game position consists of three piles with 3, 4 and 7 stones, respectively. It is your turn. What is the correct move?

14. You are playing Nim and the current game position consists of three piles with 13, 14 and 17 stones, respectively. It is your turn. What is the correct move?

15. Suppose your little brother challenges you to a game of Nim using the 20 pennies from his piggy bank. How can you divide these into three piles at the outset of the game so that you, as Player 2, have a winning strategy?

(a)

	B		
R	W		
B	G		
	R		

	W		
R	G		
B	B		
	R		

	W		
W	B		
W	R		
	W		

	W		
	G	R	
G	G		
	B		

(b)

	W		
G	W		
B	G		
	R		

	G		
G	G		
B	B		
	G		

	W		
G	B		
G	R		
	G		

	W		
	G	R	
R	B		
	B		

Figure 7-96 Two puzzles for Problem 8.

16. Suppose two people play Nim with seven piles of stones with 1, 2, 3, 4, 5, 6, and 7 stones in the respective piles. What is the correct first move for Player 1, or will she eventually lose no matter what she does (assuming Player 2 has read this section)?

17. What if in the above scenario, there is an eighth pile containing 11 stones?

18. Explain why the kernel for the two-pile Nim game outlined in Example 5 is a consequence of the general description given in Theorem 5.

19. Consider a game that begins with three piles with two, five, and seven stones, and has the rule that on any turn a player may remove up to two stones from any one pile. Find the kernel of this game, and decide which player has the winning strategy.

20. Play this game with a friend. Start with the 4 × 4 grid shown in Figure 7-97. Players alternate putting X's in the squares with the limitation that X's must be placed in only one row or column on each move, but any number may be inserted. The person to X the last available space wins. What is the strategy for the second player to win this game? (HINT. Consider the symmetry of the grid.)

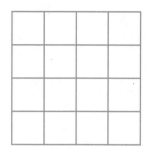

Figure 7-97 The grid game in Problem 20.

21. Draw a picture of a clock face and play the following game with a friend. On each move, a player can cross out any number not already crossed out or she can cross out *two* numbers if they are consecutive numbers (like 5 and 6). The player who crosses out the last number wins. Find the kernel of this game, and determine which player has the winning strategy. (HINT. Consider the symmetry of the clock face.)

22. Investigate some of the stone removal games we have studied if the rules are changed so that the player who removes the last stone *loses* the game.

7.6 Excursion: Binary Trees

Trees are a special type of graph first discussed formally in Section 7.2. In the current section, we will study a more specific graph structure called a *binary tree*. Although binary trees have properties in common with general trees, their most important features stem from their additional structure. This section emphasizes that additional structure and its applications. It is not necessary to have worked through the material in the earlier section to understand this one.

Even before Section 7.2, we saw "trees" in several parts of this text, primarily as a visualization or organizational aid. One important type was the decision tree used for counting and in probability. The first example like this came early in the text.

Example 1 *The tree in Figure 7-98 is the game tree for a "best-of-three sets" tennis match that we first saw in the introductory section of Chapter 1.*

Another familiar context for trees is the idea of a family tree. For example, Figure 7-99 shows a small family tree giving Brenda's descendants for several generations. An important third example comes from our experience with arithmetic and calculators. In particular, a tree can represent the way arithmetic calculations are performed.

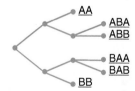

Figure 7-98 Game tree for a "best-of-three" series.

AA
ABA
ABB
BAA
BAB
BB

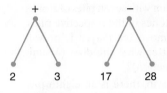

Tree for 2 + 3   Tree for 17 − 28

Figure 7-100 Representing
simple arithmetic expressions.

Figure 7-99 Brenda's
family tree.

Example 2 *We can use trees to represent arithmetic expressions as the two simple examples in Figure 7-100 illustrate. How can these examples be extended to represent more complicated expressions?*

SOLUTION   As in Figure 7-100, we place the numbers at the bottom, and the operation at the top, with edges joining the operation to the two numbers it applies to. For more complicated expressions, we simply repeat the process. For example, for the expression $(2 + 3) \times (17 - 28)$ we place another node above these two trees, with the × operation, and with edges to the two smaller trees that represent the subexpressions. This construction is shown on the left in Figure 7-101. If one of the operands is a subexpression and the other is a number, as in $(2 + 3) \times 7$, we simply connect the operation to a tree for the subexpression and a node containing the number, as shown in the tree on the right. □

Practice Problem 1 *Write the tree representation of the arithmetic expression*

$$(2 + (3 \times 6)) \div 4$$

All these examples share several important properties. As graphs, they are connected and have no cycles. Unlike many graphs, however, the way in which they are drawn is important. We could draw the family tree with Brenda at the left (as in the decision tree), or perhaps on the right, or perhaps even at the bottom, as shown in Figure 7-102 on the left. However, drawing the family tree like the diagram on the right would totally destroy the meaning of the tree.

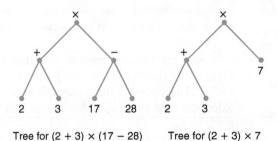

Tree for (2 + 3) × (17 − 28)   Tree for (2 + 3) × 7

Figure 7-101 More complicated arithmetic
expressions.

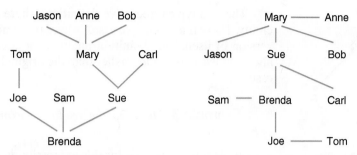

Figure 7-102 Other ways to draw Brenda's family tree.

Notice that the tree on the left in Figure 7-102 looks more like what we would call a tree (or at least a bush) in ordinary life. Much of the terminology used in connection with trees comes from the context of either family trees or trees in nature. For example, we say that Brenda is at the *root* of the tree, and that Tom, Sam, Jason, Anne, Bob, and Carl are at the *leaves*. We also say that the nodes labeled Joe, Sam, and Sue are *children* of the Brenda node, and that Mary and Carl are *siblings*. This is obviously based on the family tree analogy.

In all our examples there is a special node, called the *root* of the tree, that constitutes a "starting point" for applications. Among these *rooted* trees is a type with particular importance known as a *binary tree*. Although it would be possible to give a formal definition of *binary tree* based on graph theory, it is customary instead to give a recursive definition, because so many of the important applications are based on the recursive nature of the tree.

Basic Definitions

The terminology used in connection with binary trees is extensive and (just as with graphs) not universally agreed on. Since this is not a computer science text, we will not try to be exhaustive in our formal definitions. We will emphasize the mathematical properties of trees and restrict ourselves to definitions that will be involved in these mathematical discussions. Other terms that arise will be addressed more informally.

Definition

1. A *binary tree T* is a structure that is either:
   (a) empty, or
   (b) consists of one or more *nodes*. In this case, one of those nodes is designated as the *root* node, and the remaining nodes are separated into two disjoint sets that are in turn (possibly empty) binary trees. One of these subtrees is designated the *left subtree of T* and the other the *right subtree of T*.

2. A nonempty binary tree that has empty left and right subtrees is called a *leaf*.

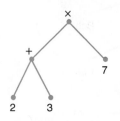

Figure 7-103 The tree for the expression $(2 + 3) \times 7$.

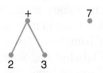

Left subtree Right subtree

Figure 7-104 The subtrees for the tree in Figure 7-104.

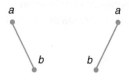

Tree T_1 Tree T_2

Figure 7-105 Similar but different binary trees.

This is a typical recursive definition. There is a "base case"—the tree may be empty. There is a "recursive case," where the concept is defined in terms of a smaller version of itself. The definition might seem circular, but because the subtrees are always at least one node smaller than the original tree, eventually you reach the base case.

Example 3 *Show that the tree for the expression $(2 + 3) \times 7$ fits the definition of binary tree.*

SOLUTION   The tree for this expression is shown in Figure 7-103. The root node is the node labeled with the $\times$ symbol. The remaining nodes are divided into two disjoint sets: the left subtree T_L containing the nodes $+$, 2, and 3, and the right subtree T_R containing the node 7. We show these subtrees in Figure 7-104.

We must now show that each of these is in turn a binary tree. The right subtree is a leaf since its root is 7 and both its subtrees are empty. The left subtree has root $+$, left subtree 2, and right subtree 3. Finally, both 2 and 3 are leaves since each of their subtrees is empty. Notice that in an expression tree such as this, the numbers are at leaf nodes and the operations are at nonleaf nodes.   □

In the formal definition of a binary tree, nodes are mentioned, but there is no mention of edges. As you can see in the examples, when we draw a binary tree as a graph, we place an edge from each node to its children (i.e., to the roots of that node's subtrees). In the preceding example, there is an edge from the root node $\times$ to the nodes $+$ and 7. In the left subtree, there is an edge from the subtree's root node $+$ to the nodes 2 and 3.

A subtlety in the definition of binary trees is the fact that the left and right subtrees are distinguished as part of the definition of the structure of the tree. As an example, consider the binary trees shown in Figure 7-105. They appear very similar, but they are, in fact, different. The tree T_1 has a root a with an empty left subtree and a nonempty right subtree, while the tree T_2 has a root a with a nonempty left subtree and an empty right subtree.

Practice Problem 2

(a) *For the tree you drew in Practice Problem 1, list the root and sketch its two subtrees. Then list all the leaves of the left subtree, list all the leaves of the right subtree, and list all the leaves for the entire tree. Finally, list all the nonleaves (sometimes called internal nodes) for the entire tree.*

(b) *The family tree shown in Figure 7-99 is not a binary tree since some nodes have too many children, but we can still make sense of most of the terms we have introduced. List the root and the leaves. List the children of the Mary node. List the siblings of the Mary node. What node do you think we would call the parent of the Mary node?*

Levels and Height

When we draw a binary tree as a graph, with the tree and all subtrees throughout the graph having their roots at the top, the tree appears to have "levels." Informally,

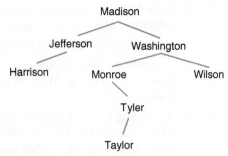

Figure 7-106 U.S. Presidents born in Virginia.

we can identify a node's level based on how far away it is from the root, which we consider to be at level 1.

Example 4 *Consider the tree in Figure 7-106, where the nodes are labeled with the names of presidents born in Virginia. Starting at the top with level 1, partition the presidents by level in the tree.*

SOLUTION   Table 7-14 shows all presidents at each level of the tree.   ☐

In order to be able to exploit the notion of "level" in induction proofs a bit later, it will be helpful to have a recursive definition that meshes with our earlier recursive definition of a binary tree.

Definition   For a node v in a (nonempty) binary tree T, v is either the root of T, or v is a node in the left or right subtree of T (called the subtree T_s).

● If v is the root of T, then the *level of v in T* is 1.
● If v is not the root of T, then the *level of v in T* is 1 more than the level of v in the subtree T_s.

Example 5 *Verify that this definition yields the anticipated result for the "Monroe" node in Figure 7-106.*

SOLUTION

1. Since Monroe is not the root of the entire Madison tree, we must first calculate its height in the Washington subtree.

Level	Nodes
1	Madison
2	Jefferson, Washington
3	Harrison, Monroe, Wilson
4	Tyler
5	Taylor

Table 7-14   Table for Example 4

2. Since Monroe is not the root of the Washington subtree, we must first calculate its height in the Monroe subtree.

3. In the Monroe subtree, Monroe is the root, so its level there is 1.

4. According to the definition, its level in the Washington subtree is 1 more than its level in the Monroe subtree, so in the Washington subtree it is at level 2.

5. According to the definition, its level in the Madison subtree is 1 more than its level in the Washington subtree, so in the Madison subtree it is at level 3. This agrees with our earlier answer.

□

Practice Problem 3   *Intuitively, at what level is Taylor in the Washington tree? At what level is it in the entire Madison tree? Does this match the formal definition?*

Once we understand this notion of level, it is possible to define the height of a tree T as the maximum level of the nodes in T. However, in order to illustrate the way these recursive definitions are used, we will define "height" recursively and *prove* the relationship between height and level.

> **Definition**   An empty binary tree has *height* 0. For a nonempty binary tree T, the *height* of T is 1 more than the larger of the heights of the left and right subtrees of T.

Proposition 1   *For all $n \geq 0$, in any binary tree T with height n, the level of any node in T is less than or equal to n.*

PROOF   (By induction on n.) Let $P(n)$ be the statement "In a binary tree T with height n, the level of any node in T is less than or equal to n." The first statement $P(0)$ states, "In a binary tree T with height 0, the level of any node in T is less than or equal to 0." According to the definition of "height," the only binary tree of height 0 is an empty binary tree T. Since an empty binary tree has no nodes, the statement "The level of any node in T is less than or equal to 0" is vacuously true. Hence, statement $P(0)$ is true.

Now let $m \geq 1$ be given such that the statements $P(0), \ldots, P(m-1)$ have been checked, and we consider the statement $P(m)$ that states, "In a binary tree T with height m, the level of any node in T is less than or equal to m," the next to be checked. Let a binary tree T of height m be given, and let v be any node in T. Either v is the root of T, or v is in either the left subtree of T or the right subtree of T.

Case 1: If v is the root of T, then by the definition of "level," the level of v in T is 1. Since we know $m \geq 1$, we can conclude that "the level of v in T is less than or equal to m."

Case 2: If v is not the root of T, then we can let T_s be the subtree of T that contains v. By the recursive definition of "height," the height k of T_s is less than or equal to $m-1$. So we know that $P(k)$ has already been checked to be true, and hence we know that the level of v in T_s is less than or equal to k. Since (by the recursive definition of "level") the level of v in T is 1 more than the level of v in T_s, and since the level of v in T_s is less than or equal to k, we conclude that the level of v in T is less than or equal to $k + 1$, which in turn is less than or equal to m.

In either case, we have established that the level of v in T is less than or equal to m, completing the induction. ∎

This establishes one relationship between height and level. Exercise 14 at the end of this section explores another. The main point here is that the recursive definitions make induction proofs about binary trees very straightforward. We end this discussion with one more example of this.

Proposition 2 *For all $n \geq 1$, any nonempty binary tree of height n has no more than 2^{n-1} leaves.*

PROOF (By induction on n) When $n = 1$, the above statement says, "A nonempty binary tree of height 1 has no more than 2^0 leaves." According to the definition of "height," a binary tree of height 1 must consist of a root node with empty left and right subtrees. Hence, the only node in such a binary tree is a leaf, so the given statement is true when $n = 1$.

Now let $m \geq 2$ be given such that the statement in the proposition has been checked for $n = 1, \ldots, m - 1$, and consider the next statement "A binary tree level of height m has no more than 2^{m-1} leaves." Let a binary tree T of height m be given. Since $m \geq 2$, we know that T is nonempty, and so we can identify the left subtree and right subtree of T by T_L and T_R, respectively. Let i and j be the heights of these subtrees with $i \geq j$. By the recursive definition of height, we know that $i = m - 1$, so both subtrees have smaller heights than T, and hence we know that the statement of the proposition has already been checked for binary trees of these heights. This means that there are at most $2^{i-1} + 2^{j-1}$ leaves combined in the left and right subtrees of T. Since $j - 1 \leq i - 1 = m - 2$, it follows that

$$2^{i-1} + 2^{j-1} \leq 2^{m-2} + 2^{m-2}$$
$$= 2 \cdot 2^{m-2} = 2^{m-1}$$

so there are at most 2^{m-1} leaves in the binary tree T. ∎

Practice Problem 4 *What is the maximum number of total nodes in a binary tree of height n?*

Searching Lists with Binary Trees

An important application of binary trees in computer science is the storing, retrieving, and sorting of data. A *binary search tree* is one way to organize a list of numbers so that retrieval is very efficient.

Definition A *binary search tree* is a binary tree where every nonleaf has these properties:

- The label on that node is greater than all the labels on the nodes (if any) of its left subtree.
- The label on that node is less than all the labels on the nodes (if any) of its right subtree.

Figure 7-107 The binary search tree for Example 6.

Example 6 *The numbers in the list $L = [3, 5, 1, 9, 11]$ can be stored in the binary search tree T shown in Figure 7-107.* □

The advantage of this structure is that it is very efficient to determine if a value is in the tree. For example, to search for 8 in the tree, we know that 8 must be either the label of the root, or it is in the left subtree if 8 is less than the root, or it is in the right subtree if 8 is greater than the root. Based on this idea, we carry out the following recursive search algorithm:

- Is T empty? If so, there is no 8 in the tree.
- If T is not empty, then

    Is 8 the root of T? If so, there is an 8 in the tree.
    If 8 is not the root of T, then

        If 8 is less than the root of T, then search the left subtree of T.
        If 8 is greater than the root of T, then search the right subtree of T.

For the particular tree in Figure 7-107, we compare 8 to 3. Since 8 is greater than 3, we search the right subtree. In that right subtree, we compare 8 to 9. Since 8 is less than 9, we search 9's left subtree. In that left subtree, we compare 8 to 5. Since 8 is greater than 5, we search 5's right subtree. But since this right subtree is empty, we conclude that there is no 8 in the tree.

Observe that we made three comparisons, one at each level of the original tree. By contrast, to search in the list L itself, we would have to make five comparisons, one at each member of the list. The difference between three comparisons and five comparisons is not worth worrying about, but if we have a list with 1,000 items organized into a tree with, say, 13 levels, the difference between 1,000 comparisons and 13 comparisons becomes quite significant.

In general, to sequentially search for an element in an n-element list requires as many as n comparisons, whereas the number of comparisons in searching a binary search tree is in the worst case just the height of the tree. Therefore, the relationship between the number of nodes in a binary tree and the height of the binary tree established in Practice Problem 4 becomes the central issue in using binary trees for searching data.

Proposition 3 *If a list of n distinct numbers is stored in a binary search tree T, then the height of T is at least $\lceil \log_2 n \rceil$.*

PROOF Let h be the height of a binary search tree T with nodes labeled with the n numbers in L. By Practice Problem 4 in Exercise 13, we know that the total number of nodes in T is at most 2^h. Since the nodes of T are labeled with the n elements in L, this means that $n \leq 2^h$. By the definition of logarithms, this is the same thing as saying $h \geq \log_2 n$. Since h is an integer value, it follows that $h \geq \lceil \log_2 n \rceil$, as desired. ∎

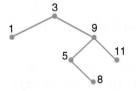

Figure 7-108 The revised binary search tree T'.

An interesting feature of this representation is that adding a number to the binary search tree consists of one additional step following an unsuccessful search for that number. In the previous example, we realized that 8 was not in the tree when we reached the empty right subtree of the 5 node. Suppose we now insert 8 into the tree as the right subtree of that 5 node. The revised tree T' is shown Figure 7-108.

In this example, the naive insertion strategy increased the height of the tree, making future searches take longer. In a computer science course, one might study algorithms for doing insertions in a manner that keeps the height from growing more than necessary. According to Proposition 3, the height must always be at least $\lceil \log_2 n \rceil$. The goal of these algorithms is to keep the height as close to the theoretical bound $\lceil \log_2 n \rceil$ as possible, without causing the insertion algorithm to be excessively slow. The details of this topic are not addressed in this text.

Sorting Lists

We will next show how binary trees tell us something of significant theoretical importance about sorting lists of numbers. It is very easy to capture the essence of ordering a list of numbers with a tree structure.

Example 7 *Given the list of three distinct numbers $[a, b, c]$, the binary tree in Figure 7-109 illustrates one strategy for determining their order from smallest to largest by comparing two numbers at a time.* □

The order of the questions in the above binary tree could certainly be changed to yield a different tree structure, but note that there is no way for a binary tree of height less than 3 to contain the six leaves corresponding to the 3! different possible orderings of a, b, and c. This is really a simple consequence of Proposition 2.

Theorem 4   *Given a list of n distinct numbers, any binary tree with leaves labeled with the n! possible orderings of these numbers has a height of at least $1 + \lceil \log_2(n!) \rceil$.*

PROOF   By the contrapositive of Proposition 2, a binary tree with $n!$ leaves must have height h such that $2^{h-1} \geq n!$. This means that $h - 1 \geq \log_2(n!)$, or $h \geq 1 + \log_2(n!)$. Since h must be an integer value, this means that, in practice, $h \geq 1 + \lceil \log_2(n!) \rceil$. ∎

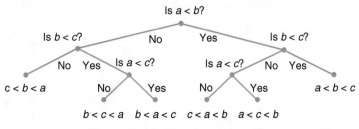

Figure 7-109   Tree for sorting a list.

The significance of the previous result is that it provides a theoretical lower bound on the number of comparisons that must be made in an algorithm intended to sort a set of data.

Corollary 5 *Any algorithm that uses a binary comparison relation to sort a list of $n \geq 3$ distinct values must use at least $\frac{n}{2} \log_2 n$ comparisons.*

PROOF Any such algorithm can be represented by a binary tree with the $n!$ possible orderings as the leaves. The number of comparisons (in the worst case) is precisely the height of this tree. By Theorem 4, this height h satisfies the following:

$$h \geq 1 + \log_2(n!)$$
$$\geq \log_2(n^{n/2})$$
$$= \frac{n}{2} \log_2(n)$$

Since the second inequality is an algebraic property of "factorial," we defer the proof of that inequality to Exercise 19. ■

There are many algorithms for sorting data that use essentially this number of comparisons, and so this result tells us that these algorithms are the best possible for generic sets of data.

Traversing Binary Trees

A *traversal* of a binary tree is any algorithm that visits each of the nodes of the tree. The term "visits" can mean any number of things. One simple type of visit consists of printing the data stored in the node. There are three common traversals known as *preorder*, *inorder*, and *postorder*, each defined recursively. Rather than give detailed formal algorithms, we will simply use examples to try to understand the differences between these traversals.

Preorder traversal

In the preorder traversal of a binary tree, we visit the root first, then we visit each of the nodes of the left subtree, and finally we visit each of the nodes of the right subtree. To visit the nodes of the subtree, we do a preorder traversal of that subtree. Hence, the definition is recursive—preorder traversal of a tree is defined in terms of preorder traversal of two smaller subtrees.

 Example 8 *Give a preorder print of the first two trees in Figure 7-110.*

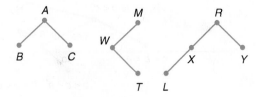

Figure 7-110 Preorder prints.

SOLUTION   For each subtree including the whole tree, we print the root, then the entire left subtree (if any), then the entire right subtree (if any). For the first tree, this gives us the straightforward result A, B, C. When we reach a leaf, we simply print that leaf since both of its subtrees are empty. For the second tree, it is obvious that the root (M) should come first. Then we print the W subtree in preorder, first the root (W), then the empty left subtree (nothing), then the right subtree (T). The final result is M, W, T.   □

Practice Problem 5   *Do a similar analysis to determine the preorder print for the third tree (on the far right) in Figure 7-110.*

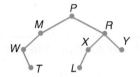

Figure 7-111   Tree for
Example 9.

Example 9   *Give the preorder traversal for the tree in Figure 7-111.*

SOLUTION   The solution consists of the root P, followed by the preorder traversal for the left subtree, followed by the preorder traversal for the right subtree. Example 8 tells us that the left subtree prints in the order M, W, T, and Practice Problem 5 tells us that the right subtree prints in the order R, X, L, Y. Therefore, the complete print order is P, M, W, T, R, X, L, Y. This illustrates how the answer for the whole tree is recursively built up from the answers for the subtrees.   □

Inorder traversal

For the *inorder traversal* of a binary tree, we first visit each of the nodes of the left subtree, then we visit the root, and finally we visit each of the nodes of the right subtree. The subtree nodes are visited by doing an inorder traversal of the subtree. Notice that this time the root of each subtree is visited in between its left and right subtrees. For the trees of Example 8, the inorder prints are shown in Figure 7-112.

Practice Problem 6   *Use the inorder prints for the trees of Example 8 to give the inorder print for the tree of Example 9.*

Postorder traversal

The postorder traversal follows the same general scheme but visits the root last. We visit each of the nodes of the left subtree by doing a postorder traversal of the subtree. Then we visit each of the nodes of the right subtree by doing a postorder traversal

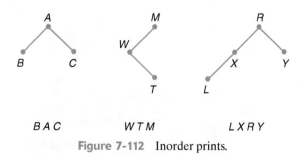

B A C   W T M   L X R Y

Figure 7-112   Inorder prints.

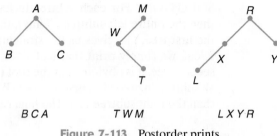

B C A   *T W M*   *L X Y R*

Figure 7-113   Postorder prints.

of the subtree. Finally, we visit the root. For the trees of Example 8, the postorder prints are shown in Figure 7-113.

Practice Problem 7   *Use the postorder prints for the trees of Example 8 to give the postorder print for the tree of Example 9.*

Traversals and Expression Trees

Binary tree traversals are of particular interest for trees that represent arithmetic expressions with binary operations. We call these *expression trees*. In general, it is not possible to reconstruct a tree from any one traversal. However, for expression trees we can determine the tree from either the preorder traversal or the postorder traversal. Expression trees have additional structure not present in all binary trees. In particular, every node is either a leaf or has exactly two children, and in the latter case the node must contain an operation, not a number.

In a way, it is unfortunate that the inorder traversal does not also have this property. The inorder traversal, with the root in between its two subtrees, would display the operation in between the two numbers. This, of course, is how we ordinarily write expression such as 3×4 by hand. However, as this example illustrates, the inorder print of an expression tree is ambiguous.

Example 10   *In Figure 7-114, the binary tree on the left represents the arithmetic expression $2 + (3 \times 4)$, and the binary tree on the right represents the arithmetic expression $(2 + 3) \times 4$. Give the inorder traversals of these two trees.*

SOLUTION   For the first tree, we print the left subtree (2), then the root (+), then the right subtree in inorder $(3, \times, 4)$. The complete result is $2, +, 3, \times, 4$. For the second tree, we print the left subtree $(2, +, 3)$, then the root $(\times)$, then

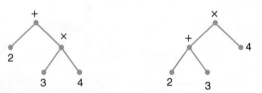

The binary tree for $2 + (3 \times 4)$ The binary tree for $(2 + 3) \times 4$

Figure 7-114   Expression trees for Example 10.

the right subtree (4). The complete result is $2, +, 3, \times, 4$. Observe that both trees give the same result.   □

As the example illustrates, the list $2, +, 3, \times, 4$ does not uniquely identify which binary tree it came from—the inorder traversal of the binary tree must be augmented with parentheses and/or precedence rules to correctly identify the expression. Because of its connection with inorder traversal, arithmetic notation (with parentheses) like $2 + (3 \times 4)$ or $(2 + 3) \times 4$ is called *infix notation*.

Prefix and Postfix Notation

The previous problem might seem insurmountable since we are all familiar with how lack of parentheses causes ambiguity in arithmetic expressions. However, it turns out that the preorder and postorder traversals *can* be unambiguously associated back to the original expression tree, as illustrated in the following examples and practice problems.

Example 11 *The preorder traversals of the binary trees in Example 10 are as follows:*

- *The preorder traversal of the binary tree representing $2 + (3 \times 4)$ lists the nodes in order $+, 2, \times, 3, 4$.*
- *The preorder traversal of the binary tree representing $(2 + 3) \times 4$ lists the nodes in order $\times, +, 2, 3, 4$.*

We use the term *prefix notation* for this way of representing arithmetic expressions. Hence, the prefix notation for $2 + (3 \times 4)$ is $+, 2, \times, 3, 4$ and the prefix notation for $(2 + 3) \times 4$ is $\times, +, 2, 3, 4$.

Practice Problem 8 *An arithmetic expression is given in prefix notation $+, -, 3, 1, \div, 8, 4$. Draw a binary expression tree representing this expression, rewrite it in infix notation (with parentheses), and evaluate it.*

Likewise, the postorder traversal uniquely determines the expression tree, as the following example and practice problem illustrate.

Example 12 *The postorder traversals of the binary trees in Example 10 are as follows:*

- *The postorder traversal of the binary tree representing $2 + (3 \times 4)$ lists the nodes in order $2, 3, 4, \times, +$.*
- *The postorder traversal of the binary tree representing $(2 + 3) \times 4$ lists the nodes in order $2, 3, +, 4, \times$.*

Just as with preorder traversal, we give the special name *postfix notation* to the notation for an arithmetic expression written as a postorder traversal of the corresponding expression tree.

Practice Problem 9 *An arithmetic expression is given in postfix notation* $3, 2, +, 5, 1, -, \times$. *Draw a binary expression tree representing this expression, rewrite it in infix notation (with parentheses), and evaluate it.*

Solutions to Practice Problems

1 The tree is given in Figure 7-115.

2 (a) The root is the node labeled $\div$. The two subtrees are given in Figure 7-116. The left tree has leaves 2, 3, 6, and the right tree has leaf 4. The entire tree has leaves 2, 3, 6, and 4. The nonleaves are $+$, $\times$, and $\div$.

   (b) The root is Brenda. The leaves are Tom, Sam, Jason, Anne, Bob, and Carl. Mary's children are Jason, Anne, and Bob. Mary's sibling is Carl, and her parent is Sue.

3 In the Washington tree, Taylor is level 4, and in the Madison tree, it is level 5. This matches the recursive definition of *level*.

4 The maximum number of nodes in a binary tree of height n is $2^n - 1$. This is proved in Exercise 13.

5 The root (R) comes first, then the X subtree in preorder (root X, left subtree L). Finally, we print R's right subtree Y. Putting it together, we get $RXLY$.

6 $WTMPLXRY$

7 $TWMLXYRP$

8 The tree is given in Figure 7-117. In infix notation the expression is $(3-1) + (8 \div 4)$, with a value of $2 + 2 = 4$.

9 The tree is given in Figure 7-118. In infix notation the expression is $(3 + 2) \times (5 - 1)$, with a value of $5 \times 4 = 20$.

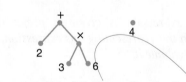

Figure 7-115 Solution to Practice Problem 1.

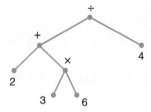

Figure 7-116 Solution to Practice Problem 2.

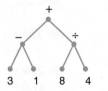

Figure 7-117 Solution to Practice Problem 8.

Figure 7-118 Solution to Practice Problem 9.

Exercises for Section 7.6

1. Draw a picture of your family tree with your father's father as a root and with his descendants below. What is the height of this tree? Is this a binary tree?

2. Draw a picture of your family tree (with three or four levels) with yourself as the root, your mother's family tree as the left subtree, and your father's family tree as the right subtree. Is this a binary tree?

3. Place the numbers $1, 4, -3, 12, 6, 13, 9$ into a binary search tree with the smallest possible height.

4. Place the numbers $11, -4, 3, 2, 16, 8, 5, -1$ into a binary search tree with the smallest possible height.

5. Give the preorder, inorder, and postorder traversals for the tree shown in Figure 7-119.

6. Give the binary tree that represents each of the following "inorder" arithmetic expressions:
   (a) $3 + (5 \times 4)$
   (b) $a \times (3 - b)$
   (c) $(((3 + 5) - 7) \times 4) \div 2$

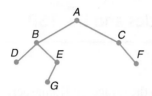

Figure 7-119 Tree for Problem 5.

(d) $((a \times 1) + (3 - 2)) - ((3 + a) \div 2)$

7. For each of the trees in Exercise 6, give the result of preorder and postorder traversals.

8. For each of the following arithmetic expressions in prefix notation, write the equivalent infix expression (with parentheses) and evaluate it:

   (a) $\div, +, \times, 5, 4, 7, 3$

   (b) $-, \times, 5, 4, +, 6, 3$

   (c) $\times, 8, -, 5, \div, 6, 2$

9. For each of the following arithmetic expressions in postfix notation, write the equivalent infix expression (with parentheses) and evaluate it:

   (a) $6, 2, 3, \times, -, 12, 4, \div, \times$

   (b) $1, 5, 12, 5, 3, -, \div, \times, +$

   (c) $1, 2, +, 3, \times, 4, 5, \times, 6, +, +$

10. List the people in your family tree in Exercise 2 resulting from inorder, preorder, and postorder traversals.

11. Give two different binary trees whose preorder print is A, B, C. Give two different trees whose postorder print is A, B, C. (This illustrates the fact that being able to reconstruct the tree from only one traversal is a special property of expression trees, and does not apply to all binary trees.)

12. Explain why if a set of data is stored as a binary search tree, then an inorder print of that tree will print the data in sorted order.

13. Prove by induction on n that the maximum number of nodes in a binary tree of height n is $2^n - 1$.

14. Prove by induction on h that in any binary tree T with height h, there is a node at level h in T.

15. Explain how Proposition 1 and Exercise 14 establish that the height of a binary tree is the largest level of any node in T.

16. A binary tree of height h is called a *complete binary tree* if it is either empty, or it has two complete subtrees both of height $h - 1$. (In the picture, all the levels from level 1 to level h have as many nodes as they can hold.) Prove by induction on $n \geq 0$ that a complete binary tree with height n has $2^n - 1$ total nodes.

17. The following algorithm allows us to think of any list of data as a binary tree. Given a list $a_1, a_2, a_3, \ldots, a_n$, build a tree by this rule: Place a_1 at the root, a_2 as the left child of a_1, a_3 as the right child of a_1, and in general:

   ● If k is even, place a_k as the left child of $a_{\lfloor k/2 \rfloor}$

   ● If k is odd, place a_k as the right child of $a_{\lfloor k/2 \rfloor}$

   (a) Apply this algorithm to these lists of words, letters, or numbers:

     i. short, above, to, well, indicate, closed, section, student, find

     ii. a, b, c, d, e, f, g, h, i, j, k, l, m, n, o, p

     iii. 16, 23, 1020, 145, 29, 17, 772

   (b) For each list in part (a), what was the height of the resulting tree?

   (c) For what size list would the resulting tree be a *complete* tree of height 2? of height 3? of height 6? of height 10?

   (d) For what sizes of list would the resulting tree be a (not necessarily complete) tree of height 2? of height 3? of height 6? of height 10?

   (e) Give a formula for calculating the height of the tree as a function of the size of the list.

18. Refer to Exercise 17. Prove by induction on n that the nth item in the list is at level $1 + \lfloor \log_2(n) \rfloor$.

19. To complete the proof of Corollary 5, show that for all $n \geq 3$, $n! > n^{n/2}$. (HINT: You may use the fact that $n^n < n^2 \cdot (n-1)^{n-1}$ for all $n \geq 3$ without proving it.)

20. Suppose a binary search tree is created with new data inserted using the simple process described after the proof of Proposition 3. Explain what happens if the data are inserted in increasing order. (For example, describe what happens if you insert the numbers 1 to 10 in order into a new binary search tree.)

7.7 Excursion: Hamiltonian Cycles and the TSP

Hamilton's Puzzle

Another problem that seems to be related to the bridges of Königsberg puzzle comes from a little puzzle that was marketed by the Irish mathematician W. R. Hamilton (1805–1865) in 1857. The puzzle consists of a wooden dodecahedron, with its 20 nodes labeled using the names of cities. The object was to find a path using the edges of the dodecahedron that visits every "city" once and then returns to the starting point. In terms of graphs, this means that the goal is to find a *cycle* in the graph in Figure 7-120 that uses every node exactly once.

Definition   Consider a simple graph G.

1. Recall from Section 7-1 that a *path* is a list of vertices $v_0, v_1, v_2, \ldots, v_n$ in which every consecutive pair of vertices is connected by an edge and no vertex appears more than once. A *cycle* is a path with the exception that there is exactly one pair of repeated nodes—namely the first and last nodes in the list are identical.

2. A *Hamiltonian cycle* in G is a cycle that uses every node of G. The graph G is called *Hamiltonian* if it contains a Hamiltonian cycle.

The basic problem generated by Hamilton's puzzle is to find a characterization for Hamiltonian graphs in the same way that we found one for Eulerian graphs and graphs with Eulerian trails. Surprisingly, no such characterization is currently known. In fact, mathematicians do not even know how to program a computer to efficiently determine whether a graph is Hamiltonian. Here "efficient" means something rather

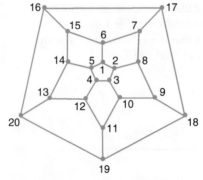

Figure 7-120   The graph of the
dodecahedron puzzle.

technical*, but suffice it to say that listing all cycles in a graph and checking whether they use all the nodes are not efficient. On the other hand, testing whether a connected graph is Eulerian by checking whether every node has even degree is very efficient. Research on this and related problems in graph theory is of the foremost interest to many mathematicians and computer scientists today. And it all started with Hamilton's wooden puzzle!

Practice Problem 1   *Spend a little time trying to find a solution to Hamilton's puzzle that begins with the nodes* 1, 5, 14, 15, ...

By completing Practice Problem 1, you should appreciate that Hamilton's dodecahedron puzzle is much harder to solve than the envelope-drawing puzzle or the bridges of Königsberg puzzle. It is a different sort of "traversal problem" altogether. In this section, we will discuss the complexity of solving this type of problem, and we will examine some efficient ways to find approximate solutions.

Traveling Salesperson Problems

The Hamilton puzzle fits into a wider class of problems called "traveling salesperson problems." We will begin our discussion with an example of this more general problem before returning to Hamilton's dodecahedron. A simple example can be described as follows. The graph in Figure 7-121 shows five cities A, B, C, D, and E, with the edges between them marked with the cost of traveling between those two cities. A salesperson lives in city A, and he needs to visit the other four cities and return home as cheaply as possible. What route should he take?

Before we attack this problem, we observe that this particular graph has a special property: For each pair of distinct nodes, there is one edge joining them. Such a graph is called a *complete* graph. In a graph with this property there will always be a

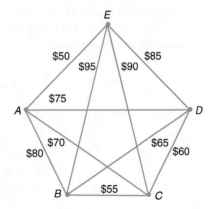

Figure 7-121   Traveling salesperson example.

---

* If you have worked through Section 4.8, you will appreciate the fact that no known algorithm for checking for Hamiltonian cycles is $O(n^p)$ for any p whatsoever. We say that such a problem has *nonpolynomial* complexity, and that is very bad news for a computer.

Hamiltonian cycle. In fact, there will be a lot of them! To find one, just list the nodes in any order, then repeat the first node listed. To get from one node to the next in your list, simply use the edge between them. Here are four Hamiltonian cycles for the graph of Figure 7-121:

- C, A, B, E, D, C
- A, D, E, B, C, A
- B, E, D, C, A, B
- B, A, C, D, E, B

Observe that the first and third are essentially the same cycle, just with different starting points. Also, the fourth is the reverse of the third.

Definition   The *complete graph* on n nodes, denoted by K_n, is the simple graph with nodes $\{1, \ldots, n\}$ and an edge between every pair of distinct nodes.

Using this terminology, we determine that the graph in Figure 7-121 is the graph K_5. (Technically, it is only isomorphic to K_5—in K_5 the node labels would be $1, 2, 3, 4, 5$ rather than A, B, C, D, E.)

We are now ready to solve the problem described earlier.

Example 1   *The graph in Figure 7-121 shows five cities with the edges between them marked with the cost of traveling between those two cities. A salesperson lives in city A, and he needs to visit the other four cities and return home as cheaply as possible. What route should he take?*

SOLUTION   In Table 7-15, we simply list all possible routes (Hamiltonian cycles) that start at node A, compute the cost of each one, and take the smallest! Notice that we only have to list half of all possible routes, since for each route, the reverse route has the same cost. For example, the route A, E, D, C, B, A is the reverse of the first route A, B, C, D, E, A, and consequently has the same cost as that route. Hence, the cheapest route is A, C, B, D, E, A, and its cost is \$325. Notice that starting the cycles at any other city would not change the costs or the final answer.   □

Number	Route	Cost	
1	A, B, C, D, E, A	\$330	
2	A, B, C, E, D, A	\$385	
3	A, B, D, C, E, A	\$345	
4	A, B, D, E, C, A	\$390	
5	A, B, E, C, D, A	\$400	
6	A, B, E, D, C, A	\$390	
7	A, C, B, D, E, A	\$325	← Cheapest
8	A, C, B, E, D, A	\$380	
9	A, C, D, B, E, A	\$340	
10	A, C, E, B, D, A	\$395	
11	A, D, B, C, E, A	\$335	
12	A, D, C, B, E, A	\$335	

Table 7-15   Table for Example 1

We can formalize this problem somewhat. The graph in Figure 7-121 is usually called a weighted graph. It is a graph with costs (generically called *weights*) associated with the edges.* Our goal then is to find a Hamiltonian cycle having the smallest total cost possible.

> **Definition**   Given a graph G with weighted edges, the problem of finding the Hamiltonian cycle of smallest possible weight is called the *traveling salesperson problem (TSP)* on G.

We can use our counting ideas to produce a formula for the total number of routes to check.

Proposition 1   *The number of Hamiltonian cycles that need to be checked in the TSP on K_n is $\frac{(n-1)!}{2}$.*

PROOF   As in the previous example, we can see that there is a one-to-one correspondence between the set of Hamiltonian cycles that begin and end with 1 and the set of permutations on $\{2, 3, 4, \ldots, n\}$. There are $(n-1)!$ such permutations, and only half of them need to be checked since the others are the reverses, and hence correspond to the same weight cycles. ∎

In one sense, the TSP is totally solved for complete graphs. It would be fairly easy to write a computer program to generate the list of all the Hamiltonian cycles, calculate the total cost for each, and find the smallest total cost. This is all very promising until we think harder about the function $\frac{(n-1)!}{2}$. Table 7-16 shows us how this function grows as the number of nodes grows. The size of these numbers is a little more than just cause for concern. The number of Hamiltonian cycles in K_{100}, for example, is far more than any current estimate of the number of elementary particles in the universe. This is unfortunate since problems involving paths and cycles in weighted graphs are very common in real-world applications. For example, any business concerned with a transportation network (like a bus line, airline, or express delivery service) has hundreds of destinations (nodes) and costs for moving between destinations (weighted edges). Telecommunication companies might consider the cost of cables (weights of edges) between thousands of switching stations (nodes). Clearly, the method of solution in these problems cannot consist of first generating the list of all Hamiltonian cycles of a graph like K_{500} or $K_{5,000}$, so we will discuss some other things we might try. Before we do that, however, we should establish how Hamilton's dodecahedron puzzle is just an example of this class of problem.

n	5	6	8	10	20	50	100
$(n-1)!/2$	12	60	2,520	181,440	6.08×10^{16}	3.04×10^{62}	4.67×10^{156}

Table 7-16   The Number of Hamiltonian Cycles in K_n

---

* Formally, a *weighted graph* is a graph G along with a function w from the set of edges of G to the set of real numbers.

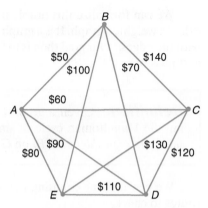

Figure 7-122   TSP graph for
Practice Problem 2.

 Example 2 *Hamilton's puzzle is a TSP problem.*

SOLUTION   Form the complete graph on 20 nodes, and weight the edges in the dodecahedron puzzle with a 0 and weight the edges not in the dodecahedron puzzle with a 1. Asking the question "Does Hamilton's puzzle have a solution?" is the same as asking, "Does the minimum cost Hamiltonian cycle in the weighted K_{20} have weight 0?" □

Practice Problem 2   *Find the minimum-weight Hamiltonian cycle in the graph shown in Figure 7-122.*

Approximate Solutions

It is a computational burden to list all possible routes, so we would like to have a condition like the one for Eulerian circuits that we can use to find an answer without listing every possibility. Unfortunately, no one knows of any such condition. In fact, the TSP is in a class of difficult problems that are computationally equivalent, called the *class of NP-complete problems*.

When an important problem cannot feasibly be solved exactly, the next best thing is an approximate solution. Because of the industrial interest in problems related to the TSP, improving on the best-known approximation schemes is one of the few ways a mathematician can get rich and famous. We will not delve into a study of these schemes in this book, but we will look at a couple of obvious choices for approximation schemes to see how well they do. These algorithms go by a variety of names. We use the adjective "greedy" in describing them because their overall strategy matches the overall strategy of a class of algorithms described using that term. In essence, a "greedy" algorithm tries to do as well as it can within small parts of the problem, hoping that will lead to a good solution for the overall problem.

Definition   Given a weighted complete graph K_n, the *vertex-greedy algorithm* starts at a designated node v_1, and uses the following rule: For each $i \geq 2$, choose v_i to be the node not yet used for which $[v_{i-1}, v_i]$ has the smallest weight. Once all the nodes are used, the edge from v_n back to v_1 completes the cycle.

Example 3 *For the graph in Example 1, find the result of the vertex-greedy algorithm, starting with vertex A.*

SOLUTION   The steps are shown in Table 7-17. The final route produced by this algorithm is A, E, D, C, B, A (or A, B, C, D, E, A), and its cost is \$330.   □

While this algorithm does not necessarily give the cheapest route, it can be completed in much less time than the process of considering every possible route in Example 1. In the example, the total number of edges listed in the "Edges to examine" column is $4 + 3 + 2 + 1$. In general, examining k edges to see which has the smallest weight can be done using $k - 1$ comparisons. So the algorithm for this example uses $3 + 2 + 1 + 0 = 6$ comparisons of weights for the sample graph. This example can be easily generalized to the following proposition.

Proposition 2   *The vertex-greedy algorithm on a weighted complete graph K_n uses $\frac{(n-1)(n-2)}{2}$ comparisons.*

PROOF   We start with any vertex v_1. The algorithm must look at each of the $n - 1$ edges leading to the other nodes of the graph. This process requires $n - 2$ comparisons. Similarly, after v_2 has been determined, each of the edges leading to the remaining $n - 2$ nodes must be considered, a process that requires $n - 3$ comparisons. In general (we could rigorously prove by induction), after vertex v_{i-1} has been determined, $n - i$ comparisons are required to determine v_i. The total number of comparisons required then is

$$(n - 2) + (n - 3) + \cdots + 2 + 1 + 0 = \sum_{i=1}^{n-2} i = \frac{(n - 1)(n - 2)}{2}$$
∎

Although this algorithm does not necessarily return the cheapest Hamiltonian cycle, its execution time is considerably smaller than the process of checking every possible cycle that we saw in Proposition 1. We can write the result as $\frac{n^2 - 3n + 2}{2}$. As n gets larger, the $-3n + 2$ portion becomes less and less significant, so the result is approximately proportional to n^2. A simple variation of this algorithm would be to repeat the algorithm for each of the n nodes of the graph, resulting in an algorithm whose complexity is approximately proportional to n^3.

Step	Current route	Edges to examine	Cheapest next edge	Current cost
1	A	$[A, B], [A, C], [A, D], [A, E]$	$[A, E]$	\$50
2	A, E	$[E, B], [E, C], [E, D]$	$[E, D]$	\$135
3	A, E, D	$[D, B], [D, C]$	$[D, C]$	\$195
4	A, E, D, C	$[C, B]$	$[C, B]$	\$250
5	A, E, D, C, B	Go back to start	$[B, A]$	\$330

Table 7-17   Steps for Example 3

Finally, we look at one other obvious method for finding an approximate solution to the TSP. The complexity of this algorithm is related to the complexity of sorting the edge weights from cheapest to most expensive, a topic that we will not take up in this book. It can be shown that this algorithm's complexity is approximately proportional to $n^2 \log_2 n$, which lies between n^2 and n^3.

Definition   Given a weighted complete graph K_n, the *edge-greedy algorithm* first sorts the edges into increasing order by weight. It then builds the Hamiltonian cycle by using the edges in this increasing order, except that it skips over any edge that would create a node of degree 3 or a cycle. It continues until only one edge can be added to complete the Hamiltonian cycle.

Example 4   *For the graph in Example 1, find the result of the edge-greedy algorithm.*

SOLUTION   We will show the steps once again in table form. We begin by listing the edges in increasing order by weight:

$$[A, E], [B, C], [C, D], [B, D], [A, C], [A, D], [A, B], [D, E], [C, E], [B, E]$$

Then we go down the rows of Table 7-18 filling in the last column. You may find it useful to follow along by highlighting the edges on the graph as they are added. The final cycle is A, E, B, C, D, A with a total cost of $50 + 95 + 55 + 60 + 75 = \$335$. Note that after step 5 we already had four edges, and it would have been possible to stop and figure out what the fifth edge would have to be. Since in practice most of the time taken doing this algorithm occurs in the sorting step, this would not significantly reduce the complexity.   □

Step	Edges and weights	Action taken	Cost so far
1	$[A, E]$, \$50	Add to result	\$50
2	$[B, C]$, \$55	Add to result	\$105
3	$[C, D]$, \$60	Add to result	\$165
4	$[B, D]$, \$65	Skip (would form cycle B, C, D, B)	
5	$[A, C]$, \$70	Skip (would make C have degree 3)	
6	$[A, D]$, \$75	Add to result	\$240
7	$[A, B]$, \$80	Skip (degree 3 and cycle problems)	
8	$[D, E]$, \$85	Skip (degree 3 and cycle problems)	
9	$[C, E]$, \$90	Skip (degree 3 and cycle problems)	
10	$[B, E]$, \$95	Add to result	\$335

Table 7-18   Steps for Example 4

Practice Problem 3 *Perform both approximation algorithms on the graph in Practice Problem 2.*

These approximate solutions can be pretty bad, as we will see in the exercises for this section, and they are presented here as simple, naive examples, not current research. Because of the importance of this type of problem in industry, a great deal of money is spent each year on research into *good* approximate solutions to the TSP. However, the problem is far from solved.

Hamiltonian Graphs Again

We end this section with two examples of classical theoretical results concerning Hamiltonian graphs. Recall that there was a very simple necessary and sufficient condition (namely, that the nodes all have even degree) to determine if a given connected graph contained an Eulerian circuit. There is no known useful necessary and sufficient condition for determining if a graph contains a Hamiltonian cycle. The theorems we will prove do, however, provide some information:

- The first theorem gives a sufficient condition. If that condition is true, the graph has a Hamiltonian cycle, but if that condition is false, we can make no conclusion.
- The second theorem gives a different necessary condition. If that condition is false, the graph does not have a Hamiltonian cycle, but if that condition is true, we can make no conclusion.

To assist your understanding of the proof of the first theorem, we first look at a few small examples.

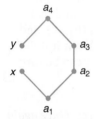

Figure 7-123
Partial graph
for Example 5.

Example 5 *Consider a graph G that contains six vertices as shown in Figure 7-123. The graph contains at least the five edges shown, but additional edges in the graph are possible.*

1. *For each part, show that, if the graph also contains the edge(s) listed, it must contain a Hamiltonian cycle:*
   (a) $[x, y]$
   (b) $[x, a_2]$ *and* $[y, a_1]$
   (c) $[x, a_3]$ *and* $[y, a_2]$
   (d) $[x, a_4]$ *and* $[y, a_3]$
2. *Now suppose you are told that the graph G does not have a Hamiltonian cycle. Explain why* $\deg(x) + \deg(y)$ *must be less than* 6.

SOLUTION

1. Refer to the graphs in Figure 7-124 for the subsequent discussion.
   (a) If we add the edge $[x, y]$ to the graph in the figure, there is a Hamiltonian cycle $x, a_1, a_2, a_3, a_4, y, x$.
   (b) If we add both $[x, a_2]$ and $[y, a_1]$, there is a Hamiltonian cycle $x, a_1, y, a_4, a_3, a_2, x$.

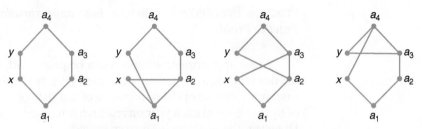

Figure 7-124   Graphs for the solution to Example 5.

(c) If we add both $[x, a_3]$ and $[y, a_2]$, there is a Hamiltonian cycle $x, a_1, a_2, y, a_4, a_3, x$.

(d) If we add both $[x, a_4]$ and $[y, a_3]$, there is a Hamiltonian cycle $x, a_1, a_2, a_3, y, a_4, x$.

2. The graph as originally shown has edges $[x, a_1]$ and $[y, a_4]$, so that $\deg(x) + \deg(y)$ is 2. If we list all the edges that could make this sum larger, we have

$$[x, y], [x, a_2], [x, a_3], [x, a_4], [y, a_1], [y, a_2], [y, a_3]$$

These are the same seven edges we looked at in the first part of this example. We showed that our graph cannot have the edge $[x, y]$ (part (a)), and that it can have no more than one of the edges listed in each of parts (b), (c), and (d). So it cannot have more than three of these seven edges. Even if it did have this maximum, we would have $\deg(x) + \deg(y) = 5$. So we can be sure that $\deg(x) + \deg(y) < 6$.   □

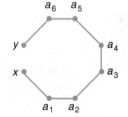

Figure 7-125
Graph for Practice
Problem 4.

Practice Problem 4   *Figure 7-125 shows a graph with eight vertices. Once again, the figure shows some of the edges of the graph, but there could be more.*

1. For each part, show that if the graph also contains both the edges listed, then there is a Hamiltonian cycle:

(a) $[x, a_3]$ *and* $[y, a_2]$

(b) $[x, a_6]$ *and* $[y, a_5]$

2. What other pairs of edges following this general pattern have this same property?

3. What is the largest value for $\deg(x) + \deg(y)$ *if the graph is not Hamiltonian?*

We are now ready for the theorem that gives a sufficient condition for a graph to have a Hamiltonian cycle. Recall that we say nodes are *adjacent* if they are joined by an edge.

Theorem 3   *(Ore) Let G be a graph with $n \geq 3$ nodes. If $(\deg x + \deg y) \geq n$ for all nonadjacent nodes x and y in G, then G is Hamiltonian.*

PROOF   We will prove the contrapositive: If G is *not* Hamiltonian, then there are nonadjacent nodes x and y in G with $(\deg x + \deg y) < n$.

Let G be a graph that is not Hamiltonian. Keep adding edges, one at a time, until the resulting graph *is* Hamiltonian. (We know that eventually this will happen because the complete graph K_n is Hamiltonian). Call the last non-Hamiltonian graph in this process H, and call the edge that was added next (the one that caused a Hamiltonian graph) $[x, y]$. Notice that H did not contain the edge $[x, y]$, so x and y are nonadjacent in H (and also in the original graph G). Also notice that the Hamiltonian cycle which results from adding the edge $[x, y]$ must contain this $[x, y]$ edge (or else H would have already had a Hamiltonian cycle).

By labeling the vertices, we can write the Hamiltonian cycle as $x, a_1, a_2, \ldots, a_{n-2}, y, x$. Since all the edges but $[x, y]$ were already in H, we know that H has a path $x, a_1, a_2, \ldots, a_{n-2}, y$, plus perhaps some other edges. (This is exactly the situation illustrated by the examples and practice problems.)

Considering just this path, we have $\deg(x) + \deg(y) = 2$, and the only edges H could have that can increase this sum are $[x, a_{i+1}]$ and $[y, a_i]$ (where i can range from 1 to $n-3$). However, for each value of i from 1 to $n-3$, H cannot contain both members of the pair $[x, a_{i+1}]$, $[y, a_i]$, since this would give a Hamiltonian cycle $x, a_1, \ldots, a_i, y, a_{n-2}, a_{n-3}, \ldots, a_{i+1}, x$ in H. Thus, no more than $n-3$ of these edges can be in H, so in H we have $\deg(x) + \deg(y) \leq 2 + (n-3) = n-1 < n$. Since H was obtained from G by adding additional edges, this same inequality must also hold in G. ∎

This gives us a sufficient condition for testing if a graph is Hamiltonian, but this condition is certainly not necessary. (See Exercise 17.) On the other hand, the following result gives us a necessary condition that alas is not sufficient. (See Exercise 18.)

Theorem 4 *If G has a Hamiltonian cycle, then removing any edge from G will result in a connected graph.*

PROOF Let a simple graph G with a Hamiltonian cycle be given. Let's name the cycle $v_1, v_2, v_3, \ldots v_n, v_1$, where n is the number of nodes in G and no node is repeated in the cycle. Now let the graph G' be formed by removing any single edge from G. To show that G' is connected, we must show that for any given pair of nodes in G', there exists a path between these nodes. Let two nodes in G' (and hence in G) be given. Since the cycle given above completely lists all nodes in G, we know that the two chosen nodes come from this list, so let's say they are v_i and v_j, where $i < j$. Since only one edge of G is missing in G', one of the following two paths must be intact in G': (i) $v_i, v_{i+1}, \ldots, v_j$ or (ii) $v_j, v_{j+1}, \ldots, v_n, v_1, \ldots, v_i$. Hence, there is a path connecting v_i and v_j in G'. ∎

Solutions to Practice Problems

1 Here is one solution to Hamilton's puzzle:

1, 5, 14, 15, 16, 17, 18, 9, 10, 11, 19, 20, 13, 12, 4, 3, 2, 8, 7, 6, 1

2 Table 7-19 shows every possibility and highlights the cheapest option.

3 Table 7-20 shows the vertex-greedy solution, and Table 7-21 shows the edge-greedy solution.

4 1. There are these Hamiltonian cycles:

(a) $x, a_1, a_2, y, a_6, a_5, a_4, a_3, x$
(b) $x, a_1, a_2, a_3, a_4, a_5, y, a_6, x$

2. $[x, a_2]$, $[y, a_1]$ and $[x, a_4]$, $[y, a_3]$ and $[x, a_5]$, $[y, a_4]$

3. The two there now along with no more than five additional means there will be no more than seven total.

Number	Route	Cost	
1	A, B, C, D, E, A	$500	
2	A, B, C, E, D, A	$520	
3	A, B, D, C, E, A	$450	
4	A, B, D, E, C, A	$420	← Cheapest
5	A, B, E, C, D, A	$490	
6	A, B, E, D, C, A	$440	
7	A, C, B, D, E, A	$460	
8	A, C, B, E, D, A	$500	
9	A, C, D, B, E, A	$430	
10	A, C, E, B, D, A	$450	
11	A, D, B, C, E, A	$510	
12	A, D, C, B, E, A	$530	

Table 7-19 Exhaustive Solution to Practice Problem 2

Step	Current route	Edges to examine	Cheapest next edge	Current cost
1	A	$[A, B], [A, C], [A, D], [A, E]$	$[A, B]$	$50
2	A, B	$[B, C], [B, D], [B, E]$	$[B, D]$	$120
3	A, B, D	$[D, C], [D, E]$	$[D, E]$	$230
4	A, B, D, E	$[E, C]$	$[E, C]$	$360
5	A, B, D, E, C	Go back to start	$[C, A]$	$420

Table 7-20 Vertex-Greedy Solution to Practice Problem 3

Step	Edges and weights	Action taken	Cost so far
1	$[A, B]$, $50	Add to result	$50
2	$[A, C]$, $60	Add to result	$110
3	$[B, D]$, $70	Add to result	$180
4	$[A, E]$, $80	Skip (would make A have degree 3)	
5	$[A, D]$, $90	Skip (would make A have degree 3)	
6	$[B, E]$, $100	Skip (would make B have degree 3)	
7	$[D, E]$, $110	Add to result	$290
8	$[C, D]$, $120	Skip (degree 3 and cycle problems)	
9	$[C, E]$, $130	Add to result	$420
10	$[B, C]$, $140	Skip (degree 3 and cycle problems)	

Table 7-21 Edge-Greedy Solution to Practice Problem 3

Exercises for Section 7.7

1. Find another solution to Hamilton's dodecahedron puzzle.

2. In Example 2, we stated that Hamilton's puzzle can be thought of as a TSP problem in an appropriately weighted K_{20} graph. How many different Hamiltonian cycles are there in K_{20}?

3. Suppose that a computer can calculate the weight of a single Hamiltonian path in K_n in 10^{-15} sec. (This means that it can compute the weights of a thousand trillion different cycles per second.) If you use this computer to check all possible cycles, how long would it take to find the shortest (in total driving miles) cycle that visits all state capitals in the lower 48 United States?

4. Suppose the entire population of the earth (which we can round to 10 billion people) has a computer like the one described in the previous exercise, and they all set about checking different possible cycles. How long would it take to check all possible cycles in K_{48}?

5. For what values of m and n is the complete bipartite graph $K_{m,n}$ a Hamiltonian graph?

6. Solve the TSP on each of the complete graphs given in Figure 7-126.

7. Provide a simple example to show that the edge-greedy algorithm does not always give the exact solution to the TSP.

8. For each of the graphs in Exercise 6, use the vertex-greedy algorithm starting with vertex a to find a cheap Hamiltonian cycle.

9. For each of the graphs in Exercise 6, use the edge-greedy algorithm to find a cheap Hamiltonian cycle.

10. Give an example of a weighted complete graph for which the cheapest Hamiltonian cycle costs $100, but the vertex-greedy algorithm finds a Hamiltonian cycle that costs $1,000.

11. Using the previous problem as a starting point, explain why the vertex-greedy algorithm can be wrong by any given amount, depending on the graph.

12. Notice that the result of the vertex-greedy algorithm depends on the starting vertex. One way to counteract this is to run the vertex-greedy algorithm starting at each vertex, and taking as our approximate solution the cheapest Hamiltonian cycle out of all those runs. (This multiplies our efforts by the number of vertices in the graph, but this increase is nothing compared to the increase in the brute force method of attack.) Try this new version on the graph shown in Figure 7-127.

13. For the graph in Exercise 12, use the edge-greedy algorithm to find a cheap Hamiltonian cycle, and compare this result to your answer to that exercise.

14. Let G_n denote the graph whose nodes are labeled with all 2^n binary sequences of length n, and which has an edge from a to b whenever sequences a and b differ in only one position. For example, G_2 has nodes $\{00, 01, 10, 11\}$ and edges $\{[00, 01], [00, 10], [01, 11], [10, 11]\}$. Show that G_2, G_3, and G_4 are Hamiltonian.

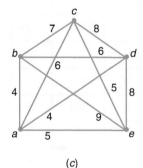

(a)

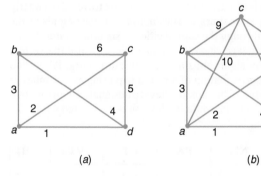

(b)

(c)

Figure 7-126 Graphs for Problem 6.

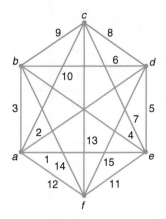

Figure 7-127 Graphs for Problem 12.

15. Ironically, Hamilton's dodecahedron puzzle (ca. 1857) is not the earliest puzzle problem involving a "Hamiltonian cycle" and solved using graph theoretical ideas. A much older puzzle is the "knight's tour problem," which was treated mathematically by Euler (in 1759) and A. T. Vandermonde (in 1771):

A knight moves in an L-shape on a standard 8×8 chessboard, "jumping" two spaces horizontally or vertically and one space in the perpendicular direction. Can a knight start at a square of the chessboard, make a sequence of legal moves, and return to its starting square having visited every other square on the chessboard exactly once?

(a) Explain why the classical knight's tour problem is solved by finding a Hamiltonian cycle in a graph. That is, explain what the meaning of the nodes and edges is in a graph so that a Hamiltonian cycle in the graph gives a solution to the puzzle.

(b) Using a smaller 4×4 chessboard, solve the knight's tour puzzle using the type of graph you described above.

(c) Explain why it is impossible to solve the knight's tour puzzle on a 5×5 chessboard. (HINT: Compare the numbers of white and black squares.)

16. How many (base ten) digits would it take to write out the number 99!? (HINT: The number of base ten digits of a positive integer x is given by $\log_{10} x$ (i.e., 1 more than the greatest integer less than or equal to $\log_{10} x$), and $\log_{10}(ab) = \log_{10} a + \log_{10} b$.)

17. Find a counterexample to the converse of Theorem 3.

18. Find a graph on five nodes that is a counterexample to the converse of Theorem 4.

19. Explain how the following derives from Ore's theorem: If G is a simple, connected graph on $n \geq 3$ nodes for which every node has a degree of at least $n/2$, then G is Hamiltonian.

20. Explain how you can tell that the graph in Figure 7-128 is not Hamiltonian. (This graph is called the *Petersen graph*, and it pops up frequently as an important example in graph theory.)

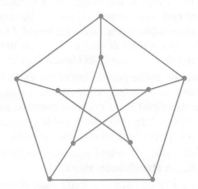

Figure 7-128 Graph for Problem 20.

21. On a recent trip to New England, we traveled round trip from Harrisburg to Burlington, VT, passing along the way through the capital cities of six other states. Given the mileage chart shown in Table 7-22, use the vertex-greedy algorithm starting with Harrisburg, PA to find an approximation of a round trip using as few miles as possible. (The capital cities are designated with the state abbreviation, while Burlington is designated with a *B*.)

	NY	MA	B	NH	PA	CT	VT	RI
NY	*	170	154	151	293	114	173	165
MA	170	*	216	68	393	101	180	50
B	154	216	*	155	446	235	39	266
NH	151	68	155	*	449	154	119	119
PA	293	393	446	449	*	296	494	358
CT	114	101	235	154	296	*	200	87
VT	173x	180	39	119	494	200	*	230
RI	165	50	266	119	358	87	230	*

Table 7-22 Mileage Chart for Exercise 21

22. Repeat the previous exercise using the vertex-greedy algorithm starting with Boston, MA.

23. Refer to the mileage chart in Exercise 21. Use the edge-greedy algorithm to find an approximation of the round trip using the fewest possible total miles.

24. Refer to the mileage chart in Exercise 21. Find the lowest mileage round trip involving just the state capitals of CT, MA, NH, RI, and VT.

25. Imagine a graph with 20 vertices, containing the path

$$x, a_1, a_2, a_3, a_4, a_5, a_6, a_7, a_8, a_9, a_{10}, a_{11}, a_{12}, a_{13}, a_{14},$$
$$a_{15}, a_{16}, a_{17}, a_{18}, y$$

Show that if it also contains the two edges $[x, a_{13}]$ and $[y, a_{12}]$, then it contains a Hamiltonian cycle that begins $x, a_1, \ldots$. Generalize. What is the largest possible value for $\deg(x) + \deg(y)$ if G is not Hamiltonian?

Chapter 7 Summary

7.1 **Graph Theory**

Terms and concepts

- You should recognize that a graph consists of *vertices* (*nodes*) and *edges*, and that each edge has one or two *endpoints*. The edge is said to be *incident with* its endpoints. Each node has a *degree* that measures how many edges are incident with that node.

- You should recognize special types of edges and nodes: *loops, multiple* or *parallel* edges, *adjacent* nodes.

- You should know what we mean by a *walk*, and be able to calculate the *length* of the walk.

- You should be able to distinguish special types of walks: *trivial walk, trail, path, circuit, trivial circuit, cycle*.

- You should understand that there are variations on the fundamental graph idea (for example, *simple graphs* and *directed graphs*) and that for certain applications one or the other variation may better model the application.

- The notation $[a, b]$ (or in some situations $\{a, b\}$) is used to indicate the edge with endpoints a and b. In a directed graph the ordered pair notation (a, b) emphasizes that the order of the endpoints is significant

- A graph may be *connected*, and in any case it will contain *subgraphs* which are *connected components* of the graph.

- You should understand the meaning of *Eulerian circuit*, *Eulerian trail*, and *Eulerian graph*.

Applying graph terminology, and Eulerian graphs

- You should be able to draw a graph with specified nodes and edges, using points for the nodes and lines for the edges.

- You should be able to calculate the degree of a node.

- Within a given graph, you should be able to identify walks, cycles, subgraphs, connected components, etc. In short, you should be able to demonstrate your compre-

hension of all the terms listed above by identifying them in a graph.

- You should recognize and be able to apply the formula $\sum_{i=1}^{n} \deg(v_i) = 2m$, where the v_i are the vertices and m is the number of edges in the graph.

- You should be able to determine if a given graph contains an Eulerian circuit or Eulerian trail. If it does, you should be able to construct that circuit or trail.

7.2 **Proofs About Graphs and Trees**

Terms and concepts

- You should know that a *tree* is a connected, simple graph with no cycles, and you should be able to identify the *leaves* of a tree.

- You should know what we mean by a *spanning tree* for a graph.

- You should be familiar with *weighted graphs*, and with the notion of a *minimal spanning tree* for a weighted graph.

Proofs

- You should be able to provide direct proofs of graph properties (by starting with the hypotheses and establishing that the conclusions must follow).

- You should be able to state the contrapositive of a given theorem, and recognize that you can prove the theorem by establishing that the contrapositive is true.

- You should be able to use induction to prove graph properties. You should recognize that, in contrast to many earlier induction proofs, it is not always $P(m - 1)$ that proves crucial in the proof.

- You should be able to illustrate steps in your proofs with concrete examples.

Spanning trees

- Given a graph, you should be able to construct a spanning tree for that graph.
- Given a weighted graph, you should be able to use *Prim's algorithm* to contruct a minimal spanning tree for that graph.

7.3 Isomorphism and Planarity

Terms, concepts, and formulas

- You should understand the terms *isomorphism* and *isomorphic graphs*.
- You should know that a *planar* graph can be drawn (on a plane) so that no edges cross, and that the drawing is called an *embedding* of the graph in the plane.
- You should know the meaning of the term *bipartite* graph.
- You should recognize the notation K_n for the *complete graph* on n nodes, and the notation $K_{n,m}$ for the *complete bipartite graph* on n, m nodes.
- You should be familiar with the *five regular polyhedra* (the *five Platonic solids*), and their connection to the question of planarity.
- You should know what is meant by a *face* of a planar graph embedded in the plane.
- You should recognize Euler's formula for planar graphs: $V + F = E + 2$.

Isomorphism

- For small isomorphic graphs, you should be able to produce the function that establishes the isomorphism.
- For small graphs, you should be able to use the idea of "tacks and elastic string" to help decide if they are isomorphic. (You may use the assistance of technology for this.)
- You should be familiar with properties that isomorphic graphs share, and be able to use the absence of those properties to demonstrate that graphs are *not* isomorphic.

Planarity

- You should know that K_5 and $K_{3,3}$ are not planar, and that any graph that is not planar contains a "copy" of one of these graphs.
- You should be able to use the idea of "tacks and elastic string" to help decide if a graph is planar. (You may use the assistance of technology for this.)
- You should be able to use Euler's formula and Theorem 7 to demonstrate that a graph is not planar.

7.4 Connections to Matrices and Relations

Terms and concepts

- You should understand the use of an *adjacency matrix* M to represent a graph, and know the meaning of the notation M_{ij}.
- You should know the meaning of a *directed graph*, where we represent the edges as *ordered pairs*, and the meaning of a *walk* in a directed graph.
- You should recognize the standard use of algebraic operations applied to matrices: for example, $A + B$, $A \cdot B$, A^2.
- You should understand the use of the *Boolean product* of matrices whose entries are 1s and 0s, and in particular the notation $M^{(k)}$ used in this context.

Matrix operations

- For given matrices A and B, you should know how to calculate $A + B$ and $A \cdot B$ (and you should know when these operations are/are not possible). Appendix B contains further information on matrix operations.
- You may have learned how to use computer software to calculate expressions involving matrices.
- You should be able to compute Boolean products of matrices; in particular, you should be able to compute $M^{(k)}$ for the adjacency matrix M of a binary relation.
- You should be able to compute Boolean sums of matrices, denoted $A \vee B$.
- You should understand the connection between the usual product M^k and the Boolean product $M^{(k)}$, and the connection between the usual sum $A + B$ and the Boolean sum $A \vee B$.

Modeling graphs with matrices

- You should know how to build the adjacency matrix M for a graph or a directed graph.
- You should be able to compute the number of walks from node to node of length 2, length 3, and so on, using matrix multiplication.
- You should be able to compute the number of walks of length k or less, using matrix multiplication and addition.
- You should be able to use Boolean matrix operations to explore the existence of walks, and connectivity of a graph.

Matrices, graphs, and binary relations

- You should understand, and be able to exploit, the connection between a binary relation, its associated graph, and its associated adjacency matrix.

- In particular, you should be able to use boolean matrix operations to model relation composition.

- You should be able to use matrices to explore transitivity and other properties of relations.

7.5 Graphs in Puzzles and Games

Terms and concepts

- In the context of the Instant Insanity puzzle, you should know what we mean by a *good subgraph* of *G*.

- You should understand what is meant by a *finitely progressive game*.

- For the directed graph of a game, you should understand the significance of the *kernel* of the graph (also referred to as the set of *good* nodes).

- In the context of the Nim games, you should know what is meant by the *direct sum* of rows and the *binary number of the game position*.

Simple puzzles and games

- You should be able to use directed graphs to represent puzzles where the nodes represent the legal states of the puzzle, and the edges represent legal moves. (An example is the *wolves, goats, and cabbages* game.)

- Similarly, you should be able to use directed graphs to represent games where the nodes represent the possible states of the game, and the edges represent legal moves. (The simple stone-removal games are examples.)

- For small puzzles and games, you should be able to use the graph to analyze the puzzle or game, answering questions such as:
  - Starting at a given state, is it possible to reach some other given state?
  - Which player has a winning strategy?

- You should know the properties that characterize the kernel of a graph, and be able to construct the kernel for games with small graphs.

Instant Insanity and Nim

- You should understand the particular application of graphs to *Instant Insanity* puzzles:

  - The nodes represent colors and the edges represent opposite faces of a cube.

  - You should know how to use a good subgraph of *G* to construct a *partial solution*, working with just the front and back columns while ignoring the other pair of columns.

  - You should know how to use a second good subgraph of *G* to move from the partial solution to a complete solution.

- You should know how to use binary numbers to analyze the classic game of Nim and variations of the game.

7.6 Excursion: Binary Trees

Terms and concepts

- You should be familiar with the terms *tree* and *binary tree*, and with a large variety of terms used to describe trees: *node, root, leaf, internal node, child, parent, sibling, subtree*.

- You should know the related concepts of *levels* within a tree, and the *height* of the tree.

- You should understand the particular type of binary tree known as a *binary search tree*.

- You should know about *traversals* of a binary tree (*preorder, inorder,* and *postorder*). For expression trees, you should be familiar with the related terms *infix notation, prefix notation,* and *postfix notation*.

Applications and proofs involving tree concepts

- You should be able to illustrate the various concepts (subtree, leaf, etc.) for specific trees.

- You should be able to illustrate the method of using binary trees to represent algebraic expressions.

- For a binary search tree, you should be able to describe and trace the algorithms for searching for a particular value, and for inserting a new value.

- You should be familiar with the connection between binary trees and bounds on the number of comparisons needed to sort a list.

- You should be able to give the preorder, inorder, and postorder prints for a given binary tree.

- You should be able to use the recursive definitions of binary tree, height, etc., to develop proofs by induction for various properties of binary trees.

7.7 Excursion: Hamiltonian Cycles and the TSP

Terms and concepts

- You should recall the terms *path* and *cycle*, and understand their use in describing *Hamiltonian circuits* and *Hamiltonian graphs*.

- You should know what is meant by a *traveling salesperson problem (TSP)* and its connection to the concept of *weighted graphs*.

- You should recognize the notation K_n for the *complete graph* on n nodes.

- You should realize that the TSP problem is an example from the *class of NP-complete problems*.

- You should understand the terminology used in the section for two algorithms for obtaining approximate solutions to the TSP: the *vertex-greedy algorithm*, and the *edge-greedy algorithm*.

Finding Hamiltonian cycles

- For small graphs, you should be able to solve the TSP by listing all possible Hamiltonian cycles and calculating the weights for each.

- You should be able to apply the *vertex-greedy algorithm* and the *edge-greedy algorithm* to particular weighted graphs.

- You should realize that there are no known conditions which are both necessary and sufficient for a graph to be Hamiltonian.

A | Rules of the Game

Throughout this book we refer to games played by people all over the world. However, since not every game is familiar to everyone, we devote this short appendix to an overview of the terms and rules associated with some of these games. Of course, more details on any of them are simply a click away on any web search engine.

Cards

A standard *deck of cards* consists of 52 cards, each of which consists of a **value** and a **suit**. The value of a card is one of, *ace, 2, 3, 4, 5, 6, 7, 8, 9, 10, jack, queen, king*, and the suit of a card is one of, *club, diamond, heart, spade*. Examples of cards are *3 of diamonds*, *queen of hearts*, and *ace of spades*.

There are four cards with each possible value and there are thirteen cards of each possible suit. Cards whose values are jack, queen or king are called *face cards*. Aces are considered the highest value or the lowest value in the deck, depending on the game or the situation. When cards are distributed in a game, we say that cards are *dealt*, and the particular set of cards that a player has is called a *hand* of cards.

A few specific card games are mentioned in problems and examples throughout the book. Although knowledge of the rules and strategies for playing these games is not necessary for completing the problems, it might be useful to know a little bit about them.

Blackjack

Blackjack is a game in which the initial hand consists of a face down card and a face up card. The value of the hand is simply the sum of the values of the two cards in the hand, with some simple interpretations: Face cards all have a value of ten. The player decides on an individual basis whether an Ace is worth one or eleven. The object of the game is to draw cards until the sum of the values equals as close as possible to but not exceeding 21. For this reason the game of blackjack is sometimes called *Twenty-one*.

Poker

In poker, a **hand** consists of five cards. There are many variations of poker that involve the different ways in which these five cards are obtained, but these are irrelevant to the basic way hands are compared.

1. A hand is said to contain a *pair* if exactly two cards have the same value and the other three cards have three different values.
2. A hand is said to contain *two pairs* if there are two values such that exactly two cards have one value and exactly two cards have the other value.
3. A hand is said to contain *three of a kind* if exactly three cards have the same value and the other two cards have two different values.
4. A hand is said to contain a *straight* if all five card values are consecutive.
5. A hand is said to contain a *flush* if all five cards are of the same suit.
6. A hand is said to contain a *full house* if two cards have one value and three cards have another value.
7. A hand is said to contain a *four of a kind* if four of the cards have the same value.
8. A hand is said to contain a *straight flush* if it contains a straight which is also a flush.

When comparing hands in poker, a hand lower in the list above beats a hand that is higher in the list.

Sports

In our study of games, we use several examples of sports played between two players or two teams. We summarize here some of the terminology used for those examples.

Series

In many situations, teams play a *series* of games to determine a winner. In tennis, the first player to win two sets wins the match in many tournaments. In the first round of professional playoffs, two teams play games until one team wins three games. In the final round of playoffs for baseball, basketball and hockey, two teams play until one team wins four games. These formats are called respectively, best of three, best of five, and best of seven series.

In less formal settings, people might decide something based on a coin flip and decide to use a best two out of three rule. Or perhaps two siblings play a travel game and decide to use a best five out of nine rule to make it more interesting. In all of these cases, the underlying structure of repeating a basic component until one player has first won a certain fixed number of times is the same.

Baseball and Softball

Few sports are as inundated with numerical data as baseball. Players are identified by their statistics in traditional categories, media frenzy follows lengthy streaks, and legends are affirmed when sacred records are broken. So common is the language among fans, it is easy to forget that it is not a common language for everyone. Hence, a short lesson in terminology is in order here.

Central to baseball and softball is the competition between pitcher and batter that occurs dozens of times in a game. We refer to one such occurrence as a *plate appearance* for the batter. To simplify the terminology, we refer to an outcome of this confrontation as taking the form of a *hit*, a *walk*, or an *out*. All less common outcomes are equivalent to one of these for the purposes of the problems we will consider. For example, a player's *batting average* is reported as the ratio of the number of hits to the total number of hits and outs. (The sum of the number of hits and outs is called the number of *at-bats*, considered to be the number of fair chances for the player to hit.) For example, a player who has 3 hits, 2 walks, and 5 outs in 10 plate appearances has a batting average of 0.375.

Tennis

In tennis, two players play *points*. A sequence of points comprises a *game*, a sequence of games make a *set*, and a sequence of sets make a *match*. In each case, there is a target number necessary for winning, and there are various rules for breaking ties.

1. To win a *game*, a player must score four points before his or her opponent does, with the extra rule that a game must be won by at least two points. This extra rule keeps games going for many points on occasion. (At the 1995 Wimbledon Finals, Steffi Graf and Arantxa Sanchez Vicario played a game that lasted 32 points before Graf won the game.)

2. To win a *set*, a player must win six games before her opponent does, also with the extra rule that a set must be won by at least two games. However, if a set is tied six games apiece, most tournaments play one game (with special rules) called a *tiebreaker* which determines the winner of the set.

3. To win a *match*, a player must be the first to win either two or three sets depending on the rules of the tournament.

Miscellaneous Games

A few other games mentioned in the book appear only once or twice, so the reader needs to know less about them. Here are brief surveys of these games:

- *Table tennis* is different from tennis. In table tennis, a game is a sequence of *points* with a target number (like 21) necessary for a win. In table tennis, it is usually required that the winner of a game must win by two. That is, if the score in the game is 20-20, then the game will not end until one player has scored two more points than his opponent.

- *Bridge* is a card game with many subtle rules, but its basis is the distribution (dealing) of the entire deck of cards into four 13-card hands. The order in which the player receives his cards is irrelevant to the subsequent play of the game.

- *Scrabble*®* is a word game played with tiles with one letter on each. The game board is a large grid of rectangles that match the dimensions of the tiles. The goal of the game is for the player to interweave her tiles with those already played on the board to form new words. Each letter has a value, and the player's score on a turn is based on the sum of the values of the letters he has used.

Exercises for Appendix A

1. If you received each of the following hands in a game of poker, what would you call it?

    (a) {8 of hearts, jack of spades, jack of clubs, 8 of spades, jack of hearts}

    (b) {2 of diamonds, 7 of diamonds, ace of diamonds, 9 of diamonds, king of diamonds}

    (c) {9 of diamonds, 10 of spades, 6 of spades, 8 of clubs, 7 of spades}

    (d) {6 of hearts, queen of clubs, 3 of hearts, 5 of diamonds, 3 of clubs}

    (e) {ace of spades, ace of diamonds, ace of clubs, 2 of clubs, ace of hearts}

    (f) {3 of hearts, 7 of clubs, 3 of diamonds, 3 of clubs, king of clubs}

    (g) {queen of spades, 4 of diamonds, 4 of hearts, 8 of hearts, queen of diamonds}

2. Choose a card that would complete the hand type given.

    (a) Flush: {_____, king of hearts, jack of hearts, 2 of hearts, 3 of hearts}

    (b) Pair: {_____, 8 of diamonds, ace of hearts, 4 of clubs, queen of clubs}

    (c) Four of a kind: {_____, 2 of clubs, 2 of diamonds, 5 of hearts, 2 of spades}

    (d) Full house: {_____, 9 of clubs, 10 of spades, 10 of hearts, 9 of diamonds}

    (e) Two pair: {_____, queen of diamonds, queen of spades, 7 of clubs, 2 of spades}

    (f) Straight: {_____, king of diamonds, 10 of diamonds, jack of hearts, queen of spades}

    (g) Three of a kind: {_____, 3 of spades, 6 of clubs, 3 of clubs, ace of hearts}

3. For each of the hands in Exercise 2, how many different cards can be put in the blanks?

4. During the first week of the baseball season, John got 5 hits, 3 walks, and 12 outs. What is his batting average?

5. How many consecutive hits would John need to raise his batting average to at least 0.400?

6. A batter's on-base percentage can be calculated using the formula (walks + hits)/(walks + at-bats). Using the information from Exercise 4, what is John's on-base percentage?

7. Kelly exactly doubled her season batting average in one game by getting 3 hits in 4 at bats. How many total hits does she have this season? (HINT: There are two solutions.)

8. Kelly improved her season batting average (which was greater than 0) by exactly 1/5 (i.e., 0.200) by getting 4 hits in 4 at-bats in today's game. How many hits does she now have this season? (HINT: There is a unique answer.)

9. We can represent a best-of-five series between teams A and B as a list that denotes the order in which the games were won. For example, in the series $AABA$, team A won games 1, 2 and 4, and team B won game 3. Which of the following lists of letters does **not** correspond to a real best-of-five series?

(a) *AABBB*   (b) *BABAB*   (c) *AABAA*
(d) *AAABB*   (e) *ABABB*

10. We can represent a tennis game between players *A* and *B* as a list that denotes the order in which the points were won. For example, in the game *AABAA*, player

B won only the third point. Which of the following lists of letters does **not** correspond to a real tennis game? (Remember the tie-breaking rules in tennis games.)
(a) *AABBBABB*   (b) *BABBB*   (c) *AABAABBB*
(d) *AAABBA*   (e) *ABAABBB*

Answers to the Exercises

1. **(a)** Full house
   (b) Flush
   (c) Straight
   (d) Pair
   (e) Four of a kind
   (f) Three of a kind
   (g) Two Pair

2. There is more than one solution for each question. Listed here is one possible answer for each.
   (a) 9 of hearts
   (b) Ace of spades
   (c) 2 of hearts
   (d) 10 of clubs
   (e) 7 of diamonds
   (f) Ace of clubs
   (g) 3 of diamonds

3. **(a)** 9
   (b) 12
   (c) 1
   (d) 4
   (e) 6
   (f) 8
   (g) 2

4. 5/17, which would normally be written as 0.294.

5. 3 consecutive hits

6. 2/5, which would normally be written as 0.400.

7. 4 or 5

8. 10

9. Series (c) should have ended after four games.

10. Game (c) should have ended after the fifth point. Game (e) is not yet over.

B | Matrices and Their Operations

In Sections 6.6 and 7.4 of this book, we present matrices and matrix operations as they pertain to specific concepts in discrete mathematics. This brief appendix serves as a refresher or a source of more practice for those students who are less familiar with these ideas.

Matrix Operations

An $m \times n$ matrix is a rectangular array with m rows and n columns. Objects in a matrix M are referenced by writing $M_{i,j}$ for the entry in Row i, Column j. Traditionally Row 1 is topmost and Column 1 is leftmost. When the objects in the matrix have a well-defined system of arithmetic (as do numbers, for example, since we know how numbers can be added, subtracted, multiplied and divided), then we can define a system of arithmetic on the matrices themselves.

 Example 1 *To compute the sum*

$$\begin{bmatrix} 1 & 1 & 3 & -1 \\ -1 & 2 & 3 & -1 \\ 2 & 0 & 0 & -1 \end{bmatrix} + \begin{bmatrix} 1 & -1 & 3 & 2 \\ -2 & 0 & 2 & -1 \\ -2 & 0 & 0 & 1 \end{bmatrix}$$

we simply add each entry in A with the entry in B in the same position:

$$\begin{bmatrix} 1 & 1 & 3 & -1 \\ -1 & 2 & 3 & -1 \\ 2 & 0 & 0 & -1 \end{bmatrix} + \begin{bmatrix} 1 & -1 & 3 & 2 \\ -2 & 0 & 2 & -1 \\ -2 & 0 & 0 & 1 \end{bmatrix} = \begin{bmatrix} 2 & 0 & 6 & 1 \\ -3 & 2 & 5 & -2 \\ 0 & 0 & 0 & 0 \end{bmatrix}$$

It is easy to turn this into a formal definition using the notation for the individual entries of a matrix.

Definition   If A and B are both $m \times n$ matrices (so A and B have exactly the same size), then the sum $A + B$ is the $m \times n$ matrix C with

$$C_{i,j} = A_{i,j} + B_{i,j}$$

 Example 2 *To compute the product*

$$\begin{bmatrix} 1 & -1 & 3 & \frac{1}{2} \\ -\frac{1}{2} & 0 & \frac{1}{2} & -1 \\ 0 & 0 & 0 & 1 \end{bmatrix} \cdot \begin{bmatrix} 1 & -1 \\ \frac{1}{2} & \frac{1}{3} \\ -\frac{2}{3} & \frac{1}{2} \\ 2 & 0 \end{bmatrix}$$

we take each row of the first matrix times each column of the second matrix using the following "row-column multiplication":

$$(\text{Row 2 of } A) \cdot (\text{Column 1 of } B) = \begin{bmatrix} -\frac{1}{2} & 0 & \frac{1}{2} & -1 \end{bmatrix} \cdot \begin{bmatrix} 1 \\ \frac{1}{2} \\ -\frac{2}{3} \\ 2 \end{bmatrix}$$

$$= \left(-\frac{1}{2}\right)(1) + (0)\left(\frac{1}{2}\right) + \left(\frac{1}{2}\right)\left(-\frac{2}{3}\right) + (-1)(2)$$

$$= -\frac{17}{6}$$

It is a little more complicated (but elegant with the use of summation notation) to turn this into a formal definition.

Definition   If A is an $l \times m$ matrix and B is a $m \times n$ matrix (so the number of **columns** in A is the same as the number of **rows** in B), then the product AB is the $l \times n$ matrix C with

$$C_{i,j} = (A_{i,1})(B_{1,j}) + (A_{i,2})(B_{2,j}) + (A_{i,3})(B_{3,j}) + \cdots + (A_{i,m})(B_{m,j})$$

$$= \sum_{k=1}^{m}(A_{i,k})(B_{k,j})$$

There are special kinds of matrices that are impervious to the effects of addition and multiplication, in much the same way that the numbers 0 and 1 are impervious to the effects of addition and multiplication (respectively) of numbers.

> **Definition** For any positive integer n, the $n \times n$ *identity matrix*, denoted $\mathbf{I}_n$, is the matrix consisting of 0's everywhere except along the main (top-left to bottom-right) diagonal where there are 1's, and the $n \times n$ *zero matrix*, denoted $\mathbf{0}_n$, is the matrix consisting of 0's everywhere. When the size of the matrix is clear, we simply use $\mathbf{I}$ and $\mathbf{0}$ for these matrices.

Example 3 *Here is an example of each type. Show that they perform as expected under multiplication and addition.*

$$\mathbf{I}_3 = \begin{bmatrix} 1 & 0 & 0 \\ 0 & 1 & 0 \\ 0 & 0 & 1 \end{bmatrix} \text{ and } \mathbf{0}_4 = \begin{bmatrix} 0 & 0 & 0 & 0 \\ 0 & 0 & 0 & 0 \\ 0 & 0 & 0 & 0 \\ 0 & 0 & 0 & 0 \end{bmatrix}$$

SOLUTION Let

$$A = \begin{bmatrix} 2 & 0 & -1 \\ 1 & 3 & 7 \\ 0 & 5 & 1 \end{bmatrix} \text{ and } B = \begin{bmatrix} 1 & 10 & 0 & 3 \\ 5 & -2 & 1 & 0 \\ 7 & -1 & 0 & 5 \\ 0 & 4 & 1 & 0 \end{bmatrix}$$

Then

$$A \cdot \mathbf{I} = \begin{bmatrix} 2 & 0 & -1 \\ 1 & 3 & 7 \\ 0 & 5 & 1 \end{bmatrix} = A$$

and

$$B + \mathbf{0} = \begin{bmatrix} 1 & 10 & 0 & 3 \\ 5 & -2 & 1 & 0 \\ 7 & -1 & 0 & 5 \\ 0 & 4 & 1 & 0 \end{bmatrix} = B$$

as expected. □

Matrix Arithmetic with Technology

We will use the TI-83 calculator for the specific instructions on how to perform matrix arithmetic with a handheld calculator. Other calculators function similarly. See your user manual for how to do this on your calculator.

Enter or Edit Matrices

● Press the [MATRIX] button. You will see a list of matrix names, [A], [B], etc., in a vertical list and the menu options [NAMES], [MATH] and [EDIT] across the top of the screen.

● Use the cursor arrows to highlight the [EDIT] menu option and then the matrix you wish to create or edit. For this example, choose the matrix [A].

- On the EDIT screen, you will now see MATRIX[A] followed by the dimensions (rows × columns) of the matrix. For this example, let's designate that A is a 3×4 matrix, and then use the cursor arrows to set the entry values to

$$A = \begin{bmatrix} 1 & 1 & 3 & -1 \\ -1 & 2 & 3 & -1 \\ 2 & 0 & 0 & -1 \end{bmatrix}$$

After editing the matrix, press [QUIT] to return to the main screen.

- Similarly you can EDIT matrix [B] to create

$$B = \begin{bmatrix} -1 & 0 & 1 \\ -1 & 2 & -3 \\ 2 & 0 & 0 \end{bmatrix}$$

After editing the matrix, press [QUIT] to return to the main screen.

Arithmetic Operations

- To perform arithmetic on matrices, you will enter the arithmetic expression normally but you will get the names of the matrices involved from the matrix menu. For example, to find the sum $B + B^2$ for the matrix above, do the following steps from the main screen:

  –Press [MATRIX] and choose the NAMES menu option with matrix [B] selected. Pressing enter returns you to the main menu with the expression [B] displayed.

  –Press '+'.

  –Press [MATRIX], choose the NAMES menu option with matrix [B] selected, and press enter. You will see the main menu with the expression [B] + [B] displayed.

  –Press '^2' to complete the expression $[B] + [B]^2$, and press enter to evaluate it.

- The identity matrix can be quickly referenced on the calculator. For example, to compute the sum $B + \mathbf{I}$, do the following steps:

  –Press [MATRIX] and choose the NAMES menu option with matrix [B] selected. Pressing [ENTER] returns you to the main menu with the expression [B] displayed.

  –Press '+'.

  –Press [MATRIX], choose the MATH menu option and select option 5 (identity). You will see the main menu with the expression "[B]+identity(" displayed. The identity function must be told the dimensions of $\mathbf{I}$ that you want.

  –Press '3)' to complete the expression $[B] + identity(3)$, and press [ENTER] to evaluate it.

- The inverse of an $n \times n$ matrix A is the $n \times n$ matrix denoted A^{-1} with the property that $A \cdot A^{-1} = \mathbf{I}_n$. Finding inverses by hand can be time consuming, so the calculator is a good tool for doing this. For example, to find the inverse of the matrix B above and check that it is correct, we can do the following steps:

  –Press [MATRIX] and choose the NAMES menu option with matrix [B] selected. Pressing [ENTER] returns you to the main menu with the expression [B] displayed.

–Press the 'x^{-1}' button. You will see the main menu with the expression $[B]^{-1}$ displayed.

–Press [ENTER] to evaluate the expression. The matrix shown is supposed to be the inverse of B.

–To check this, press [ANS] button, the multiplication sign '$\times$', and then [MATRIX] $\rightarrow$ [NAMES] $\rightarrow$ [B] $\rightarrow$ [ENTER]. You will see the main menu with the expression $Ans * [B]$ displayed. This means to multiply the previous answer (which was B^{-1}) by the matrix B.

–Press [ENTER] to evaluate the expression. Since the answer is the identity matrix, we know your answer for the inverse of B was correct.

Explore more on the Web.

There are many computer programs, both free and commercial, that will perform matrix operations. On the web page for this book, there is a tool for computing the matrix powers relevant to the Markov chains in Section 6.6 and the composition of relations in Section 7.4.

Exercises for Appendix B

1. Given matrices A and B, compute each of the following or explain why the expression makes no sense:

$$A = \begin{bmatrix} 1 & 1 & 3 & -1 \\ -1 & 2 & 3 & -1 \\ 2 & 0 & 0 & -1 \end{bmatrix} \text{ and } B = \begin{bmatrix} -1 & 0 & 1 \\ -1 & 2 & -3 \\ 2 & 0 & 0 \end{bmatrix}$$

(a) $A + B$

(b) $A \cdot B$

(c) $B \cdot A$

(d) B^2

(e) A^2

2. For each of the following equations, choose your own 2×2 matrices A and B and check if the equation is true.

(a) $A + B = B + A$

(b) $A \cdot B = B \cdot A$

(c) $A \cdot (B + I) = A \cdot B + A$

(d) $A \cdot I = I \cdot A$

3. Each of the following statements is false. Find a counterexample to each.

(a) For all 2×2 matrices A and B, $A \cdot B = \mathbf{0}$ only if $A = \mathbf{0}$ or $B = \mathbf{0}$.

(b) For every 2×2 matrix A, $A^2 = \mathbf{I}$ only if $A = \mathbf{I}$.

(c) For every 2×2 matrix $B \neq \mathbf{0}$, there exists a matrix A such that $A \cdot B = \mathbf{I}$. (We would say that A is the *inverse* of B in this case.)

4. Match each matrix on the left with a matrix on the right so that the pair can be multiplied together to equal $\mathbf{I}$. (Matrices in such a pair are said to be *inverses* of each other.)

(a) $\begin{bmatrix} 2 & 1 \\ 5 & 3 \end{bmatrix}$ (i) $\begin{bmatrix} 3 & -1 \\ -5 & 2 \end{bmatrix}$

(b) $\begin{bmatrix} 2 & 3 \\ 3 & 5 \end{bmatrix}$ (ii) $\begin{bmatrix} 5 & -3 \\ -3 & 2 \end{bmatrix}$

(c) $\begin{bmatrix} 1 & 2 & -1 \\ 2 & 4 & 1 \\ -1 & 2 & 0 \end{bmatrix}$ (iii) $\begin{bmatrix} -\frac{1}{8} & \frac{3}{8} & -\frac{1}{4} \\ -\frac{1}{16} & \frac{3}{16} & \frac{3}{8} \\ \frac{5}{16} & \frac{1}{16} & \frac{1}{8} \end{bmatrix}$

(d) $\begin{bmatrix} 0 & -1 & 3 \\ 2 & 1 & 1 \\ -1 & 2 & 0 \end{bmatrix}$ (iv) $\begin{bmatrix} \frac{1}{6} & \frac{1}{6} & -\frac{1}{2} \\ \frac{1}{12} & \frac{1}{12} & \frac{1}{4} \\ -\frac{2}{3} & \frac{1}{3} & 0 \end{bmatrix}$

5. Finding inverse matrices is a topic we will leave for another course, but there is a case where it is pretty easy to figure out what to do. This exercise will lead you through this discovery process.

(a) Through trial-and-error find the inverse of each of the following matrices. To be more formal, you can multiply each by the generic matrix $\begin{bmatrix} w & x \\ y & z \end{bmatrix}$ and determine what these four variables must be to have the product equal to $\mathbf{I}$.

$$A = \begin{bmatrix} 1 & 4 \\ 0 & 1 \end{bmatrix} \qquad B = \begin{bmatrix} 2 & 4 \\ 0 & 2 \end{bmatrix} \qquad C = \begin{bmatrix} 2 & 4 \\ 0 & 1 \end{bmatrix}$$

(b) In terms of a and c, what is the inverse of the matrix $A = \begin{bmatrix} a & c \\ 0 & a \end{bmatrix}$? What conditions on a and c are necessary for the inverse of A to exist?

(c) In terms of a, b and c, what is the inverse of the matrix $A = \begin{bmatrix} a & c \\ 0 & b \end{bmatrix}$? What conditions on a, b and c are necessary for the inverse of A to exist?

6. The *transpose of a matrix* is the matrix formed by interchanging the rows and columns of the original. We denote the transpose of A by writing A^T. For example,

$$\text{if } A = \begin{bmatrix} 1 & 3 & 4 \\ 0 & -2 & 2 \\ 2 & 1 & 5 \\ 2 & 2 & 7 \end{bmatrix}, \text{ then } A^T = \begin{bmatrix} 1 & 0 & 2 & 2 \\ 3 & -2 & 1 & 2 \\ 4 & 2 & 5 & 7 \end{bmatrix}.$$

Given the matrix A above along with matrices B and C below, compute each of the following, if possible.

$$B = \begin{bmatrix} 2 & 2 & 1 \\ 0 & 1 & 1 \\ 0 & 1 & 2 \end{bmatrix} \text{ and } C = \begin{bmatrix} 0 & 1 & 2 & 0 \\ -1 & 0 & 1 & 0 \\ 0 & 2 & 1 & 3 \\ 3 & 1 & 0 & 2 \end{bmatrix}$$

(a) $A \cdot A^T$
(b) $A^T \cdot A$
(c) $A \cdot A^T + C$
(d) $A^T \cdot A + B$

7. For the matrices A, B, and C from Exercise 6, use a calculator to complete the following computations (if possible). On the TI-83, the transpose operator can be found in the MATH menu of the MATRIX screen.
(a) $(A \cdot A^T) \cdot C$
(b) $(A^T \cdot A) \cdot B$
(c) $(A^T \cdot A)^{-1}$
(d) $(A \cdot A^T)^{-1}$

8. For the matrix B from Exercise 6, use a calculator to complete the following computations (if possible).
(a) $B + B^2 + B^3 + B^4 + B^5$
(b) $(\mathbf{I} - B)^{-1}$
(c) $(B^6 - B) \cdot (B - \mathbf{I})^{-1}$

Answers to the Exercises

1. (a) A and B do not have the same number of columns, so they cannot be added.
(b) The length of a row in A is different from the length of a column in B, so $A \cdot B$ makes no sense.
(c) $B \cdot A = \begin{bmatrix} 1 & -1 & -3 & 0 \\ -9 & 3 & 3 & 2 \\ 2 & 2 & 6 & -2 \end{bmatrix}$
(d) $B^2 = \begin{bmatrix} 3 & 0 & -1 \\ -7 & 4 & -7 \\ -2 & 0 & 2 \end{bmatrix}$
(e) Since A is not a square matrix, then $A \cdot A$ is not defined.

2. Solutions will vary. The answers below use $A = \begin{bmatrix} 1 & -2 \\ -2 & 3 \end{bmatrix}$ and $B = \begin{bmatrix} 4 & 3 \\ 5 & 2 \end{bmatrix}$.
(a) $A + B = B + A = \begin{bmatrix} 5 & 1 \\ 3 & 5 \end{bmatrix}$.
(b) $A \cdot B \neq B \cdot A$, since $A \cdot B = \begin{bmatrix} -6 & -1 \\ 7 & 0 \end{bmatrix}$ and $B \cdot A = \begin{bmatrix} -2 & 1 \\ 1 & -4 \end{bmatrix}$.
(c) $A \cdot (B + I) = \begin{bmatrix} 1 & -2 \\ -2 & 3 \end{bmatrix} \cdot \begin{bmatrix} 5 & 3 \\ 5 & 3 \end{bmatrix} = \begin{bmatrix} -5 & -3 \\ 5 & 3 \end{bmatrix}$,

and $A \cdot B + A = \begin{bmatrix} -6 & -1 \\ 7 & 0 \end{bmatrix} + \begin{bmatrix} 1 & -2 \\ -2 & 3 \end{bmatrix} = \begin{bmatrix} -5 & -3 \\ 5 & 3 \end{bmatrix}$, so they are equal.

(d) $A \cdot I = I \cdot A = \begin{bmatrix} 1 & -2 \\ -2 & 3 \end{bmatrix} = A$

3. (a) If $A = \begin{bmatrix} 0 & 2 \\ 0 & 3 \end{bmatrix}$ and $B = \begin{bmatrix} 1 & -1 \\ 0 & 0 \end{bmatrix}$, then the product $A \cdot B = \begin{bmatrix} 0 & 0 \\ 0 & 0 \end{bmatrix}$.
(b) If $A = \begin{bmatrix} 0 & 2 \\ 1/2 & 0 \end{bmatrix}$, then $A^2 = \begin{bmatrix} 1 & 0 \\ 0 & 1 \end{bmatrix}$.
(c) If $A = \begin{bmatrix} 0 & 1 \\ 0 & 0 \end{bmatrix}$ and $B = \begin{bmatrix} w & x \\ y & z \end{bmatrix}$ (that is, B is entirely generic), then $A \cdot B = \begin{bmatrix} y & z \\ 0 & 0 \end{bmatrix}$, and this product can never be $\mathbf{I}$ since it has all 0's on the bottom row.

4. (a) and (i); (b) and (ii); (c) and (iv); and (d) and (iii)

5. (a) $A^{-1} = \begin{bmatrix} 1 & -4 \\ 0 & 1 \end{bmatrix}$, $B^{-1} = \begin{bmatrix} 1/2 & -1 \\ 0 & 1/2 \end{bmatrix}$, and $C^{-1} = \begin{bmatrix} 1/2 & -2 \\ 0 & 1 \end{bmatrix}$

(b) In general, $\begin{bmatrix} a & c \\ 0 & a \end{bmatrix}^{-1} = \begin{bmatrix} 1/a & -c/a^2 \\ 0 & 1/a \end{bmatrix}$ as long as $a \neq 0$.

(c) In general, $\begin{bmatrix} a & c \\ 0 & b \end{bmatrix}^{-1} = \begin{bmatrix} 1/a & -c/ab \\ 0 & 1/b \end{bmatrix}$ as long as $a \neq 0$ and $b \neq 0$.

6. (a) $A \cdot A^T = \begin{bmatrix} 26 & 2 & 25 & 36 \\ 2 & 8 & 8 & 10 \\ 25 & 8 & 30 & 41 \\ 36 & 10 & 41 & 57 \end{bmatrix}$

(b) $A^T \cdot A = \begin{bmatrix} 9 & 9 & 28 \\ 9 & 18 & 27 \\ 28 & 27 & 94 \end{bmatrix}$

(c) $A \cdot A^T + C = \begin{bmatrix} 26 & 3 & 27 & 36 \\ 1 & 8 & 9 & 10 \\ 25 & 10 & 31 & 44 \\ 39 & 11 & 41 & 59 \end{bmatrix}$

(d) $A^T \cdot A + B = \begin{bmatrix} 10 & 11 & 29 \\ 9 & 19 & 27 \\ 28 & 28 & 96 \end{bmatrix}$

7. (a) $(A \cdot A^T) \cdot C = \begin{bmatrix} 106 & 112 & 79 & 147 \\ 22 & 28 & 20 & 44 \\ 115 & 126 & 88 & 172 \\ 161 & 175 & 123 & 237 \end{bmatrix}$

(b) $(A^T \cdot A) \cdot B = \begin{bmatrix} 18 & 55 & 74 \\ 18 & 63 & 81 \\ 56 & 177 & 243 \end{bmatrix}$

(c) $(A^T \cdot A)^{-1} = \begin{bmatrix} \frac{107}{61} & -\frac{10}{61} & -\frac{29}{61} \\ -\frac{10}{61} & \frac{62}{549} & \frac{1}{61} \\ -\frac{29}{61} & \frac{1}{61} & \frac{9}{61} \end{bmatrix}$

(d) $(A \cdot A^T)^{-1}$ does not exist.

8. (a) $B + B^2 + B^3 + B^4 + B^5 = \begin{bmatrix} 62 & 257 & 331 \\ 0 & 55 & 88 \\ 0 & 88 & 143 \end{bmatrix}$

(b) $(B - \mathbf{I})^{-1} = \begin{bmatrix} 1 & 1 & -2 \\ 0 & -1 & 1 \\ 0 & 1 & 0 \end{bmatrix}$

(c) $(B^6 - B) \cdot (B - \mathbf{I})^{-1} = \begin{bmatrix} 62 & 257 & 331 \\ 0 & 55 & 88 \\ 0 & 88 & 143 \end{bmatrix}$

Selected Answers and Hints for *Discrete Mathematics*

Section 1.1 Exercises

1. **(a)** hsdC **(c)** hsdC

3. **(a)** person 5 is last and 14 is next-to last; **(c)** person 15 is last and 7 is next-to last.

4. The game tree is shown below; 12 of 20 = 60% are five-set matches

7. **(a & b)** In the game tree, the first branch represents the nickel, the second branch the dime, and the third branch the quarter. For example, HTT means the nickel is heads and the dime and quarter are tails. The systematic list appears on the right.

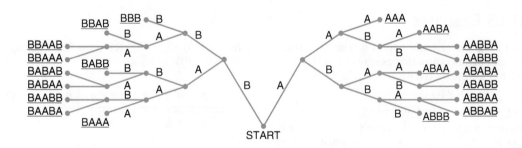

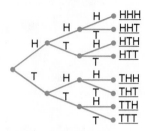

(c) Getting exactly two heads is more likely than getting three heads.

9. Relabeling H's in Problem 7 with F's makes the game tree identical to the one in Problem 8.

11. (See Figure 1-2) All ways of drawing the envelope must start at corner 1 and end at 2, or start at 2 and end at 1.

12. For game **(a)** the move shown in bold guarantees a win. In **(b)**, you will lose if your opponent knows what she is doing.

(a)

X	$\boxed{X}$	X	X
X	X	X	
X		X	X
X	X	X	X

Section 1.2 Exercises

1. (a) $a_n = a_{n-1} + 2$ with $a_1 = 2$, or $a_n = 2n$; **(c)** $a_n = a_{n-1} + (2n-1)$ with $a_1 = 2$, or $a_n = n^2 + 1$; **(e)** $a_n = 2 \cdot a_{n-1}$ with $a_1 = 1$, or $a_n = 2^{n-1}$; **(g)** $a_n = a_{n-1} + 3$ with $a_1 = 2$, or $a_n = 3n - 1$; **(i)** $a_n = 2 \cdot a_{n-1}$ with $a_1 = 5$, or $a_n = 5 \cdot 2^{n-1}$; **(k)** $a_n = a_{n-1} \cdot a_{n-2}$ with $a_1 = 2$ & $a_2 = 5$; **(m)** $a_n = a_{n-1} + 3$ with $a_1 = 3$, or $a_n = 3n$

2. (a) $a_{k-1} = 5k - 7$, and $a_{k+1} = 5k + 3$; **(c)** $a_{k-1} = 2k + 5$, and $a_{k+1} = 2k + 9$; **(e)** $a_{k-1} = 2^{3k-2} - 1$, and $a_{k+1} = 2^{3k+4} - 1$

4. (a) $a_n = 2 + a_{n-1}$ with $a_1 = 2$; **(c)** $a_n = 1 + 2a_{n-1}$ with $a_1 = 2$

5. (a) $a_n = 2n$; **(c)** $a_n = 3 \cdot 2^{n-1} - 1$

6. (a) true; **(c)** not true.

7. (a) $a_n = a_{n-1} + 3$ and $a_1 = 4$; **(c)** $a_n = a_{n-1} + 2$ and $a_1 = 9$; **(e)** $a_n = a_{n-1} + 7$ and $a_1 = 1$

8. (a) $a_n = 5^n$; **(c)** $a_n = 3 \cdot 2^{n-1}$; **(e)** $a_n = 4n - 3$

9. (a) $a_n = \frac{n(n+9)}{2}$

11. $a_n = 99 + 2n$

13. $a_n = (2n - 1)^2$

15. $1 = 2^3 - 3 \cdot 2^2 + 3 \cdot 2 - 1$

17. $a + 3(n - 1)$

19. $s_n = 2 \cdot a_n - 2$

21. (a) 84; **(c)** 36

23. (a) 15 and 30

24. (a) $\sum_{k=1}^{9} 4k$; **(c)** $\sum_{k=1}^{6} 5$

27. (a)   **i.** Anne
     ii. The person with name tag 5.

  (b)   **i.** Player 4
     ii. Players 5, 6, 1, 2, and 3 are left, in that order.
     iii. Player 5 will win.

Section 1.3 Exercises

1. (a) B is telling the truth; **(b)** only A is telling the truth for sure, and we also know that exactly one of B or C is truthful, but we cannot tell which.

3. (a) $\neg p \vee q$ **(b)** $p \wedge q$ **(c)** $p \wedge \neg q$

5. (a) $f \wedge (\neg m)$; **(c)** $(\neg f) \wedge (\neg m)$, or $\neg(f \vee m)$

6. (a) This person is a male math major. **(c)** This person is a female, and she is either over age 30 or a math major.

7. (a) $t \wedge d \wedge h$; **(c)** $(t \vee h) \wedge \neg(t \wedge h)$, or $(t \wedge (\neg h)) \vee (h \wedge (\neg t))$

8. (a) Bill is tall or dark or handsome, but not all three. **(c)** Bill is dark, but not tall and light.

10. (a) $(x > 0) \wedge (y > 0)$; **(c)** $((x > 0) \wedge (y \leq 0)) \vee ((x \leq 0) \wedge (y > 0))$

11. (a)

p	q	$\neg p$	$\neg p \vee q$	$p \wedge (\neg p \vee q)$
T	T	F	T	T
T	F	F	F	F
F	T	T	T	F
F	F	T	T	F

p	q	r	$q \vee r$	$p \wedge (q \vee r)$
T	T	T	T	T
T	T	F	T	T
T	F	T	T	T
(d) T	F	F	F	F
F	T	T	T	F
F	T	F	T	F
F	F	T	T	F
F	F	F	F	F

12. (a) Let h and t represent the same statements as in the example, and p represent "the snackbar makes a profit." Then the statement can be written, $\neg h \vee (t \wedge p)$, and it has the following truth table:

h	t	p	$t \wedge p$	$\neg h \vee (t \wedge p)$
T	T	T	T	T
T	T	F	F	F
T	F	T	F	F
T	F	F	F	F
F	T	T	T	T
F	T	F	F	T
F	F	T	F	T
F	F	F	F	T

(c) Let f represent "the staff is friendly" and p represent "the staff is very well paid." Then the statement can be written, $\neg p \wedge f$, and it has the following truth table:

f	p	$\neg p$	$\neg p \wedge f$
T	T	F	F
T	F	T	T
F	T	F	F
F	F	T	F

Section 1.4 Exercises

1. (a) $(x > 0) \wedge (y > 0)$; **(c)** $(x > 0) \vee (y > 0) \wedge \neg((x > 0) \wedge (y > 0))$

3. (a) $2, 4, 6, 8$
   (b) $6, 7, 8, 9$
   (c) $6, 8$
   (d) 4
   (e) $3, 6, 9$
   (f) $2, 5$

5. (a) $2, 4, 6, 8, 10$
   (b) None of the elements of D (The predicate $R(n)$ is true for all the elements of D.)

13. (a) $h \wedge (\neg t \vee \neg p)$. Everyone is hungry at mealtime, and either everyone is not tired or the snack bar does not make a profit. **(c)** $p \vee \neg f$. The staff is well paid, or they are not friendly.

14. $(b \geq 600) \vee (m \geq 25)$. The negation is $(b < 600) \wedge (m < 25)$.

16. The equivalence follows from the fact that the indicated columns of the truth table are identical.

p	q	$p \wedge q$	$\neg(p \wedge q)$	$\neg p$	$\neg q$	$\neg p \vee \neg q$
T	T	T	F	F	F	F
T	F	F	T	F	T	T
F	T	F	T	T	F	T
F	F	F	T	T	T	T
			↑			↑

18. The equivalence follows from the fact that the indicated columns of the truth table are identical.

p	q	r	$q \vee r$	$p \wedge (q \vee r)$	$p \wedge q$	$p \wedge r$	$(p \wedge q) \vee (p \wedge r)$
T	T	T	T	T	T	T	T
T	T	F	T	T	T	F	T
T	F	T	T	T	F	T	T
T	F	F	F	F	F	F	F
F	T	T	T	F	F	F	F
F	T	F	T	F	F	F	F
F	F	T	T	F	F	F	F
F	F	F	F	F	F	F	F
				↑			↑

21. (a) Not equivalent; **(c)** Not equivalent

23. (a) $p \wedge q$; **(b)** $\neg p$

24. (a) $(p \wedge \neg q) \vee p \equiv p \vee (p \wedge \neg q)$ by Commutative property, and $p \vee (p \wedge \neg q) = p$ by Absorption property

(c) $2, 4, 6, 8, 10, 12$ (that is, all the elements of D)
   (d) 2

7. (a) false $(x = 1)$; **(b)** true; **(c)** false $(x = 1)$; **(d)** false $(x = 8)$

8. (a) $\forall s \in B,\ G(s)$, where B is the set of biology majors and $G(s)$ is the predicate, "s is required to take geometry."

(b) $\exists s \in C,\ \neg M(s)$, where C is the set of computer science majors and $M(s)$ is the predicate, "s minors in mathematics."

(c) $\forall s \in M, \neg B(s)$, where M is the set of math majors and $B(s)$ is the predicate, "s is required to take a business course."

(d) $\exists x \in P, \neg S(x)$, where P is the set of puzzles and $S(x)$ is the predicate, "x has a solution."

10. (a) and **(d)** have the same meaning

   (a) $\forall s \in B, \neg G(s)$

   (b) $\exists s \in B, \neg G(s)$

   (c) $\forall s \in B, G(s)$

   (d) $\forall s \in B, \neg G(s)$

11. (a) Let D be the set of all friends of Alaina, and $P(x)$ the predicate, "x gets tired of playing at the beach." The two forms are "$\neg \exists x \in D, P(x)$" and "$\forall x \in D, \neg P(x)$."

   (b) Let D be the set of all friends of Alaina, and $C(x)$ the predicate, "x dislikes doing cartwheels." The two forms are "$\neg \exists x \in D, C(x)$" and "$\forall x \in D, \neg C(x)$". (or we could let $L(x)$ stand for "x likes doing cartwheels" and write "$\neg \exists x \in D, \neg L(x)$" and "$\forall x \in D, L(x)$").

(c) Let M be the set of all math courses, and $T(x)$ the predicate, "x is too hard for Jennica." The two forms are "$\neg \exists x \in M, T(x)$" and "$\forall x \in M, \neg T(x)$".

(d) Let M be the set of all the meals at the camp, and $B(x)$ the predicate, "x is too bad." The two forms are "$\neg \exists x \in M, B(x)$" and "$\forall x \in M, \neg B(x)$".

13. (a) (i) -12 **(ii)** $0, 2$ **(iii)** any even number is a counterexample

   (b) (i) $-3, 23, 3, -31$ **(ii)** use $3 - 2y$ where y is the number they have chosen

14. (a) $\exists a \in \mathbb{R}, \exists b \in \mathbb{Z}, a^2 + b \notin \mathbb{Z}$; **(c)** $\exists x \in \mathbb{Z}, \forall y \in \mathbb{R}, x \neq 2y$

17. (a) There is an integer x that is at least as big as every integer.

(b) There is a set of integers that does not have a smallest number.

(c) There is a positive integer x such that no matter how the positive integer y is chosen, either y is at least as big as x or y is not a factor of x.

18. (a) original statement; **(b)** negation; **(c)** negation

Section 1.5 Exercises

1. (a) $\neg c \to f$; **(c)** $l \wedge \neg b$; **(e)** $\neg(t \wedge s)$

2. The truth tables follow below. **(a)** False only if you do not attend the concert and you do not get an F; **(c)** True only if I ate lunch but not breakfast; **(e)** False only if this triangle has both a 30 and a 60 degree angle.

(a)

c	f	$\neg c \to f$
T	T	T
T	F	T
F	T	T
F	F	F

(c)

l	b	$l \wedge \neg b$
T	T	F
T	F	T
F	T	F
F	F	F

(e)

t	s	$\neg(t \wedge s)$
T	T	F
T	F	T
F	T	T
F	F	T

4. (a)

p	q	$p \wedge q$	$(p \wedge q) \to q$
T	T	T	T
T	F	F	T
F	T	F	T
F	F	F	T

(c)

p	q	$p \vee q$	$(p \vee q) \to q$
T	T	T	T
T	F	T	F
F	T	T	T
F	F	F	T

(e)

p	q	r	$q \to r$	$p \wedge (q \to r)$
T	T	T	T	T
T	T	F	F	F
T	F	T	T	T
T	F	F	T	T
F	T	T	T	F
F	T	F	F	F
F	F	T	T	F
F	F	F	T	F

5. In each part we highlight the columns that must be checked to see if the statements are equivalent.

p	q	$p \to q$	$q \to p$	
T	T	T	T	
T	F	F	T	
F	T	T	F	
F	F	T	T	

(a) (Not equivalent)

p	q	$p \to q$	$p\wedge (p \to q)$	$p \wedge q$
T	T	T	T	T
T	F	F	F	F
F	T	T	F	F
F	F	T	F	F

(c) (Equivalent)

7. Let b stand for "Alaina likes basketball," s for "Alaina likes swimming," and g for "Alaina likes gymnastics." From the truth tables we see that **(a)**, **(c)**, and **(d)** are equivalent.

(a) $b \to (s \wedge g)$ has truth table:

b	s	g	$s \wedge g$	$b \to (s \wedge g)$
T	T	T	T	T
T	T	F	F	F
T	F	T	F	F
T	F	F	F	F
F	T	T	T	T
F	T	F	F	T
F	F	T	F	T
F	F	F	F	T

(b) $g \to (s \wedge b)$ has truth table:

b	s	g	$s \wedge b$	$g \to (s \wedge b)$
T	T	T	T	T
T	T	F	T	T
T	F	T	F	F
T	F	F	F	T
F	T	T	F	F
F	T	F	F	T
F	F	T	F	F
F	F	F	F	T

(c) $(\neg g \vee \neg s) \to \neg b$ has truth table:

b	s	g	$\neg b$	$\neg s$	$\neg g$	$\neg g \vee \neg s$	$(\neg g \vee \neg s) \to \neg b$
T	T	T	F	F	F	F	T
T	T	F	F	F	T	T	F
T	F	T	F	T	F	T	F
T	F	F	F	T	T	T	F
F	T	T	T	F	F	F	T
F	T	F	T	F	T	T	T
F	F	T	T	T	F	T	T
F	F	F	T	T	T	T	T

(d) $\neg b \vee (s \wedge g)$ has truth table:

b	s	g	$\neg b$	$s \wedge g$	$\neg b \vee (s \wedge g)$
T	T	T	F	T	T
T	T	F	F	F	F
T	F	T	F	F	F
T	F	F	F	F	F
F	T	T	T	T	T
F	T	F	T	F	T
F	F	T	T	F	T
F	F	F	T	F	T

8. (a) $(x > 0) \to (y > 0)$; **(b)** $(x > 0) \to (y \le 0)$; **(c)** $(x \le 0) \to (y > 0)$; **(d)** $(x \le 0) \to (y \le 0)$

10. (b) is the only statement that is not true of all elements in D. A counterexample is $x = 3$.

12. (a) $P(n)$ is "n is even," and $Q(n)$ is "$n^2 + n$ is even." **(b)** $P(n)$ is "n is a multiple of 5," and $Q(n)$ is "n has ones' digit of 5." **(c)** $P(n)$ is "n is prime," and $Q(n)$ is "$2^n - 1$ is prime."

14. Use $\mathbb{Z}$ to represent the set of all integers, and X for the set of positive real numbers. Let E_3 represent, "ends in the digit 3," E_5 represent, "ends in the digit 5," D_3 represent, "is evenly divisible by 3," and D_5 represent, "is evenly divisible by 5." For part (d), let $L(x)$ represent "$x < \sqrt{2}$," and $G(x)$ represent, "$2/x > \sqrt{2}$."

(a) $\forall n \in \mathbb{Z}, E_5(n) \to D_5(n)$

(b) $\forall m \in \mathbb{Z}, E_3(m) \to D_3(m)$

(c) $\forall n \in \mathbb{Z}, D_5(n) \to D_3(n^2 - 1)$

(d) $\forall x \in X, L(x) \to G(x)$

15. Parts **(a)** and **(d)** are true. For **(b)**, 13 is a counterexample. For **(c)**, 15 is a counterexample. Notice that part **(d)** would be false if we had quantified over "non-zero real numbers" with counterexample -1.

16. Using the same sets and predicates as in Exercise 14, the negations are as follows.

(a) $\exists n \in \mathbb{Z}, E_5(n) \wedge \neg D_5(n)$

(b) $\exists m \in \mathbb{Z}, E_3(m) \wedge \neg D_3(m)$

(c) $\exists n \in \mathbb{Z}, D_5(n) \wedge \neg D_3(n^2 - 1)$

(d) $\exists x \in X, L(x) \wedge \neg G(x)$

20. (a) $n = -2$; **(c)** $x = 3\pi/2$

22. There are many equivalent answers. We present versions consistent with the way we have written statements in the previous sections.

(a) For every positive integer n, if n is even, then $\frac{1}{n} \leq 1$.

(b) For all positive integers a and b, if $a - b$ is odd, then $a^2 \neq 2b^2$.

(c) For all integers a and b, if a and b are positive, then $a/b \neq 1 + b/a$.

25. (a) If you do not get an F for the course, then you will attend the concert. **(c)** If you will not be hungry, then you eat your breakfast.

26. (a) If you will get an F for the course, then you do not attend the concert. **(c)** If you will be hungry, then you do not eat your breakfast.

27. (a) If you attend the concert, then you will not get an F for the course. **(c)** If you eat your breakfast, you will not be hungry.

28. (a) Both are telling the truth.

	p	q	*A says* $\neg q \to \neg p$	*B says* $\neg q \to \neg p$
★	T	T	T	T
	T	F	F	F
	F	T	T	T
	F	F	T	T

30. (a) $\forall s \in S, C(s) \to D(s)$. **(c)** $\forall s \in S, C(s) \to \neg D(s)$. **(e)** $\forall s \in S, C(s) \to \neg D(s)$

Section 1.6 Exercises

1. No. The poor child has fallen for the *inverse fallacy*.

2. (a) inverse fallacy; **(b)** converse fallacy; **(c)** inverse fallacy; **(d)** *modus ponens*; **(e)** converse fallacy.

4. (a) If you read the book, then you will pass the course. **(b)** If you will pass the course, then you must read the book. **(c)** If you will pass the course, then you must read the book.

6. (a) neither; **(b)** tautology; **(c)** tautology; **(d)** neither.

8. (Your examples will vary.) **(a)** John ordered pizza and Jill ordered pizza. Conclusion: John ordered pizza.

p	q	$p \wedge q$	$(p \wedge q) \to p$
T	T	T	T
T	F	F	T
F	T	F	T
F	F	F	T

(c) Bill got a haircut. Sue went shopping. Conclusion: Bill got a haircut and Sue went shopping.

p	q	$p \wedge q$	$(p \wedge q) \to (p \wedge q)$
T	T	T	T
T	F	F	T
F	T	F	T
F	F	F	T

(e) If Keith attends the party, Meg will attend. If Meg attends the party, Nancy will attend. Conclusion: If Keith attends the party, Nancy will attend. The truth table shows the following statements: **(i)** $p \to q$; **(ii)** $q \to r$; **(iii)** $(p \to q) \wedge (q \to r)$; **(iv)** $p \to r$; **(v)** $((p \to q) \wedge (q \to r)) \to (p \to r)$

p	q	r	(i)	(ii)	(iii)	(iv)	(v)
T	T	T	T	T	T	T	T
T	T	F	T	F	F	F	T
T	F	T	F	T	F	T	T
T	F	F	F	T	F	F	T
F	T	T	T	T	T	T	T
F	T	F	T	F	F	T	T
F	F	T	T	T	T	T	T
F	F	F	T	T	T	T	T

9. (a) when p and q are both false, then the premises are all true while the conclusion of the argument is false; **(c)** This is exactly like 8(e).

10. (a) N is "Newton is considered a great mathematician," L is "Leibniz work is ignored," and C is "Calculus is the center)))" piece." This argument has the structure: $(\neg N \wedge \neg L) \to \neg C, N \to L \therefore C \wedge \neg L$. The truth table shows the following statements: **(i)** $\neg N \wedge \neg L$; **(ii)** $(\neg N \wedge \neg L) \to \neg C$; **(iii)** $N \to L$; **(iv)** $C \wedge \neg L$; and **(v)** $(((\neg N \wedge \neg L) \to \neg C) \wedge (N \to L)) \to (C \wedge \neg L)$

N	L	C	(i)	(ii)	(iii)	(iv)	(v)
T	T	T	F	T	T	F	F
T	T	F	F	T	T	F	F
T	F	T	F	T	F	T	T
T	F	F	F	T	F	F	T
F	T	T	F	T	T	F	F
F	T	F	F	T	T	F	F
F	F	T	T	F	T	T	T
F	F	F	T	T	T	F	F

The first row of the table shows that if N, L and C are all true, then the hypotheses of the argument are true while the conclusion is false.

(c) r is "I have a good round of golf," c is "the wind is calm," and d is "the weather is dry." This argument has the structure, $r \to (c \vee d)$, $c \wedge d \therefore r$. The truth table shows the following statements: **(i)** $c \vee d$; **(ii)** $r \to (c \vee d)$; **(iii)** $c \wedge d$; and **(iv)** $((r \to (c \vee d)) \wedge (c \wedge d)) \to r$.

r	c	d	(i)	(ii)	(iii)	(iv)
T	T	T	T	T	T	T
T	T	F	T	T	F	T
T	F	T	T	T	F	T
T	F	F	F	F	F	T
F	T	T	T	T	T	F
F	T	F	T	T	F	T
F	F	T	T	T	F	T
F	F	F	F	T	F	T

The fifth row of the table shows that if r is false while c and d are both true, then the hypotheses of the argument are true while the conclusion is false.

11. (a) g is "we take our gas heater," e is "we take extra blankets," and m is "we take our air mattress." This argument has the structure: $g \to e$, $\neg e \to \neg m$, $\therefore (m \vee g) \to e$. The truth table shows the following statements: **(i)** $g \to e$; **(ii)** $\neg e \to \neg m$; **(iii)** $m \vee g$; **(iv)** $(m \vee g) \to e$; and **(v)** $((g \to e) \wedge (\neg e \to \neg m)) \to ((m \vee g) \to e)$.

g	e	m	(i)	(ii)	(iii)	(iv)	(v)
T	T	T	T	T	T	T	T
T	T	F	T	T	T	T	T
T	F	T	F	F	T	F	T
T	F	F	F	T	T	F	T
F	T	T	T	T	T	T	T
F	T	F	T	T	F	T	T
F	F	T	T	F	T	F	T
F	F	F	T	T	F	T	T

Section 2.1 Exercises

1. **(a)** 12 **(c)** -8 **(e)** $n^2 + 2n - 3$ **(g)** -12
2. **(a)** Counterexample $m = 3, n = 2$
   (b) Seems to be true
   (c) Counterexample $n = 9$
   (d) Counterexample $n = 3$
   (e) Seems to be true
   (f) Seems to be true
   (g) Counterexample $n = 6$
   (h) Counterexample $n = 4$
4. **(a)** Dear READER,

   If you choose any even number and call it n, then I can show you that $3n$ is also even. Here is how. Since n is even, we agree it can be written as $n = 2k$ for some integer k. Then if I calculate $3n$, I get $3n = 3(2k) = 6k = 2(3k)$. Now $3k$ is also an integer, so this takes the form of 2 times some integer, which is exactly what we mean by saying it is even.

   Your friend, AUTHOR

   (c) Dear READER,

   Let's call your even number n, and use our agreement on what even means to write $n = 2k$ (k being some integer). Now calculate $n + 1 = 2k + 1$. By our agreement on the meaning of odd, this shows that $n + 1$ is odd.

   Your friend, AUTHOR

   (e) Dear READER,

   This one is a little tougher. Let's start by naming your odd number n, and writing $n = 2k + 1$ (k some integer). (This is what n being odd means.) If we calculate $n^3 - n$, we get $(2k + 1)^3 - (2k + 1) = 8k^3 + 12k^2 + 6k + 1 - 2k - 1 = 8k^3 + 12k^2 + 4k = 4(2k^3 + 3k^2 + k)$. Since we have written $n^3 - n$ as 4 times the integer $2k^3 + 3k^2 + k$, this shows that $n^3 - n$ is divisible by 4.

   Your friend, AUTHOR

5. (a) Proposition. The sum of two odd number is even.
Proof. Let m and n be odd numbers. This means there is an integer K such that $m = 2K + 1$ and there is an integer L such that $m = 2L + 1$, and so

$$m + n = \underline{(2K + 1)} + \underline{(2L + 1)}$$
$$= 2 \cdot \underline{(K + L + 1)}$$

Since $\underline{K + L + 1}$ is an integer, this means that $m + n$ is even. ∎

(b) Proposition. If n is even, then n^2 is even.
Proof. Let n be an even number. This means that we can write $n = \underline{2m}$ for some integer m. This in turn means that $n^2 = 4m^2 = 2 \cdot \underline{(2m^2)}$, so n^2 is even, because $\underline{2m^2}$ is an integer. ∎

(c) Proposition. Every odd perfect square can be written in the form $4k + 1$ where k is an integer.
Proof. Let s be an odd perfect square. So $s = n^2$ for some integer n, and n^2 is odd. By the contrapositive of the previous exericse, n is odd. Since n is odd, there is an integer L such that $n = \underline{2L + 1}$. This means

$$s = n^2 = \underline{(2L + 1)^2}$$
$$= \underline{4L^2 + 4L + 1}$$

So $s = 4\underline{(L^2 + L)} + 1$, where $\underline{L^2 + L}$ is an integer, as desired. ∎

	x	y	K	L	$K + L$	$2 \cdot (K + L)$	$x + y$
6. (a)	18	20	9	10	19	38	38
(c)	18	18	9	9	18	36	36

9. (a) Dear Reader,

Take an example of two odd integers, and let's agree to call your integers x and y. Since x is odd, we know there must be an integer K that makes $x = 2K + 1$ true, and likewise we know we can write $y = 2L + 1$ for some integer L. Now let's see what happens when we multiply your two integers:

$$x \cdot y = (2K + 1) \cdot (2L + 1)$$
$$= 4KL + 2L + 2K + 1$$
$$= 2 \cdot (2KL + L + K) + 1$$

Since $2KL + L + K$ is an integer, we can see that we have written $x \cdot y$ as $2 \cdot (\text{integer}) + 1$, which shows us that $x \cdot y$ is odd. ∎

Your friend, AUTHOR

(c) Dear Reader,

Take an example of an even integer and an integer that is divisible by 3, and let's agree to call your integers x (for the even one) and y (for the one that is divisible by 3). Since x is even, we know there must be an integer K that makes $x = 2K$ true. Similarly, since y is divisible by 3, we know we can write $y = 3L$ for some integer L. Now let's see what happens when we multiply your two integers:

$$x \cdot y = (2K)(3L)$$
$$= 6(KL)$$

Since KL is an integer, we can see that we have written $x \cdot y$ as $6 \cdot (\text{integer})$, which is exactly what it means to say that $x \cdot y$ is divisible by 6. ∎

Your friend, AUTHOR

11. (a) *Proof.* Let x and y be odd integers. Then there is an integer K such that $x = 2K + 1$ and there is an integer L such that $y = 2L + 1$. It follows then that

$$x \cdot y = (2K + 1) \cdot (2L + 1)$$
$$= 4KL + 2L + 2K + 1$$
$$= 2 \cdot (2KL + L + K) + 1$$

Since $2KL + L + K$ is an integer, we can conclude that $x \cdot y$ is odd. ∎

12. (a) A counterexample to **(i)** requires that my brother and I both root for the Braves while a counterexample to **(ii)** requires that I do not root for the Braves while my brother does.

(c) A counterexample to each would be a student who does not do math problems every night but is good at math.

13. (a) Proposition. If $m = 0$ and $n = 0$, then $m^2 + n^2 = 0$.
Proof. If $m = 0$ and $n = 0$, then $m^2 + n^2 = 0^2 + 0^2 = 0$. ∎

(c) Proposition. If m and n are odd integers, then $m + n$ is an even integer.
Proof. Let odd integers m and n be given. This means $n = 2K + 1$ for some integer K, and $m = 2L + 1$ for some integer L. So

$$m + n = (2K + 1) + (2L + 1)$$
$$= 2K + 2L + 2$$
$$= 2(K + L + 1)$$

which means $m + n$ can be written as twice an integer, hence $m + n$ is even. ∎

Section 2.2 Exercises

1. **(a)** $73 = 6 \cdot 12 + 1$; **(c)** $-1234 = 15 \cdot (-83) + 11$; **(e)** $1000 = 7 \cdot 142 + 6$

2. **(a)** 1; **(c)** 11; **(e)** 6

3. **(a)** 1; **(c)** 6; **(e)** 5; **(g)** 1

6. **(a)** ... for some integer m. From this we can conclude that $c = m \cdot b = m \cdot (3 \cdot k) = 3 \cdot (mk)$. Since mk is an integer, ...
   (b) ... so that $n = 2m + 1$. This means $n^3 = (2m + 1)^3 = 8m^3 + 12m^2 + 6m + 1 = 2(4m^3 + 6m^2 + 3m) + 1$. Since $4m^3 + 6m^2 + 3m$ is an integer, we can see that n^3 is 1 more than twice an integer. Hence n^3 is odd, completing the proof.
   (c) Fill in the blanks as follows: $3k + 1$, $4k + 1$, $4k + 1$, and $4k + 1$.

7. **(a)** *Proof.* Let integers a, b and c be given such that a divides b and a divides c. This means that $b = k \cdot a$ for some integer k and $c = l \cdot a$ for some integer l, so $b + c = k \cdot a + l \cdot a = (k + l) \cdot a$. Since $k + l$ is an integer, this means that a divides $b + c$. ∎
   (c) *Proof.* Let integers a, b, c and d be given such that a divides b and c divides d. This means that $b = k \cdot a$ for some integer k and $d = l \cdot c$ for some integer l, so $b \cdot d = (k \cdot a) \cdot (l \cdot c) = (kl) \cdot (ac)$. Since kl is an integer, this means that ac divides bd. ∎
   (e) *Proof.* Let an integer n be given such that 9 divides $10^{n-1} - 1$. This means that $10^{n-1} - 1 = 9k$ for some integer k. We can show algebraically that

   $$10^n - 1 = 10 \cdot (10^{n-1} - 1) + 9$$
   $$= 10 \cdot (9k) + 9$$
   $$= 9 \cdot (10k + 1)$$

   Since $10k + 1$ is an integer, this means that 9 divides $10^n - 1$. ∎

8. Fill in the blanks with: a is rational; b is rational; $xw + yz$; $xw + yz$; and $w \neq 0$ and $y \neq 0$.

9. *Hint.* This begins exactly as in Exercise 8, but instead of calculating $a + b$, you calculate $a - b$.

11. *Proof.* Let x and y be rational numbers. Then there are integers a and b with $b \neq 0$ such that $x = \frac{a}{b}$, and there are integers c and d with $d \neq 0$ such that $y = \frac{c}{d}$. So $\frac{x+y}{2} = \frac{\frac{a}{b} + \frac{c}{d}}{2} = \frac{ad + bc}{2bd}$. Since $ad + bc$ and $2bd$ are integers and $2bd \neq 0$, this shows that $\frac{x+y}{2}$ is a rational number. ∎

13. **(a)** **(i)** *Proof.* Let n be given such that a_{n-1} is even. This means that $a_{n-1} = 2k$ for some integer k. Therefore, $a_n = a_{n-1} + 2n = 2k + 2n = 2(k + n)$, which is even. **(ii)** $10, 14, 20, 28, 38$. Every term is even. **(iii)**

7, 11, 17, 25, 35. It is vacuously true that every even term is followed by an even term.

14. **(a)** Fill in the blanks as follows: "... the given n is not divisible by 3"; $(3q + 1)^2 = 9q^2 + 6q + 1$; $3q^2 + 2q$; 1; $(3q + 2)^2 = 9q^2 + 12q + 4$; $3q^2 + 4q + 1$; 1.

16. *Proof.* Let an integer n which is not divisible by 3 be given. By the Division Theorem, when any integer is divided by 3 it leaves a remainder of 0, 1 or 2. That is, one of the following cases must be true:

    ● **Case 1:** It might be that $n = 3q$ for some integer q. However, for this particular integer n, we know this case does not happen, because the given n is not divisible by 3.

    ● **Case 2:** It might be that $n = 3q + 1$ for some integer q. In this case,

    $$n^2 + 2 = (3q + 1)^2 + 2$$
    $$= (9q^2 + 6q + 1) + 2$$
    $$= 3 \cdot (3q^2 + 2q + 1)$$

    ● **Case 3:** It might be that $n = 3q + 2$ for some integer q. In this case,

    $$n^2 + 2 = (3q + 2)^2 + 2$$
    $$= (9q^2 + 12q + 4) + 2$$
    $$= 3 \cdot (3q^2 + 4q + 2)$$

    Thus, in every case that satisfies the hypothesis, we see that $n^2 + 2$ is divisible by 3, completing the proof. ∎

18. *Outline of proof.* First, prove this **Proposition:** For any n, $n^3 + 2n$ is divisible by 3. (*Hint:* Write $n^3 + 2n = n(n^2 + 2)$ and consider two cases: either n is divisible by 3, or it is not.) Then, use this proposition together with previous exercises to make the desired conclusion.

20. The sum of any three consecutive perfect cubes* is divisible by 9.
    Proof. Call the three consecutive perfect cubes, $(n - 1)^3$, n^3 and $(n + 1)^3$. Then

    $$(n - 1)^3 + n^3 + (n + 1)^3 = 3n^3 + 6n$$
    $$= 3n(n^2 + 2)$$

    Now n is either divisible by 3 or it is not.

    ● **Case 1:** If n is divisible by 3, then $3n$ is divisible by 9, and hence $3n(n^2 + 2)$ is divisible by 9.

    ● **Case 2:** If n is not divisible by 3, by Exercise 16 we know that $n^2 + 2$ is divisible by 3, and hence

---

*An integer x is a perfect cube if $x = y^3$ for some integer y.

$3\left(n^2+2\right)$ is divisible by 9. It follows then that $3n(n^2+2)$ is divisible by 9.

In either case, $3n(n^2+2)$ is divisible by 9. ∎

22. For all integers n, $n^5 - n$ is divisible by 5.

Proof. Let an integer n be given. By the Division Theorem, when n is divided by 5, it leaves a remainder of 0, 1, 2, 3, or 4.

● **Case 1:** It might be that $n = 5q$ for some integer q. In this case,

$$n^5 - n = 3125q^5 - 5q$$
$$= 5 \cdot (625q^5 - q)$$

● **Case 2:** It might be that $n = 5q + 1$ for some integer q. In this case,

$$n^5 - n = (5q+1)^5 - (5q+1)$$
$$= 3125q^5 + 3125q^4 + 1250q^3 + 250q^2 + 20q$$
$$= 5 \cdot (625q^5 + 625q^4 + 250q^3 + 50q^2 + 4q)$$

● **Case 3:** It might be that $n = 5q + 2$ for some integer q. In this case,

$$n^5 - n = (5q+2)^5 - (5q+2)$$
$$= 3125q^5 + 6250q^4 + 5000q^3 + 2000q^2$$
$$+ 395q + 30$$
$$= 5 \cdot (625q^5 + 1250q^4 + 1000q^3 + 400q^2$$
$$+ 79q + 6)$$

● **Case 4:** It might be that $n = 5q + 3$ for some integer q. In this case,

$$n^5 - n = (5q+3)^5 - (5q+3)$$
$$= 3125q^5 + 9375q^4 + 11\,250q^3 + 6750q^2$$
$$+ 2020q + 240$$
$$= 5 \cdot (625q^5 + 1875q^4 + 2250q^3 + 1350q^2$$
$$+ 404q + 48)$$

● **Case 5:** It might be that $n = 5q + 4$ for some integer q. In this case,

$$n^5 - n = (5q+4)^5 - (5q+4)$$
$$= 3125q^5 + 12500q^4 + 20000q^3 + 16000q^2$$
$$+ 6395q + 1020$$
$$= 5 \cdot (625q^5 + 2500q^4 + 4000q^3$$
$$+ 3200q^2 + 1279q + 204)$$

Hence, in every possible case, $n^5 - n$ is divisible by 5. ∎

24. Fill in the blanks with: "this is the same as $5(A - C) = D - B$"; -4; 4; 0; and "$5(A - C) = D - B = 0$."

26. **(a)** The contrapositive of the statement, "If n^2 is even, then so is n" is the statement, "If n is odd, then n^2 is odd."

Proof. Let n be a given odd integer. Since n is odd, we can write $n = 2k + 1$ where k is an integer. Then $n^2 = (2k+1)^2 = 4k^2 + 4k + 1 = 2(2k^2 + 2k) + 1$. This shows that n^2 is odd. ∎

(b) This was proven in Proposition 7.

(c) **Claim.** In Pythagorean triples $a^2 + b^2 = c^2$, if c is even, then so are both a and b.

Proof. Let integers a, b, and c, with $a^2 + b^2 = c^2$ and c being even, be given. Since c is even, we can write $c = 2k$ for some integer k. There are two possibilities for a — either it is even or it is odd. We consider both cases:

● **Case 1.** a is even. Then $a = 2m$ for some integer m. In this case, the equation $a^2 + b^2 = c^2$ can be rewritten $b^2 = c^2 - a^2 = 4k^2 - 4m^2 = 2(2k^2 - 2m^2)$, from which it follows that b^2 is even, and thus b is even by part (a).

● **Case 2.** a is odd. Then $a = 2m + 1$ for some integer m. In this case, the equation $a^2 + b^2 = c^2$ can be rewritten $b^2 = c^2 - a^2 = 4k^2 - (4m^2 + 4m + 1) = 4k^2 - 4m^2 - 4m - 1 = 4(k^2 - m^2 - m - 1) + 3$. By part (b), it is impossible for a perfect square to be written in the form $4M + 3$ where M is an integer.

Since Case 2 is impossible, it must be the case that a is even, from which it follows that b is also even. ∎

28. *Outline of proof.* Let n be a perfect square integer. Write $n = m^2$, and write $m = 3q + r$. Consider the three possible cases ($r = 0$, $r = 1$, and $r = 2$).

30. *Outline of proof.* Let an integer c be given. By the Division Theorem, dividing c by 4 leaves a remainder of 0, 1, 2, or 3.

● **Case 1:** If $c = 4q$, then show that $c + 2$ is not a perfect square, using Proposition 7.

● **Case 2:** If $c = 4q + 1$, then show that $c + 2$ is not a perfect square

● **Case 3:** If $c = 4q + 2$, then show that $2c + 2$ is not a perfect square.

● **Case 4:** If $c = 4q + 3$, then show that $7c + 2$ is not a perfect square.

In every case, at least one of the values $c + 2$, $2c + 2$, or $7c + 2$ is not a perfect square.

32. *Outline of proof.* We should first establish that (*) every perfect square is of the form $8k$, $8k + 1$, or $8k + 4$. This can be done by extending the reasoning in Proposition 7. We now consider three perfect squares a^2, b^2 and c^2, and list all the possibilities for their sum, in the form $8M+??$. The only form not possible is $8M + 7$.

Section 2.3 Exercises

1. (a) $P(1)$ is "2 is prime," $P(2)$ is "5 is prime," and $P(12)$ is "145 is prime." Only $P(1)$ and $P(2)$ true. $P(m-1)$ is "$(m-1)^2 + 1$ is prime."

(b) $L(1)$ is "$1 < 2$," $L(2)$ is "$4 < 4$," $L(3)$ is "$9 < 8$," $L(4)$ is "$16 < 16$," $L(5)$ is "$25 < 32$," and $L(6)$ is "$36 < 64$." Of these, $L(1)$, $L(5)$ and $L(6)$ are true. $L(m-1)$ is "$(m-1)^2 < 2^{m-1}$."

(c) $S(1)$ is "$1^2 = \frac{1(2)(3)}{6}$," $S(2)$ is "$1^2 + 2^2 = \frac{2(3)(5)}{6}$," $S(3)$ is "$1^2 + 2^2 + 3^2 = \frac{3(4)(7)}{6}$," $L(4)$ is "$1^2 + 2^2 + 3^2 + 4^2 = \frac{4(5)(9)}{6}$," $L(5)$ is "$1^2 + 2^2 + 3^2 + 4^2 + 5^2 = \frac{5(6)(11)}{6}$," and $L(6)$ is "$1^2 + 2^2 + 3^2 + 4^2 + 5^2 + 6^2 = \frac{6(7)(13)}{6}$." Each of these is true. $S(m-1)$ is "$\sum_{i=1}^{m-1} i^2 = \frac{(m-1)(m)(2m-1)}{6}$."

3. We give either the outline of the proof or the formal proof (both for part (a).

(a) *Outline of proof.* The table shows the verification for $n = 1$, $n = 2$, $n = 3$, and $n = 4$. It also shows the last row checked ($n = m - 1$) and the next row to be checked ($n = m$).

n	a_n (recursive formula)	closed formula	equal?
1	1	$4 \cdot 1 - 3 = 1$	yes
2	$1 + 4 = 5$	$4 \cdot 2 - 3 = 5$	yes
3	$5 + 4 = 9$	$4 \cdot 3 - 3 = 9$	yes
4	$9 + 4 = 13$	$4 \cdot 4 - 3 = 13$	yes
...	...	...	...
$m-1$	$a_{m-2} + 4$ $= 4m - 7$	$4(m-1) - 3$ $= 4m - 7$	yes
m	$a_{m-1} + 4$	$4m - 3$	???

Now we simplify the recursive formula for a_m: $a_m = a_{m-1} + 4 = (4m - 7) + 4 = 4m - 3$, and observe that this is equal to the desired closed formula. *Formal proof.* Let $P(n)$ be the statement, "$a_n = 4n - 3$." Since $P(1)$ is the statement "$a_1 = 4 - 3$," we know that it is true from the given definition of $a_1 = 1$. Now let the integer $m \geq 2$ be given such

that $P(1), P(2), \ldots, P(m-1)$ have already been checked to be true. In particular, the last statement we checked was $a_{m-1} = 4(m-1) - 3$." It now follows that

$$a_m = a_{m-1} + 4 \text{ (by the recurrence relation for } a)$$
$$= (4(m-1) - 3) + 4 \text{ (by } P(m-1))$$
$$= 4m - 4 - 3 + 4$$
$$= 4m - 3$$

Thus we have shown that $a_m = 4m - 3$, which is precisely statement $P(m)$. ∎

(c) *Outline of proof.* The table shows the verification for $n = 1$, $n = 2$, $n = 3$, and $n = 4$. It also shows the last row checked ($n = m - 1$) and the next row to be checked ($n = m$).

n	a_n (recursive formula)	closed formula	equal?
1	1	$\frac{1(2)(3)}{6} = 1$	yes
2	$1 + 2^2 = 5$	$\frac{2(3)(5)}{6} = 5$	yes
3	$5 + 3^2 = 14$	$\frac{3(4)(7)}{6} = 14$	yes
4	$14 + 4^2 = 30$	$\frac{4(5)(9)}{6} = 30$	yes
...	...	...	...
$m-1$	$a_{m-2} + (m-1)^2$ $= \frac{(m-1)(m)(2m-1)}{6}$	$\frac{(m-1)(m)(2m-1)}{6}$	yes
m	$a_{m-1} + m^2$	$\frac{m(m+1)(2m+1)}{6}$	???

Now we simplify the recursive formula for a_m: $a_m = a_{m-1} + m^2 = \frac{(m-1)(m)(2m-1)}{6} + m^2 = \frac{(m-1)(m)(2m-1)+6m^2}{6} = \frac{m[(m-1)(2m-1)+6m]}{6} = \frac{m(2m^2-3m+1+6m)}{6} = \frac{m(2m^2+3m+1)}{6} = \frac{m(m+1)(2m+1)}{6}$, and observe that this is equal to the desired closed formula.

(d) *Outline of proof.* The table shows the verification for $n = 1$, $n = 2$, $n = 3$, and $n = 4$. It also shows the last row checked ($n = m - 1$) and the next row to be checked ($n = m$).

n	a_n (recursive formula)	closed formula	equal?
1	1	$2^1 - 1 = 1$	yes
2	$2 \cdot 1 + 1 = 3$	$2^2 - 1 = 3$	yes
3	$2 \cdot 3 + 1 = 7$	$2^3 - 1 = 7$	yes
4	$2 \cdot 7 + 1 = 15$	$2^4 - 1 = 15$	yes
...	...	...	...
$m-1$	$2 \cdot a_{m-2} + 1$ $= 2^{m-1} - 1$	$2^{m-1} - 1$	yes
m	$2 \cdot a_{m-1} + 1$	$2^m - 1$	???

Now we simplify the recursive formula for a_m: $a_m = 2 \cdot a_{m-1} + 1 = 2 \cdot (2^{m-1} - 1) + 1 = 2 \cdot 2^{m-1} - 2 + 1 = 2^m - 1$, and observe that this is equal to the desired closed formula.

6. $S(1)$ is "$\frac{1(2)}{2} = \frac{1(2)(3)}{6}$," $S(2)$ is "$\frac{1(2)}{2} + \frac{2(3)}{2} = \frac{2(3)(4)}{6}$," $S(3)$ is "$\frac{1(2)}{2} + \frac{2(3)}{2} + \frac{3(4)}{2} = \frac{3(4)(5)}{6}$," and $S(4)$ is "$\frac{1(2)}{2} + \frac{2(3)}{2} + \frac{3(4)}{2} + \frac{4(5)}{2} = \frac{4(5)(6)}{6}$." Each of these is true. Without sigma notation, $S(m)$ can be written,

$$\frac{1(2)}{2} + \frac{2(3)}{2} + \frac{3(4)}{2} + \cdots + \frac{m(m+1)}{2}$$
$$= \frac{m(m+1)(m+2)}{6}$$

and $S(m - 1)$ can be written,

$$\frac{1(2)}{2} + \frac{2(3)}{2} + \frac{3(4)}{2} + \cdots + \frac{(m-1)(m)}{2}$$
$$= \frac{(m-1)(m)(m+1)}{6}$$

8. (a) *Outline of proof.* The table shows the verification for $n = 1$, $n = 2$, $n = 3$, and $n = 4$. It also shows the last row checked ($n = m - 1$) and the next row to be checked ($n = m$).

n	summation	closed formula	equal?
1	1	$1^2 = 1$	yes
2	$1 + 3 = 4$	$2^2 = 4$	yes
3	$1 + 3 + 5 = 9$	$3^2 = 9$	yes
4	$1 + 3 + 5 + 7$ $= 16$	$4^2 = 16$	yes
...			...
$m-1$	$\sum_{i=1}^{m-1}(2i - 1)$ $= (m-1)^2$	$(m-1)^2$	yes
m	$\sum_{i=1}^{m}(2i - 1)$	m^2	???

Now we simplify the summation for $n = m$: $\sum_{i=1}^{m}(2i - 1) = \left(\sum_{i=1}^{m-1}(2i - 1)\right) + (2m - 1) = (m - 1)^2 + (2m - 1) = m^2 - 2m + 1 + (2m - 1) = m^2$, and observe that this is equal to the desired closed formula.

Formal proof. Let $P(n)$ be the statement, "$\sum_{i=1}^{n}(2i - 1) = n^2$". Since $P(1)$ is the statement "$1 = 1^2$", it is clearly true. Now let the integer $m \geq 2$ be given such that $P(1), P(2), \ldots, P(m - 1)$ have already been checked to be true. In particular, the last statement we checked was $P(m - 1)$, which said, "$\sum_{i=1}^{m-1}(2i - 1) = (m - 1)^2$". It now follows that

$$\sum_{i=1}^{m}(2i - 1) = \left(\sum_{i=1}^{m-1}(2i - 1)\right) + (2m - 1)$$
$$= (m - 1)^2 + (2m - 1) \text{ (by } P(m - 1))$$
$$= m^2 - 2m + 1 + (2m - 1)$$
$$= m^2$$

Thus we have shown that $\sum_{i=1}^{m}(2i - 1) = m^2$, which is precisely statement $P(m)$. ∎

(c) *Outline of proof.* The table shows the verification for $n = 1$, $n = 2$, $n = 3$, and $n = 4$. It also shows the last row checked ($n = m - 1$) and the next row to be checked ($n = m$).

n	summation	closed formula	equal?
1	1	$2^2-1-2=1$	yes
2	$1+3=4$	$2^3-2-2=4$	yes
3	$1+3+7=11$	$2^4-3-2=11$	yes
4	$1+3+7+15$ $=26$	$2^5-4-2=26$	yes
...			...
$m-1$	$\sum_{i=1}^{m-1}(2^i-1)$ $=2^m-m-1$	$2^m-(m-1)-2$ $=2^m-m-1$	yes
m	$\sum_{i=1}^{m}(2^i-1)$	$2^{m+1}-m-2$	???

Now we simplify the summation for $n=m$: $\sum_{i=1}^{m}(2^i-1)=\left(\sum_{i=1}^{m-1}(2^i-1)\right)+(2^m-1)=(2^m-m-1)+(2^m-1)=2\cdot2^m-m-1-1=2^{m+1}-m-2$, and observe that this is equal to the desired closed formula.

Formal proof. Let $P(n)$ be the statement, "$\sum_{i=1}^{n}(2^i-1)=2^{n+1}-n-2$." Since $P(1)$ is the statement "$2-1=4-1-2$," it is clearly true. Now let the integer $m\geq2$ be given such that $P(1), P(2), \ldots, P(m-1)$ have already been checked to be true. In particular, the last statement we checked was $P(m-1)$, which said, "$\sum_{i=1}^{m-1}(2^i-1)=2^m-(m-1)-2$." It now follows that

$$\sum_{i=1}^{m}(2^i-1)=\left(\sum_{i=1}^{m-1}(2^i-1)\right)+(2^m-1)$$
$$=2^m-(m-1)-2+(2^m-1)$$
$$\text{(by } P(m-1))$$
$$=2\cdot2^m-m+1-2-1$$
$$=2^{m+1}-m-2$$

Thus we have shown that $\sum_{i=1}^{m}(2^i-1)=2^{m+1}-m-2$, which is precisely statement $P(m)$. ∎

(e) *Outline of proof.* The table shows the verification for $n=1$, $n=2$, $n=3$, and $n=4$. It also shows the last row checked ($n=m-1$) and the next row to be checked ($n=m$).

n	summation	closed formula	equal?
1	$\frac{1}{2}$	$1-\frac{1}{2}=\frac{1}{2}$	yes
2	$\frac{1}{2}+\frac{1}{4}=\frac{3}{4}$	$1-\frac{1}{4}=\frac{3}{4}$	yes
3	$\frac{1}{2}+\frac{1}{4}+\frac{1}{8}=\frac{7}{8}$	$1-\frac{1}{8}=\frac{7}{8}$	yes
4	$\frac{1}{2}+\frac{1}{4}+\frac{1}{8}+\frac{1}{16}=\frac{15}{16}$	$1-\frac{1}{16}=\frac{15}{16}$	yes
...			...
$m-1$	$\sum_{i=1}^{m-1}\frac{1}{2^i}=1-\frac{1}{2^{m-1}}$	$1-\frac{1}{2^{m-1}}$	yes
m	$\sum_{i=1}^{m}\frac{1}{2^i}$	$1-\frac{1}{2^m}$	???

Now we simplify the summation for $n=m$: $\sum_{i=1}^{m}\frac{1}{2^i}=\left(\sum_{i=1}^{m-1}\frac{1}{2^i}\right)+\frac{1}{2^m}=1-\frac{1}{2^{m-1}}+\frac{1}{2^m}=1-\frac{2}{2^m}+\frac{1}{2^m}=1-\frac{1}{2^m}$, and observe that this is equal to the desired closed formula.

9. (a) *Proof.* Let $P(n)$ be the statement, "$\frac{(1)(2)}{2}+\frac{(2)(3)}{2}+\cdots+\frac{(n)(n+1)}{2}=\frac{n(n+1)(n+2)}{6}$." Since $P(1)$ is the statement "$\frac{(1)(2)}{2}=\frac{(1)(2)(3)}{6}$," we know that it is true. Now let $m\geq2$ be given such that we have already checked $P(1), P(2), \ldots, P(m-1)$ to be true. Since

$$\frac{(1)(2)}{2}+\frac{(2)(3)}{2}+\cdots+\frac{(m-1)(m)}{2}+\frac{(m)(m+1)}{2}$$
$$=\left[\frac{(1)(2)}{2}+\frac{(2)(3)}{2}+\cdots+\frac{(m-1)(m)}{2}\right]+\frac{(m)(m+1)}{2}$$
$$=\frac{(m-1)(m)(m+1)}{6}+\frac{(m)(m+1)}{2}$$
$$\text{(since } P(m-1) \text{ is true)}$$
$$=\frac{(m)(m+1)(m+2)}{6}$$

we see that $P(m)$ is also true. ∎

10. (a) $\sum_{i=1}^{n}\frac{i(i+1)}{2}=\frac{n(n+1)(n+2)}{6}$

11. Claim. For all $n\geq1$, $\sum_{i=1}^{n}(i)(2^i)=(n-1)2^{n+1}+2$

Proof. Let $S(n)$ represent the statement, "$\sum_{i=1}^{n}(i)(2^i)=(n-1)2^{n+1}+2$." Then $S(1)$ is "$(1)(2^1)=(1-1)2^2+2$," which is true. Now let $m\geq2$ be given such that $P(1), P(2), \ldots, P(m-1)$ have all been checked to be true. In particular, $P(m-1)$ is "$\sum_{i=1}^{m-1}(i)(2^i)=(m-2)2^m+2$." So

$$\sum_{i=1}^{m}(i)(2^i) = \left(\sum_{i=1}^{m-1}(i)(2^i)\right) + (m)(2^m)$$
$$= ((m-2)\,2^m + 2) + (m)(2^m) \text{ by } P(m-1)$$
$$= 2 \cdot m \cdot 2^m - 2 \cdot 2^m + 2$$

$$= m \cdot 2^{m+1} - 2^{m+1} + 2$$
$$= (m-1) \cdot 2^{m+1} + 2$$

which verifies that statement $P(m)$ is true.   ■

Section 2.4 Exercises

1. (a) $a_n = a_{n-1} + 2 \cdot 3^{n-1}$ with $a_1 = 2$; **(c)** $c_n = c_{n-1} + \frac{1}{n(n+1)}$ with $c_1 = \frac{1}{2}$

2. (a) *Proof.* It is easy to see that $a_1 = 3^1 - 1$ since the definition above gives us that $a_1 = 2$. Let $m \geq 2$ be given such that the closed formula has been checked to work for $a_1, a_2, \ldots, a_{m-1}$. In particular, it has been checked that $a_{m-1} = 3^{m-1} - 1$. Now

$$a_m = a_{m-1} + 2 \cdot 3^{m-1} \text{ by the recurrence for } a$$
$$= \left(3^{m-1} - 1\right) + 2 \cdot 3^{m-1}$$
$$= 3 \cdot 3^{m-1} - 1 = 3^m - 1$$

So $a_m = 3^m - 1$, completing the induction.   ■

(c) *Proof.* It is easy to see that $c_1 = \frac{1}{1+1}$ since the definition above gives us that $c_1 = \frac{1}{2}$. Let $m \geq 2$ be given such that the given closed formula has been checked to work for $c_1, c_2, \ldots, c_{m-1}$. In particular, it has been checked that $c_{m-1} = \frac{m-1}{m}$. Now

$$c_m = c_{m-1} + \frac{1}{m(m+1)} \text{ by the recurrence for } c$$
$$= \left(\frac{m-1}{m}\right) + \frac{1}{m(m+1)}$$
$$= \frac{(m-1)(m+1)}{m(m+1)} + \frac{1}{m(m+1)}$$
$$= \frac{m^2}{m(m+1)} = \frac{m}{m+1}$$

So $c_m = \frac{m}{m+1}$, completing the induction.   ■

3. (a) Let $g_n = n^3 + 2n$. Note that $g_{n-1} = (n-1)^3 + 2(n-1) = n^3 - 3n^2 + 5n - 3 = (n^3 + 2n) - 3n^2 + 3n - 3$, so $g_{n-1} = g_n - 3(n^2 - n + 1)$, or equivalently, $g_n = g_{n-1} + 3(n^2 - n + 1)$. **(b & c)** We prove by induction that for all $n \geq 0$, g_n is divisible by 3.
Proof. Let $P(n)$ be the statement, "g_n is divisible by 3." Since $g_0 = 0$, which is divisible by 3, statement $P(0)$ is true. Now let $m \geq 1$ be given such that statements $P(0), P(1), \ldots, P(m-1)$ have all been checked to be true. In particular, it has been checked that g_{m-1} is divisible by 3, so there is an integer K such that $g_{m-1} = 3K$.

From the (a) part,

$$g_m = g_{m-1} + 3(m^2 - m + 1)$$
$$= 3K + 3(m^2 - m + 1) \text{ by } P(m-1)$$
$$= 3(K + m^2 - m + 1)$$

This means that g_m is divisible by 3, completing the induction.   ■

4. (a) *Proof.* Let $P(n)$ be the statement, "$n^2 - n$ is divisible by 2." $P(1)$ says, "$1^2 - 1$ is divisible by 2," which is true. Let $m \geq 2$ be given such that $P(1), P(2), \ldots, P(m-1)$ have all been checked to be true. In particular, $P(m-1)$ states, "$(m-1)^2 - (m-1)$ is divisible by 2," so there is an integer K such that $(m-1)^2 - (m-1) = 2K$. Now consider the next statement $P(m)$:

$$m^2 - m = ((m-1)^2 - (m-1)) + 2m - 2$$
$$= 2K + 2m - 2$$
$$= 2(K + m - 1)$$

Hence, $m^2 - m$ is divisible by 2, verifying that statement $P(m)$ is true.   ■

5. *Proof.* Let $P(n)$ be the statement, "$10^n - 1$ is divisible by 9." Then $P(0)$ is the statement, "$10^0 - 1$ is divisible by 9," that is, "0 is divisible by 9." This is true since $0 = 0 \cdot 9$. Now let $m \geq 1$ be given such that $P(0), P(1), \ldots, P(m-1)$ have all been verified. In particular, we know that $10^{m-1} - 1$ is divisible by 9, say $10^{m-1} - 1 = 9k$ for some integer k. Using algebra, we write:

$$10^{m-1} - 1 = 9k$$
$$10^{m-1} = 9k + 1$$
$$10 \cdot 10^{m-1} = 10(9k + 1)$$
$$10^m = 90k + 10$$
$$10^m - 1 = 90k + 10 - 1$$
$$= 90k + 9$$
$$= 9(10k + 1)$$

Since $10k + 1$ is an integer, this shows that $P(m)$ is true, and the result follows by induction.   ■

8. (a) *Proof.* It is easy to check that $F_1 < 2^1$ and $F_2 < 2^2$ since these values ($F_1 = F_2 = 1$) are given in the definition of the Fibonacci numbers. Let $m \geq 3$ be given such that the inequality "$F_n < 2^n$" has been checked for the terms $F_1, F_2, \ldots, F_{m-1}$. In particular, we know that $F_{m-1} < 2^{m-1}$ and $F_{m-2} < 2^{m-2}$. Now

$$\begin{aligned} F_m &= F_{m-1} + F_{m-2} \\ &< 2^{m-1} + 2^{m-2} \\ &< 2^{m-1} + 2^{m-1} \\ &= 2^m \end{aligned}$$

Hence, $F_m < 2^m$, completing the induction. ∎

(b) *Proof.* Let $P(n)$ be the statement, "$F_2 + F_4 + \cdots + F_{2n} = F_{2n+1} - 1$." It is easy to check that $P(1)$ (which says "$F_2 = F_3 - 1$") is true since $F_2 = 1$ and $F_3 = F_2 + F_1 = 2$. Let $m \geq 2$ be given such that $P(1), P(2), \ldots, P(m-1)$ have already been checked to be true. In particular, $P(m-1)$ says, "$F_2 + F_4 + \cdots + F_{2m-2} = F_{2m-1} - 1$." Now

$$\begin{aligned} F_2 + F_4 + \cdots + F_{2m} &= (F_2 + F_4 + \cdots + F_{2m-2}) + F_{2m} \\ &= (F_{2m-1} - 1) + F_{2m} \\ &= F_{2m+1} - 1 \end{aligned}$$

This verifies that $P(m)$ is true, completing the induction. ∎

9. (a) Claim: For all $n \geq 1$, F_{4n} is divisible by 3. *Proof by induction on n.* We can compute $F_4 = F_3 + F_2 = 2 + 1 = 3$ to see that the first statement (which is "F_4 is divisible by 3") is true. Let $m \geq 2$ be given such that $F_4, F_8, F_{12}, \ldots, F_{4(m-1)}$ have all been checked to be divisible by 3. In particular, since we know that F_{4m-4} is divisible by 3, we know there is an integer K such that $F_{4m-4} = 3K$. Now using the recurrence for the Fibonacci sequence, we see that

$$\begin{aligned} F_{4m} &= F_{4m-1} + F_{4m-2} \\ &= (F_{4m-2} + F_{4m-3}) + (F_{4m-3} + F_{4m-4}) \\ &= (F_{4m-3} + F_{4m-4}) + 2F_{4m-3} + F_{4m-4} \\ &= 3F_{4m-3} + 2F_{4m-4} \\ &= 3F_{4m-3} + 2(3K) \\ &= 3(F_{4m-3} + 2K) \end{aligned}$$

From this, it follows that F_{4m} is divisible by 3. ∎

10. (a) *Proof.* Let $P(n)$ be the statement, "In the Josephus game with 2^n people, Joe's friend should stand in position $2^{n-1} + 1$." Since $P(1)$ is the statement "In the Josephus game with 2 people, Joe's friend should stand in position 2," we know that it is true. Now let $m \geq 2$ be given, and assume that we already know $P(1), P(2), \ldots, P(m-1)$ to all be true. In the play

of the game with 2^m people, the order of elimination is $2, 4, 6, \ldots, 2^m$ leaving the 2^{m-1} people labeled $1, 3, 5, 7, \ldots, 2^m - 1$ for the next round of elimination. Since $P(m-1)$ is true, this game will eliminate the person in position $2^{m-2} + 1$ next to last. Because of their labels, we determine that the last remaining person was originally labeled the $(2^{m-2} + 1)^{th}$ odd number which is $2(2^{m-2} + 1) - 1 = 2^{m-1} + 1$. This establishes the truth of $P(m)$. ∎

12. *Proof.* Let $P(n)$ be the statement, "the product of n odd integers is an odd integer." The first statement $P(1)$ states, "the product of 1 odd integer is an odd integer," which is strange to say, but certainly true. We have proved $P(2)$ before in Exercise 9(a) of Section 2.1. Now let $m \geq 3$ be given such that $P(1), P(2), \ldots, P(m-1)$ have all been verified, and let $a_1, a_2, \ldots, a_m$ be m odd integers. We must show that $a_1 a_2 \ldots a_m$ is an odd integer. Let p indicate the product $a_1 a_2 \ldots a_{m-1}$, so that $a_1 a_2 \ldots a_m = p \cdot a_m$ By $P(m-1)$, we know that p is odd. Hence by $P(2)$ we conclude that $p \cdot a_m$ is odd, and this establishes $P(m)$. The result follows by induction. ∎

14. (a) *Proof.* Let $P(n)$ be the statement, "there exist integers q and r such that $n = 3 \cdot q + r$ and $0 \leq r \leq 2$." If n is less than 3, this statement is clearly true (just use $q = 0$ and $r = n$). This establishes $P(0)$, $P(1)$, and $P(2)$. Now let $m \geq 3$ be given such that $P(0), P(1), P(2) \ldots, P(m-1)$ have all been verified. By $P(m-3)$ we know we can write $m - 3 = 3 \cdot q + r$ with $0 \leq r \leq 2$. Adding 3 to both sides, we obtain $m = 3 \cdot q + r + 3 = 3 \cdot (q + 1) + r$. Since $q + 1$ and r are integers and r still satisfies $0 \leq r \leq 2$, this establishes $P(m)$. The result follows by induction. ∎

15. *Proof by induction.* Let $P(n)$ be the statement, "One can make n-cents in postage using a combination of 3-cent and 8-cent stamps." We can check the first three statements as follows:

- $P(14)$ is true since $3 + 3 + 8 = 14$.
- $P(15)$ is true since $3 + 3 + 3 + 3 + 3 = 15$.
- $P(16)$ is true since $8 + 8 = 16$.

Now let $m \geq 17$ be given such that $P(m)$ is the first statement not yet checked. In particular, $P(m-3)$ has been checked, so we know that it is possible to make $m - 3$ cents in postage using just these types of stamps. Adding a 3-cent stamp to this postage consitutes m cents in postage, so $P(m)$ is true. ∎

17. *Proof.* Let $P(n)$ be the statement,

$$\sum_{i=2}^{2^n} \frac{1}{i} \geq \frac{n}{2}$$

For example, $P(1)$ says, "$\frac{1}{2} \geq \frac{1}{2}$" which is true. Now let $m \geq 2$ be given such that $P(1), P(2), \ldots, P(m-1)$ have already been checked to be true. In particular, $P(m-1)$ says,

$$\sum_{i=2}^{2^{m-1}} \frac{1}{i} \geq \frac{m-1}{2}$$

We now consider the next statement $P(m)$:

$$\sum_{i=2}^{2^m} \frac{1}{i} = \left(\frac{1}{2} + \frac{1}{3} + \cdots + \frac{1}{2^{m-1}}\right)$$

$$+ \left(\frac{1}{2^{m-1}+1} + \frac{1}{2^{m-1}+2} + \cdots + \frac{1}{2^m}\right)$$

$$\geq \left(\frac{m-1}{2}\right) + \frac{1}{2} = \frac{m}{2}$$

Section 2.5 Exercises

1. Fill in the blanks with: $3K + 1, 9L + 5, 3K + 1, 9L + 5$, $K - 3L$, and "no counterexample exists."

3. (a) *Proof.* Suppose a counterexample n to this statement exists. Since n makes the hypothesis true, $n^2 = 2K$ for some integer K. Since n makes the conclusion false, $n = 2L + 1$ for some integer L. Combining these equations gives us $2K = (2L+1)^2$. Using algebra, we find that $2K - 4L^2 - 4L = 1$, from which it follows that $K - 2L^2 - 2L = \frac{1}{2}$. Since $K - 2L^2 - 2L$ is an integer and $\frac{1}{2}$ is not, this contradiction tells us no counterexample exists. ∎

(b) *Proof.* Suppose a counterexample n to this statement exists. Since n makes the hypothesis true, $n^2 = 2K + 1$ for some integer K. Since n makes the conclusion false, $n = 2L$ for some integer L. Combining these equations gives us $2K + 1 = (2L)^2$. Using algebra, we find that $2K - 4L^2 = -1$, from which it follows that $K - 2L^2 = -\frac{1}{2}$. Since $K - 2L^2$ is an integer and $-\frac{1}{2}$ is not, this contradiction tells us no counterexample exists. ∎

5. *Proof by contradiction.* Suppose as a counterexample there are odd perfect squares a and b whose sum is the perfect square c. From Exericse 3(b), we know that a and b are the squares of odd integers. That is, $a = (2K+1)^2$ and $b = (2L+1)^2$ for some integers K and L. In this case,

$$c = a + b = 4K^2 + 4K + 1 + 4L^2 + 4L + 1$$

$$= 4(K^2 + K + L^2 + L) + 2$$

Hence, c is an even perfect square of the form $4M + 2$, where M is the integer $K^2 + K + L^2 + L$. But Exercise 4 tells us this is impossible, so no such counterexample exists. ∎

The key step here is the fact that

$$\frac{1}{2^{m-1}+1} + \frac{1}{2^{m-1}+2} + \cdots + \frac{1}{2^m}$$

$$\geq \underbrace{\frac{1}{2^m} + \frac{1}{2^m} + \cdots + \frac{1}{2^m}}_{2^{m-1} \text{ terms}} = \frac{1}{2}$$

This establishes that $P(m)$ is true, completing the induction. ∎

19. (a) When $m = 1$, if the Reader picks the set $S = \varnothing$, then it is impossible to follow the instruction, "choose an element $a \in S$."

7. *Proof by contradiction* Suppose there is an integer n which is of the form $5K + 3$ and of the form $5L + 1$. for some integers K and L. This means that

$$5K + 3 = 5L + 1, \text{ or}$$

$$\frac{2}{5} = L - K$$

Since $L - K$ is an integer, this contradiction shows that there is no such integer n. ∎

9. *Proof outline.* Suppose to the contrary that there *are* positive integers a and b, with no common divisors, satisfying $a^2 = 2b^2$. Since a and b have no common divisor greater than 1, they cannot both be even. Thus either: (1) a is even and b is odd; or (2) b is even and a is odd; or (3) both are odd. Proceed by cases.

- **Case 1.** If a is even and b is odd, write $a = 2m$ and $b = 2n + 1$, substitute into $a^2 = 2b^2$ and simplifying to get a contradiction.

- **Case 2.** If a is odd and b is even, write $a = 2m + 1$ and $b = 2n$ and again substitute into $a^2 = 2b^2$ to find a contradiction.

- **Case 3.** If a and b are both odd, proceed similarly to obtain a contradiction.

10. (a) Suppose a positive number x divided by a positive number y results in a negative number z. Since $x \div y = z$ implies that $x = y \cdot z$, this means that the positive number x is the product of a positive number and a negative number, contradicting "rule" (ii). Hence, z must be positive.

11. Suppose, to the contrary, that there is a rational number a such that $a + \sqrt{2}$ is rational. Then $(a + \sqrt{2}) - a$ is rational. But this means $\sqrt{2}$ is rational, a contradiction of Theorem 4. Hence, there is no such rational number a.

13. The contrapositive is, "If $a + b$ is rational, then a is irrational or b is rational."

Proof. Let a and b be given such that $a + b$ is rational. Either a is irrational or a is rational.

Case 1. If a is irrational, then the conclusion, "a is irrational or b is rational" is certainly true.

Case 2. If a is rational, then by Exercise 9 in Section 2.2, $(a + b) - a = b$ is rational, and hence the conclusion, "a is irrational or b is rational" is also true.

Since in either case the same conclusion holds, it must be true that a is irrational or b is rational. ∎

14. (a) *Proof.* Since $6 \left(\frac{5}{3} \right)^2 + 11 \left(\frac{5}{3} \right) = \frac{50}{3} + \frac{55}{3} = 35$, it follows that there exists a positive rational number r such that $6r^2 + 11r = 35$.

16. The contrapositive of Proposition 3 is, "If r is rational, then $r^2 \neq 2$."

Proof. Let the rational number r be given. By Proposition 1, $r = \frac{a}{b}$ for integer a and b having no common divisor greater than 1. This means that a and b do not have 2 as a common divisor. By the contrapositive of Proposition 2, this means that $(\frac{a}{b})^2 \neq 2$. That is, $r^2 \neq 2$, as desired. ∎

18. (a) For every integer n, n is not the largest integer.

(b) Suppose we are given an integer n. To show that n is not the largest integer, all we have to do is find a larger integer – and $n + 1$ certainly fits that description.

19. *Hint:* The contrapositive statement is, "If n is even, then $5n + 4$ is even." The proof is very straightforward. (Write $n = 2k$ and use this to write $5n + 4$ as $2 \cdot$ (some integer)).

22. The contrapositive statement is, "If each of the four children is less than ten years old, then the average age of the children is not ten years old."

Proof. Let a, b, c, d represent the ages of the four children, each of whom is less than ten years old. Since $a < 10, b < 10, c < 10$, and $d < 10$, we know that $a + b + c + d < 40$, from which it follows that $\frac{a+b+c+d}{4} < 10$. Hence, the average of a, b, c, and d is not ten. ∎

24. *Proof by contradiction.* Suppose to the contrary that there exist 10 numbers that make the hypothesis true but the conclusion false. Then we know the average of the numbers is 89.63, and we also know that one of the numbers is less than 89.63 – call that number s. Label the remaining numbers $x_1, x_2, \ldots, x_9$. Since the conclusion is false, each x_i satisfies $x_i \leq 89.63$. Thus, $x_1 + x_2 + \ldots + x_9 \leq 9(89.63)$. Now since $s < 89.63$, we know $s + x_1 + x_2 + \ldots + x_9 < 89.63 + 9(89.63)$, that is $s + x_1 + x_2 + \ldots + x_9 < 10(89.63)$. But this contradicts the statement that the average is 89.63. ∎

26. (a) Pigeonhole Principle (Basic Version): If $n + 1$ objects are distributed among n boxes, then some box must contain more than one object.

Proof. Let $n + 1$ objects be given, and label the n boxes $1, 2, 3, \ldots, n$. After the objects have been distributed among the boxes, define $x_1, x_2, \ldots, x_n$ by the following rule:

$$x_i = \text{the number of objects in box } i$$

Since each object can go into only one box, we know that

$$x_1 + x_2 + x_3 + \cdots + x_n = n + 1$$

which means the average value of the x's is

$$\frac{x_1 + x_2 + x_3 + \cdots + x_n}{n} = \frac{n + 1}{n}$$
$$= 1 + \frac{1}{n}$$

According to the Average Version, there is an x_i that is at least $1 + \frac{1}{n}$. Since x_i is an integer value, this means that $x_i > 1$, which can be interpreted as meaning, "more than 1 object is in the box labeled i." ∎

27. Statement of contrapositive. If integers x, y, and z satisfy $x < 4$, $y < 4$, and $z < 5$, then $x + y + z < 11$.

Proof. Since $x < 4$, $y < 4$, and $z < 5$, and since x, y, and z are integers, we know that $x \leq 3, y \leq 3$, and $z \leq 4$. So $x + y + z \leq 3 + 3 + 4$, that is $x + y + z \leq 10$. Since $10 < 11$, we conclude that $x + y + z < 11$. ∎

30. We can write the basic version as: Objects are distributed among n boxes. If $n + 1$ objects are distributed, then some box must contain more than one object.

Contrapositive: If no box contains more than one object, then the number of object distributed is not $n + 1$.

Proof. After the objects have been distributed among the boxes, define $x_1, x_2, \ldots, x_n$ by the following rule:

$$x_i = \text{the number of objects in box } i$$

Since no box contains more than one object, we know that each x_i satisfies $x_i \leq 1$. Thus the total number of objects distributed is given by

$$x_1 + x_2 + x_3 + \cdots + x_n \leq n$$

This shows that the total cannot be $n + 1$. ∎

31. This statement is similar to Exercise 24: "For any list of numbers, if one of the numbers is greater than the average, then at least one of the numbers must be less than the average." The proof of the statement is related to the Average Version of the Pigeonhole Principle. However, its truth is intuitively obvious to almost everyone, and that is why Garrison Keillor's sign-off is funny to almost everyone. (Some of today's bureaucrats don't get the joke.)

32. (a) *Proof.* Define Boxes 0, 1, 2, 3, 4, 5, and 6 by the following rule: "Place integer x into Box r if x is of the form $6q + r$." The Remainder Theorem tells us that every integer can be placed in this way. Let seven integers be given. By the Pigeonhole Principle, some box must contain (at least) two integers. Let's call the two integers x and y, and say they are in Box d. The rule defining the boxes tells us that $x = 6 \cdot K + d$ and $y = 6 \cdot L + d$ for some integers K and L. In this case,

$$x - y = (6K + d) - (6L + d)$$
$$= 6 \cdot (K - L).$$

Since $K - L$ is an integer, this means that $x - y$, the difference between x and y, is divisible by 6. ∎

33. *Proof.* Let five integers be given. Think of two boxes, one labeled "divisible by 3" and one labeled "not divisible by 3." By the distribution version of the Pigeonhole Principle (with $n = \underline{3}$ and $m = \underline{2}$), we conclude that there are at least three of the numbers in one box. Let's refer to these three numbers as a, b and c, and consider two cases based on which box they are in.

- **Case 1.** If a, b, c are in the box labeled "divisible by 3," then $a^2 + b^2 + c^2$ is divisible by 3 because $a = 3K$, $b = 3L$ and $c = 3J$ (where K, L, and J are integers) imply that

$$a^2 + b^2 + c^2 = (3K)^2 + (3L)^2 + (3J)^2$$
$$= 3(3K^2 + 3L^2 + 3J^2)$$

which is certainly divisible by 3.

- **Case 2.** If a, b, c are in the box labeled "not divisible by 3," then by Practice Problem 4 from Section 2.2, a^2 can be written in the form $\underline{3K + 1}$, b^2 can be written in the form $\underline{3L + 1}$, and c^2 can be written in the form $\underline{3J + 1}$. Hence, $a^2 + b^2 + c^2$ is divisible by

3 because in this case,

$$a^2 + b^2 + c^2 = (3K + 1) + (3L + 1) + (3J + 1)$$
$$= 3(K + L + J + 1)$$

In either case, $a^2 + b^2 + c^2$ is divisible by 3, completing the proof. ∎

35. (a) Considering the four triangular regions shown to be "boxes", any five points will require two to share one box by the Pigeonhole Principle.

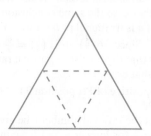

The greatest distance between two points in one of these small triangular regions is the distance between vertices, which is $\frac{1}{2}$. Hence, there must be two points within $\frac{1}{2}$ of each other.

36. (a) If $K\pi$ is in Box 1, then $K\pi$ is within $\frac{1}{n}$ of an integer.

(b) If no multiple of π is in Box 1, then all n given multiples are distributed among $n - 1$ boxes, so the Pigeonhole Principle guarantees that some box will contain (at least) two multiples.

(c) If $K\pi$ and $L\pi$ are both in the same box (where $K > L$, say), then $K\pi - L\pi$ has a fractional part between $\frac{n-1}{n}$ and 1 or between 0 and $\frac{1}{n}$.

(d) In either case described in (e), $K\pi - L\pi$ is within $\frac{1}{n}$ of an integer. Since $(K - L)\pi$ is one of the multiples originally described, this is the desired conclusion.

Section 2.6 Exercises

1. (a) 100011; **(c)** 1111011
2. (a) 120; **(c)** 443
3. (a) 43; **(c)** 173
4. (a) 29; **(c)** 10
5. For convenience in reading them, we show the binary numbers in groups of 4 bits. **(a)** $(1101\ 1010\ 1101)_2$; **(b)** $(1\ 1111\ 0000\ 1011)_2$; **(d)** $(19)_{16}$; **(e)** $(B0DE)_{16}$
6. For convenience in reading them, we show the binary numbers in groups of 3 bits. **(a)** $(1\ 111)_2$; **(b)** $(10\ 000\ 000\ 101)_2$; **(d)** $(31)_8$; **(e)** $(130336)_8$
7. (a) $(F)_{16}$; **(b)** $(2A5)_{16}$; **(d)** $(6655)_8$; **(e)** $(4210421)_8$
8. Since this is an "if and only if" proposition, there are two separate proofs to be completed.

- **Claim 1.** If x is divisible by 3, then the sum of the decimal digits of x is divisible by 3.
  Proof. Let a natural number x which is divisible by 3 be given, and let s be the sum of the decimal digits of x. Since x is divisible by 3, we know that $x = 3K$ for some integers. By Proposition 3, we know that $x - s$ is divisible by 9, so $x - s = 9L$ for some integer L. From this it follows that $3K - 9L = s$, so $s = 3(K - 3L)$. Hence, s is divisible by 3. ∎
- **Claim 2.** If the sum of the decimal digits of x is divisible by 3, then x is divisible by 3.
  Proof. Let a natural number x whose digits sum to a number s that is divisible by 3 be given. That is, $s = 3K$ for some integer K. By Proposition 3, we

know that $x - s = 9L$ for some integer L, and so $x = s + 9L = 3K + 9L = 3(K + 3L)$. Hence, x is divisible by 3. ∎

9. 0, 1 or 4

11. **(a)** and **(c)**

12. Let the integer n be represented as $d_4 d_3 d_2 d_1 d_0$ in base b. This means that

$$n = d_4 \cdot b^4 + d_3 \cdot b^3 + d_2 \cdot b^2 + d_1 \cdot b + d_0 \text{ and so}$$
$$n \cdot b = d_4 \cdot b^5 + d_3 \cdot b^4 + d_2 \cdot b^3 + d_1 \cdot b^2 + d_0 \cdot b + 0$$

Hence, $n \cdot b$ is represented as $d_4 d_3 d_2 d_1 d_0 0$ in base b.

13. Nine is a counterexample.

14. Seven is a counterexample.

17. *Outline.* First use induction to prove **Proposition 1**: For all $n \geq 0$, $8^n - 1$ is divisible by 7. The proof is similar to Exercise 5 in Section 2.4.

Next prove **Proposition 2**: If s is the sum of the digits in the octal representation of x, then $x - s$ is divisible by 7. The proof is similar to Proposition 3 in this section.

Finally, to prove the "if and only if" statement of Exercise 17, we must prove two separate things: **Claim 1**. If a natural number x is divisible by 7, then the sum of its octal digits is divisible by 7; and **Claim 2**. If a natural number x has the sum of its octal digits divisible by 7, then x is divisible by 7. These are similar to the proof of Exercise 8 in this section.

19. **(b)** #FFFF00 is yellow. **(c)** 16777216

20. We use T for ten and E for eleven. **(a)** 4321; **(c)** $3T1$

Section 2.7 Exercises

1. **(a)** $a = 2$, $b = 2$, $n = 4$; **(c)** $a = 2$, $n = 8$; **(e)** A trivial example is $n = 2$ since $1^2 \equiv_2 5$ and 5 does not divide $2 \cdot 1$. A more substantial example is $n = 19$ noting that $9^2 \equiv_{19} 5$, but 5 does not divide $19 \cdot 18$.

2. **(a)** The missing steps are: *This means $a - b = Kn$ and $c - d = Ln$ for some integers K and L. Adding these equations, we get*

$$a - b + c - d = Kn + Ln$$

which can be rearranged as $(a + c) - (b + d) = (K + L) \cdot n$. Since $K + L$ is an integer, this means that $(a + c) - (b + d)$ is divisible by n.

3. The mod 7 tables are given below.

+	0	1	2	3	4	5	6
0	0	1	2	3	4	5	6
1	1	2	3	4	5	6	0
2	2	3	4	5	6	0	1
3	3	4	5	6	0	1	2
4	4	5	6	0	1	2	3
5	5	6	0	1	2	3	4
6	6	0	1	2	3	4	5

·	0	1	2	3	4	5	6
0	0	0	0	0	0	0	0
1	0	1	2	3	4	5	6
2	0	2	4	6	1	3	5
3	0	3	6	2	5	1	4
4	0	4	1	5	2	6	3
5	0	5	3	1	6	4	2
6	0	6	5	4	3	2	1

5. In mod 11 arithmetic, we have

$$1^{-1} = 1, 2^{-1} = 6, 3^{-1} = 4, 5^{-1} = 9, 7^{-1} = 8, 10^{-1} = 10$$

8. Proposition 3 tells us that $x - s_x$ is divisible by 9. This means $x - s_x = 9k$ for some integer k, which means $x \equiv_9 s_x$.

10. **(a)** Every $x \equiv_{15} 2$ will work; **(c)** Every $x \equiv_{11} 4$ or $x \equiv_{11} 8$ will work; **(e)** Every $x \equiv_{17} 4$ or $x \equiv_{17} 13$ will work.

11. p can be any element of $\{2, 5, 13, 17, 29, 37, 41, 53, 61, 73, 89, 97\}$. Except for 2, these are the primes of the form $4K + 1$, where K is an integer.

13. *Outline of proof.* This is equivalent to the claim, "$n^5 \equiv_{10} n$ for all $n \in \mathbb{N}$." (Let $n \in \mathbb{N}$ be given. By the Remainder Theorem, $n \equiv_{10} d$ for some $d \in \{0, 1, 2, 3, 4, 5, 6, 7, 8, 9\}$ (d is the units' digit of n), so you can argue in ten cases (for example, if $n \equiv_{10} 5$, then $n^5 \equiv_{10} 3125 \equiv_{10} 5$).

14. (a) 7; **(b)** Meet at Moe's **(c)** (1) Making $n = 77$ and $e = 43$ publicly available, makes it easy to determine $p = 7$ and $q = 11$ and hence that $k = 60$. From this, it is pretty easy to find the decryption key. Hence, using a small value of n is not very secure. (2) Encrypting one character at a time means there will be a lot of repetition of numbers in your message. A long message can be easily broken using knowledge about common letters and letter patterns. Hence, encrypting one letter at a time can be broken without using RSA methods at all.

Section 3.1 Exercises

1. (a) $\{2\}$

(b) $\{1, 2, 4, 8\}$

(c) $\{3, 4, 7, 8, 9\}$

(d) $\{8\}$

(e) $\{5, 6, 10\}$

3. (a) $\{x \in \mathbb{Z} : x = 2y \text{ for some } y \in \mathbb{Z}\}$

(b) $\{x \in \mathbb{N} : x = 2^y \text{ for some } y \in \mathbb{N}\}$

(c) $\{x \in \mathbb{N} : x = 2y + 1 \text{ for some } y \in \mathbb{N}\}$

(d) $\{x \in \mathbb{N} : x = (2y + 1)^2 \text{ for some } y \in \mathbb{N}\}$

5. (a) $\{7, 13, 19, 25, 31, 37, 43, \ldots\}$

(b) $\{\ldots -17, -11, -5, 1, 4, 7, 10, 13, 16, \ldots\}$

(c) $\{3, 5, 9, 11, 15, 17, 21, 23, 27, \ldots\}$

(d) $\{4, 10, 16, 22, 28, 34, 40, 46, 52 \ldots\}$

7. (a) $\{2x : x \in \mathbb{Z}\}$

(b) $\{x^3 : x \in \mathbb{Z}\}$

(c) $\{10x + 7 : x \in \mathbb{N}\}$

(d) $\{\frac{a}{b} \in \mathbb{Q} : -b < a < b\}$

8. (a) $\not\subseteq$ because $0 \in \mathbb{N}$ and $0 \notin \mathbb{Z}^+$; **(c)** $\not\subseteq$ because $-\frac{1}{2} \in \mathbb{Q}$ and $-\frac{1}{2} \notin \mathbb{R}^+$; **(e)** $\subseteq$

9. (a) B

10. (a) $\{2x + 1 : x \in \mathbb{Z}\}$; **(c)** $\{2y^2 : y \in \mathbb{N}\}$

11. (a) $\{\frac{2}{3}, \frac{4}{5}, \frac{6}{7}, \frac{8}{9}, \frac{10}{11}, \ldots\}$, universal set $= \mathbb{Q}$; **(c)** $\{1, 2, \frac{1}{2}, 4, \frac{1}{4}, 8, \frac{1}{8}, 9, \frac{1}{9}, 27, \frac{1}{27}, \ldots\}$, universal set $= \mathbb{Q}$

12. (a) all positive odd numbers $\{2x + 1 : x \in \mathbb{N}\}$; **(c)** starting at 1, increased by 8 every time $\{8x + 1 : x \in \mathbb{N}\}$

13. (a) True.

(b) False. $A = \{1, 2\}$, $B = \{a, b\}$

(c) True.

(d) False. $A = \{1, 2\}$, $B = \{a, b, c\}$.

15. (a) $A \cap (B \cup C) = (A \cap B) \cup (A \cap C) = \{1, 3, 5\}$; **(c)** $(A \cup B)' = A' \cap B' = \{6, 7, 8, 9, 10\}$; **(e)** $A \cap (A \cup B) = A = \{1, 3, 5\}$

17. (a) A computer search yields 341, 561, and 645 as the only three possibilities; **(c)** A computer search yields 124, 217, 561, and 781 as the only four possibilities.

18. (a) Proposition.   If $n \equiv_6 5$, then $n \equiv_2 1$ and $n \equiv_3 2$. *Proof.*   Let n be given with $n \equiv_6 5$. This means that we can write $n = 6k + 5$ for some integer k. So $n = 6k + 5 = 2(3k + 2) + 1$. Since $3k + 2$ is an integer, this shows that $n \equiv_2 1$. Also, $n = 6k + 5 = 3(2k + 1) + 2$, and since $2k + 1$ is an integer this shows that $n \equiv_3 2$. ∎

19. (a) $x = 43$; **(c)** No solution; **(e)** No solution.

16. Only the final Venn diagrams illustrating the property are shown.

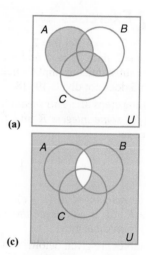

17. The Venn diagram for each side of the equation is given. For each that do not match, an example is given to illustrate the difference.

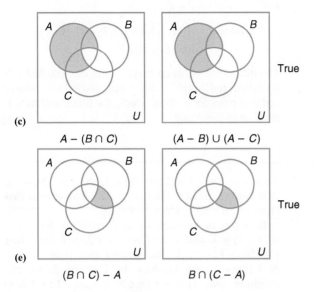

(c)

$A - (B \cap C)$

$(A - B) \cup (A - C)$ — True

(e)

$(B \cap C) - A$

$B \cap (C - A)$ — True

19. (a) Any partial sum of the harmonic series, in lowest terms, has an even denominator.

(b) Any partial sum of the harmonic series is between 1 and 2.

(c) No rational number between 1 and 2 has an even denominator (in lowest terms).

(d) $C \subseteq A$

20. (a) $\{-5, -4, -3, -2, -1, 0, 1, 2, 3\}$

(b) $[3.1, 4]$

(c) $\emptyset$

22. (a) $[0, 12]$; **(c)** $[2, 27]$

23. (a) $n(A) = 8, n(B) = 9, n(C) = 12$

(b) (i) 4, 13; **(ii)** 5, 15; **(iii)** 7, 14; **(iv)** 3; **(v)** 16

26. (a) 668

27. (a) 4200; **(c)** 560

29. As the diagram shows, there are three students that have completed all three stations.

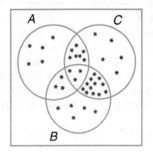

32. (a) True; **(c)** False. $A = \{1\}, B = \{2\}$

Section 3.2 Exercises

1. (a) $\{(2, 1), (2, 2), (2, 8), (4, 1), (4, 2), (4, 8)\}$;
(c) $\{\{\}, \{1\}, \{2\}, \{8\}, \{1, 2\}, \{1, 8\}, \{2, 8\}, \{1, 2, 8\}\}$

3. (a) 4; **(b)** 6; **(c)** 4

4. (a) $(\{1\}, \{1,2\}), (\{2\}, \{2,3\}), (\emptyset,\{1\}), (\{1,2\}, \{1,2,3\}), (\{2,3\}, \{1,2,3\})$

5. $A \cup B \cup C \neq \mathbb{Z}$; for example, $3 \notin A \cup B \cup C$.

7. $5 \notin A \cup B \cup C \cup D$ and so $A \cup B \cup C \cup D \neq \mathbb{Z}$.

9. We know the three sets are non-empty since $0 \in A$, $1 \in B$ and $3 \in C$. The Remainder Theorem tells us that every integer can be written in one of the forms, $4k$, $4k + 1$, $4k + 2$ or $4k + 3$. This means that any given integer is either in A (if it's of the form $4k$ or $4k + 2$) or in B (if it's in the form $4k + 1$) or in C (if it's of the form $4k + 3$). No pair of these sets overlap, as can be shown with a simple proof by contradiction. For example, to show that $A \cap B = \emptyset$, we simply assume that there *is* an $x \in A \cap B$, and argue that $x = 2K$ and $x = 4L + 1$ for some integers K and L. From this, it will follow that $K - 2L = \frac{1}{2}$, a contradiction to the closure of the integers.

10. (a) Let $A = \{1\}, B = \{2\}, C = \{3\}$; **(c)** Let $A = \{1, 2\}, B = \{2, 3\}$

13. (a) 8; **(b)** 16; **(c)** 12; **(d)** 4096; **(e)** 128

15. 10

16. (a) Not a partition; because not all elements of S are subsets of A; **(c)** Not a partition; since 6 is not an element of any part.

17. (a) $\{\{1, 2, 3, 4\}, \{5, 6, 7, 8\}\}$; **(c)** $\{\{2, 4, 6, 8\}, \{1, 3, 5, 7\}\}$

19. *Proof.* Let $S(n)$ be the statement, "If $n \geq 1$ is an integer, $A = \{1, 2, 3, 4, 5\}, B = \{1, 2, 3, \ldots, n\}$, then $n(A \times B) = 5n$." Statement $S(1)$ is true since

$$A \times B = \{1, 2, 3, 4, 5\} \times \{1\} = \{(1, 1), (2, 1), (3, 1), (4, 1), (5, 1)\}$$

Now let $m \geq 2$ be given such that statements $S(1), S(2), \ldots, S(m - 1)$ have all been verified to be true. In particular, $S(m - 1)$ states that $\{1, 2, 3, 4, 5\} \times \{1, 2, \ldots, m - 1\}$ is a set containing $5(m - 1)$ elements/ordered pairs. Now consider the next statement, $S(m)$. The set/cartesian product $\{1, 2, 3, 4, 5\} \times \{1, 2, \ldots, m - 1, m\}$ clearly contains all of the elements/ordered pairs in the set $\{1, 2, 3, 4, 5\} \times \{1, 2, \ldots, m-1\}$, as well as the five elements $(1, m), (2, m), (3, m), (4, m)$ and $(5, m)$, for a total of $5(m-1)+5 = 5m$ elements. This establishes statement $S(m)$, completing the induction. ∎

21. Let $A = \{1, 2\}, B = \{b_1, b_2\}, C = \{x\}$.
Then,

$$A \times B \times C = \{(1, b_1, x), (1, b_2, x), (2, b_1, x), (2, b_2, x)\},$$

Section 3.3 Exercises

1. (a) *Proof.* Dear READER, Recall that $N = \{0, 1, 2, 3, \ldots\}$ and that a natural number x is odd if and only if it can be written as $x = 2y + 1$, for some $y \in N$. Now, you must agree that each of $1, 3, 5, 7$ and 9 is a natural number. Also, note that $1 = 2 \cdot 0 + 1, 3 = 2 \cdot 1 + 1, 5 = 2 \cdot 2 + 1, 7 = 2 \cdot 3 + 1$ and $9 = 2 \cdot 4 + 1$, which establishes that each of $1, 3, 5, 7$ and 9 is an odd, natural number. Hence, each element of $\{1, 3, 5, 7, 9\}$ is also an element of $\{k \in N : k \text{ is odd}\}$. That is, $\{1, 3, 5, 7, 9\} \subseteq \{k \in N : k \text{ is odd}\}$. Your old friend, AUTHOR
(c) This is false since 2 is a prime number that is not odd.

2. (a) *Proof.* Let $x \in \{4m : m \in \mathbb{Z}\}$, so that $x = 4m$, for some $m \in \mathbb{Z}$. We can write $x = 4m = 2(2m)$. Since $2m \in \mathbb{Z}, x \in \{2n : n \in \mathbb{Z}\}$, and so $\{4m : m \in \mathbb{Z}\} \subseteq \{2n : n \in \mathbb{Z}\}$. ∎

(c) *Proof.* Let $x \in \mathbb{Z}$. We can write, $x = \frac{x}{1}$. Since 1 and x are both in $\mathbb{Z}$, it follows that $x \in \mathbb{Q}$, and so $\mathbb{Z} \subseteq \mathbb{Q}$. ∎
(e) *Proof.* Let $x \in \{2n+1 : n \in \mathbb{Z}\} \cap \{5m+4 : m \in \mathbb{Z}\}$, so that $x = 2n+1$ and $x = 5m+4$, for some $n, m \in \mathbb{Z}$. Now the fact that $2n+1 = 5m+4$ implies that $2n + 2 = 5m + 5 = 5(m+1)$, and so 2 divides $m+1$. Therefore, $m+1 = 2k$ or $m = 2k-1$ for some $k \in \mathbb{Z}$, and so

$$x = 5m + 4 = 5(2k-1) + 4 = 10(k-1) + 9.$$

Since $(k-1) \in \mathbb{Z}$, it follows that $x \in \{10k+9 : k \in \mathbb{Z}\}$. Thus we have established that $(\{2n+1 : n \in \mathbb{Z}\} \cap \{5m+4 : m \in \mathbb{Z}\}) \subseteq \{10k+9 : k \in \mathbb{Z}\}$. ∎

4. (a) The missing step: This means that $x \in A$ and $x \in B$.
(b) The missing step: It follows that $x \in A$ or $x \in B$.
(c) The missing steps:
- **Case 1**: If $x \in A$, then since $A \subseteq B$, we can infer that $x \in B$.
- **Case 2**: If $x \in B$, then we can also infer that $x \in B$.

5. The missing steps: Since $A \subseteq B$, we know that $x \in B$, and since $A \subseteq C$, we know that $x \in C$.

6. (a) *Proof.* Let $x \in \{10n-1 : n \in \mathbb{Z}\}$ be given. This means that $x = 10n - 1$ for some $n \in \mathbb{Z}$. We can

and

$$(A \times B) \times C = \{((1, b_1), x), ((1, b_2), x), ((2, b_1), x),$$
$$((2, b_2), x)\}.$$

The difference between these sets is that the first is a set of ordered 3-tuples, whereas the second set contains ordered pairs for each of which, the first coordinate is itself an ordered pair.

write, $x = 10n - 1 = 2(5n - 1) + 1$, and since $5n - 1 \in \mathbb{Z}$, it follows that $x \in \{2k+1 : k \in \mathbb{Z}\}$. This proves that $\{10n - 1 : n \in \mathbb{Z}\} \subseteq \{2k+1 : k \in \mathbb{Z}\}$.

In a similar manner, we can write $x = 10n - 1 = 5(2n - 1) + 4$, and since $(2n - 1) \in \mathbb{Z}$, it follows that $x \in \{5m+4 : m \in \mathbb{Z}\}$. This proves that $\{10n - 1 : n \in \mathbb{Z}\} \subseteq \{5m+4 : m \in \mathbb{Z}\}$. Now by Exercise 5, we can conclude that $\{10n - 1 : n \in \mathbb{Z}\} \subseteq (\{2k+1 : k \in \mathbb{Z}\} \cap \{5m+4 : m \in \mathbb{Z}\})$. ∎

7. The missing step for Case 1: Since $A \subseteq C$, we can infer that $x \in C$. The missing step for Case 2: Since $B \subseteq C$, we can infer that $x \in C$.

8. (a) *Proof.* Let $x \in \{4k+1 : k \in \mathbb{Z}\}$, so that $x = 4k + 1$ for some $k \in \mathbb{Z}$. Since we can write $x = 4k+1 = 2(2k) + 1$, and $2k \in \mathbb{Z}$, we have that $x \in \{2n+1 : n \in \mathbb{Z}\}$. Hence, $\{4k+1 : k \in \mathbb{Z}\} \subseteq \{2n+1) : n \in \mathbb{Z}\}$. Now, let $y \in \{4m+3 : m \in \mathbb{Z}$, so that $y = 4m+3$ for some $m \in \mathbb{Z}$. Since we can write $y = 4m+3 = 2(2m+1) + 1$, and $(2m+1) \in \mathbb{Z}$, we have that $y \in \{2n+1 : n \in \mathbb{Z}\}$. Hence $\{4m+3 : m \in \mathbb{Z} \subseteq \{2n+1 : n \in \mathbb{Z}\}$. By exercise 7, we now have that $(\{4k+1 : k \in \mathbb{Z}\} \cup \{4m+3 : m \in \mathbb{Z}\}) \subseteq \{2n+1 : n \in \mathbb{Z}\}$. ∎

9. (a) The missing steps:
Case 1. If $x \in A$, then we can truthfully say that $x \in A$ or $x \in B$, which can be written, $x \in A \cup B$. But we can also truthfully say that $x \in A$ or $x \in C$, which can be written, $x \in A \cup C$. Since both of these inferences are valid, we can conclude $x \in (A \cup B) \cap (A \cup C)$.
Case 2. If $x \in B \cap C$, then $x \in B$ and $x \in C$, so we can say that $x \in A$ or $x \in B$, which can be written, $x \in A \cup B$. We can also say that $x \in A$ or $x \in C$, which can be written $x \in A \cup C$. Since both of these inferences are valid, we can conclude $x \in (A \cup B) \cap (A \cup C)$.

11. (a) *Proof.* Let sets A and B be given such that $A \cup B = B$. To show that $A \cap B = A$, we must show that $A \cap B \subseteq A$ and $A \subseteq A \cap B$. Proposition 1 establishes that $A \cap B \subseteq A$ is always true, so we only need to establish $A \subseteq A \cap B$ using an element-wise proof.

Let $x \in A$ be given. Since Practice Problem 1 tells us that $A \subseteq A \cup B$, we can infer that $x \in A \cup B$. Since we are given that $A \cup B = B$, we know that $x \in B$. Since $x \in A$ and $x \in B$, we know that $x \in A \cap B$. ■

This establishes that $A \subseteq A \cap B$, and hence we conclude that $A = A \cap B$, as desired.

(c) *Proof.* Let sets A, B and C be given such that $A \cap B = A$ and $B \cap C = B$. To show that $A \cap C = A$, we must show that $A \cap C \subseteq A$ and $A \subseteq A \cap C$. Proposition 1 establishes that $A \cap C \subseteq A$ is always true, so we only need to establish $A \subseteq A \cap C$ using an element-wise proof.

Let $x \in A$ be given. Since we are given that $A \cap B = A$, we know that $x \in A \cap B$, which implies that $x \in B$, since $A \cap B \subseteq B$ by Proposition 1. Since we are given that $B \cap C = B$, we know that $x \in B \cap C$, which implies that $x \in C$, since $B \cap C \subseteq C$ by Proposition 1 again. Hence, $x \in A$ **and** $x \in C$, so $x \in A \cap C$. This establishes that $A \subseteq A \cap C$, and hence we conclude that $A = A \cap C$, as desired. ■

12. (a) *Proof by contradiction.* Assume that **there is** an element $a \in \mathbb{N}$ in both sets $\{2k + 1 : k \in \mathbb{N}\}$ and $\{4k : k \in \mathbb{N}\}$. This means that $a = 2K + 1$ for some $K \in \mathbb{N}$ and $a = 4L$ for some $L \in \mathbb{N}$. Combining these facts leads us to conclude that $2K + 1 = 4L$, which implies that

$$2(2L - K) = 1$$

or $2L - K = \frac{1}{2}$. We know (from closure properties of $\mathbb{Z}$) that it is impossible to subtract integers and get a result that is not an integer, so this last statement is absurd. Hence, there is no such number $a \in \mathbb{N}$. That is, $(\{2k + 1 : k \in \mathbb{N}\} \cap \{4k : k \in \mathbb{N}\}) = \emptyset$. ■

(c) *Proof by contradiction.* Assume that **there is** an element $(a, b) \in \mathbb{R} \times \mathbb{R}$ in both the set $\{(x, y) \in \mathbb{R} \times \mathbb{R} : x^2 - 2x - 3 = y\}$ and the set $\{(x, y) \in \mathbb{R} \times \mathbb{R} : x - 6 = y\}$. This means that

$$a^2 - 2a - 3 = b = a - 6$$

This in turn implies that $a^2 - 3a + 3 = 0$, which the quadratic formula tells us has no real solutions, a contradiction to the fact that $a \in \mathbb{R}$. ■

13. (a) *Proof.* Let sets A and B of elements in U be given such that $A \cap B = A$. To show that $A' \cup B = U$, we must show that $A' \cup B \subseteq U$ and $U \subseteq A' \cup B$. Since all sets are subsets of the universal set U, we know that $A' \cup B \subseteq U$ is true, so we only need to establish $U \subseteq A' \cup B$ using an element-wise proof.

Let $x \in U$ be given. By the definition of complement, we know that either $x \in A$ or $x \in A'$, so we consider each case separately to establish that the given x must be in $A' \cup B$.

● If $x \in A$, then since we are given that $A \cap B = A$, we know that $x \in A \cap B$, from which it follows that $x \in B$. But Practice Problem 1 can be used to establish that $B \subseteq A' \cup B$, so we conclude that $x \in A' \cup B$, as desired.

● If $x \in A'$, then since $A' \subseteq A' \cup B$ (again by Practice Problem 1), we know that $x \in A' \cup B$, as desired.

In either case, we see that $x \in A' \cup B$.

This establishes that $U \subseteq A' \cup B$, and hence we conclude that $A' \cup B = U$. ■

14. (a) The properties to use are distributive, identity or universal bound, and identity.

(c) The properties to use are distributive, negation, commutative, and identity.

15. (a) $(A \cap \emptyset) \cup (A \cap U) = A$

(b) $A \cup (A' \cap B) = A \cup B$

(c) $A \cap (A' \cup B) = A \cap B$

(d) $(A \cap B) \cup (B \cap C) = (A \cup C) \cap B$

(e) $(A \cap B) \cup (A' \cup C)' = A \cap (B \cup C')$

16. (a) Use properties d, e, c, e, and d for the steps.

(b) Use properties d, e, c, a, d, and e for the steps.

(c) Use properties d, c, a, i, and d for the steps.

18. (a) If $A \cup B = B$, then $A \cap B = A$.
Proof. Let sets A and B be given such that $A \cup B = B$. Then the following holds true:

$B = A \cup B$	given
$A \cap B = A \cap (A \cup B)$	substitution
$A \cap B = A$	absorption

(b) If $A \cap B = A$, then $A \cup B = B$.
Proof. Let sets A and B be given such that $A \cap B = A$. Then the following holds true:

$A = A \cap B$	given
$A \cup B = (A \cap B) \cup B$	substitution
$A \cup B = B \cup (B \cap A)$	commutative
$A \cup B = B$	absorption

(c) If $A \cap B = A$, then $A' \cup B = U$.
Proof. Let sets A and B be given such that $A \cap B = A$. Then the following holds true:

$$
\begin{array}{ll}
A \cap B = A & \text{given} \\
\hline
A' \cup (A \cap B) = A' \cup A & \text{substitution} \\
\hline
A' \cup (A \cap B) = U & \text{negation} \\
\hline
(A' \cup A) \cap (A' \cup B) = U & \text{distributive} \\
\hline
(A \cup A') \cap (A' \cup B) = U & \text{commutative} \\
\hline
U \cap (A' \cup B) = U & \text{negation} \\
\hline
(A' \cup B) \cap U = U & \text{commutative} \\
\hline
A' \cup B = U & \text{identity} \\
\hline
\end{array}
$$

21. (a) *Proof.* Suppose $A \subseteq B$, and let $(a, c) \in A \times C$, for some set C. By definition of $A \times C$, $a \in A$ and $c \in C$. Since, $A \subseteq B$, $a \in B$, and $(a, c) \in B \times C$, proving that $(A \times C) \subseteq (B \times C)$.

(b) *Proof.* Suppose $A \subseteq B$, and let $C \in \mathcal{P}(A)$; that is, let C be a subset of A, written $C \subseteq A$. Since we are assuming $A \subseteq B$, we have that $C \subseteq B$ as well. Hence, $C \in \mathcal{P}(B)$, proving that $\mathcal{P}(A) \subseteq \mathcal{P}(B)$.

23. (a) *Proof.* Let $C \in \mathcal{P}(A) \cap \mathcal{P}(B)$; that is, let $C \subseteq A$ and $C \subseteq B$. Thus, $C \subseteq A \cap B$ and so $C \in \mathcal{P}(A \cap B)$, establishing that $\mathcal{P}(A) \cap \mathcal{P}(B) \subseteq \mathcal{P}(A \cap B)$. Conversely, let $C \subseteq \mathcal{P}(A \cap B)$, so that $C \subseteq A \cap B$. Well then, $C \subseteq \mathcal{P}(B)$, from which it follows that $C \in \mathcal{P}(A) \cap \mathcal{P}(B)$. This establishes the inclusion, $\mathcal{P}(A) \cap \mathcal{P}(B) \supseteq \mathcal{P}(A \cap B)$, so that we now have that $\mathcal{P}(A) \cap \mathcal{P}(B) = \mathcal{P}(A \cap B)$.

Section 3.4 Exercises

1. (a) $(a \cdot b') + a = a$; **(c)** $(a \cdot b')' \cdot (a + b) = a$; **(e)** $(a \cdot b)' + (a + b) = 1$

2. (a) $a \cdot (b' + a) = a$; **(c)** $(a + b) \cdot (a' \cdot c) = a + (b \cdot c')$; **(e)** $(a \cdot b')' = a' + (a \cdot b)$

3. (a) Use properties i, a, d, and d. **(b)** Use properties c, e, a, and d.

4. (a) Claim: $(a + b) \cdot (b + c) = ac + b$
Proof. Version 1.

$$
\begin{array}{ll}
(a + b) \cdot (b + c) = (a + b) \cdot b & \text{Distributive(c)} \\
\quad + (a + b) \cdot c & \\
\hline
= b \cdot (b + a) & \text{Commutative(a)} \\
\quad + c \cdot (a + b) & \\
\hline
= b + c \cdot (a + b) & \text{Absorption(j)} \\
\hline
= b + c \cdot a + c \cdot b & \text{Distributive(c)} \\
\hline
= b + b \cdot c + a \cdot c & \text{Commutative(a)} \\
\hline
= b + a \cdot c & \text{Absorption(j)} \\
\hline
= a \cdot c + b & \text{Commutative(a)} \\
\hline
\end{array}
$$

Version 2.

$$
\begin{array}{ll}
a \cdot c + b = b + a \cdot c & \text{Commutative(a)} \\
\hline
= (b + a) \cdot (b + c) & \text{Distributive(c)} \\
\hline
= (a + b) \cdot (b + c) & \text{Commutative(a)} \\
\hline
\end{array}
$$

(c) Claim: $(a + b)(a'c)' = a + bc'$

Proof.

$$
\begin{array}{ll}
(a + b)(a'c)' = (a + b)(a'' + c') & \text{DeMorgan(h)} \\
\hline
= (a + b)(a + c') & \text{Double Negation(f)} \\
\hline
= a + b \cdot c' & \text{Distributive (c)} \\
\hline
\end{array}
$$

5. (a) Use properties d, e, c

6. (a) *Proof.* Let a and b be given such that $a + b = b$. Then

$$
\begin{array}{ll}
a \cdot b = a \cdot (a + b) & \text{Since } b = a + b \\
\hline
= a & \text{Absorption(j)} \\
\hline
\end{array}
$$

(c) *Proof.* Let a and b be given such that $a' + b = 1$. Then

$$
\begin{array}{ll}
ab' = (a')' \cdot b' & \text{Double Negative(f)} \\
\hline
= (a' + b)' & \text{DeMorgan(h)} \\
\hline
= (1)' & \text{Since } a' + b = 1 \\
\hline
= 0 & \text{Complements(k)} \\
\hline
\end{array}
$$

7. (a) becomes "If $A \cup B = B$, then $A \cup (B - A) = B$," and

(b) becomes "If $A \cap B = A$, then $B - (B - A) = A$." It is easier to prove with the abstract properties than using definitions of sets.

9. (a) The completed tables are given below:

•	1	2	3	5	6	10	15	30
1	1	1	1	1	1	1	1	1
2	1	2	1	1	2	2	1	2
3	1	1	3	1	3	1	3	3
5	1	1	1	5	1	5	5	5
6	1	2	3	1	6	2	3	6
10	1	2	1	5	2	10	5	10
15	1	1	3	5	3	5	15	15
30	1	2	3	5	6	10	15	30

+	1	2	3	5	6	10	15	30
1	1	2	3	5	6	10	15	30
2	2	2	6	10	6	10	30	30
3	3	6	3	15	6	30	15	30
5	5	10	15	5	30	10	15	30
6	6	6	6	30	6	30	30	30
10	10	10	30	10	30	10	30	30
15	15	30	15	15	30	30	15	30
30	30	30	30	30	30	30	30	30

(b) $u = 30$

(c) $z = 1$

(d) $1' = 30, 2' = 15, 3' = 10, 5' = 6, 6' = 5, 10' = 3, 15' = 2, 30' = 1$

11. Properties (a), (c), (d), & (e) are all true, so this structure is a Boolean algebra.

12. (a) A check mark indicates those pairs where $L(a, b)$ is true.

≤	1	2	3	5	6	10	15	30
1	✓	✓	✓	✓	✓	✓	✓	✓
2		✓			✓	✓		✓
3			✓		✓		✓	✓
5				✓		✓	✓	✓
6					✓			✓
10						✓		✓
15							✓	✓
30								✓

(b) $L(a, b)$ is true precisely when a evenly divides b.

13. (a) *Proof.*

$$
\begin{aligned}
a &= a \cdot 1 && \text{(d) Identity} \\
&= a \cdot (a + a') && \text{(e) Negation} \\
&= a \cdot a + a \cdot a' && \text{(c) Distributive} \\
&= a \cdot a + 0 && \text{(e) Negation} \\
&= a \cdot a && \text{(d) Identity}
\end{aligned}
$$

Section 3.5 Exercises

1. (a) Distributivity: The columns for $a \cdot (b + c)$ and $(a \cdot b) + (a \cdot c)$ are identical.

a	b	c	$b+c$	$a\cdot(b+c)$	$a \cdot b$	$a \cdot c$	$(a \cdot b) + (a \cdot c)$
0	0	0	0	0	0	0	0
0	0	1	1	0	0	0	0
0	1	0	1	0	0	0	0
0	1	1	1	0	0	0	0
1	0	0	0	0	0	0	0
1	0	1	1	1	0	1	1
1	1	0	1	1	1	0	1
1	1	1	1	1	1	1	1

(c)

a	b	c	ab	$ab+c'$	$b'+c$	$(ab+c')(b'+c)$
0	0	0	0	1	1	1
0	0	1	0	0	1	0
0	1	0	0	1	0	0
0	1	1	0	0	1	0
1	0	0	0	1	1	1
1	0	1	0	0	1	0
1	1	0	1	1	0	0
1	1	1	1	1	1	1

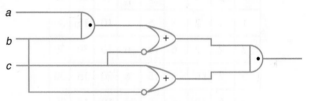

3. (a)

a	b	$a+b$	$b(a+b)$	$a+b(a+b)$
0	0	0	0	0
0	1	1	1	1
1	0	1	0	1
1	1	1	1	1

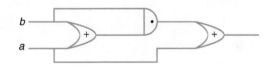

(e)

a	b	c	abc	abc'	ab'	$abc+abc'+ab'$
0	0	0	0	0	0	0
0	0	1	0	0	0	0
0	1	0	0	0	0	0
0	1	1	0	0	0	0
1	0	0	0	0	1	1
1	0	1	0	0	1	1
1	1	0	0	1	0	1
1	1	1	1	0	0	1

4. (a) $a + b$; **(c)** $c'b' + abc$; **(e)** a

5. (a) $ab'c' + abc$; **(b)** $a'b'c' + a'b'c + a'bc' + a'bc$; **(c)** $a'b'c + a'bc + ab'c + abc$

8. (a) Equivalent; **(c)** Equivalent

9. (a) $x'z'$; **(c)** z; **(e)** yz

10. (a) $x'z'w' + xy'z' + xzw' + x'y'z$; **(b)** $zw' + x'yw +$
$xy'w'$; **(c)** $x'y + yw + x'zw + xy'w' + y'z'w'$

11. (a) $xy + x'y'z' + yzw'$; **(c)** $z + xy' + x'y$

12. (a) $x'y + zw' + xy'z + xz'w$

13. (a) $y + x'$; **(c)** $y' + z'w$

Section 4.1 Exercises

1. **(a)**

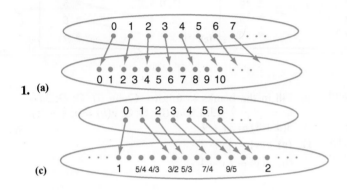

(c)

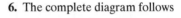

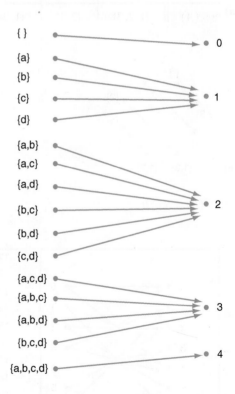

Arrow diagram for f

2. Answers will vary. **(a)** Domain $\{z \in \mathbb{R} : z \neq -1\}$,
Codomain $\mathbb{R}$; **(c)** Domain $\{x \in \mathbb{R} : x \geq -\frac{1}{2}\}$, Codomain
$\{y \in \mathbb{R} : y \geq 0\}$; **(e)** Domain $\mathbb{R}$, Codomain $\{y \in \mathbb{R} : 0 <$
$y \leq 1\}$

3. The complete diagram is shown below:

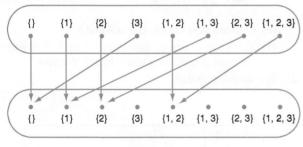

Arrow diagram for f

4. The complete diagram is shown below. Once again, a
"loop" at a value indicates that the function maps that
value to itself.

6. The complete diagram follows

8. The diagrams follow.

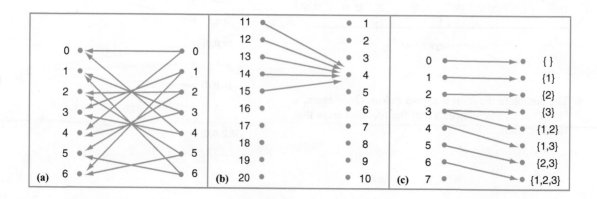

10. (a) Yes, this is a function.

(b) No. The element 1 in the domain is mapped to 2 and 5 in the codomain.

(c) No. The rational number 0.5 can be written as $\frac{1}{2}$ or $\frac{2}{4}$ (as well as many other ways), hence the domain element 0.5 is mapped to 1 and 2 (among other values) in the codomain.

12. (a) and **(c)**

14. (a)

A	$\emptyset$	$\{1\}$	$\{2\}$	$\{3\}$	$\{1, 2\}$	$\{1, 3\}$	$\{2, 3\}$	$\{1, 2, 3\}$
$c(A)$	$\{1, 2, 3\}$	$\{2, 3\}$	$\{1, 3\}$	$\{1, 2\}$	$\{3\}$	$\{2\}$	$\{1\}$	$\emptyset$

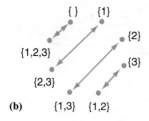

(b)

(c) $c^{-1}(B) = S - B$

16. Fill in the blanks as follows. **Claim (i):** $\mathbb{Q}$; $\mathbb{Q}$; $3a + 9$; $a = \frac{b-9}{3} = \frac{b}{3} - 3$. **Claim (ii):** $\mathbb{Q}$; $\mathbb{Q}$; $g(b) = a$; $\frac{b}{3} - 3$; $b = 3(a + 3) = 3a + 9$.

17. (a) $g(y) = \frac{1}{3}y + 2$

(b) $g(y) = \frac{1}{2}y - 4$

(c) $g(y) = 3y + 3$

(d) $g(y) = y - 5$

19. (a) Let $a \in \mathbb{Q}^{\geq 0}$ and $b \in \mathbb{R}$ be given such that $f(a) = b$. That is, $\sqrt{a} = b$. From this it follows that $a = b^2$ and hence $a = g(b)$.

(b) One counterexample is $a = 4$ and $b = -2$.

(c) There are g-arrows, like from -2 to 4, that when reversed do not correspond to any f-arrow.

21. The diagrams follow.

23. (a)

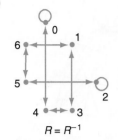

$$R = R^{-1}$$

(b)

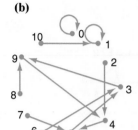

R

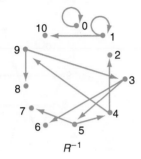

R^{-1}

(c)

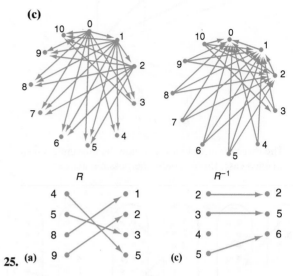

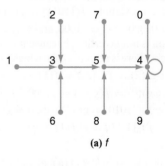

25. (a) ... **(c)**

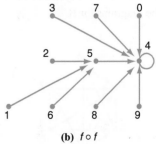

26. (a) yes; **(b)** yes; **(c)** no; **(d)** yes

Section 4.2 Exercises

1. (a) $2z^2 - 1$; **(b)** $4y + 3$; **(c)** $3x + 2$

3. (a) $f(1) = 11$ and $g(11) = 5$
   (b) The diagrams for f and g appear below:

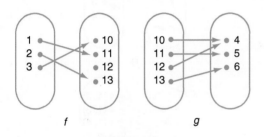

f g

(c) Only $g \circ f$ is defined.
(d) The function $(g \circ f) : \{1, 2, 3\} \to \{4, 5, 6\}$ has the following arrow diagram:

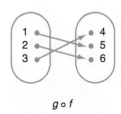

$g \circ f$

5. f is not invertible since $f(0) = f(9) = f(5) = f(4) = 4$, for example, and there is no input that has the out-

put of 9, for example. The diagrams for f, $f \circ f$, and $f \circ f \circ f$ are given below. (In each case, the "loop" at 4 indicates an arrow pointing from 4 to itself.)

(a) f

(b) $f \circ f$

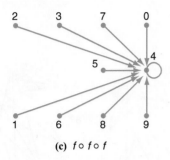

(c) *f ∘ f ∘ f*

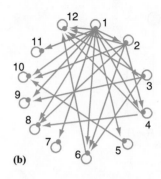

(b)

6. The three pictures below provide the missing diagrams. In each case, there is only one possible answer.

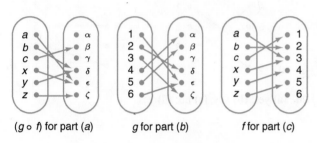

(*g ∘ f*) for part (*a*)   *g* for part (*b*)   *f* for part (*c*)

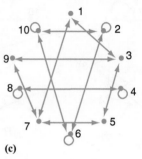

(c)

15. Here is the complete picture:

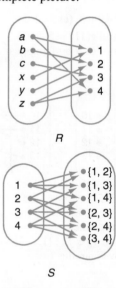

8. Parts **(b)** and **(c)** are impossible. For part **(a)**, *g ∘ f* has domain {1, 2, 3, 4}, codomain {*x, y, z*}, and rule {(1, *y*), (2, *x*), (3, *x*), (4, *z*)}.

10. (a) $f(11011) = 4$, $f(01101) = 3$, and $f(11000) = 2$. This function is not invertible because it is not one-to-one. For example, $f(11011) = f(11110)$.

(b) $g(0) = 00000$, $g(2) = 11000$, and $g(4) = 11110$. This function is not invertible because it is not onto. For example, there is no value *n* for which $g(n) = 10101$.

(c) ● $(f ∘ g)(2) = f(g(2)) = f(11000) = 2$
● $(f ∘ g)(0) = f(g(0)) = f(00000) = 0$
● $(g ∘ f)(11010) = g(f(11010)) = g(3) = 11100$
● $(g ∘ f)(11100) = g(f(11100)) = g(3) = 11100$

(d) No.

11. Fill in the blanks with: *b*; *a*; $f(a)$; *a*; *a*; *b*; *a*; *b*; *f ∘ g*.

13. Here are the diagrams for R:

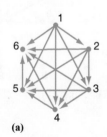

(a)

R

S

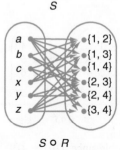

S ∘ R

17. $(x, y) \in R \circ R$ means x beat at least one team that beat y.

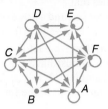

19. Student x is in instructor b's course.

21. $(u, v) \in R$ means that production facility u produces machine part v.

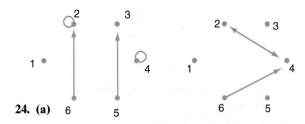

24. (a)

26. (a) $(x, y) \in R_1 \circ R_1^{-1}$ if classes x and y have a student in common, and $(x, y) \in R_1^{-1} \circ R_1$ if student x and student y have a class in common.
(c) $(x, y) \in R_3 \circ R_3^{-1}$ if integers x and y are such that $x = \pm y$, and $R_3^{-1} \circ R_3$ relates each nonnegative perfect square to itself.

Section 4.3 Exercises

1. (a) There is no $x \in \mathbb{R}$ for which $f(x) = -4$ since the equation $x^2 + 4x + 1 = -4$ has no real solutions, by the quadratic formula; **(c)** There is no $x \in [1, \infty)$ for which $h(x) = 2$ since the equation $\frac{1}{x+1} = 2$ can only have solution $-\frac{1}{2}$, which is not in $[1, \infty)$.

2. (a) -1 and -3; **(c)** 0 and $-\frac{2}{3}$

3. (a) One-to-one and onto, and hence invertible; **(c)** One-to-one but not onto

4. (a) not $1-1$ but onto

5. (a) yes; **(b)** yes; **(c)** $c^{-1} = c$

7. Fill in the blanks with: $h(x_1) = h(x_2)$; "f is one-to-one"; $2 \cdot x_1 = 2 \cdot x_2$; $x_1 = x_2$

10. (a) Consider any player of the game. Since the cards were evenly divided, the players all have at least 17 cards (they have 17, 17, and 18). Let A be the set of cards the player has, and let B be the set of possible values (i.e., $\{ace, king, \ldots, 4, 3, 2\}$). Define f by $f(x) =$ the value of the card x. Since A is size 17 or 18, and B is size 13, f is not $1-1$. So there are two cards in A of the same value.

(c) Let $A = \{a, b, c, d, e\}$ be the set of five positive integers. Let $B = \{7, 9, 3, 1\}$. Define f by $f(x) =$ the ones' digit of 7^x. Since A has size 5 and B has size 4, f is not $1-1$. Thus there are two numbers y, z in the set A for which 7^y and 7^z have the same ones' digit.

(e) Let A be the set of numbers in the phone number, and let $B = \{0, 1, 2, 3, 4, 5\}$. Define f by $f(x) = x$ mod 6 (that is, the remainder when x is divided by 6). Since A is size 7 and B is size 6, f is not $1-1$. For the two numbers x and y for which $f(x) = f(y)$, $x - y$ is divisible by 6.

11. *Proof.* Let $x_1, x_2 \in \mathbb{Z}$. Assume that $f(x_1) = f(x_2)$. Then $2x_1 + 3 = 2x_2 + 3$. This implies that $2x_1 = 2x_2$ and it follows that $x_1 = x_2$. Therefore f is $1-1$. ∎

14. (a) Let $x_1, x_2 \in [0, \infty)$ be given such that $f(x_1) = f(x_2)$. That is, $x_1^2 + 4 = x_2^2 + 4$, so $x_1^2 - x_2^2 = 0$, hence either $x_1 = x_2$ or $x_1 = -x_2$. Since x_1 and x_2 are both from $[0, \infty)$, they must have the same sign, so only the option $x_1 = x_2$ is possible.

(b) Let $y \in [4, \infty)$ be given, and set $x = \sqrt{y-4}$. We can verify with algebra that

$$g(x) = g(\sqrt{y-4})$$
$$= \left(\sqrt{y-4}\right)^2 + 4$$
$$= (y-4) + 4 = y$$

Hence, y is an output of the function g. Therefore g is onto.

(c) $f^{-1}(y) = \sqrt{y-4}$

16. (a) *Proof.* Let $x_1, x_2 \in \mathbb{R}$ be given such that $f(x_1) = f(x_2)$. That is, $x_1^3 - 2 = x_2^3 - 2$, or $x_1^3 = x_2^3$. Taking the cube root of both sides tells us that $x_1 = x_2$, as desired. ∎

(b) *Proof.* Let $y \in \mathbb{R}$ be given, and set $x = \sqrt[3]{y+2}$, a real number. Then $f(x) = x^3 - 2 = (y+2) - 2 = y$. Hence, y is an output of the function f. ∎

(c) $g(y) = \sqrt[3]{y+2}$ is the inverse of f.

18. (a) *Proof.* Let $x_1, x_2 \in (1, \infty)$ be given such that $f(x_1) = f(x_2)$. That is, $\frac{x_1}{x_1-1} = \frac{x_2}{x_2-1}$. Multiplying both sides by $(x_1-1)(x_2-1)$ gives us $x_1x_2 - x_1 = x_1x_2 - x_2$, from which it follows that $x_1 = x_2$, as desired. ∎

(b) *Proof.* Let $y \in (1, \infty)$ be given, and set $x = \frac{y}{y-1}$. Since $y > y - 1 > 0$, it follows that $\frac{y}{y-1} > 1$. Hence, $x \in (1, \infty)$, and

$$f(x) = \frac{\frac{y}{y-1}}{\frac{y}{y-1} - 1} = \frac{\frac{y}{y-1}}{\frac{y-(y-1)}{y-1}} = \frac{\frac{y}{y-1}}{\frac{1}{y-1}} = y$$

Therefore, y is an output of the function f. ∎

(c) $f^{-1}(y) = \frac{y}{y-1}$

19. (a) Let $(x_1, y_1), (x_2, y_2) \in A$ be given such that $f(x_1, y_1) = f(x_2, y_2)$. That is, $2^{x_1} \cdot 5^{y_1} = 2^{x_2} \cdot 5^{y_2}$, which implies that $2^{x_1-x_2} = 5^{y_2-y_1}$. The only number that is a power of 2 and a power of 5 is the number 1. Hence, $x_1 - x_2 = y_2 - y_1 = 0$, so $(x_1, y_1) = (x_2, y_2)$.

(b) Let n be a positive factor of 500. Since $500 = 2^2 \times 5^3$, then n must have the form $2^a \times 5^b$ where $a \in \{0, 1, 2\}$ and $b \in \{0, 1, 2, 3\}$. That is, $n = f(a, b)$. Hence, n is an output of the function f.

(c) Theorem 7

(d) $n(A) = 3 \times 4 = 12$

21. Since $f(abc) = f(acb) = ab$, f is not 1 – 1. However, f is onto. Given a word y of length 4 or less over the alphabet $\{a, b\}$, form the word x by putting the letter c at the end of the word y. (For example, if y is *abab*, then x would be *ababc*.) Then x is a word of length 5 or less over the alphabet $\{a, b, c\}$, and $f(x) = y$. The set A is bigger.

24. (a) Let $S \in \mathcal{P}(\{1, 2\})$. Set $T = S \cup \{3\}$. Since $S \subseteq \{1, 2\}$, it follows that $T \subseteq \{1, 2, 3\}$. Moreover, $f(T) = T - \{3\} = S$. Hence, S is an ouput of the function f.

(b) $f(\{1, 2\}) = f(\{1, 2, 3\}) = \{1, 2\}$

(c) Given $S \in \mathcal{P}(\{1, 2\})$, we have $f(S \cup \{3\}) = f(S) = S$.

(d) Since every element of the codomain has exactly two arrows pointing to it, we conclude that the domain has twice as many elements as the codomain.

28. The function $f : \mathbb{R}^{\geq 0} \to \mathbb{R}$ with the rule $f(x) = x$ is one-to-one, and the function $g : \mathbb{R} \to \mathbb{R}^{\geq 0}$ with the rule $g(y) = 2^y$ is one-to-one. Hence, by Theorem 9, $\mathbb{R}^{\geq 0}$ and $\mathbb{R}$ have the same size.

29. $B = \{a, c, e\}$

Section 4.4 Exercises

1. (a) This relation is transitive, antisymmetric and reflexive.

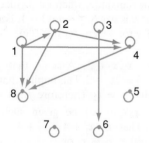

(c) This relation is transitive, but not antisymmetric since $(2, 8) \in R$ and $(8, 2) \in R$, and not reflexive since $(5, 5) \notin R$.

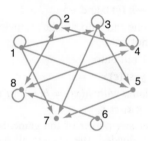

2. (a) Reflexive

3. (a) Since $(1, 3) \in R_1$ and $(3, 1) \in R_1$, R_1 is not antisymmetric.

4. (a) Yes

5. (a) The blanks might be filled in as follows.

- $R_1 = \{(a, b) \in A \times A : a \text{ is a proper factor of } b\}$, where $A = \{1, 2, 3, 4, 6, 12\}$
- $R_2 = \{(a, b) \in B \times B : \underline{a < b}\}$, where $B = \{1, 2, 3, 4, 5\}$

(b) Neither relation is reflexive. We can extend these relations to the reflexive relations R'_1 and R'_2:

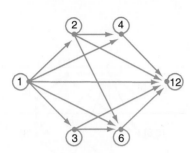

Relation R'_1

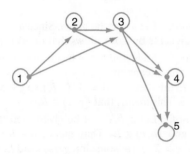

Relation R'_2

These can be interpreted as follows: $(x, y) \in R'_1$ means "x divides y"; and $(x, y) \in R'_2$ means "$x \le y$"

(c) Both of the relations are transitive.

7. (a) irreflexive, antisymmetric; **(b)** irreflexive, transitive, antisymmetric

9. (a) reflexive, transitive; **(b)** transitive, antisymmetric; **(c)** antisymmetric, transitive, irreflexive

11. transitive

13. (a) There is a loop at every point. **(b)** There are no 2-way streets. **(c)** If there's an arrow from a to b and one from b to c, then there must be an arrow from a to c.

15.

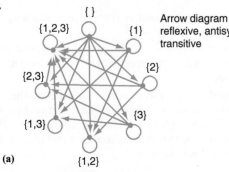

Arrow diagram
reflexive, antisymmetric
transitive

(a)

(c) Not reflexive, since $\{1\} \cap \{1\} \neq \emptyset$. Not antisymmetric, since $\{1\}R\{2\}$ and $\{2\}R\{1\}$. Not transitive, since $\{1\}R\{2\}$ and $\{2\}R\{1\}$, but not $\{1\}R\{1\}$.

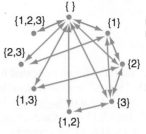

not reflexive, not antisymmetric
not transitive

17. (a) If R is reflexive, then R^{-1} is reflexive.
Proof. Let the reflexive relation R on the set A be given. Let $a \in A$ be given. Since R is reflexive, we know $(a, a) \in R$, from which it follows (by reversing the coordinates) that $(a, a) \in R^{-1}$. Hence, R^{-1} is reflexive. ∎

(b) If R is antisymmetric, then R^{-1} is antisymmetric.
Proof. Let the antisymmetric relation R on the set A be given. Let $a, b \in A$ be given such that $(a, b) \in R^{-1}$ and $(b, a) \in R^{-1}$. By definiton of R^{-1}, this means that $(b, a) \in R$ and $(a, b) \in R$, and since R is antisymmetric, it follows that $a = b$. Hence, R^{-1} is antisymmetric. ∎

(c) If R is transitive, then R^{-1} is transitive.
Proof. Let the transitive relation R on the set A be given. Let $a, b, c \in A$ be given such that $(a, b) \in R^{-1}$ and $(b, c) \in R^{-1}$. This means that $(b, a) \in R$ and $(c, b) \in R$, and the transitivity of R tells us that $(c, a) \in R$. From this, it follows that $(a, c) \in R^{-1}$. Hence, R^{-1} is transitive. ∎

20. (a) $\{(0, 1), (1, 2), (0, 3), (1, 4), (0, 0), (1, 1), (2, 1), (2, 2)\}$; **(b)** $\{(1, 2)\}$; **(c)** $\{(0, 1), (0, 3), (1, 4), (0, 0)\}$

21. (a) *Proof.* Let R_1 and R_2 be reflexive relations on the set A. Let $a \in A$ be given. Since R_1 and R_2 are reflexive, we know that $(a, a) \in R_1$ and $(a, a) \in R_2$, from which it follows that $(a, a) \in R_1 \cup R_2$. Hence, $R_1 \cup R_2$ is reflexive. ∎

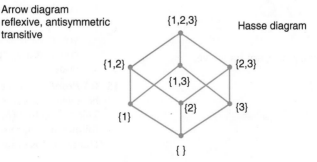

Hasse diagram

(c) This statement is false, as we can see from an example of relations on the set $A = \{1, 2, 3\}$. Let $R_1 = \{(1, 2), (1, 3), (2, 3)\}$ and $R_2 = \{(2, 1), (3, 1), (3, 2)\}$. Both of these relations are antisymmetric, while their union

$$R_1 \cup R_2 = \{(1, 2), (1, 3), (2, 3), (2, 1), (3, 1), (3, 2)\}$$

is clearly not antisymmetric, having $(1, 2)$ and $(2, 1)$ both within, for example.

Section 4.5 Exercises

1. Relation R_1 is symmetric and reflexive as indicated by the "double arrows" and the "loops," respectively, in the figure below. The figure also shows that relation R_3 is not symmetric since $(1, 3) \in R_1$ but $(3, 1) \notin R_3$ for example, and R_3 is not reflexive since $(3, 3) \notin R_3$ for example.

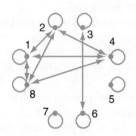   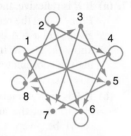

Relation R_1 Relation R_3

2. (a) Yes; **(c)** No, since $(2, 1) \in R_3$ but $(1, 2) \notin R$.

3. (a) symmetric; **(c)** symmetric

4. (a) This is not a partition since not every element is a set. **(c)** This is a partition of A.

6. $\{\{ABCD, BCDA, CDAB, DABC\}, \{ABCD, BDCA, DCAB, CABD\}, \{ACBD, CBDA, BDAC, DACB\}, \{ACDB, CDBA, DBAC, BACD\}, \{ADBC, DBCA, BCAD, CADB\}, \{ADCB, DCBA, CBAD, BADC\}\}$

8. (a) Reflexive, not symmetric, transitive.

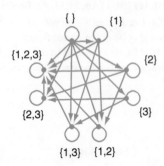

(c) Not reflexive, symmetric, not transitive.

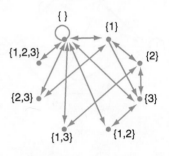

10. (a) *Proof.* Let $a \in A$ be given. Since $a - a = 0$ and 0 is divisible by 3, it follows that $(a, a) \in R$.

(b) *Proof.* Let a and $b \in A$ be given such that $(a, b) \in R$. This means that $x - y = 3 \cdot K$ for some integer K. In this case, $y - x = 3 \cdot (-K)$, so $y - x$ is divisible by 3. This means that $(y, x) \in R$.

(c) *Proof.* Let $a, b, c \in A$ be given such that $(a, b) \in R$ and $(b, c) \in R$. That means $a - b = 3 \cdot K$ and $b - c = 3 \cdot L$ for some integers K and L. In this case,

$$
\begin{aligned}
a - c &= (a - b) + (b - c) \\
&= 3K + 3L \\
&= 3(K + L)
\end{aligned}
$$

Hence, $a - c$ is divisible by 3, so $(a, c) \in R$.

(d) $\{\{0, 3, 6\}, \{1, 4\}, \{2, 5\}\}$

12. Each person who has a son is related to himself/herself, and to his/her spouse (if in the set E). This relation is not reflexive, but it is symmetric and transitive.

13. (a) A relation is reflexive if there is a loop at every node.

(b) A relation is symmetric if only double arrows are necessary because all arrows point both ways.

(c) A relation is transitive if every time you see ⟋, you must also see ⟋, and every pair of nodes connected by "double arrows" have a loop at each node.

15. (a) *Proof.* Let the symmetric relation R on the set A be given. Let $a, b \in A$ be given such that $(a, b) \in R^{-1}$. This means that $(b, a) \in R$, and since R is symmetric, it follows that $(a, b) \in R$. But this means that $(b, a) \in R^{-1}$. Hence, R^{-1} is symmetric. ∎

(c) *Proof.* Let R be any relation on A. Let $a, b \in A$ be given such that $(a, b) \in R \circ R^{-1}$. This means that, for some $c \in A$, we have $(a, c) \in R^{-1}$ and $(c, b) \in R$. This implies that $(c, a) \in R$ and $(b, c) \in R^{-1}$, from which we can conclude that $(b, a) \in R \circ R^{-1}$. Therefore, $R \circ R^{-1}$ is symmetric. ∎

16. **(a)** $\{(0,0), (0,1), (0,2), (1,2), (2,1), (3,3), (3,4), (4,3), (4,4)\}$; **(b)** $\{(1,2), (3,4), (4,3)\}$; **(c)** $\{(0,0), (2,1), (3,3), (4,4)\}$; **(d)** $\{(0,1), (1,2), (0,2), (3,4), (4,3), (3,3), (4,4)\}$; **(e)** $\{(0,0), (1,1), (1,2), (2,1), (2,2), (3,3), (3,4), (4,3), (4,4)\}$

18. **(a)** $R_1 = \{(1,1), (2,2), (3,3), (4,4), (5,5)\}$ and $R_2 = \{(1,2), (2,1)\}$. Note that the empty set is technically

a symmetric (and transitive) relation, but it is not reflexive.
(b) $R_1 = \{(1,1), (2,2), (2,3), (3,3), (4,4), (5,5)\}$ and $R_2 = \{(1,2), (2,1)\}$
(c) $R_1 = \{(1,1), (1,2), (2,1), (2,2)\}$ and $R_2 = \{(1,1), (1,3), (3,1), (3,3)\}$

19. **(a)** $\{(0,1), (1,0), (0,0), (1,2), (2,1), (0,3), (3,0), (2,2)\}$; **(b)** Yes; **(c)** If R is symmetric, then $R = R^{-1}$, so $R \cup R^{-1} = R$.

22. *Proof.* Let R be a relation on a set A with the property that $R = R^{-1}$, and let $a, b \in A$ be given such that $(a, b) \in R$. From this we can conclude that $(b, a) \in R^{-1}$. However, since $R = R^{-1}$ this tells us that $(b, a) \in R$. Therefore, R is symmetric. ∎

Section 4.6 Exercises

1. The two columns at the right are identical as predicted by the Example.

x	$-x$	$\lfloor -x \rfloor$	$-\lfloor -x \rfloor$	$\lceil x \rceil$
0.5	−0.5	−1	1	1
10.4	−10.4	−11	11	11
−5.3	5.3	5	−5	−5
4	−4	−4	4	4
−17	17	17	−17	−17

2. **(a)** False. $x = 2.3$, $y = 2.7$; **(c)** False. $x = 1.5$
4. 48
6. 955
8. 19729
10. We have
$$\frac{\log_{10}(3^n)}{\log_{10}(2^n)} = \frac{n \cdot \log_{10}(3)}{n \cdot \log_{10}(2)} = \frac{\log_{10}(3)}{\log_{10}(2)} \approx 1.58$$

so the number of digits in 3^n is roughly 1.6 times the number of digits in 2^n.

11. Let $a = \log_2 10$ and $b = \log_{10} 2$. This means $2^a = 10$ and $10^b = 2$. So $(2^a)^b = 10^b = 2$, which means $a \cdot b = 1$. Hence, $a = \frac{1}{b}$, as desired.
13. **(a)** 666; **(b)** 400; **(c)** 285
15. **(a)** 933; **(c)** 1200; **(e)** 1466
18. Since $100! = 100 \cdot 99!$, $100!$ is $99!$ with 2 more zeroes on the right.
19. 158
22. 24
24. 992
27. **(a)** 494
28. **(a)** 3768
29. 2, 3, 5, 7, 9, 11, 13, 17, 19, 23, 29 and 31
31. **(a)** True; **(c)** True
32. **(a)** $Pow(x, Sum(y, z)) = Prod(Pow(x, y), Pow(x, z))$
33. **(a)** False. $x = 5$, $y = 2$; **(c)** False. $x = 5$, $y = 3$, $z = 2$

Section 4.7 Exercises

1. **(a)** 34; 96. **(b)** For example, 1 is not in the range. For $f(n)$ to be 1, n would have to be spelled "A," and no number has that spelling. **(c)** Since $f(255) = 240$ and $f(240) = 216$, this completes the cycle. **(d)** It takes too long to compute f for one thing, but more importantly f has no fixed point so you cannot know the value at any time.

2. **(a)** If a card is in position x, then after the deal into three face up piles, the card will be in position $\lceil x/3 \rceil$ of

its particular pile. To see this, just note that cards 1, 2, and 3 will be in position 1 of their piles; cards 4, 5, and 6 will be in position 2; and so on. The formula $\lceil x/3 \rceil$ yields precisely these results. Since the magician places a pile with 11 cards on top after dealing all the cards, the card in position x before the deal is in position $11 + \lceil x/3 \rceil$ after the deal.

3. The only two cycles are 1, 6, 3, 8, 4, 2, 1 and 5, 10, 5. Formally we prove the following by induction: For every

$n \geq 1$, the iterated function sequence (i.f.s.) for g starting with n leads to one of these two cycles.

Proof. Let $P(n)$ be the statement, "The i.f.s. for g starting with n leads to one of these two cycles." The numbers in the two given cycles illustrate that $P(1), P(2), \ldots, P(6)$ are all true. Let $m \geq 7$ be given such that the statement $P(m)$ is the first one not yet checked. We argue in two cases depending on whether m is even or odd.

Case 1: If m is even, then $f(m) = m/2$, and the i.f.s. starting with m consists of m followed by the i.f.s. starting with $m/2$. Since $m/2 < m$, we know statement $P(m/2)$ has already been checked to be true, so the i.f.s. starting with $m/2$ leads to one of the two given cycles.

Case 2: If m is odd, then $f(m) = m+5$ and $f(f(m)) = \frac{m+5}{2}$, and the i.f.s. starting with m consists of m followed by $m+5$ followed by the i.f.s. starting with $\frac{m+5}{2}$. Since $\frac{m+5}{2} < m$ (since $m > 5$) we know statement $P\left(\frac{m+5}{2}\right)$ has already been checked to be true, so the i.f.s. starting with $\frac{m+5}{2}$ leads to one of the two given cycles.

5. (a) There is a fixed point at 1 and cycles $2 \to 3 \to 2$ and $4 \to 5 \to 7 \to 9 \to 6 \to 4$.

6. $\frac{1+\sqrt{5}}{2} \to \frac{1+\sqrt{5}}{2}$ and $\frac{1-\sqrt{5}}{2} \to \frac{1-\sqrt{5}}{2}$

8. $1 \to 0 \to 1$

10. $\frac{1+\sqrt{3}}{2} \to \frac{1+\sqrt{3}}{2}$ and $\frac{1-\sqrt{3}}{2} \to \frac{1-\sqrt{3}}{2}$. There are no cycles with period two.

13. $\sqrt{z} \to \sqrt{z}$ and $-\sqrt{z} \to -\sqrt{z}$ (Remember that the letter z is a fixed positive number throughout the process.) There are no cycles with period two.

15. $0 \to 0, 2/3 \to 2/3$, and $2/5 \to 4/5 \to 2/5$

17. Regardless of the starting value (between 0 and 1), the iterated function sequence in the spreadsheet will eventually be 0 even when we should see the the the cycles discussed in Exercise 16. Since computers typically use "base two" representation of numbers, the values of these sequences as fractions with odd denominators cannot be represented exactly. Hence even though we set an initial value like 2/3 in cell A1, the computer stores something very close to but not exactly equal to 2/3 in its memory. This very small initial difference gradually changes the very nature of the sequence as it converges to 0 instead of staying in a cycle.

19. (a) *Proof.* Let $n \geq 2$ be given. Since $F_n = F_{n-1} + F_{n-2}$, it follows (by dividing through by F_{n-1}) that

$$\frac{F_n}{F_{n-1}} = 1 + \frac{F_{n-2}}{F_{n-1}}$$

By definition of r_n, this is the same thing as $r_n = 1 + \frac{1}{r_{n-1}}$.

(b) $g(x) = 1 + \frac{1}{x}$

(c) $\frac{1+\sqrt{5}}{2} \to \frac{1+\sqrt{5}}{2}$ and $\frac{1-\sqrt{5}}{2} \to \frac{1-\sqrt{5}}{2}$. The values of r_n are the ratio of successive Fibonacci numbers, and these ratios "converge" to $\frac{1+\sqrt{5}}{2}$ as n is increased.

Section 4.8 Exercises

1. The completed table is shown below.

Length of list (n)	1	2	3	4	5	6	7	8	9	10
Number of comparisons	1	2	2	3	3	3	3	4	4	4
Value of $\lfloor \log_2(n) \rfloor$	0	1	1	2	2	2	2	3	3	3

2. (a) doubled; (c) roughly doubled; (e) roughly multiplied by 4; (g) roughly squared.

4. (a) $2 \cdot n \leq 2n + 1 \leq 2.5 \cdot n$ for all $n \geq 2$; (c) $1 \cdot n \leq n + \sqrt{n} \leq 2 \cdot n$ for all $n \geq 1$.

5. (b) $10n + 7 \leq 2 \cdot n^2$ for all $n \geq 6$.

6. (a) $3n - 7 \geq 2 \cdot n$ for all $n \geq 7$.

7. Taking the ratio a_n/n^2 and simplifying yields $\frac{1}{2} - \frac{1}{2n}$. We show that $\frac{1}{4} \leq \frac{1}{2} - \frac{1}{2n} \leq \frac{1}{2}$ for all $n \geq 2$. Since $n > 0$, then $\frac{1}{2n} > 0$, and hence $\frac{1}{2} - \frac{1}{2n} \leq \frac{1}{2}$. On the other hand, since $n \geq 2$, we know that $2n \geq 4$, and therefore $\frac{1}{2n} \leq \frac{1}{4}$, so $\frac{1}{2} - \frac{1}{2n} \geq \frac{1}{2} - \frac{1}{4} = \frac{1}{4}$. By Proposition 3, it follows that $a_n \in \Theta(n^2)$.

9. 17; Thus for $n \geq 1$, we have $5 \leq a_n/n^3 \leq 22$. By Proposition 3, it follows that $a_n \in \Theta(n^3)$.

11. (a) Taking the ratio a_n/n^3 and simplifying yields $5 - \frac{4}{n} - \frac{6}{n^2} - \frac{7}{n^3} = 5 - (\frac{4}{n} + \frac{6}{n^2} + \frac{7}{n^3})$. For $n \geq 12$, each of the fractions $\frac{4}{n}, \frac{6}{n^2}$, and $\frac{7}{n^3}$ is less than $\frac{1}{3}$. Thus for $n \geq 12$, we have $4 \leq a_n/n^3 \leq 5$.

(b) Taking the ratio a_n/n^3 and simplifying yields $5 + \frac{4}{n} - \frac{6}{n^2} + \frac{7}{n^3}$. This satisfies the inequality

$$5 - \frac{4}{n} - \frac{6}{n^2} - \frac{7}{n^3} \leq 5 + \frac{4}{n} - \frac{6}{n^2} + \frac{7}{n^3}$$
$$\leq 5 + \frac{4}{n} + \frac{6}{n^2} + \frac{7}{n^3}$$

From part (a), we know that for $n \geq 12$, $\frac{4}{n} + \frac{6}{n^2} + \frac{7}{n^3} \leq 1$. Hence the inequality,

$$5 - \left(\frac{4}{n} + \frac{6}{n^2} + \frac{7}{n^3} \right) \leq a_n \leq 5 + \left(\frac{4}{n} + \frac{6}{n^2} + \frac{7}{n^3} \right)$$

implies the inequality, $4 \leq a_n \leq 6$. Thus for $n \geq 12$, we have $4 \leq a_n/n^3 \leq 6$.

(c) Taking the ratio a_n/n^3 and simplifying yields $5 - \frac{4}{n} - \frac{6}{n^2} + \frac{7}{n^3}$. This satisfies the inequality

$$5 - \left(\frac{4}{n} + \frac{6}{n^2} + \frac{7}{n^3} \right) \leq 5 - \frac{4}{n} - \frac{6}{n^2} + \frac{7}{n^3} \leq 5 + \left(\frac{4}{n} + \frac{6}{n^2} + \frac{7}{n^3} \right)$$

and the result follows just as in part (b).

13. (a) 11, 21, 51, 101; **(c)** 16, 26, 56, 106

14. (a) Let $a_n = n^2$. Let positive numbers K and N be given. We must find an integer $n > N$, for which $\frac{a_n}{n} > K$. Since $\frac{a_n}{n} = n$, we may take any integer for n that is larger than both K and N. If we do, we will have $n > N$ and also $\frac{a_n}{n} > K$.

15. Answers can vary in each case.

(a) For $a_n = 2a_{n-1} + n^3$ and $a_1 = 1$, it appears in the spreadsheet that $\frac{1}{4} \cdot n^3 \leq a_n \leq \frac{1}{2} \cdot n^3$ for all $n \geq 10$.

(b) For $a_n = a_{n-1} + (\frac{1}{2}n + 17)$ and $a_1 = 10$, it appears in the spreadsheet that $\frac{1}{5} \cdot n^2 \leq a_n \leq \frac{1}{2} \cdot n^2$ for all $n \geq 69$.

(c) For $a_n = a_{n-1} + \sqrt{n}$ and $a_1 = 3$, it appears in the spreadsheet that $\frac{1}{2} \cdot n^{3/2} \leq a_n \leq \frac{3}{4} \cdot n^{3/2}$ for all $n \geq 13$.

18. (a) *Proof by induction.* We will prove that $a_n \leq 3n^2$ for all $n \geq 1$. Since $a_1 = 1$, the first statement, "$a_1 \leq 3$" is true. Let $m \geq 2$ be given such that the statement "$a_m \leq 3m^2$" is the first one not yet checked to be true. In particular, the statement, "$a_{m-1} \leq 3(m-1)^2$" has already been checked to be true. From this we can infer that

$$\begin{aligned} a_m &= a_{m-1} + 3m \\ &\leq 3(m-1)^2 + 3m \\ &= 3m^2 - 3(m-1) \\ &\leq 3m^2 \text{ since } m - 1 \geq 0 \end{aligned}$$

Hence $a_m \leq 3m^2$, which is precisely statement $P(m)$. ∎

(d) *Proof by induction.* We will prove that $d_n \geq n$ for all $n \geq 1$. Since $d_1 = 1$, the first statement, "$d_1 \geq 1$" is true. Let $m \geq 2$ be given such that the statement "$d_m \geq m$" is the first one not yet checked to be true. In particular, the statement, "$d_{m-1} \geq m - 1$" has already been checked to be true. >From this we can

infer that

$$\begin{aligned} d_m &= d_{m-1} + \lceil \sqrt{m} \rceil \\ &\geq (m-1) + \lceil \sqrt{m} \rceil \\ &= m + (\lceil \sqrt{m} \rceil - 1) \\ &\geq m \text{ since } \lceil \sqrt{m} \rceil \geq 1 \end{aligned}$$

Hence $d_m \geq m$, which is precisely statement $P(m)$. ∎

20. (a) *Proof.* Since for all $n \in \mathbb{N}$, $1 \cdot f(n) \leq f(n) \leq 1 \cdot f(n)$, it follows that $f(n) \in \Theta(f(n))$. ∎

(b) *Proof.* Let functions f and g be given such that $f(n) \in \Theta(g(n))$. This means that there is a natural number M such that for all $n \geq M$, $K \cdot g(n) \leq f(n) \leq L \cdot g(n)$ for some positive real numbers K and L. From this it follows that for all $n \geq M$, $\frac{1}{L} \cdot f(n) \leq g(n) \leq \frac{1}{K} \cdot f(n)$, and so $g(n) \in \Theta(f(n))$. ∎

(c) *Proof.* Let functions f, g and h be given such that $f(n) \in \Theta(g(n))$ and $g(n) \in \Theta(h(n))$. This means that there is a natural number M_1 such that for all $n \geq M_1$, $K_1 \cdot g(n) \leq f(n) \leq L_1 \cdot g(n)$ for some positive real numbers K_1 and L_1, and there is a natural number M_2 such that for all $n \geq M_2$, $K_2 \cdot h(n) \leq g(n) \leq L_2 \cdot h(n)$ for some positive real numbers K_2 and L_2. We choose M to be the larger of M_1 and M_2. In this case, when $n \geq M$, then

$$\begin{aligned} f(n) &\leq L_1 g(n) \text{ since } n \geq M_2 \\ &\leq L_1 L_2 h(n) \text{ since } n \geq M_1 \end{aligned}$$

and

$$\begin{aligned} f(n) &\geq K_1 g(n) \text{ since } n \geq M_2 \\ &\geq K_1 K_2 h(n) \text{ since } n \geq M_1 \end{aligned}$$

It follows from this that $f(n) \in \Theta(h(n))$. ∎

23. (a) We can show that $3^n \notin O(2^n)$ as follows: Let any $K > 0$ be given, and consider an integer n that is greater than $\log_{3/2} K$. In this case,

$$\frac{3^n}{2^n} = \left(\frac{3}{2} \right)^n > \left(\frac{3}{2} \right)^{\log_{3/2} K} = K$$

Since this can be done for any $K > 0$, this shows (by Proposition 3) that $3^n \notin O(2^n)$.

(c) $f(n) = 2n^2$ and $g(n) = n^2$

25. *Proof by induction on k.* The proposition contains the predicate we wish to prove, but it is a bit of a mouthful. Therefore, we will use the notation $W(n)$ to stand for the statement, "Algorithm COUNTERFEIT DETECTOR requires k weighings to find a counterfeit among 2^k coins." This statement is clearly true when $k = 1$ since you need one weighing (i.e., weigh either one and see if it is one ounce or not) when you have two coins. That is, $W(1)$ is true.

Now let $m \geq 2$ be given such that statements $W(1), \ldots, W(m-1)$ have all been checked to be true. To verify $W(m)$ we must envision the algorithm being performed on a collection of 2^m coins, one of which is counterfeit. In the first step of the algorithm, the coins are broken into two groups, each consisting of 2^{m-1} coins, and one of the groups is weighed to determine whether it or the other group contains the counterfeit. After this step, the algorithm operates on this remaining collection of 2^{m-1} coins, and so by the previously checked statement $W(m-1)$, it requires $m-1$ more weighings. Therefore, a total of $1 + (m-1) = m$ weighings are needed to find the counterfeit among the original 2^m coins. This is precisely statement $W(m)$ that we had hoped to verify, completing the induction. ∎

26. The completed table is shown below.

Number of coins (n)	1	2	3	4	5	6	7	8	9	10
Number of weighings	0	1	2	2	3	3	3	3	4	4
Value of $\lceil \log_2(n) \rceil$	0	1	2	2	3	3	3	3	4	4

29. (a) C; **(b)** B; **(c)** A

k	1	2	3	4	5	6	7	8	9	10
n	2	4	8	16	32	64	128	256	512	1024

32. (a)

c_n	5	19	65	211	665	2059	6305	19171	58025	175099
$n^{\log_2 3}$	3	9	27	81	243	729	2187	6561	19683	59049
Ratio	1.7	2.1	2.4	2.6	2.7	2.8	2.9	2.9	2.95	2.97

(c)

k	1	2	3	4	5	6	7	8	9	10
n	2	4	8	16	32	64	128	256	512	1024
c_n	8	56	296	1400	6248	26936	113576	471800	1939688	7916216
n^2	4	16	64	256	1024	4096	16384	65536	262144	1048576
Ratio	2	3.5	4.6	5.5	6.1	6.6	6.9	7.2	7.4	7.5

33. (a) *Proof by induction.* When $n = 1$, the inequality is "$b_1 \leq 2$," which is true since we are given that $b_1 = 1$ as part of the recursive description. Let $m \geq 2$ be given such that b_m is the first element of the sequence not yet checked to satisfy the inequality. Then

$$b_m = b_{\lfloor m/2 \rfloor} + m$$
$$\leq 2\lfloor m/2 \rfloor + m$$
$$\leq 2(m/2) + m = 2m$$

∎

Section 5.1 Exercises

1. **(a)** List two choices for candy bar in alphabetical order: {kk, kl, km, ll, lm, mm}

   (b) Since the first bar is given away, order matters this time, so we get {kk, kl, lk, km, mk, ll, lm, ml, mm}

   (c) Number the shirts from 1 to 9, and list the 3 chosen in numerical order: {123, 124, 125, . . . , 358, 359, . . . }

   (d) The set has all the entries from the previous problem, plus entries such as 111, 344, 667, etc.

   (e) Number the flavors 1 to 8, and list two choices of flavors in numerical order. Here are a few: {11, 12, 13, . . . , 18, 22, 23, . . . , 77, 78, 88}

   (f) Assuming you think chocolate on top of vanilla is different from vanilla on top of chocolate, you get all the entries from the previous problem, plus their opposites (for example, 21 along with 12).

3. In Exercise 1, we have **(a)** unordered list (bag), **(b)** ordered list, **(c)** set, **(d)** bag, **(e)** bag, **(f)** ordered list

4. **(a)** set; **(c)** unordered list (or perhaps ordered list if you care which number was on which die); **(e)** unordered list (or set if duplicate toppings are not allowed)

5. **(a)** 36; **(b)** 6; 2; 7 is more likely; **(c)** 10; $\frac{10}{36} \approx 28\%$

7. **(a)** 8; **(b)** 3; **(c)** 7; **(d)** 3

9. **(a)** 6; **(b)** 2; **(c)** 2; **(d)** A win in three games for player A must have 2 A's and 1 B, and one of the A's must come last: __, __, A. We define a correspondence from these to the ordered lists of length 2 containing 1 A by simply dropping the final A.

11. **(a)** 24; yes, by counting rows and columns; **(b)** 12; 2; **(c)** 12

12. **(a)** 6; **(c)** 81; **(e)** 24

13. **(a)** 12; **(c)** 120; **(e)** 21

14. **(a)** 8; **(c)** 8; **(e)** 8

16. **(a)** 18; **(b)** 24; **(c)** 48; **(d) i.** 8; **ii.** 16

18. **(a)** Given the set, write a sequence of I's and O's with I's in the positions given by the set.

OIIOO	OOOII	OOIOI	IOOOI
{2,3}	{4,5}	{3,5}	{1,5}

   (b)

IOOIO	OOIIO	IOIOO	OIOOI
{1,4}	{3,4}	{1,3}	{2,5}

20. **(a)** This is simply a matter of changing the A's and B's in an arrangement from (i) into 1's and 0's, respectively. This will change them (one-for-one) into the binary sequences described in (ii).

   (b) Items described by (i) look like {x, y} where x and y are two different elements from {1, 2, 3, 4, 5}. Each of these can be "linked" with the set {a, b, c} from {1, 2, 3, 4, 5} of three entries which are neither x nor y. In this way, each item described by (i) is linked (one-for-one) with an item described by (ii).

   (c) For each permutation of all 5 elements we associate the permutation of length 4 that comes from taking the first 4 elements in the same order. For example, to 13ax9 we associate 13ax. It is easy to see that this association is one-to-one since the first four entries in the list determine the fifth exactly.

   (d) If we represent a sequence of 5 coin tosses as a list of length 5 with entries taken from {H, T} then the items in (i) will look like HHTTH, THHHT, etc. To each of these we can associate the set {x, y, z} of the three positions of the "H"s. For example, HHTTH is associated with {1,2,5}, THHHT is associated with {2,3,4} and so on.

Section 5.2 Exercises

1. **(a)** 60,466,176; **(b)** 9,765,625

3. **(a)** 175,760,000; **(b)** 174,790,000

5. 24

6. **(a)** 4080; **(b)** 3840; **(c)** 3855; **(d)** 240

8. **(a)** 1776; **(b)** 8760; **(c)** 8256

9. **(a)** 216; **(b)** 120

11. **(a)** 311,875,200; **(b)** 18,345,600; **(c)** 617,760

13. **(a)** 40,320; **(b)** 10,080

14. **(a)** 16; **(b)** 384; **(c)** 51

16. **(a)** 43,680; **(b)** 3640; **(c)** 28,756; **(d)** 1800

18. **(a)** 362,880; **(b)** 2880

20. **(a)** 2,985,984; **(b)** 665,280; **(c)** 77.72%

22. **(a)** 128; **(b)** 128; **(c)** 16; **(d)** 240

23. **(a)** 5292; **(b)** 56: **(c)** 3276

25. **(a)** 159; **(b)** 960

26. **(a)** $19 \cdot 18 \cdot 17 \cdot 16 \cdot 15 = 1,395,360$; **(b)** $6 \cdot 5 \cdot 4 \cdot 3 \cdot 2 \cdot 1 = 720$; **(c)** $120 \cdot 119 \cdot 118 \cdot 117 \cdot \cdots \cdot 106 \cdot 105 \cdot 104 \approx 6.755 \times 10^{34}$

27. **(a)** $P(9, 4)$; **(c)** $P(365, 28)$; **(e)** $P(k, m+1)$

29. The problem is that, by the time you get to vice-president, there could be either 7 or 6 choices, depending on the sex of the president. You must put the more restrictive steps before the less restrictive steps in your algorithm: There are 7 choice for the vice-president, 10 for the secretary, and 15 for the president, yielding $7 \cdot 10 \cdot 15 = 1050$ possible results.

31. This algorithm "double-counts" each outcome that has a math major for both of the offices. One way to fix

it is to properly apply the Inclusion-Exclusion Principle, by subtracting the overlap. The overlap is given by this algorithm: "8 choices for president, 7 for vice-president, and 24 for secretary," so the total count is $9600 - 8 \cdot 7 \cdot 24 = 8256$. Another possible algorithm counts three non-overlapping sets: (1) president is math major and vice-president is not; (2) vice-president is math major and president is not; (3) both are math majors. [$8 \cdot 18 \cdot 24 + 18 \cdot 8 \cdot 24 + 8 \cdot 7 \cdot 24 = 8256$]. A third algorithm counts the complement (neither are

Section 5.3 Exercises

1. **(a)** $ab, ac, ad, ae, ba, bc, bd, be, ca, cb, cd, ce, da,$ $db, dc, de, ea, eb, ec, ed$; **(c)** 2 equivalence classes, each of size 10
2. **(a)** 60; **(c)** $\{ade, aed, dae, dea, ead, eda\}$; **(e)** 10
3. **(a)** $\{ACDBFE, CDBFEA, DBFEAC, BFEACD,$ $FEACDB, EACDBF\}$
5. 96
7. 144
10. 32
12. 120
14. 35
15. **(a)** 5,985; **(c)** 2,475
17. **(a)** 220; **(c)** 10
19. **(a)** 256; **(c)** 56
21. Second solution: Solve the complementary problem, giving 8,008.
23. 189,303,411,990
24. HINT: Break the committee selections up into five cases based on how many married women are chosen.
25. 886,656
27. **(a)** 74,613
    (b) 7,560
29. **(a)** 10,477,677,064,400; **(c)** About 9,626,413 to 1
32. 35 is the coefficient of t^6; the coefficient of t^5 is 0.
35. *Proof by induction.* Let $P(n)$ be the statement, "$\sum_{k=0}^{n} C(n, k) = 2^n$" It is easy to check that $P(0)$ (which says "$C(0, 0) = 2^0$") and $P(1)$ (which says "$C(1, 0) + C(1, 1) = 2^1$") are both true. Now let $m \geq 2$ be given such that statements $P(0), P(1), \ldots, P(m - 1)$ have all been checked, and $P(m)$ is the next statement to be checked. So

$$\sum_{k=0}^{m} C(m, k) = C(m, 0) + \sum_{k=1}^{m-1} (C(m - 1, k - 1)$$
$$+ C(m - 1, k)) + C(m, m)$$

math majors) and subtracts from the overall count $[26 \cdot 25 \cdot 24 - 18 \cdot 17 \cdot 24 = 8256]$.

34. 13,440
36. 3168
37. 325
39. **(a)** 375; **(b)** 303; **(c)** 216
41. 955,514,880
42. 10 subsets, which is half the number of permutations.

$$= C(m - 1, 0) + \sum_{k=1}^{m-1} (C(m - 1, k - 1)$$
$$+ C(m - 1, k)) + C(m - 1, m - 1)$$
$$= \sum_{k=0}^{m-1} C(m - 1, k - 1) + \sum_{k=0}^{m-1} C(m - 1, k)$$
$$= 2^{m-1} + 2^{m-1} = 2^m \quad \text{by statement } P(m - 1)$$

Hence statement $P(m)$ is also true. ∎

39. The 3rd entry of Row 7.
41. **(a)** The coefficient of x^2 in $x \cdot (1 + (1 + x) + (1 + x)^2 + \cdots + (1 + x)^n)$ is the same as the coefficient of x in $(1 + (1 + x) + (1 + x)^2 + \cdots + (1 + x)^n$, which can be found by summing the coefficients of x in the component terms $1, (1 + x), (1 + x)^2, \ldots, (1 + x)^n$. From the binomial theorem, this is simply

$$C(1, 1) + C(2, 1) + C(3, 1) + \cdots + C(n, 1) =$$
$$1 + 2 + 3 + \cdots + n$$

Of course, the coefficient of x^2 in $(1 + x)^{n+1}$ is $C(n + 1, 2)$.

43. This problem uses the mod 10 arithmetic from Section 2.7. Let A, B, C, D and E be the values of the original cards. The rows formed are
    ● $[A, B, C, D, E]$
    ● $[A + B, B + C, C + D, D + E]$
    ● $[A + 2B + C, B + 2C + D, C + 2D + E]$
    ● $[A + 3B + 3C + D, B + 3C + 3D + E]$
    ● $[A + 4B + 6C + 4D + E]$
    where each operation uses mod 10 arithmetic. Since $4 \equiv_{10} -6$, we know that

$$A + 4B + 6C + 4D + E \equiv_{10} A + 4B - 4C + 4D + E$$

The right-hand side of this equation is precisely the calculated value $A + E + 4(B + D - C)$ in the magic trick.

Section 5.4 Exercises

1. (a) 28; **(c)** 247

3. (a) 24; **(c)** 4,096

5. (a) 1,048,576; **(c)** 489,888

7. 4,200

9. $C(19, 2)C(18, 2)C(16, 2)\cdots C(4, 2)C(2, 2) \approx 2.13829 \times 10^{15}$

13. (a) 24 weird arrangements look like ALAAMAB once the goofines is removed.

   (b) 210

15. (a) The equations are $2 + 2 + 2 + 2 + 2 + 2 = 12$, $6 + 0 + 0 + 0 + 6 + 0 = 12$, and $0 + 0 + 0 + 12 + 0 + 0 = 12$. The binary sequences are 00100100100100100, 00000011110000001, and 11100000000000011.

17. 1,771

19. 78

21. Using cases based on the value of z, we get 161 solutions.

23. 120

25. Using four cases, we get a total of 16 solutions.

27. 1,391,975,640.

29. $C(210, 99) - C(166, 99) \approx 6.4319 \times 10^{61}$

31. Taking the number of integer solutions to the equation

$$x_1 + x_2 + x_3 + x_4 = 12$$

that have each $x_i \geq 1$ and subtracting the number of solutions that have any of the $x_i \geq 7$ leaves the number of desired solutions as 125.

Section 5.5 Exercises

1. Let a_n be the number of n-digit numbers which do not use "0" for a digit. Any such n-digit number can be made by choosing a non-zero digit as the leading digit, and then following it with an $(n-1)$-digit number with the same property. This means that

$$a_n = 9 \cdot a_{n-1}$$

This along with the fact that $a_1 = 9$ completely solves the problem.

3. *Proof by induction.* From Example 5 we know $T_1 = 0$, $T_2 = 2$ and $T_n = 2 \cdot T_{n-2}$. Let $P(n)$ be the statement, "If n is even, then $T_n = 2^{n/2}$; if n is odd, then $T_n = 0$" for $n \geq 2$. Since $T_1 = 0$ and $T_2 = 2 = 2^{2/2}$ from the example, we know that statements $P(1)$ and $P(2)$ and true. Now suppose we have checked $P(1), P(2), \ldots, P(m-1)$ have all been checked to be true for some integer $m \geq 3$. We have two cases to consider depending on whether m is even or odd.

Case 1. If m is odd, then

$$\begin{aligned} T_m &= 2 \cdot T_{m-2} \text{ from Example 5} \\ &= 2 \cdot 0 \text{ by statement } P(m-2) \text{ since } m-2 \\ &\quad \text{ is also odd} \\ &= 0 \end{aligned}$$

Case 2. If m is even, then

$$\begin{aligned} T_m &= 2 \cdot T_{m-2} \text{ from Example 5} \\ &= 2(2^{(m-2)/2}) \text{ by statement } P(m-2) \\ &\quad \text{ since } m-2 \text{ is also even} \\ &= 2(2^{m/2-1}) \\ &= 2^{m/2} \end{aligned}$$

In either case, we have confirmed the truth of statement $P(m)$, completing the induction.

5. (a) $P(n, 1) = n \cdot P(n-1, 0) = n$; **(c)** $P(n, 3) = n \cdot P(n-1, 2) = n \cdot (n-1) \cdots (n-2)$ by part (b)

7. Let c_n be the number of ways to cover a $2 \times n$ chessboard with 1×2 dominoes. For example, $c_3 = 3$ since we can cover a 2×3 chessboard in the following ways:

1	2	2
1	3	3

1	2	3
1	2	3

1	1	3
2	2	3

Similarly one can verify that $c_1 = 1$ and $c_2 = 2$. To see the recursive model, just observe that the left side of every covering looks like one of the following:

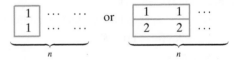

By thinking of how to complete the covering in each case, we can conclude that $c_n = c_{n-1} + c_{n-2}$.

9. *Proof by induction.* Since $\sum_{k=0}^{1}(-1)^k \frac{1}{k!} = 1 - 1 = 0$ and $d_1 = 0$, the statement is true when $n = 1$. Since $2! \sum_{k=0}^{2}(-1)^k \cdot \frac{1}{k!} = 2(1 - 1 + \frac{1}{2}) = 1$ and $d_2 = 1$, the statement is true when $n = 2$. Let $m \geq 3$ be given such that the first statement that has not yet been checked. Then

$$\begin{aligned} d_m &= (m-1) \cdot (d_{m-1} + d_{m-2}) \\ &= (m-1)(m-1)! \cdot \sum_{k=0}^{m-1}(-1)^k \frac{1}{k!} + (m-1) \end{aligned}$$

$$\cdot(m-2)!\sum_{k=0}^{m-2}(-1)^k\cdot\frac{1}{k!}$$

$$= m\cdot(m-1)!\sum_{k=0}^{m-1}(-1)^k\cdot\frac{1}{k!} - (m-1)!\sum_{k=0}^{m-1}(-1)^k\frac{1}{k!}$$

$$+(m-1)!\sum_{k=0}^{m}(-1)^k\frac{1}{k!}$$

$$= m!\sum_{k=0}^{m-1}(-1)^k\frac{1}{k!} - (m-1)!\cdot(-1)^{m-1}\cdot\frac{1}{(m-1)!}$$

$$= m!\sum_{k=0}^{m-1}(-1)^k\frac{1}{k!} + (-1)^m$$

$$= m!\sum_{k=0}^{m}(-1)^k\frac{1}{k!}$$

12. $f_{0,k}=1$ for all $k\geq 0$, $f_{n,1}=1$ for all $n\geq 1$, and

$$f_{n,k}=\sum_{m=0}^{n}f_{m,k-1}.$$

14. Let a_n be the number of positive n-digit numbers with no consecutive 1's. We observe that $a_1=9$ and $a_2=89$ (any number from 10 to 99 except 11). Assume that we understand these sorts of numbers when they have less than n-digits but we are now asked to count how many are like this with n-digits. To form these we can add $0,2,3,4,\ldots,8,9$ to the right end of an $(n-1)$-digit number without consecutive 1's, or we can add one of $\{01,21,31,41,51,61,71,81,91\}$ to the right end of an $(n-2)$-digit number with no consecutive 1's. This means that $a_n=9a_{n-1}+9a_{n-2}$.

Section 5.6 Exercises

1. (a) $a_n=7n+3$; **(c)** $a_n=1+2n+\frac{3}{2}n(n-1)$

3. (a) $n+2n(n-1)$; **(c)** $6n+\frac{15}{2}n(n-1)+\frac{4}{3}n(n-1)(n-2)$

5. $\Delta_0^0=s_0$

6. (a) $a_n=\frac{5}{2}\cdot 2^n-3$; **(c)** $a_n=\frac{5}{6}(-2)^n+\frac{5}{3}$

7. The mortgage will be paid off in the 139th month.

9. $s_0=5$ and $s_n=3\cdot s_{n-1}-2$ for $n\geq 1$ has closed formula $s_n=\frac{4}{3}\cdot 3^n+1$.

11. Since $a_n=L\cdot b^n+\frac{c}{1-b}$, if $a_0=1$, we must have $L=1-\frac{c}{1-b}$.

13. *Proof by induction.* If $n=0$, the statement is "$s_0=s_0+K\cdot 0$," which is certainly true. Let $m\geq 1$ be given such that all of the statements up to "$s_{m-1}=s_0+K\cdot(m-1)$" have been verified to be true. Now

$$s_m=s_{m-1}+(s_m-s_{m-1})$$
$$=s_{m-1}+K \text{ since } s_n \text{ has constant first difference } K$$
$$=(s_0+K\cdot(m-1))+K$$
$$=s_0+Km$$

15. (c) $a_n=7\cdot 2^n-n^2-4n-6$

17. (a) $a_n=C\cdot 3^n+K\cdot 4^n$; **(c)** $a_n=C\cdot 4^n+K\cdot(-3)^n$

18. (a) $a_n=4^n-3^n$; **(c)** $a_n=\frac{1}{7}(4^n-(-3)^n)$

19. (a) $a_n=C\cdot\left(\frac{1+\sqrt{13}}{2}\right)^n+K\cdot\left(\frac{1-\sqrt{13}}{2}\right)^n$

20. (a) $a_n=\left(\frac{13+\sqrt{13}}{26}\right)\cdot\left(\frac{1+\sqrt{13}}{2}\right)^n+\left(\frac{13-\sqrt{13}}{26}\right)\cdot\left(\frac{1-\sqrt{13}}{2}\right)^n$

21. *Proof.* Suppose that the equation $x^2=cx+d$ has distinct solutions $x=r$ and $x=s$, and consider the closed formula $a_n=C\cdot r^n+K\cdot s^n$. From this formula, it follows that $a_{n-1}=C\cdot r^{n-1}+K\cdot s^{n-1}$ and $a_{n-2}=C\cdot r^{n-2}+K\cdot s^{n-2}$. Now we can combine this information to get

$$c\cdot a_{n-1}+d\cdot a_{n-2}=c\cdot(C\cdot r^{n-1}+K\cdot s^{n-1})+d$$
$$\cdot(C\cdot r^{n-2}+K\cdot s^{n-2})$$
$$=C\cdot(c\cdot r^{n-1}+d\cdot r^{n-2})+K$$
$$\cdot(c\cdot s^{n-1}+d\cdot s^{n-2})$$
$$=r^{n-2}\cdot C\cdot(c\cdot r+d)+s^{n-2}\cdot K$$
$$\cdot(c\cdot s+d)$$
$$=r^{n-2}\cdot(C\cdot r^2)+s^{n-2}\cdot(K\cdot s^2)$$
$$=C\cdot r^n+K\cdot s^n=a_n$$

This verifies the relation $c\cdot a_{n-1}+d\cdot a_{n-2}=a_n$.

22. (a) $a_n=(C+K\cdot n)\cdot 2^n$

23. (a) $a_n=\left(1-\frac{1}{2}n\right)\cdot 2^n$

Section 6.1 Exercises

1. (a) $\frac{1}{4}$; **(c)** $\frac{2}{13}$

2. (a) $\frac{1}{5}$

3. (a) $\frac{5}{18}$; **(c)** $\frac{1}{6}$; **(e)** $\frac{5}{6}$

4. (a) $\frac{1}{2}$

5. Some have more than one legitimate answer. **(a)** The set of ordered lists of length 3 using elements from $\{1,2,3,4,5,6\}$; **(b)** The set of permutations of length 2 with elements from the set of club members (Treat the first person listed as president, the second as vice president.); **(c)** The set of ordered lists of length 6 with elements from $\{H,T\}$; **(d)** The set of sets (combinations) of size 3 with elements from the set of 52 cards, OR the set of permutations of length 3 with elements from the set of 52 cards

7. (a) $\frac{1}{13}$; **(c)** $\frac{1}{52}$; **(e)** $\frac{4}{13}$

9. There are a total of $C(23, 3) = 1771$ different committees possible. **(a)** $\frac{190}{253}$; **(c)** $\frac{20}{161}$

11. $\frac{2}{5}$

12. Our sample space will be the set of all combinations of size 5 with elements from the deck of 52 cards. **(a)** $\frac{33}{16\,660}$; **(c)** $\frac{1}{64\,974}$

15. The probability is approximately 0.753.

17. The probability that two people in a group of 500 have the same last four digits of their Social Security number is approximately 0.9999969.

19. (a) $\frac{5}{12}$; **(c)** $\frac{m-1}{2n}$ (assuming $n \geq m$)

21. $\frac{1}{22}$

23. $\frac{C(n-1,31)}{C(n+31,31)}$

25. Answers will vary but should support the theoretical probability of $\frac{1}{3}$.

26. Answers will vary but should support the theoretical probability of approximately 0.22.

27. To see why this works, imagine one half (consisting of 13 red cards and 13 black cards) came from a blue-backed deck while the other half (consisting of 13 red cards and 13 black cards) came from a green-backed deck. The final deck will alternate colors of their backs even though the faces of the cards are fairly shuffled. As the solitaire game is played, cards are removed from the deck in adjacent pairs which share the same face-color. Hence at any point in the process, (i) the number of blue-backed red cards is equal to the number of green-backed red cards, (ii) the number of blue-backed black cards is equal to the number of green-backed black cards, and (iii) the blue-backed and green-backed cards alternate. Given these three properties that remain invariant as the game is played, it is impossible that the game should ever end in a loss. This is true because a losing final position would consist of cards alternating in face colors, and property (iii) then dictates that the red cards and black cards should have different back colors contrary to properties (i) and (ii).

Section 6.2 Exercises

1. (a) A draw which includes the Ace of Clubs is in both events, so these events are not disjoint; **(c)** These events are disjoint.

2. (a) Let E_1 be the set of outcomes where the card is an Ace, and E_2 be the set of outcomes where the card is a Jack. Then since these events are disjoint,

$$Prob(E_1 \text{ or } E_2) = Prob(E_1) + Prob(E_2) = \frac{2}{13}$$

3. (a) Let E_1 be the set of outcomes where the card is an ace, and E_2 be the set of outcomes where the card is an heart. Then

$$Prob(E_1 \text{ or } E_2) = Prob(E_1) + Prob(E_2)$$
$$-Prob(E_1 \text{ and } E_2) = \frac{4}{13}$$

(c) Let E_1 be the set of outcomes where the card is a diamond or a club, and E_2 be the set of outcomes where the card is a king. Then

$$Prob(E_1 \text{ or } E_2) = Prob(E_1) + Prob(E_2)$$
$$-Prob(E_1 \text{ and } E_2) = \frac{4}{13}$$

4. (a) These events are not independent. Let E_1 be the set of outcomes where the first card is an Ace, and E_2 be the set of outcomes where the second card is a Ten, Jack, Queen or King. Then $Prob(E_1) = \frac{1}{13}$, $Prob(E_2) = \frac{4}{13}$,

but

$$Prob(E_1 \text{ and } E_2) = \frac{16}{663} \neq \frac{1}{13} \cdot \frac{4}{13}$$

(c) These events are independent.

5. Because the first card is replaced and the deck is shuffled before the second card is drawn, in each case, the events described are independent. **(a)** $\frac{1}{26}$; **(c)** $\frac{81}{169}$

7. $\frac{7}{18}$

9. $\frac{104}{216} \approx 0.48$.

11. $\frac{15}{64}$

13. $Prob(\text{sum is even}) = \frac{1}{2}$; $Prob(\text{sum is a multiple of 3}) = \frac{1}{3}$

15. $\frac{5}{324}$

17. $\frac{1,279}{216,580}$

19. (a) Let E_1 be the set of outcomes where the faceup card is an Ace, and let E_2 be the set of outcomes where the facedown card is a Ten, Jack, Queen or King. Then

$$Prob(E_2|E_1) = \frac{16}{51} \approx 0.314$$

(c) Let E_1 be the set of outcomes where the opponent's faceup card is an Ace and the next two cards (mine) are a Four and a Five, in either order, and let E_2 be the set of outcomes where the opponent's facedown card is a Ten, Jack, Queen or King. Then

$$Prob(E_2|E_1) = \frac{16}{49} \approx 0.327$$

21. (a) Let E_1 be the set of outcomes where the opponent's faceup cards are {2D,6C}, and let E_2 be the set of outcomes where the opponent's facedown cards have values {3,4,5}. Then

$$Prob(E_2|E_1) \approx 0.0112$$

(c) Let E_1 be the set of outcomes where the opponent's faceup cards are {5D,6C}, and let E_2 be the set of outcomes where the opponent's facedown cards have values {2,3,4} or {3,4,7} or {4,7,8} or {7,8,9}. Then

$$Prob(E_2|E_1) \approx 0.0449$$

23. We use the numbers already calculated in Example 9.

    (a) $Prob(P \text{ and } \overline{S})$ is the probability that the steroid result is positive and the athlete has not used steroids.

$$Prob(P \text{ and } \overline{S}) = Prob(\overline{S}) \cdot Prob(P|\overline{S})$$
$$= (1 - Prob(S)) \cdot Prob(P|\overline{S})$$
$$= 0.0006$$

(b) Events $P \cap \overline{S}$ and $P \cap S$ are disjoint and $(P \cap \overline{S}) \cup (P \cap S) = P$, so by the sum rule,

$$Prob(P) = Prob(P \text{ and } S) + Prob(P \text{ and } \overline{S})$$
$$\approx 0.02985 + 0.0006 = 0.03045$$

(c) $Prob(S|P)$ is the probability that an athlete has used steroids given that the result of the test was positive. $Prob(S|P) \approx 0.9803$.

(d) $Prob(S|\overline{P})$ is the probability that an athlete has used steroids given that the result of the test was negative. $Prob(S|\overline{P}) \approx 0.000155$.

24. $\frac{8}{27} \approx 0.296$

26. $\frac{1}{2}; \frac{65}{128} \approx 0.5078$

Section 6.3 Exercises

1. $\frac{175}{93,312} \approx 0.001875$

3. $\frac{15}{128}$

5. $\frac{9,765,625}{30,233,088} \approx 0.323$

7. $\frac{763}{3888} \approx 0.196$

9. $4\left(\frac{1}{6}\right)\left(\frac{4}{6}\right)^3 + 6\left(\frac{1}{6}\right)^2\left(\frac{4}{6}\right)^2 + 12\left(\frac{1}{6}\right)^3\left(\frac{4}{6}\right) + 4\left(\frac{1}{6}\right)^3\left(\frac{4}{6}\right) + 4\left(\frac{1}{6}\right)^4 + \left(\frac{1}{6}\right)^4 \approx 0.325$

11. $\frac{352}{833} \approx 0.4226$.

14. The probability that I win the series is at least 0.9 when $p \geq 0.8042$.

16. The following graph was made in Excel:

18. $\frac{35}{972} \approx 0.036$

19. (a) $\frac{131}{243}$. **(c)** $\frac{767}{1296} \approx 0.59$.

20. $\frac{3807}{4096} \approx 0.93$.

22. (a) $\frac{81}{125}$; **(b)** $\frac{2133}{3125}$; **(c)** $\frac{11\,097}{15\,625}$

23. HINT: This is harder because there is no obvious simple case structure. List all outcomes and find the probability of each.

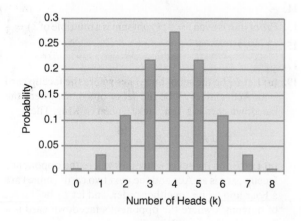

Section 6.4 Exercises

1. Letting X denote the sum of a pair of dice, we will have $E[X] = 7$, which is twice the expected value of a single die roll.

3. 1.2

5. $\frac{5}{13} \approx 0.385$

6. HINT: Let X be the sum of the values of the pair of cards. X can be any number from $\{4, 5, 6, \ldots, 22\}$. The probability of X having any particular one of these values must be calculated separately, leading to the final answer, $E[X] \approx 14.88$. (This is the value of the average starting hand in Blackjack, counting Ace as 11).

9. The expected payoff is approximately \$5.42. If 10,000 people play, the casino will take in \$100,000 and expect to pay out about \$54200, so the casino expects to make a profit of about \$45,800 per day on this game.

11. Using the hint, we recall the identity from Exercise 36 of Section 5.3:

$$\sum_{k=1}^{6} k \cdot C(6, k) = 6 \cdot 2^5$$

Now using X to denote the number of 1's we see in our tossing of six coins, we have

$$E[X] = (0)C(6, 0)\left(\frac{1}{2}\right)^6 + (1)C(6, 1)\left(\frac{1}{2}\right)^6 + \cdots$$

$$+ (6)C(6, 6)\left(\frac{1}{2}\right)^6$$

$$= \left(\frac{1}{2}\right)^6 \left(\sum_{k=0}^{6} k \cdot C(6, k)\right)$$

$$= \left(\frac{1}{2}\right)^6 (6 \cdot 2^5) = 3$$

13. Since the general identity from Exercise 37 of Section 5.3 is

$$\sum_{k=1}^{n} k \cdot C(n, k) \cdot x^k = n \cdot x \cdot (1 + x)^{n-1}$$

we can use $x = 1/5$ and $n = 10$ to get

$$\sum_{k=1}^{10} k \cdot C(10, k) \cdot (1/5)^k = 10 \cdot (1/5) \cdot \left(\frac{6}{5}\right)^9$$

and we can multiply both sides of this equation by $(5/6)^{10}$ to get

$$\sum_{k=1}^{10} k \cdot C(10, k) \cdot \left(\frac{1}{6}\right)^k \left(\frac{5}{6}\right)^{10-k} = 2 \cdot \frac{5}{6}$$

Now if X is the number of "1"s that occur in the ten rolls, then

$$E[X] = (1)C(10, 1)\left(\frac{1}{6}\right)^1 \left(\frac{5}{6}\right)^9 + \cdots$$

$$+ (10)C(10, 10)\left(\frac{1}{6}\right)^{10}\left(\frac{5}{6}\right)^0$$

$$= \sum_{k=1}^{10} k \cdot C(10, k) \cdot \left(\frac{1}{6}\right)^k \left(\frac{5}{6}\right)^{10-k}$$

$$= 2 \cdot \frac{5}{6} = \frac{5}{3}$$

by the identity above.

15. If the sample space is $\{MF, FM, MM\}$, the expected number of boys is 4/3. If the sample space is $\{MF, MM\}$, the expected number of boys is 3/2.

17. $\frac{107}{27} \approx 3.96$

19. Following the method of Example 7 with X representing the number of games in the best-of-three series, we have $E[X] = 6\,p^4 - 12\,p^3 + 3\,p^2 + 3\,p + 3$. The graph of this expression shows that the maximum length of the series occurs when $p = 1/2$.

21. The solution to Example 7 and the fact that the winner of a series must win 4 games together tell us that the average winning margin is 2.1875 games.

23. If we "zoom in" on the graph in the solution to Example 7, we can see that when p is less than about 0.215 or more than about 0.785.

Section 6.5 Exercises

1. $E[X] = 6$.

3. $E[X] = \frac{36}{11}$.

5. **(a)** Use p_n to denote the probability that the nth card is the first "Ace." Then for example, $p_1 = \frac{1}{13}$, $p_2 = \frac{48}{52} \cdot \frac{1}{13}$, $p_3 = \frac{48}{52} \cdot \frac{47}{51} \cdot \frac{1}{13}$, and in general,

$$p_n = \frac{P(48, n-1)}{P(52, n-1)} \cdot \frac{1}{13}$$

(c) The solution to Exercise 4 assumes that the trials are independent. "Drawing cards without replacement" does not have this property.

7. We have the recurrence relation $h_1 = 2$ and

$$h_n = \left(\frac{1}{2}\right)(1 + h_{n-1}) + \left(\frac{1}{2}\right)(1 + h_n)$$

or more simply, $h_n = 2 + h_{n-1}$, for all $n \geq 2$.

9. $p = 81/82$.

11. 3.2.

13. $E[X] = \frac{2}{p^2+(1-p)^2}$

15. $p = \frac{2992}{3645} \approx 0.82$

17. **(c)** $a = p = 5$ and $t = 9$

19. The expected length of this game is 48 coin tosses.

21. In the Hank and Ted game played with a fair coin, if Hank starts with X markers and Ted starts with Y markers, then the probability that Hank wins is $\frac{X}{X+Y}$, the probability that Ted wins is $\frac{Y}{X+Y}$, and the expected length of the game is $(X)(Y)$ moves.

23. Let $d_n = (n)(M - n)$. The conditions $d_0 = 0$ and $d_M = 0$ are obviously satisfied by this definition. Since

$$d_{n+1} + 1 = (n+1)(M - (n+1)) + 1 = nM - n^2 - 2n + M$$

and

$$d_{n-1} + 1 = (n-1)(M - (n-1)) + 1 = nM - n^2 + 2n + M$$

we conclude that

$$\frac{1}{2}(d_{n+1}+1) + \frac{1}{2}(d_{n-1}+1) = \frac{1}{2}((nM - n^2 - 2n + M)$$
$$+ (nM - n^2 + 2n + M))$$
$$= \frac{1}{2}(2nM - 2n^2 + 2M)$$
$$= nM - n^2 + M + 1$$
$$= n(M-n) = d_n$$

Section 6.6 Exercises

1. For each i with $1 \le i \le 7$, State i will be the game where Hank has $i - 1$ markers and Ted has $7 - i$ markers.

$$M = \begin{bmatrix} 1 & 0 & 0 & 0 & 0 & 0 & 0 \\ 2/3 & 0 & 1/3 & 0 & 0 & 0 & 0 \\ 0 & 2/3 & 0 & 1/3 & 0 & 0 & 0 \\ 0 & 0 & 2/3 & 0 & 1/3 & 0 & 0 \\ 0 & 0 & 0 & 2/3 & 0 & 1/3 & 0 \\ 0 & 0 & 0 & 0 & 2/3 & 0 & 1/3 \\ 0 & 0 & 0 & 0 & 0 & 0 & 1 \end{bmatrix}$$

3. Define states as follows:

State 1 Andre wins	State 4 Andre up 1	State 7 Pete up 2
State 2 Andre up 3	State 5 Tied	State 8 Pete up 3
State 3 Andre up 2	State 6 Pete up 1	State 9 Pete wins

which is Equation (6.2).

25. **(a)** Let $d_n = c\left(1 - \frac{1}{2^n}\right) - 3n$. Then $\frac{2}{3}(d_{n+1} + 1) + \frac{1}{3}(d_{n-1} + 1)$ can be simplifed as

$$\frac{2}{3}\left(c\left(1 - \frac{1}{2^{n+1}}\right) - 3(n+1) + 1\right)$$
$$+ \frac{1}{3}\left(c\left(1 - \frac{1}{2^{n-1}}\right) - 3(n-1) + 1\right)$$
$$= \left(\frac{2c}{3} - \frac{1}{3}\left(\frac{c}{2^n}\right) - 2n - \frac{4}{3}\right)$$
$$+ \left(\frac{c}{3} - \frac{2}{3}\left(\frac{c}{2^n}\right) - n + \frac{4}{3}\right)$$
$$= c - \frac{c}{2^n} - 3n = c\left(1 - \frac{1}{2^n}\right) - 3n$$

This final simplification is precisely the definition of d_n, so the relationship holds.

(b) If $d_n = c\left(1 - \frac{1}{2^n}\right) - 3n$, then $c \approx 60.00005722$.

(c) Using the previous result, $d_{10} = \approx 29.94$.

(d) Looking at all values of d_n for $0 \le n \le 20$, we see that the game that begins with $H = 4$, $T = 16$ is expected to last the longest.

26. No matter how many total markers Ted has, Hank will win this game with probability greater than 0.999.

$$M = \begin{bmatrix} 1 & 0 & 0 & 0 & 0 & 0 & 0 & 0 & 0 \\ 2/3 & 0 & 1/3 & 0 & 0 & 0 & 0 & 0 & 0 \\ 0 & 2/3 & 0 & 1/3 & 0 & 0 & 0 & 0 & 0 \\ 0 & 0 & 2/3 & 0 & 1/3 & 0 & 0 & 0 & 0 \\ 0 & 0 & 0 & 2/3 & 0 & 1/3 & 0 & 0 & 0 \\ 0 & 0 & 0 & 0 & 2/3 & 0 & 1/3 & 0 & 0 \\ 0 & 0 & 0 & 0 & 0 & 2/3 & 0 & 1/3 & 0 \\ 0 & 0 & 0 & 0 & 0 & 0 & 2/3 & 0 & 1/3 \\ 0 & 0 & 0 & 0 & 0 & 0 & 0 & 0 & 1 \end{bmatrix}$$

5. States 1 through 6 will refer to the game piece being on squares A through F, respectively. The transition matrix is

$$M = \begin{bmatrix} 0 & 1/4 & 1/2 & 1/4 & 0 & 0 \\ 1/4 & 0 & 1/4 & 1/4 & 1/4 & 0 \\ 1/4 & 0 & 0 & 1/2 & 1/4 & 0 \\ 1/4 & 0 & 0 & 1/4 & 1/4 & 1/4 \\ 0 & 0 & 0 & 1/4 & 0 & 3/4 \\ 0 & 0 & 0 & 0 & 0 & 1 \end{bmatrix}$$

7. Answers vary but should support Exercises 12 and 24.

11. Answers vary but should support Exercise 28.

13. The probability of going from State 5 to State 1 in no more than 16 moves is approximately 0.0642, since this is the entry in Row 5, Column 1 of the matrix M^{16}.

15. Using the following fifteen states of this game (with A-B reflecting the series score), the matrix N for the transient states $(1, 2, 3, 5, 6, 7, 9, 10$ and $11)$ is shown below.

1: 0-0	**6:** 1-1	**11:** 2-2
2: 1-0	**7:** 2-1	**12:** 3-2
3: 2-0	**8:** 3-1	**13:** 0-3
4: 3-0	**9:** 0-2	**14:** 1-3
5: 0-1	**10:** 1-2	**15:** 2-3

$$N = \begin{bmatrix} 0 & \frac{1}{2} & 0 & \frac{1}{2} & 0 & 0 & 0 & 0 & 0 \\ 0 & 0 & \frac{1}{2} & 0 & \frac{1}{2} & 0 & 0 & 0 & 0 \\ 0 & 0 & 0 & 0 & 0 & \frac{1}{2} & 0 & 0 & 0 \\ 0 & 0 & 0 & 0 & \frac{1}{2} & 0 & \frac{1}{2} & 0 & 0 \\ 0 & 0 & 0 & 0 & 0 & \frac{1}{2} & 0 & \frac{1}{2} & 0 \\ 0 & 0 & 0 & 0 & 0 & 0 & 0 & 0 & \frac{1}{2} \\ 0 & 0 & 0 & 0 & 0 & 0 & 0 & \frac{1}{2} & 0 \\ 0 & 0 & 0 & 0 & 0 & 0 & 0 & 0 & \frac{1}{2} \\ 0 & 0 & 0 & 0 & 0 & 0 & 0 & 0 & 0 \end{bmatrix}$$

We can add 1 to the sum of the entries in the first row (corresponding to those games beginning in state 1) of $N + N^2 + N^3 + N^4 + N^5$, giving us $33/8 \approx 4.125$ games expected in this series.

17. Emulating Exercise 15, there are $104/27 \approx 3.85$ games.

19. *Proof by induction.* The first statement is, "$M^1 = \begin{bmatrix} 1/2 & 0 \\ 0 & 1/2 \end{bmatrix}$," which is true by the given definition of M. Let $m \geq 2$ be given such that the first statement *not* yet proven is the one involving M^m. In particular, the previous statement,

$$M^{m-1} = \begin{bmatrix} (1/2)^{m-1} & 0 \\ 0 & (1/2)^{m-1} \end{bmatrix}$$

has already been checked to be true. In this case, we know

$$M^m = M \cdot M^{m-1} = \begin{bmatrix} 1/2 & 0 \\ 0 & 1/2 \end{bmatrix} \cdot \begin{bmatrix} (1/2)^{m-1} & 0 \\ 0 & (1/2)^{m-1} \end{bmatrix} = \begin{bmatrix} (1/2)^m & 0 \\ 0 & (1/2)^m \end{bmatrix}$$

which verifies the next statement, completing the induction.∎

21. *Proof by induction.* The first statement is, "$M^1 = \begin{bmatrix} 0 & 2(1/4)^1 \\ 2(1/4)^1 & 0 \end{bmatrix}$ and $M^2 = \begin{bmatrix} (1/4)^1 & 0 \\ 0 & (1/4)^1 \end{bmatrix}$," which

is true by the given definition of M and the computation

$$M^2 = M \cdot M = \begin{bmatrix} 0 & 1/2 \\ 1/2 & 0 \end{bmatrix} \cdot \begin{bmatrix} 0 & 1/2 \\ 1/2 & 0 \end{bmatrix} = \begin{bmatrix} (1/2)(1/2) & 0 \\ 0 & (1/2)(1/2) \end{bmatrix} = \begin{bmatrix} 1/4 & 0 \\ 0 & 1/4 \end{bmatrix}$$

Let $m \geq 3$ be given such that the first statement *not* yet proven is the one involving M^{2m-1} and M^{2m}. From this we know

$$M^{2m-1} = M \cdot M^{2m-2} = \begin{bmatrix} 0 & 1/2 \\ 1/2 & 0 \end{bmatrix} \cdot \begin{bmatrix} (1/4)^{m-1} & 0 \\ 0 & (1/4)^{m-1} \end{bmatrix} = \begin{bmatrix} 0 & (1/2)(1/4)^{m-1} \\ (1/2)(1/4)^{m-1} & 0 \end{bmatrix} = \begin{bmatrix} 0 & 2(1/4)^m \\ 2(1/4)^m & 0 \end{bmatrix}$$

and so

$$M^{2m} = M \cdot M^{2m-1} = \begin{bmatrix} 0 & 1/2 \\ 1/2 & 0 \end{bmatrix} \cdot \begin{bmatrix} 0 & 2(1/4)^m \\ 2(1/4)^m & 0 \end{bmatrix} = \begin{bmatrix} (1/4)^m & 0 \\ 0 & (1/4)^m \end{bmatrix}$$

which verifies the next statement, completing the induction.∎

23. *Proof by induction.* Let M be an $n \times n$ transition matrix reflecting the one-move transition probabilities for States 1 through n of a game, and let $P(k)$ be the statement, "The entry in Row i, Column j of the matrix M^k is the probability of the game moving from State i to State j in k moves." The statement $P(1)$ is true by the definition of the transition matrix M. Let $m \geq 2$ be given such that all plays of the game consisting of less than m moves have been checked, and we are now considering m moves of the game. Let States i and j be given, and let's compute the probability that the game goes from State i to State j in exactly m moves.

In order to go from State i to State j, the game must go from State i to some State l in $m - 1$ moves and then from that State l to State j in one move. By the induction hypothesis, the probability of going from State i to State l in $m - 1$ moves is $M_{i,l}^{m-1}$, the entry in Row i, Column l of M^{m-1}. The probability of going from State l to State j in one move is $M_{i,j}$.

Hence by the product rule, the probability of going from State i to State l in $m - 1$ moves followed by one

move to State j is

$$M_{i,l}^{k-1} \cdot M_{l,j}$$

By the sum rule, we add all possible cases of which state is State l, to get the probability of going from State i to State j in m moves as

$$\sum_{l=1}^{n} M_{i,l}^{k-1} \cdot M_{l,j}$$

which is precisely the entry in Row i, Column j of

$$M^{k-1} \cdot M = M^k$$

This completes the induction step.   ■

Section 7.1 Exercises

1. (a) Nine nodes and eleven edges; **(c)** The sum of the degrees is 22, which is twice the number of edges; **(d)** *ABDA, DEFIHGD, FIHF, DEFHGD*

4. (a) 12346781 and 12643781; **(c)** 3487645; **(e)** 28732

5. (a) $deg(1) = 4, deg(2) = 4, deg(3) = 2, deg(4) = 2,$ $deg(5) = 6, deg(6) = 6, deg(7) = 2, deg(8) = 2, deg(9) = 4, deg(10) = 4$; **(c)** One possibility is $1, 3, 5, 2, 6, 10, 5, 7, 9, 6, 1$

7. (a) Here are two possibilities:

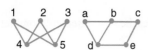

(c) Here are two possibilities:

(d) If there are 7 nodes in a simple graph, the largest degree is six. **(e)** There cannot be an odd number of vertices of odd degree.

8. (a) Add edges [2, 7] and [3, 6], and parallel edge [4, 8]; **(c)** Add edges [1, 7], [2, 8], [3, 9], [4, 10], [5, 6]

9. (a) 3, 4, 8, 2, 6, 4, 5, 6, 7, 8, 1, 2, 3, 7 is an Eulerian trail. **(c)** 1, 2, 3, 4, 5, 1, 6, 2, 7, 3, 8, 4, 9, 5, 10, 6, 7, 8, 9, 10, 1 is an Eulerian circuit.

11. In this graph, there is an Eulerian trail since the "In" and "Out" nodes are the only ones with odd degree.

25. There will are approximately 13.21 moves expected in this game.

27. There are approximately 8.33 moves expected in this game.

29. We use the matrix

$$N = \begin{bmatrix} 5/6 \end{bmatrix}$$

corresponding to the transient states from the matrix given in Exercise 6, and compute

$$N + N^2 + \cdots + N^{1000} \approx \begin{bmatrix} 5.0 \end{bmatrix}$$

Adding 1 to the sum of the entries in the first row (corresponding to games beginning in State 1) gives us approximately 6.0 moves expected in this game.

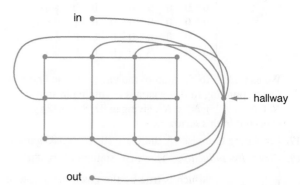

12. (a) There is only one connected graph with 0 edges with every vertex having even degree; namely, a single node with no edges. The trivial circuit consisting of just this node is an Eulerian circuit. **(c)** Here are both connected graphs with 2 edges with every vertex having even degree. The graph on the left has the Eulerian circuit $a, 1, a, 2, a$, and the graph on the right has the Eulerian circuit $a, 1, b, 2, a$.

13. Yes, it is possible. This will be possible in any connected graph with no more than four nodes of odd degree.

15. (a) 2; **(c)** 2

17. Define the graph G having nodes labeled 0, 1, 2, 3, 4, 5, 6 (for the number of spots on the dominoes) and an edge for each domino with endpoints reflecting the number of spots on that domino. In this graph, all nodes have degree 6. When one domino (edge) is removed, the nodes labeled with the two numbers on that domino have degree 5. The resulting graph has an Eulerian path starting and ending at these two nodes of odd degree. The trick works because of this.

Section 7.2 Exercises

1. Refer to the three graphs in the figure below.

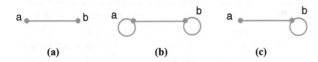

(a) **(b)** **(c)**

(a) The walk $W = a, 1, b, 1, a$ is closed but does not contain a cycle.

(b) Every vertex is of degree 3 but the only cycles have size 1.

(c) There are two vertices, one of degree 1 and one of degree 3.

2. The three blanks should be filled in as follows: G is connected; W itself; W with the edge $[v_0, v_1]$ replaced by the walk $v_0, v_n, \ldots, v_2, v_1$

3. (a) $1, 2, 4$; **(c)** $9, 3, 2, 4$; **(e)** $3, 5, 4, 8, 9, 3, 2, 8, 7 \rightarrow 3, 2, 8, 7$

7. Let G be a simple graph with n vertices, where $n \geq 2$. Since G is simple, the degree of each vertex must be no more than $n - 1$. Moreover, if there is a vertex of degree $n - 1$, then that vertex is adjacent to every other vertex in the graph, making it impossible for there to also be a degree 0 vertex. Create boxes labeled 1 to $n - 2$ and another box labeled "0 or $n - 1$," and assign each vertex of the graph to a box based on its degree. Since there are $n - 1$ boxes and n vertices, the Pigeonhole Principle tells us that at least one box contains two vertices. These two vertices must have the same degree.

9. Let G be a graph on n nodes, and let P be a path in G. Since a path cannot use the same node twice, P contains no more than n nodes. Since for every path the number of edges is one less than the number of nodes, the number of edges in P cannot be more than $n - 1$.

10. HINT: Consider the contrapositive statement, "For every connected graph G, if there exists a pair of vertices a, b in G with two (or more) paths from a to b, then G has at least one cycle."

11. Let G be a tree with at least one edge, let $e = [v_0, v_1]$ denote the deleted edge, and let G' denote the resulting graph. By the previous exercise, the only path from v_0

to v_1 in G is v_0 to v_1. Thus in G' there is no path from v_0 to v_1, and hence no walk from v_0 to v_1.

Define $H_0 = \{v \in V \mid$ there is a walk from v to v_0 in $G'\}$ and $H_1 = \{v \in V \mid$ there is a walk from v to v_1 in $G'\}$. Since for every vertex v in G there is a walk in G from v to v_0 and a walk in G from v to v_1, it follows that H_0 and H_1 are each connected and that every vertex is in one of the two sets. Also note that no vertex can be in both sets since otherwise we could build a path from v_0 to v_1 in G'. It follows that H_0 and H_1 are the connected components of G'.

14. The blanks should be filled in as follows: Any tree G with 0 edges has one node. The graph $\bullet$ has one vertex and no edges. Proposition 4; $m - 1$; $m - 1$; m; G has one more node that G'; $m + 1$

17. All relevant graphs with no edges, one edge, two edges or three edges are given in the figure below.

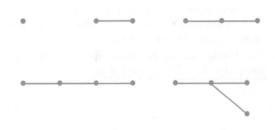

20. *Proof by induction.* Let $P(k)$ stand for, "A simple graph with n vertices, k connected components, and no cycles has $n - k$ edges." The first statement $P(1)$ is true by Theorem 7 since if $k = 1$, such a graph is connected and therefore a tree.

Now let $m > 1$ be given such that statements $P(1), \ldots, P(m-1)$ have been checked. Let G be a graph with n vertices, m connected components, and no cycles. Choose any component H, and let G' be the graph with H removed. Let n_1 be the number of vertices in H, and let n_2 be the number of vertices in G'. Observe that $n_1 + n_2 = n$.

By $P(1)$, the component H has $n_1 - 1$ edges, and by statement $P(m-1)$, the graph G' has $n_2 - (m-1)$

edges. Thus G has a total of

$$(n_1 - 1) + (n_2 - (m - 1)) = n_1 - 1 + n_2 - m + 1$$
$$= n_1 + n_2 - m = n - m$$

edges. This establishes statement $P(m)$, and the result follows by induction. ∎

21. The blanks should be filled in as follows: Exercise 20; 1; $n-1$; 1; $-n$; $\frac{n^2-n}{2}$; $\frac{n}{2}$; 1; $n-1$; $\frac{(n-1)(n-2)}{2}$

23. **(a)** There is only one, with edges $[a, d], [a, c], [a, b]$ and a total weight of 6. **(c)** There is only one, with edges $[a, b], [a, d], [a, e], [c, e]$ and a total weight of 18.

25. **(a)** We list the edges in the order [a, d],[a, c],[a, b], [b, d],[c, d],[b, c]. Edges are added at steps 1, 2 and 3, giving the minimal spanning tree with edges $[a, d], [a, c], [a, b]$ and total weight 6. **(c)** The algorithm adds $[a, b]$ and $[a, d]$ in the first two steps, and $[a, e]$ and $[c, e]$ in the next two steps.

26. *Outline of the proof.* Let $P(n)$ be the statement, "T_n is included in a minimal spanning tree of G." In the induction step, we know from the algorithm that T_m is formed by adding the edge e_m to the tree T_{m-1}. By the inductive hypothesis $P(m-1)$, we know that T_{m-1} is a subgraph of a minimal spanning tree of G. Call this minimal spanning tree **T**. The edge e_m either is in **T** or it is not, so you can argue by the following cases: **Case 1.** If the edge e_m is actually in **T**, then ...; **Case 2.** If e_m is *not* an edge in **T**, then form the new graph H by adding e_m to **T**. Graph H will have a cycle so we can let f be the edge on this cycle with the smallest weight. Show that the weight of f must be equal to the weight of e_m, and conclude that the tree obtained by removing e_m from H is a minimal spanning tree of G including T_m.

27. The blanks should be filled in as follows: $n+1$; $2(n+1) = 2n + 2$; $2n$; $2n + 2 > 2n$

29. Using Kruskal's algorithm, we add the edges in this order: $ae, fj, ld, bc, bf, gj, ij, im, kl, ab, hi, jn, ad$. The total cost is $3(\$100) + 6(\$200) + 3(\$300) + \$400 = \$2800$.

31. **(a)** One solution is to remove cd from $acda$, then bd from $abda$, and finally bc from $abca$. This leaves the spanning tree ab, ac, ad with a total weight of 6. **(c)** One solution is to remove bc from $abca$, then de from $cdec$, then cd from $acda$, then ac from $acea$, then bd from $abda$, and finally be from $abea$. This leaves the spanning tree ab, ad, ae, ce with a total weight of 18.

33. Let $P(n)$ denote the statement, "In the algorithm of Exercise 28, graph G_n contains a minimal spanning tree of G." Since G_0 is G and G is given as being connected, $P(0)$ is true. Now let $m \geq 1$ be given such that statements $P(0), \ldots, P(m-1)$ have all been checked to be true. Recall that graph G_m is formed from G_{m-1} be removing the most expensive edge e from some cycle C in G_{m-1}. By statement $P(m-1)$, graph G_{m-1} contains a spanning tree T that is a minimal spanning tree for G. Since trees don't have cycles, the cycle C must include some edge f that is not in T. There are two cases to consider:

Case 1. If $e = f$, then all edges of T are included in G_m, so G_m contains the minimal spanning tree T of G.

Case 2. If $e \neq f$, then edge e must weigh at least as much edge f (since the algorithm deleted e to form G_m), so the new tree T' formed by adding f to T and removing e will weigh no more overall than T. Hence T' is a different minimal spanning tree for G, and T' is contained in G_m.

Since in either case, G_m contains a minimal spanning tree for G, this establishes statement $P(m)$, completing the induction.

Section 7.3 Exercises

1. **(a)** Graph G_2 has a nodes of degree 4 and graph G_1 does not.

   (b) Graph H_1 has a cycle of length 5 and graph H_2 does not.

3. If we write down the degrees of each node, we find that graph (B) has two nodes of degree 2, and the others have only one. Hence (B) is not isomorphic to any of the others. Since all the graphs have a unique node of degree 1, these would have to correspond to each other under an isomorphism. Note that the degree 1 node in graph (D) is adjacent to a degree 3 node, while in graphs (A) and (C), the degree 1 node is adjacent to a degree

4 node. Thus (D) is not isomorphic to (A) or (C). This leaves (A) and (C) as the only candidates. To see the isomorphism, imagine moving the bottom left node in (C) halfway toward the bottom right node, then dragging the node in the middle in (C) down to the bottom left.

5. The graph on the left has no cycles of length 3 while the graph on the right does. Another difference is that the graph on the left is $K_{3,3}$ so it is non-planar, while the graph on the right has no edge-crossings.

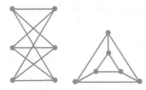

7. The three graphs are $K_{5,5}$ and the two graphs shown in the figure below.

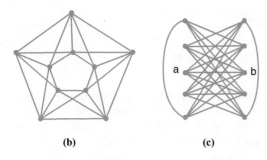

(b)   (c)

All cycles in $K_{5,5}$ have an even number of edges. Graphs (B) and (C) contain cycles of length 3, so neither is isomorphic to $K_{5,5}$. Graph (B) has the property that every edge is on some cycle of length 3. In graph (C), edge $[a, b]$ is not on such a cycle, hence graphs (B) and (C) are not isomporphic.

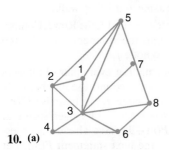

10. (a)

Section 7.4 Exercises

1. (a) $\begin{bmatrix} 0 & 1 & 0 & 1 \\ 1 & 0 & 1 & 1 \\ 0 & 1 & 0 & 0 \\ 1 & 1 & 0 & 0 \end{bmatrix}$; **(c)** $\begin{bmatrix} 0 & 1 & 0 & 1 & 1 \\ 0 & 0 & 0 & 0 & 0 \\ 0 & 0 & 1 & 1 & 0 \\ 0 & 1 & 0 & 0 & 0 \\ 0 & 0 & 0 & 1 & 0 \end{bmatrix}$;

(e) $\begin{bmatrix} 0 & 1 & 0 & 0 & 0 \\ 1 & 0 & 0 & 0 & 0 \\ 0 & 0 & 0 & 0 & 0 \\ 0 & 0 & 1 & 0 & 1 \\ 0 & 0 & 1 & 0 & 0 \end{bmatrix}$

11. (a) Suppose (A) is planar. (A) has 8 vertices, 24 edges, every edge is on a cycle, and the smallest cycle is length 3. By Euler's formula, (A) has exactly $24 + 2 - 8 = 18$ faces. By Theorem 7, the number of edges is at least $\frac{3}{2}$ times the number of faces, that is $24 \geq \frac{3}{2} \cdot 18$, or $24 \geq 27$. Since this is a contradiction, we conclude that (A) is not planar. **(c)** Suppose (C) is planar. (C) has 10 vertices, 15 edges, every edge is on a cycle, and the smallest cycle is length 5. By Euler's formula, (C) has exactly $15 + 2 - 8 = 9$ faces. By Theorem 7, the number of edges is at least $\frac{5}{2}$ times the number of faces, that is $15 \geq \frac{5}{2} \cdot 9$, or $15 \geq 22.5$. Since this is a contradiction, we conclude that (A) is not planar.

12. (a) This graph is non-planar because it contains $K_{3,3}$. Every node in the set $\{1, 2, 6\}$ is connected to every node in $\{3, 4, 5\}$. The additional edges $[4, 5]$ and $[3, 4]$ only make it worse.

13. (a) For the given illustration, $V = 12, E = 18, F = 8$ and $12 + 8 = 18 + 2$. In general, $V = 2k, E = 3k, F = k + 2; 2k + (k + 2) = 3k + 2$. **(c)** For the given illustration, $V = 5, E = 8, F = 5$ and $5 + 5 = 8 + 2$. In general, $V = k + 1, E = 2K, F = K + 1; (k + 1) + (k + 1) = 2K + 2$.

15. *Proof.* By Theorem 7, $m \geq \frac{3}{2}F$, hence $F \leq \frac{2}{3}m$. By Euler's formula, $n + F = m + 2$. So $m + 2 = n + F \leq n + \frac{2}{3}m$. Solving the inequality $m + 2 \leq n + \frac{2}{3}m$, we get $m \leq 3n - 6$. ∎

17. In the figure below, the dashed line indicates a line on the opposite side of the surface. Each of the houses A, B and C are connected to every utility 1, 2 and 3.

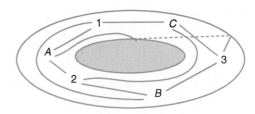

3. (a) The degree of the ith node is the sum of the ith row. **(b)** The degree of the ith node is the sum of the ith row plus the value of the (i, i) entry. (This means the (i, i) entry is counted twice.)

4. (a) The outdegree for the ith node is the sum of the ith row. The indegree is the sum of the ith column.

5. (a) $M = \begin{bmatrix} 0 & 0 & 1 & 1 & 0 & 0 \\ 1 & 0 & 0 & 0 & 1 & 0 \\ 0 & 1 & 0 & 1 & 0 & 0 \\ 0 & 1 & 0 & 0 & 1 & 1 \\ 1 & 0 & 1 & 0 & 0 & 1 \\ 1 & 1 & 1 & 0 & 0 & 0 \end{bmatrix};$

$M^2 = \begin{bmatrix} 0 & 2 & 0 & 1 & 1 & 1 \\ 1 & 0 & 2 & 1 & 0 & 1 \\ 1 & 1 & 0 & 0 & 2 & 1 \\ 3 & 1 & 2 & 0 & 1 & 1 \\ 1 & 2 & 2 & 2 & 0 & 0 \\ 1 & 1 & 1 & 2 & 1 & 0 \end{bmatrix};$

$M^3 = \begin{bmatrix} 4 & 2 & 2 & 0 & 3 & 2 \\ 1 & 4 & 2 & 3 & 1 & 1 \\ 4 & 1 & 4 & 1 & 1 & 2 \\ 3 & 3 & 5 & 5 & 1 & 1 \\ 2 & 4 & 1 & 3 & 4 & 2 \\ 2 & 3 & 2 & 2 & 3 & 3 \end{bmatrix}$

$M + M^2 + M^3 = \begin{bmatrix} 4 & 4 & 3 & 2 & 4 & 3 \\ 3 & 4 & 4 & 4 & 2 & 2 \\ 5 & 3 & 4 & 2 & 3 & 3 \\ 6 & 5 & 7 & 5 & 3 & 3 \\ 4 & 6 & 4 & 5 & 4 & 3 \\ 4 & 5 & 4 & 4 & 4 & 3 \end{bmatrix}$

So there are three walks of length 3 or less from node 1 to node 6. They are 146, 1346 and 1456.

6. (a) $\{(1, 3), (1, 4), (2, 1), (2, 5), (3, 2), (3, 4), (4, 2), (4, 5), (4, 6), (5, 1), (5, 3), (5, 6), (6, 1), (6, 2), (6, 3)\}$

7. (a) The matrices are $\begin{bmatrix} 1 & 1 & 1 & 0 & 1 & 1 \\ 1 & 1 & 1 & 1 & 1 & 1 \\ 1 & 1 & 1 & 1 & 1 & 1 \\ 1 & 1 & 1 & 1 & 1 & 1 \\ 1 & 1 & 1 & 1 & 1 & 1 \\ 1 & 1 & 1 & 1 & 1 & 1 \end{bmatrix}$,

$\begin{bmatrix} 0 & 0 & 0 & 1 & 1 & 1 \\ 0 & 0 & 0 & 0 & 1 & 1 \\ 0 & 0 & 0 & 0 & 0 & 1 \\ 0 & 0 & 0 & 0 & 0 & 0 \\ 0 & 0 & 0 & 0 & 0 & 0 \\ 0 & 0 & 0 & 0 & 0 & 0 \end{bmatrix}$

and $\begin{bmatrix} 0 & 0 & 0 & 0 & 1 & 0 \\ 0 & 0 & 0 & 0 & 0 & 0 \\ 0 & 0 & 0 & 0 & 0 & 0 \\ 0 & 0 & 0 & 0 & 0 & 0 \\ 0 & 0 & 0 & 0 & 0 & 0 \\ 0 & 0 & 0 & 0 & 0 & 0 \end{bmatrix}$

(b) For the first graph, let $A = \{1, 2, 3, 4, 5, 6\}$. Then $R \circ R \circ R$ is $(A \times A) - \{(1, 4)\}$. For the second graph $R \circ R \circ R = \{(1, 4), (1, 5), (1, 6), (2, 5), (2, 6), (3, 6)\}$. For the third graph, $R \circ R \circ R = \{(1, 5)\}$.

8. (a) The matrices for R and $R \circ R$, respectively, are shown below, followed by the graphs of each.

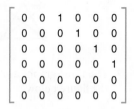

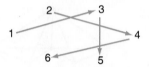

9. (a) It determines all pairs of nodes between which there is a walk of length 5 or less. **(c)** $M^{(1)} \vee M^{(2)} \vee M^{(3)} \vee M^{(4)} \vee M^{(5)} \vee M^{(6)} \vee M^{(7)} \vee M^{(8)} \vee M^{(9)}$

10. (a) $2 \cdot 5 = 10$; **(c)** $1 \cdot 2 = 2$; **(e)** 30

11. (a) There are $a_{61} \cdot b_{13}$ walks from 6 to 3 that begin by going to node 1, $a_{62} \cdot b_{23}$ that begin by going to node 2, and so on, for a total of $\sum_{t=1}^{9} a_{6,t} \cdot b_{t,3}$. **(c)** $\sum_{t=1}^{9} a_{i,t} \cdot b_{t,j}$

12. (a) For every pair of integers i and j, the Row i, Column j entry of M^1 counts the number of 1-step walks from node i to node j.

(b) M^1 is the same as M. The Row i, Column j entry of M is the number of edges from node i to node j, and an edge is the same as a 1-step walk.

(c) For every pair of integers i and j, the Row i, Column j entry of M^{k-1} counts the number of $(k-1)$-step walks from node i to node j.

13. Let $P(k)$ be the statement, "For every pair of integers i and j, the (i, j) entry of M^k counts the number of k-step walks from node i to node j." In Exercise 12(b), we established $P(1)$. Now let $m \geq 2$ be given such that statements $P(1), \ldots, P(m - 1)$ have already been established, and consider the next statement $P(m)$. Let i and j be given, and let w_t denote the number of the m-step walks from node i to node j whose first edge goes from node i to node t. Clearly there are $\sum_{t=1}^{n} w_t$ m-step walks from node i to node j, since every such walk must contain a first edge leading to exactly one of the nodes in the graph. Moreover by the product rule for counting, w_t is the product of the number of 1-step walks from i to t and the number of $(m-1)$-step walks from t to j. Using the induction hypothesis $P(m - 1)$, this is the same as the product of the (i, t) element from M times the (t, j) element from M^{m-1}. In this case, $\sum_{t=1}^{n} w_t = \sum_{i=1}^{n} M_{i,t} \cdot (M^{m-1})_{t,j}$, which is precisely the $((i, j)$ entry in M^m by the definition of matrix multiplication. This establishes $P(m)$, completing the induction proof. ∎

15. In each graph, an edge without arrows indicates an edge in both directions.

(a) $\begin{bmatrix} 1 & 0 & 0 & 0 \\ 0 & 1 & 0 & 0 \\ 0 & 0 & 1 & 0 \\ 0 & 0 & 0 & 1 \end{bmatrix}$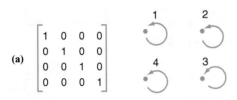

(c) $\begin{bmatrix} 0 & 1 & 1 & 0 \\ 1 & 0 & 1 & 0 \\ 1 & 1 & 0 & 0 \\ 0 & 0 & 0 & 0 \end{bmatrix}$

(e) $\begin{bmatrix} 1 & 1 & 1 & 0 \\ 1 & 1 & 0 & 0 \\ 1 & 0 & 1 & 0 \\ 0 & 0 & 0 & 0 \end{bmatrix}$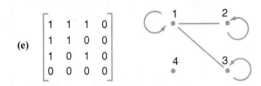

16. (a) For the M in part (a) of Exercise 15, $I \leq M$ is obviously true. **(c)** For the M in part (c) of Exercise 15, $I \not\leq M$ since $m_{1,1} = m_{2,2} = m_{3,3} = m_{4,4} = 0$. **(e)** For the M in part (e) of Exercise 15, $I \not\leq M$ since $m_{4,4} = 0$.

17. For every subscript t, $I_{t,t} = 1$. Therefore the (t, t) entry of $(I \vee M)$ is always 1, and hence $I \leq (I \vee M)$. Therefore, $I \vee M$ is reflexive.

18. (a) This relation is the same as R in Exercise 15(a).

(b) This relation is $\{(1, 1), (2, 2), (3, 3), (4, 4), (2, 3), (3, 2), (4, 1)\}$. The matrix and graph are shown below.

$\begin{bmatrix} 1 & 0 & 0 & 0 \\ 0 & 1 & 1 & 0 \\ 0 & 1 & 1 & 0 \\ 1 & 0 & 0 & 1 \end{bmatrix}$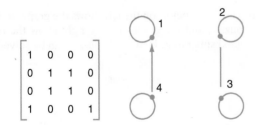

(c) This relation is $\{(1, 1), (2, 2), (3, 3), (4, 4), (1, 2), (2, 3), (1, 3), (3, 1), (3, 2), (2, 1)\}$. The matrix and graph are shown below.

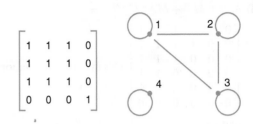

$\begin{bmatrix} 1 & 1 & 1 & 0 \\ 1 & 1 & 1 & 0 \\ 1 & 1 & 1 & 0 \\ 0 & 0 & 0 & 1 \end{bmatrix}$

19. (a) $M = M^T$, so this relation is symmetric.

(b) $M^T \neq M$, since $M_{14}^T = 1$ but $M_{14} = 0$.

(c) $M = M^T$, so this relation is symmetric.

21. For parts (a),(c),(e), and (g) the relations (and hence the graphs and matrices) are the same as in Exercise 15.

(b) This relation is $\{(1,1),(1,4),(2,3),(3,2),(4,1), (4,4)\}$. The matrix and graph are shown below.

$\begin{bmatrix} 1 & 0 & 0 & 1 \\ 0 & 0 & 1 & 0 \\ 0 & 1 & 0 & 0 \\ 1 & 0 & 0 & 1 \end{bmatrix}$

22. (a) $\begin{bmatrix} 1 & 0 & 0 & 0 \\ 0 & 1 & 0 & 0 \\ 0 & 0 & 1 & 0 \\ 0 & 0 & 0 & 1 \end{bmatrix} \leq \begin{bmatrix} 1 & 0 & 0 & 0 \\ 0 & 1 & 0 & 0 \\ 0 & 0 & 1 & 0 \\ 0 & 0 & 0 & 1 \end{bmatrix}$ is true.

(b) $\begin{bmatrix} 1 & 0 & 0 & 0 \\ 0 & 1 & 0 & 0 \\ 0 & 0 & 1 & 0 \\ 1 & 0 & 0 & 1 \end{bmatrix} \not\leq \begin{bmatrix} 1 & 0 & 0 & 0 \\ 0 & 0 & 1 & 0 \\ 0 & 1 & 0 & 0 \\ 1 & 0 & 0 & 1 \end{bmatrix}$ since $2R3$ and $3R2$, but not $2R2$.

23. (a) $\{(1, 2), (1, 3), (1, 4), (1, 5), (1, 6), (2, 3), (2, 4), (2, 5), (2, 6), (3, 4), (3, 5), (3, 6), (4, 5), (4, 6), (5, 6)\}$

(b) The set in the part (a) solution is the transitive closure.

(c) $M = \begin{bmatrix} 0 & 1 & 0 & 0 & 0 & 0 \\ 0 & 0 & 1 & 0 & 0 & 0 \\ 0 & 0 & 0 & 1 & 0 & 0 \\ 0 & 0 & 0 & 0 & 1 & 0 \\ 0 & 0 & 0 & 0 & 0 & 1 \\ 0 & 0 & 0 & 0 & 0 & 0 \end{bmatrix}$

(d) $M^{(1)} + M^{(2)} + M^{(3)+M^{(4)}+M^{(5)}} =$

$$\begin{bmatrix} 0 & 1 & 1 & 1 & 1 & 1 \\ 0 & 0 & 1 & 1 & 1 & 1 \\ 0 & 0 & 0 & 1 & 1 & 1 \\ 0 & 0 & 0 & 0 & 1 & 1 \\ 0 & 0 & 0 & 0 & 0 & 1 \\ 0 & 0 & 0 & 0 & 0 & 0 \end{bmatrix}$$ This matches the relation in

part (b).

25. (a) The relation is already transitive, so the transitive closure is the same relation.

(c) This relation is $\{(1, 1), (1, 2), (1, 3), (2, 1), (2, 2), (2, 3), (3, 1), (3, 2), (3, 3)\}$. The matrix and graph are shown below.

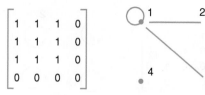

Section 7.5 Exercises

1. A solution corresponds to a path from 8, 0, 0 to 4, 4, 0 in the graph below. Note that in the graph, every node within the square also has edges pointing to two of the four corners of the square, but we have left these edges out for clarity. Note that the puzzle has two solutions, one slightly shorter than the other.

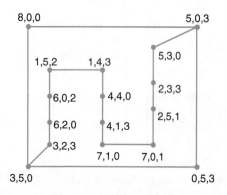

3. The graph model below shows that there is no path from $(10, 0, 0)$ to a node labeled $(5, 5, 0)$. Since the only transitions involve even numbers of quarts and the beginning state has an even number of quarts, we can never measure any odd number of quarts with these containers.

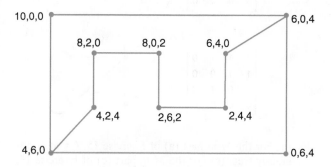

5. Cubes 1, 2, 3 and 4 are shown left-to-right below.

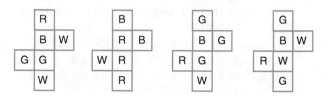

7. (a) The figure below on the left shows the graph of this puzzle, and the graphs on the right show the two good subgraphs. Hence this puzzle can be solved.

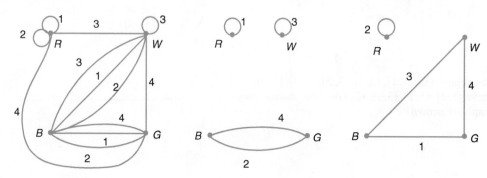

(b) See the figure below. We can argue that the puzzle has no solution as follows. If there *are* two good subgraphs then one will have to include the loop from cube 4 at vertex R while the other must include the edges $[W, R]$ from cube 3 and $[R, G]$ from cube 1. Let's focus on the first of these. The only ways for this subgraph to have every node of degree 2 is to either have B, G, W, B 3-cycle or a B, W, B 2-cycle along with a loop at G. The first of these is impossible since we would be forced to take the $[B, W]$ edge from cube 1 (since cube 4 was already used with the loop at R) leaving us no possible edge $[W, G]$. The second is impossible since there are only two edges of the form $[W, B]$ so we would need to use them both but one is labeled 4 which has already been used with the loop at R. Note that the latter subgraph originally described *is* possible, so there is a partial solution to this puzzle but not a complete solution.

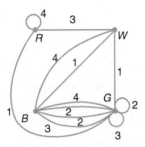

9. The kernel is $K = \{(4, 7), (3, 5), (5, 3), (1, 2), (2, 1), (0, 0)\}$, so Player 1 has the winning strategy. On her first move, she should either remove 1 stone from the smaller pile or 4 stones from the larger pile.

11. (a) This game is identical to the stone removal game played in Exercise 9 but played on the grid of states (X, Y) where X and Y range from 0 to 8. Generalizing the process in that solution, we will have a kernel in this game of $K = \{(7, 4), (4, 7), (3, 5), (5, 3), (1, 2), (2, 1), (0, 0)\}$.

Since the beginning position has the form $(8, X)$ or $(X, 8)$, it cannnot be in the kernel, so Player 1 will have a winning strategy regardless of the queen's starting spot.

(b) This is like a stone removal game combining the rules of Exercises 9 and 10 and played on the grid shown in the solution to part (a). We can follow the process as before to find that

$$K = \{(0, 0), (1, 3), (3, 1), (2, 6), (4, 5), (5, 4), (6, 2)\}$$

Since the beginning position has the form $(8, X)$ or $(X, 8)$, it cannnot be in the kernel, so Player 1 will have a winning strategy regardless of the queen's starting spot.

13. The game is already in a winning position. This means Player 1 will not have a winning strategy assuming that Player 2 has read this section.

3	=	0	1	1
4	=	1	0	0
7	=	1	1	1

Direct Sum	=	0	0	0

15. One possibility is piles of 2, 8 and 10 as shown below, but any division of 20 pennies into three piles so that the direct sum is zero will give you the winning strategy.

10	=	1	0	1	0
8	=	1	0	0	0
2	=	0	0	1	0

Direct Sum	=	0	0	0	0

17. Player 1 should remove all 11 stones in the 8th pile. This will make the direct sum 0.

19. The complete kernel is

$$K = \{\, 000, 003, 006, 011, 014, 017, 022, 025, 030, 033,$$
$$036, 041, 044, 047, 052, 055, 101, 104, 107, 110,$$
$$113, 116, 131, 134, 137, 140, 143, 146, 202, 205,$$
$$220, 223, 226, 232, 235, 250, 253, 256 \,\}$$

21. The strategy for Player 2 to win is to mirror the moves of Player 1 using the 180° rotational symmetry of the clock face. That is, for each number Player 1 crosses out, Player 2 crosses out the number that is 6 hours later. On each move, Player 1 destroys the 180° symmetry of the picture and Player 2 restores it. Since the winning position has this symmetry, Player 2 must win.

Section 7.6 Exercises

2. Answers will obviously vary, but they should be binary trees as long as only biological parents are listed. Here is one possibility:

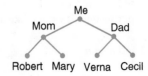

4. Since a complete binary tree with height 3 has $2^3 - 1 = 7$ nodes, we will need to use a tree with height 4 to store 8 values. The following diagram shows one of several possibilities:

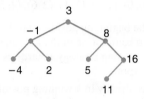

5. Preorder: $ABDEGCF$; Inorder: $DBGEACF$; Postorder: $DGEBAFC$

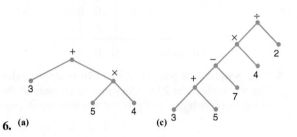

6. (a)   (c)

7. (a) Prefix: $+, 3, \times, 5, 4$; Postfix: $3, 5, 4, \times, +$;
(c) Prefix: $\div, \times, -, +, 3, 5, 7, 4, 2$; Postfix: $3, 5, +, 7, -, 4, \times, 2, \div$

8. (a) $((5 \times 4) + 7) \div 3 = 9$

9. (a) $(6 - (2 \times 3)) \times (12 \div 4) = 0$

10. Inorder: Robert, Mom, Mary, Me, Verna, Dad, Cecil; Preorder: Me, Mom, Robert, Mary, Dad, Verna, Cecil; Postorder: Robert, Mary, Mom, Verna, Cecil, Dad, Me

13. *Proof by induction.* Let $P(n)$ be the statement, "The maximum number of nodes in a binary tree of height n is $2^n - 1$." Statement $P(1)$ refers to a binary tree of height 1, of which there is only one: a single root with empty left and right subtrees. Since such a tree has $2^1 - 1 = 1$ node, we conclude that the first statement $P(1)$ is true.

Now let $m \geq 2$ be given such that statements $P(1), P(2), \ldots, P(m-1)$ have all been checked to be true. Let T be a binary tree with height m. By definition of "height," each of the left subtree T_L and right subtree T_R of T have height less than or equal to $m - 1$. Letting h_L and h_R denote the respective heights of T_L and T_R, we can cite statements $P(h_L)$ and $P(h_R)$, which tell us that tree T_L has a maximum of $2^{h_L} - 1$ nodes and tree T_R has a maximum of $2^{h_R} - 1$ nodes. Hence, the original tree T has a maximum of

$$1 + \left(2^{h_L} - 1\right) + \left(2^{h_R} - 1\right) \leq 1 + \left(2^{m-1} - 1\right)$$
$$+ \left(2^{m-1} - 1\right) = 2^m - 1$$

nodes. This confirms statement $P(m)$, completing the induction. ∎

15. *Proof.* Let T be any binary tree. Let L denote the maximum level of any node in T, and let H denote the height of T. Proposition 1 tells us that $L \leq H$, and Exercise 14 tells us that $H \leq L$. Therefore, $L = H$. ∎

17. (a) Here is one of the trees:

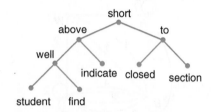

(c) $3; 7; 63; 1023$; (e) $f(n) = 1 + \lfloor \log_2(n) \rfloor$.

19. *Proof by induction.* Let $P(n)$ be the statement, "$(n!)^2 > n^n$." Since $(3!)^2 = 36$ and $3^3 = 27$, it is clear that statement $P(3)$ is true. Now let $m \geq 4$ be given such that statements $P(3), P(4), \ldots, P(m-1)$ have all been checked to be true. In particular, it has been checked that $((m-1)!)^2 > (m-1)^{m-1}$ is true. From this it follows that

$$(m!)^2 = m^2 \cdot ((m-1)!)^2$$
$$> m^2 \cdot (m-1)^{m-1} \text{ by statement } P(m-1)$$
$$= m^m \text{ by the given fact}$$

∎

Section 7.7 Exercises

1. $1, 2, 8, 7, 17, 16, 20, 13, 12, 11, 19, 18, 9, 10, 3, 4, 5, 14, 15, 6, 1$

3. There are $\frac{47!}{2} \approx 1.3 \times 10^{60}$ different cycles to check. At a rate of 10^{15} per second, it would take about 10^{45} seconds.

5. $m \geq 2, n \geq 2$ and $m = n$

6. (a) a, d, b, c, a has weight 13

7. If we start with vertex e, the Hamiltonian cycle generated by the algorithm is e, c, d, a, b, f, e and has weight 21, but the cycle a, b, c, d, e, f, a has weight 14.

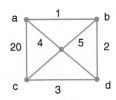

8. (a) a, d, b, c, a has weight 13

9. (a) $[a, d], [a, c], [b, d], [b, c]$ yields a, d, b, c, a with weight 13; **(c)** $[a, b], [a, d], [c, e], [b, c], [d, e]$ yields a, b, c, e, d, a with weight 28

13. $[a, e], [a, d], [e, b], [c, d], [c, f], [b, f]$ yields a, e, b, f, c, d, a with weight 42.

14. (b) 000, 100, 101, 111, 110, 010, 011, 001, 000 is a Hamiltonian cycle in G_3 shown in the figure below.

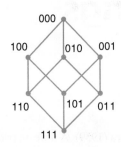

15. (b) The graph on the right below shows the representation of the Knight's Tour on the 4×4 chessboard shown on the left.

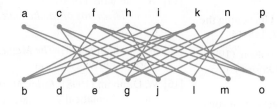

17. The graph shown below has a Hamiltonian cycle, but $\deg(a) + \deg(c) = 4 < 5$.

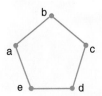

19. *Proof.* Let G be a simple connected graph on $n \geq 3$ nodes in which each node has degree at least $\frac{n}{2}$. This means that for every pair of non-adjacent vertices u and v, we can be sure that

$$\deg(u) + \deg(v) \geq \frac{n}{2} + \frac{n}{2} = n$$

By Ore's Theorem, G is Hamiltonian. ∎

20. HINT: Assume that the Petersen graph has a Hamiltonian cycle, and argue that a contradiction must result. If you consider there are five "outside edges," five "star edges," and five "connecting edges," then you can argue that there must be an even number of edges of each type in the Hamiltonian cycle. Each possible case will lead to a contradiction.

21. The cycle PA, NY, CT, RI, MA, NH, VT, B, PA has weight 1216 miles.

23. $[B, VT], [RI, MA], [MA, NH], [RI, CT], [CT, NY], [NH, VT], [PA, B], [PA, NY]$ yields the cycle $PA, B, VT, NH, MA, RI, CT, NY, PA$ with weight 1216 miles.

References and Further Readings

[1] E. Akin and M. Davis (1985). Bulgarian Solitaire. *The Amer. Math. Monthly*, **92**, 237–250.

[2] M. A. Ascher (1990). A river-crossing problem in cross-cultural perspective. *Math. Magazine*, **63**, 26–29.

[3] E. T. Bell (1937). *Men of Mathematics*. New York: Simon & Schuster.

[4] A. T. Benjamin and J. J. Quinn (1999). Unevening the odds of 'Even Up'. *Math. Magazine*, **72**, 145–146.

[5] N. L. Biggs, E. K. Lloyd, and R. J. Wilson (1999). *Graph Theory, 1736–1936*. New York: Oxford University Press.

[6] L. Blaine (1991). Theory versus computation in some very simple dynamical systems. *College Math Journal*, **22**, 39–41.

[7] C. Bouton (1902). Nim: A game with a complete mathematical solution. *Ann. of Math.*, **3**, 35–39.

[8] C. B. Boyer (1989). *A History of Mathematics, 2nd ed.* New York: John Wiley & Sons.

[9] R. Calinger, ed. (1994). *Classics of Mathematics*. Upper Saddle River: Prentice Hall.

[10] P. J. Davis and R. Hersh (1981). *The Mathematical Experience*. Boston: Birkhauser.

[11] P. J. Davis and R. Hersh (1986). *Descartes' Dream: The World According to Mathematics*. San Diego: Harcourt Brace Jovanovich.

[12] E. W. Dijkstra (1959). A note on two problems in connexion with graphs. *Numer. Math.*, **1**, 269–271.

[13] S. E. Dreyfus and A. M. Law (1977). *The Art and Theory of Dynamic Programming*. New York: Academic Press.

[14] W. Dunham (1990). *Journey Through Genius: The Great Theorems of Mathematics*. New York: John Wiley & Sons.

[15] W. Dunham (1994). *The Mathematical Universe*. New York: John Wiley & Sons.

[16] W. Dunham (1998). *Euler, Master of Us All*. Washington: Mathematical Association of America.

[17] P. Eisele and K. P. Hadeler (1990). Game of cards, dynamical systems, and a characterization of the floor and ceiling functions. *Amer. Math. Monthly*, **97**, 466–477.

[18] R. Engel (1992). *Exploring Mathematics with your Computer*. Washington: Mathematical Association of America.

[19] D. E. Ensley (1999). Invariance under group actions to amaze your friends. *Mathematics Magazine*, **72**, 383–387.

[20] D. E. Ensley (2003). Fibonacci's triangle and other abominations. *Math Horizons*, **11** (September), 10–14.

[21] D. E. Ensley (2004). Unshuffling for the imperfect mathemagician. *Math Horizons*, **11** (February), 13–21.

[22] S. Epp (2004). *Discrete Mathematics with Applications*, Third Edition. Pacific Grove, CA: Brooks/Cole.

[23] W. Feller (1950). *An Introduction to Probability Theory and Its Applications*, Volume I. New York: John Wiley & Sons.

[24] M. Gardner (1956). *Mathematics, Magic and Mystery*. New York: Dover Publications.

[25] M. Gardner (1977). *Mathematical Carnival*. New York: Vintage Books.

[26] M. Gardner (1979). *Mathematical Circus*. New York: Vintage Books.

[27] M. Gardner (1989). *Penrose Tiles to Trapdoor Ciphers*. New York: Freeman.

[28] P. Gerdes (1999). *Geometry from Africa*. Washington: Mathematical Association of America.

[29] R. Graham, D. Knuth and O. Patashnik (1994). *Concrete Mathematics*. Reading, MA: Addison-Wesley.

[30] R. A. Griggs and J. R. Cox (1984). The elusive thematic-materials effect on Wason's Selection Task. *British Journal of Psychology*, **73**, 407–420.

[31] R. V. Heath (1953). *Mathemagic: Magic, Puzzles, and Games with Numbers*. New York: Dover Publications.

[32] I. N. Herstein & I. Kaplansky (1974). *Matters Mathematical*. New York: Chelsea Publishing Company.

[33] D. Hofstadter (1979). *Gödel, Escher, and Bach: An Eternal Golden Braid*. New York: Basic Books.

[34] E. Kasner & J. Newman (1940). *Mathematics and the Imagination*. New York: Simon & Schuster.

[35] D. Knuth (1997). *The Art of Computer Programming, Volume I: Fundamental Algorithms*, Third Edition. Reading, MA: Addison-Wesley.

[36] J. G. Kemeny & J. L. Snell (1960). *Finite Markov Chains*. Princeton: Van Nostrand.

[37] J. B. Kruskal, Jr. (1956). On the shortest spanning subtree of a graph and the traveling salesman problem. *Proc. Amer. Math. Soc.*, **7**, 48–50.

[38] J. C. Lagarias (1985). The $3x + 1$ Problem and its generalization. *Amer. Math. Monthly*, **92**, 3–23.

[39] S. B. Morris (1997). *Magic Tricks, Card Shuffling and Dynamic Computer Memories*. Washington: Mathematical Association of America.

[40] J. R. Newman (1956). *The World of Mathematics*. New York: Simon & Schuster.

[41] C. S. Peirce (1877). The fixation of belief, in *Popular Science Monthly*, **12**, 1–15.

[42] R. G. Prim (1957). Shortest connection networks and some generalizations. *Bell Systems Technical Journal*, **36**, 1389–1401.

[43] L. Sigler (2003). *Fibonacci's Liber Abacci*. New York: Springer.

[44] W. Simon (1993). *Mathematical Magic*. New York: Dover Publications.

[45] R. Smullyan (1978). *What is the Name of this Book?*. New York: Simon & Schuster.

[46] P. D. Straffin (1993). *Game Theory and Strategy*. Washington: Mathematical Association of America.

[47] A. Tucker (2002). *Applied Combinatorics*, Fourth Edition. New York: John Wiley & Sons.

[48] P. C. Wason (1968). Reasoning about a rule. *Quaterly Journal of Experimental Psychology*, **20**, 273–281.

[49] R. Young (1992). *Excursions in Calculus*. Washington: Mathematical Association of America.

Index

Special symbols and notation

2^A, 207
$\forall$, 43
$\wedge$, 27
$C(n, k)$, 372, 399
$\lceil x \rceil$, 327
$\times$, 198
$\deg(v)$, 510, 513
Δ_n, 424
Δ_n^k, 425
$\in$, 43, 183
$E[X]$, 467, 469
$\exists$, 43
f^{-1}, 261
!, 16, 330, 385, 393
$\lfloor x \rfloor$, 327
gcd, 228
$\mathbf{I}_n$ (I), 497, 620
$\leftrightarrow$, 62
$\rightarrow$, 53
$\cap$, 186
ι, 272
K_n, 536, 598
$K_{m,n}$, 536, 607
lcm, 228
$\leq$, 559
M^{-1}, 497, 621
$M^{(k)}$, 560
M^T, 623
M_∞, 487

$M_{i,j}$, 482, 547, 618
mod (%), 105, 166
$n(A)$, 191, 443
$\mathbb{N}$, 182
$\neg$, 27
O, 349
$\circ$, 268, 274
Ω, 349
$\vee$, 27, 561
$\mathcal{P}(A)$, 203
$P(n)$, 116
$P(n, k)$, 372, 393, 419
$Prob(E_1|E_2)$, 452
$Prob(E)$, 443
$\mathbb{Q}$, 182
$\mathbb{Q}^+$, 182
$\mathbb{Q}^{\geq 0}$, 182
R^{-1}, 257
$\mathbb{R}$, 45, 182
$\mathbb{R}^+$, 182
$\mathbb{R}^{\geq 0}$, 182
Σ, 19, 114
$\subseteq$, 183
$\subset$, 195
Θ, 349
$\cup$, 186
$\mathbb{Z}$, 45, 182
$\mathbb{Z}^+$, 182
$\mathbb{Z}^{\geq 0}$, 182
$\mathbf{0}_n$, 620

Absorbing state, 483, 487, 494
Absorption property
    Boolean algebra, 225
    logic, 35, 39, 222
    sets, 195, 215, 222
Abstract algebra, 166
Abstraction, 221, 507
Addition of matrices, 491, 551, 619
Adjacency matrix, 547
    binary relation, 553
    Boolean, 555
    directed graph, 549
Adjacent nodes, 508, 547, 604
Adleman, L., 172
Alabama, 370, 411, 417
Alcuin of York, 567
Algorithm
    base two conversion, 154
    binary search, 353
    complexity of, 346, 601
    counterfeit detection, 354, 363
    for counting, 391
    dictionary search, 346
    division, 103
    edge greedy (TSP), 602
    efficient, 352
    Euclidean, 173
    finding kernel, 574
    graph, 525
    greedy, 600

Algorithm (*Contd.*)
  Kruskal's, 532
  merge sort, 357
  minimal spanning tree, 527, 532
  Prim's, 527–529
  recursive, 356
  selection sort complexity, 350
  sorting, 589
  spanning tree, 526
  vertex greedy (TSP), 600
Almost Eulerian, 518
Alphabet, 300
Always, 42
And ($\wedge$), 27, 38
And gate, 230
Antiprism, 545
Antisymmetric relation, 302, 306, 321
Apples, 382, 448
Argument, 71
  conclusion, 74
  fallacy, 72
  invalid, 74
  premises, 74
  valid, 74
Argument structure, 74
  modus ponens, 72, 76
  modus tollens, 72, 76
Aristotle, 68
Arithmetic-logic unit, 230
Arithmetic operations, 331
Arithmetic progression, 22
Arithmetic triangle, 403, 428
Arrow diagram, 252, 505
Associative property
  Boolean algebra, 225
  logic, 35, 222
  sets, 188, 215, 222
  AUTHOR, 87
Average value, 467, 469

Babylonians, 344
Backgammon, 449, 465
Bacon, K., 276
Bag, 372, 411, 448
Barnsley, M., 342
Base b numeral, 156
Base ten, 150
Base two, 152
  conversion algorithm, 154
Baseball, 5, 393, 455, 615
  batting average, 466, 615
Basketball, 127
Batting average, 466, 615
Bernoulli, J., 455, 460
Bernoulli trial, 455, 460, 469
Bernstein, F., 295
Biconditional ($\leftrightarrow$), 62
Binary magic cards, 165
Binary number of a game position, 577

Binary numeral, 152, 326, 333
  conversion algorithm, 154
Binary relation (see Relation)
Binary search algorithm, 353
Binary search tree, 587
Binary sequence, 383, 393, 408, 412, 420, 475, 607
Binary string, 266, 280, 300
Binary tree, 583
  child in, 583
  complete, 595
  empty, 583
  height, 586
  inorder traversal, 591
  internal node, 584
  leaf in, 583
  left subtree, 583
  level, 585
  node in, 583
  parent in, 584
  postorder traversal, 591
  preorder traversal, 590
  right subtree, 583
  root of, 583
  sibling in, 583
  traversal, 590
Binomial, 402
Binomial coefficient, 403
Binomial probability distribution, 465
Binomial theorem, 403
Bipartite graph, 536
Birthday problem, 396, 444, 447
Bit, 152, 327
Blackjack, 373, 388, 396, 447, 454, 459, 614
Board games, 65
Boole, G., 223
Boolean adjacency matrix, 555
Boolean algebra, 223
Boolean algebra properties, 225
  DeMorgan's Law, 229
Boolean arithmetic, 555
Boolean expression, 224
  sum of products, 234
Boolean matrix operations, 554
Boolean power, 560
Boolean product of matrices, 555
Boolean sum of matrices, 561
Bouton, C., 576
Bridge, 372, 382, 616
Bridges of Königsberg, 506, 564
Bulgarian solitaire, 345
Butterfly effect, 341

$C(n, r)$, 372, 399
Cabbage, 567
Calculus, 77
Calinger, R., 140
Candy, 381, 448
Cantor, G., 295

Cantor's Theorem, 296
Cardano, G., 441
Cards, 613
  deck, 161, 294, 315, 344, 345, 388, 399, 405, 443, 445, 450, 613
  face, 458, 613
  hand, 294, 315, 372, 396, 447, 613
  suit, 294, 395, 400, 447, 454, 613
  tricks with, 2, 163, 344, 405
  value, 299, 613
Cardinality of a set, 292, 295
Carmichael numbers, 177
Cartesian graph, 198
Cartesian product, 198, 201
Cases, proof by, 101
Catfish, 432
Cauchy, A., 539
Ceiling function, 327
Chaff, 3
Chaos, 338
Chaos game, 342
Charlemagne, 567
Chessboard games, 580
Child in a tree, 583
Chinese remainder theorem, 178
Chokwe, 514
Circuit
  Eulerian, 508, 511
  graph, 508, 511
  logic, 229
  simplification, 231, 235
Circular arrangements, 400
Clock game, 581
Closed formula, 11, 111, 426
  summation, 114
Closed interval, 196
Closed trail, 508
Closure
  property of integers, 91
  property of rationals, 100
  reflexive, 566
  symmetric, 566
  transitive, 323, 557, 566
Codomain, 249
Collatz, L., 344
Collatz Problem, 338
Combination, 372
  number of ($C(n, k)$), 399
Combinatorial equivalence, 376
Combinatorics, 368
Commutative property
  Boolean algebra, 225
  logic, 35, 222
  sets, 188, 215, 222
Complement, 187
Complementary event, 444
Complementary problem, 390
Complete binary tree, 595
Complete bipartite graph ($K_{m,n}$), 536, 607

Complete graph (K_n), 536, 598
Complexity
  algorithm, 346, 601
  Master theorem, 358
  non-polynomial, 597
  O, 349
  Ω, 349
  selection sort, 350
  Θ, 349
Composition of functions, 268
Composition of relations, 274, 554
Conclusion
  of an argument, 74
  of an implication, 53, 57, 83
Conditional probability, 452
Congruent mod n, 166
Connected component, 511
Connected graph, 511, 560–562
Contradiction, 35
Contrapositive, 60, 84
  proof by, 94, 100, 132, 287, 520
Converse, 60, 68
Converse fallacy, 73
Correspondence, 286, 376
Countable set, 297
Counterexample, 43, 51, 57, 83, 190
Counterfeit detection algorithm, 354, 363
Counting algorithm, 391
Cross ($\times$), 198
Cryptography, 172
Cube, 539
Cycle
  in a graph, 508
  Hamiltonian, 596, 603
  in a sequence, 336

Deal, 613
Decimal numeral, 150
Decision tree, 457, 475, 581
Deck of cards, 161, 294, 315, 344, 345, 388,
  399, 405, 443, 445, 450, 613
Decryption, 172
Degree (deg(v)), 510, 513
Degree sequence, 516, 544
DeMéré, C., 441
Democrat, 78, 401
DeMorgan, A., 440
DeMorgan's Law
  Boolean algebra, 225, 229
  logic, 32, 35, 46, 222
  sets, 215, 222
Derangement numbers, 423
Descartes, R., 77, 198
Dessert, 69
Deuce, 421, 476, 489
Diagram
  arrow, 252
  Hasse, 302
  Venn, 189

Dice, 65, 414, 419, 452, 467
Difference of sets, 186
Difference table, 424–425
Differences, 424–425
Direct sum, 577
Directed edge, 549
Directed graph, 505, 509, 548
  adjacency matrix, 549
  in-degree, 564
  out-degree, 564
  of a relation, 553
  walk, 549, 551
Disjoint events, 448
Disjoint sets, 186
Distributive property
  Boolean algebra, 225
  logic, 35, 222
  sets, 188, 215, 222
Divide-and-conquer, 356, 364
Dividend, 104
Divisible, 82, 96
  by 4, 90
  by k, 98
Division theorem, 103
Divisor, 104
Divisor lattice, 229
Dodecahedron, 539
Dodecahedron puzzle, 596, 600
Domain
  function, 249
  predicate, 42, 47, 116
  relation, 253
Dominoes, 118, 514
Double arrow, 254, 317
Double negative property
  Boolean algebra, 225
  logic, 33, 35, 46, 222
  sets, 215, 222
Drug testing, 454
Dual expression, 216, 225
Duality principle
  Boolean algebra, 226
  sets, 216

$E[X]$, 467, 469
Edge, 506
  directed, 510, 548
  multiple, 508
  parallel, 508
Edge crossing, 536
Edge greedy algorithm, 602
Eeny-meeny-miney-more, 3
Element of a set, 183
Element-wise set proof, 205
Embedding, 536
Empty binary tree, 583
Empty set ($\emptyset$ or { }), 183, 211
Encryption, 172
Endpoint, 508

Enumeration, 369
Envelope puzzle, 6, 9, 513, 564
Equality of sets ($=$), 183, 213
Equally likely outcomes, 441
Equivalence
  class, 398
  combinatorial, 376
  logical, 32, 41, 60
Equivalence relation, 313, 319, 398
  induced by partition, 315
Euclid, 68
  Elements, 140
Euclidean algorithm, 173
Euler, L., 6, 77, 423, 506, 538–541, 608
Eulerian, 508
  circuit, 506, 511
  graph, 511, 525
  trail, 506, 513
Euler's formula, 538, 541, 545
Even number, 44, 90
Event, 441
  disjoint, 448
  independent, 451
  mutually exclusive, 448
Exclusive or, 29
Existence proof, 90, 100, 137
Exists ($\exists$), 43
Expected value, 467, 469
Experiment, 441
Expression tree, 582, 592

Face of planar graph, 540, 542
Face card, 458, 613
Factor, 98
Factorial, 16, 330, 385, 393, 589
Failed trial, 460
Fallacy, 72
  converse, 73
  inverse, 73
False advertising, 68
Family tree, 581, 594
Fermat, P., 441
Fermat numbers, 110, 332
Fermat's little theorem, 171, 174
Fibonacci, L., 16
  Liber Abacci, 150
Fibonacci numbers, 16, 125, 131, 345, 433
Finite set, 295
Finite structures, 324, 369, 415
Finitely progressive game, 572
First differences (Δ_n), 424
Floor function, 327
Flush, 447, 459, 614
For all ($\forall$), 43
Formal proposition, 27
Four of a kind, 474, 614
Fractal, 342
Fraction, 100
Full house, 465, 614

Function, 11, 249
  arithmetic operations, 331
  arrow diagram, 252
  as a binary relation, 255
  ceiling, 327
  codomain, 249
  composition ($\circ$), 268
  domain, 249
  floor, 327
  growth hierarchy, 356
  growth of, 345
  identity (ι), 272
  increasing, 349
  inverse ($^{-1}$), 261
  invertible, 283, 286
  iterated, 336
  linear, 363
  map, 249
  one-to-one, 286
  one-to-one correspondence, 286, 376
  onto, 286
  rule, 249
  sequence, 347
  sublinear, 363
  superlinear, 363
Fundamental theorem of arithmetic, 129
Fundamental theorem of sums, 425

Galilei, G., 441
Gambler's ruin problem, 478
Gambling, 441, 474
Game tree, 7, 25
Games
  backgammon, 449, 465
  baseball, 5, 393, 455
  basketball,
  blackjack, 373, 388, 396, 447, 454, 459, 614
  board games, 65, 500
  bridge, 372, 382, 616
  Bulgarian solitaire, 345
  card, 614
  chaos, 342
  chessboard, 580
  children's, 485, 495, 498, 500
  clock game, 581
  finitely progressive, 572
  on a graph, 572
  grid game, 6, 581
  Hank and Ted, 478–481, 482
  tournament, 127, 419
  Josephus game, 3, 8, 22, 121, 128, 131, 159
  kernel, 573, 577
  Kevin Bacon, 276, 282
  Nim, 576, 580
  poker, 399, 447, 459, 614
  Scrabble, 335, 616
  series of, 461, 471, 501

softball, 470, 615
solitaire, 445
state, 482
stone removal, 549, 572
table tennis, 480, 491, 500, 616
tennis, 5, 9, 382, 421, 463, 476, 480, 489, 501, 615
twenty-one, 614
Gardner, M., 580
Gate, 230
  and, 230
  inverter, 230
  not, 230
  or, 230
Gauss, C., 373, 418
General sum rule, 391, 450
Geometry, 539
Geometry of position, 506
Gergonne pile trick, 344
Goat, 567
Gödel, K., 296
Goldbach, C., 539
Golden ratio, 139, 182
Goldilocks, 54, 82
Golf, 72
Goose, 394
Graph, 505
  acyclic, 521
  adjacency matrix, 547, 549
  algorithm, 525
  almost Eulerian, 518
  bipartite, 536
  circuit, 508
  closed walk, 508
  complete (K_n), 536, 598
  complete bipartite ($K_{m,n}$), 536, 607
  connected, 511, 519, 560–562
  connected component, 511
  cycle, 508
  directed, 509, 548
  edge, 506
  edge-crossing, 536
  embedding, 536
  Eulerian, 506, 511
  face, 540
  of a game, 573
  Hamiltonian, 596, 603
  Hamiltonian cycle, 596
  induction proofs, 522
  isomorphism, 534
  kernel, 573, 577
  longest path, 530, 566
  loop, 508
  minimal spanning tree, 527
  multiple edge, 508
  node, 508, 583
  parallel edge, 508
  path, 508, 568, 596
  Petersen, 545

planar, 536, 538
of a relation, 553
simple, 510
spanning tree, 526
subgraph, 511, 569
trail, 508
traversal, 590
trivial circuit, 508
trivial walk, 508
tree, 505, 521
vertex, 506
walk, 506, 551
weighted, 527, 599
Graph theory, 505
Greatest common divisor, 228
Greedy algorithm, 600
Greek, 129, 538
Grid game, 6, 581

Hall, M., 4
Hamilton, W., 596
Hamiltonian cycle, 596
Hamiltonian graph, 596, 603
Hand of cards, 294, 315, 372, 396, 447, 613
Hank and Ted game, 478–481, 482
Harmonic series, 131
Hasse, H., 302
Hasse diagram, 302
Height of binary tree, 586
Herring, 394
Hexadecimal numeral, 158
  binary conversion, 159
Hockey stick identity, 428
Home field advantage, 465
Hypothesis, 53, 57, 83

Icosahedron, 539
Idempotent property
  Boolean algebra, 225
  logic, 35, 222
  sets, 215, 222
Identity function, 272
Identity matrix (I), 497, 566, 620
Identity property
  Boolean algebra, 225
  logic, 35, 222
  sets, 215, 222
If and only if ($\leftrightarrow$), 62
If-then statements, 53, 83
Implication, 53, 82
  conclusion, 53, 57, 83
  contrapositive, 60, 84
  converse, 60, 69
  counterexample, 57, 83
  hypothesis, 53, 57, 83
  inverse, 60, 69
  language of, 61
  negation of, 58
Implies ($\rightarrow$), 53

In-degree, 564
In-shuffle, 162
Incident, 510
Inclusion-exclusion principle, 192, 415
Incomparable, 309
Increasing function, 349
Independent, 401
Independent events, 451
Induced equivalence relation, 315
Induction, 110, 117
   as a game, 111
Inductive reasoning, 110
Infinite set, 295
Infix notation, 254, 593
Injection (See one-to-one function)
Inorder traversal, 591
Input, 249
Instant Insanity, 568, 580
   good subgraph, 569, 571
   partial solution, 569
Integer solutions, 411, 413, 448
Integers ($\mathbb{Z}$), 45, 182
Integrated circuit, 537
Internal node, 584
Intersection ($\cap$), 186
Interval notation, 184, 196, 290
Invalid argument, 74
Inverse, 60
Inverse fallacy, 73
Inverse function, 261
Inverse matrix, 497, 621
Inverse relation, 257
Inverter, 230
Invertible function, 283, 286
Iota, 272
Irrational number, 100, 138
Irreflexive relation, 308
Isomorphic graphs, 534–535
Isomorphism, 534
Iterated function sequence, 336

Jail, 65
Jones, T., 54, 82
Josephus, F., 3
Josephus game, 3, 8, 22, 121, 128, 131, 159
Josephus permutation, 4

Kabjlie, 579
Karnaugh, M., 235
Karnaugh map, 235
Keillor, G., 149
Kernel, 573
Kernel algorithm, 574
King, S., 77
Knight's Tour puzzle, 608
Königsberg, 506
Kruskal's algorithm, 532
Kuratowski, K., 537
Kuratowski's Theorem, 538

Lagrange, J., 110
Lattice, 229
Leaf (leaves), 521, 583
Least common multiple, 228
Left subtree, 583
Leibniz, G., 77, 506
Length of walk, 508
Level, 585
Liars, 24
Libertarian, 78
Linear algebra, 485, 496, 498
Linear first order recurrence, 430
Linear function, 363
List
   ordered, 371, 393
   unordered, 371, 413, 510
Logarithm, 324, 356, 588
   base b, 326, 330
   base ten, 324, 608
   base two, 355
   properties, 325
Logarithmic growth, 356
Logic
   and, 27, 38
   circuit, 229
   exclusive or, 29
   if and only if, 62
   implies, 53
   not, 27, 38
   or, 27, 38
   predicate, 40, 116
   propositional, 27, 68
   substitution rule, 36
Logic properties, 35
   absorption, 39, 222
   double negative, 33
Logic puzzles, 26–30, 37, 63, 67
Logically equivalent, 32, 60
Longest path, 530, 562
Loop, 508, 510
Lorayne, H., 405
Lorenz, E., 340
Lottery, 369, 407, 447, 474

Magic
   binary magic cards, 165
   club trick, 2, 128, 250
   domino trick, 514
   Gergonne pile trick, 344
   number spelling trick, 334
   ordinal trick, 344
   perfect shuffle trick, 163
   Scrabble trick, 335
   triangle card trick, 405
Map, 249
Markov chain, 482, 498
Master theorem, 358
Mathematical induction, 110, 117
Matrix

addition, 491, 551, 619
adjacency, 547
arithmetic, 550, 619
   Boolean adjacency, 555
   Boolean operations, 554
   Boolean power, 560, 566
   Boolean product, 555
   Boolean sum, 561, 566
   entry ($M_{i,j}$), 482, 547, 618
   identity (I), 497, 566, 620
   inverse (M^{-1}), 497, 621
   multiplication, 484, 550, 555, 619
   row-column multiplication, 484, 555, 619
   transition, 482
   transpose (M^T), 566, 623
   zero ($\mathbf{0}_n$), 620
Meaningless, 373
Member of a set, 41
Merge sort algorithm, 357
Minimal spanning tree, 527
   Kruskal's algorithm, 532
   Prim's algorithm, 527–529
Mississippi, 410
mod (%) notation, 105, 166
Modular arithmetic, 168, 176
Modus ponens, 72, 76
Modus tollens, 72, 76
Morris, B., 162
Morton, M., 40
Multiple, 54, 98, 328
Multiple edge, 508
Multiple quantifiers, 46
Mutually exclusive events, 448
Mutually exclusive sets, 186

$n(A)$, 191, 443
n-tuple, 201
Natural numbers ($\mathbb{N}$), 182
Necessary, 70
Negation
   of propositions, 27, 31, 39, 58
   quantified statements, 45–46
Negation property
   Boolean algebra, 225
   logic, 35
   sets, 215
Networks, 6
Never, 42
New York City, 299
Newton, I., 77
Nicholson, J., 276
Nim, 576, 580
Node, 508, 583
Non-absorbing states, 494
Non-existence proof, 137
Non-planar graph, 537
Not ($\neg$), 27, 38
Not gate, 230
NP-complete, 600

Number puzzles, 9, 13, 23
Number of set elements, 191
Number theory, 140, 145, 171
Numbers
  Bell, 423
  binary, 152, 326
  Carmichael, 177
  derangement, 423
  even, 42, 90
  Fermat, 110, 332
  Fibonacci, 16, 125, 131, 345, 433
  fraction, 100
  Hindu-Arabic, 150
  integers ($\mathbb{Z}$), 45, 182
  irrational, 100, 138
  natural ($\mathbb{N}$), 182
  odd, 42, 90
  Pell, 131
  perfect cube, 98
  perfect square, 15, 88, 164, 176
  prime, 52, 82, 140
  rational ($\mathbb{Q}$), 100, 182
  real ($\mathbb{R}$), 45, 182
  relatively prime, 138
  twin primes, 109
Numeral, 150
  base b, 156, 330
  binary, 152, 326
  decimal, 150
  hexadecimal, 158
  octal, 164
  Roman, 150

O notation, 349
Octahedron, 539
Octal, 164
Odd number, 10, 90
Omega notation (Ω), 349
One-set arrow diagram, 254
One-to-one correspondence, 286, 376
One-to-one function, 286
Only if, 70
Onto function, 286
Operations research, 369
Optimization, 369
Or ($\vee$), 27, 38
Or gate, 230
Oracle of Bacon, 276, 282
Order
  of function growth, 350, 356
  partial, 302
  relation, 301
  strict partial, 308
  strict total, 309
  total, 309
  well-ordering, 140
Order relation, 301
Ordered list, 371, 393
Ordered n-tuple, 201

Ordered pair, 198, 510, 548
Ordered quadruple, 201
Ordered triple, 201
Ore's Theorem, 604
Oresme, N., 131
Out-degree, 564
Out-shuffle, 162
Output, 249

$\mathcal{P}(A)$, 203
$P(n)$, 116
$P(n, r)$, 372, 393, 419
Pair, 614
Paradox, 197
Parallel edge, 508
Parent in a tree, 584
Parity, 314
Part, 205, 315
Partial order, 302
  strict, 308
Partition, 205, 314, 423
  equivalence relation of, 315
Pascal, B., 403, 441
Pascal's triangle, 403, 428
Path, 508, 568, 596
Pedantic, 95
Peirce, C., 68
Pell sequence, 131
Pennsylvania, 70, 83
Perfect cube, 98, 109
Perfect shuffle, 161
  how to fake, 163
Perfect square, 15, 88, 164, 176
Period of cycle, 336
Permutation, 300, 372
  derangements, 423
  Josephus permutation, 4
  number of ($P(n, k)$), 393, 419
  perfect shuffle, 161
Petersen graph, 545, 608
Pigeonhole principle, 142, 530
  average version, 148
  basic version, 144, 293
  general version, 143
Pittsburgh, 70
Pizza, 38
Planar graph, 536, 538
  application, 537
  Euler's formula, 538, 541, 545
  face, 540
  face size, 542
Platonic solids, 538
Poker, 399, 447, 614
Polyhedra
  antiprism, 545
  prism, 545
  pyramid, 545
  regular, 538
Population model, 432

Postfix notation, 593
Postorder traversal, 591
Potato, 3
Power set, 203
Precedence, 28
Predicate, 40
  domain, 42, 47
  induction, 116
  quantified, 43
Prefix notation, 593
Premise, 77
Preorder traversal, 590
Prerequisite, 308
Prim's algorithm, 527–529
Prime interest rate, 78
Prime number, 52, 88, 140
  relatively, 138
Principle of Inclusion/Exclusion, 192, 415
Prism, 545
$Prob(E)$, 443
Probability, 440, 443
  conditional ($Prob(E_1|E_2)$), 452
  experiment, 441
  general product rule, 453
  general sum rule, 450
  product rule, 452
  sample space, 441
  sum rule, 449
Product of matrices, 484, 550, 619
Product rule
  for counting, 387
  for probability, 452, 453
Proof, 85
  of antisymmetric relation, 307
  by cases, 101
  by contradiction, 133
  by contrapositive, 94, 100, 132, 287, 520
  direct, 85, 89, 132
  element-wise set, 205
  existence, 90, 100, 137
  about functions, 287
  as a game, 87
  about graphs, 519
  by induction, 110, 117, 132
  induction on graphs, 522–524
  of invertible function, 292
  as a letter, 89
  non-existence, 137
  about numbers, 90, 96
  of one-to-one function, 289
  of onto function, 292
  of reflexive relation, 305
  set equality, 213
  of subsets, 210
  of symmetric relation, 318
  tracing, 89
  of transitive relation, 306
Proper subset ($\subset$), 195
Proposition, 27

Propositional connective, 30, 51, 65
Propositional variable, 27
Pseudocode, 391
Pseudoprime, 177
Psychology, 54
Puzzles
  coin weighing, 354, 363
  envelope puzzle, 6, 9, 513, 564
  Hamilton's dodecahedron, 596, 600, 608
  higher or lower, 363
  Instant Insanity, 568, 580
  knight's tour, 608
  line tracing, 514
  logic puzzles, 24–30, 37
  number puzzles, 9, 13, 23
  river-crossing puzzle, 567, 579
  towers of Hanoi, 420
  utilities puzzle, 536, 543, 546
  water puzzle, 579
Pyramid, 545

$\mathbb{Q}$, 182
Quadrilateral, 54
Quantified predicate, 43
Quantifier, 43
Quotient, 103

Random variable, 466
Rational numbers, 100, 182
READER, 87
Real numbers ($\mathbb{R}$), 45, 182
Reciprocal, 170
Rectangle, golden, 139
Recurrence relation, 11, 423
  constant differences, 426
  divide-and-conquer, 356, 364
  linear first order, 430
  master theorem, 358
  second order, 432, 436
Recursion, 475
Recursive algorithm, 356
Recursive formula, 11, 111
Recursive model, 4, 418, 475, 511
Reflexive closure, 566
Reflexive relation, 302, 566
Red herring (See wild goose)
Regular polyhedra, 538
Relation, 253
  adjacency matrix for, 553
  antisymmetric, 302, 306, 321
  binary, 253
  codomain, 253
  composition ($\circ$), 274, 554
  domain, 253
  equivalence, 315, 319, 398
  function, 255
  graph of, 553
  identity (ι), 272
  infix notation, 254

inverse, 257
irreflexive, 308
order, 301
partial order, 302
reflexive, 302, 566
rule, 253
strict partial order, 308
symmetric, 317, 566
transitive, 302, 323, 557
Remainder, 103
Republican, 78, 401
Reverse, 257
RGB color values, 165
Right subtree, 583
River-crossing puzzle, 567, 579
Rivest, R., 172
Roman numerals, 150
Rook, 580
Root of a tree, 583
Roster method, 183
Round robin tournament, 419, 435
Row-column multiplication, 484, 619
RSA cryptosystem, 172
Rule, 249, 253
Rule of complements, 390
Rule of products, 387
Rule of sums, 389, 391
Russell, B., 197
Russell's paradox, 182, 197

Sample space, 441
Schoolhouse Rock, 40
Scrabble, 335
Search tree, 587
Second differences, 425
Second order recurrence relation, 432, 436
Selection sort complexity, 350
Selection tree, 388
Self-similarity, 342
Senate, 399
Sensitivity to initial conditions, 340, 345
Sequence, 9
  binary, 383, 393, 408, 412, 420, 475, 607
  closed formula, 11, 111, 426
  differences, 424–425
  function, 347
  iterated function, 336
  notation, 11, 199
  recursive formula, 11, 111
Series of games, 461, 471, 501
Set, 181, 372
  cardinality, 295
  Cartesian product, 198, 201
  complement, 187
  countable, 297
  difference, 186
  disjoint, 186
  duality principle, 216
  element, 41, 183

empty, 183
equality, 183, 213
finite, 295
form description, 185
infinite, 295
integers ($\mathbb{Z}$), 182
intersection, 186
membership ($\in$), 43, 183
natural numbers ($\mathbb{N}$), 182
number of elements, 191
rational numbers ($\mathbb{Q}$), 182
real numbers ($\mathbb{R}$), 182
roster method, 183
partition, 205, 314, 423
power set, 203
property description, 184
subset, 183
uncountable, 297
union, 186
universal, 183
Set builder notation, 184
Set operation properties, 215, 222
  absorption, 195, 220
  associative, 188
  commutative, 188, 220
  DeMorgan's law, 220
  distributive, 188, 220
Shakespeare, W., xiii
Shamir, A., 172
Shuffle
  in-shuffle, 162
  out-shuffle, 162
  perfect, 161
Sibling in a tree, 583
Sierpinski Triangle, 343
Sigma notation, 19, 114
Simple graph, 510
Simple interest, 431
Single elimination tournament, 127
Six Degrees of Separation, 276
Smarty pants, 418
Smullyan, R., 24
Smullyan's Island, 24–30, 34, 37, 63, 67, 199
Softball, 470, 615
Solitaire, 345, 445
Sometimes, 42
Sorting algorithm, 350, 357, 589
Spanning tree, 526
  algorithm, 527
  minimal, 526
Spreadsheet, 345, 351, 360–364
Stamps, 131
State of a game, 482
  absorbing, 483
  transient, 494
Statement for induction, 116
Statistics, 454
Steroids, 454
Stone removal game, 549, 572

Straight, 447, 459, 614
Straight flush, 447, 614
Strategy, 7, 572
Strict partial order, 308
Strict total order, 309
Strong induction, 119
Subgraph, 511, 569
Subset ($\subseteq$), 183
Substitution rule, 36
Subtree, 583
Successful outcomes, 443
Successful trial, 460
Sufficient, 70
Suit of a card, 294, 395, 400, 447, 454, 613
Sum of matrices, 491, 551, 619
Sum notation, 19, 114
Sum of products, 234
Sum rule
    for counting, 389, 391
    for probability, 449, 450
Summation, 19
Super Bowl, 150
Surjection (see onto function)
Symmetric closure, 566
Symmetric relation, 317, 566
Syracuse Problem, 338

Table tennis, 480, 491, 500, 616
Tartaglia's triangle, 403
Tautology, 35, 74
Telecommunications, 599
Tennis, 5, 9, 382, 421, 463, 476, 480, 489, 501, 615
Ternary sequence, 417
Tetrahedron, 539
There exists ($\exists$), 43
Theta notation, 349
Third differences, 425
Thoreau, H., xi
Three of a kind, 465, 614
TI-83 calculator, 326, 447, 559, 620, 623
Time complexity, 346
Topology, 538
Total order, 309
Tournament
    first round, 423
    round robin, 419, 435
    single elimination, 127
Towers of Hanoi, 420

Tracing a proof, 89
Trail, 508, 513
Transient states, 494
Transition matrix, 482
Transitive closure, 323, 566
Transitive relation, 302, 323, 557, 566
Transportation network, 599
Transpose of a matrix, 566, 623
Traveling Salesperson Problem, 597
Traversal of a binary tree, 590
Traversal problem, 597
Tree, 505, 521
    binary, 581
    binary search, 587
    child in, 583
    decision, 457, 475, 581
    expression, 582, 592
    family, 581, 594
    game, 7, 9, 25
    height, 586
    inorder traversal, 591
    internal node, 584
    leaf (leaves), 521, 583
    level, 585
    minimal spanning, 527
    parent in, 584
    postorder traversal, 591
    preorder traversal, 590
    root of, 583
    selection, 388
    sibling in, 583
    spanning, 526
    subtree, 583
    traversal, 590
Trial, 455
Triangle, 52
Trick-or-treat, 373
Trivial circuit, 508
Trivial walk, 508
Truth table, 24
    and ($\wedge$), 30
    for circuit, 230
    as function, 250
    implication ($\rightarrow$), 55
    not ($\neg$), 30
    or ($\vee$), 30
Truth-tellers, 24
TSP, 597

Tukey, J., 152
Tuple, 201
Twenty-one game, 614
Twin primes, 109
Two pair, 465, 614
Two-set arrow diagram, 254

U.S. presidents, 585
Ulam's Problem, 338
Uncountable set, 297
Unemployment, 78
Union ($\cup$), 186
Universal bound property
    Boolean algebra, 225
    logic, 35, 222
    sets, 215, 222
Universal set, 183
Universe, 183
Unordered list, 371, 413, 510
Utilities puzzle, 536, 543, 546

Valid argument structure, 74
Validity of arguments, 71
Value of card, 613
Vandermonde, A., 608
Variable, 27, 40, 466
Venn diagram, 189
Vertex (vertices), 506
    degree, 510, 513, 564
Vertex-greedy algorithm, 600
Virginia, 276, 585

Walk, 506
    length, 508
    trivial, 508
Wason's selection task, 84
Water puzzle, 579
Weak induction, 120
Weighted graph, 527, 599
Well-ordering principle, 140
Welles, O., 276
Wild goose chase (See red herring)
Wolf, 567
Word, 300
World Series, 462

$\mathbb{Z}$, 45, 182
Zahlen, 45
Zero matrix ($\mathbf{0}_n$), 620